MCP Mathematics

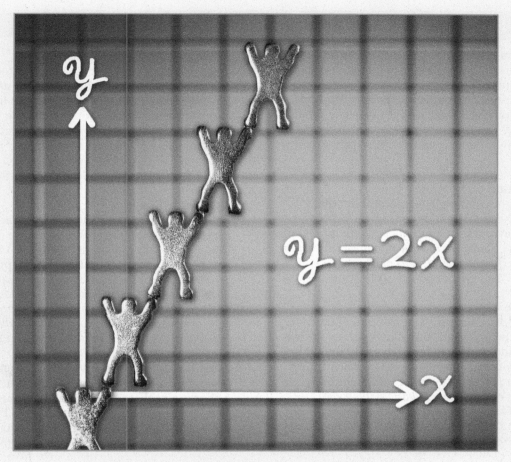

$$y = 2x$$

Richard Monnard • Royce Hargrove

TEACH THE COVER: What equation can be used to describe the line of people on the graph?

Project Staff

Art & Design: Robert Dobaczewski, Kathleen Ellison, Senja Lauderdale, David Mager, Jim O'Shea, Angel Weyant

Editorial: Stephanie P. Cahill, Gina Dalessio, Phyllis Dunsay, Mary Ellen Gilbert, Dena Kennedy, Theresa McCarthy, Marilyn Sarch

Marketing: Doug Falk, Clare Harrison

Production/ Manufacturing: Irene Belinsky, Lawrence Berkowitz, Louis Campos, Diane Fristachi, Pamela Gallo, Leslie Greenberg, Suellen Leavy, Ruth Leine, Karyn Mueller, Michele Uhl

Publishing Operations: Carolyn Coyle, Richetta Lobban

SAVVAS
LEARNING COMPANY

ISBN-13: 978-0-7652-6067-3
ISBN-10: 0-7652-6067-0
15   20

MCP Mathematics

About the Program

This comprehensive math program will help students in Grades K to 6 develop a solid mathematics background. It is designed to encourage critical-thinking skills, active participation, and mastery of skills within the context of problem-solving situations. The program's developmental sequence introduces and extends skills taught in most mathematics curricula, such as number sense, operations, algebra, geometry, data collection and analysis, logic, and probability.

A Research-Based Approach

The program offers a strong pedagogical approach that is research based. First, students are provided a developed model that introduces the lesson concept. Then, students are given guided practice opportunities to get started. Next, abundant practice is provided for mastery. Finally, students can apply their skills to problem-solving and enrichment activities. An overview of this approach follows.

● Direct Instruction

Each lesson begins with a developed model that demonstrates the algorithm or concept in a problem-solving situation and gets students actively involved in the situation. . . . *Students taught using direct instruction were found to perform better on tests of computation and year-end tests (Crawford and Snider 2000).**

● Guided Practice

A Getting Started section provides samples of the concept or skill being taught and allows you to guide and observe students as they begin to apply the skills learned. *When low-achieving students were taught using such direct instruction methods as . . . guided practice, . . . they showed improved mastery of basic skills, solved computation and word problems correctly in less time, and had higher self-ratings of academic motivation (Kame´enui, Carnine, Darch, and Stein 1986; Din 1998; Ginsburg-Block and Fantuzzo 1998).**

● Independent Practice

The Practice section can be used to develop and master basic skills by allowing students to independently practice the algorithms and to apply learning from the lesson or from previous lessons. *Research indicates that providing students with extended practice appears to serve two purposes: re-teaching of the skill for students who had not yet mastered it and relating of the previously learned skill to new skills, resulting in the formation of interrelationships among concepts that improved retention and yielded higher achievement test results (Hett 1989).**

● Higher-Order Thinking Skills

A collection of real-life word problems in the Problem Solving section provides application opportunities related to the lesson concepts. *Frequent practice with word problems is associated with higher-order skill development (Coy 2001). This finding is especially true when the word problems present familiar real life situations (Coy 2001).**

● Problem-Solving Strategies

Once per chapter, a special lesson introduces students to the techniques of problem solving using Polya's four-step model. The Apply activities in these lessons allow students to use problem-solving strategies in everyday situations. *Students who received instruction in problem-solving processes showed better performance on tests of skills, tasks, and problem solving, as well as on a measure of the transfer of learning (Durnin et al. 1997).**

● Calculator Usage

Calculator lessons in Grades 3 to 6 teach students the functions and the skills needed to use calculators intelligently after they've developed a foundation of competence in pencil-and-paper computation. *. . . Students who used calculators in mathematics instruction had more positive attitudes toward mathematics and a higher math self-concept. . . . A special curriculum developed around calculators has been shown to improve mathematics achievement (Hembree and Dessart 1986).**

● Frequent and Cumulative Assessment

Chapter Tests provide both students and teachers with a checkpoint that tests all the skills taught in a chapter. You will find Alternate Chapter Tests based on the same objectives in the Teacher's Edition. *Assessment has been found to be most effective when it is a frequent and well-integrated aspect of mathematics instruction (Brosnan and Hartog 1993).* Cumulative Assessments maintain skills that have been taught in all previous chapters. A standardized test format is used starting at the middle of the second-grade text. The Teacher's Edition also contains Alternate Cumulative Assessments. *. . . Frequent cumulative tests result in higher levels of achievement than do infrequent tests or tests related only to content since the last test (Dempster 1991).**

● Remedying Common Errors

The comprehensive Teacher's Edition provides abundant additional help for teachers, including a four-step lesson plan that walks the teacher through the lesson, pointing out common errors and providing intervention strategies. *Curricula with an error correction component were found to result in higher scores for computation, math concepts, and problem solving (Stefanich and Rokusek 1992; Crawford and Snider 2000).**

*Research compiled by PRES Associates, Inc. (2004). Research Support for *MCP Mathematics* (unpublished).

Using the Student Edition

Use the First Page of a Lesson for Direct Instruction

Each two-page lesson focuses on one main objective. The first page begins with a developed model that students can actively work on as you walk them through it.

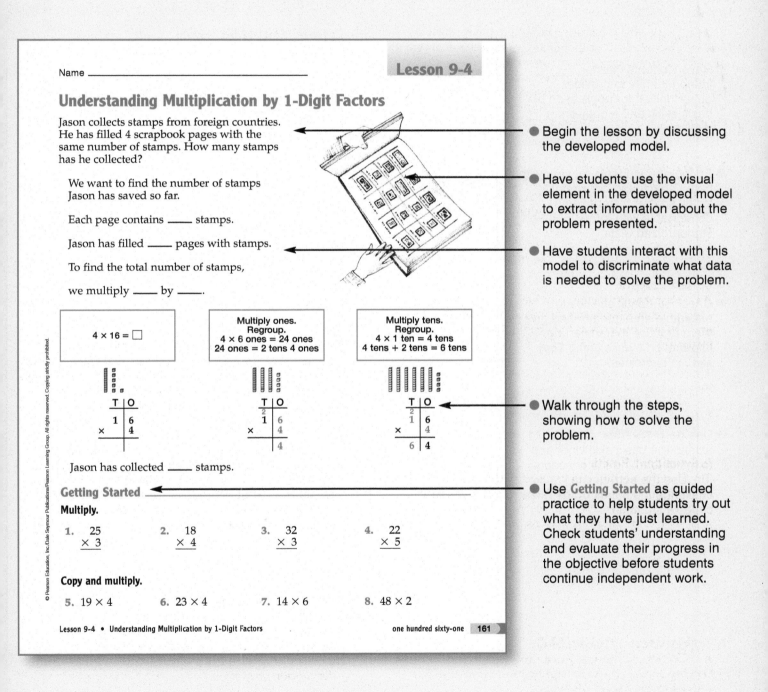

Name _____

Lesson 9-4

Understanding Multiplication by 1-Digit Factors

Jason collects stamps from foreign countries. He has filled 4 scrapbook pages with the same number of stamps. How many stamps has he collected?

We want to find the number of stamps Jason has saved so far.

Each page contains _____ stamps.

Jason has filled _____ pages with stamps.

To find the total number of stamps,

we multiply _____ by _____.

$4 \times 16 = \square$	Multiply ones. Regroup. 4×6 ones = 24 ones 24 ones = 2 tens 4 ones	Multiply tens. Regroup. 4×1 ten = 4 tens 4 tens + 2 tens = 6 tens

```
    T | O              T | O              T | O
                        2                  2
    1 | 6              1 | 6              1 | 6
  × |   4            × |   4            × |   4
                          4              6 | 4
```

Jason has collected _____ stamps.

Getting Started

Multiply.

1. 25
   × 3

2. 18
   × 4

3. 32
   × 3

4. 22
   × 5

Copy and multiply.

5. 19 × 4   6. 23 × 4   7. 14 × 6   8. 48 × 2

Lesson 9-4 • Understanding Multiplication by 1-Digit Factors   one hundred sixty-one   161

- Begin the lesson by discussing the developed model.

- Have students use the visual element in the developed model to extract information about the problem presented.

- Have students interact with this model to discriminate what data is needed to solve the problem.

- Walk through the steps, showing how to solve the problem.

- Use Getting Started as guided practice to help students try out what they have just learned. Check students' understanding and evaluate their progress in the objective before students continue independent work.

Using the Student Edition

Assign Practice from the Second Page of a Lesson

The second page of a lesson provides practice and extension of the lesson's objective. You can begin the process of individual mastery by assigning Practice exercises that students can work on independently.

- Have students Practice independently to provide opportunities for application of basic skills and higher-order thinking.

- Encourage students to work with both vertical and horizontal forms so that they become comfortable with forms found on standardized tests.

- Check students' abilities to assemble an algorithm and give them practice in transferring information by assigning **Copy and . . .** exercises.

- Use Now Try This! activities to extend the basic skill work and to make learning the concepts fun. Use the activities to challenge the minds of the more capable students.

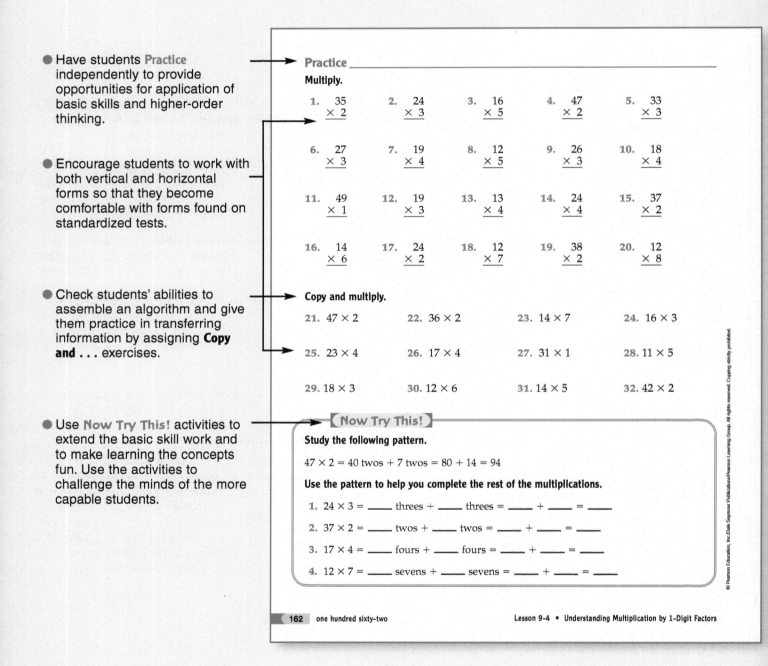

Practice

Multiply.

| 1. | 35 × 2 | 2. | 24 × 3 | 3. | 16 × 5 | 4. | 47 × 2 | 5. | 33 × 3 |

| 6. | 27 × 3 | 7. | 19 × 4 | 8. | 12 × 5 | 9. | 26 × 3 | 10. | 18 × 4 |

| 11. | 49 × 1 | 12. | 19 × 3 | 13. | 13 × 4 | 14. | 24 × 4 | 15. | 37 × 2 |

| 16. | 14 × 6 | 17. | 24 × 2 | 18. | 12 × 7 | 19. | 38 × 2 | 20. | 12 × 8 |

Copy and multiply.

21. 47×2 22. 36×2 23. 14×7 24. 16×3

25. 23×4 26. 17×4 27. 31×1 28. 11×5

29. 18×3 30. 12×6 31. 14×5 32. 42×2

Now Try This!

Study the following pattern.

$47 \times 2 = 40$ twos $+ 7$ twos $= 80 + 14 = 94$

Use the pattern to help you complete the rest of the multiplications.

1. $24 \times 3 = $ _____ threes $+$ _____ threes $= $ _____ $+$ _____ $= $ _____

2. $37 \times 2 = $ _____ twos $+$ _____ twos $= $ _____ $+$ _____ $= $ _____

3. $17 \times 4 = $ _____ fours $+$ _____ fours $= $ _____ $+$ _____ $= $ _____

4. $12 \times 7 = $ _____ sevens $+$ _____ sevens $= $ _____ $+$ _____ $= $ _____

Lesson 9-4 • Understanding Multiplication by 1-Digit Factors

Using the Student Edition

Teach Problem Solving and Calculators in Every Chapter

Problem Solving lessons focus on different problem-solving strategies using the chapter concepts and skills. Calculator lessons teach students to use the technology while reinforcing chapter content.

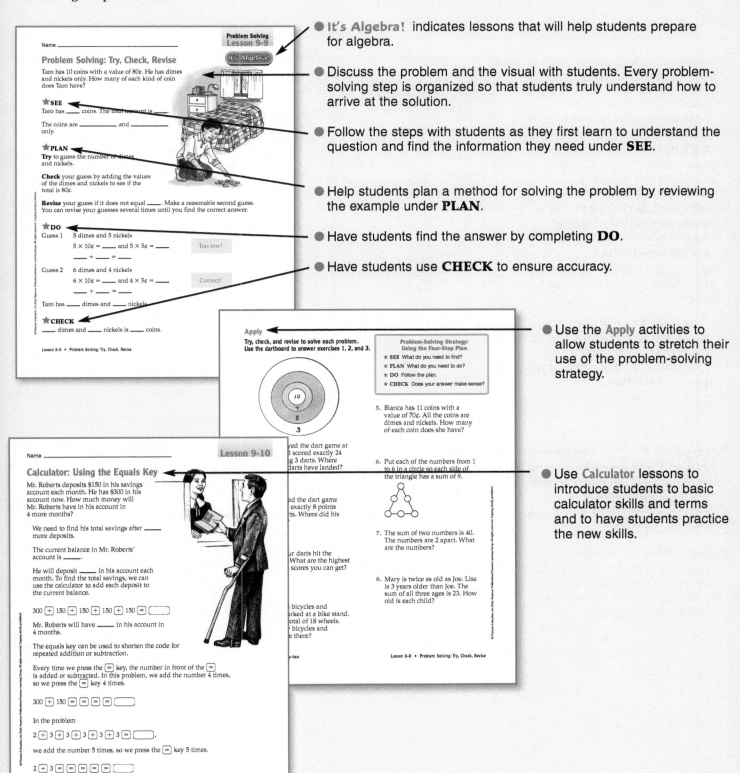

- **It's Algebra!** indicates lessons that will help students prepare for algebra.

- Discuss the problem and the visual with students. Every problem-solving step is organized so that students truly understand how to arrive at the solution.

- Follow the steps with students as they first learn to understand the question and find the information they need under **SEE**.

- Help students plan a method for solving the problem by reviewing the example under **PLAN**.

- Have students find the answer by completing **DO**.

- Have students use **CHECK** to ensure accuracy.

- Use the Apply activities to allow students to stretch their use of the problem-solving strategy.

- Use Calculator lessons to introduce students to basic calculator skills and terms and to have students practice the new skills.

v

Using the Teacher's Edition

Instruct Using a Four-Step Lesson Plan

The Teacher's Edition is designed and organized with you in mind. Consistent four-step lesson plans on two pages will help you make efficient use of your planning time and will help you deliver effective instruction.

- Reduced student pages provide point-of-use information.

- Use Getting Started to review the Objectives and to set a clear course for the lesson goal.

- Reduce class preparation time by gathering Materials early.

- Use the Warm Up exercises to help students brush up on skills at the beginning of each day's lesson.

- Teach gives practical suggestions for introducing the problem and developing the skill. You will find specific suggestions for an effective presentation of the model in Introduce the Lesson.

- Develop Skills and Concepts gives suggestions for presenting and developing the algorithm, skill, or concept. Some include ideas for the use of manipulatives.

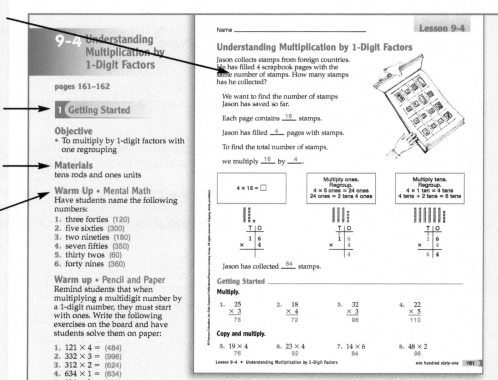

9-4 Understanding Multiplication by 1-Digit Factors

pages 161–162

1 Getting Started

Objective
- To multiply by 1-digit factors with one regrouping

Materials
tens rods and ones units

Warm Up • Mental Math
Have students name the following numbers:
1. three forties (120)
2. five sixties (300)
3. two nineties (180)
4. seven fifties (350)
5. thirty twos (60)
6. forty nines (360)

Warm up • Pencil and Paper
Remind students that when multiplying a multidigit number by a 1-digit number, they must start with ones. Write the following exercises on the board and have students solve them on paper:
1. $121 \times 4 =$ (484)
2. $332 \times 3 =$ (996)
3. $312 \times 2 =$ (624)
4. $634 \times 1 =$ (634)
5. $114 \times 2 =$ (228)

2 Teach

Introduce the Lesson Have students read and identify the problem. Ask students to read and complete the information sentences (16; 4) and the plan sentence. (16, 4)

- Make a quick sketch from Jason's stamp collection book on the board. Point out the *rows* and the *columns* and ask students to repeat. Then draw pages with different numbers of rows and columns and ask: *How many rows (columns) are there?*
- Have students examine the first step of the model. Ask a student to multiply 6 ones by 4. (24) Tell students to write the number of ones—in this case 4—in the ones column, and write the number of tens (2) over the tens column.
- Now, have them look at the next step and ask a student to tell the product of 4×1 ten. (4) Point out that the 2 tens left from the one's multiplication must be added

to the 4 tens. (4 tens + 2 tens = 6 tens) Have students complete the solution sentence. (64)

Develop Skills and Concepts Write the following on the board:

$$24 \times 3$$

Ask the class where to start multiplying. (in the ones column) Ask a volunteer to multiply ones. (3×4 ones = 12 ones) Ask what number to write in the ones column of the answer (2) and what number will be carried over to the tens column to be added later. (1) Use the base-ten blocks to clarify the regrouping. Display 12 ones units and have a student exchange 10 ones for 1 ten rod. Ask students how many ones are remaining. (2) Let another student multiply tens and add the 1 ten to complete the problem. (72) Repeat with several similar problems.

T161

Practice

Multiply.

1. 35 ×2 = 70	2. 24 ×3 = 72	3. 16 ×5 = 80	4. 47 ×2 = 94	5. 33 ×3 = 99
6. 27 ×3 = 81	7. 19 ×4 = 76	8. 12 ×5 = 60	9. 26 ×3 = 78	10. 18 ×4 = 72
11. 49 ×1 = 49	12. 19 ×3 = 57	13. 13 ×4 = 52	14. 24 ×4 = 96	15. 37 ×2 = 74
16. 14 ×6 = 84	17. 24 ×2 = 48	18. 12 ×7 = 84	19. 38 ×2 = 76	20. 12 ×8 = 96

Copy and multiply.

21. 47 × 2 = 94	22. 36 × 2 = 72	23. 14 × 7 = 98	24. 16 × 3 = 48
25. 23 × 4 = 92	26. 17 × 4 = 68	27. 31 × 1 = 31	28. 11 × 5 = 55
29. 18 × 3 = 54	30. 12 × 6 = 72	31. 14 × 5 = 70	32. 42 × 2 = 84

[Now Try This!]

Study the following pattern.

47 × 2 = 40 twos + 7 twos = 80 + 14 = 94

Use the pattern to help you complete the rest of the multiplications.

1. 24 × 3 = _20_ threes + _4_ threes = _60_ + _12_ = _72_
2. 37 × 2 = _30_ twos + _7_ twos = _60_ + _14_ = _74_
3. 17 × 4 = _10_ fours + _7_ fours = _40_ + _28_ = _68_
4. 12 × 7 = _10_ sevens + _2_ sevens = _70_ + _14_ = _84_

162   one hundred sixty-two   Lesson 9-4 • Understanding Multiplication by 1-Digit Factors

---

For Mixed Abilities

Common Errors • Intervention
When students multiply the ones, they may not regroup but write both digits in the answer.

INCORRECT	CORRECT
38	1
× 2	38
616	× 2
	76

Have them work with partners and base-ten blocks to model the problem.

Enrichment • Measurement
Have students think about the 8- and 6-hour clocks they made in the Enrichment on page T160. Ask them to explain the problem with these clocks. (The clock is in the same configuration 3 or 4 times a day.) Ask each student to devise a different way of distinguishing one part of the day from another, as our use of A.M. and P.M. does.

More to Explore •
Measurement
Have the students make a simple rain gauge. Ask them to mark off a glass or plastic jar in quarter-inches. Have students place the jar in a safe but open place at home or school. Ask the students to take measurements each time it rains or snows and record the results for a two-week period. Have students make a bar graph to show the amount of rainfall for the two weeks. Have the class compare their results with the official measurements given by the TV weather report or the newspaper.

ESL/ELL STRATEGIES

Make a quick sketch on the board of a page from Jason's stamp collection book. Point out the *rows* and *columns* and ask students to repeat the words. Then, draw pages with different numbers of rows and columns and ask: *How many rows (columns) are there?*

T162

---

3 Practice

Have students solve all the exercises on the page. Remind them to start multiplying in the ones column, and write any tens over the tens column and add them in last.

Now Try This! Discuss with the students how multiplying by decades helps to do problems mentally. Have students complete the exercises.

4 Assess

Ask students where they must begin multiplying a two-digit number. (at the ones)

● Use the **Common Errors • Intervention** feature to explore a common error pattern and to provide remediation to struggling students. Collectively, all the Common Errors features in any chapter constitute a complete set of the common errors that students are likely to make when working in that chapter.

● **Enrichment** activities are a direct extension of the skills being taught. Some students may do these activities on their own while you work with those students who need more help.

● **More to Explore** activities are challenging and independent, expanding the mathematical experiences of the students. The More to Explore section encompasses a wide variety of activities and projects and introduces and extends skills taught in the normal curriculum—including data collection and analysis, logic, and probability.

● **ESL/ELL Strategies** are offered twice per chapter. These activities will help you provide insights into English vocabulary and increase comprehension of mathematical concepts. Specific techniques cited ensure that learning is taking place. The techniques also remove potential language barriers for English-language learners at beginning levels of proficiency.

● **Practice** offers you guidelines to assist your students as they practice independently.

● **Assess** provides you with a short question or activity that can be used to quickly evaluate if students have grasped the main objective of the lesson.

Assessment

Assess Often With Chapter and Cumulative Tests

A variety of assessments help you track your students' mastery of algorithms, basic skills, and problem solving.

- Use **CHAPTER TESTS** to help you assess your students' mastery of all the skills taught in the chapters.

- Also, **ALTERNATE CHAPTER TESTS** are included in the Teacher's Edition.

- A variety of problems give students a better chance to score well on the tests and cover all concepts.

- A **CUMULATIVE ASSESSMENT** is provided in each chapter to maintain prior learning by reviewing all skills taught up to that point.

- Use the standardized test format of the Cumulative Assessments to prepare students for high-stakes testing.

- Use the **ALTERNATE CUMULATIVE ASSESSMENT** provided in the Teacher's Edition.

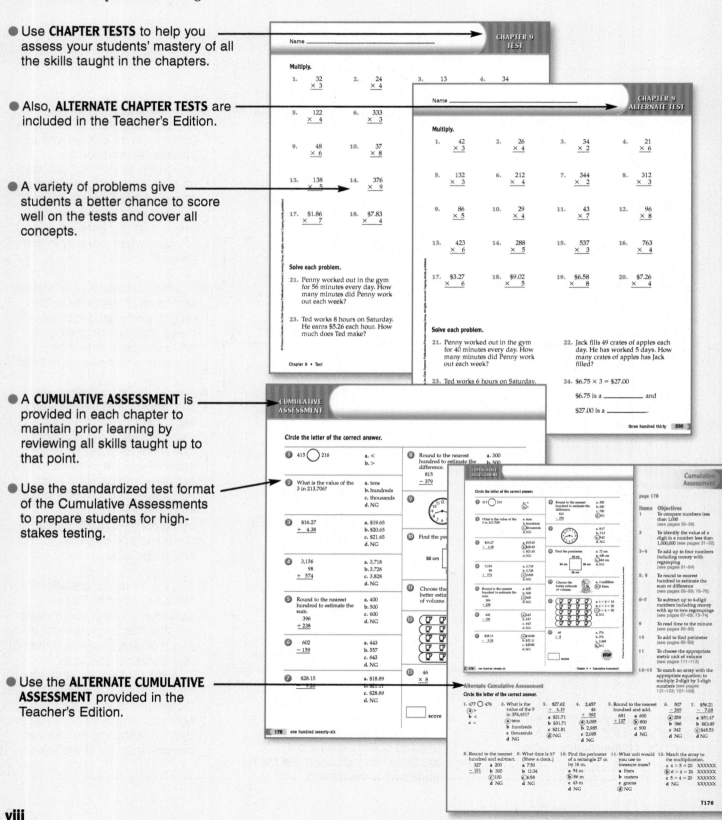

Scope and Sequence

Levels	K	A	B	C	D	E	F
Readiness							
Using attributes of size, shape, and color	●	●	●				
Sorting and classifying	●	●					
Spatial relationships	●						
Numeration							
One-to-one correspondence	●						
Understanding numbers	●	●	●	●	●	●	●
Writing numbers	●	●					
Counting objects	●	●	●				
Sequencing numbers	●	●	●	●	●		
Numbers before and after	●	●	●	●	●		
Ordering numbers	●	●	●	●	●	●	●
Comparing numbers	●	●	●	●	●	●	●
Grouping numbers	●	●	●	●	●		
Ordinal numbers	●	●	●	●			
Number words		●	●	●	●	●	●
Expanded numbers				●	●	●	●
Place value	●	●	●	●	●	●	●
Skip-counting	●	●	●	●	●		
Roman numerals				●	●		●
Rounding numbers				●	●	●	●
Squares and square roots							●
Primes and composites				●	●	●	●
Multiples			●	●	●	●	●
Least common multiples						●	●
Greatest common factors						●	●
Exponents						●	●
Addition							
Addition facts	●	●	●	●	●	●	●
Fact families		●	●	●	●	●	●
Missing addends		●	●	●	●	●	●
Adding money	●	●	●	●	●	●	●
Column addition		●	●	●	●	●	●
Two-digit addends		●	●	●	●	●	●
Multidigit addends			●	●	●	●	●
Addition with regrouping		●	●	●	●	●	●
Basic properties of addition				●	●	●	●
Estimating sums			●	●	●	●	●
Addition of fractions				●	●	●	●

Scope and Sequence

Levels	K	A	B	C	D	E	F
Addition (continued)							
Addition of mixed numbers				●	●	●	●
Addition of decimals				●	●	●	●
Addition of customary measures						●	●
Addition of integers						●	●
Subtraction							
Subtraction facts	●	●	●	●	●	●	●
Fact families		●	●	●	●	●	●
Missing subtrahends		●	●	●	●		
Subtracting money	●	●	●	●	●	●	●
Two-digit numbers		●	●	●	●	●	●
Multidigit numbers			●	●	●	●	●
Subtraction with regrouping		●	●	●	●	●	●
Zeros in the minuend			●	●	●	●	●
Basic properties of subtraction				●	●	●	●
Estimating differences			●	●	●	●	●
Subtraction of fractions				●	●	●	●
Subtraction of mixed numbers						●	●
Subtraction of decimals				●	●	●	●
Subtraction of customary measures						●	●
Subtraction of integers						●	●
Multiplication							
Multiplication facts			●	●	●	●	●
Fact families			●	●	●	●	●
Missing factors				●	●	●	●
Multiplying money				●	●	●	●
Multiplication by powers of ten				●	●	●	●
Multidigit factors				●	●	●	●
Multiplication with regrouping				●	●	●	●
Basic properties of multiplication			●	●	●	●	●
Estimating products				●	●	●	●
Multiples				●	●		
Least common multiples						●	●
Multiplication of fractions						●	●
Factorization						●	●
Multiplication of mixed numbers						●	●
Multiplication of decimals					●	●	●
Exponents						●	●
Multiplication of integers						●	●

Scope and Sequence

Levels	K	A	B	C	D	E	F
Division							
Division facts			●	●	●	●	●
Fact families			●	●	●	●	●
Divisibility rules				●	●	●	●
Two-digit quotients				●	●	●	●
Remainders				●	●	●	●
Multidigit quotients					●	●	●
Zeros in quotients					●	●	●
Division by multiples of ten					●	●	●
Two-digit divisors					●	●	●
Properties of division					●	●	●
Averages					●	●	●
Greatest common factors						●	●
Division of fractions						●	●
Division of mixed numbers						●	●
Division of decimals						●	●
Division of integers							●
Money							
Counting pennies	●	●	●	●	●		
Counting nickels	●	●	●	●	●		
Counting dimes	●	●	●	●	●		
Counting quarters		●	●	●	●		
Counting half-dollars		●	●	●	●		
Counting dollar bills		●	●	●	●		
Writing dollar and cent signs			●	●	●	●	●
Matching money with prices	●	●	●				
Determining amount of change			●	●	●	●	
Determining sufficient amount		●	●				
Determining which coins to use		●	●				
Addition	●	●	●	●	●	●	●
Subtraction	●	●	●	●	●	●	●
Multiplication				●	●	●	●
Division				●	●	●	●
Estimating amounts of money			●	●	●	●	●
Rounding amounts of money			●	●	●	●	●
Buying from a table of prices		●	●	●	●	●	●
Fractions							
Understanding equal parts	●	●	●	●			
One-half	●	●	●	●	●		

Scope and Sequence

Levels	K	A	B	C	D	E	F
Fractions (continued)							
One-fourth	●	●	●	●	●		
One-third	●	●	●	●	●		
Identifying fractional parts of figures	●	●	●	●	●	●	●
Identifying fractional parts of sets		●	●	●	●	●	●
Finding unit fractions of numbers				●	●		
Equivalent fractions				●	●	●	●
Comparing and ordering fractions				●	●	●	●
Simplifying fractions				●	●	●	●
Mixed numbers				●	●	●	●
Addition of fractions				●	●	●	●
Subtraction of fractions				●	●	●	●
Addition of mixed numbers					●	●	●
Subtraction of mixed numbers						●	●
Multiplication of fractions						●	●
Multiplication of mixed numbers						●	●
Division of fractions						●	●
Division of mixed numbers						●	●
Renaming fractions as decimals				●	●	●	●
Renaming fractions as percents						●	●
Decimals							
Place value				●	●	●	●
Reading decimals				●	●	●	●
Writing decimals				●	●	●	●
Converting fractions to decimals					●	●	●
Writing parts of sets as decimals				●	●	●	●
Comparing decimals				●	●	●	●
Ordering decimals				●	●	●	●
Addition of decimals				●	●	●	●
Subtraction of decimals				●	●	●	●
Rounding decimals					●	●	●
Multiplication of decimals					●	●	●
Division of decimals						●	●
Renaming decimals as percents						●	●
Geometry							
Polygons	●	●	●	●	●	●	●
Sides and vertices of polygons		●	●	●	●	●	●
Faces, edges, and vertices		●	●	●	●	●	●
Solid figures	●	●	●	●	●	●	●

Scope and Sequence

Levels	K	A	B	C	D	E	F
Geometry (continued)							
Symmetry	●	●	●	●	●	●	●
Lines and line segments				●	●	●	●
Rays and angles				●	●	●	●
Measuring angles						●	●
Transformations			●	●	●	●	●
Congruency				●	●	●	●
Similar figures				●	●	●	
Circles				●	●	●	
Triangles				●	●	●	●
Quadrilaterals				●	●	●	●
Measurement							
Nonstandard units of measure	●	●					
Customary units of measure		●	●	●	●	●	●
Metric units of measure		●	●	●	●	●	●
Renaming customary measures				●	●	●	●
Renaming metric measures				●	●	●	●
Selecting appropriate units		●	●	●	●		
Estimating measures		●	●	●	●		
Perimeter by counting		●	●				
Perimeter by formula				●	●		●
Area of polygons by counting		●	●	●	●		
Area of polygons by formula					●	●	●
Volume by counting				●	●	●	
Volume by formula					●	●	●
Addition of measures						●	●
Subtraction of measures						●	●
Circumference of circles							●
Area of circles							●
Surface area of space figures							●
Estimating temperatures				●	●	●	●
Reading temperature scales		●	●	●	●	●	●
Time							
Ordering events	●						
Calendars	●	●	●	●	●		
Telling time to the hour	●	●	●	●	●		
Telling time to the half-hour		●	●	●	●		
Telling time to the five minutes		●	●	●	●		
Telling time to the minute			●	●	●		

Scope and Sequence

Levels	K	A	B	C	D	E	F
Time (*continued*)							
Understanding A.M. and P.M.				●	●	●	●
Elapsed time			●	●	●	●	●
Graphing							
Tallies	●	●	●	●	●	●	
Bar graphs		●	●	●	●	●	●
Picture graphs	●	●	●	●	●		
Line graphs				●	●	●	●
Circle graphs						●	
Line plots			●			●	
Stem-and-leaf plots						●	●
Histograms							●
Ordered pairs			●	●	●	●	●
Statistics and Probability							
Understanding probability			●	●	●	●	●
Listing outcomes					●	●	●
Mean, median, and mode				●	●	●	●
Writing probabilities				●	●	●	●
Compound probability							●
Making predictions							●
Tree diagrams					●	●	●
Ratios and Percents							
Understanding ratios					●	●	●
Equal ratios						●	●
Proportions						●	●
Scale drawings						●	●
Ratios as percents						●	●
Percents as fractions						●	●
Fractions as percents						●	●
Finding the percents of numbers						●	●
Integers							
Understanding integers						●	●
Addition of integers						●	●
Subtraction of integers						●	●
Multiplication of integers							●
Division of integers							●
Graphing integers on coordinate planes							●

Scope and Sequence

Levels	K	A	B	C	D	E	F
Problem Solving							
Act it out	●	●	●	●	●	●	●
Choose a strategy	●	●	●	●	●		
Choose the correct operation	●	●	●	●	●		
Collect and use data		●	●	●	●	●	●
Determine missing or extra data		●	●	●	●		
Draw a picture or diagram	●	●	●	●	●	●	●
Identify a subgoal						●	●
Look for a pattern	●	●	●	●	●	●	●
Make a graph	●	●	●	●	●	●	●
Make a list		●	●	●	●	●	●
Make a model					●		
Make a table		●	●	●		●	●
Make a tally graph						●	
Restate the problem					●	●	●
Solve a simpler but related problem				●	●	●	●
Solve multistep problems				●	●	●	●
Try, check, and revise	●	●	●	●	●	●	●
Use a formula						●	●
Use a four-step plan				●	●	●	●
Use an exact answer or an estimate				●	●		
Use logical reasoning		●	●	●	●		●
Work backward				●	●	●	●
Write a number sentence		●	●	●	●		
Calculators							
Calculator codes				●	●	●	●
Equals key				●	●	●	●
Addition/subtraction keys				●	●	●	●
Multiplication/division keys				●	●	●	●
Clear key				●	●		
Calculators: Real-World Applications							
Averages						●	●
Formulas						●	●
Money					●	●	●
Percents						●	●
Repeating decimals						●	●
Statistics						●	●

Scope and Sequence

Levels	K	A	B	C	D	E	F
Algebra							
Patterns	●	●	●	●	●	●	●
Completing number sentences	●	●	●	●	●	●	●
Properties of numbers				●	●	●	●
Numerical expressions				●	●	●	●
Evaluating numerical expressions					●	●	●
Algebraic expressions						●	●
Evaluating algebraic expressions						●	●
Order of operations				●	●	●	●
Integers						●	●
Addition and subtraction of integers						●	●
Multiplication and division of integers							●
Ordered pairs			●	●	●	●	●
Function tables				●	●	●	●
Graphing a rule or an equation						●	●
Variables				●	●	●	●
Equations				●	●	●	●
Solve addition and subtraction equations						●	●
Solve multiplication and division equations						●	●
Model problem situations with equations						●	●
Solve inequalities							●
Formulas						●	●
Properties of equality						●	●

Contents

Chapter ⑤ Number Theory and Fractions

Chapter ⑥ Add and Subtract Fractions

Chapter ⑦ Multiply and Divide Fractions

1-1 Addition Facts and Properties

pages 1–2

1 Getting Started

Objective
• To review addition facts and properties

Vocabulary
addend, sum, Cummutative Property, Associative Property, Identity Property

Materials
addition fact cards; counters

Warm Up • Mental Math
Have students find each sum.

1. $3 + 4$ (7)
2. $2 + 7$ (9)
3. $5 + 1$ (6)
4. $8 + 2$ (10)
5. $3 + 2$ (5)
6. $6 + 2$ (8)
7. 4 and 4 (8)
8. 4 and 2 plus 1 (7)

Warm Up • Activity
Use addition fact cards to review basic addition facts through $9 + 9$. Divide the class into two teams. A team's score is the number of facts answered correctly in three minutes.

Lesson 1-1

It's Algebra!

Addition Facts and Properties

The Cubs won the Little League District Championship. How many games did they play?

	Won	Lost
Cubs	9	6
Pirates	8	7
Cards	5	9
Giants	6	8

We want to know how many games the Cubs played.

We know the Cubs won __9__ games and lost __6__ games.

To find the total games played, we add the games won and the games lost. We add __9__ and __6__.

$$\underset{\text{addends}}{\underbrace{9 \quad + \quad 6}} = \underset{\text{sum}}{15}$$

The Cubs played __15__ games in all.

Understanding the basic properties of addition makes it easier to find sums.

Commutative Property	Associative Property	Identity Property
Two numbers can be added in any order without affecting the sum. Addends can be grouped in any order without affecting the sum.	Addends can be grouped in any order without affecting the sum.	When zero is one of two addends, the sum is the other addend.

Getting Started

Find each sum.

1. $6 + 4 = $ __10__
2. $8 + 8 = $ __16__
3. $9 + 0 = $ __9__
4. $7 + 1 = $ __8__

Add. Check by adding in the reverse order.

5. $\begin{array}{r} 3 \\ + 8 \\ \hline 11 \end{array}$
6. $\begin{array}{r} 5 \\ + 4 \\ \hline 9 \end{array}$
7. $\begin{array}{r} 9 \\ + 2 \\ \hline 11 \end{array}$
8. $\begin{array}{r} 7 \\ 1 \\ + 8 \\ \hline 16 \end{array}$
9. $\begin{array}{r} 3 \\ 6 \\ + 5 \\ \hline 14 \end{array}$
10. $\begin{array}{r} 7 \\ 8 \\ + 0 \\ \hline 15 \end{array}$

Lesson 1-1 • Addition Facts and Properties

1

---

2 Teach

Introduce the Lesson Ask a student to read the problem aloud and tell what is to be solved. (how many games the Cubs played) Have students tell what they need to know to solve the problem. (number of games won and lost) Ask where this information can be found. (in the chart) Have students tell what facts are known. (The Cubs won 9 games and lost 6 games.) Have students complete the sentences as they read aloud with you to solve the problem.

Develop Skills and Concepts Recall that addition involves joining together two or more groups to find a total. Stress that the groups being joined are called *addends* and that the total is the *sum*. The groups need not be the same size. Write an addition fact in both vertical and horizontal form. Discuss each of the addition properties. Have students write examples of each property on the board. Reinforce the properties by having students use counters to demonstrate these properties:

$2 + 4 = 4 + 2$
$7 + 0 = 7$
$(3 + 1) + 5 = 3 + (1 + 5)$

It's Algebra! The concepts in this lesson prepare students for algebra.

3 Practice

Have students complete all the exercises. Before they begin, be sure students understand the directions for each section. Remind them to include the units with their answers to each word problem. Then, have students complete the page independently.

Practice

Find each sum.

1. $4 + 4 = \underline{8}$
2. $7 + 5 = \underline{12}$
3. $3 + 7 = \underline{10}$
4. $9 + 3 = \underline{12}$

5. $2 + 8 = \underline{10}$
6. $8 + 6 = \underline{14}$
7. $7 + 6 = \underline{13}$
8. $0 + 5 = \underline{5}$

9. $6 + 3 = \underline{9}$
10. $8 + 2 = \underline{10}$
11. $8 + 9 = \underline{17}$
12. $3 + 6 = \underline{9}$

13. $9 + 7 = \underline{16}$
14. $3 + 2 = \underline{5}$
15. $9 + 4 = \underline{13}$
16. $7 + 7 = \underline{14}$

17. $1 + 7 = \underline{8}$
18. $4 + 8 = \underline{12}$
19. $5 + 6 = \underline{11}$
20. $7 + 8 = \underline{15}$

Add. Check by adding in the reverse order.

21. 9 $+5$ 14	22. 6 $+6$ 12	23. 0 $+0$ 0	24. 7 $+4$ 11	25. 8 $+6$ 14	26. 5 $+2$ 7
27. 4 3 $+2$ 9	28. 1 6 $+8$ 15	29. 6 3 $+5$ 14	30. 8 1 $+3$ 12	31. 5 0 $+8$ 13	32. 5 2 $+5$ 12
33. 0 $+9$ 9	34. 5 $+1$ 6	35. 2 $+4$ 6	36. 8 $+3$ 11	37. 4 $+6$ 10	38. 8 $+8$ 16

Problem Solving

Solve each problem.

39. Chris paid $6 to see a football game. He paid $2 to park his car. How much did he pay altogether?
$8

40. Ellie scored 3 soccer goals in the first half and 2 soccer goals in the second half. How many goals did Ellie score?
5 goals

41. In a football game, Walt scored a field goal for 3 points. Hal ran for a touchdown and kicked the extra point for 7 points altogether. How many points did both boys score?
10 points

42. Mickey ran 5 kilometers on Monday, 3 kilometers on Tuesday, and 8 kilometers on Friday. How far did he run during the week?
16 kilometers

43. Annie earned $5 babysitting on Friday and $9 on Saturday. How much did she earn in all?
$14

44. Ryan's ski class met for 2 hours before lunch and 1 hour after lunch. How many hours did he ski with his class?
3 hours

Lesson 1-1 • Addition Facts and Properties

4 Assess

Use addition fact cards to assess students on the more difficult teen facts.

For Mixed Abilities

Common Errors • Intervention

Some students may not have mastered all of the addition facts. Have them work with partners to make their own fact cards for those facts that give them trouble. Have partners take turns practicing with these fact cards. Then collect the fact cards from each pair of students and use them with the whole class, suggesting that these are facts that are frequently missed.

Enrichment • Number Sense

Have students work in pairs. Provide 3 number cubes for each pair of students. One student rolls two number cubes and gives the sum. The other student rolls a number cube and adds that number to the sum given by the first student.

More to Explore • Sets

Illustrate on the board or on a transparency an example of a Venn diagram using overlapping circles. For example, label one circle M and one circle F. Tell students to use a Venn diagram to illustrate their family including siblings, grandparents, aunts, uncles, and cousins, using their first names. Tell students circle M represents the members of their mother's side of the family, and is written M = { }. Circle F represents their father's side of the family, and is written F = { }.

Ask which members belong to both Sets M and F. (the student and siblings) Point out that whenever elements belong to both sets, it is the *intersection* of the two sets. Whenever the elements of both sets are combined, it is the *union* of both sets. Have students list the names of those family members that represent the intersection of the two sets and the names of those family members that represent the union of the two sets.

pages 3–4

1 Getting Started

Objective
- To review subtraction facts and related addition facts

Vocabulary
minuend, subtrahend, difference, inverse operation, fact family

Materials
*subtraction fact cards; sets of strips with dots on them to represent each number 1–9

Warm Up • Mental Math
Have students find each difference.

1. $6 - 4$ (2)
2. $8 - 5$ (3)
3. $9 - 7$ (2)
4. $12 - 4$ (8)
5. $7 - 3$ (4)
6. $10 - 5$ (5)
7. the difference between 11 and 6 (5)
8. the difference between 13 and 7 minus 2 (4)

Warm Up • Activity
Use subtraction fact cards to review basic subtraction facts. Divide the class into two teams. A team's score is the number of facts correctly answered in three minutes.

2 Teach

Introduce the Lesson Have a student read the problem aloud and tell what is to be solved. (how much change Lynn receives) Call attention to the picture and ask what information it provides. (the cost of the calculator) Have students complete the sentences to solve the problem. Emphasize the vocabulary and recall the use of the minus sign to indicate subtraction. Ask students how they might check the answer. (add $3 to $7) Discuss the Inverse Property and explain that addition can be used to check subtraction. Have students name other members of the fact family for 9 + 8. (8 + 9; 17 − 8; 17 − 9)

Develop Skills and Concepts Write **14 − 6** both vertically and horizontally on the board. Have students identify the *minuend* and *subtrahend,* and name the *difference.*

*indicates teacher demonstration materials

T3

Subtraction Facts and Properties

Lynn is buying a new calculator. She pays for it with a ten-dollar bill. How much change will she receive?

We want to know how much change Lynn will receive.

We know she gives the clerk __$10__ and the calculator costs __$7__.

To find the difference, we subtract the cost of the calculator from the amount Lynn has. We subtract __$7__ from __$10__.

$$\underset{\text{minuend}}{\$10} - \underset{\text{subtrahend}}{\$7} = \underset{\text{difference}}{\$3}$$

Lynn receives __$3__ in change.

Understanding the relationship between addition and subtraction makes it easier to find sums and differences.

Addition and subtraction check each other. They are called **inverse operations**.

Any three numbers can be used to write four related facts called a **fact family**.

6	13	13	7
+ 7	− 7	− 6	+ 6
13	6	7	13

Using related facts helps you find missing numbers in equations.

$7 + ? = 10$ $10 - 7 = \underline{3}$

Getting Started

Find each difference.

1. $11 - 3 = \underline{8}$ 2. $16 - 8 = \underline{8}$ 3. $4 - 0 = \underline{4}$ 4. $7 - 7 = \underline{0}$

Solve. Check by using the inverse operation.   Write the missing addend.

5.   8   6.   14   7.   0
   − 5   − 6   − 0
    3   8   0

8. $6 + \underline{3} = 9$ 9. $0 + \underline{6} = 6$

Lesson 1-2 • Subtraction Facts and Properties   3

Recall that subtraction is used to determine how much is left or to compare two groups. Develop skill with the Inverse Property by having students complete the following problems:

$4 + (2) = 6$ $(6) - 2 = 4$
$(2) + 4 = 6$ $6 - (4) = 2$

It's Algebra! The concepts in this lesson prepare students for algebra.

3 Practice

Have students complete all the exercises. Before they begin, be sure students understand the directions for each section. Make certain that students can recognize the inverse of a given operation. Then, have students complete the page independently.

Practice

Find each difference.

1. $9 - 6 = \underline{3}$ 2. $14 - 9 = \underline{5}$ 3. $9 - 5 = \underline{4}$ 4. $15 - 6 = \underline{9}$

5. $7 - 2 = \underline{5}$ 6. $13 - 7 = \underline{6}$ 7. $17 - 9 = \underline{8}$ 8. $5 - 2 = \underline{3}$

9. $10 - 9 = \underline{1}$ 10. $18 - 9 = \underline{9}$ 11. $3 - 3 = \underline{0}$ 12. $10 - 4 = \underline{6}$

13. $7 - 0 = \underline{7}$ 14. $13 - 6 = \underline{7}$ 15. $15 - 7 = \underline{8}$ 16. $11 - 7 = \underline{4}$

Solve. Check by using the inverse operation.

17. $\begin{array}{r} 6 \\ -4 \\ \hline 2 \end{array}$ 18. $\begin{array}{r} 11 \\ -7 \\ \hline 4 \end{array}$ 19. $\begin{array}{r} 14 \\ -7 \\ \hline 7 \end{array}$ 20. $\begin{array}{r} 9 \\ -3 \\ \hline 6 \end{array}$ 21. $\begin{array}{r} 8 \\ -0 \\ \hline 8 \end{array}$ 22. $\begin{array}{r} 12 \\ -3 \\ \hline 9 \end{array}$

23. $\begin{array}{r} 8 \\ -1 \\ \hline 7 \end{array}$ 24. $\begin{array}{r} 14 \\ -5 \\ \hline 9 \end{array}$ 25. $\begin{array}{r} 6 \\ -0 \\ \hline 6 \end{array}$ 26. $\begin{array}{r} 12 \\ -5 \\ \hline 7 \end{array}$ 27. $\begin{array}{r} 10 \\ -8 \\ \hline 2 \end{array}$ 28. $\begin{array}{r} 5 \\ -2 \\ \hline 3 \end{array}$

29. $\begin{array}{r} 8 \\ -3 \\ \hline 5 \end{array}$ 30. $\begin{array}{r} 16 \\ -7 \\ \hline 9 \end{array}$ 31. $\begin{array}{r} 14 \\ -8 \\ \hline 6 \end{array}$ 32. $\begin{array}{r} 12 \\ -6 \\ \hline 6 \end{array}$ 33. $\begin{array}{r} 18 \\ -9 \\ \hline 9 \end{array}$ 34. $\begin{array}{r} 13 \\ -8 \\ \hline 5 \end{array}$

Write each missing addend.

35. $\underline{9} + 7 = 16$ 36. $\underline{6} + 9 = 15$ 37. $5 + \underline{7} = 12$ 38. $\underline{7} + 9 = 16$

39. $8 + \underline{9} = 17$ 40. $5 + \underline{8} = 13$ 41. $\underline{4} + 0 = 4$ 42. $7 + \underline{4} = 11$

43. $\underline{8} + 8 = 16$ 44. $1 + \underline{6} = 7$ 45. $\underline{9} + 2 = 11$ 46. $2 + \underline{8} = 10$

47. $6 + \underline{3} = 9$ 48. $\underline{7} + 8 = 15$ 49. $3 + \underline{3} = 6$ 50. $\underline{2} + 1 = 3$

Problem Solving

Solve each problem.

51. Alan bought one tape for $9. He bought a poster for $17. How much more did the poster cost?
$8

52. Paula bought a belt for $8 and a pair of socks for $3. How much more did the belt cost?
$5

53. Sandi had 16 records in her collection. She gave 7 records to her brother. How many records did Sandi have left?
9 records

54. Robert had 12 pictures to take. He took 3 pictures on Friday. How many pictures did Robert still have left to take?
9 pictures

Lesson 1-2 • Subtraction Facts and Properties

4

4 Assess

Write

$$8 + 6 =$$

on the board. Have a student complete the addition sentence and write the two related subtraction sentences.
(14; $14 - 6 = 8$; $14 - 8 = 6$)

For Mixed Abilities

Common Errors • Intervention

Have students who have not mastered the subtraction facts work with partners. Give each pair two sets of strips with dots on them to represent 1 through 9. Have students choose two strips and place them side by side. See the example below.

■ ■ ■ ■ ■   ■ ■ ■ ■

Have students write an addition sentence to describe the strips: $5 + 4 = 9$. Then, have them take away the strip showing the fewer number of dots, 4, and write the related subtraction sentence, $9 - 4 = 5$.

Enrichment • Number Sense

Have students complete the following patterns:

1. 13, 12, 11, (10), (9), (8)

2. 16, 14, 12, (10), (8), (6)

3. 11, 8, (5), (2)

4. 15, 11, (7), (3)

5. 18, 15, (12), (9), (6), (3), (0)

More to Explore • Application

Have students bring in the front page from the daily newspaper. Tell them to scan the page and write down the first 10 numbers they find that are 4 digits or larger. Have them list the numbers from least to greatest, and round each to the nearest tens, hundreds, thousands, and ten thousands. Have them exchange papers to check each other's work.

ESL/ELL STRATEGIES

Review the pronunciation and meaning of all key mathematical terms. Write them on the board capitalizing the stressed syllables. For example: MINuend, SUBtrahend, DIFFerence. Help students explain each term in their own words. For example: *The minuend is the smaller number.*

T4

pages 5–6

1 Getting Started

Objective
- To review multiplication facts and properties

Vocabulary
factor, product, Commutative Property, Associative Property, Zero Property, Identity Property

Warm Up • Mental Math
Have students find each missing number.

1. $9 + (6) = 15$
2. $17 - (8) = 9$
3. $4 + 7 = (11)$
4. $8 + (8) = 16$
5. $(15) - 8 = 7$
6. $(14) - (6) = 8$
7. $13 - (8) = 5$
8. $(6) + 3 = 9$

Warm Up • Activity
Call on individual students to complete each number sentence on the board. Have each student explain how he or she found the answer.

1. $5 + 5 + 5 + 5 = (20)$
2. $3 + 5 + 4 = (12)$
3. $3 + 3 + 3 + 3 + 3 = (15)$
4. $4 + 2 + 9 = (15)$
5. $6 + 6 = (12)$

2 Teach

Introduce the Lesson Ask what problem is to be solved. (how many students go on the field trip) Ask what facts are already known. (There are 7 cars, and 5 students go in each car.) Have students name two different ways to find the total number of students that can go on the field trip. (multiply 5 by 7; add $5 + 5 + 5 + 5 + 5 + 5 + 5$) Stress that multiplication can be used instead of addition only when the groups are of equal size. Have students complete the sentences as they read aloud to solve the problem.

Develop Skills and Concepts Emphasize the use of the words *factors* and *product*. Write a basic multiplication fact both vertically and horizontally on the board. Call on students to read each problem, identify the factors, and name the product. Discuss each of the properties

presented on page 5. Compare the Commutative and Associative properties for multiplication with those for addition. Have students give examples of the Zero and One properties.

It's Algebra! The concepts in this lesson prepare students for algebra.

3 Practice

Have students complete all the exercises. Before they begin, be sure students understand the directions for each section. Point out that some word problems require more than one operation to solve. Then, have students complete the page independently.

Multiplication Facts and Properties

The sixth-grade class is going on a field trip to the Natural History Museum. Five students can ride in each car. How many students can go on the field trip?

We need to know how many students can go on the trip.

There are __7__ cars and __5__ students can go in each car.

To find the total number of students who can go on the trip, we multiply the number of students in each car by the number of cars. We multiply __5__ times __7__.

$$\underset{\text{factors}}{\underline{5} \times \underline{7}} = \underset{\text{product}}{\underline{35}}$$

__35__ students can go on the field trip.

Understanding the basic properties of multiplication makes it easier to find products.

Commutative Property Two numbers can be multiplied in any order.	**Associative Property** Factors can be grouped in any way.
$6 \times 8 = \underline{48}$ $8 \times 6 = \underline{48}$	$(4 \times 2) \times 5 = \underline{40}$ $4 \times (2 \times 5) = \underline{40}$

Zero Property When zero is one of the factors the product is zero.	**Identity Property** When 1 is one of the two factors, the product is the other factor.
$5 \times 0 = \underline{0}$ $0 \times 5 = \underline{0}$	$7 \times 1 = \underline{7}$ $1 \times 7 = \underline{7}$

Getting Started

Find each product.

1. $7 \times 8 = \underline{56}$
2. $9 \times 0 = \underline{0}$
3. $1 \times 6 = \underline{6}$
4. $5 \times (2 \times 3) = \underline{30}$

Multiply. Check by multiplying in the reverse order.

5. $\begin{array}{r} 6 \\ \times 8 \\ \hline 48 \end{array}$
6. $\begin{array}{r} 7 \\ \times 3 \\ \hline 21 \end{array}$
7. $\begin{array}{r} 8 \\ \times 9 \\ \hline 72 \end{array}$

Practice

Find each product.

1. $7 \times 1 =$ ___7___
2. $4 \times 4 =$ ___16___
3. $2 \times 1 =$ ___2___
4. $0 \times 0 =$ ___0___

5. $2 \times 9 =$ ___18___
6. $7 \times 6 =$ ___42___
7. $3 \times 8 =$ ___24___
8. $5 \times 4 =$ ___20___

9. $4 \times 8 =$ ___32___
10. $9 \times 6 =$ ___54___
11. $7 \times 8 =$ ___56___
12. $5 \times 9 =$ ___45___

13. $(3 \times 3) \times 2 =$ ___18___
14. $3 \times (3 \times 2) =$ ___18___
15. $4 \times (1 \times 6)$ ___24___
16. $(4 \times 1) \times 6 =$ ___24___

17. $1 \times (8 \times 2) =$ ___16___
18. $(2 \times 2) \times 2 =$ ___8___
19. $2 \times (2 \times 3)$ ___12___
20. $(0 \times 9) \times 8 =$ ___0___

Multiply. Check by multiplying in the reverse order.

21. $\begin{array}{r} 2 \\ \times 1 \\ \hline 2 \end{array}$
22. $\begin{array}{r} 6 \\ \times 8 \\ \hline 48 \end{array}$
23. $\begin{array}{r} 7 \\ \times 3 \\ \hline 21 \end{array}$
24. $\begin{array}{r} 8 \\ \times 9 \\ \hline 72 \end{array}$
25. $\begin{array}{r} 9 \\ \times 4 \\ \hline 36 \end{array}$
26. $\begin{array}{r} 0 \\ \times 6 \\ \hline 0 \end{array}$

27. $\begin{array}{r} 5 \\ \times 3 \\ \hline 15 \end{array}$
28. $\begin{array}{r} 6 \\ \times 2 \\ \hline 12 \end{array}$
29. $\begin{array}{r} 5 \\ \times 5 \\ \hline 25 \end{array}$
30. $\begin{array}{r} 8 \\ \times 7 \\ \hline 56 \end{array}$
31. $\begin{array}{r} 9 \\ \times 9 \\ \hline 81 \end{array}$
32. $\begin{array}{r} 4 \\ \times 2 \\ \hline 8 \end{array}$

33. $\begin{array}{r} 9 \\ \times 6 \\ \hline 54 \end{array}$
34. $\begin{array}{r} 0 \\ \times 7 \\ \hline 0 \end{array}$
35. $\begin{array}{r} 4 \\ \times 3 \\ \hline 12 \end{array}$
36. $\begin{array}{r} 9 \\ \times 7 \\ \hline 63 \end{array}$
37. $\begin{array}{r} 6 \\ \times 1 \\ \hline 6 \end{array}$
38. $\begin{array}{r} 3 \\ \times 3 \\ \hline 9 \end{array}$

39. $\begin{array}{r} 4 \\ \times 6 \\ \hline 24 \end{array}$
40. $\begin{array}{r} 6 \\ \times 3 \\ \hline 18 \end{array}$
41. $\begin{array}{r} 4 \\ \times 7 \\ \hline 28 \end{array}$
42. $\begin{array}{r} 3 \\ \times 2 \\ \hline 6 \end{array}$
43. $\begin{array}{r} 8 \\ \times 8 \\ \hline 64 \end{array}$
44. $\begin{array}{r} 0 \\ \times 5 \\ \hline 0 \end{array}$

Problem Solving

Solve each problem.

45. One bag contains 9 apples. How many apples are there in 7 bags?
    63 apples

46. Lonnie earns $3 an hour. One week he worked 4 hours on Monday, 6 on Friday, and 8 on Saturday. How much did Lonnie earn that week?
    $54

47. Daphne made 50 sandwiches for her party. Seven guests ate 2 sandwiches each. Nine guests ate 1 sandwich each. Eight guests ate no sandwiches. How many sandwiches were left over from the party?
    27 sandwiches

48. Charley bought 7 cheeseburgers at $3 each. He gave the clerk 2 ten-dollar bills and 1 five-dollar bill. How much change did Charley get back?
    $4

6

Lesson 1-3 • Multiplication Facts and Properties

4 Assess

Have students explain how they might use the properties to find the product of $2 \times 7 \times 5$ easily. (Possible answer: Change the order of the factors. Then, group to find a product of 10. Thus, $2 \times 7 \times 5 = 2 \times 5 \times 7 = 10 \times 7 = 70$.)

For Mixed Abilities

Common Errors • Intervention

If students have trouble with basic multiplication facts, have them practice with arrays. For example, draw the array below on the board or on an overhead transparency.

```
x   x   x   x
x   x   x   x
x   x   x   x
```

Circle 3 rows of 4 X's and ask a student to write the repeated addition sentence $(4 + 4 + 4 = 12)$ and the multiplication sentence $(3 \times 4 = 12)$ for this picture.

Draw the X's again and circle 4 rows of 3. Ask for the addition sentence and the multiplication sentence for this picture.
$(3 + 3 + 3 + 3 = 12$ and $4 \times 3 = 12)$

Enrichment • Number Sense

1. Prepare multiplication puzzles for students to solve such as the following:

6	3	(18)
7	5	(35)

(42) (15)

8	9	(72)
6	4	(24)

(48) (36)

Then, have students make their own multiplication puzzles for classmates to solve.

2. Have students write a word problem that involves a two-step multiplication problem with money.

More to Explore • Application

Have students work in groups of two to research the ten states that have the largest populations. Encourage them to find the most current data available. Have them rank the states in order from largest to smallest. Then, have students write the number words for each population and read the numbers aloud. As an extension, encourage groups to rank all 50 states according to population.

1-4 Division Facts and Properties

pages 7–8

1 Getting Started

Objective
• To review division facts and related multiplication facts

Vocabulary
dividend, divisor, quotient

Materials
counters

Warm Up • Mental Math
Have students calculate each of the following:

1. $4 + 2 + 7$ (13)
2. $3 \times 9 \times 0$ (0)
3. $8 \times 1 \times 6$ (48)
4. $9 + 0 + 7$ (16)
5. $2 \times 2 \times 7$ (28)
6. $11 - 3 - 2$ (6)
7. $6 \times 9 \times 1$ (54)
8. $1 \times 2 \times 3 \times 4$ (24)

Warm Up • Activity
Have students work in small groups with a set of counters to review the division facts. Have students follow these directions:

1. Take 16 counters. Make 8 groups with the same number of counters in each. How many counters in each group? (2)
2. Take 20 counters. Make groups of 5. How many groups of 5 are there? (4)

2 Teach

Introduce the Lesson Call on a student to read the problem aloud and tell what is to be found. (how many cartons Rick needs to store 48 bottles) Have students identify the facts that are given. (There are 48 bottles, and 6 bottles fit into each carton.) Have students complete the sentences and solve the problem. Remind students they can write a basic division fact in two ways. Write $27 \div 3$ and $3)\overline{27}$ on the board. Have students read each fact, identify the *dividend* and *divisor*, and name the quotient.

Develop Skills and Concepts Explain that just as multiplication joins equal-sized groups, division separates equal-sized groups. Division answers the

---

Name _____

Division Facts and Properties

It's Algebra!

Rick is storing juice bottles in cartons. Each carton holds 6 bottles. How many cartons does Rick need to store 48 bottles?

We want to know how many cartons Rick needs.

We know there are __48__ bottles and that __6__ bottles fit into each carton.

To find the number of cartons, we divide the total number of bottles, by the number of bottles that fit into one carton. We divide __48__ by __6__.

$$\underset{\text{dividend}}{48} \div \underset{\text{divisor}}{6} = \underset{\text{quotient}}{8}$$

Rick needs __8__ cartons.

Understanding the relationship between multiplication and division makes it easier to find solutions to some equations.

Multiplication and division check each other. They are called inverse operations. We can use the inverse operation to find the missing number in a multiplication or division equation.

$6 \times ? = 18$ $18 \div 6 =$ __3__ $? \times 3 = 18$ $18 \div 3 =$ __6__

The Inverse Property explains why we cannot divide by 0.

$? \times 0 = 3$ Impossible $3 \div 0 = ?$ Cannot be solved

Getting Started

Find each quotient.

1. $4 \div 1 =$ __4__
2. $0 \div 3 =$ __0__
3. $40 \div 5 =$ __8__
4. $14 \div 7 =$ __2__

Divide. Check using the inverse operation.

5. $9)\overline{18}$ → 2
6. $3)\overline{12}$ → 4
7. $6)\overline{42}$ → 7
8. $4)\overline{12}$ → 3
9. $7)\overline{35}$ → 5
10. $4)\overline{4}$ → 1

Write each missing factor.

11. $6 \times$ __9__ $= 54$
12. __3__ $\times 7 = 21$
13. $9 \times$ __1__ $= 9$
14. __3__ $\times 8 = 24$

Lesson 1-4 • Division Facts and Properties

---

questions *How many in each group?* or *How many groups?* Refer to the introductory problem and have students distinguish which question is being answered. (how many groups) Discuss the Inverse Property. Point out that division is the inverse of multiplication. Have students name the missing factor or quotient in these problems:

1. if $24 \times 6 = 144$, then $144 \div 24 =$ (6)
2. if $192 \div 24 = 8$, then $8 \times$ (24) $= 192$

Have a student explain why zero cannot be a divisor. (Possible answer: If $8 \div 0 = ?$, then $? \times 0 = 8$. This is impossible since 0 times any number is 0.)

Remind students that zero can be a dividend; zero divided by any number is zero.

It's Algebra! The concepts in this lesson prepare students for algebra.

Practice

Find each quotient.

1. $18 \div 6 = \underline{3}$ 2. $0 \div 8 = \underline{0}$ 3. $56 \div 8 = \underline{7}$ 4. $27 \div 9 = \underline{3}$

5. $40 \div 5 = \underline{8}$ 6. $15 \div 3 = \underline{5}$ 7. $72 \div 8 = \underline{9}$ 8. $6 \div 3 = \underline{2}$

9. $28 \div 4 = \underline{7}$ 10. $36 \div 9 = \underline{4}$ 11. $2 \div 1 = \underline{2}$ 12. $18 \div 2 = \underline{9}$

13. $20 \div 4 = \underline{5}$ 14. $36 \div 6 = \underline{6}$ 15. $54 \div 9 = \underline{6}$ 16. $72 \div 9 = \underline{8}$

Divide. Check using the inverse operation.

17. $6\overline{)54}$ — 9 18. $7\overline{)49}$ — 7 19. $8\overline{)24}$ — 3 20. $4\overline{)32}$ — 8 21. $2\overline{)12}$ — 6

22. $9\overline{)81}$ — 9 23. $8\overline{)56}$ — 7 24. $5\overline{)45}$ — 9 25. $3\overline{)3}$ — 1 26. $7\overline{)21}$ — 3

Write each missing factor.

27. $3 \times \underline{9} = 27$ 28. $\underline{6} \times 5 = 30$ 29. $9 \times \underline{0} = 0$ 30. $\underline{8} \times 8 = 64$

31. $\underline{9} \times 6 = 54$ 32. $5 \times \underline{5} = 25$ 33. $7 \times \underline{4} = 28$ 34. $\underline{8} \times 6 = 48$

Problem Solving

Solve each problem.

35. Mike bought 4 bike tires for $36. How much did each tire cost?

$9

36. There are 49 days until Ann's birthday. How many weeks is that?

7 weeks

Now Try This!

Find each pattern. Write the missing number.

It's Algebra!

1.
In	Out
3	12
2	8
5	20
6	24

2.
In	Out
3	8
6	11
4	9
2	7

3.
In	Out
6	13
2	5
8	17
3	7

4.
In	Out
4	15
7	24
6	21
2	9

8

Lesson 1-4 • Division Facts and Properties

---

3 Practice

Have students complete all the exercises. Before they begin, be sure students understand the directions for each section.

Now Try This! Each set of numbers presents a different pattern: Set 1, multiply by 4; Set 2, add 5; Set 3, multiply by 2, add 1; Set 4, multiply by 3, add 3.

4 Assess

Use division fact cards to assess students on facts with 7, 8, and 9 as a divisor.

For Mixed Abilities

Common Errors • Intervention

If students have trouble mastering their division facts, have them practice by writing two multiplication sentences and two division sentences for a set of numbers such as 7, 2, 14. ($7 \times 2 = 14$, $2 \times 7 = 14$, $14 \div 7 = 2$, $14 \div 2 = 7$)

Enrichment• Number Sense

Challenge students to do the following mentally:

1. $(4 \times 6) \div 3 = (8)$

2. $(8 \times 5) \div 8 = (5)$

3. $(7 + 2) \div 3 = (3)$

4. $(16 - 8) \div 8 = (1)$

5. $(36 \div 6) \times 4 = (24)$

6. $(42 \div 7) + 6 = (12)$

7. $(81 \div 9) - 9 = (0)$

Have students write problems to illustrate each of the statements. (Possible answers are provided.)

8. A number divided by itself is 1. ($15 \div 15 = 1$)

9. A number divided by 1 is itself. ($25 \div 1 = 25$)

10. Division is not commutative. ($8 \div 2 = 4$, $2 \div 8 \neq 4$)

11. Division is not associative. $[(10 \div 2) + 3 = 8$, $10 \div (2 + 3) \neq 8]$

More to Explore • Geometry

Give each student drinking straws. Tell them to cut one straw into 3 pieces, any length, and form a triangle. Before they cut the straw, ask them to write down the probability of their successfully forming a triangle with the 3 pieces. Then have them proceed to test their prediction. Have them repeat the activity 6 times, cutting straws into different length pieces each time. Have them write a concluding report about their findings. (Students should find that the sum of the length of 2 sides of a triangle must exceed the length of the third side.)

T8

1-5 Exponents

pages 9–10

1 Getting Started

Objectives
- To write repeated multiplication in exponent form
- To write a number in exponent form as a standard number

Vocabulary
exponent, base, power

Warm Up • Mental Math
Have students complete each sequence.
1. 2, 4, 6, (8), (10), (12)
2. 5, 10, (15), (20), 25, (30)
3. 21, 18, (15), (12), 9, (6)
4. 10, (20), 30, 40, (50), (60)
5. 4, 8, (12), (16), 20, (24)
6. 81, 72, 63, (54), (45), (36)
7. 16, 24, 32, (40), (48), (56)
8. 63, 56, (49), 42, (35), (28)

Warm Up • Mental Math
Have students write each product.
1. $3 \times 3 =$ (9)
2. $2 \times 2 =$ (4)
3. $3 \times 3 \times 3 =$ (27)
4. $2 \times 2 \times 2 =$ (8)
5. $3 \times 3 \times 3 \times 3 =$ (81)
6. $2 \times 2 \times 2 \times 2 =$ (16)
7. Have students write a problem in which 4 is used as a factor two times (16); 4 is used as a factor three times (64).

2 Teach

Introduce the Lesson Have a student read the definition of *exponent* and direct their attention to the chart showing powers of 10. Talk through the factoring with 2 below the chart and have students fill in the missing numbers. Have a student read the rule about squared and cubed numbers, and fill in the missing numbers for writing 60,000 using exponents.

Develop Skills and Concepts Remind students that the exponent of any number is the *power* of that number. The exponent tells how many times the number is used as a factor. Write 5^4 on the board and have a student write the factored form and standard form.
$(5^4 = 5 \times 5 \times 5 \times 5 = 625)$

Exponents

An **exponent** tells how many times a number is used as a factor. In $10^3 = 1{,}000$, 10 is called the **base** and 3 is called the **exponent**, or **power** of ten.

Number		Factored Form		Exponent Form
10	=	10	=	10^1
100	=	10×10	=	10^2
1,000	=	$10 \times 10 \times 10$	=	10^3
10,000	=	$10 \times 10 \times 10 \times 10$	=	10^4
100,000	=	$10 \times 10 \times 10 \times 10 \times 10$	=	10^5

Complete the following:

Number	Factored Form	Exponent Form	Power of 2
2	2	2^1	first
4	2×2	2^2	second
8	$2 \times 2 \times 2$	2^3	third
16	$2 \times 2 \times 2 \times 2$	2^4	fourth
32	$2 \times 2 \times 2 \times 2 \times 2$	2^5	fifth

The second power of a number is called the number squared. The third power of a number is called the number cubed. We can write standard numbers in exponent form.

$$400 = 4 \times 100 = 4 \times 10 \times 10 = 4 \times 10^2$$

$$7{,}000 = 7 \times 1{,}000 = 7 \times 10 \times 10 \times 10 = 7 \times 10^3$$

$$60{,}000 = 6 \times \underline{10{,}000} = 6 \times \underline{10} \times \underline{10} \times \underline{10} \times \underline{10} = \underline{6 \times 10^4}$$

Getting Started

Write in exponent form.

1. $4 \times 4 = \underline{4^2}$
2. the third power of 2 $\underline{2^3}$
3. 6 squared $= \underline{6^2}$
4. $5 \times 5 \times 5 \times 5 \times 5 = \underline{5^5}$
5. 9 cubed $= \underline{9^3}$
6. the fourth power of 6 $= \underline{6^4}$
7. $3 \times 3 \times 3 \times 3 = \underline{3^4}$

Write as standard numbers.

8. $3^3 = \underline{27}$
9. $8^2 = \underline{64}$
10. $7 \times 10^4 = \underline{70{,}000}$
11. $6 \times 10 \times 10 = \underline{600}$
12. $10 \times 10 \times 10 \times 10 = \underline{10{,}000}$
13. $4 \times 4 \times 4 = \underline{64}$

Write the standard number **40,000** on the board and have a student work through the factoring to reach the exponent form.
$(40{,}000 = 4 \times 10{,}000 = 4 \times 10 \times 10 \times 10 \times 10 = 4 \times 10^4)$

It's Algebra! The concepts in this lesson prepare students for algebra.

3 Practice

Have students complete all the exercises. Be sure students write each exponent in its proper position so that 2^3 is not written as 23. Watch for students who confuse 7^2 with 7×2, or 9^3 with 9×3.

Practice

Write in exponent form.

1. $2 \times 2 \times 2 = \underline{2^3}$

2. $3 \times 3 \ \underline{3^2}$

3. $5 \times 5 \times 5 = \underline{5^3}$

4. $7 \times 7 = \underline{7^2}$

5. $8 \times 8 \times 8 \ \underline{8^3}$

6. $10 \times 10 \times 10 \ \underline{10^3}$

7. $2 \times 2 \times 2 \times 2 = \underline{2^4}$

8. 7 squared $\underline{7^2}$

9. 3 cubed $= \underline{3^3}$

10. $4 \times 4 \times 4 \times 4 = \underline{4^4}$

11. the fifth power of 10 $= \underline{10^5}$

12. the second power of 9 $= \underline{9^2}$

13. $6 \times 6 \times 6 \times 6 \times 6 = \underline{6^5}$

Write as standard numbers.

14. $2^3 = \underline{8}$

15. $4 \times 10^3 \ \underline{4{,}000}$

16. $3 \times 3 \times 3 = \underline{27}$

17. $5^2 = \underline{25}$

18. $3 \times 10 \times 10 \ \underline{300}$

19. $5 \times 10^3 \ \underline{5{,}000}$

20. $8^2 = \underline{64}$

21. $3 \times 10^1 \ \underline{30}$

22. $9^2 = \underline{81}$

23. $8 \times 10^3 = \underline{8{,}000}$

24. $5 \times 10^4 \ \underline{50{,}000}$

25. $2 \times 10^2 = \underline{200}$

[Now Try This!]

There is a pattern when you multiply or divide numbers in exponent form.

It's Algebra!

$$2^2 \times 2^3 = (2 \times 2) \times (2 \times 2 \times 2) = 2^5 = 32$$
$$2^4 \div 2^1 = (2 \times 2 \times 2 \times 2) \div (2) = 2^3 = 8$$

Use the chart and the pattern of exponents to find each answer.

Powers of 2
$2^1 = 2$
$2^2 = 4$
$2^3 = 8$
$2^4 = 16$
$2^5 = 32$
$2^6 = 64$
$2^7 = 128$
$2^8 = 256$
$2^9 = 512$

1. $2^4 \times 2^3 = \underline{2^7 = 128}$

2. $2^2 \times 2^1 = \underline{2^3 = 8}$

3. $2^8 \div 2^5 = \underline{2^3 = 8}$

4. $2^7 \div 2^4 = \underline{2^3 = 8}$

5. $2^3 \times 2^6 = \underline{2^9 = 512}$

6. $2^2 \times 2^5 \times 2^1 = \underline{2^8 = 256}$

7. $(2^6 \div 2^4) \times 2^3 = \underline{2^5 = 32}$

8. $(2^5 \times 2^4) \div 2^6 = \underline{2^3 = 8}$

9. $32 \times 16 = 2^5 \times 2^4 = \underline{2^9 = 512}$

10. $64 \times (256 \div 32) = \underline{2^6 \times (2^8 \div 2^5) = 2^9 = 512}$

10

Lesson 1-5 • Exponents

Now Try This! Ask students what the chart and examples show. (Possible answer: The chart shows the powers of 2; The examples show a shortcut for multiplying and dividing numbers written in exponent form.) Encourage students to test the shortcut by creating additional multiplication and division examples with powers of 2.

4 Assess

Have students write the standard number for each of the following:

2^3 (8) 3^2 (9) 2×10^3 (2,000)

For Mixed Abilities

Common Errors • Intervention

Some students may multiply by the exponent instead of using it to tell them how many times the base is used as a factor. Have them work with partners to complete a chart like the following:

Factored Form	Exponent Form	Number
5×5	5^2	(25)
3×3	3^2	(9)
$2 \times 2 \times 2$	2^3	(8)

Enrichment • Number Sense

1. Have students write the exponent form for each of the following numbers:

   49 (7^2) 100 (10^2)

   125 (5^3) 1,000 (10^3)

   81 (9^2) 10,000 (10^4)

   10 (10^1) 729 $(9^3$ or $3^6)$

2. Have students solve the following:

   $3^3 - 2^2 + 4^2 = $ (39)

   $9^2 - 8^2 + 7^2 - 6^2 = $ (30)

   $2^3 \times 3^2 + 2^2 = $ (76)

More to Explore • Applications

Give students the following problem: There are three links with which to make a chain:

5¢	3¢	7¢

Have students draw a chain that would cost 55¢. (Answers will vary.) Next, challenge them to construct as many different chains as possible that would cost $1.00.

ESL/ELL STRATEGIES

To help students understand the different ways we use to talk about exponential expressions, write an exponent example on the board followed by three related sentences: *$10^2 = 100$. Ten squared is 100. Ten to the power of two is 100. The second power of ten is 100.* Have students read and repeat. Then, write examples with different exponents and help them describe each one in words.

T10

1-6 Order of Operations

pages 11–12

1 Getting Started

Objective
• To use the Order of Operations to evaluate numerical expressions

Vocabulary
Order of Operations

Materials
fact cards for basic facts; counters or chips

Warm Up • Mental Math
Have students find each of the following:
1. $7 + 6$ (13)
2. 4×8 (32)
3. $63 \div 7$ (9)
4. $6 + 5$ (11)
5. $(4 + 6) + 3$ (13)
6. $(7 \times 1) \times 8$ (56)
7. $7 + (2 + 5)$ (14)
8. $4 \times (3 \times 2)$ (24)

Warm Up • Activity
Use addition, subtraction, multiplication, and division fact cards to drill students in basic facts. Keep score of the number of facts correctly answered in one minute.

Name _____

Order of Operations

Bud and Barbara have the same problem to solve, but they do not get the same answer. Who is correct?

$5 + 3 \times 4 = ?$
$8 \times 4 = 32$

$5 + 3 \times 4 = ?$
$5 + 12 = 17$

To make sure there is only one right answer to a problem, we must work operations in the same order. All mathematicians agree to use the Order of Operations.

The **Order of Operations** is a set of rules that tell you the order in which operations must be done.

• First, evaluate the powers.   There are no powers in $5 + 3 \times 4$.

• Then, do the operations inside parentheses.   There are no parentheses in $5 + 3 \times 4$.

• Then, multiply and divide in order from left to right.
$5 + \underline{3 \times 4}$
$5 + \underline{\quad 12 \quad}$

• Last, add and subtract in order from left to right.
$5 + \underline{\quad 12 \quad} = \underline{\quad 17 \quad}$

The person who found the correct answer is __Barbara__.

Getting Started

Use the Order of Operations to simplify each expression.

1. $(3 + 6) \times 4 = \underline{\ 36\ }$ 2. $9 \times 4 \div 6^2 \ \underline{\ 1\ }$ 3. $12 - 3 \times 4 = \underline{\ 0\ }$

Put the parentheses in the correct place.

4. $5 \div (4 + 1) = 1$ 5. $(3 \times 2) + 7 = 13$ 6. $(4 + 5) \times (3 - 2) = 9$

Write the correct sign.

7. $(3 \boxed{\times} 3) - 1 = 8$ 8. $9 - (12 \boxed{\div} 3) = 5$ 9. $(3 \boxed{\times} 5) - (4 \times 2) = 7$

Lesson 1-6 • Order of Operations

---

2 Teach

Introduce the Lesson Have students study the picture and discuss how Bud and Barbara arrived at their answers. (Bud added first, then multiplied; Barbara multiplied first, then added.) Have students read the four rules aloud and complete the sentences. Point out that addition, subtraction, multiplication, and division can involve only two numbers at a time.

Develop Skills and Concepts Emphasize the importance of the Order of Operations. At whatever level the student approaches the problem, they must first evaluate the powers, do all operations in parentheses, multiply or divide from left to right, and then, add or subtract from left to right. Have students apply these rules as they work these problems together:

$(6 + 2) \times 3 = (24)$ $2 \times 3 \times 6 = (36)$
$15 - 8 - 4 = (3)$ $7 + 2 \times 2 = (11)$
$8 \times 2 \div 4 = (4)$ $17 - (2 \times 4) = (9)$

It's Algebra! The concepts in this lesson prepare students for algebra.

3 Practice

Have students complete all the exercises. Have students write the rules for the Order of Operations on the exercise page to use as a reference.

T11

Practice

Use the Order of Operations to simplify each expression.

1. $(4 + 3) \times 6 = \underline{42}$ 　　2. $9 \div 3 \times 4 = \underline{12}$ 　　3. $15 \div (3 + 2) = \underline{3}$

4. $(2^2 + 5) \times 7 = \underline{63}$ 　　5. $15 - 8 + 1 = \underline{7}$ 　　6. $6 + 4^3 \div 8 = \underline{14}$

7. $6 + 9 \div 3 = \underline{9}$ 　　8. $(16 - 3^2) \times 5 = \underline{35}$ 　　9. $(5 + 3) \times 4 = \underline{32}$

10. $6 \times 4 \div 3 = \underline{8}$ 　　11. $(7 + 2) \times 8 = \underline{72}$ 　　12. $48 \div (9 - 3) = \underline{8}$

13. $5 \times (4 - 3) \times 2^2 = \underline{20}$ 　　14. $(9 - 1) \div 2 = \underline{4}$ 　　15. $(15 - 2^3) \times 6 = \underline{42}$

16. $17 - 3^3 \div 9 = \underline{14}$ 　　17. $7 \times (6 \div 6) = \underline{7}$ 　　18. $16 \div 2^2 \div 4 = \underline{1}$

Put the parentheses in the correct place.

19. $6 \div (5 + 1) = 1$ 　　20. $(4 \times 2) + 9 = 17$ 　　21. $(24 \div 6) + 5 = 9$

22. $7 + (3 \times 2) = 13$ 　　23. $54 \div (6 + 3) = 6$ 　　24. $(7 + 2) \times (5 - 3) = 18$

25. $9 + (2 \times 4) = 17$ 　　26. $36 \div (4 + 2) = 6$ 　　27. $(8 \times 3) \div 6 = 4$

Write the correct sign.

28. $(11 \boxed{-} 3) \times 2 = 16$ 　　29. $6^2 \div (2 \boxed{+} 4) = 6$ 　　30. $(3 \boxed{\times} 2) + 4 = 10$

31. $4 \times (4 \boxed{+} 4) = 32$ 　　32. $(4 + 9) - (8 \boxed{-} 2) = 7$ 　　33. $(3 \times 4) \boxed{-} (2 \times 4) = 4$

34. $(12 - 8) \boxed{-} (6 \div 2) = 1$ 　　35. $(36 \div 4) - (18 \boxed{\div} 9) = 7$ 　　36. $3 \boxed{\times} (3 \div 3) = 3$

Now Try This!

Play this game with a partner. Place a row of chips or counters on a desk. Use as many chips as you wish in each row. Take turns picking up one, two, or three chips at a time. The object is to leave one chip for your partner to pick up. There is a way to win every time.　**See Solution Notes.**

Now Try This! Distribute chips or counters randomly so that pairs of students may receive an odd or even number. After students have had a chance to experiment, discuss strategies for winning. (always leave an odd number of chips for your opponent)

4 Assess

Write **$36 \div 9 - 3 = 6$** on the board. Have students place parentheses to make this a true statement.
[$36 \div (9 - 3) = 6$]

For Mixed Abilities

Common Errors • Intervention

Some students may do the operations in order from left to right instead of using the Order of Operations. Have students work in pairs to solve sets of number sentences such as the following.

$5 + (7 \times 6) = \square$ (47)
$(5 + 7) \times 6 = \square$ (72)
$5 + 7 \times 6 = \square$ (47)

After they find the answers, have them discuss how this activity illustrates that a rule for an order of operations is necessary so that each sentence has a unique answer.

Enrichment • Number Sense

Have students find n in each number sentence.

1. $15 - (5 \times n) = 0$ (3)
2. $6 + n \div 2 = 8$ (4)
3. $3 \times (8 \div n) = 12$ (2)
4. $63 \div 9 + n = 14$ (7)
5. Challenge students to write five problems with an answer of 20. They are to use two operations and at least 3 numbers in each problem. (Check students' work.)

More to Explore • Sets

Duplicate the following for students:

On the water planet Aqua, life requires careful planning. Since all of the buildings float, care must always be taken to keep an equal number of beings in all sectors of the building. The sole exception to this precaution is at the center of the structure.

Below is a diagram of a typical Aquan building in which 33 beings work. Place a number of workers in each section, keeping the building balanced and afloat. (Answers may vary.)

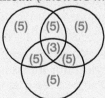

T12

pages 13–14

1 Getting Started

Objective
- To solve problems by using a four-step plan

Warm Up • Mental Math
Ask students to name each expression in exponential form and then name the product.

1. $3 \times 3 \times 3 \times 3 = (3^4, 81)$
2. $4 = (4^1, 4)$
3. $6 \times 6 \times 6 \times 6 \times 6 = (6^5, 7,776)$
4. $8 \times 8 = (8^2, 64)$
5. $10 \times 10 \times 10 \times 10 = (10^4, 10,000)$

Name _____

Problem Solving:
Use a Four-Step Plan

Marcia entered the Fly-a-Kite contest. She came in first by keeping her kite flying for 305 minutes. How much money did she win?

⭐ **SEE**

We want to know how much money Marcia won in the contest.

Her kite flew for _305_ minutes.

The prize was _$3_ for each minute in the air over _5_ hours.

⭐ **PLAN**

To find out how many minutes there are in five hours, we multiply _60_ by 5. To find out how much longer than five hours Marcia's kite was in the air, we subtract the number of minutes in 5 hours from _305_.

To figure her prize money, we multiply the prize money per minute by the number of minutes over 5 hours.

We multiply _60_ by _5_.

⭐ **DO**

60	305	$3
× 5	−300	× 5
300	5	$15

Marcia won _$15_ in the kite-flying contest.

⭐ **CHECK**

We can check by reviewing the plan, reworking the steps of the solution, and considering if the answer makes sense.

---

2 Teach

Introduce the Lesson Explain that the See, Plan, Do, Check outline gives us a set of guidelines for solving any kind of problem.

- In SEE, we state the problem again in our own way and restate the important facts that we know.

- In PLAN, we devise a strategy for using the known facts to arrive at the solution. Sometimes the plan will have more than one step.

- In DO, we execute the plan as we have outlined it in the previous step. We finish this step by restating the problem along with our solution.

- In CHECK, we re-examine the entire process to see how logical our thinking was. We check each stage in the DO step to see that we made no mechanical errors. Finally, we see if our solution makes sense based on what we know.

Develop Skills and Concepts Walk students through each of the steps, having them complete the information sentences. Encourage them to use this plan in every problem they have to solve.

3 Apply

Solution Notes

1. Alex $2.44
   Lois   + 1.85
           ─────
            $4.29

2. There are three possible solutions.

Apples	Bananas
18	1
12	8
6	15

Apply

Use the four-step plan to help solve each problem.

1. After the football game, Alex and Lois stopped for lunch. Alex ordered a fish sandwich for $3, cole slaw for $2, a piece of pie for $2, and a glass of milk for $1. Lois ordered a fish sandwich, cole slaw, and a glass of milk. What was the total cost of their lunch?

   $14

2. Larry and Gary went to the store to buy apples and bananas for their lunch the following week. The 3 apples cost $1 and 2 bananas cost $1. They spent a total of $5. How many of each did they buy? (Hint: There is more than one correct answer.)

   12 apples and 2 bananas or
   9 apples and 4 bananas or
   6 apples and 6 bananas or
   3 apples and 8 bananas

3. A box of cereal contains 19 ounces of flakes. The four-member Johnson family has planned they will each have a 2-ounce serving for breakfast, when they are camping. If they open a box of cereal on Monday, on what day will they need to open a new box?

   Wednesday

4. Mr. Moser visited his mother on Mother's Day. He drove 80 miles to his sister's house and then drove 30 miles farther to his mother's house. After his visit, Mr. Moser drove home by the most direct route that was 100 miles long. If Mr. Moser's car gets 30 miles per gallon of gasoline, how many gallons did he use on the trip?

   7 gallons

> **Problem-Solving Strategy:**
> **Using the Four-Step Plan**
> ★ **SEE** What do you need to find?
> ★ **PLAN** What do you need to do?
> ★ **DO** Follow the plan.
> ★ **CHECK** Does your answer make sense?

5. Read Problem 4 again. What if Mr. Moser gets only 15 miles per gallon of gasoline? Then how many gallons of gasoline did he use on the trip?

   14 gallons

6. Estes had to estimate the cost of 9 items costing 49¢ each. Which way would get him closer to the actual cost, 9×50 or 10×49? Explain.

   9×50

7. Rewrite Problem 1 so that the total amount of money that Alex spent for his lunch is the same but each item that he bought cost an equal amount. Then how much did Lois spend for her lunch?

   $6

8. Some friends went to the carnival. Tickets for rides were $1.25 each, 3 for $3.50, or 10 for $10.00. Which is the best buy? Describe a situation where buying this many tickets might not be the best thing to do.

   Best buy is 10 for $10.00;
   Buying this many tickets would not be the best buy if the friends go on 8 or fewer rides.

Lesson 1-7 • Problem Solving: Use a Four-Step Plan

For Mixed Abilities

More to Explore •
Number Sense

Have each student prepare equivalent decimal equation cards to use in the game Concentration. Each student must write 10 matching pairs of equations, using any operation. (for example: 4×17.5 (70) and $42.7 + 27.3$ (70)) Students work with a partner, shuffle their cards, and place them facedown. Each partner takes a turn drawing 2 cards, students keep any pair of cards with matching answers. The opponent must be responsible for checking the partner's mathematics. The player returns, facedown, any incorrect matches. Play ends when all cards have been matched. The winner has the most cards.

---

3. 2 days + 1 serving + 1 ounce
    $8)\overline{19}$
    The family uses four servings per day or 8 ounces. On Wednesday they will have only one complete serving plus 1 ounce left over.

4. $80 + 30 + 100 = 210$ miles
    $\frac{210}{30} = 7$ gallons

Higher-Order Thinking Skills

5. **Analysis:** Students should recognize that if Mr. Moser gets half the mileage, then he will need twice as much gasoline. Since the answer to the original problem is 7 gallons, the answer to the changed problem is 14 gallons.

6. **Synthesis:** 9×50 adds only one 9 to the actual product whereas 10×49 adds one 49 to the product.

7. **Synthesis:** The problems may vary but the price of each of the 4 items is the average cost of the 4, or $0.61. This means that Lois spent $3 \times \$0.61$.

8. **Evaluation:** Answers to the second part may vary; sample answers might be where the friends altogether do not have $10 or where they do not altogether want to go on 10 rides.

4 Assess

Ask students to name the four steps they learned for solving problems. Have them explain each step. (See, Plan, Do, Check; Possible answer: Read the problem; Decide on how you should solve the problem; Do the calculations outlined in the plan; Check your work.)

pages 15–16

1 Getting Started

Objective
• To review the use of calculator keys and operations

Materials
calculators

Warm Up • Mental Math
Have students evaluate each expression.

1. 6^2 (36)
2. 10^1 (10)
3. 5^2 (25)
4. $3^2 + 2^2$ (13)
5. $2^3 + 3^2$ (17)
6. $10^2 + 10^2$ (200)
7. $8^2 - 2^2$ (60)
8. $9^2 - 8^2$ (17)

Warm Up • Calculator
Have students evaluate each expression.

1. 5 dozen ÷ 12 (5)
2. (55 + 72 + 89) ÷ 27 (8)
3. $0.45 + $1.52 + $0.49 + $0.95 ($3.41)
4. 35 × 42 × 56 (82,320)
5. ($5.50 × 5) + ($6.75 × 6)($68.00)
6. 1,512 ÷ 27 (56)
7. $6.50 ÷ $0.25 (26)

Name _____

Calculator: The Basic Keys

The calculator can help make work with numbers easier. To use the calculator, we enter numbers and operations by pressing the **keys** on the **keyboard**. The answer shows up on the **screen**. We use a **calculator code** to give the order in which to press the keys. Turn your calculator on, follow the codes, and write the results in the empty screen.

9 + 3 = ⬚ 12

8 − 5 = ⬚ 3

6 − 4 + 11 = ⬚ 13

Sometimes a mistake is made, and an entry must be changed. The C key clears the calculator. The CE key cancels the most recent entry, but remembers the earlier entries.

Complete these codes.

1. 7 × 3 C = ⬚ 0
2. 15 − 8 CE 3 = ⬚ 12
3. 3 × 3 × 3 + 3 CE = ⬚ 27
4. 64 ÷ 2 ÷ 2 ÷ 2 C = ⬚ 0
5. 2 + 3 + 5 CE + 4 = ⬚ 9
6. 2 + 3 + 4 + 5 C = ⬚ 0

The = key is also called a **constant key**. It remembers the last operation and number entered. Each time the = key is pressed, the calculator continues to follow the last instructions.

Try these codes. Answers on some calculators differ.

7. 4 + 4 = = = = ⬚ 20
8. 48 − 6 = = = = ⬚ 30
9. 3 × 3 = = = ⬚ 81
10. 64 ÷ 2 = = = = ⬚ 4
11. 90 ÷ 3 = = ⬚ 10
12. 6 × 5 = = = = = ⬚ 3,750

15

---

2 Teach

Introduce the Lesson Discuss how the calculator is used in everyday life. Draw a large calculator on the board or the overhead, and display a real calculator. Emphasize that a calculator is useful for working with large numbers, for checking accuracy, and for doing long computations rapidly.

• Have a student read the introductory paragraph aloud and complete the codes.

• Point out that the chance for error occurs with human handling. The operator has to give the calculator correct information. If a mistake occurs and an entry must be changed, the operator can use the Clear or the Clear Entry key.

• Have a student read the second paragraph aloud and complete the codes.

• Show students how the = key can be used as a constant key.

• Have a student read the third paragraph aloud and complete the codes.

Develop Skills and Concepts Have students turn their calculators on and off. Have them identify the special keys: C and CE, the numbers keys, the screen and the operation keys. Tell students that if their calculators do not have CE keys, they must start each computation from the beginning if they make a mistake. If calculators do not have a constant key, students must punch in the last instruction. Have students work the following codes:

2 + 2 = = = (8)
54 − 6 = = = (36)
4 × 4 = = = (256)
36 ÷ 3 = = (4)

Practice

Complete the codes. Write the results on the screen.

1. 6 + 9 = | 15 |
2. 16 − 7 = | 9 |
3. 8 × 7 = | 56 |
4. 54 ÷ 9 = | 6 |
5. 9 + 9 + 9 C = | 0 |
6. 7 + 7 − 7 CE = | 14 |
7. 2 × 5 ÷ 10 = | 1 |
8. 36 ÷ 6 × 9 ÷ 6 = | 9 |
9. 64 ÷ 8 − 3 + 5 CE = | 5 |
10. 5 − 3 + 2 CE + 8 = | 10 |
11. 6 × 9 ÷ 3 − 9 = | 9 |
12. 63 ÷ 7 + 3 ÷ 6 C = | 0 |
13. 2 × 3 = = ÷ 3 = | 6 |
14. 6 × 4 ÷ 2 = = | 6 |
15. 81 ÷ 9 + 3 = = ÷ 5 = | 3 |
16. 49 ÷ 7 = + 6 × 7 = | 91 |

Apply

Use your calculator to solve each problem.

17. Tomas bought 6 records for $54. His brother broke one, but Tomas replaced it for the same price. How much did he spend on records altogether?

$63

18. Alicia jogged 8 miles in one week. Pat jogged 6 miles and Gwen jogged 9 miles the same week. How many more miles did Pat and Gwen together jog than Alicia?

7 miles

19. JoAnn read 3 books in 4 days. If she keeps reading at the same rate, how many books will JoAnn read in 24 days?

18 books

20. Suppose someone promises to give you 1¢ on Monday and double the amount she gives you each day for one week. How much will you have altogether on Sunday? Complete the table to find out. Remember, double means multiply by 2.

21. A touchdown is worth 6 points and a field goal is worth 3 points. Alan scored 48 points. If Alan kicked 2 field goals, how many touchdowns did he score?

7 touchdowns

	Amount Given	Total Amount
Monday	1¢	1¢
Tuesday	2¢	3¢
Wednesday	4¢	7¢
Thursday	8¢	15¢
Friday	16¢	31¢
Saturday	32¢	63¢
Sunday	64¢	$1.27

16

Lesson 1-8 • Calculator: The Basic Keys

---

3 Practice

Have students do all the problems on the page. Remind them to check each number as they enter it.

4 Assess

Ask students to explain the difference between the C key and the CE key. (Possible answer: The C key clears the calculator. The CE key cancels the most recent entry.)

For Mixed Abilities

Common Errors • Intervention

Some students may enter numbers and operations on their calculator incorrectly. Have them work with partners to write the codes for the following set of problems and then use the calculator to solve them.

1. 4 + 6 = (10)
2. 13 − 5 = (8)
3. 7 × 8 = (56)
4. 72 ÷ 9 = (8)

Enrichment • Number Sense

Direct students to create a set of number sequences. Have students exchange their sequences with a partner. Then, use their calculator to find the next five numbers in each sequence.

More to Explore • Application

Have students comparison shop for groceries. Have students bring in newspaper grocery ads. Give them a list of items to be purchased. Make sure there are some items that are common to different store ads.

- Have the students find the prices of each item and compare the total price from each store.

- Ask students to list the best buys from each store and compile the lowest possible total price.

- You may extend the activity by asking how much change they would receive from a specified amount or by introducing discount coupons for certain items.

Chapter 1
Test

page 17

Items	Objectives
1–20	To review basic addition and subtraction facts (see pages 1–4)
21–40	To review basic multiplication and division facts (see pages 5–8)
41–44	To write exponential expressions as standard numbers (see pages 9–10)
45–50	To use Order of Operations to simplify expressions (see pages 11–12)

Alternate Chapter Test

You may wish to use the Alternate Chapter Test on page 370 of this book for further review and assessment.

T17

Add or subtract.

1. $5 + 4 = \underline{9}$ 2. $9 + 6 = \underline{15}$ 3. $7 + 8 = \underline{15}$ 4. $6 + 7 = \underline{13}$

5. $16 - 7 = \underline{9}$ 6. $13 - 8 = \underline{5}$ 7. $15 - 7 = \underline{8}$ 8. $12 - 5 = \underline{7}$

9. $\begin{array}{r} 4 \\ +8 \\ \hline 12 \end{array}$ 10. $\begin{array}{r} 7 \\ +9 \\ \hline 16 \end{array}$ 11. $\begin{array}{r} 8 \\ +8 \\ \hline 16 \end{array}$ 12. $\begin{array}{r} 3 \\ +7 \\ \hline 10 \end{array}$ 13. $\begin{array}{r} 5 \\ +8 \\ \hline 13 \end{array}$ 14. $\begin{array}{r} 9 \\ +2 \\ \hline 11 \end{array}$

15. $\begin{array}{r} 11 \\ -7 \\ \hline 4 \end{array}$ 16. $\begin{array}{r} 17 \\ -9 \\ \hline 8 \end{array}$ 17. $\begin{array}{r} 15 \\ -8 \\ \hline 7 \end{array}$ 18. $\begin{array}{r} 14 \\ -6 \\ \hline 8 \end{array}$ 19. $\begin{array}{r} 16 \\ -9 \\ \hline 7 \end{array}$ 20. $\begin{array}{r} 14 \\ -8 \\ \hline 6 \end{array}$

Multiply or divide.

21. $3 \times 7 = \underline{21}$ 22. $8 \times 5 = \underline{40}$ 23. $9 \times 1 = \underline{9}$ 24. $6 \times 8 = \underline{48}$

25. $16 \div 4 = \underline{4}$ 26. $21 \div 3 = \underline{7}$ 27. $18 \div 2 = \underline{9}$ 28. $42 \div 6 = \underline{7}$

29. $\begin{array}{r} 9 \\ \times 5 \\ \hline 45 \end{array}$ 30. $\begin{array}{r} 6 \\ \times 7 \\ \hline 42 \end{array}$ 31. $\begin{array}{r} 7 \\ \times 5 \\ \hline 35 \end{array}$ 32. $\begin{array}{r} 6 \\ \times 9 \\ \hline 54 \end{array}$ 33. $\begin{array}{r} 7 \\ \times 8 \\ \hline 56 \end{array}$ 34. $\begin{array}{r} 9 \\ \times 0 \\ \hline 0 \end{array}$

35. $6\overline{)54}$ → 9 36. $3\overline{)24}$ → 8 37. $9\overline{)36}$ → 4 38. $5\overline{)40}$ → 8 39. $6\overline{)24}$ → 4 40. $4\overline{)28}$ → 7

Write as standard numbers.

41. $6^2 = \underline{36}$ 42. $5 \times 10^3 = \underline{5,000}$ 43. $2^4 = \underline{16}$ 44. $3 \times 10^2 = \underline{300}$

Use the Order of Operations to simplify each expression.

45. $3 + 9 \times 1 = \underline{12}$ 46. $18 \div 3 + 6 = \underline{12}$

47. $(3 + 5) \times 3 = \underline{24}$ 48. $(27 \div 3^2) + 6 = \underline{9}$

49. $6 + 14 \div 2 - 5 = \underline{8}$ 50. $2^4 - (5 + 3) \times 2 = \underline{0}$

Circle the letter of the correct answer.

1. 5 + 4
- a. 7
- b. 8
- **c. 9**
- d. NG

2. 9 + 7
- **a. 16**
- b. 17
- c. 18
- d. NG

3. 2 3 +4
- a. 5
- b. 7
- **c. 9**
- d. NG

4. 8 + 6 = n, n = ?
- a. 13
- **b. 14**
- c. 15
- d. NG

5. 14 − 7
- **a. 7**
- b. 8
- c. 9
- d. NG

6. 16 − 9
- a. 6
- **b. 7**
- c. 8
- d. NG

7. 3 + n = 8, n = ?
- a. 11
- b. 6
- **c. 5**
- d. NG

8. 9 × 4
- a. 32
- **b. 36**
- c. 40
- d. NG

9. 9 × 6
- a. 45
- **b. 54**
- c. 56
- d. NG

10. n × 3 = 6, n = ?
- a. 0
- **b. 2**
- c. 18
- d. NG

11. 28 ÷ 7
- a. 3
- b. 5
- c. 7
- **d. NG**

12. 8)0
- **a. 0**
- b. 8
- c. impossible
- d. NG

13. n × 6 = 42, n = ?
- a. 5
- b. 6
- **c. 7**
- d. NG

score ☐   STOP

18   Chapter 1 • Cumulative Assessment

page 18

Items Objectives
1–4 To review basic addition facts (see pages 1–2)
5–7 To review basic subtraction facts (see pages 3–4)
8–9 To review basic multiplication facts (see pages 5–6)
10–13 To review basic division facts (see pages 7–8)

NG stands for "Not Given"

Alternate Cumulative Assessment

Circle the letter of the correct answer.

1. 3 + 7
- a 9
- **b 10**
- c 11
- d NG

2. 6 + 8
- **a 14**
- b 15
- c 16
- d NG

3. 4 5 +7
- a 15
- **b 16**
- c 17
- d NG

4. 9 + 6 = n, n = ?
- a 14
- **b 15**
- c 16
- d NG

5. 16 − 8
- a 6
- b 7
- **c 8**
- d NG

6. 13 − 4
- a 7
- b 8
- **c 9**
- d NG

7. 4 + n = 9, n = ?
- a 13
- b 6
- **c 5**
- d NG

8. 7 × 9
- a 56
- **b 63**
- c 70
- d NG

9. 8 × 5
- a 35
- **b 40**
- c 45
- d NG

10. n × 2 = 8, n = ?
- a 2
- **b 4**
- c 16
- d NG

11. 6)36
- a 5
- b 7
- c 9
- **d NG**

12. 0)7
- a 0
- b 7
- **c impossible**
- d NG

13. n × 5 = 25, n = ?
- **a 5**
- b 6
- c 7
- d NG

T18

2-1 Place Value to Hundred Thousands

pages 19–20

1 Getting Started

Objectives
• To read and write numbers through 999,999
• To identify place value

Warm Up • Mental Math
Have students simplify each expression.
1. $5^2 + 5^2$ (50)
2. $6^2 - 2^2$ (32)
3. $9^2 - 1^2$ (80)
4. $2^2 \times 2^2$ (16)
5. $3^3 - 3^2$ (18)
6. $4^2 - 2^2$ (12)
7. $3^2 \times 2^2$ (36)
8. $8^2 + 1^3$ (65)

Warm Up • Pencil and Paper
Ask students to write the value of the underlined digit in each number.
1. 15$\underline{2}$ (2)
2. 3$\underline{2}$7 (20)
3. $\underline{2}$16 (200)
4. 9$\underline{2}$,423 (2,000)
5. $\underline{5}$6 (50)
6. $\underline{2}$9,453 (20,000)
7. 14$\underline{7}$,651 (7,000)

Name _____

Add and Subtract Whole Numbers

Lesson 2-1

Place Value to Hundred Thousands

A wide-body jet can weigh as much as 510,000 pounds at take-off. This kind of jet cruises at 553 miles per hour, at an altitude of 35,000 feet, for greatest fuel efficiency. Write its minimum and maximum range in expanded form.

We want to understand the meaning of the numbers that describe the range of the jet.

We know the minimum range is <u>6,154</u> miles.

The maximum range is <u>7,013</u> miles.

WIDE BODY JET
CAPACITY: 330
RANGE: 6,154
to 7,013 miles

thousands			ones		
hundred thousands	ten thousands	thousands	hundreds	tens	ones
		↑	↑	↑	↑

Standard Form {
| | | 6, | 1 | 5 | 4 |
| | | 7, | 0 | 1 | 3 |

Expanded Form
6,000 + 100 + 50 + 4
7,000 + 10 + 3

Getting Started

Write the place value of each green digit.

1. 8,246   2. 159   3. 16,240   4. 127,396
   tens   hundreds   thousands   hundred
                                                          thousands

Write in standard form.

5. 500 + 30 + 7 <u>537</u>   6. 4,000 + 30 + 6 <u>4,036</u>

7. one hundred twenty-seven thousand, two hundred nine <u>127,209</u>

Write in expanded form.   Write in words.

8. 130,058 <u>100,000 + 30,000 + 50 + 8</u>   9. 300,009 <u>three hundred thousand, nine</u>

---

2 Teach

Introduce the Lesson Have a student read the first paragraph and the numbers for the minimum and maximum range of the jet. Call attention to the place-value chart and ask students to identify the two periods shown. (ones and thousands) Point out that each period has 3 places—ones, tens, hundreds. Have students complete the model, writing the number in expanded form.

Develop Skills and Concepts Stress that the value of a digit is based on its *face* value, 1, 2, 3 . . . 9; and its *place* value, or position within a number. Emphasize that each place value in a whole number is 10 times greater than the place value to its right. Note that commas are used to separate periods when numbers are written in standard form. Review the use of zero as a placeholder. Dictate these numbers and have students write them on the board: 5,012; 3,501; 7,002.

3 Practice

Have students complete all the exercises. Students who have difficulty translating words from numbers may need to read these exercises aloud.

4 Assess

Have students read the numbers 508,639 and 580,639. (five hundred eight thousand, six hundred thirty-nine; five hundred eighty thousand, six hundred thirty-nine) Then, have them identify the place value of the digit 8 in each number. (thousands; ten thousands)

Practice

Write the place value of each green digit.

1. 6,715

   _____tens_____

2. 23,610

   _____ten thousands_____

3. 212

   _____ones_____

4. 346,158

   _____hundred thousands_____

5. 116,250

   _____thousands_____

6. 308

   _____hundreds_____

7. 5,750

   _____thousands_____

8. 14,729

   _____tens_____

Write in standard form.

9. $5,000 + 20 + 6 =$ _5,026_

10. $4,000 + 800 + 20 + 3 =$ _4,823_

11. $10,000 + 3,000 + 200 + 70 + 4 =$ _13,274_

12. $700,000 + 2,000 + 600 + 6 =$ _702,606_

13. five thousand, six hundred thirty-five _5,635_

14. twenty-seven thousand, two hundred nine _27,209_

15. four hundred eighty-seven thousand, four hundred fourteen _487,414_

Write in expanded form.

16. $2,743 =$ _2,000 + 700 + 40 + 3_

17. $6,428 =$ _6,000 + 400 + 20 + 8_

18. $9,005 =$ _9,000 + 5_

19. $14,276 =$ _10,000 + 4,000 + 200 + 70 + 6_

20. $127,382 =$ _100,000 + 20,000 + 7,000 + 300 + 80 + 2_

Write in words.

21. 318 _three hundred eighteen_

22. 6,247 _six thousand, two hundred forty-seven_

23. 126,205 _one hundred twenty-six thousand, two hundred five_

24. 14,389 _fourteen thousand, three hundred eighty-nine_

20

Lesson 2-1 • Place Value to Hundred Thousands

5 Mixed Review

1. _____ $+ 9 = 17$ (8)
2. 5^3 (125)
3. $(2 + 3) \times 5$ (25)
4. 9×8 (72)
5. $72 \div 8 - 6$ (3)
6. 6^4 (1,296)
7. $12 - 4 \times 3$ (0)
8. $63 \div 9$ (7)
9. $7 + 6$ (13)
10. $28 \div$ _____ $= 7$ (4)

For Mixed Abilities

Common Errors • Intervention

Some students may name the place value of a particular digit incorrectly. Have them work with partners to write the number in expanded form and then name the place value of each digit.

Enrichment • Number Sense

Have students write the next number in each group.

1. 3,675; 3,775; 3,875 (3,975)
2. 45,801; 45,811; 45,821 (45,831)
3. 7,958; 8,958; 9,958 (10,958)
4. 72,342; 74,342; 76,342 (78,342)
5. 64,008; 74,008; 84,008 (94,008)
6. 501,997; 501,998; 501,999 (502,000)

More to Explore • Statistics

Ask students to take a population survey. Have them count and record the number of students in their sixth-grade class. Then, have them estimate what percent of the total sixth-grade population is boys and what percentage is girls, and record their estimates. Then, have them find the total population of the school and determine what percent of the school population is actually represented by each of these groups. Have them compare this data to their estimates.

ESL/ELL STRATEGIES

When introducing the addition and subtraction material, use problems on the board to point out the following elements and review the meaning of each term: *align* (put in a row), *column* (a vertical row), *addends* (numbers to be added together), and *place value* (tens, hundreds, etc.).

T20

2-2 Comparing and Ordering Numbers

pages 21–22

1 Getting Started

Objective
• To compare and order numbers

Warm Up • Mental Math
Have students find each of the following:

1. 4 + 8 (12)
2. 17 − 9 (8)
3. 7 × 5 (35)
4. 6 + 5 (11)
5. 32 ÷ 8 (4)
6. 15 − 6 (9)
7. 9 + 3 (12)
8. 6 + 4 (10)
9. 8 − 3 (5)
10. 6 × 7 (42)
11. 56 ÷ 8 (7)
12. 5 × 6 (30)

Warm Up • Pencil and Paper
Have students write the value of the underlined digit in each number.

1. 7<u>4</u>3,980 (40,000)
2. 59,6<u>6</u>4 (60)
3. <u>9</u>56,381 (900,000)
4. 137,02<u>5</u> (5)
5. 64<u>3</u>,726 (3,000)
6. 444,<u>4</u>44 (400)

Comparing and Ordering Numbers

It's Algebra!

There are six U.S. national parks between 200,000 and 250,000 acres in size. Which is larger, Capitol Reef Park or Channel Islands Park?

U.S. National Parks	Acres
Capitol Reef (Utah)	241,904
Channel Islands (Calif.)	249,354
Volcanoes (Hawaii)	209,695
Mt. Rainier (Wash.)	235,612
Rocky Mountain (Colo.)	265,727
Voyageurs (Minn.)	218,035

We want to name the larger of the two national parks.

We know that Capitol Reef Park has ___241,904___ acres and Channel Islands Park has ___249,354___ acres.

To compare numbers, write them in a column so that the ones digits are aligned.

    24**1**,904

    24**9**,354

Start at the left and compare digits. The hundred thousands are the same. The ten thousands are the same. But 9 thousands are greater than 1 thousand.

24**9**,354 > 24**1**,904

___Channel Islands___ Park is larger than ___Capitol Reef___ Park.

Getting Started

Write < or > in each circle.

1. 364 ⟨<⟩ 374
2. 3,296 ⟨>⟩ 3,293
3. 7,243 ⟨>⟩ 7,234
4. 23,015 ⟨<⟩ 23,150
5. 16,894 ⟨<⟩ 17,894
6. 8,674 ⟨>⟩ 986

Order the numbers from least to greatest.

7. 396, 387, 390
   ___387, 390, 396___
8. 9,500; 9,275; 9,268
   ___9,268, 9,275, 9,500___
9. 459, 470, 453, 460
   ___453, 459, 460, 470___
10. 74,601; 74,106; 74,016
    ___74,016, 74,106, 74,601___
11. 952,650; 592,065; 529,056
    ___529,056, 592,065, 952,650___

Lesson 2-2 • Comparing and Ordering Numbers

21

---

2 Teach

Introduce the Lesson Have students read the paragraph and state the problem to be solved. (Which is larger, Capitol Reef Park or Channel Islands Park?) Have a student identify the facts given. (Capitol Reef Park has 241,904 acres; Channel Islands Park has 249,354 acres.) As students work through the model on the page, emphasize the importance of aligning columns carefully and comparing digits left to right.

Develop Skills and Concepts Review the symbols of equality and inequality. Put the following sentences on the board for students to read:

7 > 4 (7 is greater than 4)   3 < 6 (3 is less than 6)
4 + 7 = 11 (4 + 7 equals 11)   8 − 4 = 4 (8 − 4 equals 4)
16 < 20 (16 is less than 20)   18 > 9 (18 is greater than 9)

Explain that when we order a group of numbers, we actually compare two numbers at a time. Stress the importance of following directions, so students will know whether to order from least to greatest, or from greatest to least. Write these two number groups on the board and have students order them from least to greatest:

946,367   942,367   948,367 (942,367; 946,367; 948,367)
8,259   8,651   8,269 (8,259; 8,269; 8,651)

It's Algebra! The concepts in this lesson prepare students for algebra.

3 Practice

Have students complete all the exercises. If necessary, write on the board: > **is greater than**, < **is less than**, and = **is equal to**.

Practice

Write < or > in each circle.

1. 47 $<$ 51
2. 139 $>$ 137
3. 420 $<$ 520
4. 3,275 $>$ 3,272
5. 9,546 $>$ 9,462
6. 11,950 $<$ 12,000
7. 82,426 $<$ 82,462
8. 71,999 $<$ 80,000
9. 722,710 $>$ 722,170

Order the numbers from least to greatest.

10. 96, 76, 84
    76, 84, 96
11. 127, 375, 136, 480, 130, 250
    127, 130, 136, 250, 375, 480
12. 1,036; 1,041; 1,040
    1,036; 1,040; 1,041
13. 6,259; 6,300; 6,240
    6,240; 6,259; 6,300

Problem Solving

Use the chart to solve each problem.

Bridge	Length in Meters
George Washington	1,067
Golden Gate	1,280
MacKenzie Straits	1,158
Tacoma Narrows	852

14. Which bridge is longest? __Golden Gate__

15. Which bridge is second shortest? __George Washington__

16. How many bridges are longer than 1,100 meters? __2__

17. List the lengths of the bridges from longest to shortest.
    __1,280__, __1,158__, __1,067__, __852__

Now Try This!

These are the seven symbols used in all Roman numerals.

I = 1   V = 5   X = 10   L = 50   C = 100   D = 500   M = 1,000

These are some of the basic additions and subtractions that the Romans used.

IV = 4	VI = 6	IX = 9	XI = 11	
5 − 1	5 + 1	10 − 1	10 + 1	
LX = 60	XC = 90	CX = 110	CM = 900	MC = 1,100
50 + 10	100 − 10	100 + 10	1,000 − 100	1,000 + 100

Write < or > in each circle.

1. CD $>$ LX
2. VI $<$ IX
3. XLII $<$ LXV
4. MCDXX $>$ MCDVII
5. MCMLXXXVI $>$ MCMXLV

Lesson 2-2 • Comparing and Ordering Numbers

Now Try This! Have students compare the system of Roman numerals with our place-value system. Point out that there is no symbol for zero in Roman numerals and for most Roman numerals, the symbols are written from left to right in order of decreasing value. Note these exceptions: 4 = IV; 9 = IX; 40 = XL; 90 = XC; 400 = CD; 900 = CM.

4 Assess

Ask students to write the largest and smallest four-digit number that can be written using the digits 0, 1, 2, and 3. (3,210; 1,023)

For Mixed Abilities

Common Errors • Intervention

Some students may have difficulty comparing numbers in a side-by-side position. Have them write the numbers on a place-value chart, one under the other, and compare the places from left to right.

Enrichment • Number Sense

Have students work in groups. Ask each student to write five numbers, each with three to five digits, on separate index cards. Collect the cards, shuffle them, and place them face down. Then, have each student select three cards and place the cards in order from least to greatest or from greatest to least.

More to Explore • Number Sense

Have students construct a *Sieve of Eratosthenes*, explaining Eratosthenes was the Greek mathematician who created this method for identifying prime numbers. Have students construct a number chart for 1 through 100, consisting of 10 rows of 10 numbers each. Have them scan the first row, circle the first prime they find (2) and continue through the rows, drawing an X through all multiples of 2. Have them return to the beginning, and repeat the process for the next prime (3), and so on. The circled numbers in the chart will be primes through 100.

2-3 Rounding Numbers

pages 23–24

1 Getting Started

Objectives
- To round numbers through hundred thousands
- To round money to the nearest dollar

Materials
*three large paper number lines—1,000 through 10,000; 10,000 through 100,000; 100,000 through 1,000,000; blank index cards

Warm Up • Mental Math
Have students identify the larger number.
1. 467; 476 (476)
2. 5,392; 5,293 (5,392)
3. 6,012; 6,210 (6,210)
4. 256,256; 265,256 (265,256)
5. 90,050; 50,090 (90,050)
6. 1,026; 1,206 (1,206)

Warm Up • Pencil and Paper
Have students order each set of numbers from least to greatest.
1. 5,642   5,600   5,700
   (5,600; 5,642; 5,700)
2. 343   350   340
   (340; 343; 350)
3. 47,900   47,800   47,801
   (47,800; 47,801; 47,900)
4. 500,000   429,867   400,000
   (400,000; 429,867; 500,000)

Have students read their answers aloud.

2 Teach

Introduce the Lesson Have a student read the problem and identify what is to be solved. (how far the planes flew to the nearest thousand miles) Ask what information is given. (The planes flew exactly 26,103 miles.) As students continue reading through the model, point out that the steps for rounding numbers involve skills they already know: identifying place value and comparing numbers.

Develop Skills and Concepts Have students suggest reasons for rounding numbers. Explain that some quantities cannot be given exactly because of incomplete records or because measurements themselves are not exact.

Name _____

Rounding Numbers

One of the first around-the-world flights started on April 6, 1924, from Seattle, Washington. The two army airplanes returned to that city 35 days later. To the nearest thousand miles, how far did the planes fly?

We want to know the approximate distance that the planes flew.

We know they flew exactly __26,103__ miles.

To find about how far the planes flew, we can round the exact number of miles to the nearest thousand.

To round a number, follow these steps:

Circle the digit in the place to which you want to round.

2(6)1 0 3

Look at the digit to the right of the one circled.

2(6)1 0 3   1 < 5

If the digit is less than 5, keep the circled number and replace all the digits to its right with zeros.

If the digit is 5 or greater, add 1 to the circled number and replace all digits to its right with zeros.

The planes flew about __26,000__ miles.

DISTANCE FLOWN AROUND THE WORLD 26,103 MILES

Getting Started

Round to the nearest ten and the nearest hundred.

1. 948	2. 3,650
950	3,650
900	3,700

Round to the nearest hundred and the nearest thousand.

3. 6,257	4. 8,520
6,300	8,500
6,000	9,000

Round to the nearest ten thousand and the nearest hundred thousand.

5. 349,236	6. 657,456	7. 275,136	8. 534,999
350,000	660,000	280,000	530,000
300,000	700,000	300,000	500,000

Lesson 2-3 • Rounding Numbers

23

Write various 4-, 5-, and 6-digit numbers on index cards. Have students pick a number card. Then, round the number to the nearest thousand, ten thousand, or hundred thousand. Have students follow these steps:

1. Place the number card in the appropriate place on the correct number line.
2. Identify the values that the number is between.
3. Determine which of those values the card number is closer to.
4. Write the rounded number on the board.

Practice

Round to the nearest ten and the nearest hundred.

1. 3,275	2. 4,618	3. 10,371	4. 86,254
3,280	4,620	10,370	86,250
3,300	4,600	10,400	86,300

Round to the nearest hundred and the nearest thousand.

5. 4,372	6. 729,500	7. 16,246	8. 29,743
4,400	729,500	16,200	29,700
4,000	730,000	16,000	30,000

Round to the nearest ten thousand and the nearest hundred thousand.

9. 512,658	10. 636,248	11. 858,653	12. 299,512
510,000	640,000	860,000	300,000
500,000	600,000	900,000	300,000

Round to the nearest dollar.

13. $8.76	14. $19.85	15. $96.37	16. $146.85
$9	$20	$96	$147

17. $749.50	18. $806.12	19. $2,240.74	20. $9,210.49
$750	$806	$2,241	$9,210

Problem Solving

Round each distance.

Earthly Facts

Distance	Miles	Nearest 100	Nearest 1,000
21. circumference at the poles	24,860	24,900	25,000
22. between poles	7,899	7,900	8,000
23. circumference at the equator	24,901	24,900	25,000
24. thickness at the equator	7,926	7,900	8,000

Lesson 2-3 • Rounding Numbers

For Mixed Abilities

Common Errors • Intervention

Some students may round to the wrong place. Have students first draw an arrow above the place to which they are rounding. Then, have them look at the digit to the right to determine whether the digit under the arrow should stay the same or be increased by 1. Finally, have them replace all the digits to the right of the arrow with zeros.

Enrichment • Number Sense

Have students round each of the following numbers to the nearest hundred thousand:

1. 4,376,459 (4,400,000)

2. 58,615,987 (58,600,000)

3. 138,659,804 (138,700,000)

4. 7,051,673 (7,100,000)

5. 62,128,761 (62,100,000)

6. 8,100,795 (8,100,000)

More to Explore • Logic

Duplicate the following for students:

While examining the region around Sirius, the Dog Star, the famous astronomer I.C. Cosmos made an amazing discovery. The area contained many black holes. More amazing was that the number of black holes had a numerical correlation to some aspects of a real dog.

Professor Cosmos determined that the number of black holes was divisible by a dog's number of legs with a remainder of 3. The number was also evenly divisible by the remainder. He noticed, too, that the number would leave a remainder of 2 when divided by the total of a dog's legs, ears, and tail.

Knowing all this, how many black holes are around Sirius? (51)

---

3 Practice

Have students complete all the exercises. Make sure that students read each set of directions carefully.

4 Assess

Ask students to write three numbers that would round to 700 when rounded to the nearest hundred. (Possible answer: 729; 658; 693)

T24

2-4 Place Value to Trillions

pages 25–26

1 Getting Started

Objectives
- To identify place value through trillions
- To compare two numbers through millions

Materials
small paper rectangles, each with a number 1 through 5

Warm Up • Mental Math
Have students identify the number that is one more than:
1. 909 (910)
2. 5,899 (5,900)
3. 172,919 (172,920)
4. 9,999 (10,000)

one less than:
5. 100,000 (99,999)
6. 11,099 (11,098)
7. 1,990 (1,989)
8. 10,009 (10,008)

Warm Up • Pencil and Paper
Have each student make a place-value chart through the hundred thousands place. Have them write the following numbers in the chart:
1. 49,624
2. 845
3. 6,861
4. 39
5. 421.892

2 Teach

Introduce the Lesson Read the problem aloud. Have students identify what is being asked and tell where the needed facts are found. (the meaning of the number on the bill shown) Review the term *standard form*. Extend the place-value chart through the trillions period and work through the model with students.

Develop Skills and Concepts Study the place-value chart. Call on students to name the periods and point out the repetition of hundreds, tens, and ones within each period. To help students read large numbers, have them follow these steps:

1. Determine the name of the greatest period in the number. Read the digits in that period as any 1-, 2-, or 3-digit number, but with the "last name" of the period.

2. Repeat the procedure with the next greatest period.

3. Continue, giving each group of digits within each period its proper "last name."

Tell students 52,000,000,000,000 is read 52 (first name) trillion (last name); 120,516,000,000 is read 120 (first name) billion (last name), 516 (first name) million (last name).

3 Practice

Have students complete all the exercises. Because the large numbers may be threatening to some students, allow them to do the first few exercises with a partner. Assign the remaining exercises as independent work.

T25

Place Value to Trillions

In 2004, the U.S. national debt passed the four trillion dollar mark. What is the meaning of the number on Uncle Sam's bill?

$7,176,395,702,319

We want to understand the meaning of the number on the bill. We know the standard form for the number is $ 7,176,395,702,319 .

trillions			billions			millions			thousands			ones		
hundred trillions	ten trillions	trillions	hundred billions	ten billions	billions	hundred millions	ten millions	millions	hundred thousands	ten thousands	thousands	hundreds	tens	ones

We write: 7   1   7   6   3   9   5   7   0   2   3   1   9

We say: 7 _trillion_, 176 _billion_, 395 _million_, 702 _thousand_, 319 dollars.

Getting Started

Write the place value of each green digit.

1. 396,274,385,921
   hundred million

2. 2,709,437,209,561
   hundred billion

3. 13,076,976,146,000
   trillion

4. 27,659,437,403
   million

Write in standard form.

5. 6 billion, 19 million, 626 thousand, 529 _6,019,626,529_

6. 117 trillion, 29 billion, 6 million, 17 thousand, 26 _117,029,006,017,026_

Write < or > in each circle.

7. 3,294,276 (>) 3,249,276

8. 15,609,342 (>) 15,609,243

Round to the nearest million.

9. 126,048,759 _126,000,000_

10. 6,429,843,207 _6,430,000,000_

Lesson 2-4 • Place Value to Trillions

25

Practice

Write the place value of each green digit.

1. 29,937,207

   _____million_____

2. 126,759,483

   _____ten million_____

3. 796,500,000,000

   _____hundred billion_____

4. 4,723,156,439

   _____hundred million_____

5. 7,483,761,805,648

   _____trillion_____

6. 12,746,249,421

   _____billion_____

Write in standard form.

7. 9 billion, 161 million, 340 thousand, 17 __9,161,340,017__

8. 105 billion, 656 thousand, 729 __105,000,656,729__

9. 4 trillion, 296 billion, 512 million, 626 thousand, 250 __4,296,512,626,250__

Write < or > between the numbers.

10. 3,762,428 (>) 3,762,284

11. 9,296,410 (<) 9,364,410

12. 14,456,785 (>) 14,456,385

13. 129,476,321 (>) 129,473,510

Round to the nearest million.

14. 76,753,215 __77,000,000__

15. 14,239,452 __14,000,000__

16. 527,500,000 __528,000,000__

17. 138,472,496 __138,000,000__

[Now Try This!]

Both you and a partner need five slips of paper with the numbers 1 through 5 written on them.

| 1 | 2 | 3 | 4 | 5 |

Take turns turning over the slips of paper one at a time. Record each number as it appears beginning in the ones place. The player with the greater number wins a point.

Answers will vary.

	ten thousands	thousands	hundreds	tens	ones
First	☐	☐	☐	☐	☐
Second	☐	☐	☐	☐	☐

Lesson 2-4 • Place Value to Trillions

For Mixed Abilities

Common Errors • Intervention

Some students may omit zeros as placeholders or write too many zeros when they are writing numbers in the millions, billions, and trillions. Have them write the number on a place-value chart using period names to guide them. Every place on the chart to the right of the left-most digit must also have a digit, and no more than three digits should appear in any one period.

Enrichment • Applications

1. Have students bring in articles from newspapers and magazines containing large numbers. Begin a bulletin-board display that may be used for creating problems involving comparing and rounding numbers.

2. Tell students to write the number that is: 1,000 more than 165,214,318,723,585 (165,214,318,724,585); and 1,000 less than 2,691,745,382,016 (2,691,745,381,016).

More to Explore • Measurement

Ask students how to keep track of their mileage in a car by recording the mileage from the odometer at the beginning of a car trip and at the end. Tell students to keep track of their daily mileage for every car trip they take for one week to the nearest tenth of a mile. Have them chart each daily total, calculate a weekly total, and display their charts for students to compare.

- Randomly group students to add their mileage together, to see which group traveled the farthest.

- As an extension, have students calculate the cost of gas for their weekly mileage at the current price.

Now Try This! Direct students to record each number as it appears. Have students read their numbers and then compare them. After each student has had 3 turns, have them order the 6 numbers from least to greatest. You might also ask them to round their numbers to the nearest hundred thousand.

---

4 Assess

Write a number that is between 29 trillion and 30 trillion.
(Possible answer: 29,500,000,000,000)

2-5 Adding up to 6-Digit Numbers

pages 27–28

pages 27–28

1 Getting Started

Objectives
• To find the sum of two numbers

Materials
3 addition flashcards

Warm Up • Mental Math
Have students find each sum.

1. 43 + 24 (67)
2. 72 + 16 (88)
3. 35 + 20 (55)
4. 55 + 34 (89)
5. 39 + 60 (99)
6. 72 + 25 (97)
7. 97 + 2 (99)

Warm Up • Pencil and Paper
Have students use the symbols I, V, X, L, and C only once, to name the largest and the smallest Roman numeral possible. (CLXVI; CXLIV)

Name _____

Adding up to 6-Digit Numbers

The continental United States shares its borders with Canada and Mexico. What is the total length of border shared by the United States and Canada?

We want to know the length of the U.S.-Canadian border. We know the border between the lower 48 states and Canada is __3,987__ miles. The border between Alaska and Canada is another __1,538__ miles. To find the total length of the common border, we add these two partial lengths. We add __3,987__ and __1,538__.

Align the addends in columns by place value and add the columns from right to left.

Rename when necessary.

```
  1 11
  3,987
+ 1,538
  5,525
```

The total length of the common border is __5,525__ miles.

Getting Started

Add.

1.	2.	3.	4.
56 + 38 94	159 + 75 234	479 + 584 1,063	6,210 + 459 6,669

5.	6.	7.	8.
7,295 + 9,488 16,783	226,258 + 18,912 245,170	$386.15 + 273.94 $660.09	143,785 + 875,352 1,019,137

Copy and add.

9. 3,964 + 16,246
   20,210
10. 154,280 + 346,596
   500,876
11. $4,276.50 + $3,726.38
   $8,002.88

---

2 Teach

Introduce the Lesson Have a student read the problem aloud. Have a student describe the map and tell what information is shown. (The map gives the length of the border between the lower 48 states and Canada, and the length of the border between Alaska and Canada.) Tell students to continue to read the model, complete the information sentences, and solve the problem.

Develop Skills and Concepts Ask students why it is important to align the digits properly before adding vertically. (Possible answer: Ones must be added to ones, tens to tens, and so on.)

• Review the process of renaming: 10 ones = 1 ten; 10 tens = 1 hundred; 10 hundreds = 1 thousand; 10 thousands = 1 ten thousand; and 10 ten thousands = 1 hundred thousand.

• Write **4,370 + 395** and **869,746 + 649,787** on the board. Have two students work the problems on the board as the class works them at their seats. (4,765; 1,519,533)

3 Practice

Have students complete all the exercises. Remind them to write the dollar and cent signs when they work with money.

Practice

Add.

1. $\begin{array}{r} 29 \\ + 58 \\ \hline 87 \end{array}$	2. $\begin{array}{r} 768 \\ + 59 \\ \hline 827 \end{array}$	3. $\begin{array}{r} 8,206 \\ + 657 \\ \hline 8,863 \end{array}$	4. $\begin{array}{r} 7,953 \\ + 1,034 \\ \hline 8,987 \end{array}$
5. $\begin{array}{r} 9,742 \\ + 5,151 \\ \hline 14,893 \end{array}$	6. $\begin{array}{r} \$762.58 \\ + 863.96 \\ \hline \$1,626.54 \end{array}$	7. $\begin{array}{r} 147,546 \\ + 59,375 \\ \hline 206,921 \end{array}$	8. $\begin{array}{r} 696,458 \\ + 838,297 \\ \hline 1,534,755 \end{array}$

Copy and add.

9. 396 + 857
   1,253

10. 3,260 + 396
    3,656

11. 4,748 + 5,862
    10,610

12. $84.78 + $39.57
    $124.35

13. 7,065 + 8,396
    15,461

14. 14,158 + 7,369
    21,527

15. 27,473 + 82,925
    110,398

16. $210.56 + $396.87
    $607.43

17. 427,153 + 81,296
    508,449

18. 175,240 + 329,594
    504,834

19. 637,456 + 258,962
    896,418

20. 758,736 + 859,878
    1,618,614

Problem Solving

Solve each problem.

21. The Atlantic Ocean is 17,821 feet deeper than the Gulf of Mexico. If the Gulf of Mexico is 12,425 feet deep, how deep is the Atlantic Ocean?
    30,246 feet

22. The largest lake in North America is Lake Superior. It is 8,810 square miles larger than Lake Huron. Lake Huron is 23,010 square miles. How large is Lake Superior?
    31,820 square miles

Use the line graph to solve each problem.

23. Which state had the greatest population growth?
    California

24. Which state was second in population growth during this time?
    Texas

25. What was the total population increase in these 5 leading states?
    26,938 thousand or 26,938,000

Crowded States

More than half the population growth in the U.S. from 1980 to 2000 occurred in just 5 states:

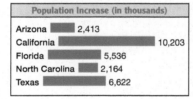

Population Increase (in thousands)

Arizona 2,413
California 10,203
Florida 5,536
North Carolina 2,164
Texas 6,622

4 Assess

Have students determine the error in this calculation.

$\begin{array}{r} 98,652 \\ + 6,428 \\ \hline 162,932 \end{array}$

(Possible answer: The digits in each addend are not correctly lined up by place value.)

Then, have students redo the addition correctly.

$\left(\begin{array}{r} 98,652 \\ + 6,428 \\ \hline 105,080 \end{array}\right)$

For Mixed Abilities

Common Errors • Intervention

Some students may be renaming incorrectly because they are adding from left to right.

Incorrect	Correct
$\begin{array}{r} \overset{11}{} \\ 7,380 \\ + 967 \\ \hline 7,258 \end{array}$	$\begin{array}{r} \overset{1\ 1}{} \\ 7,380 \\ + 967 \\ \hline 8,347 \end{array}$

Have students use counters and a place-value chart to show the two addends. Then, have them join the counters in each place, from right to left, renaming as they go.

Enrichment • Number Sense

Write the following addition square on the board. Challenge students to complete it by adding across and down. If they add correctly, the sum of A + B equals the sum of C + D.

	C	D	
A	456	5,742	(6,198)
B	38,764	59,406	(98,170)
	(39,220)	(65,148)	(104,368)

Encourage students to create original addition puzzles, exchange their puzzles with classmates, and solve the puzzles they receive.

More to Explore • Measurement

Tell students to measure to the nearest centimeter the height of each appliance in their homes. Have them record the results. Then, have them calculate the mean height, the median height, and identify the tallest and shortest appliance.

- Have students compare their lists by charting the results on the board.

- Extend the activity by having students find the average height of the common appliances listed for the class, such as all refrigerators or all washing machines.

T28

2-6 Column Addition

pages 29-30

1 Getting Started

Objectives
• To add three or more numbers

Warm Up • Mental Math
Have students compare each number pair.

1. 49,684 ($<$) 49,864
2. 3,961 ($=$) 3,961
3. 120,446 ($<$) 120,464
4. 734,609 ($>$) 73,609
5. 4,826,386 ($<$) 4,921,001
6. 36,491,851 ($=$) 36,491,851
7. 245,563,981 ($>$) 245,562,999
8. 5,801,607,802 ($>$) 5,801,802

Warm Up • Pencil and Paper
Have students find each sum.

1. 6,953 + 8,529 (15,482)
2. 153,492 + 89,677 (243,169)
3. 949,670 + 7,890,658 (8,840,328)
4. 59,328 + 674,919 (734,247)
5. 137,137 + 69,584,867 (69,722,004)
6. forty-three million, plus two hundred seventy-six thousand, twelve (43,276,012)

Name _____

Column Addition

Mrs. Rameriz had dinner at the Greenwood Inn. She ordered crabmeat cocktail, chicken kiev, and coffee. The tax on the meal was $1.41. What was the cost of Mrs. Rameriz's dinner?

We want to know how much Mrs. Rameriz spent for dinner. We know the appetizer cost __$5.95__, the entree __$12.95__, the beverage __$1.25__, and the tax __$1.41__. To find the total bill, we add these amounts together. We add __$5.95__, __$12.95__, __$1.25__, and __$1.41__.

REMEMBER to align the addends in a column by place value. Add two digits at a time. Then add that sum to the next digit in the column. Rename when necessary.

$ 5.9 5	$5 + 5 = \underline{10}$	
12.9 5		
1.2 5	$\underline{10} + 5 = \underline{15}$	
+ 1.4 1	$\underline{15} + 1 = \underline{16}$	
$21.56		

Mrs. Rameriz paid __$21.56__ for dinner.

Appetizers	
Crabmeat Cocktail	$ 5.95
Shrimp Cocktail	$ 6.50
Salads	
Fruit Salad	$ 2.50
Spinach Salad	$ 2.50
Entrees	
Fresh Fish of the day	$ 17.95
Sirloin Steak	$ 16.50
Chicken Kiev	$ 12.95
Beverages	
Coffee	$ 1.25
Milk	$ 0.75

Getting Started

Add.

1.
```
   179
    84
 + 215
   478
```

2.
```
 $12.57
  18.35
 + 0.76
 $31.68
```

3.
```
  548 million
   96 million
 + 374 million
 1,018 million
```

4.
```
  629,156
  358,712
 + 413,095
 1,400,963
```

Copy and add vertically.

5. $182.47 + $810.75 + $82.75
$1,075.97

6. 126,350 + 75,975 + 276,453
478,778

---

2 Teach

Introduce the Lesson Have a student read the problem aloud. Call attention to the menu and ask what facts students will need to solve the problem (the cost of the appetizer, $5.95; the cost of the entree, $12.95; the cost of the beverage, $1.25), and what facts they can find in the problem. (The tax is $1.41.) Have students complete the information and plan sentences and solve the problem. Point out that addition is indicated when the problem calls for a *total* of numbers or amounts.

Develop Skills and Concepts Review the Associative Property of Addition. Write on the board:

```
(15 + 12) + 35 =     15 + (12 + 35) =
   27 + 35               15 + 47
      62                    62
```

Emphasize that changing the grouping of the addends does not change the sum. Point out that students will be combining two addends at a time, no matter how many addends are in the problem. Stress that the process of finding the sum remains the same.

Rewrite the problem on the board vertically. Remind students to align the digits carefully when they write an addition problem in vertical form. Point out to the students that they will add the ones, the tens, the hundreds, the thousands, and so on.

It's Algebra! The concepts in this lesson prepare students for algebra.

3 Practice

Have students complete all the exercises. For exercises containing addends with different numbers of digits, stress the importance of aligning the digits carefully. Call attention to the Problem Solving section. Explain that a *commission* is a kind of bonus based on the amount of sales. Make certain students understand the meaning of *accumulates* in the context of the problem.

Practice

Add.

1.	2.	3.	4.
973	619	2,396	6,756
426	47	563	4,392
+ 185	+ 523	+ 1,483	+ 1,485
1,584	1,189	4,442	12,633

5.	6.	7.	8.
$86.47	272 million	17,859	126,285
7.85	159 million	9,327	415,788
+ 13.49	+ 16 million	+ 78,515	+ 907,853
$107.81	447 million	105,701	1,449,926

Copy and add.

9. 6,458 + 4,973 + 2,638
   14,069

10. $89.59 + $127.48 + $47.58
   $264.65

11. 32,474 + 12,965 + 6,478
   51,917

12. 596 + 14,273 + 6,456
   21,325

13. 128,275 + 38,975 + 16,850
   184,100

14. 453,752 + 593,856 + 323,485
   1,371,093

15. $2,346.58 + $759.48 + $3,705.46
   $6,811.52

16. 3,275,495 + 8,796 + 16,753 + 391,743
   3,692,787

Problem Solving

Mr. Ray's sales commission accumulates daily.

Complete the table to show his total for each day.

17.

	Monday	Tuesday	Wednesday	Thursday	Friday	Saturday
Daily Commission	$43.58	$21.39	$37.18	$41.60	$23.75	$17.29
Total Earned	$43.58	$64.97	$102.15	$143.75	$167.50	$184.79

Now Try This!

Find the missing digit in each problem.

1.	2.	3.	4.
6 7	1 <u>2</u> 8	2 4 5	1, <u>3</u> 6 7
2 <u>8</u>	6 <u>4</u>	<u>1</u> 3 9	2, 9 <u>4</u> 2
+ <u>3</u> 2	+ <u>3</u> 3 4	+ 2 7 <u>5</u>	+ 3, 7 6 <u>1</u>
1 2 7	5 2 6	6 5 9	8, 0 7 0

30   Lesson 2-6 • Column Addition

Now Try This! Direct students to find the missing digits in each problem. Remind them that they can use subtraction to find missing digits. Review the inverse relationship between addition and subtraction.

$$4 + (7) = 11 \longrightarrow 11 - 4 = (7)$$
$$(9) + 6 = 15 \longrightarrow 15 - 6 = (9)$$

4 Assess

Write **158 + 12,639 + 5,874** on the board. Ask students to use pencil and paper to find the sum. (18,671)

For Mixed Abilities

Common Errors • Intervention

Some students may get incorrect sums in column addition because they do not align the digits correctly under one another. Have these students write the problems on graph paper, using the vertical lines to keep the digits placed in their proper columns.

Enrichment • Number Sense

Challenge students to complete the following puzzle by finding the path to the given sum. They should begin with one of the top numbers and proceed vertically or horizontally, adding numbers until the sum of 623 is reached. Have them draw arrows indicating the path followed.

78	213	156
97	197	89
324	181	217
sum	623	

$(156 \longrightarrow 89 \longrightarrow 197 \longrightarrow 181 \longrightarrow 623)$

More to Explore • Geometry

Have students place their right hand on a blank piece of paper, spreading their fingers as far as possible. Have them trace the angles formed by their fingers and then measure and label the angles. Have them repeat the procedure for their left hands. Ask students to compare the angles formed by their two hands. Then, have students compare angles formed by their hands with angles formed by others in the class. Students can list the class frequency for each of the angles formed.

ESL/ELL STRATEGIES

Practice saying large numbers aloud with students. Start with four-digit numbers and work up to 15-digit numbers. Point out that the word *and* is not used to separate thousands, millions, etc. Example: *Nine million, ~~and~~ two hundred fifty-six thousand, ~~and~~ four hundred ~~and~~ twenty-four.*

pages 31–32

1 Getting Started

Objective
• To subtract whole numbers through hundred thousands

Materials
*subtraction fact cards

Warm Up • Mental Math
Have students find the missing number in each number sentence.

1. 3 + (8) = 11
2. (4) + 5 = 9
3. 7 + (6) = 13
4. 9 + (8) = 18 − 1
5. 6 + (4) = 5 × 2
6. (6) + 8 = 14
7. 4 + 3 + (2) = 9
8. 2 + 7 + (6) = 15

Warm Up • Pencil and Paper
Have students find each difference.

1.
   74   56   85   93
− 23 − 13 − 60 − 23
(51)   (43)   (25)   (70)

Subtracting up to 6-Digit Numbers

Mrs. Andrews flies regularly between these four cities. How much farther is the flight from Los Angeles to New York than the flight from Los Angeles to Denver?

Air Distances in Kilometers				
	Denver	Los Angeles	New York	Seattle
Denver	—	1,186	2,274	1,424
Los Angeles	1,186	—	3,951	1,548
New York	2,274	3,951	—	3,887
Seattle	1,424	1,548	3,887	—

We want to compare the distance between Los Angeles and New York to the distance between Los Angeles and Denver.

We know it is __3,951__ kilometers from Los Angeles to New York, and __1,186__ kilometers from Los Angeles to Denver.

To find the difference in kilometers, we subtract the shorter distance from the longer distance. We subtract __1,186__ from __3,951__.

Remember to align the numbers in columns by place value. Subtract the columns from right to left. Rename when necessary.

```
  8 14 11
  3,9̸5̸1̸
− 1,186
  2,765
```

It is __2,765__ kilometers farther.

Getting Started

Subtract.

1.
  795
− 273
  522

2.
  859
− 296
  563

3.
 39,884
− 16,948
 22,936

4.
126,758
− 89,962
 36,796

Copy and subtract.

5. $146.75 − $97.85
  $48.90

6. 27,437 − 9,575
  17,862

7. 746,229 − 257,863
  488,366

---

2 Teach

Introduce the Lesson Have students look at the picture and tell what is to be solved. (how much greater the distance is between Los Angeles and New York than between Los Angeles and Denver) Ask what facts are given that will help students solve the problem. (The distance from Los Angeles to Denver is 1,186 km. The distance from Los Angeles to New York is 3,951 km.) Point out that the chart provides more information than is needed for this problem. Work through the model with the students.

Develop Skills and Concepts Ask why it is necessary to align the digits properly before subtracting. (so that ones will be subtracted from ones, tens from tens, and so on) Review the process of renaming. Have students summarize the place-value relationships: 1 ten = 10 ones; 1 hundred = 10 tens; 1 thousand = 10 hundreds;

1 ten thousand = 10 thousands; and 1 hundred thousand = 10 ten thousands. Provide practice by renaming the following:

1. 31 = 2 tens + (11) ones
2. 406 = 3 hundreds + (10) tens + 6 ones
3. 2,135 = 1 thousand + (11) hundreds + 2 tens + (15) ones

Remind students to subtract from right to left.

3 Practice

Have students complete all the exercises. Remind students to align digits properly when copying problems.

Practice

Subtract.

1.	437 − 159 278	2.	1,679 − 857 822	3.	9,271 − 4,859 4,412	4.	12,246 − 9,758 2,488
5.	46,439 − 17,296 29,143	6.	75,314 − 57,965 17,349	7.	139,485 − 97,859 41,626	8.	$4,682.56 − 2,939.47 $1,743.09
9.	35,728 − 14,374 21,354	10.	58,213 − 23,556 34,657	11.	247,362 − 52,718 194,644	12.	$7,543.47 − 3,816.39 $3,727.08

Copy and subtract.

13. 1,678 − 437
1,241

14. $68.43 − $49.67
$18.76

15. 13,951 − 7,295
6,656

16. 26,214 − 7,956
18,258

17. 47,213 − 28,475
18,738

18. $785.39 − $291.95
$493.44

19. 67,425 − 48,698
18,727

20. 212,415 − 63,827
148,588

21. 596,254 − 123,975
472,279

22. $3,752.31 − $1,816.85
$1,935.46

23. 381,426 − 198,759
182,667

24. 821,236 − 457,346
363,890

Problem Solving

Solve each problem.

25. Kerry is driving from San Francisco to Boston, a distance of 3,095 miles. The odometer in Kerry's car shows he has driven 2,142 miles when he gets to Chicago. How much farther does he still have to drive?
953 miles

26. The robots on an assembly line are programmed to produce 25,231 widgets in a 5-day week. On the first day, they make 5,193 widgets. On the second day, they make 7,641. How many more widgets will the robots make by the end of the fifth day?
12,397 widgets

32

Lesson 2-7 • Subtracting up to 6-Digit Numbers

4 Assess

Have students subtract 562,974 from 807,139. (244,165)

5 Mixed Review

1. (8) × 7 = 56
2. 273 + 4,176 + 27 (4,476)
3. 24 ÷ 6 × 8 (32)
4. 3^3 (27)
5. 6 × 9 (54)
6. 5 + 8 + 4 + 7 (24)
7. $23.56 + $127.93 ($151.49)
8. 62,146 + 38 + 1,005 (63,189)
9. 10^5 (100,000)
10. 15 − 9 (6)

For Mixed Abilities

Common Errors • Intervention

Some students may forget to add the digit in the minuend after they rename and before they subtract.

Incorrect	Correct
6 10 8 10	6 14 8 16
7,496	7,496
− 3,589	− 3,589
3,501	3,907

Have students work in pairs with place-value materials to do subtraction exercises such as 56 − 39. Help them to recognize that, when they trade for subtraction, 5 tens 6 ones become 4 tens 16 ones.

Enrichment • Number Sense

Challenge students to fill in the missing digits for each problem, using 0 through 9 only once.

1.
9	7	5	3	1
− (8)	− (6)	− (4)	− (2)	− (0)
1	1	1	1	1

2.
9	7	5	3	1
− (8)	− (6)	− (4)	− (2)	− (0)
1	1	1	1	1

3.
9	(7)	6	5	(4)
− (8)	− 3	− (2)	− (1)	− 0
1	4	4	4	4

4.
(8)	2	(6)	5	9
− 7	− (0)	− 3	− (1)	− (4)
1	2	3	4	5

More to Explore • Applications

Have students bring in an atlas of road maps of the United States. Tell them to plan a trip to the city or area of their choice. They are to write down their starting point and destination. Using the most direct route, have them measure the distances from one point to another along the highways, remembering to refer to the map's scale. After they have calculated their mileage, have them figure the amount of gasoline they will need for the trip if their car gets 25 miles to a gallon. Have them figure the cost of the gasoline, using current local prices.

T32

2-8 Subtracting From Zeros

pages 33–34

1 Getting Started

Objective
• To subtract with renaming over zeros

Materials
counters or chips;
place-value charts

Warm Up • Mental Math
Have students simplify each expression.
1. 6 × 7 − 2 (40)
2. 63 ÷ 9 × 6 (42)
3. 9 + 8 − 7 (10)
4. 2 × 4 × 7 (56)
5. 64 ÷ 8 × 4 (32)
6. 6 + 7 − 4 (9)
7. 72 ÷ 8 × 5 (45)
8. 3 × 7 + 9 (30)

Warm Up • Pencil and Paper
Have students find each difference.
1. 914 − 365 (549)
2. 643 − 187 (456)
3. 5,914 − 2,567 (3,347)
4. $235.16 − $87.49 ($147.67)
5. $6,372.15 − $1,596.03 ($4,776.12)
6. two thousand, six hundred twenty-one minus eight hundred seventy-four (1,747)

Name _____

Subtracting From Zeros

Mr. Chen bought a used car at Carl's Cars. His monthly payments are $159. What is Mr. Chen's balance due after he makes the first payment?

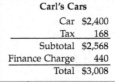

Carl's Cars	
Car	$2,400
Tax	168
Subtotal	$2,568
Finance Charge	440
Total	$3,008

We want to know how much Mr. Chen owes after his first payment. We know he owed a total of __$3,008__, and his first payment was __$159__.

To find how much Mr. Chen still owes, we subtract his first payment from the total owed. We subtract __$159__ from __$3,008__.

REMEMBER Align the money amounts in columns by place value, and subtract from right to left.

Subtract the ones.	Subtract the tens.	Subtract the hundreds.	Subtract the thousands.
2 9 9 18 $3,008 − 159 9	2 9 9 $3,008 − 159 49	2 9 $3,008 − 159 849	2 $3,008 − 159 $2,849

Mr. Chen still owes __$2,849__.

Getting Started

Subtract.

1.	2.	3.	4.
703 − 296 407	800 − 245 555	3,601 − 1,965 1,636	$50.03 − 7.58 $42.45

5.	6.	7.	8.
40,001 − 23,162 16,839	13,000 − 7,485 5,515	36,005 − 9,329 26,676	$700.07 − 248.09 $451.98

Copy and subtract.

9. 100,000 − 67,596
  32,404
10. 1,000,000 − 267,385
  732,615
11. 907,039 − 214,658
  692,381

Lesson 2-8 • Subtracting From Zeros

---

2 Teach

Introduce the Lesson Have a student read the problem aloud. Discuss the picture and have students identify the problem to be solved. (how much Mr. Chen owes after his first payment) As students read through the model and name the facts they know, point out that pictures and other visual material often contain important information. Verify that students recognize they must subtract to find out how much is left. Work through each step of the model problem with students.

Develop Skills and Concepts Write the following expanded form for 508 on the board:

5 hundreds	0 tens	8 ones
4 hundreds	10 tens	8 ones
4 hundreds	9 tens	18 ones

Explain that these are three ways of thinking about the number 508.

Have students rename 3,008 in four different ways.

3 thousands	0 hundreds	0 tens	8 ones
2 thousands	10 hundreds	0 tens	8 ones
2 thousands	9 hundreds	10 tens	8 ones
2 thousands	9 hundreds	9 tens	18 ones

Continue developing this renaming skill with the following numbers: 809; 3,005; 60,004; 32,001.

Remind students they can check subtraction by adding the difference to the subtrahend.

3 Practice

Have students complete all the exercises. Remind students to align the digits carefully in Exercises 9–20. Continue to encourage students to check their answers with addition.

Practice

Subtract.

1. 908
   − 309
   599

2. 750
   − 236
   514

3. 4,076
   − 2,785
   1,291

4. $21.08
   − 9.29
   $11.79

5. 30,006
   − 5,278
   24,728

6. 60,702
   − 4,965
   55,737

7. $420. 05
   − 139. 79
   $280.26

8. 840,039
   − 565,752
   274,287

Copy and subtract.

9. 9,005 − 3,778
   5,227

10. $140.06 − $79.58
    $60.48

11. $260 − $39.59
    $220.41

12. 13,075 − 6,782
    6,293

13. 50,072 − 37,888
    12,184

14. 967,006 − 208,448
    758,558

15. $6,005.29 − $3,758.59
    $2,246.73

16. 500,000 − 372,653
    127,347

17. 800,006 − 352,977
    447,029

18. 3,000,000 − 275,839
    2,724,161

19. 6,707,005 − 2,968,476
    3,738,529

20. 18,006,403 − 9,779,586
    8,226,817

Problem Solving

Solve each problem.

21. In 1982, the population of the metropolitan New York area was 17,539,344. Of these, 10,467,705 lived outside the city. What was the population of New York City at that time? 7,071,639

22. Complete Tania's checkbook record for the first half of September.

Check Number	Date	Description	Amount	Balance $606.08
101	9/7	Bike Repair	$17.56	$588.52
102	9/8	Gift	$29.37	$559.15
	9/9	Deposit	$140.87	$700.02
103	9/14	Shoes	$56.40	$643.62
	9/15	Deposit	$92.74	$736.36

34

Lesson 2-8 • Subtracting From Zeros

4 Assess

On the board, write

700,603
− 563,928
136,675

Have students show how they would use addition to check this subtraction.

5 Mixed Review

1. 614,208 + 25,196 (639,404)
2. $20.00 − $14.72 ($5.28)
3. 4 + 3 × 5 (19)
4. 7×10^4 (70,000)
5. 6,209 − 3,478 (2,731)
6. $729.65 + $47.48 + $106.56 ($883.69)
7. 7 + (6) = 13
8. 16 + 25 + 18 + 9 (68)
9. 500,200 − 39,176 (461,024)
10. 8^3 (512)

For Mixed Abilities

Common Errors • Intervention

Some students may bring down the numbers that are being subtracted when there are zeros in the minuend.

Incorrect
4,008
− 1,962
3,966

Correct
4,008
− 1,962
2,046

Have students work in pairs and use play money to model a problem such as $500 − $246, where they see that they must trade 5 hundreds for 4 hundreds, 9 tens, and 10 ones before they can subtract.

Enrichment • Number Sense

Write the following numbers on the board:

5,004   4,704 37,008 70,006
8,040 64,001 40,069   3,906

Challenge students to make up subtraction problems using these numbers. Have students exchange and solve each other's problems.

More to Explore • Logic

Present the following problem.

Nine of the runners in Mr. Shirley's sixth-grade class challenged each other to a final race, to see who really was the fastest in the class. Each runner wore a number 1 through 9. Read the following clues to help you decide who came in first, second, and third:

• The sum of all the numbers who entered the race was three times the sum of the numbers of first, second, and third places.

• Adding the numbers for second and third places, you will have a number that is $\frac{2}{3}$ of first place.

• Adding the number for first and second places, you have a number that is twice the third place number.

Which numbers came in first, second, and third place? (9, 1, 5)

T34

2-9 Estimating Sums and Differences

pages 35–36

1 Getting Started

Objective
• To estimate a sum or difference

Warm Up • Mental Math
Have students use the Order of Operations to simplify each expression.

1. $(5 + 6 - 1) \div 2 \times 5$ (25)
2. $(3 \times 8) \div 4 \times 9$ (54)
3. $(4 \times 3 + 12) \div 8$ (3)
4. $10 \times 10 \times 10 - 10$ (990)
5. $6 + 7 + 7 - 11$ (9)
6. $(45 - 10) \div 7 \times 6$ (30)
7. $56 \div 7 \div 8 + 0$ (1)
8. $(37 + 49 - 16) \times 0$ (0)

Warm Up • Mental Math
Have students round each number to the nearest ten:

1. 34 (30)   2. 425 (430)
3. 4,592 (4,590) 4. 62,957 (62,960)

to the nearest hundred:

5. 461 (500)   6. 7,828 (7,800)
7. 64,554 (64,600) 8. 3,864 (3,900)

to the nearest thousand:

9. 7,428 (7,000)   10. 82,650 (83,000)
11. 9,498 (9,000) 12. 65,581 (66,000)

Name _____

Estimating Sums and Differences

Ron and Phyllis are doing a vehicle count to see if a traffic light is needed at a freeway entrance. About how many vehicles used this entrance on Thursday and Friday? About how many more vehicles used the entrance ramp on Friday than on Thursday?

DAY	TRAFFIC COUNT
Thursday	8,576
Friday	18,735

We need to estimate the total number of vehicles that used the ramp on both days, and the difference between the traffic count for the two days.

We know that __8,576__ vehicles used the entrance on Thursday, and __18,735__ vehicles used it on Friday.

To estimate the total number of vehicles, we round to the nearest thousand and add.

$$18,735 \longrightarrow 19,000$$
$$8,576 \longrightarrow \underline{+\ \ 9,000}$$
$$28,000$$

About __28,000__ vehicles used the ramp.

To estimate how many more vehicles entered on Friday than on Thursday, we round to the nearest thousand and subtract.

$$18,735 \longrightarrow 19,000$$
$$8,576 \longrightarrow \underline{-\ \ 9,000}$$
$$10,000$$

About __10,000__ more vehicles entered on Friday than on Thursday.

Getting Started

Estimate by rounding to the nearest thousand.

1.	2.	3.	4.
7,693	8,653	32,450	94,271
+ 3,256	− 4,965	+ 65,825	− 15,372
11,000	4,000	100,000	70,000

Estimate by rounding to the nearest ten thousand. (for 3 and 4)

Copy and add or subtract. Estimate to check each answer.

5. $46,914 + 7,842$
  54,756
6. $601,426 - 178,969$
  422,457
7. $704,123 - 66,574$
  637,549

---

2 Teach

Introduce the Lesson Have students read the problem. Call attention to the art and have students discuss how it relates to the problems to be solved. (Art shows the number of vehicles that use the freeway entrance on Thursday and Friday.) Emphasize that the problem does not call for an exact answer and involves two different estimations. Have students read and complete the information sentences and solve the problem. Point out the differences between estimating and computing an exact answer.

Develop Skills and Concepts Emphasize that the method shown on page 35 is to round first and then compute. Explain that the advantage of working in this order is to simplify the numbers so that computation can be done mentally and easily. Remind students to look only at the digit at the right of the place to which they are rounding, and to round down if the digit is less than five. Have

students round each of the following numbers to the nearest hundred, thousand, and ten thousand:

1. 45,208   (45,200; 45,000; 50,000)
2. 64,971   (65,000; 65,000; 60,000)
3. 38,569   (38,600; 39,000; 40,000)
4. 704,829   (704,800; 705,000; 700,000)

Point out that rounding to the nearest hundred before computing provides a closer estimate than rounding to the nearest ten thousand.

3 Practice

Have students complete all the exercises. Point out that estimation can be used to predict an answer and to check whether an answer is reasonable.

T35

Practice

Estimate by rounding to the nearest thousand.

1.	4,575 − 1,836 3,000	2.	7,150 + 6,416 13,000	3.	12,739 − 5,895 7,000

4.	21,239 − 14,710 6,000	5.	39,758 + 21,837 62,000	6.	145,795 + 86,439 232,000

Estimate by rounding to the nearest ten thousand.

7.	43,286 + 11,723 50,000	8.	29,841 − 16,372 10,000	9.	74,185 + 16,347 90,000

10.	139,216 − 83,950 60,000	11.	437,615 − 168,212 270,000	12.	512,817 + 273,548 780,000

Copy and add or subtract. Estimate to check each answer.

13. 33,458 + 9,748
    43,206

14. 71,485 − 19,285
    52,200

15. 76,936 + 54,636
    131,572

16. 137,459 − 86,653
    50,806

17. 326,085 − 119,408
    206,677

18. 296,485 + 175,817
    472,302

19. 246,327 − 75,385
    170,942

20. 215,294 + 318,467
    533,761

21. 684,372 − 431,729
    252,643

(Now Try This!)

Another way to estimate a sum or a difference is to use **front-end estimation**. Add or subtract the front-end digits. Then, adjust your answer to get a closer estimate.

19,841
+ 17,750
20,000   9,841 + 7,750 is about 16,000

731,903
− 429,716
300,000

31,903 > 29,716
more than
300,000

Estimate. Use front-end estimation with adjusting.

1.	568,157 + 124,275 600,000	2.	91,495 − 10,970 80,000	3.	746,178 + 245,513 900,000	4.	264,396 − 29,812 240,000

See Solution Notes for adjustments.

36   Lesson 2-9 • Estimating Sums and Differences

---

4 Assess

Ask students to describe a situation in which estimating the sum or difference is appropriate. (Answers will vary.)

For Mixed Abilities

Common Errors • Intervention

Students may estimate incorrectly because they round incorrectly. Have students round each of the following numbers to the nearest ten thousand.

47,564 (50,000)   25,398 (30,000)
72,348 (70,000)   14,586 (10,000)

Then, have students use the rounded numbers to estimate.

47,564 + 14,586 (60,000)	72,348 − 25,398 (40,000)
25,398 − 14,586 (20,000)	25,398 + 47,564 (80,000)

Enrichment • Number Sense

Challenge students to match each exercise with its answer.

4,982 − 1,529 (b)	49,641 + 3,248 (d)	a. 44,244 b. 3,453
3,564 + 5,208 (c)	72,065 − 27,821 (a)	c. 8,772 d. 52,889

More to Explore • Applications

Tell students to draw a Venn diagram to represent this information:

- The Universal set is all 300 students in school.

- A is the set of all 60 students in the band.

- B is the set of all 15 brass instrument players.

- C is the set of all 7 percussion instrument players.

- D is the set of 27 students in drama club, 3 of whom play trumpet in the band.

Have students calculate how many students are not in the drama club or band. (216 students)

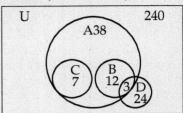

T36

pages 37–38

1 Getting Started

Objective
• To solve a problem by drawing a diagram

Materials
35 objects with perpendicular sides

Warm Up • Mental Math
Have students find the number that represents each term.

1. dozen (12)
2. score (20)
3. gross (144)
4. century (100)
5. quadruplets (4)
6. decade (10)
7. bicentennial (200)
8. baker's dozen (13)

Warm Up • Pencil and Paper
Have students find the following:

1. 5 right angles in the classroom
2. $5 \times 3 \times \$0.25$ ($3.75)
3. a rectangle divided into 5 parts with the fewest lines
4. the area of a 5 by 7 centimeter rectangle (35 sq cm)
5. 250×3 in. (750 in.)
6. $3 \times 9 \times \$0.50$ ($13.50)
7. $1 \times 1 \times 1$ (1)
8. the next number: 6, 5, 7, 6, 8, (7)

Name _____

Problem Solving: Draw a Diagram

A fireman was standing on the middle rung of a ladder. He climbed up 3 rungs. A sudden flare sent him down 7 rungs. When the fire died down again, he climbed up 9 rungs. When the fire was finally out, he climbed up the remaining 4 rungs to the top of the ladder. How many rungs were on the ladder?

★ SEE
We want to know the number of rungs on the ladder.

The fireman starts on the __middle__ rung. He climbed up __3__ rungs. He stepped down __7__ rungs. He then moved up __9__ rungs. Finally, he climbed up __4__ rungs to the top of the ladder.

★ PLAN
We can use all of the facts to draw a diagram. Remember that there must always be an equal number of rungs above and below the middle rung.

★ DO
1. He climbed up 3 rungs. There are also __3__ rungs below the middle rung.

2. He climbed down 7 rungs. We need to go __4__ rungs below the middle rung.

3. He climbed up 9 rungs. We need to go __5__ rungs above the middle rung.

4. He climbed 4 rungs to the top. We need to go __4__ more rungs above the middle rung.

5. We need to go as many rungs below the middle rung as we have above it.

There were __19__ rungs on the ladder.

★ CHECK
We can check the diagrams with the facts of the problem.

Lesson 2-10 • Problem Solving: Draw a Diagram

37

---

2 Teach

Introduce the Lesson Review the function of each of the four steps of the SEE, PLAN, DO, CHECK outline with your class. Assign a student to read the entire problem aloud. As you walk the class through the four steps, be sure the students are writing the input answers. Be sure students understand that their diagrams do not have to be realistic. Symbolic drawings are easier to make, require no artistic talent, and are easier to understand.

Develop Skills and Concepts Select one of the problems on page 38 to be worked together by the class. After they have reached a solution, have students put the drawings that helped them arrive at the solution on the board.

3 Apply

Solution Notes
1. Draw a clock face and mark the position of the hour hand each time right angles occur.
2. The end of the train must travel 2 miles.
3. Note the following diagram:
4. 8 steps to advance.
5. First open the links. Then, join the links.

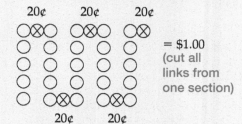

T37

Apply

Draw a diagram to help solve each problem.

1. Consider the movements of the hour and minute hands of a clock for 12 hours. How many times do they form right angles?
   22

2. A train, one mile long, travels at a rate of one mile per minute through a tunnel one mile long. How long will it take the train to pass completely through the tunnel?
   2 minutes

3. Seven monkeys living in a big cage at the zoo are always throwing banana peels at each other. What is the smallest number of straight wire fences that the zookeeper can put up in the large cage, so that each monkey will be in a cage by himself?
   3 fences

4. On May 29, 1953, Hunt, Hillary, and Tenzing were very close to reaching the peak of Mount Everest. They had just 5 more steps to go before reaching the peak when they met bad weather. For every 2 steps forward, the driving wind forced them back 1 step. How many steps did they have to actually take to get to the peak of Mount Everest?
   8 steps

5. You have 6 small bracelets, each made up of 5 links. You decide to make one necklace by joining them together. It costs $0.10 to open a link and $0.10 to weld a link. How much would it cost to have the necklace made?
   $1.20

6. A person has 6 sections of chain, each consisting of 4 links, and wants to form one continuous long chain. A metal shop charges $0.20 to open a link and $0.50 to weld a link. What would be the cost of the least expensive method for creating the chain?
   $3.50

Problem-Solving Strategy: Using the Four-Step Plan

★ **SEE** What do you need to find?

★ **PLAN** What do you need to do?

★ **DO** Follow the plan.

★ **CHECK** Does your answer make sense?

---

7. Doreen Director was told how many people attended the play on Friday night and how many attended on Saturday night. To estimate the difference between these two numbers, she rounded the higher number down and the lower number up. Did she underestimate or overestimate the actual difference?
   She underestimated.

8. Jack and Jill are grocery shopping. They have only $20.00 to spend. They have selected 7 items and, just before checking out, they review the costs. They discover that the prices are $0.70, $7.38, $5.90, $2.34, $1.10, $1.78, and $3.58. This is more than $20.00. What items should they put back? How would you decide?
   See Solution Notes.

9. The Crazy Cat Calendar Club meets on the 10th of the month. But it only meets every two months. How many days are there between the 10th of a month and the 10th two months later?
   58, 59, 60 or 61 days

10. Sam Clemente said, "If I add 4 equal addends, I will get a sum less than 25." Sam Antonio says, "If I add 3 equal addends, I will get the same sum." What is the sum?
    12 or 24

38

Lesson 2-10 • Problem Solving: Draw a Diagram

More to Explore • Applications

Have students bring in the business section of the newspaper and focus on the stock page. Explain the layout of the data given about the New York Stock Exchange and the American Stock Exchange. Point out that the closing price is indicated by decimals rather than dollars and cents, and the amount of price change for each stock is shown by a plus or minus sign. Have students scan either the NYSE or AMEX for the following:

1. the most expensive stock per share

2. the least expensive stock per share

3. the most active stock (showing the most change)

4. how no stock activity is shown

5. the initials representing well-known companies you randomly select

---

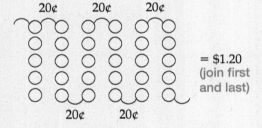

= $1.20
(join first and last)

6. See problem #5.

Higher-Order Thinking Skills

7. **Analysis:** Students should try it with different numbers until they are convinced.

8. **Evaluation:** The total cost of the 7 items is $22.78. So, they must return items worth at least $2.78. Choices range from returning the 2 items that cost $1.10 and $1.78, which leaves them the closest to $20.00 as possible, to returning everything. Reasons

for returning any items will vary, and this problem lends itself nicely to cooperative-group work.

9. **Analysis:** There are four possible answers, depending upon the two months of the year and whether or not it is leap year.

10. **Analysis:** There are two possible answers: 12 (3 + 3 + 3 + 3 and 4 + 4 + 4) and 24 (6 + 6 + 6 + 6 and 8 + 8 + 8).

4 Assess

Write this problem on the board.

A farmer wants to build a pen using the side of the barn as one side of the pen. He has 12 feet of fencing. What are the dimensions of the pen with the greatest area? Hint: Area is length times width. (3 ft by 6 ft)

Alternate Chapter Test

You may wish to use the Alternate Chapter Test on page 372 of this book for further review and assessment.

Write the place value of each green digit.

1. 39,476

   hundreds

2. 8,276

   thousands

3. 128,586

   ten thousands

4. 247,968

   hundred thousands

Write < or > in each circle.

5. 574 $<$ 745

6. 1,296 $>$ 1,295

7. 15,536 $>$ 15,529

8. 8,425 $<$ 8,524

Round to the nearest thousand and to the nearest ten thousand.

9. 18,750

   19,000

   20,000

10. 26,500

    27,000

    30,000

11. 136,926

    137,000

    140,000

12. 586,151

    586,000

    590,000

Write the place value of each green digit.

13. 426,396,247

    ten millions

14. 3,960,481

    ten thousands

15. 1,392,461,571

    billions

16. 13,750,185

    millions

Add.

17.
```
   9,275
+ 14,981
  24,256
```

18.
```
  $476.35
+ 129.29
 $605.64
```

19.
```
  375,248
+ 859,188
1,234,436
```

20.
```
  339,275
  148,458
+ 829,874
1,317,607
```

Subtract.

21.
```
 11,276
− 4,864
  6,412
```

22.
```
 16,006
− 9,859
  6,147
```

23.
```
 676,385
−297,196
 379,189
```

24.
```
 $6,246.25
− 1,457.83
$ 4,788.42
```

Estimate by rounding to the nearest thousand.

25.
```
  9,728   10,000
+ 3,256 + 3,000
         13,000
```

26.
```
 49,500   50,000
−26,438 −26,000
         24,000
```

27.
```
 49,500   50,000
+26,481 +26,000
         76,000
```

28.
```
  8,513    9,000
− 3,198 − 3,000
          6,000
```

29.
```
 36,279   36,000
+29,701 +30,000
         66,000
```

30.
```
 74,937   75,000
−26,481 −26,000
         49,000
```

Circle the letter of the correct answer.

❶
$$\begin{array}{r} 6 \\ + 3 \\ \hline \end{array}$$
a. 7
b. 8
c. 9 (circled)
d. NG

❷
$$\begin{array}{r} 5 \\ 2 \\ + 4 \\ \hline \end{array}$$
a. 10
b. 11 (circled)
c. 12
d. NG

❸ $8 + 9 = n$
$n = ?$
a. 17 (circled)
b. 18
c. 19
d. NG

❹
$$\begin{array}{r} 9 \\ - 9 \\ \hline \end{array}$$
a. 0 (circled)
b. 1
c. 2
d. NG

❺
$$\begin{array}{r} 15 \\ - 8 \\ \hline \end{array}$$
a. 6
b. 8
c. 10
d. NG (circled)

❻ $n + 6 = 11$
$n = ?$
a. 5 (circled)
b. 6
c. 9
d. NG

❼
$$\begin{array}{r} 8 \\ \times 7 \\ \hline \end{array}$$
a. 54
b. 56 (circled)
c. 63
d. NG

❽ $6 \times 4 = n$
$n = ?$
a. 18
b. 24 (circled)
c. 30
d. NG

❾ $3\overline{)18}$
a. 0
b. 6 (circled)
c. 9
d. NG

❿ $n \times 6 = 42$
$n = ?$
a. 5
b. 6
c. 7 (circled)
d. NG

⓫ $8 - 3 \times 2 = n$
$n = ?$
a. 2 (circled)
b. 10
c. 12
d. NG

⓬ 5×10^3
a. 50
b. 500
c. 5,000 (circled)
d. NG

⓭ Give the place value for the 6 in 946,321.
a. ones
b. tens
c. hundreds
d. NG (circled)

STOP

☐ score

Chapter 2 • Cumulative Assessment

page 40

Items	Objectives
1–3	To review basic facts and properties of addition (see pages 1–2)
4–6	To review basic facts and properties of subtraction (see pages 3–4)
7–8	To review basic facts and properties of multiplication (see pages 5–6)
9–10	To review basic facts and properties of division (see pages 7–8)
11	To simplify an expression having two or more operations (see pages 11–12)
12	To simplify numbers written in exponential form (see pages 9–10)
13	To identify place value through hundred thousands (see pages 19–20)

Alternate Cumulative Assessment

Circle the letter of the correct answer.

1.
$$\begin{array}{r} 9 \\ + 2 \\ \hline \end{array}$$
a 11 (circled)
b 12
c 13
d NG

2.
$$\begin{array}{r} 2 \\ 6 \\ + 7 \\ \hline \end{array}$$
a 14
b 15 (circled)
c 16
d NG

3. $4 + 8 = n$
$n = ?$
a 10
b 11
c 12 (circled)
d NG

4.
$$\begin{array}{r} 14 \\ - 7 \\ \hline \end{array}$$
a 6
b 7 (circled)
c 8
d NG

5.
$$\begin{array}{r} 9 \\ - 1 \\ \hline \end{array}$$
a 8 (circled)
b 9
c 10
d NG

6. $n + 4 = 11$
$n = ?$
a 6
b 7 (circled)
c 8
d NG

7.
$$\begin{array}{r} 7 \\ \times 4 \\ \hline \end{array}$$
a 24
b 28 (circled)
c 32
d NG

8. $3 \times 8 = n$
$n = ?$
a 18
b 21
c 24 (circled)
d NG

9. $9\overline{)72}$
a 7
b 8 (circled)
c 9
d NG

10. $n \times 5 = 45$
$n = ?$
a 7
b 8
c 9 (circled)
d NG

11. $15 - 6 \times 2 = n$
$n = ?$
a 3 (circled)
b 7
c 18
d NG

12. 9×10^2
a 90
b 900 (circled)
c 9,000
d NG

T40

3-1 Multiplying by Multiples of 10

pages 41–42

1 Getting Started

Objective
- To multiply by multiples of 10; 100; 1,000; and 10,000

Materials
*multiplication fact cards

Warm Up • Mental Math
Have students find each product.
1. $5 \times 2 \times 4$ (40)
2. $7 \times 7 \times 1$ (49)
3. $9 \times 1 \times 7$ (63)
4. $2 \times 4 \times 6$ (48)
5. $2 \times 3 \times 3$ (18)
6. $4 \times 4 \times 2$ (32)

Warm Up • Pencil and Paper
Have students write all whole numbers less than 999,999,999 whose three greatest place-value digits are 3, 0, 8, in that order, and whose other digits, if any, are zero.
(308,000,000; 30,800,000; 3,080,000; 308,000; 30,800; 3,080; 308)

Multiply Whole Numbers

CHAPTER **3**

Lesson 3-1

Multiplying by Multiples of 10

It is time for Kiku to trade in the pennies she has saved. Each box holds 300 pennies. How many pennies does Kiku have to wrap?

We want to know how many pennies Kiku has.

We know that Kiku has __6__ boxes of pennies and each box contains __300__ pennies. To find the total number of pennies, we multiply the number in each box by the number of boxes.

We multiply __300__ by __6__.

Consider the pattern in these multiplication sentences.

$6 \times 1 = 6$ $6 \times 3 = $ __18__

$6 \times 10 = 60$ $6 \times 30 = $ __180__

$6 \times 100 = 600$ $6 \times 300 = $ __1,800__

Kiku has __1,800__ pennies to wrap.

We can use multiplication facts and properties to find the product of pairs of numbers such as 60 and 30 or 20 and 800.

$60 \times 30 = 6 \times 10 \times 3 \times 10$ $20 \times 800 = 2 \times 10 \times 8 \times 100$

$\qquad = 18 \times 100$ $\qquad = 16 \times 1,000$

$\qquad = $ __1,800__ $\qquad = $ __16,000__

Getting Started

Complete the pattern.

1. $4 \times 3 = $ __12__
   $4 \times 30 = $ __120__
   $4 \times 300 = $ __1,200__
   $4 \times 3,000 = $ __12,000__
   $4 \times 30,000 = $ __120,000__

Multiply.

2. $20 \times 30 = $ __600__

3. $9 \times 200 = $ __1,800__

4. $6 \times 5,000 = $ __30,000__

5. $200 \times 500 = $ __100,000__

6. $9,000 \times 3 = $ __27,000__

7. $30 \times 4,000 = $ __120,000__

8. $50 \times 600 = $ __30,000__

9. $7,000 \times 600 = $ __4,200,000__

---

2 Teach

Introduce the Lesson Have a student read the problem and tell what is to be solved. (the total number of pennies Kiku has to wrap) Ask what information is given in the problem and picture. (Kiku has 6 boxes of pennies and each box contains 300 pennies.) Have students complete the sentences as they read aloud with you to solve the problem. Discuss using multiplication facts, multiplication properties, and patterns to find products.

Develop Skills and Concepts Write the following exercises on the board:

$7 \times 1 = 7$	$4 \times 6 = 24$
$7 \times 10 = 70$	$4 \times 60 = 240$
$7 \times 100 = 700$	$4 \times 600 = 2,400$
$7 \times 1,000 = 7,000$	$4 \times 6,000 = 24,000$

As you work through the problems, ask students to compare the number of zeros in the second factor to the number of zeros in the product. Remind students that:

- When multiplying by 10, write 0 in the ones place, and multiply by the number of tens

- When multiplying by 100, write 0 in the ones and tens place, and multiply by the number of hundreds

- When multiplying by thousands, write in the ones, tens, and hundreds place, and multiply by the number of thousands

3 Practice

Have students complete all the exercises. Remind students to watch the place value of digits.

Practice

Complete each pattern.

1. $7 \times 2 = \underline{14}$
$7 \times 20 = \underline{140}$
$7 \times 200 = \underline{1,400}$
$7 \times 2,000 = \underline{14,000}$
$7 \times 20,000 = \underline{140,000}$

2. $8 \times 5 = \underline{40}$
$8 \times 50 = \underline{400}$
$8 \times 500 = \underline{4,000}$
$8 \times 5,000 = \underline{40,000}$
$8 \times 50,000 = \underline{400,000}$

3. $5 \times 5 = \underline{25}$
$5 \times 50 = \underline{250}$
$50 \times 50 = \underline{2,500}$
$500 \times 50 = \underline{25,000}$
$5,000 \times 50 = \underline{250,000}$

Multiply.

4. $400 \times 7 = \underline{2,800}$

5. $30 \times 50 = \underline{1,500}$

6. $90 \times 200 = \underline{18,000}$

7. $3,000 \times 30 = \underline{90,000}$

8. $6 \times 10,000 = \underline{60,000}$

9. $400 \times 800 = \underline{320,000}$

10. $90 \times 900 = \underline{81,000}$

11. $200 \times 3,000 = \underline{600,000}$

12. $5,000 \times 200 = \underline{1,000,000}$

13. $800 \times 9 = \underline{7,200}$

14. $80 \times 60 = \underline{4,800}$

15. $90 \times 2,000 = \underline{180,000}$

Problem Solving

Solve each problem.

16. A fuel truck can hold 5,000 gallons of gasoline. How many gallons can be delivered by the truck in 3 trips?
15,000 gallons

17. A book has 20 pages with 30 lines each. Each line contains 60 characters of type. How many characters of type are there in the book?
36,000 characters

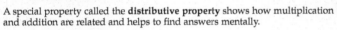

Now Try This!

A special property called the **distributive property** shows how multiplication and addition are related and helps to find answers mentally. **It's Algebra!**

$$\begin{array}{l} 7 \text{ groups of } 4 \ (7 \times 4) \\ \underline{+ \ 3 \text{ groups of } 4 \ (3 \times 4)} \\ 10 \text{ groups of } 4 \ (10 \times 4) \end{array} \quad \text{or} \quad 10 \times 4 = (7 \times 4) + (3 \times 4)$$

Solve each equation.

1. $7 \times 10 = (7 \times 2) + (7 \times \underline{8})$

2. $(9 \times 2) + (9 \times 8) = \underline{9} \times 10$

3. $10 \times 6 = (3 \times \underline{6}) + (7 \times \underline{6})$

4. $(4 \times 5) + (6 \times 5) = 10 \times \underline{5}$

5. $8 \times 10 = (8 \times 1) + (8 \times \underline{9})$

6. $10 \times \underline{3} = (4 \times 3) + (6 \times 3)$

42

Lesson 3-1 • Multiplying by Multiples of 10

Now Try This! Have students use the Distributive Property to find the product of $6 \times (40 + 6)$.

$6 \times (40 + 6) = (6 \times 40) + (6 \times 6) = 240 + 36 = 276$

Then, have students use the Distributive Property to complete Exercises 1 through 6.

4 Assess

Have students find the product of $4,000 \times 500$. (2,000,000) Ask students to explain why there are six 0s in this product instead of five 0s. (Possible answer: The product of 4×5 is 20. Then, you need 5 more 0s in the product for the thousand and the hundreds in the factors.)

For Mixed Abilities

Common Errors • Intervention

Some students may write an incorrect number of zeros when multiplying multiples of ten times multiples of ten. Have them rewrite a problem such as $300 \times 7,000$ in the following manner to help them see what the correct number of zeros should be.

$$300 \times 7,000$$
$$(3 \times 100) \times (7 \times 1,000)$$
$$(3 \times 7) \times (100 \times 1,000)$$
$$21 \times 100,000$$
$$= 2,100,000$$

Enrichment • Number Sense

Have students name 2 different pairs of factors for each product. (Answers may vary.)

1. 6,300,000
($9 \times 700,000$; $90 \times 70,000$)

2. 490,000
(700×700; $7 \times 70,000$)

3. 2,400,000
($8,000 \times 300$; $40 \times 60,000$)

4. 36,000,000
($900 \times 40,000$; $6,000 \times 6,000$)

More to Explore • Number Sense

Have students follow these steps to find the secret number.

1. Start with any number.
2. Add the next larger whole number.
3. Add 9 to that sum.
4. Divide that sum by 2.
5. Subtract the original number.

What is the resulting number? (5)

ESL/ELL STRATEGIES

Explain that the word *property* has a special meaning in the field of math. A mathematical property describes how a certain operation works. One property of multiplication, the Commutative Property, is that numbers can be multiplied in any order and the answer will be the same.

T42

3-2 Estimating Products

pages 43–44

1 Getting Started

Objective
• To estimate products

Warm Up • Mental Math
Have students find each product.

1. 30×50 (1,500)
2. 700×8 (5,600)
3. 90×600 (54,000)
4. $10 \times 4,000$ (40,000)
5. 600×50 (30,000)
6. 37×100 (3,700)
7. $20 \times 50 \times 80$ (80,000)
8. $400 \times 5 \times 90$ (180,000)

Warm Up • Pencil and Paper
Have students round each number to the nearest ten.

1. 43 (40) 2. 38 (40)
3. 65 (70) 4. 82 (80)

Have students round each number to the nearest hundred.

1. 395 (400) 2. 754 (800)
3. 615 (600) 4. 462 (500)

Have students round each number to the nearest thousand.

1. 2,417 (2,000) 2. 5,721 (6,000)
3. 4,843 (5,000) 4. 7,526 (8,000)

Name _____

Estimating Products

During the big holiday sale, Brewster's Shirt Shop sold 876 shirts. Estimate the amount of money taken in from the sale of shirts.

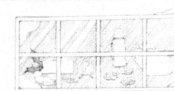

All Shirts $11.79

We need to estimate how much money the store took in from the sale of shirts.

We know there were __876__ shirts sold, and each cost __$11.79__.

To find the amount of money received, we multiply the number of shirts sold by the cost of each shirt. Since we want an estimated product, we round each factor to its greatest place value. Then we multiply the rounded numbers.

$$\underset{\text{number sold}}{876} \times \underset{\text{cost of shirt}}{\$11.79}$$

$$\underset{\text{rounded}}{900} \times \underset{\text{rounded}}{\$10} = \underline{\$9,000}$$

The shirt shop took in about __$9,000__.

Getting Started

Round the factors and estimate each product.

1. 486×33
   $500 \times 30 = 15,000$
2. 615×295
   $600 \times 300 = 180,000$
3. $79 \times 3,415$
   $80 \times 3,000 = 240,000$
4. $350 \times 7,968$
   $400 \times 8,000 = 320,000$
5. $415 \times \$18.75$
   $400 \times \$20 = \$8,000$
6. $563 \times \$203.40$
   $600 \times \$200 = \$120,000$

Copy and multiply. Then, estimate to check.

7. $375 \times \$86$
   $32,250
8. 916×493
   451,588
9. $56 \times 5,042$
   282,352

2 Teach

Introduce the Lesson Have a student read the problem and tell what is to be solved. (about how much money the store makes from the sale of 876 shirts) Ask what information is given in the problem and the picture. (There are 876 shirts sold at $11.79 per shirt.)

Have students complete the sentences as they read aloud with you to solve the problem. Tell students that one way to estimate a product is to round and then multiply. Point out that this simplifies the problem so students can multiply by multiples of 10, 100, and 1,000.

Develop Skills and Concepts Review rounding to the nearest ten, hundred, and thousand. Remind students that the purpose of rounding each factor is to get numbers that can be multiplied easily to provide an estimate. Point out that in this lesson, students will round each factor to its greatest place value. Emphasize

that the estimated product will be close enough to the exact answer to use as a check to determine whether or not a computed result is reasonable.

3 Practice

Have students complete all the exercises. Remind students to include units as part of their answers for the word problems.

T43

Practice

Round the factors and estimate each product.

1. 476×83
   $500 \times 80 = 40,000$

2. 42×356
   $40 \times 400 = 16,000$

3. 58×472
   $60 \times 500 = 30,000$

4. 396×67
   $400 \times 70 = 28,000$

5. $83 \times \$9.68$
   $80 \times \$10 = \$800$

6. $\$6.87 \times 412$
   $\$7 \times 400 = \$2,800$

7. $95 \times 6,296$
   $100 \times 6,000 = 600,000$

8. $5,065 \times 128$
   $5,000 \times 100 = 500,000$

9. $7,967 \times 684$
   $8,000 \times 700 = 5,600,000$

10. $396 \times \$41.56$
    $400 \times \$40 = \$16,000$

11. $787 \times 4,973$
    $800 \times 5,000 = 4,000,000$

12. $850 \times \$78.45$
    $900 \times \$80 = \$72,000$

13. $3,907 \times 218$
    $4,000 \times 200 = 800,000$

14. $\$25.83 \times 896$
    $\$30 \times 900 = \$27,000$

15. $\$286.36 \times 476$
    $\$300 \times 500 = \$150,000$

16. $526 \times 9,738$
    $500 \times 10,000 = 5,000,000$

17. $\$321.15 \times 610$
    $\$300 \times 600 = \$180,000$

18. $721 \times 8,801$
    $700 \times 9,000 = 6,300,000$

Copy and multiply. Then, estimate to check.

19. $\$85.63 \times 48$
    $\$4,110.24$

20. 681×757
    $515,517$

21. $9,008 \times 62$
    $558,496$

22. $7,042 \times 508$
    $3,577,336$

23. $\$4,983 \times 78$
    $\$388,674$

24. $2,389 \times 236$
    $563,804$

25. $98 \times 6,473$
    $634,354$

26. 746×815
    $607,990$

27. $\$589.56 \times 36$
    $\$21,224.16$

Problem Solving

Solve each problem.

28. One square unit on the map at the planetarium shows 2,315 stars. Estimate how many stars 89 square units will show.
    180,000 stars

29. Each share of Ace Airlines stock sells for $7.87. Estimate how much 48,976 shares will cost.
    $400,000

30. Each container on a cargo ship can hold up to 28,475 pounds of freight. Estimate the maximum weight the cargo ship can carry in 985 containers.
    30,000,000 pounds

31. In March, the record store sold 3,885 sale albums at $11.79 each. The shop also sold 2,956 full-price albums at $13.49 each. About how much did the shop make from the sale of these albums?
    $87,000

44

Lesson 3-2 • Estimating Products

---

4 Assess

Have students estimate the cost of 359 shirts at $17.39 each. ($8,000)

For Mixed Abilities

Common Errors • Intervention

Some students may estimate products incorrectly because they round incorrectly. First, have them work with partners to round each of the following numbers to its greatest place.

47 (50)   729 (700)   4,523 (5,000)
72 (70)   351 (400)   2,065 (2,000)
69 (70)   917 (900)   3,531 (4,000)

Then, have them take turns estimating the following products.

47	351
$\times 72$	$\times 69$
(3,500)	(28,000)

729	4,523
$\times 917$	$\times 2,065$
(630,000)	(10,000,000)

Enrichment • Estimation

Have students estimate:

1. the number of hours in a year
   $(24 \times 365 = 20 \times 400 = 8,000)$

2. the number of minutes in a year
   $(60 \times 24 \times 365 = 60 \times 20 \times 400 = 480,000)$

3. the number of minutes they sleep in a year (Answers will vary.)

More to Explore • Number Sense

Have students complete the following prime number square, where all the numbers are prime except 1. Remind them that the sums must be the same across, down, and diagonally.

67	1	43
(13)	(37)	(61)
31	(73)	(7)

ESL/ELL STRATEGIES

When introducing estimating, explain in simple terms the instruction ...round each factor to its greatest place value. For example: If you have a hundreds number like 562, your estimate is the closest whole hundreds number, which is 600.

T44

3-3 Multiplying by a 1-Digit Factor

pages 45–46

1 Getting Started

Objective
• To multiply 2- or 3-digit numbers by 1-digit factors

Materials
*multiplication fact cards

Warm Up • Mental Math
Have students:

1. count backward 4 times by 11s, starting at 53 (53, 42, 31, 20, 9)
2. count forward 5 times by 13s, starting at 36 (36, 49, 62, 75, 88, 101)
3. count backward 5 times by 20s, starting at 401 (401, 381, 361, 341, 321, 301)
4. count forward 4 times by 7s, starting at 516 (516, 523, 530, 537, 544)

Warm Up • Pencil and Paper
Have students write >, <, or = to complete each statement.

1. 3 × 80 (=) 40 × 6
2. 50 × 40 (<) 200 × 100
3. 600 × 50 (>) 30 × 100
4. 400 × 400 (=) 80 × 2,000
5. 90 × 300 (>) 270 × 10
6. 67 × 1,000 (=) 670 × 100
7. 500 × 100 (>) 250 × 20
8. 30 × 4 × 25 (<) 3,000 × 100

Name _____

Multiplying by a 1-Digit Factor

On Saturday, there are 5 shows at the movie theater. The ushers clear the theater after each show. Last Saturday, the theater was full for all performances. How many people saw the movie?

We want to find the total movie attendance for Saturday.

We know that __348__ people saw the movie at each showing, and there were __5__ shows.

To find the total attendance, we multiply the number of viewers of each showing by the number of times the film was shown. We multiply __348__ by __5__.

Multiply the ones. Rename if needed.	Multiply the tens. Add any extra tens. Rename if needed.	Multiply the hundreds. Add any extra tens. Rename if needed.
$\begin{array}{r} 4 \\ 348 \\ \times\ 5 \\ \hline 0 \end{array}$	$\begin{array}{r} 2\ 4 \\ 348 \\ \times\ 5 \\ \hline 40 \end{array}$	$\begin{array}{r} 2 \\ 348 \\ \times\ 5 \\ \hline 1{,}740 \end{array}$

The total Saturday attendance was __1,740__ people.

Getting Started

Multiply.

1. $\begin{array}{r} 94 \\ \times\ 7 \\ \hline 658 \end{array}$
2. $\begin{array}{r} 467 \\ \times\ 4 \\ \hline 1{,}868 \end{array}$
3. $\begin{array}{r} 307 \\ \times\ 6 \\ \hline 1{,}842 \end{array}$
4. $\begin{array}{r} 59 \\ \times\ 8 \\ \hline 472 \end{array}$
5. $\begin{array}{r} 915 \\ \times\ 3 \\ \hline 2{,}745 \end{array}$

Write >, <, or = in each circle.

6. 79 × 3 (<) 248
7. 8 × 576 (>) 4,600
8. 7 × 340 (=) 2,380

Copy and multiply.

9. 39 × 4
156
10. 6 × 705
4,230
11. 657 × 8
5,256

---

2 Teach

Introduce the Lesson Have a student read the problem and tell what is to be solved. (the total number of people who saw the movie) Ask what information is given in the problem and the picture. (348 people saw the movie each time; the movie was shown 5 times.) Have students complete the sentences as they read aloud with you. Work through the renaming needed to compute the product. Remind students that multiplication is called for when we want to find the total number of equal-sized groups.

Develop Skills and Concepts As you discuss the model, be sure students understand they should begin by multiplying the ones and showing 4 tens and 0 ones. Then, they should multiply the tens (5 × 4 tens = 20 tens) and add the 4 tens. Finally, students should multiply the hundreds (5 × 3 hundreds) and add the 2 hundreds. Emphasize that students should multiply the digit in the

tens or hundreds column before they add. Continue developing this skill with these problems worked vertically on the board:

1. 65 × 3 (195)
2. 39 × 6 (234)
3. 149 × 5 (745)
4. 360 × 70 (25,200)

3 Practice

Have students complete all the exercises. Point out that students need to provide information to solve Problem 40.

T45

Practice

Multiply.

1. $\times\ \underline{\begin{array}{r}73\\6\end{array}}$ 438	2. $\times\ \underline{\begin{array}{r}118\\7\end{array}}$ 826	3. $\times\ \underline{\begin{array}{r}296\\3\end{array}}$ 888	4. $\times\ \underline{\begin{array}{r}56\\4\end{array}}$ 224	5. $\times\ \underline{\begin{array}{r}708\\5\end{array}}$ $3,540$
6. $\times\ \underline{\begin{array}{r}296\\7\end{array}}$ $2,072$	7. $\times\ \underline{\begin{array}{r}49\\9\end{array}}$ 441	8. $\times\ \underline{\begin{array}{r}375\\8\end{array}}$ $3,000$	9. $\times\ \underline{\begin{array}{r}961\\2\end{array}}$ $1,922$	10. $\times\ \underline{\begin{array}{r}90\\8\end{array}}$ 720
11. $\times\ \underline{\begin{array}{r}352\\6\end{array}}$ $2,112$	12. $\times\ \underline{\begin{array}{r}58\\7\end{array}}$ 406	13. $\times\ \underline{\begin{array}{r}496\\8\end{array}}$ $3,968$	14. $\times\ \underline{\begin{array}{r}749\\3\end{array}}$ $2,247$	15. $\times\ \underline{\begin{array}{r}80\\9\end{array}}$ 720

Write >, <, or = in each circle.

16. $59 \times 3 \; \textcircled{<} \; 180$

17. $6 \times 184 \; \textcircled{=} \; 1,104$

18. $159 \; \textcircled{>} \; 22 \times 7$

19. $9 \times 16 \; \textcircled{=} \; 18 \times 8$

20. $7 \times 125 \; \textcircled{>} \; 109 \times 8$

21. $9 \times 189 \; \textcircled{<} \; 4 \times 426$

22. $6 \times 47 \; \textcircled{<} \; 32 \times 9$

23. $4 \times 156 \; \textcircled{<} \; 140 \times 5$

24. $8 \times 282 \; \textcircled{>} \; 5 \times 415$

Copy and multiply.

25. 15×7
105

26. 8×139
$1,112$

27. 6×47
282

28. 573×3
$1,719$

29. 219×8
$1,752$

30. 9×307
$2,763$

31. 6×98
588

32. 43×2
86

33. 439×4
$1,756$

34. 518×5
$2,590$

35. 6×847
$5,082$

36. 8×78
624

Problem Solving

Solve each problem.

37. Harry collects stamps. Each stamp book holds 248 stamps. How many stamps does Harry have in 6 books?
1,488 stamps

38. Josie works in a biology laboratory making microscope slides. She has 695 slides to make for one project. She makes 85 slides a week. By the end of 6 weeks, how many slides does she have left to make?
185 slides

39. Mr. Li's office is 26 miles from his house. How many miles does Mr. Li drive to and from work, in a 5-day week?
260 miles

40. A newspaper columnist wrote a weekly article for 8 years. How many columns did she write?
416 columns

Lesson 3-3 • Multiplying by a 1-Digit Factor

---

4 Assess

Without doing the algorithm, have students give the units digit in the product for each exercise below.

1. 6×847 (2)   2. 345×7 (5)   3. 538×8 (4)

For Mixed Abilities

Common Errors • Intervention

Some students may rename correctly but then forget to add the renamed numbers after multiplying.

Incorrect	Correct
$\begin{array}{r}\,^{2\ 4}\\148\\\times\ \ 5\\\hline 500\end{array}$	$\begin{array}{r}\,^{2\ 4}\\148\\\times\ \ 5\\\hline 740\end{array}$

Have students use a place-value form like the following to chart the steps.

	100s	10s	1s
5×8 ones		4	0
5×4 tens	2	0	0
5×1 hundred	5	0	0
Total product	7	4	0

Enrichment • Number Sense

Have students solve each problem.

1. $5^4 \times 7$ (4,375)

2. $7^3 \times 9$ (3,087)

3. $8^3 \times 4$ (2,048)

4. $6^4 \times 3$ (3,888)

5. $4^5 \times 6$ (6,144)

6. $7^4 \times 2$ (4,802)

More to Explore • Logic

Duplicate the following for students to solve. Remind them to look beyond an obvious answer.

Every day of each year at 12:00 noon a bus leaves Washington, D.C., for Seattle, Washington. At the same time, a bus leaves Seattle bound for Washington. The buses travel the same route. Each trip takes exactly seven days. (168 hours)

If you are a passenger on Monday's noon bus leaving Washington, D.C., how many buses from Seattle will you have met by the time you arrive in Seattle? (15 buses—1 for each of the 7 days before you leave, 1 for each of the days of your trip, and 1 that is arriving as you are leaving.)

T46

pages 47–48

1 Getting Started

Objective
• To multiply whole numbers or money by 1-digit factors

Materials
index cards

Warm Up • Mental Math
Have students tell how many minutes are in each period of time.

1. 3 hours (180 min)
2. 5 hours (300 min)
3. 9 hours (540 min)
4. 4 hours (240 min)
5. $2\frac{1}{2}$ hours (150 min)
6. 8 hours (480 min)
7. $1\frac{1}{2}$ hours (90 min)
8. 12 hours (720 min)

Warm Up • Pencil and Paper
Have students find each product.

1. 7 × 246 (1,722)
2. 6 × 158 (948)
3. 9 × 328 (2,952)
4. 4 × 677 (2,708)
5. 8 × 539 (4,312)
6. 3 × 987 (2,961)

Name _____

Multiplying up to 6-Digit Numbers

The students at Hamilton High School sold 3,416 raffle ticket books to raise money for band instruments. Each ticket cost $2.35, and there are 6 tickets in each book. How many raffle tickets were sold? How much did each book cost?

We want to know the number of raffle tickets sold altogether. We know the students sold __3,416__ books; there are __6__ tickets in each book, and each ticket costs __$2.35__.

To find the total number of tickets, we multiply the number of tickets in each book by the number of books. We multiply __3,416__ by __6__.

Multiply the ones.	Multiply the tens.	Multiply the hundreds.	Multiply the thousands.
$\begin{array}{r} 3 \\ 3,416 \\ \times \quad 6 \\ \hline 6 \end{array}$	$\begin{array}{r} 3 \\ 3,416 \\ \times \quad 6 \\ \hline 96 \end{array}$	$\begin{array}{r} 2 \\ 3,416 \\ \times \quad 6 \\ \hline 496 \end{array}$	$\begin{array}{r} 2 \\ 3,416 \\ \times \quad 6 \\ \hline 20,496 \end{array}$

The students sold __20,496__ raffle tickets.

To find the total cost of one book, we multiply the cost of one ticket by the number of tickets in the book. We multiply __$2.35__ by __6__.

Multiply money the same way you multiply whole numbers. Remember to place the dollar sign and decimal point in the product.

$\begin{array}{r} \$2.35 \\ \times \quad 6 \\ \hline \$14.10 \end{array}$

Each book of tickets costs __$14.10__.

Getting Started

Multiply.

1. $\begin{array}{r} \$9.27 \\ \times \quad 5 \\ \hline \$46.35 \end{array}$

2. $\begin{array}{r} \$567.25 \\ \times \quad 4 \\ \hline \$2,269.00 \end{array}$

Copy and multiply.

3. 9 × 592,403
   5,331,627

4. 6 × $19.56
   $117.36

---

2 Teach

Introduce the Lesson Have students read the problem and identify the questions being asked. (the number of raffle tickets sold and the cost of each book of tickets) Ask students to identify the known facts. (The students sold 3,416 raffle books; there are 6 tickets in each book; and each ticket costs $2.35.) As students read aloud with you to solve the problem, explain that there are two separate operations necessary. Remind students to include the dollar sign and decimal point in their answer.

Develop Skills and Concepts Explain to students that in the model they were multiplying to the thousands place, renaming as they have been doing. Explain that multiplying money is just like multiplying whole numbers with the inclusion of the dollar sign and decimal (cents) point in the product. Have students work the following problems at the board:

1. $\begin{array}{r} 7,364 \\ \times 5 \\ \hline (36,820) \end{array}$

2. $\begin{array}{r} \$4.85 \\ \times 7 \\ \hline (\$33.95) \end{array}$

3. $\begin{array}{r} \$35.07 \\ \times 6 \\ \hline (\$210.42) \end{array}$

4. $\begin{array}{r} 4,291 \\ \times 8 \\ \hline (34,328) \end{array}$

3 Practice

Have students complete all the exercises. Remind students to align the dollars and cents when necessary.

Practice

Multiply.

1. 1,738 × 4 6,952	2. $9.57 × 7 $66.99	3. 8,406 × 3 25,218	4. $46.37 × 9 $417.33	5. $125.36 × 6 $752.16
6. 52,640 × 8 421,120	7. $247.48 × 2 $494.96	8. 87,396 × 5 436,980	9. $312.60 × 7 $2,188.20	10. 43,257 × 4 173,028
11. 126,348 × 9 1,137,132	12. $3,274.13 × 8 $26,193.04	13. 509,412 × 2 1,018,824	14. $421,212 × 6 $2,527,272	15. $900.48 × 5 $4,502.40
16. 521,465 × 8 4,171,720	17. $9,215.63 × 3 $27,646.89	18. 757,462 × 7 5,302,234	19. $983,415 × 5 $4,917,075	20. $803.75 × 6 $4,822.50

Copy and multiply.

21. 5 × 7,961
39,805
22. $43.46 × 8
$347.68
23. 3,975 × 2
7,950
24. 8,658 × 4
34,632
25. 3 × 6,006
18,018
26. 9 × $63.47
$571.23
27. 5 × 16,743
83,715
28. 7 × $148.57
$1,039.99
29. 83,925 × 8
671,400
30. $423.15 × 3
$1,269.45
31. 72,456 × 2
144,912
32. 6 × $5,246.15
$31,476.90
33. 8 × 626,248
5,009,984
34. 4 × $6,247.12
$24,988.48
35. 7 × 327,248
2,290,736
36. 9 × 528,751
4,758,759
37. 5 × $8,493.49
$42,467.45
38. 6 × 857,489
5,144,934

Problem Solving

Solve each problem.

39. Wimona earns $4.27 an hour. On Saturday, Wimona worked 6 hours. How much did she earn that day?
$25.62

40. A truck carries 11,746 kilograms of freight each trip. How many kilograms of freight does the truck carry on 6 trips?
70,476 kilograms

41. Sonar waves travel at 1,410 meters per second in water. It takes 9 seconds for sonar waves to travel between two submarines. How far apart are the submarines?
12,690 meters

42. Boneless chicken costs $2.16 a pound. Mr. Irons bought 6 pounds of boneless chicken. He gave the clerk a twenty-dollar bill. How much change did Mr. Irons receive?
$7.04

Lesson 3-4 • Multiplying up to 6-Digit Numbers

4 Assess

Write

419,235
× 4

on the board. Without multiplying, ask students to predict where regrouping will occur. (4 × 5 ones; 4 × 3 tens; 4 × 9 thousands; 4 × 4 hundred thousands)

For Mixed Abilities

Common Errors • Intervention
Some students may answer incorrectly because they have difficulty keeping the digits aligned. Have these students do their work on grid paper or lined paper turned sideways so the lines are vertical and help students keep the digits properly aligned.

Enrichment • Number Sense
Have students write the missing numbers in each multiplication problem.

1. 4(5)2
× 6
2,712

2. 3(9)56
× 8
31,648

3. 26(3)7
× 5
13,185

4. 2(9)(9)2
× 6
17,952

5. 3(8)5(2)
× 7
26,964

6. (3)4(0)3
× 9
30,627

More to Explore • Measurement
Have students bring measuring tapes to class. Have them work in pairs to find each other's body measurements for the following: size of foot, size of hand-span, circumference of head, length of arm, length of lower leg, and length of spine from base of neck to waist.

- List the measurements on the board and have students calculate the averages for each body part.

- Then, have students draw "The Average Sixth Grader" using the average body measurements of the class, and filling in proportionate parts for the body measurements not taken. Give students large sheets of paper, to allow them to draw life-sized pictures.

- Have them draw faces and clothing on their figures and display their creations around the classroom.

3-5 Multiplying by Tens

pages 49–50

1 Getting Started

Objective
- To multiply a whole number by tens

Warm Up • Mental Math
Have students simplify each expression.
1. $16 - 5 + 1$ (12)
2. $39 - 2 + 5$ (42)
3. $41 - 4 + 7$ (44)
4. $25 - 6 + 2$ (21)
5. $12 - 5 + 1$ (8)
6. $9 - 2 + 5$ (12)
7. $30 - 4 + 7$ (33)
8. $15 - 6 + 2$ (11)

Warm Up • Pencil and Paper
Have students write =, <, or > to compare each pair of factors.
1. 6×572 (>) 5×576
2. 7×643 (<) 9×613
3. 4×856 (=) 8×428
4. 5×717 (>) 3×919
5. 8×435 (<) 9×392

Multiplying by Tens

Mr. Schwarz is a buyer for a large electronics store. One month he bought 30 television sets. How much did Mr. Schwarz's company pay for the sets?

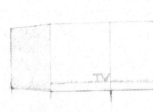

We want to know the total cost of the television sets.

We know each set costs __$215__, and Mr. Schwarz bought __30__ sets.

To find the total money spent, we multiply the cost of each set by the number of sets.

We multiply __$215__ by __30__.

Write 0 in the ones place.	Multiply by the digit in the tens place.
$215 × 30 0	$215 × 30 $6,450

Mr. Schwarz's company paid __$6,450__ for the television sets.

Getting Started

Multiply.

1. 76
× 20
1,520
2. 84
× 30
2,520
3. 176
× 70
12,320
4. $2.20
× 50
$110.00
5. 74,368
× 40
2,974,720

Copy and multiply.

6. 93×60
5,580
7. $40 \times 9,127$
365,080
8. 12.06×90
$1,085.40

Lesson 3-5 • Multiplying by Tens

49

---

2 Teach

Introduce the Lesson Have students read the problem, tell what is to be solved (the total cost of 30 television sets), and what information is provided. (Each television costs $215 and Mr. Schwarz bought 30 sets.) Have students continue reading, complete the sentences, and work through the model problem.

Develop Skills and Concepts Be sure students understand that when multiplying by a multiple of 10, they first write 0 in the ones place of the product and then multiply by the number of tens. Continue developing this skill by having students work the following problems at the board:

1. 87
× 50
(4,350)
2. 426
× 40
(17,040)
3. 5,629
× 70
(394,030)

3 Practice

Have students complete all the exercises. Remind them to use the dollar sign and decimal point in the product where necessary.

Practice

Multiply.

1.	2.	3.	4.	5.
27 × 40 1,080	89 × 70 6,230	217 × 30 6,510	$8.67 × 90 $780.30	615 × 20 12,300

6.	7.	8.	9.	10.
276 × 50 13,800	$9.08 × 80 $726.40	630 × 60 37,800	3,728 × 30 111,840	$49.26 × 90 $4,433.40

11.	12.	13.	14.	15.
526 × 40 21,040	$8.57 × 20 $171.40	750 × 70 52,500	4,382 × 50 219,100	$92.64 × 80 $7,411.20

Copy and multiply.

16. 30 × 68
2,040

17. 57 × 80
4,560

18. 39 × 20
780

19. 156 × 60
9,360

20. 50 × 214
10,700

21. 90 × $6.15
$553.50

22. 856 × 40
34,240

23. $8.58 × 70
$600.60

24. 324 × 60
19,440

Problem Solving

Solve each problem.

25. Mr. James earns $9.26 per hour for the first 40 hours he works each week. For every additional hour, he earns $14.45. Last week he worked 46 hours. How much did he earn?
$457.10

26. A box contains 16 shirts. Each shirt costs $20. How much will 5 boxes cost?
$1,600

(Now Try This!)

Start at *A*. Find all paths to *Z*. You cannot retrace a line.

1. *ABZ*
2. _____

Start at *A*. Find all paths to *Z*.

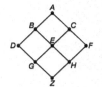

1. *ABDGZ*
2. *ABEGZ*
3. *ABEHZ*
4. *ACEHZ*
5. *ACFHZ*
6. *ACEGZ*
7. *ABECFHZ*
8. *ACEBDGZ*
9. *ACFHEBDGZ*
10. *ABDGECFHZ*

Lesson 3-5 • Multiplying by Tens

Now Try This! Have students trace the various paths with a pencil. Some students might enjoy putting this activity on the overhead projector to share with the class.

4 Assess

Have students find each product.

1.	2.	3.
25 × 30 (750)	25 × 40 (1,000)	25 × 80 (2,000)

For Mixed Abilities

Common Errors • Intervention

When multiplying by a multiple of 10, some students may forget to write the zero in the ones place.

Incorrect	Correct
83	83
× 20	× 20
166	1,660

Correct by having students circle the zero in the multiple and write the zero in the ones place before they multiply by the number of tens.

Enrichment • Number Sense

Divide students into small groups. Write ten 4- and 5-digit numbers on index cards. Place the cards facedown in a box. Then, write the numbers 10, 20, 30, 40, 50, 60, 70, 80, and 90 on other index cards. Place these cards facedown in another box. Have students choose a card from each box, and find and record the product of the two numbers. After five rounds, have students add their products to determine which group has the greatest total.

More to Explore • Number Sense

Have pairs of students prepare a set of cards, each showing numbers $\frac{1}{2}$ through $\frac{11}{12}$. Have students shuffle the cards and place them face down in two piles to play "War." Students each turn over a card, and the one having the greater number keeps the cards. If equivalent cards are turned over, another draw is made and the larger number wins all cards. Have students play until all cards are won or a specified time limit is reached.

pages 51–52

1 Getting Started

Objective
• To multiply whole numbers by 2-digit factors

Warm Up • Mental Math
Have students simplify each expression.

1. $7 + 0$ (7)
2. $15 - 0$ (15)
3. 9×0 (0)
4. 789×1 (789)
5. $64 \div 1$ (64)
6. $0 + 91$ (91)
7. $429 - 0$ (429)
8. $15 \div 1$ (15)
9. 1×46 (46)
10. 0×58 (0)

Warm Up • Pencil and Paper
Have students find each product.

1. 50×30 (1,500)
2. 800×20 (16,000)
3. 700×900 (630,000)
4. 6×49 (294)
5. $9 \times 2,087$ (18,783)
6. 40×72 (2,880)
7. $70 \times 1,368$ (95,760)
8. $3 \times 52,763$ (158,289)

Name _____

Multiplying by a 2-Digit Factor

Jumbo jets make 26 daily flights between Los Angeles and Chicago. If all the seats are occupied on each flight, how many passengers fly between the two cities each day?

We want to know the total number of passengers flying between Chicago and Los Angeles each day.

We know each jumbo jet has a capacity of ___387___ passengers, and there are ___26___ daily flights.

To find the total number of passengers, we multiply the number of passengers on each flight by the number of flights. We multiply ___387___ by ___26___.

Multiply by ones.	Multiply by tens.	Add the partial products.
$\begin{array}{r} {}^{5\,4}387 \\ \times\ \ 26 \\ \hline 2322 \end{array}$ ← 6×387	$\begin{array}{r} {}^{1\,1}387 \\ \times\ \ 26 \\ \hline 2322 \\ 7740 \end{array}$ ← 20×387	$\begin{array}{r} 387 \\ \times\ \ 26 \\ \hline 2\ 322 \\ 7\ 740 \\ \hline 10{,}062 \end{array}$ ← 6×387 ← 20×387 ← 26×387

If all the seats are occupied, ___10,062___ passengers fly between the two cities.

Getting Started

Multiply.

1. $\begin{array}{r} 57 \\ \times\ 34 \\ \hline 1{,}938 \end{array}$
2. $\begin{array}{r} 80 \\ \times\ 47 \\ \hline 3{,}760 \end{array}$
3. $\begin{array}{r} 309 \\ \times\ 74 \\ \hline 22{,}866 \end{array}$
4. $\begin{array}{r} \$4.80 \\ \times\ \ 68 \\ \hline \$326.40 \end{array}$

5. $\begin{array}{r} 3{,}724 \\ \times\ \ 53 \\ \hline 197{,}372 \end{array}$
6. $\begin{array}{r} 6{,}839 \\ \times\ \ 72 \\ \hline 492{,}408 \end{array}$
7. $\begin{array}{r} 4{,}786 \\ \times\ \ 29 \\ \hline 138{,}794 \end{array}$
8. $\begin{array}{r} \$137.12 \\ \times\ \ \ \ 28 \\ \hline \$3{,}839.36 \end{array}$

Copy and multiply.

9. 37×428
   15,836
10. $46 \times \$9.51$
    $437.46
11. $29 \times 53,267$
    1,544,743

Lesson 3-6 • Multiplying by a 2-Digit Factor

---

2 Teach

Introduce the Lesson Have a student read the problem and tell what is to be solved. (total number of passengers who fly between Los Angeles and Chicago each day) Ask what information is given in the problem and the picture. (Each jumbo jet can carry 387 passengers and there are 26 daily flights.) Have students complete the sentences to solve the problem. Work through the renaming in the three steps in the model.

Develop Skills and Concepts Explain to students that the multiplication is based on the Distributive Property. The model shows that the product of 26×387 is the sum of the product of $(6 \times 387) + (20 \times 387)$.

Write on the board:

$\begin{array}{r} 387 \\ \times\ \ 6 \\ \hline 2322 \end{array}$ $\begin{array}{r} 387 \\ \times\ 20 \\ \hline 7740 \end{array}$ $\begin{array}{r} 2322 \\ +\ 7740 \\ \hline 10{,}062 \end{array}$ ← 6×387 ← 20×387 ← 26×387

Have students distribute one factor over the other to find the products of:

1. 86×42 $[(86 \times 2) + (86 \times 40) = 172 + 3{,}440 = 3{,}612]$

2. 408×65 $[(408 \times 5) + (408 \times 60) = 2{,}040 + 24{,}480 = 26{,}520]$

3. $3{,}902 \times 27$ $[(3{,}902 \times 7) + (3{,}902 \times 20) = 27{,}314 + 78{,}040 = 105{,}354]$

Using Estimation Have students estimate each product to get a sense of the magnitude of a product before multiplying.

3 Practice

Have students complete all the exercises. Point out they will be using Robin's 18-day travel budget to solve the word problems.

Practice

Multiply.

1. 79
    × 53
    ———
    4,187

2. 96
    × 48
    ———
    4,608

3. 783
    × 86
    ———
    67,338

4. $8.09
    × 72
    ———
    $582.48

5. $6.53
    × 28
    ———
    $182.84

6. 385
    × 36
    ———
    13,860

7. 1,215
    × 66
    ———
    80,190

8. $15.35
    × 92
    ———
    $1,412.20

9. 4,763
    × 81
    ———
    385,803

10. 9,009
    × 45
    ———
    405,405

11. $212.17
    × 73
    ———
    $15,488.41

12. 9,228
    × 59
    ———
    544,452

Copy and multiply.

13. 29 × 48
    1,392

14. 67 × 36
    2,412

15. 52 × 88
    4,576

16. 436 × 29
    12,644

17. 39 × $2.12
    $82.68

18. 476 × 76
    36,176

19. 51 × 285
    14,535

20. 709 × 96
    68,064

21. $9.37 × 68
    $637.16

22. 93 × 1,475
    137,175

23. 28 × 6,208
    173,824

24. 4,800 × 43
    206,400

25. $49.56 × 82
    $4,063.92

26. 75 × 8,272
    620,400

27. 37 × $29.86
    $1,104.82

Problem Solving

Use the chart to solve each problem.

28. How much did Robin budget for recreation altogether?
    $360

29. How much did she budget for motels and meals altogether?
    $1,044

30. At the end of 12 days, Robin had spent $252 on miscellaneous items. According to her budget, how much did she have left in this category?
    $72

Robin's 18-Day Travel Budget	
Air Fare	$425
Motel	$43 per day
Meals	$15 per day
Recreation	$20 per day
Miscellaneous	$18 per day

52   Lesson 3-6 • Multiplying by a 2-Digit Factor

4 Assess

Without multiplying, have students determine the magnitude of each product.

1. 29 × 48 (thousands)

2. 476 × 59 (ten thousands)

3. 3,592 × 78 (hundred thousands)

Have students use a calculator to check.

5 Mixed Review

1. 4^4 (256)
2. 900 × 7,000 (6,300,000)
3. $1,276.27 + $893.27 ($2,169.54)
4. 4,050,382 − 2,175,150 (1,875,232)
5. 25 − 5 × 4 (5)
6. 6,048 × 9 (54,432)
7. 6,056,205 − 948,096 (5,108,109)
8. 6×10^3 (6,000)
9. $8.75 × 5 ($43.75)
10. 42 ÷ 7 (6)

For Mixed Abilities

Common Errors • Intervention

Some students may forget to write a zero in the ones place of the second partial product when multiplying by the tens. Correct by having them rewrite the problem as shown below.

$$41 \times 23 = (40 \times 23) + (1 \times 23)$$
$$= 920 + 23$$
$$= 943$$

Enrichment • Number Sense

Have students find the missing digit to make the products correct.

1. (6)(8)
    × 52
    ———
    3,536

2. 76
    × (4)(9)
    ———
    3,724

3. (8)(8)
    × 37
    ———
    3,256

4. 99
    × (4)(4)
    ———
    4,356

More to Explore • Logic

Duplicate the following for students to solve:

The emperor of China had a problem. He thought one of his 10 official coin makers had been stealing gold by making coins having 1 gram less of gold. But according to law, the emperor could only test the coins of his coin makers in one weighing a year.

The emperor went to his wise advisor for help. The advisor logically arrived at a way of finding the guilty official. The advisor called all 10 coin makers to bring 1 bag of their coins to the palace. In one weighing, the advisor was able to tell the emperor who the thief was. How did he find the thief?

(The advisor took 1 coin from the first coin maker's bag, 2 coins from the second, and so on through 10 coins from the 10th bag. He then weighed all 55 coins at once. If the total was 5 grams less than it should have been, the 5th coin maker was the thief; 3 grams less and the 3rd coin maker was the thief, etc.)

T52

3-7 Multiplying by Hundreds

pages 53–54

1 Getting Started

Objective
• To multiply whole numbers by hundreds

Materials
index cards

Warm Up • Mental Math
Have students name the next three numbers in each number sequence.

1. 528, 530, 532, . . . (534, 536, 538)
2. 1, 2, 4, 8, . . . (16, 32, 64)
3. 12, 24, 36, 48, . . . (60, 72, 84)
4. 1,234; 2,341; 3,412; 4,123, . . . (1,234; 2,341; 3,412)
5. 4, 5, 7, 10, . . . (14, 19, 25)

Warm Up • Pencil and Paper
Have students find each product.

1. 4 × 300 (1,200)
2. 6 × 700 (4,200)
3. 2 × 400 (800)
4. 9 × 600 (5,400)
5. 30 × 247 (7,410)
6. 50 × 85 (4,250)
7. 20 × 706 (14,120)
8. 60 × 3,175 (190,500)

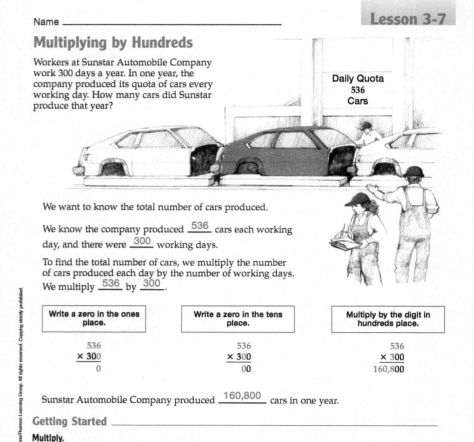

Name _____

Multiplying by Hundreds

Workers at Sunstar Automobile Company work 300 days a year. In one year, the company produced its quota of cars every working day. How many cars did Sunstar produce that year?

Daily Quota
536
Cars

We want to know the total number of cars produced.

We know the company produced __536__ cars each working day, and there were __300__ working days.

To find the total number of cars, we multiply the number of cars produced each day by the number of working days. We multiply __536__ by __300__.

Write a zero in the ones place.	Write a zero in the tens place.	Multiply by the digit in hundreds place.
536 × 300 0	536 × 300 00	536 × 300 160,800

Sunstar Automobile Company produced __160,800__ cars in one year.

Getting Started
Multiply.

1. 239
 × 400
 95,600
2. $6.87
 × 600
 $4,122.00
3. 57,138
 × 700
 39,996,600
4. $16.50
 × 300
 $4,950.00

Copy and multiply.

5. 428 × 200
 85,600
6. 500 × $157.86
 $78,930.00
7. 800 × 3,975
 3,180,000
8. 356 × 700
 249,200
9. 400 × $189.78
 $75,912
10. 600 × 4,539
 2,723,400

Lesson 3-7 • Multiplying by Hundreds

53

2 Teach

Introduce the Lesson Explain the meaning of a quota. Have students read the problem and identify the question being asked. (What is the total number of cars produced in one year?) Ask what information is given. (the number of cars produced each working day; the number of working days in the year) Work through the model problem with students, discussing each step in the multiplication. Remind students of the rules for multiplying by hundreds.

Develop Skills and Concepts Write **200 × 444 = 88,800** vertically on the board. Have students compare the number of zeros in the factor that is a multiple of one hundred (two) to the number of zeros in the product. (two) Students should recognize that when they multiply by hundreds, there would always be zeros in the ones and tens places in the product.

Have students find each product.

1. 76 × 400 (30,400)
2. 329 × 500 (164,500)
3. 4,607 × 200 (921,400)
4. 19,856 × 700 (13,899,200)

Compare this with multiplying by tens. Ask students how many zeros are in the product when multiplying by tens. (at least one) Caution students that there may be more than one zero in the product. For example, the product of 25 × 40 has three zeros. There is a zero in the ones place for 40, and two more zeros because 25 × 4 is 100. 25 × 40 = 1,000.

3 Practice

Have students complete all the exercises. Remind them to include the dollar sign and decimal point when they work with money.

Practice

Multiply.

1. $\begin{array}{r}84\\ \times\ 200\\ \hline 16{,}800\end{array}$	2. $\begin{array}{r}156\\ \times\ 700\\ \hline 109{,}200\end{array}$	3. $\begin{array}{r}782\\ \times\ 500\\ \hline 391{,}000\end{array}$
4. $\begin{array}{r}\$6.37\\ \times\ \ \ 300\\ \hline \$1{,}911.00\end{array}$	5. $\begin{array}{r}2{,}475\\ \times\ \ \ 400\\ \hline 990{,}000\end{array}$	6. $\begin{array}{r}\$73.49\\ \times\ \ \ 600\\ \hline \$44{,}094.00\end{array}$
7. $\begin{array}{r}8{,}400\\ \times\ \ \ 500\\ \hline 4{,}200{,}000\end{array}$	8. $\begin{array}{r}11{,}365\\ \times\ \ \ \ 200\\ \hline 2{,}273{,}000\end{array}$	9. $\begin{array}{r}\$458.73\\ \times\ \ \ \ 600\\ \hline \$275{,}238.00\end{array}$

Copy and multiply.

10. 400×96
38,400

11. 54×700
37,800

12. 138×900
124,200

13. 465×500
232,500

14. $200 \times \$3.76$
$752.00

15. $\$14.56 \times 600$
$8,736.00

Problem Solving

Solve each problem.

16. The music store pays $6.25 for each cassette ordered. How much does the store pay for 600 cassettes?
$3,750

17. A plane ticket for a peak time flight is $426.50. The same ticket for a night flight is $283.25. What is the price difference if 300 passengers exchange the more expensive ticket for the less expensive one?
$42,975.00

(Now Try This!)

Lattice multiplication is a special way to multiply two numbers. Study this lattice multiplication of 785 by 84.

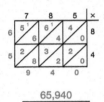

65,940

Use lattice multiplication to multiply 693 by 75.

51,975

Common Errors • Intervention

Some students may forget the zeros in the ones and tens places when they are multiplying by hundreds. Make a set of pattern cards similar to those shown below.

$621 \times 2 = 1{,}242$
$621 \times 20 = (12{,}420)$
$621 \times 200 = (124{,}200)$

Have students work with partners, taking turns to name the products that complete each pattern.

Enrichment • Number Sense

1. Direct students to write two money word problems that involve multiplying a whole number by hundreds. Then, have students exchange their word problems with a friend and solve.

2. Have students find each product.
$50{,}000 \times 412$ (20,600,000)
$9{,}000 \times 596$ (5,364,000)
$3 \times 10^5 \times 83$ (24,900,000)
$8 \times 10^4 \times 677$ (54,160,000)
$(7 \times 10^3) \times (3 \times 10^3) + (2 \times 10^2) + (5 \times 10)$ (21,000,250)

More to Explore • Statistics

Divide the class into groups and ask students to choose a current events topic that concerns them. This issue can be on a national, local, or school level. After groups have chosen their topic, have them create an opinion poll of 10 survey questions about this issue. Have them conduct an opinion poll of classmates, teachers, parents, and so on. When they have gathered their data, tell them to graph the results, choosing the most appropriate type of graph for their data. Have them report their findings to the rest of the class.

Now Try This! Tell students to follow these steps: first, multiply each digit of the right-hand vertical number by each digit of the horizontal number, and write the product in each square. (ex. $8 \times 5 = 40$) Then, add all digits on the diagonal, starting in the bottom right-hand corner (0; $0 + 2 + 2 = 4$; $4 + 4 + 3 + 8 = 19$; etc.), regrouping each time as needed. Write each sum at the end of each diagonal. Then, combine the left-hand digits (65) with the horizontal digits (940) to write the product 65,940.

4 Assess

Have students find each product, highlighting the zeros in each product that are a result of multiplying by hundreds.

1. $\begin{array}{r}25\\ \times\ 300\\ \hline (7{,}500)\end{array}$

2. $\begin{array}{r}25\\ \times\ 400\\ \hline (10{,}000)\end{array}$

3. $\begin{array}{r}25\\ \times\ 800\\ \hline (20{,}000)\end{array}$

3-8 Multiplying by a 3-Digit Factor

pages 55–56

1 Getting Started

Objective
- To multiply whole numbers by 3-digit factors

Warm Up • Mental Math
Have students simplify each expression.

1. $40 \times 20 + 5$ (805)
2. $60 \times 80 - 10$ (4,790)
3. $30 \times 40 + 6$ (1,206)
4. $70 \times 90 - 9$ (6,291)
5. $20 \times 50 - 20$ (980)
6. $60 \times 50 + 100$ (3,100)

Warm Up • Pencil and Paper
Have students find each product.

1. 4×386 (1,544)
2. 2×726 (1,452)
3. 30×437 (13,110)
4. $50 \times 3,642$ (182,100)
5. $40 \times 25,901$ (1,036,040)
6. 400×815 (326,000)
7. 700×562 (393,400)
8. $500 \times 3,774$ (1,887,000)

Name _____

Lesson 3-8

Multiplying by a 3-Digit Factor

National Discount Company keeps track of sales of cameras, computers, and cassettes in all of its stores. If each of these items continues to sell at its average rate, how many cassettes should be sold in one year? The store is closed for 53 days during the year.

National Discount Company	
Item	**Daily Sales Average**
Cameras	210
Cassettes	859
Computers	767

We want to know how many cassettes the company can expect to sell in one year.

We know the daily sales average of cassettes is $\underline{859}$, and there are $\underline{365}$ days in a year.

To find the number sold in a year, we must first subtract the days the store is closed from the number of days in a year. Then, we multiply the number of cassettes sold in one day, by the number of days the store is open.

We subtract $\underline{53}$ from $\underline{365}$. Then, we multiply $\underline{859}$ by $\underline{312}$.

Multiply by the ones.	Multiply by the tens.	Multiply by the hundreds.	Add the partial products.
859	859	859	859
× 312	× 312	× 312	× 312
1718	1718	1718	1 718 ← 2 × 859
	8590	8590	8 590 ← 10 × 859
		257700	257 700 ← 300 × 859
			268,008 ← 312 × 859

The company expects to sell $\underline{268,008}$ cassettes in one year.

Getting Started
Multiply.

1. 586
   × 349
   204,514

2. $18.26
   × 730
   $13,329.80

3. 42,659
   ×   508
   21,670,772

Copy and multiply.

4. 901×756
   681,156

5. $\$76.14 \times 523$
   $39,821.22

6. $56,312 \times 764$
   43,022,368

Lesson 3-8 • Multiplying by a 3-Digit Factor

55

---

2 Teach

Introduce the Lesson Have students read the problem, identify the question, and state the known facts. (About how many cassettes will be sold in one year; about 859 cassettes are sold every day.) Point out that students also need a fact that is not given in the problem. Have students identify the needed fact. (365 days in a year) Have students complete the sentences as they read aloud with you.

As students work through the model, emphasize that multiplying by 3 digits merely extends what they already know. Remind students first to multiply by the ones, by the tens, by the hundreds, and add the three partial products.

Develop Skills and Concepts Stress the use of the Distributive Property in the multiplication steps. Emphasize that finding each partial product is a skill students already know. The two most common errors when multiplying 3- or more digit numbers are place value and renaming errors. Encourage students to align the partial products carefully and to write the renamed numbers in the proper columns. Have students work these problems on the board:

1. 429
   × 315
   (135,135)

2. 783
   × 246
   (192,618)

3. 631
   × 186
   (117,366)

3 Practice

Have students complete all the exercises. Remind them to write the dollar sign and decimal point when multiplying money.

T55

Practice

Multiply.

1. 647 × 453 293,091	2. 981 × 276 270,756	3. $14.29 × 855 $12,217.95	4. $2,585.50 × 693 $1,791,751.50
5. 3,708 × 436 1,616,688	6. 9,512 × 693 6,591,816	7. 16,245 × 781 12,687,345	8. 365,812 × 215 78,649,580
9. 5,649 × 843 4,762,107	10. 8,933 × 745 6,655,085	11. 18,302 × 912 16,691,424	12. 655,308 × 163 106,815,204

Copy and multiply.

13. 429 × 587
 251,823

14. 321 × 853
 273,813

15. 750 × 926
 694,500

16. 398 × $7.36
 $2,929.28

17. 854 × 576
 491,904

18. 973 × 187
 181,951

19. 2,349 × 147
 345,303

20. $44.36 × 210
 $9,315.60

21. 666 × 7,394
 4,924,404

22. 861 × 2,764
 2,379,804

23. 6,208 × 427
 2,650,816

24. 468 × $82.39
 $38,558.52

25. 3,770 × 289
 1,089,530

26. 727 × $39.95
 $29,043.65

27. 6,943 × 806
 5,596,058

28. 15,362 × 639
 9,816,318

29. $394.57 × 950
 $374,841.50

30. 525 × $615.05
 $322,901.25

Problem Solving

Solve each problem.

31. An average of 384 people shop in Metroland each day. How many people will shop in Metroland in 185 days?
 71,040 people

32. Al's Burgerie sold 3,814 hamburgers last month. Each cost $3.95. Across the street, McTavish's sold 5,956 hamburgers at $3.80 apiece. How much more did McTavish's make from the sale of hamburgers than did Al's Burgerie?
 $7,567.50

33. Mr. Ryan earns $12.85 each hour. In December, Mr. Ryan worked 168 hours. How much did he earn?
 $2,158.80

34. A factory buys small wheels for $8.67 and large wheels for $13.35. How much will the factory manager pay for 256 small wheels and 483 large wheels?
 $8,667.57

56

Lesson 3-8 • Multiplying by a 3-Digit Factor

---

4 Assess

Have students use a calculator to find the partial products for 835 × 264. (835 × 4 = 3,340; 835 × 60 = 50,100; 835 × 200 = 167,000)

5 Mixed Review

1. 60,208 − 43,197 (17,011)
2. $362.15 × 37 ($13,399.55)
3. 36 ÷ (2 × 3) (6)
4. $4,398.76 + $12,196.50 ($16,595.26)
5. 73 × 48 (3,504)
6. 70 × 200 (14,000)
7. 2^4 (16)
8. 50 × 24 (1,200)
9. 195 + 1,270 + 495 (1,960)
10. $9 × 10^3$ (9,000)

For Mixed Abilities

Common Errors • Intervention

Students may forget to write zeros in the ones place of the second partial product and the tens and ones place of the third partial product when multiplying by a 3-digit number. Correct by having them rewrite the problem as shown in the example below for 245 × 317.

$$(200 × 317) + (40 × 317) + (5 × 317)$$
$$63,400 \quad + \quad 12,680 \quad + \quad 1,585$$
$$77,665$$

Enrichment • Number Sense

Have students use lattice multiplication to check some of the exercises on pages 55 and 56. For example:

586
 × 349
 204,514

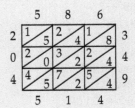

More to Explore • Number Sense

Divide students into groups, giving each $\frac{1}{2}$ in.-graph paper. Tell them to create a crossword-style puzzle, using numbers as answers. Tell them to use whole number problems involving addition, subtraction, multiplication, or division for the clues. Ask students to try to keep an equal number of across and down clues, and put their answers on a separate sheet of paper. Then, have groups exchange puzzles to solve.

3-9 Problem Solving: Look for a Pattern

pages 57–58

1 Getting Started

Objective
- To solve problems by drawing patterns

Materials
calculators

Warm Up • Mental Math
Dictate the following slowly, for students to solve:

1. $\sqrt{64} \times 4 + 3 \div 5 - 2$ (5)
2. $3 \times 40 \div 2 + 3 \div 9 \times$ itself (49)
3. $\sqrt{49} + 2 \times$ itself $- 1 \times 10$ (800)
4. $10 \times 42 \times \frac{1}{2} \div 7 + 2 \div 4$ (8)
5. $\sqrt{36} - 1 \div 5 \times$ itself $+ 1 \div 2$ (1)
6. $100 \times 15 \div 5 \div 10 \times 2$ (60)
7. $\sqrt{1} \times 69 \div 3 + 2 \div 5$ (5)
8. $7 \times 1,000 \times \frac{1}{2} \div 100 \div 5$ (7)

Warm Up • Pencil and Paper
Have students find:

1. 16×2 (32)
2. $100 - (65 + 32)$ (3)
3. the next number: 3, 7, 11, 15 (19)
4. $1 + 3 + 5 + 7 + 9$ (25)
5. the next number: 1, 2, 3; 2, 3; 1, 3 (2, 3)
6. the next number: 18, 16, 14, 12 (10)

Name _____

Problem Solving: Look for a Pattern

Simon is designing a number triangle to quiz his classmates. If he continues the pattern, what will be the sum of the numbers in the tenth row?

⭐ **SEE**

We want to find the sum of the numbers in the tenth row. By studying the diagram, we can see that consecutive odd numbers are used, and each row has __1__ more number than the preceding row.

⭐ **PLAN**

We can add the numbers on the part of the triangle that is shown. By putting the results of our work in a table, we can look for a pattern. We can extend the pattern to find the sum of the numbers in the tenth row.

⭐ **DO**

Row	1	2	3	4	5	6	7	8	9	10
Sum	1	8	27	64	125	216	343	512	729	1,000

The sum of the numbers in the tenth row is __1,000__.

⭐ **CHECK**

We can check by drawing 6 more rows of the triangle to find the numbers that will appear in the tenth row. We can then find the sum of these numbers.

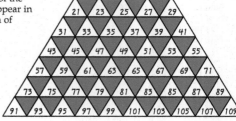

Lesson 3-9 • Problem Solving: Look for a Pattern

57

---

2 Teach

Introduce the Lesson Review the function of each of the SEE, PLAN, DO, CHECK steps. After a student has read the problem aloud, work through the steps with the class as they insert input responses.

Develop Skills and Concepts Discuss the various clues regarding the developing pattern in Simon's triangle. (The numbers used are the odd numbers from 1 to 19. Each row has one more number.) Encourage them to generalize from these clues. (If Simon used 1 number in row 1, and 2 in row 2, then it is probable he used 10 in row 10.) Tell students a list or table is frequently helpful in arranging the data in a pattern. Encourage students to actually complete the triangle as a check of the solution. They can use a calculator to help them with the computations.

It's Algebra! The concepts in this lesson prepare students for algebra.

3 Apply

Solution Notes

1. Answers may vary. A possible answer is that on the 14th turn, the numbers that are uncovered have the sum 14. (2, 4, and 8) This problem is related to the way computers use the binary numbers.
2. Each turn adds five additional marbles.
3. $20 + 18 + 16 + 14 + 12 + 10 = 90$; 6 weeks
4.

Number of stairs	Number of markers needed
1	1
2	3
3	6
⋮	⋮
10	55

Apply

Look for patterns to help you solve each problem.

1. Consider the four numbers 1, 2, 4, and 8. A game is played in which on the first turn the 1 is the only number not covered. On the second turn only the 2 is uncovered. On the third turn both the 1 and 2 are uncovered. On the fourth turn the 4 is uncovered. What numbers are uncovered on the fourteenth turn?

   2, 4, 8

2. John is lining up marbles in the following patterns:

   First turn: OOO

   Second turn: OOOOOOO

   Third turn: OOOOOOOOOOO

   How many marbles should be in the eighth turn?

   52 marbles

3. Dumbo the elephant went on a diet and wanted to lose 90 pounds. Dumbo lost 20 pounds the first week, 18 pounds the second week and 16 pounds the third week. If Dumbo continued to lose weight in this fashion, how many weeks would it take for Dumbo to lose the 90 pounds?

   6 weeks

4. The pattern for building a 3-step staircase is shown below:

   0

   00

   000

   How many markers would be required to build a pattern for a 10-step staircase?

   55 markers

Problem-Solving Strategy: Using the Four-Step Plan

★ **SEE** What do you need to find?

★ **PLAN** What do you need to do?

★ **DO** Follow the plan.

★ **CHECK** Does your answer make sense?

---

5. Tom Encherri wants to rent a car. One choice is to pay $159.95 for 5 days plus $39.95 for each additional day. Another choice is to pay $39.95 a day. What conditions would make one choice better than the other?

   Answers will vary.

6. Alexis Acura uses her calculator to find this sum: 235 + 457 + 518. However, she discovers that her [5] button does not work. How might she use her calculator anyway to find the sum?

   Answers will vary.

7. Greg's Great Grocery sells Coo-Coo Cola at 3 cans for $1.00. If you buy fewer than 3 cans, they charge you 34¢ for 1 can and 67¢ for 2 cans. Explain why the store does this.

   Answers will vary.

8. Barney Bubble's Bike Shop sells bicycles and bicycles with training wheels. Write a problem where there are 56 wheels and the answer is 9 bikes with training wheels.

   Answers will vary.

For Mixed Abilities

More to Explore • Applications

Have students conduct a poll of the class to find the group's favorite restaurant. After students have tallied these results, have them write an invitation to a chef or cook from that restaurant to speak to the class about their job and how math skills play a part in it. Have students prepare questions to ask following the talk. Beforehand, you may want to ask the speaker to discuss such things as preparing large recipes, ordering in large quantities, quantity control of serving size, and so on. Suggest they could bring copies of a week's meat or produce order for students.

Higher-Order Thinking Skills

5. **Evaluation:** This problem is appropriate for cooperative-group discussions. For example, students might note that 4 days at $39.95 is $159.80, only 15¢ less than the 5-day rate. Thus, if there is a possibility that something might occur to make a 4-day trip into a 5-day trip, it would be wise to take the first choice. However, if the trip is definitely less than 4 days, the second choice is better.

6. **Synthesis:** A possible answer is:

   234 [+] 1 [+] 447 [+]10 [+] 418 [+]100 [=]

7. **Synthesis:** Stores always will round to the next whole cent because they do not want to lose money; for example, if they charge 33¢ for 1 can, then for every 3 cans they sell, they lose a whole cent. In large supermarkets, a cent here and a cent there adds up to thousands of dollars.

8. **Synthesis:** A sample problem is, "The store has 19 bicycles on display." With this information and the data that there are 56 wheels, the reader can reason that there are 10 bicycles without training wheels and 9 bicycles with training wheels.

4 Assess

Draw the following pattern on the board:

Have students find the pattern that tells how many dots are in each display. (12, 22, 32, 42) Then ask, *How many dots will be in the tenth display?* (100)

T58

Name _____

page 59

Items	Objectives
1–8	To multiply 2-, or 3-digit factors by 1-digit factors (see pages 45–46)
9, 12–13, 16	To multiply 4-, 5-, or 6-digit factors by 1-digit factors (see pages 47–48)
10–11, 14–15	To multiply money by 1-digit factors (see pages 47–48)
17–24	To multiply whole numbers or money by 2-digit factors (see pages 51–52)
25–32	To multiply whole numbers or money by 3-digit factors (see pages 55–56)

Multiply. Use estimation to check your answers.

1. 89
    × 6
    534

2. 53
    × 9
    477

3. 136
    × 4
    544

4. 809
    × 7
    5,663

5. 72
    × 8
    576

6. 69
    × 5
    345

7. 195
    × 6
    1,170

8. 307
    × 3
    921

9. 3,924
    × 8
    31,392

10. $68.96
    × 2
    $137.92

11. $379.58
    × 5
    $1,897.90

12. 692,453
    × 3
    2,077,359

13. 4,839
    × 6
    29,034

14. $82.14
    × 3
    $246.42

15. $525.72
    × 4
    $2,102.88

16. 897,465
    × 2
    1,794,930

17. 75
    × 24
    1,800

18. $7.46
    × 78
    $581.88

19. 358
    × 67
    23,986

20. 3,849
    × 43
    165,507

21. 63
    × 82
    5,166

22. $9.56
    × 87
    $831.72

23. 477
    × 54
    25,758

24. 6,589
    × 75
    494,175

25. 639
    × 276
    176,364

26. 1,276
    × 638
    814,088

27. 36,758
    × 439
    16,136,762

28. $708.15
    × 743
    $526,155.45

29. 498
    × 651
    324,198

30. 3,492
    × 563
    1,965,996

31. 75,847
    × 698
    52,941,206

32. $903.47
    × 843
    $761,625.21

Alternate Chapter Test

You may wish to use the Alternate Chapter Test on page 374 of this book for further review and assessment.

Circle the letter of the correct answer.

1.
$$\begin{array}{r} 7 \\ 2 \\ + 4 \end{array}$$
a. 13 ✓
b. 14
c. 15
d. NG

2. $7 + 8 = n$
$n = ?$
a. 13
b. 14
c. 15 ✓
d. NG

3. $17 - 9$
a. 7
b. 8 ✓
c. 9
d. NG

4. $n + 7 = 13$
$n = ?$
a. 5
b. 9
c. 21
d. NG ✓

5. 8×8
a. 54
b. 56
c. 64 ✓
d. NG

6. $7 \times 0 = n$
$n = ?$
a. 0 ✓
b. 1
c. 7
d. NG

7. $6\overline{)48}$
a. 7
b. 8 ✓
c. 9
d. NG

8. $n \times 9 = 36$
$n = ?$
a. 3
b. 4 ✓
c. 5
d. NG

9. $15 + 10 \div 5 = n$
$n = ?$
a. 5
b. 17 ✓
c. 30
d. NG

10. What is the place value of the 0 in 4,702,931?
a. hundreds
b. thousands
c. ten thousands ✓
d. NG

11.
$$\begin{array}{r} 42,698 \\ + 46,853 \end{array}$$
a. 88,451
b. 89,551 ✓
c. 90,551
d. NG

12.
$$\begin{array}{r} 5,086 \\ - 3,994 \end{array}$$
a. 1,092 ✓
b. 1,192
c. 2,112
d. NG

13. 397×5
a. 1,455
b. 1,585
c. 1,985 ✓
d. NG

☐ score

STOP

page 60

Items	Objectives
1–2	To review basic facts of addition (see pages 1–2)
3–4	To review basic facts of subtraction (see pages 3–4)
5–6	To review basic facts of multiplication (see pages 5–6)
7–8	To review basic facts of division (see pages 7–8)
9	To simplify an expression having two or more operations (see pages 11–12)
10	To identify place value through hundred thousands (see pages 19–20)
11	To add two 5-digit numbers (see pages 27–28)
12	To subtract, renaming over zeros (see pages 33–34)
13	To multiply 3-digit factors by 1-digit factors (see pages 45–46)

Alternate Cumulative Assessment

Circle the letter of the correct answer.

1.
$$\begin{array}{r} 8 \\ 1 \\ + 6 \end{array}$$
a 14
b 15 ✓
c 16
d NG

2. $5 + 7 = n$
$n = ?$
a 12 ✓
b 13
c 14
d NG

3.
$$\begin{array}{r} 16 \\ - 7 \end{array}$$
a 7
b 8
c 9 ✓
d NG

4. $n + 6 = 12$
$n = ?$
a 5
b 6 ✓
c 7
d NG

5. $7 \times 6 = ?$
a 36
b 43
c 48
d NG ✓

6. $1 \times 2 = n$
$n = ?$
a 0
b 1
c 2 ✓
d NG

7. $4\overline{)32}$
a 7
b 8 ✓
c 9
d NG

8. $n \times 9 = 27$
$n = ?$
a 2
b 3 ✓
c 4
d NG

9. $24 + 16 \div 8 = n$
$n = ?$
a 5
b 19
c 26 ✓
d NG

10. Give the place value of the 2 in 5,092,746.
a tens
b hundreds
c thousands ✓
d NG

11.
$$\begin{array}{r} 39,475 \\ + 47,814 \end{array}$$
a 77,289
b 86,289
c 87,289 ✓
d NG

12.
$$\begin{array}{r} 8,054 \\ - 7,169 \end{array}$$
a 885 ✓
b 895
c 985
d NG

13.
$$\begin{array}{r} 485 \\ \times 7 \end{array}$$
a 3,365
b 3,395 ✓
c 3,465
d NG

4-1 Dividing With Multiples of 10

pages 61–62

1 Getting Started

Objective
- To divide multiples of 10, 100, or 1,000 by a multiple of 10

Materials
*division fact cards

Warm Up • Mental Math
Use division fact cards to challenge students in fact drills. A student's score is the number of facts correctly completed in one minute.

Warm Up • Mental Math
Have students find each product.

1. 5 × 9 (45)
   50 × 9 (450)
   50 × 90 (4,500)
   500 × 90 (45,000)
2. 4 × 6 (24)
   4 × 60 (240)
   40 × 60 (2,400)
   400 × 600 (240,000)
3. 3 × 17 (51)
   3 × 170 (510)
   30 × 170 (5,100)
   300 × 170 (51,000)
4. 8 × 46 (368)
   80 × 460 (36,800)
   800 × 46 (36,800)
   80 × 4,600 (368,000)

2 Teach

Introduce the Lesson Have students read the problem, tell what is to be solved and what information is given. (the total number of jumps; there are 5,400 steps, and a Numerarian jumps 6 steps at a time) Have students explain why they divide to find the answer. (separating a total into equal-sized groups) Continue to read the text and discuss the pattern shown. Review the terms: *dividend, divisor,* and *quotient.* Point out that multiplication is used to check division.

Develop Skills and Concepts Write division in these forms on the board:

$$A \div B = C \qquad B\overline{)A}$$

Have students identify the dividend, divisor, and quotient in each form above. (A is the dividend, B is the divisor, C is the quotient.)

Divide Whole Numbers

Lesson 4-1

Dividing With Multiples of 10

On the planet Numera, the entrance to the capitol has 5,400 steps. Numerarians can jump 6 steps at a time. How many jumps would a Numerarian take to get to the top of the steps?

We want to know how many jumps it would take for a Numerarian to climb all the steps.

We know that there are __5,400__ steps, and a Numerarian can jump __6__ steps at a time.

To find the number of jumps needed, we divide the total number of steps by the number of steps climbed in each jump. We divide __5,400__ by __6__.

Consider the pattern in these division sentences.

54 ÷ 6 = 9

540 ÷ 6 = 90   **CHECK**

5,400 ÷ 6 = __900__   __900__ × 6 = 5,400

A Numerarian could reach the top of the steps in __900__ jumps.

Getting Started

Complete the pattern.

1. 35 ÷ 7 = __5__
    350 ÷ 70 = __5__
    3,500 ÷ 70 = __50__
    35,000 ÷ 70 = __500__

Write each quotient. Check using mental math.

2. 800 ÷ 4 = __200__

3. 6,300 ÷ 90 = __70__

4. 21,000 ÷ 70 = __300__

Lesson 4-1 • Dividing With Multiples of 10   61

Then, have students find each quotient.

1. 36 ÷ 4 = (9)   2. 16 ÷ 8 = (2)
   360 ÷ 40 = (9)   160 ÷ 80 = (2)
   3,600 ÷ 40 = (90)   1,600 ÷ 80 = (20)
   36,000 ÷ 40 = (900)   16,000 ÷ 80 = (200)

As students work through each problem, discuss the relationship between the number of zeros in the dividend, divisor, and quotient. Point out that division with multiples of 10 is an application of the basic division facts.

3 Practice

Have students complete all the exercises.

Practice

Complete each pattern.

1. $48 \div 6 = \underline{8}$
   $480 \div 60 = \underline{8}$
   $4,800 \div 60 = \underline{80}$
   $48,000 \div 60 = \underline{800}$

2. $45 \div 9 = \underline{5}$
   $450 \div 90 = \underline{5}$
   $4,500 \div 90 = \underline{50}$
   $45,000 \div 90 = \underline{500}$

3. $40 \div 8 = \underline{5}$
   $400 \div 80 = \underline{5}$
   $4,000 \div 80 = \underline{50}$
   $40,000 \div 80 = \underline{500}$

Write each quotient. Check using mental math.

4. $350 \div 5 = \underline{70}$

5. $3,200 \div 4 = \underline{800}$

6. $540 \div 9 = \underline{60}$

7. $3,600 \div 40 = \underline{90}$

8. $2,800 \div 70 = \underline{40}$

9. $1,800 \div 90 = \underline{20}$

10. $4,200 \div 60 = \underline{70}$

11. $45,000 \div 50 = \underline{900}$

12. $32,000 \div 40 = \underline{800}$

Problem Solving

Solve each problem.

13. Mrs. Santana drove 270 miles on Sunday. She used 9 gallons of gasoline. How many miles did Mrs. Santana travel on each gallon of gas?
   30 miles

14. Bob said that he has worked for 36,000 seconds. How many hours has Bob worked?
   10 hours

(Now Try This!)

It's Algebra!

Multiplication and division are **inverse operations**.
To solve the equation $n \times 9 = 18$, we divide the product by the known factor.

$18 \div 9 = n$ $n = \underline{2}$

To solve the equation $n \div 4 = 9$, we multiply the quotient by the divisor.

$9 \times 4 = n$ $n = \underline{36}$

Use the inverse operation to solve these problems.

1. $n \times 6 = 36$ $n = \underline{6}$

2. $n \div 8 = 7$ $n = \underline{56}$

3. $n \times 30 = 2,100$ $n = \underline{70}$

4. $n \div 40 = 800$ $n = \underline{32,000}$

62   Lesson 4-1 • Dividing With Multiples of 10

For Mixed Abilities

Common Errors • Intervention

If students have difficulty dividing by multiples of 10, have them work with partners to complete the following.

Dividend	÷ 4	÷ 40
120	30	(3)
240	(60)	(6)
360	(90)	(9)
1,600	(400)	(40)
2,800	(700)	(70)
8,000	(2,000)	(200)
32,000	(8,000)	(800)

Enrichment • Number Sense

Have students use the following numbers to write three division exercises.

dividends: 6,400   45,000   42,000
divisors: 60   80   50
quotients: 700   80   900

(Possible answers: $42,000 \div 60 = 700$; $6,400 \div 80 = 80$; $45,000 \div 50 = 900$)

Encourage students to use multiplication to check each division.

More to Explore • Number Sense

Using the seven basic symbols of Roman numerals, (I, V, X, L, C, D, and M) have students write ten problems requiring a variety of operations: addition, subtraction, multiplication, and division. Have them exchange papers and solve the problems. Tell them all answers must be in Roman numerals.

(ESL/ELL) STRATEGIES

To support students' comprehension and pronunciation of the words *divisor, dividend, quotient,* and *remainder,* write the words on the board, capitalizing the stressed syllables: deVISor, DIVidend, QUOtient, reMAINder. Explain the meaning of each in simple English and have students repeat the words.

Now Try This! Remind students that multiplication and division can be used to check each other. Give students the problem $25 \div 5 = 5$. Ask them what the multiplication check would be. ($5 \times 5 = 25$) Have students use this property to find *n* in the equation $n \div 5 = 5$. ($n = 25$) Have students complete the exercise.

4 Assess

Ask students why the quotient of $20,000 \div 400$ is 50 instead of 500. (Possible answer: Because $20 \div 4$ is 5, you do not count the 0 in 20.)

4-2 Dividing by a 1-Digit Divisor

pages 63–64

1 Getting Started

Objective
• To divide 2- or 3-digit numbers by a 1-digit number

Warm Up • Mental Math
Have students simplify each expression.

1. $10^3 + 10^2$ (1,100)
2. $10^4 - 10^2$ (9,900)
3. $10^3 + 10^2 - 10$ (1,090)
4. $10^3 - 10$ (990)
5. $10^3 + 10^3$ (2,000)
6. $10^4 + 10^2 - 10$ (10,090)
7. $10^3 - 10^2$ (900)
8. $10^4 - 10^3$ (9,000)
9. $10^4 + 10^4 - 10$ (19,990)
10. $10^3 + 10$ (1,010)

Warm Up • Pencil and Paper
Have students write the missing factor in each equation.

1. $8 \times (40) = 320$
2. $90 \times (70) = 6,300$
3. $6 \times (800) = 4,800$
4. $7 \times (70) = 490$
5. $80 \times (7) = 560$
6. $9 \times (500) = 4,500$
7. $400 \times (90) = 36,000$

Name _____

Dividing by a 1-Digit Divisor

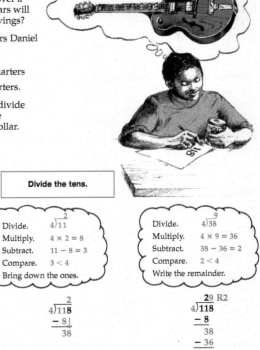

Daniel has been saving quarters for over a year to buy a guitar. How many dollars will Daniel receive in exchange for his savings?

We want to know how many dollars Daniel will get for his quarters.

We know that Daniel has __118__ quarters and one dollar is worth __4__ quarters.

To find the number of dollars, we divide the total number of quarters by the number of quarters equal to one dollar. We divide __118__ by __4__.

Divide the hundreds.	Divide the tens.	
There are not enough hundreds. We move to the tens place.	Divide. $4\overline{)11}$ — 2	Divide. $4\overline{)38}$ — 9
	Multiply. $4 \times 2 = 8$	Multiply. $4 \times 9 = 36$
	Subtract. $11 - 8 = 3$	Subtract. $38 - 36 = 2$
	Compare. $3 < 4$	Compare. $2 < 4$
	Bring down the ones.	Write the remainder.

$$4\overline{)118}$$

$$\begin{array}{r} 2 \\ 4\overline{)118} \\ -8\downarrow \\ \hline 38 \end{array}$$

$$\begin{array}{r} 29 \text{ R2} \\ 4\overline{)118} \\ -8 \\ \hline 38 \\ -36 \\ \hline 2 \end{array}$$

Daniel will receive __29__ dollars, and have __2__ quarters left over.

CHECK

$$\begin{array}{r} 29 \\ \times\ 4 \\ \hline 116 \end{array} \qquad \begin{array}{r} 116 \\ +\ 2 \\ \hline 118 \end{array}$$

Getting Started

Divide and check.

1. $3\overline{)72}$ — 24
2. $8\overline{)969}$ — 121 R1

Copy and divide.

3. $309 \div 4$ — 77 R1
4. $203 \div 9$ — 22 R5

---

2 Teach

Introduce the Lesson Have a student read the problem and tell what is to be solved. (total number of dollars Daniel will get in exchange for his quarters) Ask what additional information is not given. (4 quarters in $1.00) Have students complete the sentences as they read aloud with you.

Explain that in this problem, division answers the question: How many equal-sized groups? Then, work through the steps in the model with the students.

Develop Skills and Concepts Put $7\overline{)420}$ and 7×60 on the board. Compare the division steps with the multiplication steps. For example, in multiplying, they start with the ones place; in division, they start with the largest place, in this case the hundreds place. If there are not enough hundreds, they should divide the tens place.

Have students tell how many digits will be in each quotient.

1. $6\overline{)743}$ (3)
2. $4\overline{)389}$ (2)
3. $5\overline{)613}$ (3)

Then, have students find each quotient. (123 R5; 97 R1; 122 R3) Remind students first to *divide*, then to *multiply*, and then to *subtract*. Tell students to compare the remainder with the divisor before deciding if the division is complete. Ask what the greatest possible remainder could be if the divisors were 9 (8), 14 (13), or 106 (105).

Review the method for checking division.

divisor × quotient + remainder = dividend

3 Practice

Have students complete all the exercises. Remind students to divide and check each problem.

T63

Practice

Divide and check.

1. $3)\overline{93}$ → 31
2. $7)\overline{115}$ → 16 R3
3. $2)\overline{326}$ → 163
4. $9)\overline{585}$ → 65

5. $6)\overline{848}$ → 141 R2
6. $5)\overline{935}$ → 187
7. $8)\overline{889}$ → 111 R1
8. $9)\overline{672}$ → 74 R6

9. $3)\overline{956}$ → 318 R2
10. $8)\overline{734}$ → 91 R6
11. $4)\overline{734}$ → 183 R2
12. $6)\overline{834}$ → 139

13. $2)\overline{847}$ → 423 R1
14. $7)\overline{359}$ → 51 R2
15. $5)\overline{275}$ → 55
16. $3)\overline{537}$ → 179

Copy and divide.

17. $417 \div 5$ — 83 R2
18. $912 \div 6$ — 152
19. $653 \div 2$ — 326 R1
20. $721 \div 9$ — 80 R1

21. $95 \div 4$ — 23 R3
22. $108 \div 8$ — 13 R4
23. $536 \div 3$ — 178 R2
24. $815 \div 7$ — 116 R3

25. $715 \div 6$ — 119 R1
26. $963 \div 3$ — 321
27. $847 \div 5$ — 169 R2
28. $235 \div 8$ — 29 R3

Problem Solving

Solve each problem.

29. Roberta ran 784 meters in 4 minutes. How far did Roberta run in 1 minute?
196 meters

30. Mr. Davis works 6 hours each day. One month he worked 162 hours. How many days did Mr. Davis work?
27 days

31. The sixth grade is renting vans for a science field trip. The charge for each van is $35.25. Each van can carry 9 students. How much will it cost to rent enough vans to take 132 students on the field trip?
$528.75

32. Mark bought 192 marbles. They were in bags of 8 each. If each bag of marbles cost $1.19, how much did Mark pay?
$28.56

64

Lesson 4-2 • Dividing by a 1-Digit Divisor

4 Assess

Ask student to determine where they would begin dividing for each division example.

$7)\overline{458}$ (tens place) $8)\overline{972}$ (hundreds place) $6)\overline{58}$ (ones place)

5 Mixed Review

1. $2,542.98 + $4,938.64$ ($7,481.62)
2. 497×30 (14,910)
3. $3,207 + 41,956 + 8,219$ (53,382)
4. 503×275 (138,325)
5. 6×10^6 (6,000,000)
6. $27 + 4 \times 0$ (27)
7. $400.08 - 176.39 ($223.69)
8. $1,258 \times 600$ (754,800)
9. 500×600 (300,000)
10. $15,173 - 8,394$ (6,779)

For Mixed Abilities

Common Errors • Intervention

Some students may give incorrect answers by forgetting to give the remainder as part of the answer. Ask them to use multiplication and addition to check their answers. Students who continue to have difficulty with long division should work in small cooperative groups with place-value materials, such as hundred-flats, ten-strips, and unit squares to model the problems.

Enrichment • Application

Have students solve the following problem:

> There are 675 bumblebees in 9 hives. There are the same number of bumblebees in each hive. How many bumblebees are in each hive? (75)

Have students write a division story problem for their classmates to solve.

More to Explore • Number Sense

Tell students to find the secret number in this puzzle by following the steps:

1. Start with any whole number.

2. Multiply the number by 2.

3. Add 18 to the product.

4. Divide this sum by 2.

5. Subtract the original number.

(The answer will always be nine.)

Challenge students to devise their own formula in which the answer will always be the same number.

T64

pages 65–66

1 Getting Started

Objective
- To divide a number through 7 digits by a 1-digit divisor

Warm Up • Mental Math
Have students find the missing number in each equation.

1. $4 \times \underline{(8)} = 32$
2. $\underline{(9)} + 6 = 15$
3. $64 \div \underline{(8)} = 8$
4. $8 + \underline{(6)} = 14$
5. $3 \times \underline{(9)} = 27$
6. $56 \div \underline{(7)} = 8$
7. $12 - \underline{(5)} = 7$
8. $\underline{(7)} \times 7 = 49$
9. $32 \div \underline{(4)} = 8$
10. $\underline{(35)} \div 5 = 7$

Warm Up • Pencil and Paper
Have students write the missing term in each problem.

1. divisor = 6
   quotient = 34 R2
   dividend = (206)
2. dividend = 84
   divisor = 9
   quotient = (9 R3)
3. dividend = 57
   quotient = 8 R1
   divisor = (7)
4. divisor = 8
   quotient = 117 R5
   dividend = (941)

2 Teach

Introduce the Lesson Have students read the problem, identify the problem question, and define the given information. As students work through the division in the model, emphasize the 4 steps and the order in which they are done. Point out the remainder needs special interpretation in a real-life situation.

Develop Skills and Concepts To help students determine the number to write in the quotient, suggest that they ask themselves, *What number times the divisor will be close to, but not greater than, the dividend number?*

Write the following division exercise on the board:

$5)\overline{7,286}$ (1,457 R1)

Name _____

Dividing up to 7-Digit Numbers

The Wild Water Ride boats each carry 4 passengers and are filled every time they leave the dock. On Saturday, 9,838 people took the Wild Water Ride. How many trips did the boats make?

We want to know how many trips the boats made.

We know there were ___9,838___ passengers on the boats and each boat held ___4___ people.

To find how many trips were made, we divide the total number of passengers by the number of passengers in each boat. We divide ___9,838___ by ___4___.

REMEMBER Divide, multiply, subtract, compare, and bring down.

Divide the thousands.	Divide the hundreds.	Divide the tens.	Divide the ones.
$\begin{array}{r} 2 \\ 4\overline{)9{,}838} \\ -8 \\ \hline 1 \end{array}$	$\begin{array}{r} 2\,4 \\ 4\overline{)9{,}838} \\ -8\downarrow \\ \hline 1\,8 \\ -1\,6 \\ \hline 2 \end{array}$	$\begin{array}{r} 2\,4\,5 \\ 4\overline{)9{,}838} \\ -8 \\ \hline 1\,8\downarrow \\ -1\,6\downarrow \\ \hline 2\,3 \\ -2\,0 \\ \hline 3 \end{array}$	$\begin{array}{r} 2\,4\,5\,9\ \text{R2} \\ 4\overline{)9{,}838} \\ -8 \\ \hline 1\,8 \\ -1\,6 \\ \hline 2\,3 \\ -2\,0 \\ \hline 3\,8 \\ -3\,6 \\ \hline 2 \end{array}$

Because the remainder is ___2___, another trip was needed. The boats made ___2,460___ trips.

Getting Started

Divide and check.

1. $\overset{\text{394 R4}}{6)\overline{2,368}}$ 2. $\overset{\text{39,295 R1}}{3)\overline{117,886}}$

Copy and divide.

3. $9,631 \div 7$ 4. $503,246 \div 4$
   1,375 R6   125,811 R2

Have a student do the division on the board, discussing each step. Ask another student to check the division. ($5 \times 1,457 + 1 = 7,286$) Point out that students must remember to add in the remainder when they are checking the answer to a division problem.

Repeat this procedure with the following problems:

1. $\overset{\text{(497 R3)}}{7)\overline{3,482}}$ 2. $\overset{\text{(17,455 R1)}}{4)\overline{69,821}}$ 3. $\overset{\text{(92,027)}}{3)\overline{276,081}}$

3 Practice

Have students complete all the exercises. To check their answer, remind them to multiply the quotient times the divisor and add the remainder. The resulting number will be the dividend if their answer is correct.

Practice

Divide and check.

1. $7\overline{)2{,}494}$ 356 R2
2. $2\overline{)3{,}658}$ 1,829
3. $5\overline{)6{,}786}$ 1,357 R1

4. $8\overline{)7{,}490}$ 936 R2
5. $4\overline{)26{,}288}$ 6,572
6. $9\overline{)32{,}859}$ 3,651

7. $6\overline{)37{,}734}$ 6,289
8. $5\overline{)327{,}263}$ 65,452 R3
9. $3\overline{)652{,}702}$ 217,567 R1

Copy and divide.

10. $3{,}869 \div 4$
    967 R1
11. $8{,}346 \div 8$
    1,043 R2
12. $9{,}375 \div 2$
    4,687 R1
13. $2{,}023 \div 7$
    289
14. $3{,}196 \div 5$
    639 R1
15. $7{,}593 \div 6$
    1,265 R3
16. $37{,}791 \div 3$
    12,597
17. $56{,}241 \div 9$
    6,249
18. $50{,}620 \div 8$
    6,327 R4
19. $52{,}344 \div 6$
    8,724
20. $93{,}857 \div 4$
    23,464 R1
21. $81{,}396 \div 5$
    16,279 R1
22. $307{,}185 \div 2$
    153,592 R1
23. $836{,}459 \div 7$
    119,494 R1
24. $1{,}763{,}829 \div 9$
    195,981

Problem Solving

Solve each problem.

25. A machine sorts apples into bags of 5 apples each. How many bags does it take to store 2,643 apples?
    529 bags

26. The Jogging Club held an 11,168-meter relay on Saturday morning. Each of 8 joggers ran the same distance. How far did each jogger run?
    1,396 meters

27. Six new movie theaters opened at the shopping mall. Each theater has 242 seats. For one show, half the theaters were full, and half the theaters were half full. How many people were at the movies?
    1,089 people

28. A jet plane flew 4,575 kilometers from New York to Los Angeles in 5 hours. What was the plane's average speed per hour?
    915 kilometers per hour

66

Lesson 4-3 • Dividing up to 7-Digit Numbers

4 Assess

Have students show you how they would check the answer to the following division problem.

$6\overline{)79{,}390}$ 13,231 R4 $(13{,}231 \times 6 + 4)$

5 Mixed Review

1. 4×10^8 (400,000,000)
2. $35{,}000 \div 70$ (500)
3. $\$65.30 \times 8$ ($522.40)
4. 65×27 (1,755)
5. $400{,}200 - 275{,}074$ (125,126)
6. $276{,}129 + 308{,}175$ (584,304)
7. $1{,}958 \div 9$ (217 R5)
8. 963×400 (385,200)
9. $652 \div 8$ (81 R4)
10. $\$25.09 \times 317$ ($7,953.53)

For Mixed Abilities

Common Errors • Intervention

When the first digit in the dividend cannot be divided by the divisor, some students simply ignore it.

Incorrect	Correct
211	461
$4\overline{)1{,}844}$	$4\overline{)1{,}844}$

Have students use place-value materials to model the problem. The first step is to trade 1 thousand for 10 hundreds, giving a total of 18 hundreds. With such experiences, students soon will recognize the 18 hundreds right away without having to rename or work with manipulatives.

Enrichment • Number Sense

Have students complete the cross number puzzle.

1. (8)	2. (3)	3. (6)		
	4. (4)	(5)	5. (1)	(1)
	6. (7)	(1)	(9)	
			5. (1)	
			(2)	

Across	Down
1. $9\overline{)7{,}524}$	2. $9\overline{)3{,}123}$
4. $8\overline{)36{,}088}$	3. $7\overline{)4{,}557}$
6. $9\overline{)6{,}471}$	5. $8\overline{)15{,}296}$

More to Explore • Applications

Ask students to bring in their favorite cookie recipes. Have them figure the amount of each ingredient they would need if they doubled the recipe. Then, have them triple it. Then, have them find how much of each ingredient they would need if they cut the original recipe in half or into thirds. You might suggest that students make their recipe in the appropriate amount at home, and share the cookies with the class.

T66

1 Getting Started

Objective
- To divide with up to 5-digit dividends and with zeros in the quotient

Materials
*division fact cards

Warm Up • Mental Math
Have students name:

1. all even numbers from 151 through 165 (152, 154, 156, 158, 160, 162, 164)
2. all non-zero multiples of 6 less than 25 (6, 12, 18, 24)
3. all powers of 10 from 10 through 1,000 (10, 100, 1,000)
4. all possible remainders when the divisor is 5 (0, 1, 2, 3, 4)
5. all whole number pairs of addends that produce a sum of 6 (6 + 0, 5 + 1, 4 + 2, 3 + 3)
6. all odd numbers less than 6 (5, 3, 1)

Warm Up • Pencil and Paper
Have students find each quotient.

1. (1,388 R5) 7)9,721
2. ($81.22) 4)$324.88

Name _____

Lesson 4-4

Zeros in the Quotient

Nancy bought new tires for her bicycle. She spent $21.50. How much did Nancy pay for each tire?

We want to know the cost of each tire.

We know that Nancy bought __2__ tires and she paid a total of $21.50.

To find the cost of each tire, we divide the total cost by the number of tires. We divide $21.50 by __2__.

Divide the dollars.	Place the decimal point above the one in the dividend.	Divide the dimes and cents. Write the dollar sign.
10 2)$21.50 −2↓ 1	10. 2)$21.50 −2↓ 1	$10.75 2)$21.50 −2↓ 1 5 −1 4↓ 10 −10 0

REMEMBER For every digit brought down, there must be a digit in the quotient.

Each tire costs $10.75.

Getting Started

Divide and check.

1. $17.04 5)$85.20
2. $10.04 9)$90.36
3. 2,007 R3 4)8,031

4. $30.61 6)$183.66
5. 30,509 R2 3)91,529
6. 120,021 7)840,147

Copy and divide.

7. 407,315 ÷ 4 101,828 R3
8. 363,230 ÷ 7 51,890
9. $636.18 ÷ 6 $106.03

Lesson 4-4 • Zeros in the Quotient

67

---

2 Teach

Introduce the Lesson Have students read the problem and explain why they should divide. (to find how much each tire costs) Work through the model with students, discussing each of the three steps shown. Point out that they are first to divide the dollars, then the dimes, and then the pennies. Remind them to write the dollar sign and decimal point in the quotient.

Develop Skills and Concepts Write 3)$21.66 on the board. Point out that in dividing they divide from left to right, starting with the greatest place value. Have students cover the dollar sign so they can see the similarities between dividing whole numbers and dividing with money.

Write the 4 steps in the division process on the board:

divide, multiply, subtract, compare and bring down

Explain that after the first digit is recorded in the quotient, a digit must be recorded in each place to the right, ending with the ones place. Sometimes they will need to put a 0 in the quotient as a placeholder.

Have students tell how many digits will be in each quotient:

1. 6)$36.12 (3)
2. 7)14,357 (4)
3. 8)2,408 (3)

Have students work these problems at the board:

1. (2,090 R2) 4)8,362
2. (23,001 R1) 4)92,005
3. ($20.31) 9)$182.79

3 Practice

Have students complete all the exercises with a partner. Have them exchange their papers and check their answers together.

Practice

Divide and check.

1. 6)6,248 — 1,041 R2
2. 4)8,032 — 2,008
3. 8)9,653 — 1,206 R5
4. 2)7,810 — 3,905
5. 9)18,541 — 2,060 R1
6. 5)$952.35 — $190.47

Copy and divide.

7. 3,623 ÷ 6
   603 R5
8. 6,539 ÷ 5
   1,307 R4
9. 9,360 ÷ 8
   1,170
10. $45.09 ÷ 9
    $5.01
11. 9,017 ÷ 3
    3,005 R2
12. 8,205 ÷ 2
    4,102 R1
13. 15,238 ÷ 4
    3,809 R2
14. 49,035 ÷ 7
    7,005
15. $515.25 ÷ 5
    $103.05

Problem Solving

Solve each problem.

16. Mr. Shima bought 4 new tires for his car. He paid $563.40 for the tires. How much did each one cost?
    $140.85

17. The greatest known depth of the Pacific Ocean is 35,820 feet. A computerized bathysphere descended to this depth at a speed of 3 feet per second. How many seconds did it take to get to the bottom?
    11,940 seconds

⟮ Now Try This! ⟯

Divisibility rules for 2, 5, 3, 6, and 9 state that a number is

- divisible by 2 if it ends in 0, 2, 4, 6, or 8.
- divisible by 5 if it ends in 0 or 5.
- divisible by 3 if the sum of the digits is divisible by 3.
- divisible by 6 if it is divisible by 2 and 3.
- divisible by 9 if the sum of the digits is divisible by 9.

Apply the divisibility rules for 2, 5, 3, 6, or 9 by writing which of these numbers is a divisor for the given number.

1. 80 _2, 5_
2. 75 _5, 3_
3. 342 _2, 3, 6, 9_
4. 207 _3, 9_
5. 711 _3, 9_
6. 113 _none_
7. 435 _5, 3_
8. 876 _2, 3, 6_

Lesson 4-4 • Zeros in the Quotient

Now Try This! Encourage students to memorize the divisibility rules so that they can decide quickly upon inspection whether a number is evenly divisible by 2, 3, 5, 6, or 9.

4 Assess

Write the following on the board:

$103.01
8)$824.08

Ask students to explain why there are two 0s in the quotient. (Possible answer: When dividing, there were not enough tens and not enough dimes to divide by 8. The 0s are used as placeholders in the quotient.)

For Mixed Abilities

Common Errors • Intervention

When dividing with zeros in the quotient, some students will omit the zero or write it in the incorrect place.

Incorrect	Correct
153	1,053
3)3,159	3)3,159

Have these students use place-value materials so they recognize that 3 thousands can be divided into 3 groups of 1 thousand each, but that 1 hundred cannot be divided into 3 groups of any number of hundreds, so a zero must be written in the hundreds place of the quotient.

Enrichment • Number Sense

Have students complete the cross number puzzle.

		1. (4)			
	2. (7)	(8)	(0)	5. (6)	
3. (4)	(5)	(0)		(0)	
	4. (6)	(3)	(2)	(5)	
	(2)			(3)	

Across

2. 4)31,224
3. 8)3,600
4. 5)31,625

Down

1. 4)192,128
5. 7)42,371

More to Explore • Logic

Give each pair of students 25 toothpicks. Have each student take turns drawing one, two, three, or four toothpicks at a time from the pile. The student who draws the last toothpick wins. Have them continue playing until one of the pair discovers a strategy to consistently win. Ask students if their win strategy would change if the number of toothpicks changed.

T68

4-5 Short Division

pages 69–70

1 Getting Started

Objective
- To divide whole numbers using short division

Warm Up • Mental Math
Have students find the mystery number.

1. 3 times the number is 12 (4)
2. 5 times the number plus 2 is 17 (3)
3. the number minus 4 is 6 (10)
4. the number doubled is 100 (50)
5. half the number is 11 (22)
6. twice the number plus 5 is 15 (5)

Warm Up • Pencil and Paper
Have students find each quotient.

1. (989 R1)
$6)\overline{5,935}$

2. (228)
$7)\overline{1,596}$

3. (24,141 R2)
$3)\overline{72,425}$

4. (628)
$5)\overline{3,140}$

5. (8,028 R5)
$7)\overline{56,201}$

6. (4,138)
$9)\overline{37,242}$

2 Teach

Introduce the Lesson Have students read the problem and decide what is to be solved. (how much Bjorn earns in one hour) Ask students what information is given in the problem. (the amount Bjorn earned in a week and the number of hours he worked in that week) Ask why we use division to solve the problem. (to find the amount of money he earns each hour)

Work through the model with students. Explain that with short division, students do more mental and less written computation. Remind students multiplication is a check for division.

Develop Skills and Concepts Point out that the steps for long and short division are almost the same. Write the following on the board:

short form

```
      634              6 3 4
6)3,804          6)3,8²0²4
 − 36
  ────
   20
 − 18
  ────
   24
 − 24
```

T69

Name _____

Short Division

Bjorn works in the school cafeteria. His check for one week was $38.92. How much does Bjorn earn per hour?

Bjorn's Timecard	
Monday	2 hours
Wednesday	2 hours
Friday	3 hours
Total	7 hours

We want to know Bjorn's hourly wage.

We know that he worked ___7___ hours last week, and he earned ___$38.92___.

To find the amount he earned each hour, we divide his total wages by the number of hours he worked.

We divide ___$38.92___ by ___7___.

Divide the dollars. Place the decimal point. Think of the remainder next to the dimes.	Divide the dimes. Think of the remainder next to the cents.	Divide the cents. Place the dollar sign.
5. (39 dimes) $7)\overline{\$38.92}$	5.5 (42 cents) $7)\overline{\$38.92}$	$ 5.56 $7)\overline{\$38.92}$

Bjorn earns ___$5.56___ per hour

CHECK
```
  $5.56
×     7
───────
 $38.92
```

Getting Started

Divide and check.

1. 235 $7)\overline{1,645}$
2. 2,421 R3 $4)\overline{9,687}$
3. $247.73 $2)\overline{\$495.46}$
4. 9,432 R7 $8)\overline{75,463}$

5. 18,257 R2 $6)\overline{109,544}$
6. 36,275 $9)\overline{326,475}$
7. 248,735 $5)\overline{1,243,675}$
8. $1,748.70 $3)\overline{\$5,246.10}$

Copy and divide.

9. $291.12 ÷ 6
$48.52

10. 729,562 ÷ 8
91,195 R2

11. $4,921.95 ÷ 5
$984.39

Lesson 4-5 • Short Division

Work through this problem with the students using both the long and short forms of division. Show how the partial dividends are written out in the long form, and how in the short form, they are indicated by writing the hundreds and tens digits to the left of the appropriate digits in the dividend.

Have students work these exercises at the board. Have one student use long division and have another student use short division, explaining each step as they work.

1. (752 R3)
$5)\overline{3,763}$

2. (8,306)
$6)\overline{49,836}$

3. (49,287)
$3)\overline{147,861}$

3 Practice

Have students complete all the exercises. Encourage students to estimate before they do the computation. Then, have them check their answers.

Practice

Divide and check.

1. 8)2,368 — 296
2. 5)3,646 — 729 R1
3. 7)$73.15 — $10.45

4. 6)15,651 — 2,608 R3
5. 4)$256.84 — $64.21
6. 2)39,753 — 19,876 R1

7. 3)$3,316.53 — $1,105.51
8. 5)689,243 — 137,848 R3
9. 8)2,139,728 — 267,466

Copy and divide.

10. 5,295 ÷ 7
    756 R3
11. 56,032 ÷ 8
    7,004
12. 28,645 ÷ 3
    9,548 R1
13. $421.38 ÷ 9
    $46.82
14. $152.94 ÷ 6
    $25.49
15. 676,885 ÷ 4
    169,221 R1
16. 445,731 ÷ 2
    222,865 R1
17. $3,276.35 ÷ 5
    $655.27

Problem Solving

Solve each problem.

18. Raffi is a serious baseball fan who has collected 1,140 baseball cards. He displays the cards in a 150-page book, putting 8 cards on each page. How many more cards does Raffi need to fill the book?
    60 cards

19. Marcia buys furniture for the E-Z Chair Store. She paid $2,282.40 for 9 chairs. How much did each chair cost?
    $253.60

[Now Try This!]

Here are two more divisibility rules.

- A number is divisible by 4 if the last 2 digits are divisible by 4.
- A number is divisible by 8 if the last 3 digits are divisible by 8.

Apply the divisibility rules for 4 and 8 by writing which of these numbers is a divisor for the given number.

1. 9,704 ___4, 8___
2. 6,400 ___4, 8___
3. 13,120 ___4, 8___
4. 10,250 ___neither___

5. 26,215 ___neither___
6. 31,884 ___4___
7. 57,684 ___4___
8. 95,008 ___4, 8___

Lesson 4-5 • Short Division

For Mixed Abilities

Common Errors • Intervention

Some students may divide incorrectly because they mentally cannot keep track of the multiplications and subtractions. Have them work with partners to do each problem using the long form and then the short form, comparing the procedures.

Enrichment • Logic

Tell students to follow the steps in the following flowchart several times. Have them choose a different number each time they do the START step.

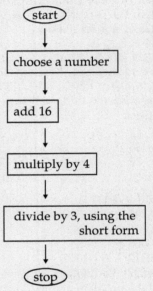

More to Explore • Applications

Have students plan a day at an amusement park of their own creation. Ask students to make a chart of prices and activities they would have in the amusement park. Have them include prices for rides, admission, show tickets, food, and other items.

Divide the class into small groups and have them exchange park price charts. Tell each group that they have $80 to spend. After some discussion have each group draw a pie graph using fractions or percents to show how they plan to spend their money on food, shows, games, tickets, and other expenses for the day.

Now Try This! Encourage students to memorize the rules so that they can decide quickly whether a number is evenly divisible by 4 or 8.

4 Assess

Write the following on the board, leaving ample space between the digits in the dividend:

```
    1 8 7, 4 3 6
3)5 6 2, 3 0 8
```

Have students insert the digits in the dividend that show each partial dividend that has been divided by 3.

```
(   1  8 7,  4 3  6   )
 3)5 26 22, 13 10 18
```

T70

4-6 Finding the Average

pages 71–72

1 Getting Started

Objective
• To find and interpret averages

Warm Up • Mental Math
Have students find each product.

1. $2 \times 5 \times 13$ (130)
2. $5 \times 4 \times 3$ (60)
3. $6 \times 5 \times 8$ (240)
4. $10 \times 6 \times 10$ (600)
5. $2 \times 50 \times 337$ (33,700)
6. $18 \times 5 \times 2$ (180)
7. $46 \times 37 \times 0$ (0)
8. $25 \times 4 \times 68$ (6,800)
9. $4 \times 7 \times 10$ (280)
10. $5 \times 8 \times 8$ (320)

Warm Up • Pencil and Paper
Have students find each sum or quotient.

1.	42	2.	83	3.	34
	16		27		36
	+ 75		+ 16		+ 37
	(133)		(126)		(107)

4. $513 \div 3$ (171)
5. $7,745 \div 5$ (1,549)
6. $2,688 \div 7$ (384)
7. $1,608 \div 4$ (402)
8. $6,312 \div 6$ (1,052)
9. $1,632 \div 8$ (204)

Finding the Average

Onida took 4 spelling tests in March to prepare for the national spelling bee. What was her average score?

MARCH – AVERAGE SCORES
ONIDA 96 90 87 91

We want to know Onida's average score.

We know that she took ___4___ tests and her test scores were: __96__, __90__, __87__, and __91__.

To find the **average**, we add the scores, and divide by the number of tests. We add __96__, __90__, __87__, and __91__ and divide the sum by __4__.

Add the numbers.	Divide by the number of addends.
96	
90	91
87	4)364
+ 91	
364	

Onida's average spelling score was __91__.

Getting Started

Find the average for each set of numbers.

1. 521, 672, 970, 153
   579
2. 83, 75, 92, 106, 94
   90
3. 1,952, 1,683, 2,068
   1,901
4. 13,725, 16,490, 12,862
   14,359

---

2 Teach

Introduce the Lesson Have students list instances in daily life when averages are used. Have a student read the problem and tell what is to be solved. (Onida's average score for 4 spelling tests) Ask what information is given in the illustration. (the test scores) Have students complete the sentences and work through the model to solve the problem.

Develop Skills and Concepts Explain that an average is a number or value around which other numbers gather. There are different ways of finding the average for a set of data. The average in this lesson is sometimes called an *arithmetic mean*. Explain to students that the average must lie between the greatest and least number. Point out that in the model example, 91 lies between 87 and 96. Have students find the average of each of the following groups of numbers.

1. 65, 64, 60, 71 (65)
2. 55, 30, 80, 60, 75 (60)
3. 364, 285, 350, 297 (324)

3 Practice

Have students complete all the exercises. Explain that in the Problem Solving section they will be using the charts to solve the word problems.

T71

Practice

Find the average for each set of numbers.

1. 646, 783, 914, 853
   799

2. 324, 278, 96, 153, 524
   275

3. 15,987, 18,273, 17,856
   17,372

4. 87, 68, 83, 74, 75, 81
   78

5. 26, 35, 33, 44, 57
   39

6. 519, 231, 117, 86, 57
   202

7. 864, 651, 865, 372
   688

8. 54, 44, 34, 38, 36, 40
   41

9. 14, 12, 6, 2, 31
   13

10. 5,015, 1,196, 7,343
    4,518

11. 65, 75, 73, 71
    71

12. 1,015, 362, 1,428, 207
    753

Problem Solving

Use the chart to solve Problems 13 and 14.

City Basketball League Total Points				
Game	Nanda	Peg	Carla	Bev
1	11	7	15	12
2	9	14	18	Abs.
3	8	13	9	15
4	6	17	12	14
5	12	9	16	19
6	8	12	14	15

13. How much higher is Peg's average than Nanda's?
    3

14. Bev missed one of the games. Compare her average points with Carla's average.
    Bev's average is 1 point higher.

Use the chart to solve Problems 15 and 16.

Robert Jones Weekly Sales Expense	
Day	Amount
Monday	$36.84
Tuesday	$29.16
Wednesday	$24.30
Thursday	$31.82
Friday	$24.53

15. What is Mr. Jones's average daily expense?
    $29.33

16. Which days did Mr. Jones spend more than his average daily expense?
    Monday and Thursday

72   Lesson 4-6 • Finding the Average

4 Assess

Write the following set of numbers on the board:

147, 132, 165, 152

From the following, have students choose the average for this set of numbers and explain why the other two numbers could not be the average.

a. 166   b. 149   c. 128

(149; Possible answer: 166 is greater than the greatest number in the set. 128 is less than the least number in the set.)

5 Mixed Review

1. 81 × 45 (3,645)
2. $21.53 × 456 ($9,817.68)
3. 478,026 ÷ 5 (95,605 R1)
4. 24 − 42 ÷ 6 (17)
5. 3,000 × 400 (1,200,000)
6. 65,103 − 28,292 (36,811)
7. 3,208 + 19,656 + 17,309 (40,173)
8. 6,152 ÷ 3 (2,050 R2)
9. 6² (36)
10. $25.09 × 317 ($7,953.53)

For Mixed Abilities

Common Errors • Intervention
Some students may always divide by 2 or some other incorrect number when they are finding averages. Have them work in small cooperative groups to discuss why the divisor must be the number of addends. They should recognize that, if the average represents each addend, then the number of addends times the average should have the same total as the sum of the addends.

Enrichment • Number Sense
Duplicate the following score chart for a dart game competition. Tell students that the average score for each player is 90. Ask students to find the score for the fifth game.

Game	1	2	3	4	5	6
Jeffrey	95	100	98	88	(85)	74
Olanda	90	81	86	97	(90)	96
Jose	87	88	89	84	(98)	94
Susan	77	85	92	98	(91)	97

More to Explore• Measurement
Explain to students that earlier civilizations did not have standard rulers, so they measured items by using a system of body units. Duplicate or write these units of measure on the board:

hands the width of the back of the hand

span the distance from the end of the little finger to the thumb with fingers spread

cubit the distance from the end of the elbow to the end of the middle finger with arm bent

foot the length of the foot

Give students a list of various classroom objects to be measured using these four body units, have the students measure each and construct a chart showing their findings. Have students compare charts and discuss why there are differences. Discuss the necessity for standard units of measure.

T72

4-7 Dividing by Multiples of 10

pages 73–74

1 Getting Started

Objective
- To divide whole numbers by a multiple of 10

Warm Up • Mental Math
Have students use basic facts to determine each answer.

1. $32 \div 4$ (8)
2. 8×6 (48)
3. $7 + 6$ (13)
4. $45 \div 5$ (9)
5. 9×0 (0)
6. $13 - 7$ (6)
7. $15 - 9$ (6)
8. $72 \div 9$ (8)
9. $17 - 8$ (9)
10. 6×7 (42)
11. $64 \div 8$ (8)

Warm Up • Mental Math
Have students write each product.

1. 3×5 (15)
2. 3×50 (150)
3. 5×7 (35)
4. 5×70 (350)
5. 6×3 (18)
6. 6×30 (180)
7. 7×2 (14)
8. 7×20 (140)
9. 9×4 (36)
10. 9×40 (360)
11. 4×7 (28)
12. 4×70 (280)

2 Teach

Introduce the Lesson Have a student read the problem and tell what is to be solved. (the amount Mr. Ellis will pay each month for the entertainment system) Explain the terms *down payment, balance,* and *monthly payment*. Ask what information is given. (the cost of the entertainment system, the amount of the down payment, and the number of months he has to pay) Ask what needs to be done to answer the question. (The $1,350 balance must be divided by 30 months.)

Have students complete the information sentences. Then, work through the division steps in the model and have a student read the answer.

Develop Skills and Concepts Point out how basic division facts are used to divide by multiples of 10. Write the following on the board for students to solve:

Name _____

Dividing by Multiples of 10

Mr. Ellis bought an entertainment system for $1,450. He paid $100 down, and promised to pay the balance in 30 months. How much will Mr. Ellis pay each month?

We want to know the amount of Mr. Ellis's monthly payments.

We know that after his down payment, he must pay a balance of $\underline{\$1,350}$ in $\underline{30}$ months.

To find the monthly payments, we divide the amount to be paid by the number of months. We divide $\underline{\$1,350}$ by $\underline{30}$.

Divide the thousands. 30 > 1	Divide the hundreds. 30 > 13	Divide the tens.	Divide the ones.
$30\overline{)\$1,350}$	$30\overline{)\$1,350}$	$\begin{array}{r} 4 \\ 30\overline{)\$1,350} \\ -1\,20 \\ \hline 15 \end{array}$	$\begin{array}{r} \$45 \\ 30\overline{)\$1,350} \\ -1\,20 \\ \hline 150 \\ -150 \\ \hline 0 \end{array}$

REMEMBER The partial dividend must be larger than the divisor for a division to take place.

Mr. Ellis will pay $\underline{\$45}$ each month.

Getting Started

Divide and check.

1. $\begin{array}{r} 116 \text{ R}19 \\ 40\overline{)4,659} \end{array}$
2. $\begin{array}{r} 73 \text{ R}25 \\ 50\overline{)3,675} \end{array}$
3. $\begin{array}{r} \$3.02 \\ 30\overline{)\$90.60} \end{array}$

4. $\begin{array}{r} 309 \text{ R}27 \\ 70\overline{)21,657} \end{array}$
5. $\begin{array}{r} 834 \text{ R}15 \\ 90\overline{)75,075} \end{array}$
6. $\begin{array}{r} 12,008 \text{ R}6 \\ 60\overline{)720,486} \end{array}$

Copy and divide.

7. $37,968 \div 80$
   474 R48
8. $\$114.40 \div 20$
   $5.72
9. $293,680 \div 70$
   4,195 R30

$\begin{array}{r}(4)\\3\overline{)12}\end{array}$ $\begin{array}{r}(4)\\30\overline{)120}\end{array}$ $\begin{array}{r}(40)\\30\overline{)1,200}\end{array}$ $\begin{array}{r}(400)\\30\overline{)12,000}\end{array}$

Then, have students determine what the first digit in the quotient should be in each of the following:

1. $\begin{array}{r}(8)\\6\overline{)50}\end{array}$ $\begin{array}{r}(8)\\60\overline{)500}\end{array}$
2. $\begin{array}{r}(5)\\8\overline{)43}\end{array}$ $\begin{array}{r}(5)\\80\overline{)430}\end{array}$

Continue developing this skill with problems like the following:

1. $1,632 \div 30$ (54 R12)
2. $8,046 \div 40$ (201 R6)
3. $44,417 \div 20$ (2,220 R17)

3 Practice

Have students complete all the exercises. Remind students to use multiplication to check division.

Practice

Divide and check.

1. $50\overline{)3{,}984}$ 79 R34

2. $70\overline{)4{,}765}$ 68 R5

3. $90\overline{)8{,}263}$ 91 R73

4. $60\overline{)8{,}455}$ 140 R55

5. $20\overline{)12{,}685}$ 634 R5

6. $80\overline{)39{,}756}$ 496 R76

7. $20\overline{)\$735.40}$ $36.77

8. $40\overline{)729{,}653}$ 18,241 R13

9. $30\overline{)695{,}406}$ 23,180 R6

Copy and divide.

10. $7{,}365 \div 40$
    184 R5
11. $8{,}296 \div 70$
    118 R36
12. $10{,}375 \div 20$
    518 R15
13. $4{,}283 \div 60$
    71 R23
14. $9{,}217 \div 30$
    307 R7
15. $15{,}408 \div 80$
    192 R48
16. $16{,}751 \div 90$
    186 R11
17. $\$217.50 \div 50$
    $4.35
18. $37{,}219 \div 60$
    620 R19
19. $75{,}215 \div 70$
    1,074 R35
20. $60{,}351 \div 30$
    2,011 R21
21. $95{,}477 \div 40$
    2,386 R37
22. $216{,}436 \div 80$
    2,705 R36
23. $785{,}381 \div 90$
    8,726 R41
24. $675{,}215 \div 50$
    13,504 R15

Problem Solving

Solve each problem.

25. The countdown began 12,780 seconds before blast-off. How many minutes before blast-off is this?
    213 minutes

26. Mrs. Murphy earns $489.60. She works a 40-hour week. What is her hourly wage?
    $12.24

27. Mr. Ikeda had 30 car payments to pay the balance of $9,150 on his new car. How much was each payment?
    $305

28. A box of 50 tacks costs $1.25. How much will 16,750 tacks cost?
    $418.75

74   Lesson 4-7 • Dividing by Multiples of 10

4 Assess

Present the following information:

Chaya earns $346.50 for a 30-hour week.

Ask, *Does Chaya earn $1.15 an hour, $11.55 an hour, or $115.50 an hour? Explain.* ($11.55 an hour; Possible answer: When you begin dividing, you begin by dividing 34 tens by 3 tens, which is about 11 tens.)

For Mixed Abilities

Common Errors • Intervention

Some students have difficulty identifying the basic fact when they are dividing by multiples of 10. Before they divide, have them circle the numbers that would identify the basic fact, helping them to determine the first digit of the quotient.

$$\underset{30\overline{)6\,6\,7}}{2} \qquad \underset{50\overline{)3\,3{,}9\,5\,0}}{6}$$

Enrichment • Number Sense

Have students use the following to write two division problems:

Divisors	40	70
Dividends	96,587	38,296
Quotients	547	2,414
Remainders	6	27

$$\left(\underset{40\overline{)96{,}587}}{2{,}424\ R27} \qquad \underset{70\overline{)38{,}296}}{547\ R6} \right)$$

More to Explore • Applications

Divide the class into small groups. Pass around a paper bag containing slips of paper that have hourly babysitting rates written on them. (for example, $8.25 through $12.75 per hour) Each group draws a slip of paper and figures the fee they would earn for working these hours:

1. 12:00 to 5:00   4. 4:15 to 6:45

2. 7:00 to 9:00   5. 1:00 to 5:15

3. 6:30 to 10:00   6. 3:00 to 7:45

Have students repeat the exercise drawing different rates from the bag.

ESL/ELL STRATEGIES

Explain that the *multiple* of a number is that number multiplied by a second number. For example, the multiples of ten are 10×2, 10×3, 10×4, or twenty, thirty, forty, and so on. Have students say the first six multiples of ten beginning with 10×1. (ten, twenty, thirty, forty, fifty, sixty)

© Pearson Education, Inc./Dale Seymour Publications/Pearson Learning Group. All rights reserved. Copying strictly prohibited.

1 Getting Started

Objective
• To divide 2- or 3-digit numbers by a 2-digit divisor

Materials
*10 index cards

Warm Up • Mental Math
Have students round each number to the nearest ten.

1. 42 (40)
2. 25 (30)
3. 457 (460)
4. 545 (550)
5. 918 (920)
6. 76 (80)
7. 64 (60)
8. 729 (730)
9. 603 (600)
10. 89 (90)

Warm Up • Pencil and Paper
Have students find each quotient.

1. (14) 8)112
2. (442 R2) 6)2,654
3. (8 R34) 90)754
4. (56 R56) 60)3,416
5. (6,549) 3)19,647
6. (316) 40)12,640

Name _____

Dividing by a 2-Digit Divisor

During its special sale, ABC Appliance stayed open around-the-clock. How many days without closing was the store open for the sale?

We want to know how many sale days the store was open.

We know ABC Appliances was open __98__ hours, and there are __24__ hours in one day.

To find the number of days, we divide the total number of hours by the number of hours in one day.

We divide __98__ by __24__.

Divide the tens. 24 > 9	Not enough tens. Think of the dividend as 98 ones.	Round the divisor to the nearest 10. Estimate how many 20s in 98 ones. Think: How many 2s in 9?
24)98	24)98	4 R2 / 24)98 / − 96 / 2

CHECK
24
× 4
——
96

96
+ 2
——
98

ABC Appliance was open about __4__ days without closing.

Getting Started

Divide and check.

1. 3 R11 / 18)65
2. 4 R1 / 12)49
3. 4 R9 / 39)165
4. 6 / 44)264
5. 5 R3 / 54)273
6. 7 R33 / 36)285
7. 9 R60 / 62)618
8. 8 R12 / 17)148

Copy and divide.

9. 300 ÷ 43 — 6 R42
10. 536 ÷ 77 — 6 R74
11. 581 ÷ 83 — 7

---

2 Teach

Introduce the Lesson Have a student read the problem and tell what is to be solved. (the number of days ABC Appliance Store was open for the sale) Ask what information is needed to solve the problem. (the total number of hours the store was open and the number of hours in a day) Point out that students must supply some of the needed information. (the number of hours in a day, 24)

Have students read and complete the information sentences.

Emphasize that students must think through each step of the division before they begin to divide. Work through the model on the board or on an overhead projector as students follow in their text. Ask a student to show how to check the quotient using multiplication and addition.

Develop Skills and Concepts Help students decide where to start dividing a 3-digit dividend. Suggest they use a piece of paper to cover all but the greatest place-value digit in the dividend. Then, have students ask themselves, *Are there enough hundreds (or tens)?* If the answer is no, students should realize there will be one digit in the quotient and they will be dividing ones. The quotient digit will be in the ones place. Write these problems on the board:

(4 R87) 87)435   (6 R10) 21)136   (4 R25) 52)233

(9 R18) 65)603   (9 R8) 39)359   (7 R7) 71)504

Have volunteers work at the board. Remind students that they can round the divisor to estimate the quotient. This will help them to determine where to place the first digit in the quotient and what that digit should be.

Practice

Divide and check.

1. $29\overline{)75}$ — 2 R17
2. $24\overline{)75}$ — 3 R3
3. $43\overline{)235}$ — 5 R20
4. $58\overline{)575}$ — 9 R53

5. $72\overline{)432}$ — 6
6. $37\overline{)163}$ — 4 R15
7. $68\overline{)308}$ — 4 R36
8. $53\overline{)437}$ — 8 R13

9. $49\overline{)304}$ — 6 R10
10. $39\overline{)250}$ — 6 R16
11. $62\overline{)535}$ — 8 R39
12. $85\overline{)743}$ — 8 R63

Copy and divide.

13. $163 \div 29$ 5 R18
14. $235 \div 33$ 7 R4
15. $204 \div 56$ 3 R36
16. $396 \div 44$ 9

17. $658 \div 72$ 9 R10
18. $304 \div 34$ 8 R32
19. $319 \div 52$ 6 R7
20. $500 \div 68$ 7 R24

21. $372 \div 41$ 9 R3
22. $437 \div 87$ 5 R2
23. $205 \div 34$ 6 R1
24. $260 \div 47$ 5 R25

25. $568 \div 94$ 6 R4
26. $346 \div 89$ 3 R79
27. $361 \div 51$ 7 R4
28. $565 \div 88$ 6 R37

Problem Solving

Solve each problem.

29. The basketball team earned $215 by washing cars. They want to buy basketballs that cost $39 each. How many can the team buy?
5 basketballs

30. Mr. Lyons used 248 gallons of hot water during December. How many gallons of hot water did Mr. Lyons use on an average day?
8 gallons

Now Try This!

Find the missing digits.

```
      8 R2 0          6 R 2          7 R 4          9 R 2 2
63)5 2 4        8 9)5 3 6      33)2 3 5      5 1)4 8 1
 -5 0 4          -5 3 4         -2 3 1         -4 5 9
    2 0               2              4             2 2
```

Lesson 4-8 • Dividing by 2-Digit Divisors

For Mixed Abilities

Common Errors • Intervention

Watch for students who have difficulty deciding where to place the first digit of the quotient. Have students multiply the divisor by 1, by 10, and by 100 to determine the number of digits in the quotient. For example,

$$68\overline{)457}$$
$$1 \times 68 = 68$$
$$10 \times 68 = 680$$
$$100 \times 68 = 6{,}800$$

Since the dividend is between 1×68 and 10×68, the quotient begins in the ones place. The quotient will be a 1-digit number. Remind students that the quotient could have a remainder.

Enrichment • Estimation

Write the following division problems on the board:

1. $78\overline{)469}$ (6 R1)
2. $34\overline{)308}$ (9 R2)
3. $65\overline{)605}$ (9 R20)
4. $23\overline{)146}$ (6 R8)
5. $45\overline{)370}$ (8 R10)
6. $58\overline{)465}$ (8 R1)

Have students copy each one on a separate index card and estimate the quotients. Have them sort their cards into stacks according to the quotient: one stack for 6, one for 8, and one for 9. They should then divide to check their estimates.

More to Explore • Number Sense

Tell students to use the digits 1 through 9 and arrange them on the three sides of a triangle so that the sum of the digits on each side is 17. Then, have them arrange the digits so that they equal 20.

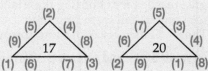

---

3 Practice

Have students complete all the exercises. Remind them to round the divisor to the nearest ten to help estimate quotients.

Now Try This! Students can use trial and error as well as basic facts to find the missing digits.

4 Assess

Write $44\overline{)396}$ on the board.

Ask, *How do you know that the quotient is a one-digit number?* (Possible answer: There are not enough tens in the dividend to be divided by 44.)

4-9 Estimated Quotients

pages 77–78

1 Getting Started

Objective
- To adjust estimated quotients to determine the actual quotient

Warm Up • Mental Math
Have students round each divisor. Then, determine the first digit in the quotient.

1. 667 ÷ 82 (8)
2. 471 ÷ 74 (6)
3. 165 ÷ 18 (8)
4. 647 ÷ 71 (9)
5. 364 ÷ 52 (7)
6. 201 ÷ 58 (3)
7. 443 ÷ 83 (5)
8. 309 ÷ 67 (4)
9. 363 ÷ 46 (7)
10. 734 ÷ 85 (8)

Warm Up • Pencil and Paper
Have students find each quotient.

1. (4 R19) 94)395
2. (5 R8) 51)263
3. (4 R22) 38)174
4. (5 R60) 76)440
5. (8 R4) 39)316
6. (8 R16) 25)216

Name _____

Estimating Quotients

A Mercury year is the same as 59 Earth days. If you lived on Mercury, how much older in Mercury years would you be for each Earth year?

We want to find how many Mercury years are in one Earth year.

We know that one Mercury year is __59__ days long, and one Earth year is __365__ days long.

To find the Mercury equivalent of an Earth year, we divide the number of days in one Earth year by the number of days in one Mercury year.

We divide __365__ by __59__.

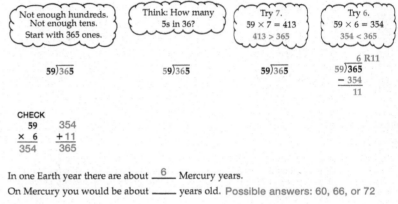

Not enough hundreds. Not enough tens. Start with 365 ones.

59)365

Think: How many 5s in 36?

59)365

Try 7.
59 × 7 = 413
413 > 365

59)365

Try 6.
59 × 6 = 354
354 < 365

     6 R11
59)365
   − 354
     11

CHECK

59	354
× 6	+ 11
354	365

In one Earth year there are about __6__ Mercury years.
On Mercury you would be about _____ years old. **Possible answers: 60, 66, or 72**

Getting Started

Divide and check.

1. 6 R7 — 18)115
2. 4 R80 — 93)452
3. 6 R57 — 73)495
4. 5 R18 — 36)198

Copy and divide.

5. 417 ÷ 87 — 4 R69
6. 252 ÷ 69 — 3 R45
7. 175 ÷ 45 — 3 R40
8. 297 ÷ 35 — 8 R17

Lesson 4-9 • Estimating Quotients

77

---

2 Teach

Introduce the Lesson Have students determine what needs to be found. (the number of Mercury years in one Earth year) Then, have them state the given facts. (a Mercury year is about 59 Earth days) Have students complete the information sentences.

Work through each step of the model. Have students give reasons for choosing the trial quotient of 7 and then rejecting it. Tell them rounding the divisor will help. They will then try 6 and find this recalls the fact 36 ÷ 6.

Develop Skills and Concepts Correcting estimates to adjust a quotient digit is difficult for many students. Rounding the divisor is very helpful, because it puts students on the familiar ground of multiplying or dividing by multiples of 10. To help students gain skill in adjusting trial quotients, write these exercises on the board:

(7 R30) 53)401   (6 R1) 47)283   (8 R9) 68)553

Have volunteers do the division on the board as their classmates work at their disks.

Remind students that a quotient digit may also be too small. If the remainder is as large as or larger than the divisor, then the quotient should be the next higher number. Have students determine which quotient digits are too small:

6 49)345 (too small)   6 58)353 (correct)   5 48)296 (too small)

3 Practice

Have students complete all the exercises. Emphasize that when the first estimate for a quotient digit is too large, students must try the next smaller number.

T77

Practice

Divide and check.

1. 12)86   **7 R2**
2. 16)148   **9 R4**
3. 49)345   **7 R2**
4. 23)180   **7 R19**

5. 36)218   **6 R2**
6. 75)530   **7 R5**
7. 12)110   **9 R2**
8. 48)296   **6 R6**

9. 86)364   **4 R20**
10. 58)355   **6 R7**
11. 22)125   **5 R15**
12. 28)239   **8 R15**

Copy and divide.

13. 327 ÷ 42   **7 R33**
14. 288 ÷ 48   **6**
15. 305 ÷ 37   **8 R9**
16. 225 ÷ 43   **5 R10**

17. 515 ÷ 66   **7 R53**
18. 418 ÷ 44   **9 R22**
19. 771 ÷ 84   **9 R15**
20. 216 ÷ 36   **6**

21. 402 ÷ 83   **4 R70**
22. 596 ÷ 71   **8 R28**
23. 261 ÷ 63   **4 R9**
24. 299 ÷ 35   **8 R19**

25. 456 ÷ 56   **8 R8**
26. 705 ÷ 86   **8 R17**
27. 423 ÷ 57   **7 R24**
28. 390 ÷ 78   **5**

Problem Solving

Solve each problem.

29. Chef George is baking muffins. He needs to bake 204 muffins for Sunday brunch. How many dozen is this?

    17 dozen

30. Kathryn is packing 328 glasses. She puts them in boxes that hold 68 glasses each. How many boxes does she need?

    5 boxes

⟨ Now Try This! ⟩

Use each of the digits 5, 6, 7, 8, and 9 only once to form a division problem.

Find the smallest possible quotient.   **5 R77**
                                         98)567

Find the largest possible quotient.   **1,975 R1**
                                        5)9,876

78   Lesson 4-9 • Estimating Quotients

Now Try This! Discuss the relationship between the dividend and the size of the divisor. Students should be able to generalize that for a given dividend, the smaller the divisor, the larger the quotient; the larger the divisor, the smaller the quotient.

4 Assess

Write the following on the board:

        4   5   6
78)390   78)390   78)390

Have students determine if each quotient is too small, correct, or too large. Then, have them multiply to check. (too small, 78 × 4 = 312; correct, 78 × 5 = 390; too large, 78 × 6 = 468)

For Mixed Abilities

Common Errors • Intervention

Watch for students who fail to notice that the first estimated quotient is not enough. Have them compare the difference in the subtraction step with the divisor. If the difference is greater than the divisor, the quotient digit is too small.

Enrichment • Number Sense

Have students play Tic Tac Quotient. Ask them to draw a 3 by 3 grid and write the following answers in the boxes randomly: 8, 3 R1, 5 R2, 4 R1, 7, 9 R11, 7 R5, 8 R9, 8 R7. Have them work these division problems on paper:

369 ÷ 45 (8 R9)   713 ÷ 78 (9 R11)

342 ÷ 68 (5 R2)   341 ÷ 85 (4 R1)

304 ÷ 38 (8)   413 ÷ 59 (7)

238 ÷ 79 (3 R1)   447 ÷ 55 (8 R7)

341 ÷ 48 (7 R5)

As students complete each problem, have them cross out the answer on their grids. The first student to cross out three answers in a row, column, or diagonal wins the game.

More to Explore • Applications

Show students two different sizes of cereal boxes with prices marked, and ask them to figure out which size is the better buy. Discuss unit pricing with the class and then have students compare the unit prices of other items. A field trip to a grocery store is an ideal way to do this. Ask, *Are bigger boxes always a better buy?*

Have students compare the sizes of boxes for a particular product. Ask, *Does the bigger box always contain more?* Suggest that students bring in some examples of products whose packaging is misleading compared to a similar product. Tell them to make sure they have the calculations to back up their example.

4-10 Dividing up to 6-Digit Numbers

pages 79–80

1 Getting Started

Objective
• To divide a number through 6 digits by a 2-digit divisor

Warm Up • Mental Math
Have students find each product.

1. 75 × 10 (750)
2. 48 × 100 (4,800)
3. 64 × 100 (6,400)
4. 31 × 10 (310)
5. 27 × 1,000 (27,000)
6. 56 × 10,000 (560,000)
7. 18 × 1,000 (18,000)
8. 93 × 10,000 (930,000)
9. 62 × 1,000 (62,000)
10. 84 × 10,000 (840,000)

Warm Up • Pencil and Paper
Have students find each quotient.

1. $48\overline{)341}$ (7 R5)
2. $53\overline{)462}$ (8 R38)
3. $37\overline{)233}$ (6 R11)
4. $62\overline{)598}$ (9 R40)
5. $25\overline{)237}$ (9 R12)
6. $75\overline{)423}$ (5 R48)

2 Teach

Introduce the Lesson Call on a student to read the problem aloud. After students identify the problem question and the facts that are given, have them complete the information sentences. Point out that in the model, the steps in the division process remain the same, regardless of the size of the dividend and the divisor. Write the steps on the board: **divide, multiply, subtract, compare, and bring down.** Have a student identify each step in the model.

Develop Skills and Concepts Remind students that the first step in any division problem is to decide where to put the first quotient digit. Point out that even though the dividend in the model problem contains ten thousands and thousands, the quotient cannot begin at the ten thousands or thousands place. Write the

---

Dividing up to 6-Digit Numbers

Computers store information in bytes. One byte is the same as one character on a typewriter. Marcia's computer can store 65,536 bytes. How many lines can Marcia store in her computer?

> I can get 78 bytes on one line.

We want to know the number of lines that can be stored.

We know that the computer can store __65,536__ bytes, and one line is __78__ bytes long.

To find how many lines can be stored, we divide the total number of bytes by the number of bytes in one line. We divide __65,536__ by __78__.

$$78\overline{)65,536}$$

> Not enough ten thousands.
> Not enough thousands.
> Start with 655 hundreds.

Divide the hundreds.	Divide the tens.	Divide the ones. Write the remainder.
$\begin{array}{r} 8 \\ 78\overline{)65,536} \\ -\ 62\ 4 \\ \hline 3\ 1 \end{array}$	$\begin{array}{r} 84 \\ 78\overline{)65,536} \\ -\ 62\ 4\downarrow \\ \hline 3\ 13 \\ -\ 3\ 12 \\ \hline 1 \end{array}$	$\begin{array}{r} 840\ \text{R16} \\ 78\overline{)65,536} \\ -\ 62\ 4 \\ \hline 3\ 13 \\ -\ 3\ 12\downarrow \\ \hline 16 \end{array}$

Marcia can store just over __840__ lines in her computer.

Getting Started

Divide and check.

1. $43\overline{)45,639}$ 1,061 R16
2. $72\overline{)\$264.24}$ \$3.67
3. $68\overline{)216,450}$ 3,183 R6

Copy and divide.

4. 537,646 ÷ 48   11,200 R46
5. \$1,776.32 ÷ 32   \$55.51
6. 82,678 ÷ 67   1,234

---

following pattern on the board and have students compare each product with the dividend, 65,536:

$$10,000 × 78 = 780,000 \text{ (too big)}$$
$$1,000 × 78 = 78,000 \text{ (too big)}$$
$$100 × 78 = 7,800 \text{ (too small)}$$

Since 7,800 < 65,536 and 78,000 > 65,536, the quotient must be between 100 and 999. Because it will have three digits, it must start in the hundreds place.

3 Practice

Have students complete all the exercises. Point out that dividing money is like dividing whole numbers, except that quotients must contain the dollar sign and decimal point.

T79

Practice

Divide and check.

1. 27)8,699   **322 R5**
2. 36)5,047   **140 R7**
3. 18)2,340   **130**

4. 57)$440.04   **$7.72**
5. 86)96,475   **1,121 R69**
6. 73)126,475   **1,732 R39**

Copy and divide.

7. 5,758 ÷ 46
   125 R8
8. 6,794 ÷ 37
   183 R23
9. 77,659 ÷ 77
   1,008 R43

10. 6,039 ÷ 12
    503 R3
11. 15,276 ÷ 23
    664 R4
12. 46,745 ÷ 66
    708 R17

13. 25,000 ÷ 16
    1,562 R8
14. 57,405 ÷ 84
    683 R33
15. 96,214 ÷ 92
    1,045 R74

16. 146,556 ÷ 58
    2,526 R48
17. 207,950 ÷ 63
    3,300 R50
18. $1,995.80 ÷ 85
    $23.48

19. 216,743 ÷ 29
    7,473 R26
20. 816,321 ÷ 42
    19,436 R9
21. 567,245 ÷ 73
    7,770 R35

Problem Solving

Use the chart to solve each problem.

The 1860 Election Results

Candidate	Popular Vote	Electoral Vote
Lincoln	1,865,593	180
Douglas	1,382,713	12
Breckenridge	848,356	72

22. How many times more electoral votes did Lincoln receive than Douglas?
    15

23. How many more popular votes did Lincoln receive than Douglas?
    482,880

24. How many times more electoral votes did Breckenridge receive than Douglas?
    6

25. Which of the 3 candidates won a majority of the total popular vote? Of the total electoral vote?
    none; Lincoln

80

Lesson 4-10 • Dividing up to 6-Digit Numbers

For Mixed Abilities

Common Errors • Intervention

Some students may have difficulty keeping all the digits aligned and in the proper places. Have them do their work on grid paper so that the squares and lines can help them divide correctly.

Enrichment • Number Sense

Have students complete the cross number puzzle.

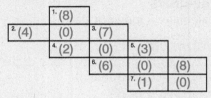

Across		Down	
2.	26,455 ÷ 65	1.	37,694 ÷ 47
4.	9,338 ÷ 46	3.	20,474 ÷ 29
6.	31,616 ÷ 52	5.	23,177 ÷ 77
7.	930 ÷ 93		

More to Explore • Measurement

Ask students what an odometer is and how it works. Tell the class they can find out how far they have traveled on their bicycles without the use of an odometer. Have students make suggestions on how this can be done. Then, have them follow these steps:

1. Measure the circumference of their bicycle wheel.

2. Tie a piece of string to the rim of the wheel or mark the rim with a piece of chalk.

3. Count the number of revolutions made by the wheel as they ride the bike a measured distance.

4. Multiply the number of revolutions by the circumference to find the distance traveled.

Extend the activity by challenging students to find the length of the school playground, sidewalk, driveway, and so on, using the circumference method.

4 Assess

Write the following on the board:

56)498,344   93)965,154   79)12,877

For each quotient, have students identify where to place the first digit of the quotient. (thousands place; ten thousands place; hundreds place)

5 Mixed Review

1. $300.50 − $179.48 ($121.02)
2. 6 × 10⁴ (60,000)
3. 37 + 143 + 98 (278)
4. 6,472 ÷ 80 (80 R72)
5. 54 × 30 (1,620)
6. 5⁴ (625)
7. 4,208 × 3 (12,624)
8. 445 ÷ 63 (7 R4)
9. 62,398 + 597,406 (659,804)
10. 60 × 7,000 (420,000)

T80

1 Getting Started

Objective
• To divide a number through 6 digits, by a 3-digit divisor

Materials
calculators; index cards

Warm Up • Mental Math
Have students tell how many digits will be in each quotient.

1. 5,916 ÷ 67 (2)
2. 8,699 ÷ 35 (3)
3. 6,012 ÷ 56 (3)
4. 25,651 ÷ 49 (3)
5. 46,102 ÷ 21 (4)
6. 37,864 ÷ 92 (3)
7. 60,535 ÷ 54 (4)
8. 33,156 ÷ 33 (4)
9. 19,495 ÷ 12 (4)
10. 593,124 ÷ 36 (5)

Warm Up • Pencil and Paper
Have students find each quotient.

1. 30,196 ÷ 37 (816 R4)
2. 1,246 ÷ 39 (31 R37)
3. 67,851 ÷ 42 (1,615 R21)
4. 8,532 ÷ 76 (112 R20)
5. 19,780 ÷ 61 (324 R16)

Name _____

Lesson 4-11

Dividing by a 3-Digit Divisor

The speed of sound depends upon the material through which the sound waves pass. About how many seconds does it take for sound to travel 2,685 meters through the air?

We want to know how long it takes sound waves to travel a certain distance through the air.

We know sound travels at about __335__ meters per second through air, and this sound traveled __2,685__ meters.

To find the number of seconds it took the sound to travel, we divide the distance the sound traveled by the speed per second. We divide __2,685__ by __335__.

$335\overline{)2,685}$

Not enough thousands.
Not enough hundreds.
Not enough tens.
Start with 2,685 ones.

$$335\overline{)2,685}$$
$$\begin{array}{r} 8 \\ -2\,680 \\ \hline 5 \end{array}$$

It takes about __8__ seconds for sound to travel 2,685 meters through the air.

⟨─ 335 mps

Getting Started

Divide and check.

1. $237\overline{)1,459}$ 6 R37

2. $118\overline{)\$271.40}$ $2.30

3. $705\overline{)4,210}$ 5 R685

4. $625\overline{)53,750}$ 86

5. $872\overline{)142,136}$ 163

6. $593\overline{)156,000}$ 263 R41

Copy and divide.

7. $272.16 ÷ 486
$0.56

8. 29,675 ÷ 342
86 R263

9. 535,684 ÷ 701
764 R120

Lesson 4-11 • Dividing by a 3-Digit Divisor

81

---

2 Teach

Introduce the Lesson Have students read the problem and note the sign in the illustration that tells how fast sound travels. Have students tell what is to be solved and what information is given. Have students complete the information sentences and work through the model. On an overhead projector, demonstrate uncovering each place value to determine the placement of the first quotient digit.

Develop Skills and Concepts Explain that division by a 3-digit divisor follows the same steps as division by 2-digit divisors: divide, multiply, subtract, compare, and bring down. Emphasize that these steps are repeated until there are no more digits to bring down.

Encourage students to round divisors to the nearest hundred to estimate trial quotients.

Continue developing division skills with these problems:

1. 23,220 ÷ 215 (108)
2. 46,802 ÷ 341 (137 R85)
3. 564,815 ÷ 645 (875 R440)

3 Practice

Have students complete all the exercises. You may want to allow them to use calculators to check each quotient.

T81

Practice

Divide and check.

1. $216\overline{)1,846}$ **8 R118**

2. $489\overline{)\$136.92}$ **$0.28**

3. $703\overline{)45,896}$ **65 R201**

4. $514\overline{)18,236}$ **35 R246**

5. $651\overline{)171,513}$ **263 R300**

6. $867\overline{)505,260}$ **582 R666**

Copy and divide.

7. $128.31 \div 329$
   $0.39

8. $58,953 \div 636$
   92 R441

9. $26,412 \div 408$
   64 R300

10. $51,634 \div 750$
    68 R634

11. $620,834 \div 987$
    629 R11

12. $2,109.80 \div 548$
    $3.85

Problem Solving

Solve each problem.

13. A jumbo jet flies about 596 miles per hour. How many hours will it take for the jet to fly 3,576 miles?
    6 hours

14. The wingspan of a jumbo jet is 198 feet. If 5 jets are parked wingtip to wingtip, what is the total distance across?
    990 feet

15. The 342 pieces of freight on a cargo ship weighed in at 59,850 pounds. What is the average weight of each piece of freight?
    175 pounds

16. A test pilot earns $171,900 a year. He makes 15 test flights each month. How much does he earn for each flight?
    $955

(Now Try This!)

Study this problem. It shows how to use a calculator to divide large numbers.

$6015 \div 235 = 25.595745$ $235 \times 25 = 5875$ $6015 - 5875 = 140$

$$235\overline{)6,015} \quad 25$$

$$235\overline{)6,015} \quad 25 \\ -5,875 \\ \hline$$

$$235\overline{)6,015} \quad 25 \text{ R}140 \\ -5,875 \\ \hline 140$$

Use a calculator to find each quotient.

1. $478\overline{)9,708}$ **20 R148**

2. $672\overline{)90,883}$ **135 R163**

3. $928\overline{)79,485}$ **85 R605**

82   Lesson 4-11 • Dividing by a 3-Digit Divisor

For Mixed Abilities

Common Errors • Intervention

Some students may stop dividing before the division process has been completed. Encourage them to draw an arrow as they bring down each digit from the dividend. The arrows will help show them that every digit in the dividend must be used before the division is complete.

$$180\overline{)171,900} \quad 955 \\ 162\ 0 \\ \hline 990 \\ 900 \\ \hline 900 \\ 900 \\ \hline 0$$

Enrichment • Number Sense

Have students find the missing numbers in the table below.

Dividend	954,629	695,067	997,097
Divisor	613	329	(142)
Quotient	(1,557)	2,112	7,021
Remainder	188	(219)	115

More to Explore • Biography

Christian Huygens, the 17th-century Dutch mathematician and astronomer, is best known for the invention of the pendulum clock. This invention greatly increased the accuracy of time measurement.

Huygens' lifelong interest was lens grinding. One of his lenses enabled him to see the first satellite of Saturn. A year later, Huygens discovered that Saturn was surrounded by a flattened ring. Huygens explained the phases and changes in the ring's shape in one of his works, *Systema Saturnium*.

Huygens worked both at The Hague, in relative solitude, and in Paris, where he mixed with some of the greatest scientists of the age and shared ideas.

In the final years of his life, Huygens presented one of the earliest discussions of the possibility of extraterrestrial life. Three hundred years later, the existence of such life is still only a possibility and an open question.

Now Try This! Be sure students have calculators to do this exercise. After the first division, they can drop all numbers to the right of the decimal point, work the multiplication, and use subtraction to come up with the correct remainder. Tell students to follow the same steps in the next three division problems.

4 Assess

Using 638 as a divisor, have students determine the number of digits in the quotient for each dividend.

752 (one-digit quotient)

1,239 (one-digit quotient)

8,460 (two-digit quotient)

369,125 (three-digit quotient)

T82

4-12 Problem Solving: Act It Out

pages 83–84

1 Getting Started

Objective
- To solve problems by acting them out

Materials
rectangular piece of paper

Warm Up • Mental Math
Have students identify the amount of time represented by each measurement.

1. week (7 days)
2. silver anniversary (25 years)
3. biweekly (2 weeks)
4. golden anniversary (50 years)
5. semi-annually (6 months)
6. biyearly (2 years)
7. millisecond ($\frac{1}{1,000}$ of a second)
8. millennium (1,000 years)

Warm Up • Number Sense
Have students:

1. count the number of 1s in their classmates' addresses.
2. count the number of 0s in their classmates' addresses.
3. $0.05 + $0.06 + $0.07 ($0.18)
4. 14 ÷ 5 (2.8)
5. 56 ÷ 5 (11.2)
6. fold a rectangle in half as many times as they can.

Problem Solving: Act It Out

An explorer wants to cross a desert with the help of porters. It takes six days to make the crossing. Each person can only carry enough food for four days. How many porters does the explorer need to take in order to complete the trip?

★ SEE
We want to know how many porters the explorer will need to take on the trip.

It takes ___6___ days to cross the desert.

Each person can carry enough food to last ___4___ days.

★ PLAN
Acting the problem out will help us to think of possible solutions. We can begin by using one student to act as the explorer and one student to act as a porter. If necessary we can add more porters and try again until we reach a solution.

★ DO
Act the problem out. Answer these questions to help reach a solution.

How much food does each person have at the end of the first day?

How many more days will it take to complete the trip? Is there any way for the explorer to complete the trip with the help of the porter or porters he presently has? If a solution cannot yet be reached, continue acting out the trip adding another porter each time. At the end of each day go back to answer the same questions until you find a solution.

The explorer will need to take ___2___ porters with him.

★ CHECK
A good way to check the solution of this problem is to explain what will happen on the trip with the required number of porters.

---

2 Teach

Introduce the Lesson Tell students some problems can be solved by actually acting out the process. Have a student read the problem aloud. Work through the SEE and PLAN stages together, allowing time for input and discussion.

Develop Skills and Concepts Divide the class into groups of 5 students each. Have each group select an explorer, with the others serving as porters. Allow sufficient time for each group to act out a possible solution. Have the explorers tell how many porters they used to cross the desert. Select a group with the correct solution to act it out for the entire class. Ask students to describe each step of the process. If time permits, select one of the solutions that used more porters than needed, have the students act it out, and have students explain why the process was defective.

3 Apply

Solution Notes
1. 8 folds would mean an increase of 256 thicknesses. (2^8)
2. Complete the computation that $\frac{1}{2}$ of $\frac{1}{2}$ of $\frac{1}{2}$ ... of $\frac{1}{2}$ of $\frac{1}{2} = \frac{1}{256}$.
3. 5 days $0.25 $\longrightarrow$ 1 roll; you will need to open 1 roll of pennies even though the amount earned is less than $0.50.
   10 days $1.00 $\longrightarrow$ 2 rolls
4. 5 days $0.31 $\longrightarrow$ 1 roll
   10 days $10.23 $\longrightarrow$ 21 rolls

Higher-Order Thinking Skills
5. **Analysis:** Both will earn the same amount of money: Income = number of hours × hourly rate, or $I = h \times r$; double either h or r and the results are the same: i.e., $(h \times 2) \times r = h \times (2 \times r)$.

Apply

Act out these situations to help you solve each problem.

1. It has been said that a rectangular piece of paper cannot be folded in half more than eight times. Using several different sizes of paper, attempt to create more than eight such folds. Good luck.
   See Solution Notes.

2. Repeat the experiment in Problem 1 by cutting a rectangular piece of paper in half. Take one of the one-half sheets and cut that in half. Take one of those half sheets and cut that one in half. Make eight cuts altogether. Approximately how much larger is the original sheet than the final piece of paper?
   The original will be 256 times larger.

3. A roll of pennies is worth $0.50. If you agree to pay an employee $0.01 the first day of work, $0.03 the second day, $0.05 the third day, $0.07 the fourth day, and so forth, how many rolls of pennies would you need to meet your payroll for five days of work? For 10 days of work?
   $\frac{1}{2}$ roll for 5 days
   2 rolls for 10 days

4. If you agree to pay an employee $0.01 for the first day of work, $0.02 the second day of work, $0.04 the third day, $0.08 the fourth day, and so forth, how many rolls of pennies would you need to meet your payroll after 5 days? After 10 days?
   1 roll for 5 days
   21 rolls for 10 days

> **Problem-Solving Strategy:**
> **Using the Four-Step Plan**
> ★ **SEE** What do you need to find?
> ★ **PLAN** What do you need to do?
> ★ **DO** Follow the plan.
> ★ **CHECK** Does your answer make sense?

---

5. Nip and Tuck get paid by the hour. They usually get the same amount per hour and work the same number of hours each week. Next week, however, Nip has a special project and will earn twice as much per hour as Tuck, but Tuck will work twice as many hours as Nip. Which one will make more money next week?
   neither

6. The Greek Drama Group is discussing how much to charge for tickets to their next play. Members believe that if they charge $14.95 per ticket, they will sell 375 tickets. But if they charge only $11.95 per ticket, they will sell 450 tickets. If you were a member of the group, what would you say to do and why?
   See Solution Notes.

7. If m and n represent two whole numbers, which of the following represents the greatest number?
   a. $m \div n$
   b. $(10 \times m) \div n$
   c. $m \div (10 \times n)$
   b

---

6. **Evaluation:** 375 at $14.95 is $5,606.25 and 450 at $11.95 is $5,377.50. So, if the only purpose is to make money, charge $14.95. However, if the purpose is to get as many people as possible to see the play, then charge $11.95.

7. **Analysis and Evaluation:** The division problem with the largest numerator will have the greatest quotient.

4 Assess

Have students write a word problem for which acting it out would be a good problem-solving strategy. Have students exchange their problems to solve.

For Mixed Abilities

More To Explore • Applications

Have students bring in the business section of the newspaper containing the stock reports. Have them start a brokerage account for themselves. Tell them they may buy 100 shares of one NYSE stock and one AMEX stock. Have students keep a log of the following data:

- initial investment
- daily fluctuation of stock prices for two weeks
- closing figure difference from initial opening figure for each stock

Have students compare results to see who made the most profitable investments.

Chapter 4
Test

page 85

Items	Objectives
1–4	To divide 2- or 3-digit numbers by 1-digit numbers (see pages 63–64)
5–6	To divide 4-, 5-, or 6-digit numbers by 1-digit numbers (see pages 65–66)
7–8	To divide by 1-digit numbers, zeros in the quotient (see pages 67–68)
9–16	To divide 2- or 3-digit numbers by 2-digit numbers (see pages 75–76)
17–19, 23–25	To divide 4-, 5-, or 6-digit numbers by 2-digit numbers (see pages 79–80)
20–22	To divide by 3-digit numbers (see pages 81–82)

Alternate Chapter Test
You may wish to use the Alternate Chapter Test on page 376 of this book for further review and assessment.

T85

Name _____

Divide and check.

1. $7\overline{)648}$ 92 R4

2. $5\overline{)315}$ 63

3. $8\overline{)896}$ 112

4. $2\overline{)\$8.50}$ \$4.25

5. $7\overline{)5{,}273}$ 753 R2

6. $4\overline{)8{,}925}$ 2,231 R1

7. $6\overline{)\$61.86}$ \$10.31

8. $9\overline{)16{,}475}$ 1,830 R5

9. $28\overline{)230}$ 8 R6

10. $46\overline{)230}$ 5

11. $75\overline{)545}$ 7 R20

12. $86\overline{)643}$ 7 R41

13. $31\overline{)127}$ 4 R3

14. $37\overline{)222}$ 6

15. $62\overline{)191}$ 3 R5

16. $72\overline{)321}$ 4 R33

17. $12\overline{)2{,}678}$ 223 R2

18. $96\overline{)\$18{,}500.16}$ \$192.71

19. $83\overline{)216{,}317}$ 2,606 R19

20. $911\overline{)1{,}958}$ 2 R136

21. $315\overline{)\$6{,}914.25}$ \$21.95

22. $723\overline{)582{,}151}$ 805 R136

23. $49\overline{)8{,}267}$ 168 R35

24. $53\overline{)15{,}286}$ 288 R22

25. $67\overline{)116{,}395}$ 1,737 R16

Circle the letter of the correct answer.

1 $3 + n = 9$
$n = ?$
a. 3
b. 6 ⊙
c. 9
d. NG

2 $8 - n = 3$
$n = ?$
a. 5 ⊙
b. 8
c. 11
d. NG

3 $7 \times 9 = n$
$n = ?$
a. 54
b. 56
c. 63 ⊙
d. NG

4 $5 \times n = 35$
$n = ?$
a. 5
b. 7 ⊙
c. 9
d. NG

5 $48 \div 6 = n$
$n = ?$
a. 5
b. 7
c. 8 ⊙
d. NG

6 $5 + 3 \times 4 = n$
$n = ?$
a. 17 ⊙
b. 27
c. 32
d. NG

7 What is the place value of the 6 in 3,248,653?
a. millions
b. thousands
c. tens
d. NG ⊙

8
27,396
+ 42,875
a. 69,161
b. 70,271 ⊙
c. 71,271
d. NG

9
83,029
− 47,853
a. 35,176 ⊙
b. 35,276
c. 44,836
d. NG

10 365×7
a. 2,125
b. 2,555 ⊙
c. 2,655
d. NG

11
$4.96
× 53
a. $39.88
b. $252.88
c. $262.88 ⊙
d. NG

12
4,659
× 468
a. 335,448
b. 2,180,412 ⊙
c. 2,190,412
d. NG

13 $379 \div 6$
a. 63 R1 ⊙
b. 6 R19
c. 631
d. NG

score

STOP

Chapter 4 • Cumulative Assessment

page 86

Items	Objectives
1–2	To review basic facts of addition and subtraction (see pages 1–4)
3	To review basic facts of multiplication (see pages 5–6)
4–5	To review basic facts of division (see pages 7–8)
6	To simplify an expression having 2 or more operations (see pages 11–12)
7	To identify place value through trillions (see pages 25–26)
8	To add two 5-digit numbers (see pages 27–28)
9	To subtract, renaming over zeros (see pages 33–34)
10	To multiply 3-digit factors by 1-digit factors (see pages 45–46)
11	To multiply money by 2-digit factors (see pages 51–52)
12	To multiply 4-digit numbers by 3-digit factors (see pages 55–56)
13	To divide 3-digit numbers by 1-digit numbers (see pages 63–64)

Alternate Cumulative Assessment

Circle the letter of the correct answer.

1. $8 + n = 15$
$n = ?$
a 6
b 7 ⊙
c 8
d NG

2. $7 - n = 5$
$n = ?$
a 2 ⊙
b 6
c 12
d NG

3. $6 \times 8 = n$
$n = ?$
a 36
b 42
c 48 ⊙
d NG

4. $2 \times n = 18$
$n = ?$
a 7
b 8
c 9 ⊙
d NG

5. $21 \div 3 = n$
$n = ?$
a 6
b 7 ⊙
c 8
d NG

6. $4 + 2 \times 7 = n$
$n = ?$
a 18 ⊙
b 36
c 42
d NG

7. Give the place value of the 0 in 802,735,946.
a ten thousands
b hundred thousands
c millions
d NG ⊙

8.
67,258
+ 29,764
a 96,922
b 97,022 ⊙
c 97,922
d NG

9.
95,037
− 67,294
a 27,743 ⊙
b 27,753
c 27,843
d NG

10. 792
× 6
a 4,252
b 4,752 ⊙
c 4,852
d NG

11. $5.83
× 49
a $255.67
b $278.67
c $285.67 ⊙
d NG

T86

5-1 Least Common Multiple

pages 87–88

1 Getting Started

Objective
- To find the least common multiple of two or more numbers

Vocabulary
least common multiple (LCM)

Warm Up • Mental Math
Have students name the next three numbers in each sequence.

1. 2, 4, 6, (8, 10, 12)
2. 3, 6, 9, (12, 15, 18)
3. 4, 8, 12, (16, 20, 24)
4. 5, 10, 15, (20, 25, 30)
5. 8, 16, 24, (32, 40, 48)
6. 10, 20, 30, (40, 50, 60)

Warm Up • Pencil and Paper
Have students write:

1. the first 5 nonzero multiples of 3 (3, 6, 9, 12, 15)
2. the first four multiples of 5 greater than 55 (60, 65, 70, 75)
3. the first 5 nonzero multiples of 14 (14, 28, 42, 56, 70)
4. the first four multiples of 100 greater than 1,000 (1,100, 1,200, 1,300, 1,400)
5. the first 5 nonzero multiples of 12 (12, 24, 36, 48, 60)

2 Teach

Introduce the Lesson Work through the model with students to find the least common multiple of 6 and 9. Have students explain the meaning of the terms *multiples*, *common multiple*, and *least common multiple* (LCM).

Provide examples to clarify the difference between factors and multiples. Point out that factors are the numbers being multiplied and multiples are the products. Explain that although 0 is a multiple of every number, we usually work with the nonzero multiples of numbers.

Develop Skills and Concepts Review the idea that in a multiplication equation, the product is also called a multiple.

- Have students name the first six nonzero multiples of 2. (2, 4, 6, 8, 10, 12) Point out that each of these numbers is a multiple of 2, because 2 is a factor of each.

Number Theory and Fractions

Lesson 5-1

Least Common Multiple

In a multiplication equation, the product is a multiple of each factor. In the multiplication equation $6 \times 9 = 54$, the product 54 is a multiple of 6 and also a multiple of 9. Sometimes, we need to find the least common multiple of two numbers. What is the least common multiple of 6 and 9?

To find the least common multiple (LCM) of 6 and 9, we identify the multiples the two numbers have in common.

Multiples of 6:
  6, 12, 18, 24, 30, 36, . . .

Multiples of 9:
  9, 18, 27, 36, . . .

The first two common multiples of 6 and 9 are __18__ and __36__.

The **LCM** of 6 and 9 is __18__.

Getting Started

Write the first multiples of each number. Circle the common multiples.

1. 4   4, 8, ⑫ 16, 20, ㉔ 28, 32, ㊱ 40

2. 6   6, ⑫ 18, ㉔ 30, ㊱ 42, 48, 54, 60

Write the first three common multiples for each pair of numbers.

3. 4 and 6   4. 2 and 5   5. 8 and 12   6. 10 and 15
   12, 24, 36   10, 20, 30   24, 48, 72   30, 60, 90

Write the LCM for each set of numbers.

7. 4 and 6   8. 5 and 7   9. 10 and 12   10. 6, 9, and 12
   12   35   60   36

- Then, have students name the first six nonzero multiples of 4. (4, 8, 12, 16, 20, 24)

- List the multiples of 2 and 4 on the board. Point out that 4, 8, and 12 are common multiples of 2 and 4. The least common multiple of 2 and 4 is 4.

Continue developing this skill by having students write the LCM of each of the following pairs of numbers: 3 and 5 (15), 5 and 4 (20), and 5 and 10 (10).

3 Practice

Have students complete all the exercises. Tell students to first determine the common multiples of each number, then determine the LCM.

Practice

Write the first four multiples for each number.

1. 2
   2, 4, 6, 8
2. 7
   7, 14, 21, 28
3. 12
   12, 24, 36, 48
4. 11
   11, 22, 33, 44
5. 4
   4, 8, 12, 16

6. 8
   8, 16, 24, 32
7. 6
   6, 12, 18, 24
8. 15
   15, 30, 45, 60
9. 20
   20, 40, 60, 80
10. 16
    16, 32, 48, 64

Write the first three common multiples for each pair of numbers.

11. 4, 6
    12, 24, 36
12. 6, 9
    18, 36, 54
13. 4, 8
    8, 16, 24
14. 6, 10
    30, 60, 90
15. 2, 5
    10, 20, 30

16. 9, 12
    36, 72, 108
17. 6, 12
    12, 24, 36
18. 5, 8
    40, 80, 120
19. 3, 10
    30, 60, 90
20. 6, 16
    48, 96, 144

Write the LCM for each set of numbers.

21. 4, 8
    8
22. 5, 7
    35
23. 4, 5
    20
24. 9, 15
    45
25. 7, 8
    56

26. 2, 3, 4
    12
27. 3, 4, 5
    60
28. 4, 6, 8
    24
29. 3, 6, 9
    18
30. 8, 9, 12
    72

[Now Try This!]

Multiples of 2 are called **even numbers**. All other counting numbers are called **odd numbers**.

Use the word *even* or *odd* to complete each equation.
Write an example of each equation.

Examples will vary.

1. even number + even number = __even__ number   Example: _____

2. even number + odd number = __odd__ number   Example: _____

3. odd number + odd number = __even__ number   Example: _____

4. even number × even number = __even__ number   Example: _____

5. even number × odd number = __even__ number   Example: _____

6. odd number × odd number = __odd__ number   Example: _____

Lesson 5-1 • Least Common Multiple

Now Try This! Have students substitute several even or odd numbers in each equation to determine if the answer to the equation will be even or odd.

4 Assess

Ask, *Does the least common multiple of 4 and 6 contain 4 and 6 as factors? Explain.* (Yes; Possible answer: The least common multiple of 4 and 6 is 12. 4 is a factor of 12 and 6 is a factor of 12.)

For Mixed Abilities

Common Errors • Intervention

Some students may have difficulty finding the least common multiple of two numbers such as 5 and 9. Have them choose one of the numbers, 9, and list multiples until they identify a number that is also a multiple of the other number, 5. As they list 9, 18, 27, 36, 45, they see that 45 is also a multiple of 5 and is the least common multiple of both numbers, 5 and 9.

Enrichment • Number Sense

Challenge students to complete this table of LCMs.

LCM of and	4	6	10	20
3	(12)	(6)	(30)	(60)
5	(20)	(30)	(10)	(20)
8	(8)	(24)	(40)	(40)
16	(16)	(48)	(80)	(80)

More to Explore • Number Sense

As a class, have students research the Egyptian system of numeration called hieroglyphics. Have them create a bulletin-board display depicting the symbols and information about the system's history, development, and application. Then, have students list common numbers they use frequently, such as telephone numbers, addresses, and Zip Codes, in hieroglyphics.

ESL/ELL STRATEGIES

Define separately each word in *least common multiple*. Explain that *least* means "smallest," *common* means "shared by both numbers," and a *multiple* is "one of two numbers that are multiplied together." Repeat the process for *greatest common factor*. Answer any questions students may have.

5-2 Greatest Common Factor

pages 89–90

1 Getting Started

Objective
- To find the greatest common factor of a pair of numbers

Vocabulary
greatest common factor (GCF)

Warm Up • Mental Math
Have students give an example of each mathematical concept.
1. Commutative Property for multiplication
2. Zero Property for addition
3. inverse operation
4. fact family
5. Property of 1 for multiplication
6. Associative Property for addition
7. least common multiple
8. divisibility test
(Answers will vary.)

Warm Up • Pencil and Paper
Have students write multiplication facts that have each of these numbers as a product.
1. 8 (1 × 8, 2 × 4)
2. 12 (1 × 12, 2 × 6, 3 × 4)
3. 10 (1 × 10, 2 × 5)
4. 16 (1 × 16, 2 × 8, 4 × 4)
5. 30 (1 × 30, 2 × 15, 3 × 10, 5 × 6)

Greatest Common Factor

We can use multiplication to find the factors of a number.

To find the factors of 12 or of 18, we think of the multiplication facts that have 12 or have 18 as a product.

__1__ × 12 = 12	__1__ × 18 = 18
__2__ × 6 = 12	__2__ × 9 = 18
__3__ × 4 = 12	__3__ × 6 = 18

The factors of 12 are 1, 2, 3, __4__, __6__, __12__.
The factors of 18 are 1, 2, 3, __6__, __9__, __18__.

To find the **greatest common factor (GCF)** of 12 and 18, we identify the factors the two numbers have in common.

Factors of 12: 1, 2, 3, **4**, **6**, **12**

Factors of 18: 1, 2, 3, 6, **9**, **18**

The common factors of 12 and 18 are
__1__, __2__, __3__, and __6__.

The GCF of 12 and 18 is __6__.

Getting Started

Write the factors for each number.

1. 8	2. 9	3. 15	4. 24	5. 36
1, 2, 4, 8	1, 3, 9	1, 3, 5, 15	1, 2, 3, 4, 6, 8, 12, 24	1, 2, 3, 4, 6, 9, 12, 18, 36

Write the common factors for each pair of numbers.

6. 4, 6	7. 5, 15	8. 10, 15	9. 12, 24	10. 15, 35
1, 2	1, 5	1, 5	1, 2, 3, 4, 6, 12	1, 5

Write the GCF for each pair of numbers.

11. 15, 27	12. 9, 12	13. 6, 24	14. 36, 48	15. 15, 35
3	3	6	12	5

---

2 Teach

Introduce the Lesson Work through the model with students to find the greatest common factor of 12 and 18. Ask students to explain the terms *factor, common factors,* and *greatest common factor* (GCF). Point out that 1 is a factor of every number. Discuss the following steps for finding the GCF of two numbers:

1. Find all of the factors of each number.
2. Identify the numbers that are factors of both numbers. These are common factors.
3. Identify the greatest number in the list of common factors. This is the greatest common factor.

Develop Skills and Concepts Write: **factor × factor = product** on the board. Explain that one way to find the factor of a number is to write the multiplication facts that have the number as a product. On the board, write the multiplication facts that have 16 as a product.

1 × 16 = 16   2 × 8 = 16   4 × 4 = 16

Call on a student to name the factors of 16. (1, 2, 4, 8, and 16)

- Have a student write the multiplication facts and the factors of 36 on the board. (1 × 36 = 36, 2 × 18 = 36, 3 × 12 = 36, 4 × 9 = 36; factors of 36; 1, 2, 3, 4, 9, 12, 18, 36)

- Point out that the common factors of 16 and 36 are 1, 2, and 4 and the GCF of 16 and 36 is 4.

Have students find the GCF of each of the following pairs of numbers: 8 and 12 (4), 14 and 21 (7), 3 and 5 (1).

3 Practice

Have students complete all the exercises. Be sure students use the three steps previously discussed to find the factors, common factors, and greatest common factor of each pair of numbers.

T89

Practice

Write the factors for each number.

1. 6
   1, 2, 3, 6
2. 10
   1, 2, 5, 10
3. 16
   1, 2, 4, 8, 16
4. 20
   1, 2, 4, 5, 10, 20
5. 26
   1, 2, 13, 26

6. 30
   1, 2, 3, 5, 6, 10, 15, 30
7. 32
   1, 2, 4, 8, 16, 32
8. 45
   1, 3, 5, 9, 15, 45
9. 50
   1, 2, 5, 10, 25, 50
10. 56
    1, 2, 4, 7, 8, 14, 28, 56

Write the common factors for each pair of numbers.

11. 6, 8
    1, 2
12. 10, 15
    1, 5
13. 7, 28
    1, 7
14. 18, 24
    1, 2, 3, 6
15. 20, 50
    1, 2, 5, 10

16. 25, 40
    1, 5
17. 16, 36
    1, 2, 4
18. 36, 48
    1, 2, 3, 4, 6, 12
19. 25, 45
    1, 5
20. 36, 81
    1, 3, 9

Write the GCF for each pair of numbers.

21. 12, 15
    3
22. 8, 36
    4
23. 10, 25
    5
24. 20, 30
    10
25. 25, 50
    25

26. 27, 48
    3
27. 35, 49
    7
28. 36, 96
    12
29. 24, 30
    6
30. 60, 90
    30

Now Try This!

The Greek mathematician Euclid discovered a process for finding the GCF by division. The **Euclidean Algorithm** can be used to find the GCF of 12 and 20.

Step 1: Divide the numbers.

$$\begin{array}{r} 1\ R8 \\ 12\overline{)20} \\ -12 \\ \hline 8 \end{array}$$

Step 2: Divide the divisor by the remainder.

$$\begin{array}{r} 1\ R4 \\ 8\overline{)12} \\ -8 \\ \hline 4 \end{array}$$

Step 3: Repeat step 2.

$$\begin{array}{r} 2 \\ 4\overline{)8} \\ -8 \\ \hline 0 \end{array}$$

When the remainder is zero, the divisor is the GCF. The GCF of 8 and 12 is 4.

Use the Euclidean Algorithm to find the GCF for each pair of numbers.

1. 24, 42
   6
2. 16, 100
   4
3. 17, 46
   1

Lesson 5-2 • Greatest Common Factor

For Mixed Abilities

Common Errors • Intervention

For students who have difficulty determining the greatest common factor of two numbers, have them continue to list the factors of each number, circling those that are common and underlining the greatest.

Enrichment • Number Sense

1. Challenge students to find the greatest common factor for each of the following sets of numbers:

   4, 12, 16 (4)   13, 52, 65 (13)

   8, 16, 36 (4)   48, 14, 27 (1)

   15, 25, 45 (5)   14, 49, 28 (7)

2. Have students complete the following table, using calculators if they wish.

Number Pair	Product of Number Pair	GCF	LCM	GCF × LCM
12, 16	(192)	(4)	(48)	(192)
9, (15)	135	(3)	(45)	(135)
(18), 48	864	(6)	(144)	(864)
35, 14	(490)	(7)	(70)	(490)
16, (18)	288	(2)	(144)	(288)

More to Explore • Logic

Introduce this strategy game to students.

• Have them line up 10 pencils in a straight row.

• Have pairs of students take turns picking up either one pencil or two adjoining pencils per turn.

• The student who picks up the last pencil is the winner.

After they have played the game, tell students they should be able to discover a strategy for consistent winning, considering they take the first turn. (Pick the third pencil from the end, leave either 2 or 4 pencils together, or mirror each move the opponent makes.)

Now Try This! This formula is an alternative method for finding the GCF. Tell students they must work the problem through to reach a remainder of 0 before finding the divisor that is the GCF.

4 Assess

Tell students that the greatest common factor of two numbers is 6 and that one of the numbers is 18. Ask, *Could the other number be 20? Why or why not?* (No; Possible answer: 6 is not a factor of 20.)

T90

5-3 Prime and Composite Numbers

pages 91–92

1 Getting Started

Objective
• To identify prime and composite numbers

Vocabulary
prime number, composite number

Materials
index cards; graph paper

Warm Up • Mental Math
Write 1732 on the board and identify it as the year George Washington was born. Ask students:

1. In what year did Washington celebrate his 5th birthday? (1737)
2. his 10th birthday? (1742)
3. his 37th birthday? (1769)

Then ask how old Washington would have been

4. in 1802? (70)
5. in 1892? (160)
6. in 1799? (67)

Warm Up • Pencil and Paper
Have students list the factors for each of the following numbers. Remind them to use divisibility tests.

1. 64 (1, 2, 4, 8, 16, 32, 64)
2. 125 (1, 5, 25, 125)
3. 87 (1, 3, 29, 87)
4. 110 (1, 2, 5, 10, 11, 22, 55, 110)
5. 198 (1, 2, 3, 6, 9, 11, 18, 22, 33, 66, 99, 198)

2 Teach

Introduce the Lesson Work through the model with students, naming the factors of 18. Then, have students name the factors of 5. Have a student read the definition of *prime number*. Stress that a prime number must be greater than 1. Then, have students read the definition of *composite number*.

Develop Skills and Concepts Refer to the model on page 91. Stress that the first step in determining whether a number is prime or composite is to list its factors.

• Explain that a prime number has exactly two factors: itself and 1.

• A composite number has more than two different factors.

T91

Name _____

Lesson 5-3

Prime and Composite Numbers

A number must be divisible by all of its factors. The number 18 has exactly six factors: 1, 2, 3, 6, 9, and 18. What are the factors of 18?

To name the factors of 18, we can use multiplication facts.

$\underline{1} \times 18 = 18$
$\underline{2} \times 9 = 18$
$\underline{3} \times 6 = 18$

The factors of 18 are
$\underline{1}$, $\underline{2}$, $\underline{3}$, $\underline{6}$, $\underline{9}$, and $\underline{18}$.
The number 18 has $\underline{six}$ factors.

What are the factors of 5? The factors of 5 are $\underline{1}$ and $\underline{5}$. The number 5 is a prime number. A **prime number** has exactly two different factors: the number 1 and itself. The number 1 is not prime because it has only one factor.

A number greater than 1 that is not prime is a **composite number**. Since 18 has six factors, 18 is a $\underline{composite}$ number.

Getting Started

Write all factors for each number.

1. 6
   1, 2, 3, 6
2. 12
   1, 2, 3, 4, 6, 12
3. 20
   1, 2, 4, 5, 10, 20
4. 23
   1, 23
5. 35
   1, 5, 7, 35

Identify each number as *prime* or *composite*.

6. 7
   prime
7. 9
   composite
8. 32
   composite
9. 47
   prime
10. 154
    composite

Lesson 5-3 • Prime and Composite Numbers

91

© Pearson Education, Inc./Dale Seymour Publications/Pearson Learning Group. All rights reserved. Copying strictly prohibited.

• Point out that 0 and 1 are neither prime nor composite numbers.

Have students list all the factors for each of the following numbers and then state whether the number is prime or composite:

14 (1, 2, 7, 14; composite)
59 (1, 59; prime)
63 (1, 3, 7, 9, 21, 63; composite)
11 (1, 11; prime)
46 (1, 2, 23, 46; composite)
67 (1, 67; prime)

3 Practice

Have students complete all of the exercises. Be sure students understand how to determine if a number is prime or composite.

Practice

Write all factors for each number.

1. 8
   1, 2, 4, 8

2. 10
   1, 2, 5, 10

3. 21
   1, 3, 7, 21

4. 25
   1, 5, 25

5. 31
   1, 31

6. 36
   1, 2, 3, 4, 6,
   9, 12, 18, 36

7. 40
   1, 2, 4, 5, 8,
   10, 20, 40

8. 45
   1, 3, 5, 9, 15,
   45

9. 57
   1, 3, 19, 57

10. 67
    1, 67

11. 28
    1, 2, 4, 7, 14,
    28

12. 42
    1, 2, 3, 6, 7,
    14, 21, 42

13. 16
    1, 2, 4, 8, 16

14. 15
    1, 3, 5, 15

15. 29
    1, 29

16. 91
    1, 7, 13, 91

17. 63
    1, 3, 7, 9, 21,
    63

18. 21
    1, 3, 7, 21

19. 9
    1, 3, 9

20. 53
    1, 53

Identify each number as *prime* or *composite*.

21. 17
    prime

22. 27
    composite

23. 49
    composite

24. 63
    composite

25. 75
    composite

26. 87
    composite

27. 96
    composite

28. 100
    composite

29. 120
    composite

30. 131
    prime

(Now Try This!)

A Greek mathematician developed a way to find prime
numbers. This method is called the **Sieve of Eratosthenes**.

Make a sieve to find all primes less than 201.

- Draw an array from 1 through 200 with 10 numbers in each row.
- Cross out 1. Circle 2. Cross out every second number after 2.
- Circle 3. Cross out every third number after 3.
- Circle the next prime number, 5. Cross out every fifth number after 5.
- Circle the next prime number, 7. Cross out every seventh number after 7.
- Continue this process for 11, 13, and 17.
- Circle all numbers that have not been crossed out.

The circled numbers are the primes less than 201.

Lesson 5-3 • Prime and Composite Numbers

Now Try This! Explain that the Sieve of Eratosthenes
is one way to find prime numbers. Have students
work on graph paper and write one number in each
square to form the sieve. Then, have students follow
the steps in this section to find all the prime numbers
less than 201.

4 Assess

Point out to the students that the number 2 is an even
prime number. Ask, *Are there other even prime numbers?
Explain.* (No; Possible answer: All other even numbers
have 2 as a factor as well as 1 and the number itself. All
other even numbers are composite numbers.)

For Mixed Abilities

Common Errors • Intervention

If students have difficulty finding
all the factors of a number, have
them play a game to practice.
Write each of the numbers 1 to 24
on a set of index cards. Provide a
set for each group of players.
Have them shuffle the cards and
place them face down. Then, have
each player choose a card and list
the factors of the number shown.
The player with the greatest
number of factors scores 2 points.
If two or more numbers have the
same number of factors, each
player scores 1 point. The first
player to score 10 points wins.

Enrichment • Number Sense

1. Explain that each even
   number can be expressed as
   the sum of two prime
   numbers. Have students write
   ten even numbers and then
   show each as the sum of two
   primes. (Possible answers:
   $4 = 2 + 2$; $6 = 3 + 3$;
   $14 = 3 + 11$ or $7 + 7$)

2. Explain that mathematicians
   believe that every odd number
   greater than 9 can be expressed
   as the sum of three odd
   primes. For example, $9 = 3 + 3
   + 3$ and $27 = 3 + 7 + 17$. Tell
   students to choose some odd
   composite numbers less than
   100 and demonstrate this.

More to Explore • Number Sense

There are a total of 293 ways to
arrange American coins from a
penny through a half-dollar to
equal $1.00. Challenge students to
find as many ways as they can to
do this. A bulletin board or chart
could be made for students to
display their findings.

This project could be continued
over a week, and could be done in
group competition.

5-4 Factoring for Primes

pages 93–94

1 Getting Started

Objective

- To write the prime factorization of a composite number using exponents

Vocabulary

factor tree, prime factorization

Warm Up • Mental Math

Have students determine the missing number in each pattern.

1. $\dfrac{18}{(6)\,|\,3}$ 4. $\dfrac{(45)}{9\,|\,5}$

2. $\dfrac{(30)}{6\,|\,5}$ 5. $\dfrac{(36)}{(6)\,|\,6}$

3. $\dfrac{54}{6\,|\,(9)}$ 6. $\dfrac{48}{(8)\,|\,6}$

Warm Up • Pencil and Paper

Have students identify each number as *prime* or *composite*. If composite, have them write the number as a product of 2 different factors, other than 1 and itself.

1. 38 (composite, 2 × 19)
2. 144 (composite, 6 × 24)
3. 84 (composite, 12 × 7)
4. 101 (prime)
5. 155 (composite, 5 × 31)
6. 201 (composite, 3 × 67)
7. 97 (prime)
8. 500 (composite, 5 × 100)

Factoring for Primes

Any whole number greater than 1 can be written as a product of prime number factors. One way to find prime number factors is to make a **factor tree**. There may be different ways to start a factor tree, but the final set of prime factors will always be the same. Use a factor tree to find the prime factors of 24. Use exponents to write this **prime factorization**.

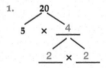
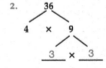

2 × 2 × 2 × 3

$2^3 \times 3$

$\underline{\ 3\ } \times \underline{\ 2\ } \times \underline{\ 2\ } \times \underline{\ 2\ }$

$\underline{\ 3\ } \times \underline{\ 2^3\ }$

REMEMBER The exponent tells how many times to use the base number as a factor.
$2^3 = 2 \times 2 \times 2$

Getting Started

Complete each factor tree.

1. 20
    5 × 4
        2 × 2

2. 36
    4 × 9
        3 × 3

3. 50
    2 × 25
         5 × 5

Write the prime factorization of each number using exponents.

4. 8   5. 35   6. 48   7. 72   8. 400
   2^3 5×7 $2^4 \times 3$ $2^3 \times 3^2$ $2^4 \times 5^2$

---

2 Teach

Introduce the Lesson Explain that every composite number can be expressed as a product of prime factors. Tell students that a factor tree is used to find the prime factors of a composite number. Have students read the paragraph aloud, noting how 24 is factored in the tree on the left. Then, have students complete the factor tree on the right. Compare the factor trees, emphasizing that a number can be factored in any order, but the final set of prime factors will be the same.

Write the following number sentences on the board. Have students circle the prime factors on the right side of each number sentence.

24 = 4 × 6   24 = ②× ③× 4
24 = ②× ②× 6   24 = ②× ③× ②× ②
24 = ②× 12

T93

Point out that only the last number sentence shows the number as the product of prime factors. Have students write the prime factors in ascending order. Then, write the prime factorization using exponents. (2 × 2 × 2 × 3; $2^3 \times 3$)

Develop Skills and Concepts Remind students that when a prime number is used more than once in the prime factorization, the expression can be written using exponents. Remind students that 2^3 is the same as $2 \times 2 \times 2$. Stress that the 3 in 2^3 is an exponent and that this exponent tells the number of times 2 is used as a factor. Have students complete each of the following factor trees and then write each prime factorization using exponents:

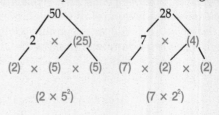

50
2 × (25)
(2) × (5) × (5)

28
7 × (4)
(7) × (2) × (2)

(2×5^2) (7×2^2)

Complete each factor tree.

1.
```
        18
       /  \
      9  ×  2
     / \
    3 × 3
```

2.
```
        20
       /  \
      4  ×  5
     / \
    2 × 2
```

3.
```
        28
       /  \
      7  ×  4
           / \
          2 × 2
```

4.
```
        50
       /  \
      2  ×  25
           / \
          5 × 5
```

5.
```
        60
       /  \
      6  ×  10
     / \   / \
    2 × 3 2 × 5
```

6.
```
        75
       /  \
      3  ×  25
           / \
          5 × 5
```

Write each prime factorization using exponents if possible.

7. 10
2×5

8. 28
$2^2 \times 7$

9. 55
5×11

10. 64
2^6

11. 66
$2 \times 3 \times 11$

12. 84
$2^2 \times 3 \times 7$

13. 100
$2^2 \times 5^2$

14. 125
5^3

15. 180
$2^2 \times 3^2 \times 5$

16. 225
$3^2 \times 5^2$

(Now Try This!)

Primes that differ by 2 are called **twin primes**. 5 and 7 are twin primes.

1. **List all the twin primes less than 100.**

3 and 5, 5 and 7, 11 and 13, 17 and 19, 29 and 31,
41 and 43, 59 and 61, 71 and 73

Primes that have digits that are reversed are called **mirror primes**.
13 and 31 are mirror primes.

2. **List all the mirror primes less than 100.**

13 and 31, 17 and 71, 37 and 73, 79 and 97

A **perfect number** is a number that is the sum of all of its factors except
itself. 6 = 1 + 2 + 3, therefore 6 is a perfect number.

3. **List all the perfect numbers less than 50.**

6

Lesson 5-4 • Factoring for Primes

For Mixed Abilities

Common Errors • Intervention

Some students do not write the prime factorization of a number correctly because they cannot identify the prime factors. Have them work with partners to name all the prime numbers from 1 to 50 and to write them on an index card. The students can use these cards as a guide for this work.

Enrichment • Number Sense

Provide this alternative method of dividing to find prime factorization of a number. Tell students they must always divide by a prime number.

```
2 |36     36 = 2² × 3²
2 |18
3 |_9
    3
```

Have students use the method above to find the prime factorization of each number. Remind students that they should write the prime factors in ascending order.

1. 120 ($2^3 \times 3 \times 5$)

2. 250 (2×5^3)

3. 1,000 ($2^3 \times 5^3$)

4. 72 ($2^3 \times 3^2$)

More to Explore • Measurement

Introduce drawing floor plans to scale. Tell students to measure their bedrooms and reproduce the measurements on $\frac{1}{4}$-inch graph paper, using the scale of $\frac{1}{4}$ in. = 1 foot. Explain that this is how architects and interior designers plan their work.

Tell them to choose major pieces of their bedroom furniture, measure them, and draw them to scale in the correct position on the floor plan of their room. Have students continue the project by making their floor plan as complete as possible, including closets, doors, windows, wastebasket, and so on.

Display their scale drawings.

---

3 Practice

Have students complete all the exercises. Students must draw their own factor trees for Exercises 7–16.

Now Try This! Encourage students to refer to the Sieve of Eratosthenes as they search for twin primes, mirror primes, and perfect numbers. You may wish to have students work in pairs to complete this activity.

4 Assess

Ask, *Why will every prime factorization of 24 be the same?*
(Possible answer: You can change the order of the factors, the product will be the same, 2 × 2 × 2 × 3 = 3 × 2 × 2 × 2.)

T94

5-5 Fractions

pages 95–96

1 Getting Started

Objective
• To write a fraction to represent a part of a region or set

Vocabulary
numerator, denominator

Materials
crayons; graph paper; inch rulers, marked off in $\frac{1}{8}''$

Warm Up • Mental Math
Have students name the next three numbers in each pattern and the rule they used to continue the pattern.

1. 7, 13, 19, (25), (31), (37) (rule: +6)
2. 94, 84, 74, (64), (54), (44) (rule: −10)
3. 4, 8, 16, (32), (64), (128) (rule: ×2)
4. 75, 71, 67, (63), (59), (55) (rule: − 4)
5. 26, 35, 44, (53), (62), (71) (rule: +9)
6. 1, 3, 7, (15), (31), (63) (rule: ×2 + 1)

Warm Up • Activity
Have students use an inch ruler to answer each question.

1. How many eighths between 1 and 2? (8)
2. How many eighths between 0 and $\frac{1}{2}$? (4)
3. What does each $\frac{1}{8}$ mark mean? (divides 1 inch into 8 equal parts)
4. Which is a larger part of 1 inch, $\frac{1}{8}$ or $\frac{1}{4}$? ($\frac{1}{4}$)
5. What is another name for $\frac{8}{8}$ inches? (1 inch)

2 Teach

Introduce the Lesson Have a student read the problem and tell what is to be solved. (what part of the dart board is green and what part of the set of boxes is filled with darts) Ask what information is given in the problem and the picture. Have students read and complete the information sentences to solve the problem.

Develop Skills and Concepts Draw the following on the board and ask students to write a fraction to tell what part is shaded and what part is not shaded:

1. a circle divided into thirds, $\frac{1}{3}$ shaded ($\frac{1}{3}$, $\frac{2}{3}$)
2. 6 squares, 2 squares shaded ($\frac{2}{6}$, $\frac{4}{6}$)

Name _____

Fractions

Sol is playing a game of darts. What part of the dartboard is green? What part of the set of boxes is filled with darts?

We know that the dart board has __4__ equal regions, and __1__ region is green.

To find what part of the dart board is green, we compare the number of green parts to the total number of parts.

$$\frac{\text{number of green parts}}{\text{number of equal parts}} = \frac{1}{4} = \frac{\text{numerator}}{\text{denominator}}$$

One-fourth or $\frac{1}{4}$ of the dart board is green.

We know there are __3__ boxes in the set and __2__ of the boxes are full.

To find what part of the set of boxes is filled, we compare the number of full boxes to the total number of boxes.

$$\frac{\text{number of full boxes}}{\text{number of boxes}} = \frac{2}{3} = \frac{\text{numerator}}{\text{denominator}}$$

Two-thirds or $\frac{2}{3}$ of the boxes are full.

Getting Started

Name each fraction.

1. $\frac{3}{4}$ __three-fourths__
2. $\frac{1}{5}$ __one-fifth__
3. $\frac{2}{8}$ __two-eighths__

Write each fraction.

4. five-eighths $\frac{5}{8}$
5. one-half $\frac{1}{2}$
6. three-fourths $\frac{3}{4}$

Write the fraction that names the shaded part.

7. $\frac{3}{8}$
8. $\frac{2}{5}$
9. $\frac{2}{6}$

---

3. a triangle divided in half, $\frac{1}{2}$ shaded ($\frac{1}{2}$, $\frac{1}{2}$)
4. 8 circles, 6 circles shaded ($\frac{6}{8}$, $\frac{2}{8}$)

Relate the numerator to the number of parts to be considered and the denominator to the number of parts in all.

Have students draw rectangles divided and shaded, on the board, to show $\frac{2}{5}$, $\frac{3}{6}$, $\frac{5}{6}$, $\frac{1}{9}$, $\frac{5}{6}$, and $\frac{7}{10}$.

3 Practice

Have students complete all the exercises.

Practice

Name each fraction.

1. $\frac{5}{8}$ ___five-eighths___
2. $\frac{1}{3}$ ___one-third___
3. $\frac{1}{10}$ ___one-tenth___
4. $\frac{3}{12}$ ___three-twelfths___
5. $\frac{7}{8}$ ___seven-eighths___
6. $\frac{5}{6}$ ___five-sixths___

Write each fraction.

7. two-thirds $\frac{2}{3}$
8. four-fifths $\frac{4}{5}$
9. nine-tenths $\frac{9}{10}$
10. five-twelfths $\frac{5}{12}$
11. zero-ninths $\frac{0}{9}$
12. two-halves $\frac{2}{2}$
13. three-sixteenths $\frac{3}{16}$
14. one-fourth $\frac{1}{4}$
15. three-sevenths $\frac{3}{7}$

Write the fraction that names the shaded part.

16.  $\frac{2}{8}$
17. $\frac{4}{5}$
18.  $\frac{2}{6}$

19. $\frac{6}{10}$
20. $\frac{1}{5}$
21.  $\frac{9}{12}$

Write the fraction that shows each situation.

22. the part of the set of glasses that is full
 $\frac{4}{5}$

23. the part of the set of apples that is *not* green
 $\frac{4}{6}$

Problem Solving

Solve each problem.

24. There are 9 players on the softball team. Three players are sixth graders. What part of the team is sixth graders?
$\frac{3}{9}$

25. Gerry has 12 coins for a total of 83¢. Seven of the coins are dimes. What part of the coins are nickels?
$\frac{2}{12}$

26. The swim team practices 3 hours a day, 6 days a week. Mort attended practice 7 hours last week. What part of the total practice did he miss?
$\frac{11}{18}$

27. Martin baked a dozen muffins and ate 3 of them. What part of the muffins are left?
$\frac{9}{12}$

96

Lesson 5-5 • Fractions

4 Assess

Ask, *Are the fractions $\frac{3}{4}$ and $\frac{3}{8}$ the same? Explain.* (No; Possible answer: $\frac{3}{4}$ means 3 out of 4 parts are being considered. $\frac{3}{8}$ means 3 out of 8 parts are being considered.)

5 Mixed Review

1. $6{,}256 \times 8$ (50,048)
2. $3 \times 8 + 7$ (31)
3. $85{,}176 - 38{,}058$ (47,118)
4. $21{,}295 \div 30$ (709 R25)
5. 265×103 (27,295)
6. 8^2 (64)
7. $\$278.93 + \$65.41 + \$493.38$ ($837.72)
8. $324 \div 48$ (6 R36)
9. 68×45 (3,060)
10. $400 \times 5{,}000$ (2,000,000)

For Mixed Abilities

Common Errors • Intervention

Some students may confuse the numerator and denominator when writing a fraction. Have these students work with partners and the following fractions: $\frac{3}{8}, \frac{4}{5}, \frac{5}{6}, \frac{1}{4},$ and $\frac{4}{9}$. Have students take turns identifying the numerator and denominator of each fraction and drawing a model to illustrate it.

Enrichment • Applications

Have students outline 8 by 8, 9 by 9, and 10 by 10 squares on graph paper. Within each outline, have them color some of the small squares red, some blue, and some yellow. Have students write three word problems about the colored squares. Suggest that the problems be as tricky as possible. Have students exchange problems and outlined squares and solve.

More to Explore • Graphing

Remind students how to make and interpret a pie graph. Have students use a paper plate, divided into sections representing blocks of hours, to construct a graph showing how they spent their time for one 24-hour day.

Help students translate the time allocations into fractional pieces of the pie. Remind them to account for all hours of their day. Ask students to answer the following questions:

1. *What conclusions can be drawn from the chart?* (Students will be able to see which activities require the most time, the least amount of time.)

2. *What is the value of this graph?* (Students can see how they could reallocate their time if they wish to make changes.)

3. *How could a chart like this one help a business or factory worker?* (A similar chart could be made to show how the worker uses his or her time during the workday. Adjustments could then be made as needed.)

T96

5-6 Equivalent Fractions

pages 97–98

1 Getting Started

Objective
• To find equivalent fractions

Vocabulary
equivalent fractions

Materials
index cards

Warm Up • Mental Math
Have students name:
1. a composite number whose prime factorization is 3^2 (9)
2. a composite number whose prime factorization is $2^2 \times 3^2$ (36)
3. a composite number that is a multiple of 5 and 6 (30, 60, 90 . . .)
4. a composite number whose prime factorization is 3×17 (51)

Warm Up • Activity
Draw each pair of regions on the board and have students write the fraction for the shaded part.
1. square divided into halves, $\frac{1}{2}$ shaded ($\frac{1}{2}$); square divided into quarters, $\frac{2}{4}$ shaded ($\frac{2}{4}$)
2. rectangle divided into eighths, $\frac{2}{8}$ shaded ($\frac{2}{8}$); rectangle divided into quarters, $\frac{1}{4}$ shaded ($\frac{1}{4}$)
3. rectangle divided into sixths, $\frac{4}{6}$ shaded ($\frac{4}{6}$); rectangle divided into thirds, $\frac{2}{3}$ shaded ($\frac{2}{3}$)

2 Teach

Introduce the Lesson Have students discuss the picture, read the problem, and tell what is to be solved. (the fraction of the group that are violin players) Ask what needed information is given. (the number of musicians in the group and the number of violinists) Have students complete the sentences as they work to solve the problem. Be sure students understand that they must multiply or divide the numerator and denominator by the same number to find an equivalent fraction.

Develop Skills and Concepts Remind students of the property of 1 for multiplication and division: any number times 1 or any number divided by 1 equals that number.

• Explain that a fraction with the same number in the numerator and denominator is a name for 1. Have students give some fractional names for 1, such as $\frac{2}{2}$ or $\frac{3}{3}$.

Name _____

Equivalent Fractions

Some members of the orchestra arrived early for rehearsal. What fraction of the group are the violinists?

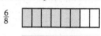

We want to write a fraction that tells what part of the group are violin players.

We know that there are __6__ musicians in the group, and __2__ of them are violinists. We compare the number of violinists to the total number of musicians. We say __$\frac{2}{6}$__ of the musicians are violinists.

We also know there are __3__ pairs of musicians, and __1__ pair of the musicians are violinists. We can say, __$\frac{1}{3}$__ of the musicians are violinists.

$\frac{2}{6}$ and $\frac{1}{3}$ are called **equivalent fractions** because they name the same number.

We find equivalent fractions by multiplying or dividing.

Multiply the numerator and the denominator by the same number.

$\frac{1}{2}$

$\frac{2}{4}$

$$\frac{1}{2} = \frac{1 \times 2}{2 \times 2} = \frac{2}{4}$$

Divide the numerator and the denominator by the same number.

$\frac{6}{8}$

$\frac{3}{4}$

$$\frac{6}{8} = \frac{6 \div 2}{8 \div 2} = \frac{3}{4}$$

Getting Started
Write a pair of equivalent fractions for the shaded part.

1. $\frac{2}{3}, \frac{6}{9}$

2. $\frac{4}{8}, \frac{1}{2}$

3. $\frac{8}{12}, \frac{2}{3}$

Complete the equations.

4. $\frac{3}{4} = \frac{3 \times 3}{4 \times 3} = \frac{9}{12}$

5. $\frac{6}{9} = \frac{6 \div 3}{9 \div 3} = \frac{2}{3}$

Complete the equivalent fraction.

6. $\frac{14}{20} = \frac{7}{10}$

7. $\frac{3}{8} = \frac{9}{24}$

Lesson 5-6 • Equivalent Fractions

97

• Point out that a fraction is equivalent to a given fraction because both the numerator and denominator are multiplied or divided by the fractional name for 1.

• Multiplying or dividing a number by 1 does not change the value of the original fraction.

Emphasize that equivalent fractions name the same number. Have students find the missing numerator or denominator. Then, write the equivalent fraction.

1. $\frac{1}{2} = \frac{1}{2} \times \frac{5}{(5)} = (\frac{5}{10})$
2. $\frac{12}{15} = \frac{12}{15} \div \frac{(3)}{3} = (\frac{4}{5})$
3. $\frac{2}{6} = \frac{2}{6} \div \frac{2}{(2)} = (\frac{1}{3})$
4. $\frac{3}{4} = \frac{3}{4} \times \frac{(5)}{5} = (\frac{15}{20})$

3 Practice

Have students complete all the exercises. Remind students that to find equivalent fractions they must multiply or divide the numerator and denominator by the same number.

T97

Practice

Write a pair of equivalent fractions for the shaded part.   Possible answers.

1. 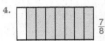 $\frac{1}{3}, \frac{2}{6}$

2. $\frac{5}{10}, \frac{1}{2}$

3.  $\frac{4}{6}, \frac{2}{3}$

4. $\frac{7}{8}, \frac{14}{16}$

5. 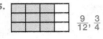 $\frac{9}{12}, \frac{3}{4}$

6. 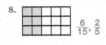 $\frac{1}{4}, \frac{2}{8}$

7. $\frac{3}{7}, \frac{6}{14}, \frac{9}{21}$

8. $\frac{6}{15}, \frac{2}{5}$

9. $\frac{5}{5}, \frac{1}{1}$

10. $\frac{8}{16}, \frac{1}{2}$

11. $\frac{3}{3}, \frac{1}{1}$

12. $\frac{8}{9}, \frac{16}{18}$

Complete the equation.

13. $\frac{1}{4} = \frac{1 \times 5}{4 \times 5} = \frac{5}{20}$

14. $\frac{3}{9} = \frac{3 \div 3}{9 \div 3} = \frac{1}{3}$

15. $\frac{5}{10} = \frac{5 \times 2}{10 \times 2} = \frac{10}{20}$

16. $\frac{10}{16} = \frac{10 \div 2}{16 \div 2} = \frac{5}{8}$

17. $\frac{2}{5} = \frac{2 \times 6}{5 \times 6} = \frac{12}{30}$

18. $\frac{4}{8} = \frac{4 \div 2}{8 \div 2} = \frac{2}{4}$

19. $\frac{14}{21} = \frac{14 \div 7}{21 \div 7} = \frac{2}{3}$

20. $\frac{5}{8} = \frac{5 \times 5}{8 \times 5} = \frac{25}{40}$

21. $\frac{4}{11} = \frac{4 \times 7}{11 \times 7} = \frac{28}{77}$

Complete the equivalent fraction.

22. $\frac{5}{12} = \frac{10}{24}$

23. $\frac{4}{8} = \frac{12}{24}$

24. $\frac{6}{18} = \frac{3}{9}$

25. $\frac{10}{15} = \frac{2}{3}$

26. $\frac{1}{3} = \frac{6}{18}$

27. $\frac{12}{16} = \frac{6}{8}$

28. $\frac{60}{100} = \frac{6}{10}$

29. $\frac{30}{48} = \frac{5}{8}$

30. $\frac{15}{24} = \frac{5}{8}$

31. $\frac{27}{45} = \frac{3}{5}$

32. $\frac{8}{12} = \frac{24}{36}$

33. $\frac{16}{20} = \frac{80}{100}$

34. $\frac{12}{32} = \frac{3}{8}$

35. $\frac{6}{21} = \frac{18}{63}$

36. $\frac{18}{27} = \frac{6}{9}$

37. $\frac{5}{12} = \frac{40}{96}$

Problem Solving

Solve each problem.

38. Zack measured a board to be $\frac{5}{8}$ of an inch thick. How many sixteenths of an inch thick is the board?

$\frac{10}{16}$ of an inch

39. Sun Li filled $\frac{9}{12}$ of a container with water for a science experiment. How many fourths full is the container?

$\frac{3}{4}$

Lesson 5-6 • Equivalent Fractions

4 Assess

Have students write the numbers 2, 4, 5, and 10 in the boxes below to create equivalent fractions.

$\frac{\square}{\square} = \frac{\square}{\square}$

$\left(\frac{2}{5} = \frac{4}{10}\right)$

5 Mixed Review

1. 9×10^5 (900,000)

2. $173 \div 31$ (5 R18)

3. $658,276 + 5,273,809$ (5,932,085)

4. $654,072 \div 6$ (109,012)

5. $31 + 4 \times 3$ (43)

6. $160,000 - 127,308$ (32,692)

7. 4^2 (16)

8. 106×427 (45,262)

9. $61,376 \div 50$ (1,227 R26)

10. $\$3.91 \times 23$ (\$89.93)

For Mixed Abilities

Common Errors • Intervention

Some students may not multiply the numerator by the same number used for the denominator. Remind them that to find an equivalent fraction, they must multiply the fraction by 1; that is, they must multiply the numerator and the denominator by the same number. When students work with a problem such as $\frac{3}{7} = \frac{n}{14}$, have them factor the numbers so that they multiply the numerator of the first fraction by the correct number.

$$\frac{3}{7} = \frac{n}{7} \times 2 \longrightarrow 3 \times \frac{2}{7} \times 2 = \frac{6}{14}$$

Enrichment • Number Sense

Prepare a set of index cards for each group of 4 students. On each of 12 cards write one of the following fractions: $\frac{3}{4}, \frac{5}{6}, \frac{1}{3}, \frac{2}{5}, \frac{7}{10}, \frac{11}{12}, \frac{3}{5}, \frac{1}{6}, \frac{3}{4}, \frac{7}{8}, \frac{2}{9}, \frac{7}{12}$. On each of 12 more cards write one of these fractions: $\frac{9}{12}, \frac{25}{30}, \frac{7}{21}, \frac{18}{45}, \frac{70}{100}, \frac{33}{36}, \frac{12}{20}, \frac{4}{24}, \frac{24}{32}, \frac{21}{24}, \frac{4}{18}, \frac{35}{60}$.

Have students shuffle the cards and place them facedown in a 6 by 4 array. Students take turns choosing 2 cards at a time. If the cards name equivalent fractions, the student keeps them and wins a point. If not, the student replaces them facedown.

Play continues until all cards have been matched. The player with the most points wins.

More to Explore • Applications

Have a student write the primary United States time zones across the board. Discuss how this pattern continues around the world. Divide students into groups and provide them with globes or flat maps. Have students choose various cities in the United States and elsewhere in the world, and determine what time it would be in those cities when it is 6:00 A.M. in their home city. Have students make another list of cities without times indicated to exchange with classmates to figure time comparisons.

T98

5-7 Simplifying Fractions

pages 99–100

1 Getting Started

Objective
- To write fractions in simplest form

Vocabulary
simplest form

Warm Up • Mental Math
Have students tell whether each fraction pair is equivalent or not equivalent.

1. $\frac{24}{32}$ (=) $\frac{3}{4}$
2. $\frac{4}{5}$ (≠) $\frac{15}{25}$
3. $\frac{1}{3}$ (≠) $\frac{3}{8}$
4. $\frac{3}{10}$ (=) $\frac{15}{50}$
5. $\frac{12}{20}$ (=) $\frac{3}{5}$
6. $\frac{14}{16}$ (≠) $\frac{5}{8}$
7. $\frac{6}{15}$ (=) $\frac{2}{5}$
8. $\frac{3}{5}$ (≠) $\frac{6}{8}$
9. $\frac{2}{4}$ (≠) $\frac{8}{12}$
10. $\frac{3}{12}$ (≠) $\frac{1}{3}$

Warm Up • Pencil and Paper
Have students write the greatest common factor for each pair of numbers.

1. 12 and 24 (12)
2. 16 and 4 (4)
3. 8 and 10 (2)
4. 5 and 20 (5)
5. 21 and 35 (7)
6. 20 and 70 (10)

2 Teach

Introduce the Lesson Call on a student to read the first part of the model and determine what fractional part of its games for the season the basketball team has already played. ($\frac{20}{25}$)

Read the definition for simplest form aloud. Then, explain that a fraction is in simplest form when 1 is the only whole number that can be evenly divided into both the numerator and the denominator.

Tell students that an efficient method for expressing a fraction in simplest form is to divide the numerator and denominator by the GCF. Work through the rest of the model to name $\frac{20}{25}$ in simplest form.

Name _____

Simplifying Fractions

The basketball team is scheduled to play 25 games this season. The team has already played 20 games. What fractional part of the basketball season has the team already played?

We want to name the fractional part of the season that the team has already played.

We know the team will play a total of __25__ games, and it has already played __20__ games.

We compare the number of games played to the total number of games.

The team has played $\frac{20}{25}$ of its games.

What is this fraction in simplest form?

A fraction is in **simplest form** when the greatest common factor of the numerator and denominator is 1.

To write $\frac{20}{25}$ in simplest form:

Find the GCF of 20 and 25.

Factors of 20: **1, 2, 4, 5, 10, 20**

Factors of 25: **1, 5, 25**

The GCF is __5__.

Divide the numerator and denominator by their GCF.

$$\frac{20}{25} = \frac{20 \div 5}{25 \div 5} = \frac{4}{5}$$

$\frac{20}{25}$ named in simplest form is __$\frac{4}{5}$__.

Getting Started

Rename each fraction in simplest form.

1. $\frac{6}{10} = \frac{3}{5}$
2. $\frac{10}{25} = \frac{2}{5}$
3. $\frac{10}{40} = \frac{1}{4}$
4. $\frac{8}{12} = \frac{2}{3}$
5. $\frac{6}{18} = \frac{1}{3}$
6. $\frac{9}{12} = \frac{3}{4}$
7. $\frac{18}{54} = \frac{1}{3}$
8. $\frac{180}{200} = \frac{9}{10}$

Lesson 5-7 • Simplifying Fractions

99

Develop Skills and Concepts Tell students that when the numerator and denominator of a fraction are different prime numbers, the fraction is already in simplest form. Have students suggest a quick way to determine whether or not a fraction is in simplest form when one term is a prime number and one is composite. (If the prime number is not a factor of the composite number, the fraction is in simplest form.)

Point out that unit fractions, those with 1 in the numerator, are in simplest form.

Review finding the GCF. Remind students that dividing both terms of a fraction by the same number is the same as dividing by 1. Point out that the equivalent fraction in simplest form has the same value as the original fraction.

Have students name each fraction in simplest form.

1. $\frac{15}{20} = \left(\frac{3}{4}\right)$
2. $\frac{21}{27} = \left(\frac{7}{9}\right)$
3. $\frac{9}{15} = \left(\frac{3}{5}\right)$

Practice

Rename each fraction in simplest form.

1. $\frac{6}{8} = \frac{3}{4}$ 2. $\frac{12}{18} = \frac{2}{3}$ 3. $\frac{9}{15} = \frac{3}{5}$ 4. $\frac{6}{12} = \frac{1}{2}$

5. $\frac{8}{20} = \frac{2}{5}$ 6. $\frac{25}{30} = \frac{5}{6}$ 7. $\frac{25}{150} = \frac{1}{6}$ 8. $\frac{4}{28} = \frac{1}{7}$

9. $\frac{14}{16} = \frac{7}{8}$ 10. $\frac{15}{18} = \frac{5}{6}$ 11. $\frac{40}{60} = \frac{2}{3}$ 12. $\frac{16}{20} = \frac{4}{5}$

13. $\frac{9}{24} = \frac{3}{8}$ 14. $\frac{14}{21} = \frac{2}{3}$ 15. $\frac{36}{48} = \frac{3}{4}$ 16. $\frac{9}{30} = \frac{3}{10}$

17. $\frac{18}{27} = \frac{2}{3}$ 18. $\frac{25}{75} = \frac{1}{3}$ 19. $\frac{24}{30} = \frac{4}{5}$ 20. $\frac{32}{40} = \frac{4}{5}$

21. $\frac{16}{24} = \frac{2}{3}$ 22. $\frac{42}{63} = \frac{2}{3}$ 23. $\frac{24}{60} = \frac{2}{5}$ 24. $\frac{64}{72} = \frac{8}{9}$

25. $\frac{12}{144} = \frac{1}{12}$ 26. $\frac{32}{120} = \frac{4}{15}$ 27. $\frac{75}{100} = \frac{3}{4}$ 28. $\frac{120}{160} = \frac{3}{4}$

Problem Solving

Use the chart to solve each problem. Write your answer in simplest form.

29. What part of the distance has Mel run?
$\frac{2}{3}$

30. What part of the distance must Nippi still run?
$\frac{7}{12}$

31. What part of the distance has Trudy run?
$\frac{1}{2}$

32. Mo has run twice as many miles as Nippi.
What part of the distance has he run?
$\frac{5}{6}$

Runners Club Run 24 Miles to Join	
Name	**Miles Run**
Mel	16
Nippi	10
Trudy	12

(Now Try This!)

Prime factors can be used to write $\frac{12}{72}$ in simplest form.

First, write the prime factors.

Then, circle the common factors.
Multiply what is left.

$\frac{12}{72} = \frac{2 \times 2 \times 3}{2 \times 2 \times 2 \times 3 \times 3}$

$\frac{2 \times 2 \times 3}{2 \times 2 \times 2 \times 3 \times 3} = \frac{1}{6}$

Use prime factors to rename each fraction in simplest form.

1. $\frac{15}{50} = \frac{3 \times 5}{2 \times 5 \times 5} = \frac{3}{10}$ 2. $\frac{27}{30} = \frac{3 \times 3 \times 3}{2 \times 3 \times 5} = \frac{9}{10}$ 3. $\frac{30}{100} = \frac{2 \times 3 \times 5}{2 \times 2 \times 5 \times 5} = \frac{3}{10}$

100

Lesson 5-7 • Simplifying Fractions

---

3 Practice

Have students complete all the exercises. Students will need to use the chart to solve the word problems.

Now Try This! This activity provides an opportunity for students to use prime factors to find the simplest form of a fraction. Have students study the model carefully. Point out that the circled numbers ($\frac{2}{2}$, $\frac{2}{2}$, and $\frac{3}{3}$) equal 1, and can be discarded. The factors that are left should be multiplied. The result will be the fraction in simplest form.

4 Assess

On the board, write $\frac{8}{24}$ $\frac{5}{9}$ $\frac{12}{30}$

Have students identify the fraction that is in simplest form. ($\frac{5}{9}$)
Then, have them simplify the other two fractions. ($\frac{1}{3}$, $\frac{2}{5}$)

For Mixed Abilities

Common Errors • Intervention

Some students may not write the fraction in simplest form because they do not divide by the largest divisor possible. Have students work in pairs with fractions not in lowest terms. For each denominator, ask them to make a list of all the numbers by which the denominator is divisible. Then, have them identify the largest number that is also a divisor of the numerator, and use this number to divide both the numerator and denominator to write the fraction in simplest form.

Enrichment • Number Sense

Provide 2 spinners for each group. Have students write a different even number between 26 and 98 in each section of the spinners. Have students take turns spinning both spinners and writing the two numbers as a simplified fraction. If the student correctly writes the fraction in simplest form, the student scores a point. The first player to score 5 points wins the game.

More to Explore • Logic

Duplicate the following for students to solve:

You are a detective called to solve the murder of the richest woman in town. The suspects are Renee, Louis, Benny, Sylvia, and Tony. They are the butler, maid, husband, cook, and gardener of the victim, not necessarily in that order. As you enter you hear the butler say, "Renee killed her!" and then you overhear "Liar, it was you!"

You question the suspects and take these statements:

Husband—Benny did the killing. Butler—Renee did it. Cook—I repeat, Tony did it. Maid—I didn't kill her. Gardener—Don't believe her.

You notice all the suspects except one had been accused and decide the others must be trying to protect the one who is really guilty. Whom do you arrest?
(Louis, the husband)

T100

5-8 Renaming Mixed Numbers

pages 101–102

1 Getting Started

Objective
• To rename whole numbers and mixed numbers as fractions

Vocabulary
mixed number

Materials
inch rulers marked off in $\frac{1}{8}$ inches; index cards

Warm Up • Mental Math
Read each product. Have students name possible factors. (Answers may vary.)

1. 12 (4, 3)
2. 75 (3, 25)
3. 80 (8, 10)
4. 36 (4, 9)
5. 72 (9, 8)
6. 100 (10, 10)
7. 550 (55, 10)
8. 39 (3, 13)
9. 95 (19, 5)
10. 67 (67, 1)

Warm Up • Activity
Have students use inch rulers to tell how many $\frac{1}{8}$-inch segments are in each measure.

1. 1 inch (8)
2. 2 inches (16)
3. $\frac{1}{2}$ inch (4)
4. $1\frac{1}{2}$ inches (12)
5. 5 inches (40)
6. $\frac{1}{4}$ inch (2)

2 Teach

Introduce the Lesson Have a student read the problem and identify what is being asked. (how many quarter inches long the mast is)

Have a student read the paragraph that discusses mixed numbers. Write examples of other mixed numbers on the board. Stress that a mixed number represents a value greater than 1.

Have students read the algorithm for renaming a mixed number as a fraction. Then, work through the model with students.

T101

Name _____

Renaming Mixed Numbers

Casey's hobby is building model ships. Today he cut a mast for a new model schooner. The mast measures $12\frac{1}{4}$ inches. How many quarter inches long is the mast?

We want to know the number of quarter inches in $12\frac{1}{4}$ inches. Numbers such as $12\frac{1}{4}$ are called **mixed numbers**. We need to rename the mixed number as a fraction.

To rename a mixed number, multiply the whole number by the denominator and add the numerator, keeping the denominator the same.

$$12\frac{1}{4} = \frac{(12 \times 4) + 1}{4} = \frac{48 + 1}{4} = \frac{49}{4}$$

The mast is $\underline{49}$ quarter inches or $\frac{49}{4}$ inches. We can also rename a whole number as a fraction by writing the whole number over 1.

$$2 = \frac{2}{1} \qquad 5 = \frac{5}{1}$$

Getting Started

Rename each whole or mixed number as a fraction.

1. $3\frac{1}{2} = \frac{7}{2}$
2. $6 = \frac{6}{1}$
3. $4\frac{2}{3} = \frac{14}{3}$
4. $11\frac{1}{9} = \frac{100}{9}$
5. $6\frac{2}{3} = \frac{20}{3}$
6. $5\frac{4}{5} = \frac{29}{5}$
7. $3 = \frac{3}{1}$
8. $109\frac{1}{5} = \frac{546}{5}$
9. $7 = \frac{7}{1}$
10. $3\frac{2}{9} = \frac{29}{9}$
11. $6\frac{7}{10} = \frac{67}{10}$
12. $15\frac{3}{5} = \frac{78}{5}$
13. $7\frac{3}{4} = \frac{31}{4}$
14. $10 = \frac{10}{1}$
15. $103\frac{3}{4} = \frac{415}{4}$

Have a student answer the question for the problem.

Develop Skills and Concepts Draw rectangles on the board, divided and shaded, for students to identify: $\frac{1}{3}$, $\frac{2}{3}$, $\frac{3}{3}$, $\frac{4}{3}$, $\frac{5}{3}$. Have students classify the 5 fractions into two groups. In one group, the numerators are smaller than the denominators. In the other group, the numerators are greater than or equal to the denominators.

Draw two circles on the board, divided into thirds, the first circle all shaded and the second circle with $\frac{1}{3}$ shaded. Have students name the shaded region in two ways. ($\frac{4}{3}$, $1\frac{1}{3}$) Discuss the method shown on page 101 for renaming a mixed number as a fraction.

Have students rename each mixed number as a fraction.

1. $1\frac{5}{8}$ $\left(\frac{13}{8}\right)$
2. $3\frac{1}{4}$ $\left(\frac{13}{4}\right)$
3. $4\frac{2}{3}$ $\left(\frac{14}{3}\right)$
4. $12\frac{1}{2}$ $\left(\frac{25}{2}\right)$
5. $10\frac{2}{5}$ $\left(\frac{52}{5}\right)$
6. $6\frac{5}{6}$ $\left(\frac{41}{6}\right)$

Explain that to write any whole number as a fraction, students should write the whole number as the numerator and write a 1 for the denominator.

Practice

Rename each whole or mixed number as a fraction.

1. $3\frac{1}{4} = \frac{13}{4}$ 2. $4 = \frac{4}{1}$ 3. $5\frac{1}{2} = \frac{11}{2}$

4. $7\frac{1}{3} = \frac{22}{3}$ 5. $9 = \frac{9}{1}$ 6. $9\frac{3}{4} = \frac{39}{4}$

7. $2\frac{4}{5} = \frac{14}{5}$ 8. $3\frac{4}{9} = \frac{31}{9}$ 9. $8\frac{2}{3} = \frac{26}{3}$

10. $9\frac{5}{8} = \frac{77}{8}$ 11. $1\frac{7}{12} = \frac{19}{12}$ 12. $3\frac{5}{8} = \frac{29}{8}$

13. $7\frac{4}{7} = \frac{53}{7}$ 14. $2\frac{5}{9} = \frac{23}{9}$ 15. $6\frac{3}{8} = \frac{51}{8}$

16. $5 = \frac{5}{1}$ 17. $9\frac{1}{9} = \frac{82}{9}$ 18. $4\frac{5}{12} = \frac{53}{12}$

19. $1\frac{7}{16} = \frac{23}{16}$ 20. $10\frac{2}{3} = \frac{32}{3}$ 21. $15\frac{1}{2} = \frac{31}{2}$

22. $16\frac{3}{4} = \frac{67}{4}$ 23. $12\frac{7}{10} = \frac{127}{10}$ 24. $100\frac{5}{8} = \frac{805}{8}$

Problem Solving

Solve each problem.

25. Are $5\frac{2}{3}$ pizzas enough for 16 people if each person eats $\frac{1}{3}$ of a pizza?

Yes

26. A gift basket holds 3 grapefruit and four times as many oranges. In simplest form, what fraction of the fruit is oranges?

$\frac{4}{5}$

(Now Try This!)

We can round a mixed number to the nearest whole number. Use a number line to help.

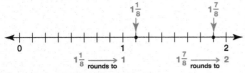

$1\frac{1}{8}$ rounds to 1 $1\frac{7}{8}$ rounds to 2

Round each mixed number to a whole number.

1. $1\frac{3}{4}$ 2   2. $3\frac{4}{5}$ 4   3. $2\frac{1}{3}$ 2   4. $5\frac{2}{7}$ 5

5. $1\frac{3}{8}$ 1   6. $4\frac{5}{6}$ 5   7. $3\frac{2}{9}$ 3   8. $9\frac{9}{10}$ 10

Lesson 5-8 • Renaming Mixed Numbers

For Mixed Abilities

Common Errors • Intervention

Some students may multiply the whole number by the denominator but forget to add the numerator when changing mixed numbers to fractions. Have them work with partners to write a problem such as $3\frac{1}{2}$ as shown here.

$$3\frac{1}{2} = 3 + \frac{1}{2} = \frac{6}{2} + \frac{1}{2} = \frac{7}{2}$$

More to Explore • Applications

Ask students to bring in several grocery ads from newspapers. Using a calculator, show them how to divide the total price of an item by the number of units in the item, to find the unit price.

Then, have students make up a shopping list for a class picnic. Have one student write on the board as the others dictate. Divide the class into small groups having ads and a calculator.

More to Explore • Statistics

Help the class make a model of *Galton's Quincunx*. Take a square piece of plywood and glue strips of balsa wood across the bottom to make a series of 7 channels. Arrange a grid of nails above the channels as illustrated. Put two sloping ramps at the top, positioned so that the top nail is exactly halfway between the ends of the two. Cover the board with a lucite panel.

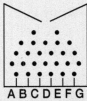

A B C D E F G

Let students roll marbles down the ramps. When each marble hits the first nail it will fall right or left. On the next row it will again fall right or left, and so on until it lands in one of the channels. Students will notice that more marbles fall in the center, fewer to the outside. Explain this is because there are more possible pathways to the center channels. Have students roll 50 marbles, recording the channels into which each marble falls. Have students show their results on a bar graph.

---

3 Practice

Have students complete all the exercises. Encourage students who have difficulty with any problems to draw a picture to help them.

> **Now Try This!** Show students how they can use a number line to round a mixed number to the nearest whole number. If needed, provide blank number lines for students to use as they do the exercises.

4 Assess

Have students describe the steps needed to change $5\frac{2}{3}$ to a fraction. (Possible answer: Multiply 5 by 3, and add 2 to the result. This is the numerator. 3 is the denominator.)

5-9 Renaming Improper Fractions

pages 103–104

1 Getting Started

Objective
• To rename improper fractions as whole or mixed numbers.

Vocabulary
improper fraction

Materials
rulers

Warm Up • Mental Math
Have students compare each number expression.

1. $10^2 \;(>)\; 2 \times 10$
2. $5 \times 4 \;(>)\; 3 \times 6$
3. $6 + 7 \;(<)\; 20 - 6$
4. $50 - 30 \;(<)\; 5 \times 5$
5. $2^3 \;(>)\; 3 \times 2$
6. $100 \times 5 \;(=)\; 50 \times 10$
7. $4 \times 3 \;(<)\; 11 + 5$
8. $7 - 3 \;(=)\; 28 \div 7$

Warm Up • Pencil and Paper
Have students write each fraction in simplest form.

1. $\frac{6}{8}\;(\frac{3}{4})$ 2. $\frac{12}{15}\;(\frac{4}{5})$
3. $\frac{21}{42}\;(\frac{1}{2})$ 4. $\frac{21}{30}\;(\frac{7}{10})$
5. $\frac{10}{32}\;(\frac{5}{16})$ 6. $\frac{4}{10}\;(\frac{2}{5})$

Have students rename each mixed number as a fraction.

1. $3\frac{4}{5}\;(\frac{19}{5})$ 2. $6\frac{1}{8}\;(\frac{49}{8})$
3. $3\frac{5}{8}\;(\frac{29}{8})$ 4. $10\frac{1}{3}\;(\frac{31}{3})$
5. $8\frac{5}{6}\;(\frac{53}{6})$ 6. $5\frac{1}{2}\;(\frac{11}{2})$

2 Teach

Introduce the Lesson Have a student read the problem and tell what is to be solved. (the number of cartons Elaine can fill) Ask what information is given in the problem and the picture. (Elaine has 14 bottles and each carton holds 4 bottles.) Have students complete the information sentences.

Point out that a fraction is a division. Write other examples of improper fractions on the board. Have students generalize that a fraction is improper if the numerator is greater than the denominator.

Work through the steps for renaming an improper fraction as a mixed number.

T103

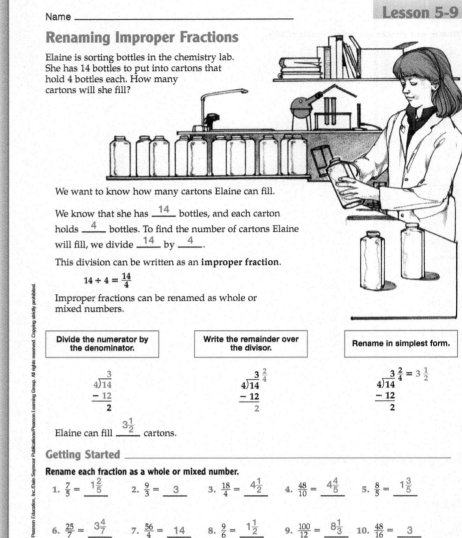

Name _____

Renaming Improper Fractions

Elaine is sorting bottles in the chemistry lab. She has 14 bottles to put into cartons that hold 4 bottles each. How many cartons will she fill?

We want to know how many cartons Elaine can fill.

We know that she has __14__ bottles, and each carton holds __4__ bottles. To find the number of cartons Elaine will fill, we divide __14__ by __4__.

This division can be written as an **improper fraction**.

$$14 \div 4 = \frac{14}{4}$$

Improper fractions can be renamed as whole or mixed numbers.

Divide the numerator by the denominator.	Write the remainder over the divisor.	Rename in simplest form.
$\begin{array}{r} 3 \\ 4\overline{)14} \\ -12 \\ \hline 2 \end{array}$	$\begin{array}{r} 3\frac{2}{4} \\ 4\overline{)14} \\ -12 \\ \hline 2 \end{array}$	$\begin{array}{r} 3\frac{2}{4} = 3\frac{1}{2} \\ 4\overline{)14} \\ -12 \\ \hline 2 \end{array}$

Elaine can fill __$3\frac{1}{2}$__ cartons.

Getting Started

Rename each fraction as a whole or mixed number.

1. $\frac{7}{5} = 1\frac{2}{5}$ 2. $\frac{9}{3} = 3$ 3. $\frac{18}{4} = 4\frac{1}{2}$ 4. $\frac{48}{10} = 4\frac{4}{5}$ 5. $\frac{8}{5} = 1\frac{3}{5}$

6. $\frac{25}{7} = 3\frac{4}{7}$ 7. $\frac{56}{4} = 14$ 8. $\frac{9}{6} = 1\frac{1}{2}$ 9. $\frac{100}{12} = 8\frac{1}{3}$ 10. $\frac{48}{16} = 3$

Have a student answer the problem question.

Develop Skills and Concepts Point out that division is used to rename an improper fraction as a mixed number. Emphasize that an improper fraction represents a number greater than 1.

Provide practice by having students rename each fraction as a whole number or mixed number.

1. $\frac{8}{5}\;(1\frac{3}{5})$ 2. $\frac{16}{4}\;(4)$ 3. $\frac{15}{12}\;(1\frac{1}{4})$ 4. $\frac{42}{6}\;(7)$
5. $\frac{6}{4}\;(1\frac{1}{2})$ 6. $\frac{56}{9}\;(6\frac{2}{9})$ 7. $\frac{83}{8}\;(10\frac{3}{8})$ 8. $\frac{45}{10}\;(4\frac{1}{2})$

3 Practice

Have students complete all the exercises. Watch for students who simply subtract the denominator from the numerator to find the whole number when renaming improper fractions as mixed numbers.

Practice

Rename each fraction as a whole or mixed number.

1. $\frac{8}{3} = 2\frac{2}{3}$
2. $\frac{9}{5} = 1\frac{4}{5}$
3. $\frac{11}{3} = 3\frac{2}{3}$
4. $\frac{6}{5} = 1\frac{1}{5}$

5. $\frac{12}{4} = 3$
6. $\frac{10}{8} = 1\frac{1}{4}$
7. $\frac{15}{6} = 2\frac{1}{2}$
8. $\frac{16}{2} = 8$

9. $\frac{17}{7} = 2\frac{3}{7}$
10. $\frac{18}{4} = 4\frac{1}{2}$
11. $\frac{36}{5} = 7\frac{1}{5}$
12. $\frac{48}{10} = 4\frac{4}{5}$

13. $\frac{180}{50} = 3\frac{3}{5}$
14. $\frac{100}{48} = 2\frac{1}{12}$
15. $\frac{90}{18} = 5$
16. $\frac{125}{25} = 5$

Problem Solving

Solve each problem.

17. Roy is helping to serve at the school picnic. He has 35 hamburgers and is putting 6 hamburgers on a tray. How many trays will Roy need?

    6 trays

18. Mrs. Mencini has 48 liters of lemonade. She wants to put the same amount of liquid in each of 9 containers. How many liters will go into each container?

    $5\frac{1}{3}$ liters

(Now Try This!)

We can always find a fraction between two fractions.

We can use a number line and equivalent fractions to find a fraction between $\frac{2}{5}$ and $\frac{3}{5}$, for example.

Rename the fractions and find their midpoint.

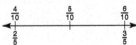

$\frac{5}{10}$ is the midpoint between $\frac{2}{5}$ and $\frac{3}{5}$.

Find a fraction between the given fractions.   Midpoints only are given. Other answers are possible.

1. $\frac{5}{7}$ and $\frac{6}{7}$

    $\frac{11}{14}$

2. $\frac{4}{9}$ and $\frac{5}{9}$

    $\frac{9}{18}$ or $\frac{1}{2}$

3. $\frac{1}{3}$ and $\frac{2}{3}$

    $\frac{3}{6}$ or $\frac{1}{2}$

4. $\frac{3}{4}$ and $\frac{4}{4}$

    $\frac{7}{8}$

5. $\frac{6}{10}$ and $\frac{7}{10}$

    $\frac{13}{20}$

6. $\frac{14}{9}$ and $\frac{15}{9}$

    $\frac{29}{18}$

7. $\frac{3}{49}$ and $\frac{4}{49}$

    $\frac{7}{98}$

8. $\frac{1}{64}$ and $\frac{2}{64}$

    $\frac{3}{128}$

104   Lesson 5-9 • Renaming Improper Fractions

Now Try This! Suggest that students draw number lines to help them solve the Now Try This! exercises. The midpoint is only one of the many possible fractions that are between any two fractions.

As an extension ask students to change $\frac{2}{5}$ to $\frac{6}{15}$ and $\frac{3}{5}$ to $\frac{9}{15}$. Students will see that there are two fractions between $\frac{6}{15}$ and $\frac{9}{15}$. ($\frac{7}{15}$; $\frac{8}{15}$)

---

4 Assess

Have students explain how to rename $\frac{16}{3}$ as a mixed number. (Possible answer: Divide 16 by 3 and write the remainder as a fraction; $5\frac{1}{3}$.)

For Mixed Abilities

Common Errors • Intervention

When students write a fraction as a mixed number, they may write the whole-number part and forget the fraction part. Have these students work in pairs rewriting an improper fraction such as $\frac{8}{7}$ in the following way.

$$\frac{8}{7} = \frac{7}{7} + \frac{1}{7} = 1 + \frac{1}{7} = 1\frac{1}{7}$$

Enrichment • Number Sense

Challenge students to use the numbers 2, 3, 6, and 9 to write fractions equivalent to $2\frac{4}{7}$, $2\frac{1}{3}$, $7\frac{1}{2}$, $3\frac{2}{3}$, $6\frac{3}{4}$, and $5\frac{2}{3}$. For each mixed number, students may use some or all of the given numbers but they may not use any number more than once. They may use the symbols $+$, $-$, and $\times$.

($2\frac{4}{7} = \frac{3 \times 6}{9 - 2}$; $2\frac{1}{3} = \frac{9 - 2}{3}$;

$7\frac{1}{2} = \frac{9 + 6}{2}$; $3\frac{2}{3} = \frac{9 + 2}{3}$;

$6\frac{3}{4} = \frac{9 \times 3}{6 - 2}$; $5\frac{2}{3} = \frac{2 + 6 + 9}{3}$)

Enrichment • Number Sense

Have each pair of students write the digits 1 through 9 on small pieces of paper, fold the papers, and put them in a paper bag. Have students take turns choosing 4 slips of paper from the bag, using the digits to write 2 fractions and comparing the fractions.

ESL/ELL STRATEGIES

Review the vocabulary of fractions by putting a variety of examples on the board. Ask students to describe each one using as many of the following terms as apply: *numerator*, *denominator*, *simplest form*, *mixed number*, and *improper fraction*.

T104

Comparing and Ordering Fractions

1 Getting Started

Objective
• To compare or order a set of fractions or mixed numbers

Vocabulary
least common denominator (LCD)

Warm Up • Mental Math
Have students estimate each product and explain how they arrived at their estimates.

1. 623×591 (360,000)
2. 135×763 (80,000)
3. 82×601 (48,000)
4. 933×810 (720,000)
5. 453×767 (400,000)
6. 65×984 (70,000)

Warm Up • Pencil and Paper
Have students order each set of numbers from least to greatest.

1. 1,011; 1,101; 1,110; 1,010 (1,010, 1,011, 1,101, 1,110)
2. $\frac{3}{8}, \frac{1}{8}, \frac{5}{8}, \frac{11}{8}$ ($\frac{1}{8}, \frac{3}{8}, \frac{5}{8}, \frac{11}{8}$)
3. 6,705; 6,507; 6,750; 6,075 (6,075; 6,507; 6,705; 6,750)
4. $3\frac{1}{4}; 4\frac{1}{4}; 2\frac{1}{4}; 7\frac{1}{4}$ ($2\frac{1}{4}; 3\frac{1}{4}; 4\frac{1}{4}; 7\frac{1}{4}$)
5. $\frac{2}{3}; 3\frac{2}{3}; 1\frac{1}{3}; \frac{13}{3}$ ($\frac{2}{3}; 1\frac{1}{3}; 3\frac{2}{3}; \frac{13}{3}$)
6. $\frac{10}{7}; 2\frac{1}{7}; \frac{17}{7}; 1\frac{2}{7}$ ($1\frac{2}{7}; \frac{10}{7}; 2\frac{1}{7}; \frac{17}{7}$)

Name _____

Comparing and Ordering Fractions

It's Algebra!

Ronald and Earle ride unicycles. One day, Ronald rode $\frac{3}{4}$ of a mile and Earle rode $\frac{2}{3}$ of a mile. Who rode the farthest?

We want to compare the distances the boys rode.

To compare fractions, we rename them as equivalent fractions with the **least common denominator (LCD)**.

$$\frac{3}{4} = \frac{9}{12} \qquad \frac{2}{3} = \frac{8}{12}$$

Compare the numerators.

$$\frac{9}{12} \; \bigcirc\!\!> \; \frac{8}{12}$$

If the denominators are the same, the fraction with the greater numerator is the greater fraction.

$$\frac{3}{4} \; \bigcirc\!\!> \; \frac{2}{3}$$

__Ronald__ rode farther than __Earle__ on his unicycle.

Mixed numbers can be compared in the same way.

First, compare the whole numbers. If they are equal, compare the fractions.

$2\frac{1}{2} > 2\frac{1}{5}$ because $2\frac{5}{10} > 2\frac{2}{10}$.

Getting Started

Find the LCD for each pair of fractions.

1. $\frac{2}{3}, \frac{1}{2}$ ___6___
2. $\frac{1}{2}, \frac{5}{8}$ ___8___
3. $\frac{7}{12}, \frac{5}{6}$ ___12___
4. $\frac{5}{6}, \frac{3}{8}$ ___24___

Write >, =, or < in each circle.

5. $\frac{3}{4}$ Ⓖ $\frac{1}{2}$
6. $5\frac{6}{8}$ Ⓖ $5\frac{8}{12}$
7. $3\frac{2}{3}$ Ⓖ $3\frac{3}{5}$
8. $\frac{1}{2}$ Ⓔ $\frac{4}{8}$

Order the fractions from least to greatest.

9. $\frac{1}{5}, \frac{2}{3}, \frac{1}{2}$ $\frac{1}{5}, \frac{1}{2}, \frac{2}{3}$
10. $7\frac{3}{8}, 7\frac{1}{4}, 8\frac{5}{6}$ $7\frac{1}{4}, 7\frac{3}{8}, 8\frac{5}{6}$

---

2 Teach

Introduce the Lesson Have a student read the problem and tell what is to be solved. (who rode his bike farther, Ronald or Earle) Ask what needed information is given. (Ronald rode his bike $\frac{3}{4}$ of a mile and Earle rode his bike $\frac{2}{3}$ of a mile.) Explain that to solve this problem students will need to compare the distances the boys rode.

Work through the model with students. Explain that, to compare fractions, the fractions must have a common denominator.

• The first step in comparing $\frac{3}{4}$ and $\frac{2}{3}$ is to rename the fractions as equivalent fractions with the least common denominator. Because the least common multiple of 4 and 3 is 12, 12 is the least common denominator.

• The second step is to write the equivalent fractions, using 12 as the denominator.

• The third step is to compare the numerators.

Repeat the three steps to compare mixed numbers.

Develop Skills and Concepts Pose this problem to students, *Harry has a bag with 6 oranges and 5 apples. Larry has a bag with 3 pears and 9 peaches. How would you compare what Larry has with what Harry has?* Students might suggest that they name the contents of the bags of fruit. Then, Larry has more fruit than Harry. Point out that students had to rename the contents of the bags before the total contents of each bag could be compared. The new name had to be common to all the items.

Using the three steps listed in Introduce the Lesson, compare $\frac{3}{5}$ and $\frac{2}{3}$ on the board. ($\frac{3}{5} < \frac{2}{3}$)

It's Algebra! The concepts in this lesson prepare students for algebra.

Practice

Find the LCD for each pair of fractions.

1. $\frac{3}{4}, \frac{1}{8}$ ___8___
2. $\frac{2}{3}, \frac{1}{4}$ ___12___
3. $\frac{3}{4}, \frac{5}{6}$ ___12___
4. $\frac{3}{5}, \frac{1}{2}$ ___10___

5. $\frac{3}{8}, \frac{1}{3}$ ___24___
6. $\frac{5}{6}, \frac{7}{8}$ ___24___
7. $\frac{2}{3}, \frac{5}{6}$ ___6___
8. $\frac{3}{4}, \frac{7}{10}$ ___20___

9. $\frac{3}{7}, \frac{5}{14}$ ___14___
10. $\frac{3}{8}, \frac{5}{12}$ ___24___
11. $\frac{5}{6}, \frac{4}{9}$ ___18___
12. $\frac{1}{4}, \frac{2}{5}$ ___20___

13. $\frac{3}{8}, \frac{7}{10}$ ___40___
14. $\frac{5}{16}, \frac{1}{6}$ ___48___
15. $\frac{7}{10}, \frac{2}{5}$ ___10___
16. $\frac{1}{11}, \frac{1}{2}$ ___22___

Write >, =, or < in each circle.

17. $\frac{3}{4}$ ⊜> $\frac{1}{8}$
18. $7\frac{2}{3}$ ⊜> $7\frac{1}{4}$
19. $\frac{3}{4}$ ⊜< $\frac{5}{6}$
20. $\frac{3}{5}$ ⊜> $\frac{1}{2}$

21. $\frac{5}{10}$ ⊜= $\frac{6}{12}$
22. $\frac{1}{8}$ ⊜< $\frac{2}{3}$
23. $9\frac{5}{7}$ ⊜< $9\frac{3}{4}$
24. $\frac{3}{4}$ ⊜> $\frac{5}{8}$

25. $6\frac{2}{3}$ ⊜> $6\frac{5}{8}$
26. $\frac{5}{9}$ ⊜> $\frac{1}{2}$
27. $\frac{4}{5}$ ⊜= $\frac{8}{10}$
28. $\frac{7}{10}$ ⊜> $\frac{2}{3}$

29. $\frac{1}{7}$ ⊜< $\frac{1}{3}$
30. $\frac{9}{12}$ ⊜> $\frac{10}{15}$
31. $1\frac{9}{30}$ ⊜> $1\frac{8}{40}$
32. $\frac{3}{15}$ ⊜< $\frac{4}{10}$

Order the fractions from least to greatest.

33. $\frac{3}{5}, \frac{2}{3}, \frac{1}{6}$
$\frac{1}{6}, \frac{3}{5}, \frac{2}{3}$

34. $4\frac{1}{4}, 14\frac{5}{8}, 4\frac{2}{3}$
$4\frac{1}{4}, 4\frac{2}{3}, 14\frac{5}{8}$

35. $\frac{4}{5}, \frac{3}{5}, \frac{1}{2}$
$\frac{1}{2}, \frac{2}{3}, \frac{4}{5}$

36. $\frac{5}{8}, \frac{5}{6}, \frac{3}{4}$
$\frac{5}{8}, \frac{3}{4}, \frac{5}{6}$

37. $2\frac{1}{2}, 1\frac{1}{3}, 2\frac{1}{9}$
$1\frac{1}{3}, 2\frac{1}{9}, 2\frac{1}{2}$

38. $\frac{2}{3}, \frac{4}{5}, \frac{3}{4}$
$\frac{2}{3}, \frac{3}{4}, \frac{4}{5}$

39. $6\frac{3}{4}, 6\frac{1}{2}, 6\frac{2}{3}$
$6\frac{1}{2}, 6\frac{2}{3}, 6\frac{3}{4}$

40. $8\frac{4}{5}, 8\frac{5}{6}, 8\frac{2}{3}$
$8\frac{2}{3}, 8\frac{4}{5}, 8\frac{5}{6}$

41. $3\frac{3}{10}, 3\frac{4}{15}, 3\frac{1}{6}$
$3\frac{1}{6}, 3\frac{4}{15}, 3\frac{3}{10}$

Problem Solving

Solve each problem.

42. Columbus got $\frac{3}{4}$ of the items correct on a math test. His friend Devin got $\frac{4}{5}$ of the items correct. Who had the better score?

Devin

43. Diane said she could finish a job in $\frac{5}{8}$ of a day. Gene said he could finish the same job in $\frac{3}{5}$ of a day. Who can finish the job faster?

Gene

44. Coleen and Howie ride the bike trail. She rests every 4 miles while he rests every 5 miles. At what point will they stop together?

20 miles

45. Which bond increased more in value; the city bond up $4\frac{1}{2}$ points or the county bond up $4\frac{2}{7}$?

city bond

Lesson 5-10 • Comparing and Ordering Fractions

106

For Mixed Abilities

Common Errors • Intervention

Some students may compare just the numerators or just the denominators when they are comparing fractions. Have them work with partners comparing two fractions, such as $\frac{5}{6}$ and $\frac{4}{5}$. Have the students write a set of equivalent fractions for each fraction until they find one from each set with the same denominator. When they compare these, they can see that $\frac{25}{30} > \frac{24}{30}$; therefore, $\frac{5}{6} > \frac{4}{5}$.

Enrichment • Number Sense

Have students work in groups to prepare two sets of index cards, a set of mixed numbers and a set of the improper fractions.

$3\frac{2}{3}, 1\frac{4}{5}, 4\frac{5}{8}, 8\frac{1}{2}, 9\frac{1}{4}, 6\frac{7}{12}, 7\frac{3}{10},$
$5\frac{5}{6}, 8\frac{7}{8}$

$\frac{11}{3}, \frac{9}{5}, \frac{37}{8}, \frac{17}{2}, \frac{37}{4}, \frac{79}{12}, \frac{73}{10}, \frac{35}{6}, \frac{71}{8}$

Have students shuffle each set separately. Tell them to place the first set of 9 cards face up in an array.

Then, have students match the other set of 9 cards to the face-up cards to form equivalent pairs.

More to Explore • Geometry

Tell students that a pattern of polygons, like the following drawn on the board, can be folded to create a polyhedron. Duplicate the pattern on paper and demonstrate the folding for students.

Duplicate the following patterns for students to cut out and fold, to identify which are cube patterns and which are not.

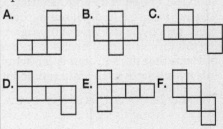

(Students will find patterns A and B will not form a cube.)

---

3 Practice

Have students complete all the exercises. Remind students that > means "is greater than," and < means "is less than."

4 Assess

Ask students why they would rename $\frac{2}{5}$ and $\frac{1}{2}$ to compare the fractions. (Possible answer: If you rename the fractions as equivalent fractions with a common denominator, you only have to compare the numerators.)

T106

5-11 Problem Solving: Make a Table

pages 107–108

1 Getting Started

Objective
• To solve problems by making a table

Materials
pentagon

Warm Up • Mental Math
Have students simplify each expression.

1. 6 + (6 + 2) + (8 + 2) (24)
2. $0.35 × 5 ($1.75)
3. 32 × 4 (128)
5. $\frac{1}{2}$ of 72 (36)
6. 3 × 15 (45)

Warm Up • Pencil and Paper
Have students simplify each expression. Encourage students to use the properties to make the work easier.

1. $6^2 - 3 ÷ 11 + 1 ×$ itself $- 1 ÷ 5$ (3)
2. $2^3 + 1 ×$ itself $- 1 ÷ 10$ (8)
3. $45 ÷ 9 + 3 × 5 × \frac{1}{2} + 4 ÷ 12$ (2)
4. $1 × 1 + 3 ×$ itself $- 1 × 2 - 2 ÷ 4$ (7)
5. $3^3 + 1 ÷ 4 × 5 + 1 ÷ 4$ (9)
6. $18 ÷ 3 + 5 - 2 × 7 + 2 ÷ 5$ (13)
7. $7^2 - 1 ÷ 6 + 2 × 10 - 5 ÷ 5$ (19)
8. $4^3 ÷ 8 - 2 × 5 + 2 ÷ 4$ (8)

Name _____

Problem Solving: Make a Table

The publisher will use exactly 201 digits in numbering the pages of a book. The book begins with page 1. How many pages will be in the book?

★ SEE
We want to know how many pages will be in the book.

Exactly __201__ digits will be used in numbering the pages.

★ PLAN
In order to solve this problem, we can make a table to record the number of digits used in numbering a chosen quantity of pages. This will help break the problem into smaller steps.

★ DO

Pages of a Book				
Pages	Number of Pages	Number of Digits Per Page	Number of Digits Used	Total Digits in Book
1–9	9	1	9	9
10–49	40	2	80	89
50–99	50	2	100	189
100	1	3	3	192
101	1	3	3	195
102	1	3	3	198
103	1	3	3	201

There will be __103__ pages in the book.

★ CHECK
We can check our solution by working backwards through the table.

Pages 100–103   4 pages   3 digits each   __12__ digits
Pages 10–99   90 pages   2 digits each   __180__ digits
Pages 1–9   9 pages   1 digit each   __9__ digits

__12__ + __180__ + __9__ = __201__ digits.

Lesson 5-11 • Problem Solving: Make a Table

2 Teach

Introduce the Lesson Lead the students through the problem and the first two steps of the solution. Remind students that making a table is a helpful strategy in many types of problems, by itself or with other strategies.

Develop Skills and Concepts Tell students the secret of a successful table is its organization. Most tables will involve two sets of data, one written across the top, the other down the left side. Although it is usually easier to list data in numerical order, there are times when it is obvious that the solution is far away and it is best to use intervals where some data can be left out.

Working backward, as shown in the CHECK section, involves using the solution to see if it fits.

3 Apply

Solution Notes

1. Every year Ann's annual salary is increased by $2,040 and Beth's by $1,000 every 6 months. Although Ann begins with a greater salary, Beth's semi-annual increases allow her to earn more than Ann by year.

Year	Ann	Beth	
0.5		$8,200	
1	$20,400	$9,200	$17,400
1.5		$10,200	
2	$22,440	$11,200	$22,400
2.5		$12,200	
3	$24,480	$13,200	$25,400

2. grade 6, 5, . . ., 1
   no. > 80% 96, 48, . . ., 3

T107

Apply

Solve each problem.

1. Ann is offered a job at a starting salary of $20,400 per year. She is told that if her work is satisfactory, her salary will be increased by $2,040 at the end of each year for the next three years. Beth is offered a job starting at $8,200 for a half year. If her work is satisfactory her half-year salary will be increased by $1,000 per half year at the end of each half year for the next three years. Who will earn more money after 3 years?

   Beth

2. In grade 6 there were 96 students who scored above 80% in the district spelling bee competition. Each grade had only half as many students above 80% as the next grade. That is, the fifth grade had only 48 students score above 80%. How many students altogether scored above 80% in grades 1 through 6?

   189 students

3. While watching a circus parade, I saw some girls and some horses come by. I counted 46 legs and 15 heads. How many horses and how many girls were in the parade?

   7 girls and 8 horses

4. A triangle has no diagonals, and a rectangle has two diagonals. How many diagonals can be drawn on a STOP sign of eight sides?

   20 diagonals

5. Rosemary was buying supplies for school. Pencils cost $0.25 and erasers cost $0.15. She spent $1.45. How many pencils and erasers did she buy?

   4 pencils and 3 erasers
   or 1 pencil and 8 erasers

6. Jennifer is preparing for her track meet. She runs one lap around the track the first day, and increases her laps by two per day. On her final day of practice she runs 27 laps. How many days did Jennifer practice?

   14 days

7. What if the pencils in Problem 5 cost 15¢ and the erasers cost 25¢? Then how many pencils and erasers did Rosemary buy for $1.45?

   4 erasers and 3 pencils or
   1 eraser and 8 pencils

8. Read Problem 6 again. How would you rewrite the exercise so that the number of days Jennifer practiced was 16 days?

   Answers may vary.

9. Tracy has some nickels, dimes, and quarters. She has $2.25 in all. She has twice as many dimes as nickels and 3 more quarters than nickels. What would be the first step you would take to find how many of each coin she has?

   Answers may vary.

10. Mattie has twice as many dimes as she has nickels. She has $1.75 in all. Ken says, "I think that Mattie has 14 nickels." Arthur says, "No, I think that she has 7 nickels." Tell who is correct, Ken or Arthur, and prove it.

    Arthur

Lesson 5-11 • Problem Solving: Make a Table

For Mixed Abilities

More to Explore • Number Sense

Have students devise their own coded number system, using a different symbol for numbers 1 through 9. Have them make a distinction for multiples of 10, including tenths through hundredths. Have them write five decimal multiplication or division problems, using their code. Then, have them exchange their work with a partner to solve each other's problems.

---

3. *girls* 0, 1, . . ., 7
   horses 15, 14, . . ., 8
   legs 60, 58, . . ., 46

5. *pencils* 0, 1, . . ., 34
   erasers 11, 10, . . . 3
   $ cost 1.65, 1.75, . . ., 1.45

4. *no. sides* 3, 4, 5, 6, 7, 8
   no. diagonals 0, 2, 5, 9, 14, 20

6. *day* 1, 2, . . ., 314
   laps 1, 3, . . ., 27

Higher-Order Thinking Skills

7. **Analysis:** The problem is the same as the original problem except that the item names have been interchanged.

8. **Synthesis:** Since Jennifer increases her laps by 2 per day and the answer to the original problem was 14 days, they need only add two 2s to 27 and say she runs 31 laps on the final day.

9. **Synthesis:** Most students probably would make a table to show the number of nickels, dimes, and quarters; e.g., if n represents the number of nickels, then $2 \times n$ represents the number of dimes, and $n + 3$ represents the number of quarters.

nickels—n			
dimes—$2n$			
quarters—$n + 3$			

10. **Evaluation:** Mattie has 7 nickels and 14 dimes, which together are worth $1.75. Students can find the answer by guessing and testing or by making a table similar to the table below.

number of nickels	1	2	3	4	5	6	7
number of dimes	2	4	6	8	10	12	14
value of nickels	$0.05	$0.10	$0.15	$0.20	$0.25	$0.30	$0.35
value of dimes	$0.20	$0.40	$0.60	$0.80	$1.00	$1.20	$1.40
total value	$0.25	$0.50	$0.75	$1.00	$1.25	$1.50	$1.75

 Assess

Ask students why making a table is a good problem-solving strategy. (Possible answer: Making a table breaks the problem into smaller steps.)

T108

5-12 Calculator: Cross Products

pages 109–110

1 Getting Started

Objective
• To use a calculator to find cross products

Materials
calculators

Warm Up • Mental Math
Ask students to name the least common multiple of each pair of numbers.

1. 4, 10 (20)   6. 25, 10 (50)
2. 8, 10 (40)   7. 10, 15 (30)
3. 2, 3, (6)   8. 4, 6 (12)
4. 6, 8 (24)   9. 8, 12 (24)
5. 12, 9 (36)   10. 5, 7 (35)

Warm Up • Pencil and Paper
Have students write the missing term in each equivalent fraction.

1. $\frac{1}{3} = \frac{(2)}{6}$ 7. $\frac{4}{5} = \frac{16}{(20)}$
2. $\frac{3}{5} = \frac{6}{(10)}$ 8. $\frac{4}{7} = \frac{(28)}{49}$
3. $\frac{42}{48} = \frac{7}{(8)}$ 9. $\frac{1}{11} = \frac{8}{(88)}$
4. $\frac{3}{8} = \frac{(27)}{72}$ 10. $\frac{2}{3} = \frac{10}{(15)}$
5. $\frac{2}{5} = \frac{(6)}{15}$ 11. $\frac{3}{4} = \frac{(36)}{48}$
6. $\frac{3}{7} = \frac{(15)}{35}$ 12. $\frac{9}{4} = \frac{36}{(16)}$

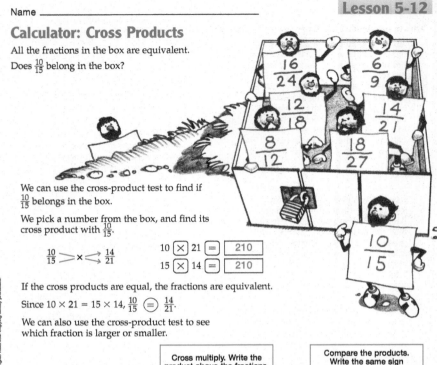

Calculator: Cross Products

All the fractions in the box are equivalent.
Does $\frac{10}{15}$ belong in the box?

We can use the cross-product test to find if $\frac{10}{15}$ belongs in the box.

We pick a number from the box, and find its cross product with $\frac{10}{15}$.

$$\frac{10}{15} \diagdown\times\diagup \frac{14}{21}$$

10 ✕ 21 = 210
15 ✕ 14 = 210

If the cross products are equal, the fractions are equivalent.

Since $10 \times 21 = 15 \times 14$, $\frac{10}{15} = \frac{14}{21}$.

We can also use the cross-product test to see which fraction is larger or smaller.

Cross multiply. Write the product above the fractions.	Compare the products. Write the same sign between the fractions.
$\frac{2}{3}$? $\frac{3}{4}$	$8 < 9$
$\begin{matrix}8 & 9\\ \frac{2}{3}\diagdown\times\diagup\frac{3}{4}\end{matrix}$	$\frac{2}{3} < \frac{3}{4}$

REMEMBER Always multiply the first numerator by the second denominator, and write that product above the first fraction.

Getting Started

Use the cross-product test to compare the fractions.
Write <, =, or > in each circle.

1. $\frac{5}{6}$ (>) $\frac{2}{3}$ 2. $\frac{12}{15}$ (=) $\frac{20}{25}$ 3. $\frac{6}{13}$ (<) $\frac{24}{45}$

4. $\frac{1}{5}$ (<) $\frac{7}{12}$ 5. $\frac{3}{4}$ (<) $\frac{7}{8}$ 6. $\frac{2}{3}$ (=) $\frac{16}{24}$

2 Teach

Introduce the Lesson Have students define the term *equivalent fractions*. (fractions that name the same numbers) Review the technique presented in the lesson on pages 105–106 for comparing fractions.

1. Find the least common denominator of the fractions.
2. Rename the fractions as equivalent fractions with the LCD.
3. Compare the numerators.

Explain that in this lesson students will learn another way to identify equivalent fractions and compare two fractions. Work through the model with students using the cross-product test.

Develop Skills and Concepts Point out that the calculator makes it easy to find large products quickly. Provide practice by having students use the cross-product test to compare each pair of fractions.

1. $\frac{5}{8}$ (>) $\frac{1}{2}$ 2. $\frac{5}{6}$ (=) $\frac{15}{18}$ 3. $\frac{14}{25}$ (<) $\frac{90}{155}$

3 Practice

Have students complete all the exercises. Remind students to compare fractions carefully.

T109

Practice

Use the cross-product test to compare the fractions. Write <, =, or > in each circle.

1. $\frac{3}{4}$ ⬤< $\frac{15}{18}$ 2. $\frac{5}{7}$ ⬤< $\frac{11}{15}$ 3. $\frac{2}{12}$ ⬤= $\frac{5}{30}$

4. $\frac{3}{8}$ ⬤> $\frac{5}{14}$ 5. $\frac{5}{9}$ ⬤> $\frac{8}{15}$ 6. $\frac{7}{8}$ ⬤> $\frac{14}{17}$

7. $\frac{2}{11}$ ⬤> $\frac{1}{6}$ 8. $\frac{3}{7}$ ⬤< $\frac{5}{11}$ 9. $\frac{5}{8}$ ⬤> $\frac{7}{12}$

10. $\frac{9}{16}$ ⬤< $\frac{15}{24}$ 11. $\frac{9}{20}$ ⬤> $\frac{13}{30}$ 12. $\frac{25}{35}$ ⬤= $\frac{15}{21}$

13. $\frac{9}{20}$ ⬤> $\frac{7}{16}$ 14. $\frac{37}{45}$ ⬤< $\frac{21}{25}$ 15. $\frac{17}{32}$ ⬤< $\frac{24}{45}$

16. $\frac{21}{37}$ ⬤< $\frac{11}{19}$ 17. $\frac{11}{56}$ ⬤> $\frac{19}{100}$ 18. $\frac{43}{56}$ ⬤< $\frac{37}{48}$

Problem Solving

Solve each problem.

19. Bill made 17 baskets in 25 tries. Ron made 13 baskets in 20 tries. Which player made the greater part of his tries?
Bill

20. Marilyn walked $\frac{2}{3}$ of a mile. Betty walked $\frac{7}{12}$ of a mile and Rita walked $\frac{5}{8}$ of a mile. Who walked the farthest?
Marilyn

21. Plunkett High won 3 out of 5 of their games this year. Last year they won 7 out of 10. Which year was better?
last year

22. Which piece is larger; $\frac{4}{32}$ of a pie or $\frac{5}{40}$ of it?
They are the same.

(Now Try This!)

The calculator makes it easy to find equivalent fractions. Complete the following.

1. $\frac{3}{9} = \frac{6}{18}$ 9 ☒ 6 ➗ 18 ＝ [3]

2. $\frac{8}{12} = \frac{26}{39}$ 8 ☒ 39 ➗ 12 ＝ [26]

3. $\frac{15}{40} = \frac{24}{64}$ _40_ ☒ _24_ ➗ _15_ ＝ [64]

110   Lesson 5-12 • Calculator: Cross Products

Now Try This! This activity focuses students' attention on the usefulness of the calculator in finding equivalent fractions. It also presents a pattern for solving equations that students will use in later work with percents and ratio.

4 Assess

Have students use a calculator to compare $\frac{3}{8}$ and $\frac{108}{288}$. (=)
Have students change 108 so that the first fraction is greater than the second fraction. (Change 108 to any number less than 108.)

For Mixed Abilities

Common Errors • Intervention

When students use the cross-product test, make sure that they write the products above the fractions and not below them. Point out to students that you really are comparing numerators of fractions whose common denominator would be the product of the two denominators.

$$35 \qquad 32$$
$$\frac{7}{8} \diagdown\diagup \frac{4}{5}$$
$$\frac{35}{40} > \frac{32}{40}$$
$$\frac{7}{8} > \frac{4}{5}$$

Enrichment • Number Sense

1. Have students use calculators to find the missing term in each equivalent fraction using the method presented in the Now Try This! on page 110.

$\frac{3}{5} = \frac{(51)}{85}$ $\frac{5}{9} = \frac{(120)}{216}$

$\frac{8}{9} = \frac{112}{(126)}$ $\frac{4}{22} = \frac{172}{(946)}$

$\frac{18}{27} = \frac{(2)}{3}$ $\frac{(38)}{95} = \frac{2}{5}$

2. Have students use the cross-product test to find the greatest fraction in each set.

$\frac{4}{5}, \frac{7}{8}, \frac{3}{4}$ $(\frac{7}{8})$

$\frac{1}{3}, \frac{3}{8}, \frac{2}{5}$ $(\frac{2}{5})$

More to Explore • Applications

Explain to students how to keep score in bowling. Define and illustrate on a sample scorecard the following terms: *frame* (each of the ten chances to knock down all the pins by rolling 2 balls); *spare* (the diagonal in the small box, count 10 plus the first ball in the next frame); *strike* (the x in the small box, count 10 plus the next 2 balls rolled).

Duplicate the following incomplete scorecard for students to fill in:

1	2	3	4	5	6
7 ⬦	6 ⎣3⎦	☒	4 ⎣3⎦	1 ⬦	☒
(16)	(25)	(42)	(49)	(69)	(86)

7	8	9	10	Total
3 ⎣4⎦	2 ⎣6⎦	☒	6 ⎣2⎦	
(93)	(101)	(119)	(127)	(127)

T110

page 111

Items Objectives

1–4 To find the least common multiple and the greatest common factor of a pair of numbers (see pages 87–90)

5–12 To write prime factorization of composite numbers, using exponents (see pages 93–94)

13–20 To write equivalent fractions (see pages 97–98)

21–28 To write fractions in simplest form (see pages 99–100)

29–36 To compare fractions (see pages 105–106)

37–44 To rename mixed numbers as fractions (see pages 101–102)

45–52 To rename improper fractions as whole and mixed numbers (see pages 103–104)

Alternate Chapter Test

You may wish to use the Alternate Chapter Test on page 378 of this book for further review and assessment.

Name _____

Write the LCM and GCF for each pair of numbers.

1. 6, 9

LCM __18__

GCF __3__

2. 4, 8

LCM __8__

GCF __4__

3. 5, 7

LCM __35__

GCF __1__

4. 8, 12

LCM __24__

GCF __4__

Write the prime factorization of each number using exponents.

5. 12 __$2^2 \times 3$__

6. 18 __2×3^2__

7. 50 __2×5^2__

8. 36 __$2^2 \times 3^2$__

9. 32 __2^5__

10. 49 __7^2__

11. 35 __5×7__

12. 54 __2×3^3__

Complete the equivalent fraction.

13. $\frac{3}{5} = \frac{9}{15}$

14. $\frac{3}{8} = \frac{12}{32}$

15. $\frac{5}{9} = \frac{20}{36}$

16. $\frac{14}{21} = \frac{2}{3}$

17. $\frac{12}{16} = \frac{3}{4}$

18. $\frac{2}{9} = \frac{12}{54}$

19. $\frac{3}{14} = \frac{9}{42}$

20. $\frac{4}{6} = \frac{32}{48}$

Rename each fraction in simplest form.

21. $\frac{15}{18} = \frac{5}{6}$

22. $\frac{25}{35} = \frac{5}{7}$

23. $\frac{10}{12} = \frac{5}{6}$

24. $\frac{24}{48} = \frac{1}{2}$

25. $\frac{36}{72} = \frac{1}{2}$

26. $\frac{35}{60} = \frac{7}{12}$

27. $\frac{14}{42} = \frac{1}{3}$

28. $\frac{50}{75} = \frac{2}{3}$

Write <, =, or > in each circle.

29. $\frac{5}{9} \; < \; \frac{3}{4}$

30. $\frac{2}{3} \; > \; \frac{5}{8}$

31. $\frac{1}{2} \; < \; \frac{7}{12}$

32. $\frac{9}{12} \; = \; \frac{6}{8}$

33. $\frac{4}{7} \; > \; \frac{2}{9}$

34. $\frac{12}{16} \; = \; \frac{3}{4}$

35. $\frac{3}{8} \; < \; \frac{4}{5}$

36. $\frac{3}{14} \; > \; \frac{1}{5}$

Rename each mixed number as a fraction.

37. $4\frac{3}{5} = \frac{23}{5}$

38. $5\frac{1}{8} = \frac{41}{8}$

39. $1\frac{3}{4} = \frac{7}{4}$

40. $9\frac{7}{10} = \frac{97}{10}$

41. $6\frac{2}{7} = \frac{44}{7}$

42. $10\frac{1}{9} = \frac{91}{9}$

43. $5\frac{4}{5} = \frac{29}{5}$

44. $15\frac{2}{3} = \frac{47}{3}$

Rename each improper fraction as a whole or mixed number.

45. $\frac{11}{3} = 3\frac{2}{3}$

46. $\frac{9}{6} = 1\frac{1}{2}$

47. $\frac{20}{5} = 4$

48. $\frac{37}{4} = 9\frac{1}{4}$

49. $\frac{54}{9} = 6$

50. $\frac{18}{12} = 1\frac{1}{2}$

51. $\frac{90}{10} = 9$

52. $\frac{75}{6} = 12\frac{1}{2}$

Circle the letter of the correct answer.

1. $n + 6 = 13$
   $n = ?$
   a. 6
   (b) 7
   c. 19
   d. NG

2. $12 - n = 8$
   $n = ?$
   (a) 4
   b. 8
   c. 12
   d. NG

3. $8 \times 7 = n$
   $n = ?$
   a. 49
   b. 54
   (c) 56
   d. NG

4. $8 \times n = 32$
   $n = ?$
   (a) 4
   b. 5
   c. 6
   d. NG

5. $6 + 8 \div 2 = n$
   $n = ?$
   a. 7
   (b) 10
   c. 12
   d. NG

6. What is the place value of the 0 in 427,063?
   a. thousands
   (b) hundreds
   c. tens
   d. NG

7. $\begin{array}{r} \$136.85 \\ + \quad 97.48 \end{array}$
   a. $134.33
   b. $233.33
   (c) $234.33
   d. NG

8. $\begin{array}{r} 74,086 \\ - 15,994 \end{array}$
   a. 58,192
   b. $57,192
   (c) 58,092
   d. NG

9. 458×6
   a. 2,708
   (b) 2,748
   c. 2,848
   d. NG

10. $\begin{array}{r} \$6.29 \\ \times \quad 47 \end{array}$
    a. $69.19
    b. $285.63
    (c) $295.63
    d. NG

11. $439 \div 7$
    a. 6 R25
    b. 62
    (c) 62 R5
    d. NG

12. $56\overline{)395}$
    a. 7
    (b) 7 R3
    c. 73
    d. NG

13. Complete the equivalent fraction.
    $\frac{6}{8} = \frac{?}{24}$
    a. 2
    b. 3
    (c) 18
    d. NG

STOP

[ ] score

Chapter 5 • Cumulative Assessment

112

page 112

Items	Objectives
1–4	To review basic addition, subtraction, multiplication, and division facts (see pages 1–8)
5	To simplify an expression (see pages 11–12)
6	To identify place value (see pages 19–20)
7	To add two numbers up to 6-digits (see pages 27–28)
8	To subtract with renaming over zeros (see pages 33–34)
9–10	To multiply 3-digit factors by up to 2-digit factors including money (see pages 45–46; 51–52)
11–12	To divide 3-digit numbers by up to 2-digit numbers (see pages 63–64; 75–76)
13	To write equivalent fractions (see pages 97–98)

Alternate Cumulative Assessment

Circle the letter of the correct answer.

1. $n + 8 = 11$
   $n = ?$
   a 2
   (b) 3
   c 4
   d NG

2. $18 - n = 95$
   $n = ?$
   a 7
   b 8
   c 9
   (d) NG

3. $4 \times 9 = n$
   $n = ?$
   a 32
   (b) 36
   c 38
   d NG

4. $3 \times n = 24$
   $n = ?$
   a 32
   (b) 8
   c 9
   d NG

5. $9 + 9 \div 3 = n$
   $n = ?$
   a 3
   b 6
   (c) 12
   d NG

6. Give the place value of the 7 in 684,702
   a tens
   (b) hundreds
   c thousands
   d NG

7. $\begin{array}{r} \$357.80 \\ + \quad 68.97 \end{array}$
   a $416.77
   b $426.70
   (c) $426.77
   d NG

8. $\begin{array}{r} 86,083 \\ - 27,597 \end{array}$
   (a) 58,486
   b 58,586
   c 59,486
   d NG

9. $\begin{array}{r} 347 \\ \times \quad 5 \end{array}$
   a 1,535
   b 1,705
   (c) 1,735
   d NG

10. $\begin{array}{r} \$4.96 \\ \times \quad 84 \end{array}$
    (a) $416.64
    b $415.64
    c $406.64
    d NG

11. $8\overline{)751}$
    a 9 R37
    b 93
    (c) 93 R7
    d NG

12. $61\overline{)325}$
    a 5 R2
    (b) 5 R20
    c 50 R2
    d NG

6-1 Estimating Sums and Differences of Mixed Numbers

pages 113–114

1 Getting Started

Objective
• To estimate sums and differences of mixed numbers

Materials
fractional number lines

Warm Up • Mental Math
Have students estimate each product.
1. 18×8 (160)
2. 22×4 (80)
3. 41×7 (280)
4. 126×6 (600)
5. 187×9 (1,800)

Warm Up • Pencil and Paper
Have students find the missing term in each equivalent fraction.
1. $\frac{2}{3} = \frac{?}{12}$ (8) 5. $\frac{7}{12} = \frac{?}{36}$ (21)
2. $\frac{3}{4} = \frac{?}{28}$ (21) 6. $\frac{3}{8} = \frac{?}{24}$ (9)
3. $\frac{3}{5} = \frac{?}{15}$ (9) 7. $\frac{3}{10} = \frac{?}{20}$ (6)
4. $\frac{1}{2} = \frac{?}{16}$ (8) 8. $\frac{5}{8} = \frac{?}{40}$ (25)

Name _____

Add and Subtract Fractions

Lesson 6-1

Estimating Sums and Differences of Mixed Numbers

Jason is following a diagram to make a picture frame. About how much longer is side A than side B?

The word *about* tells that we need to find an estimate.

We know that side A is $7\frac{3}{8}$ in. long.

We know that side B is $5\frac{6}{8}$ in. long.

We need to estimate the difference in length.

To estimate with mixed numbers, we round them to whole numbers first. Then, we add or subtract.

$7\frac{3}{8}$ Since $\frac{3}{8}$ is less than $\frac{4}{8}$ or $\frac{1}{2}$, round down. → 7

$-5\frac{6}{8}$ Since $\frac{6}{8}$ is greater than $\frac{4}{8}$ or $\frac{1}{2}$, round up. → $\frac{-6}{1}$

Side A is about ___1___ inch longer than side B.

Getting Started

Write *greater than* $\frac{1}{2}$ or *less than* $\frac{1}{2}$ for each fraction.

1. $\frac{6}{10}$ greater than $\frac{1}{2}$
2. $\frac{3}{5}$ greater than $\frac{1}{2}$
3. $\frac{5}{6}$ greater than $\frac{1}{2}$
4. $\frac{1}{3}$ less than $\frac{1}{2}$

Round each mixed number to the nearest whole number.

5. $4\frac{6}{10}$ ___5___
6. $2\frac{1}{3}$ ___2___
7. $6\frac{3}{4}$ ___7___
8. $3\frac{4}{5}$ ___4___

Estimate each sum or difference.

9. $6\frac{3}{4}$
   $+2\frac{3}{5}$ $\frac{7}{+3}$
   about ___10___

10. $4\frac{6}{10}$
    $-2\frac{1}{4}$ $\frac{5}{-2}$
    about ___3___

11. $3\frac{4}{5}$
    $+4\frac{3}{8}$ $\frac{4}{+4}$
    about ___8___

12. $15\frac{1}{3}$
    $-8\frac{3}{10}$ $\frac{15}{-8}$
    about ___7___

Lesson 6-1 • Estimating Sums and Differences of Mixed Numbers 113

---

2 Teach

Introduce the Lesson Read the problem aloud. Have a student read and compute the information sentences.

• Explain that to round a mixed number, students need to compare the fractional part to $\frac{1}{2}$. If the fractional part is $\frac{1}{2}$ or greater, the mixed number rounds up to the next whole number. If the fractional part is less than $\frac{1}{2}$, round down to the whole number in the mixed number.

• Students can use a fractional number line to help them see to which whole number the mixed number is closer.

• Ask a student to read the solution sentence.

Develop Skills and Concepts Provide another example to illustrate estimating the sum of two mixed numbers. Write $4\frac{1}{3} + 9\frac{5}{8}$ in vertical form on the board. Give students two number lines.

The mixed number $4\frac{1}{3}$ will round down to 4 or up to 5. Have students divide the section of the number line between 4 and 5, into thirds. Ask, *Is $4\frac{1}{3}$ closer to 4 or 5?* (4)

The mixed number $9\frac{5}{8}$ will round down to 9 or up to 10. Have students divide the section of the number line between 9 and 10, into eighths. Ask, *Is $9\frac{5}{8}$ closer to 9 or 10?* (10)

Now, have students find the sum $4 + 10$. (14)

Have students use the same technique to estimate the difference $12\frac{2}{5} - 8\frac{5}{6}$. (3)

Using Estimation On the board, write

$9\frac{1}{8} -$ _____ is about 3

Ask, *What does the mixed number in the blank space need to round to?* (6) Have students fill in the blank with an appropriate mixed number. (Possible answer: $5\frac{3}{4}$)

Estimate each sum or difference.

1. $6\frac{4}{5}$ 7
$+7\frac{3}{4}$ $+8$
about __15__

2. $4\frac{1}{3}$ 4
$+3\frac{7}{8}$ $+4$
about __8__

3. $2\frac{2}{5}$ 2
$+6\frac{1}{3}$ $+6$
about __8__

4. $14\frac{2}{3}$ 15
$+5\frac{4}{5}$ $+6$
about __21__

5. $12\frac{1}{6}$ 12
$+5\frac{4}{7}$ $+6$
about __18__

6. $9\frac{1}{3}$ 9
$-2\frac{4}{9}$ -2
about __7__

7. $10\frac{2}{5}$ 10
$-8\frac{1}{4}$ -8
about __2__

8. $16\frac{3}{4}$ 17
$-5\frac{7}{8}$ -6
about __11__

9. $12\frac{3}{5}$ 13
$-4\frac{1}{3}$ -4
about __9__

10. $6\frac{5}{7}$ 7
$-2\frac{1}{3}$ -2
about __5__

11. $3\frac{1}{3}$ 3
$+12\frac{4}{11}$ $+12$
about __15__

12. $18\frac{9}{12}$ 19
$-7\frac{2}{5}$ -7
about __12__

13. $6\frac{4}{10}$ 6
$+11\frac{1}{3}$ $+12$
about __18__

14. $7\frac{1}{4}$ 7
$-2\frac{5}{10}$ -3
about __4__

15. $12\frac{4}{7}$ 13
$+9\frac{8}{9}$ $+10$
about __23__

16. $17\frac{1}{3}$ 17
$-5\frac{5}{6}$ -6
about __11__

17. $3\frac{6}{7}$ 4
$+12\frac{4}{9}$ $+12$
about __16__

18. $10\frac{8}{9}$ 11
$-3\frac{4}{7}$ -4
about __7__

19. $12\frac{1}{4}$ 12
$+3\frac{4}{10}$ $+3$
about __15__

20. $6\frac{2}{5}$ 6
$-1\frac{8}{11}$ -2
about __4__

Copy. Then, estimate each sum or difference.

21. $3\frac{4}{5} + 2\frac{1}{6}$ 6

22. $15\frac{1}{3} - 4\frac{3}{5}$ 10

23. $5\frac{1}{9} + 10\frac{6}{7}$ 16

24. $13\frac{4}{7} - 2\frac{2}{9}$ 12

25. $16\frac{4}{9} - 12\frac{2}{3}$ 3

26. $17\frac{2}{3} + 4\frac{3}{5}$ 23

27. $13\frac{1}{5} - 2\frac{5}{6}$ 10

28. $6\frac{7}{10} + 4\frac{1}{6}$ 11

Problem Solving

Solve each problem.

29. Ann bought $5\frac{1}{2}$ yards of fabric to make a costume. She cut off $3\frac{1}{8}$ yards of fabric. About how much fabric was left?
3 yards

30. Sarah bought $4\frac{3}{4}$ yards of blue fabric, $6\frac{1}{3}$ yards of green fabric, and $7\frac{4}{7}$ yards of red fabric. About how much fabric did she buy?
19 yards

Lesson 6-1 • Estimating Sums and Differences of Mixed Numbers

For Mixed Abilities

Common Errors • Intervention

Some students may have difficulty deciding whether fractions such as $\frac{4}{9}$, $\frac{5}{9}$, or $\frac{4}{7}$ are greater or less than $\frac{1}{2}$. Have students work in pairs. Have each pair write the fractions $\frac{4}{7}$, $\frac{5}{7}$, $\frac{3}{8}$, $\frac{5}{8}$, $\frac{4}{9}$, $\frac{5}{9}$, $\frac{5}{12}$, and $\frac{7}{12}$ on a sheet of paper. For each fraction, the pair should decide whether it is greater than $\frac{1}{2}$ or less than $\frac{1}{2}$. Have students circle the fractions that are greater than $\frac{1}{2}$, and underline the fractions that are less than $\frac{1}{2}$.

Enrichment • Number Sense

Have students work in pairs. One student makes up a story involving mixed numbers and the other student estimates the answer. For example, the first student says, "I have $1\frac{3}{4}$ cups of flour in a bowl. Then I added $3\frac{1}{8}$ more cups. About how much flour did I have total?" (5 cups) Students reverse roles, one providing an addition problem, the other a subtraction problem.

More to Explore • Application

Challenge students to provide examples of mixed numbers in everyday life (recipes, gasoline prices, for example). Have them find examples of mixed numbers in magazines, newspapers, and other textbooks. Discuss some of their findings. How are mixed numbers being used in their examples? Create short story problems with the mixed numbers they found. Students can estimate the answers.

3 Practice

Have students complete all the exercises. Students may use number lines to help them round each mixed number to a whole number.

4 Assess

Ask students what the estimated difference would be in Exercise 14 if you round $2\frac{5}{10}$ to 2. (The estimated difference would be 5 instead of 4.) Remind students that if the fractional part of the mixed number equals $\frac{1}{2}$, you round up to the next whole number.

T114

6-2 Adding Fractions and Mixed Numbers

pages 115–116

1 Getting Started

Objective
• To add fractions with common denominators

Warm Up • Mental Math
Have students find each sum.

1. $6 + 3$ (9)
2. $7 + 6$ (13)
3. $4 + 9$ (13)
4. $3 + 5$ (8)
5. $8 + 6$ (14)
6. $5 + 6$ (11)
7. $3 + 4 + 7$ (14)
8. $8 + 2 + 6$ (16)
9. $2 + 5 + 8$ (15)
10. $6 + 1 + 6$ (13)
11. $8 + 0 + 8$ (16)
12. $2 + 4 + 6$ (12)

Warm Up • Pencil and Paper
Have students write each fraction or mixed number in simplest form.

1. $\frac{15}{35}$ $\left(\frac{3}{7}\right)$
2. $\frac{32}{24}$ $\left(1\frac{1}{3}\right)$
3. $\frac{16}{9}$ $\left(1\frac{7}{9}\right)$
4. $\frac{19}{3}$ $\left(6\frac{1}{3}\right)$
5. $\frac{14}{16}$ $\left(\frac{7}{8}\right)$
6. $\frac{30}{4}$ $\left(7\frac{1}{2}\right)$
7. $2\frac{2}{10}$ $\left(2\frac{1}{5}\right)$
8. $3\frac{12}{16}$ $\left(3\frac{3}{4}\right)$
9. $\frac{45}{25}$ $\left(1\frac{4}{5}\right)$
10. $\frac{42}{9}$ $\left(4\frac{2}{3}\right)$
11. $9\frac{6}{10}$ $\left(9\frac{3}{5}\right)$
12. $\frac{78}{10}$ $\left(7\frac{4}{5}\right)$

2 Teach

Introduce the Lesson Have a student read the problem and tell what is to be solved. Ask what information is given on the map to help them solve the problem. Have students complete the information sentences and work through the model to solve the problem. Emphasize that the fractions being added, $\frac{3}{10}$ and $\frac{5}{10}$, have the same, or a common, denominator.

Develop Skills and Concepts Explain that to add fractions with the same denominators, students should add the numerators, 3 and 5, and write that sum over the common denominator. Point out that to add mixed numbers, students should add the fraction part first and then add the whole number part. Be sure students understand that they need to simplify the fraction if necessary. Write these problems on the board to provide immediate practice and reinforcement:

$\frac{5}{12} + \frac{1}{12}$ $\left(\frac{1}{2}\right)$ $\qquad$ $\frac{3}{8} + \frac{3}{8}$ $\left(\frac{3}{4}\right)$ $\qquad$ $\frac{2}{10} + \frac{5}{10}$ $\left(\frac{7}{10}\right)$

$\begin{array}{r} 5\frac{1}{5} \\ + 3\frac{2}{5} \\ \hline \left(8\frac{3}{5}\right) \end{array}$ $\qquad$ $\begin{array}{r} 1\frac{7}{16} \\ + 3\frac{3}{16} \\ \hline \left(4\frac{5}{8}\right) \end{array}$ $\qquad$ $\begin{array}{r} 4\frac{7}{12} \\ + 3\frac{2}{12} \\ \hline \left(7\frac{3}{4}\right) \end{array}$ $\qquad$ $\begin{array}{r} 2\frac{3}{15} \\ + 4\frac{6}{15} \\ \hline \left(6\frac{3}{5}\right) \end{array}$

3 Practice

Have students complete all the exercises. Watch for students who add the denominators as well as the numerators.

T115

Name _____

Lesson 6-2

Adding Fractions and Mixed Numbers

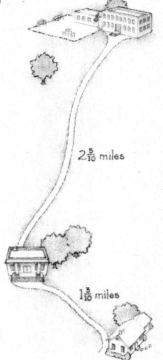

Guido has an overdue book. He decides to drop it off at the library before he bikes to school. How far does Guido bike altogether?

We want to know the distance Guido bikes altogether.

We know that it is $1\frac{3}{10}$ miles from his house to the library, and $2\frac{5}{10}$ miles from the library to school.

To find the total distance, we add the distance from the house to the library and the distance from the library to school. We add $1\frac{3}{10}$ and $2\frac{5}{10}$.

$2\frac{5}{10}$ miles

$1\frac{3}{10}$ miles

Add the fractions. Write the sum of the numerators over the common denominator.	Add the whole numbers. Simplify the fraction if necessary.

$\begin{array}{r} 1\frac{3}{10} \\ + 2\frac{5}{10} \\ \hline \frac{8}{10} \end{array}$ 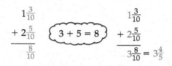 $\qquad$ $\begin{array}{r} 1\frac{3}{10} \\ + 2\frac{5}{10} \\ \hline 3\frac{8}{10} = 3\frac{4}{5} \end{array}$

$3 + 5 = 8$

Guido bikes $3\frac{4}{5}$ miles from home to school.

Getting Started

Add. Simplify answers if necessary.

1. $\frac{3}{5} + \frac{1}{5} = \frac{4}{5}$ $\qquad$ 2. $\frac{4}{9} + \frac{2}{9} = \frac{2}{3}$ $\qquad$ 3. $\frac{7}{16} + \frac{3}{16} + \frac{5}{16} = \frac{15}{16}$

4. $\begin{array}{r} 4\frac{1}{3} \\ + 5\frac{1}{3} \\ \hline 9\frac{2}{3} \end{array}$ $\quad$ 5. $\begin{array}{r} 7\frac{5}{12} \\ + 3\frac{1}{12} \\ \hline 10\frac{1}{2} \end{array}$ $\quad$ 6. $\begin{array}{r} 28\frac{1}{6} \\ + 17\frac{3}{6} \\ \hline 45\frac{2}{3} \end{array}$ $\quad$ 7. $\begin{array}{r} 522\frac{2}{15} \\ 115\frac{4}{15} \\ + 269\frac{6}{15} \\ \hline 906\frac{4}{5} \end{array}$

Lesson 6-2 • Adding Fractions and Mixed Numbers

115

Practice

Add. Simplify answers if necessary.

1. $\frac{5}{8} + \frac{2}{8} = \underline{\frac{7}{8}}$

2. $\frac{3}{7} + \frac{1}{7} = \underline{\frac{4}{7}}$

3. $\frac{5}{9} + \frac{1}{9} = \underline{\frac{2}{3}}$

4. $\frac{3}{10} + \frac{1}{10} + \frac{5}{10} = \underline{\frac{9}{10}}$

5. $\frac{3}{20} + \frac{9}{20} + \frac{3}{20} = \underline{\frac{15}{20} = \frac{3}{4}}$

6. $\frac{11}{25} + \frac{6}{25} + \frac{4}{25} = \underline{\frac{21}{25}}$

7. $5\frac{1}{8}$
$+ 2\frac{3}{8}$
$\overline{7\frac{1}{2}}$

8. $6\frac{7}{15}$
$+ 6\frac{3}{15}$
$\overline{12\frac{2}{3}}$

9. $9\frac{7}{16}$
$+ 7\frac{5}{16}$
$\overline{16\frac{3}{4}}$

10. $21\frac{7}{24}$
$+ 13\frac{9}{24}$
$\overline{34\frac{2}{3}}$

11. $3\frac{3}{6}$
$+ 8\frac{2}{6}$
$\overline{11\frac{2}{3}}$

12. $5\frac{7}{15}$
$+ 2\frac{3}{15}$
$\overline{7\frac{2}{3}}$

13. $6\frac{12}{25}$
$+ 4\frac{11}{25}$
$\overline{10\frac{23}{25}}$

14. $17\frac{7}{18}$
$+ 22\frac{4}{18}$
$\overline{39\frac{1}{3}}$

15. $3\frac{1}{12}$
$5\frac{3}{12}$
$+ 4\frac{5}{12}$
$\overline{12\frac{3}{4}}$

16. $15\frac{3}{10}$
$7\frac{1}{10}$
$+ 13\frac{1}{10}$
$\overline{35\frac{1}{2}}$

17. $17\frac{3}{18}$
$21\frac{1}{18}$
$+ 18\frac{7}{18}$
$\overline{56\frac{2}{3}}$

18. $135\frac{3}{20}$
$226\frac{7}{20}$
$+ 148\frac{5}{20}$
$\overline{509\frac{3}{4}}$

19. $2\frac{2}{7}$
$1\frac{3}{7}$
$+ 6\frac{1}{7}$
$\overline{9\frac{6}{7}}$

20. $21\frac{1}{12}$
$7\frac{4}{12}$
$+ 34\frac{1}{12}$
$\overline{62\frac{1}{2}}$

21. $13\frac{3}{16}$
$35\frac{3}{16}$
$+ 23\frac{6}{16}$
$\overline{71\frac{3}{4}}$

22. $307\frac{7}{30}$
$156\frac{13}{30}$
$+ 212\frac{5}{30}$
$\overline{675\frac{5}{6}}$

Problem Solving

Solve each problem.

23. Paula fastened a board $5\frac{3}{16}$ inches long to a board $\frac{3}{16}$ inches long. How long is the new board?
$5\frac{3}{8}$ inches

24. Bill sleeps 9 hours, goes to school 7 hours, plays 3 hours, and does homework for 2 hours. What part of a day has Bill used in these activities?
$\frac{7}{8}$ of a day

25. A snack recipe calls for $1\frac{1}{8}$ cups of raisins, $1\frac{1}{8}$ cups of unsalted peanuts, and $2\frac{5}{8}$ cups of sunflower seeds. How many cups of ingredients are needed?
$4\frac{7}{8}$ cups

26. A ribbon is $6\frac{1}{2}$ inches long. Is it possible to cut three ribbons each $2\frac{3}{16}$ inches long from the ribbon?
no

116

Lesson 6-2 • Adding Fractions and Mixed Numbers

4 Assess

Have students use colored chalk, markers, or pencils to identify each of the three steps as they find the sum of $3\frac{5}{8}$ and $9\frac{1}{8}$. ($12\frac{3}{4}$)

5 Mixed Review

1. 656×26 (17,056)
2. $3,827 \div 107$ (35 R82)
3. Simplify: $\frac{51}{7}$ ($7\frac{2}{7}$)
4. $20,156 + 8,164 + 13,298$ (41,618)
5. Simplify: $\frac{56}{64}$ ($\frac{7}{8}$)
6. $40,000 - 283$ (39,717)
7. 308×72 (22,176)
8. $3\frac{7}{12} = \frac{?}{12}$ (43)
9. 6^3 (216)
10. $\$574.00 \div 28$ ($\$20.50$)

For Mixed Abilities

Common Errors • Intervention

Some students may add or subtract both the numerator and the denominator. Have students work with partners and practice with concrete models or word names such as the following:
2 sevenths + 4 sevenths = 6 sevenths; 7 eighths − 2 eighths = 5 eighths

Enrichment • Number Sense

Write the following fractions on the board or duplicate for students:

$\frac{1}{12}, \frac{2}{12}, \frac{3}{12}, \frac{4}{12}, \frac{5}{12}, \frac{6}{12}, \frac{7}{12}, \frac{8}{12}$

Have students find which two fractions have these sums: $\frac{3}{4}, \frac{2}{3}, \frac{1}{2},$ 1, $\frac{7}{12}$. (Answers may vary.)

More to Explore • Probability

Put this table on the board:

	Colorblind	Normal color vision
Male	$2\frac{1}{2}$%	$47\frac{1}{2}$%
Female	$\frac{1}{2}$%	$49\frac{1}{2}$%

Explain that the table shows the distribution of colorblindness in a population. Duplicate the following questions:

1. Are men or women more likely to be colorblind? (men)

2. What is the total percentage of the four types of vision?
($2\frac{1}{2}$% + $\frac{1}{2}$% + $47\frac{1}{2}$% + $49\frac{1}{2}$% = 100%)

3. What does 100% mean? (the whole population)

4. How much more likely is it that a baby boy will be colorblind than if he had been born a girl?
($2\frac{1}{2} \div \frac{1}{2} = \frac{5}{2} \times \frac{2}{1} = 5$; 5 times more likely)

5. How much more likely is it that a baby girl will have normal vision than be colorblind?
($49\frac{1}{2} \div \frac{1}{2} = \frac{99}{2} \times \frac{2}{1} = 99$; 99 times)

T116

6-3 Adding Fractions With Unlike Denominators

pages 117–118

1 Getting Started

Objective
• To add fractions with unlike denominators

Warm Up • Mental Math
Have students name 6 problems each having four 4s, using any operation, and yielding the answers 0, 1, 2, 3, 4, 5.

[Possible answers:

1. $(4 \div 4) - (4 \div 4) = 0$
2. $(4 + 4) \div (4 + 4) = 1$
3. $(4 \times 4) \div (4 + 4) = 2$
4. $(4 + 4 + 4) \div 4 = 3$
5. $\frac{(4 - 4)}{4} + 4 = 4$
6. $\frac{(4 \times 4) + 4}{4} = 5$

Warm Up • Pencil and Paper
Have students rename the fractions in each group as equivalent fractions with the least common denominator.

1. $\frac{1}{3}, \frac{1}{2}$ $\left(\frac{2}{6}, \frac{3}{6}\right)$
2. $\frac{1}{3}, \frac{2}{5}$ $\left(\frac{5}{15}, \frac{6}{15}\right)$
3. $\frac{5}{6}, \frac{3}{8}$ $\left(\frac{20}{24}, \frac{9}{24}\right)$
4. $\frac{1}{6}, \frac{5}{9}$ $\left(\frac{3}{18}, \frac{10}{18}\right)$
5. $\frac{7}{12}, \frac{1}{3}$ $\left(\frac{7}{12}, \frac{4}{12}\right)$
6. $\frac{2}{3}, \frac{2}{9}, \frac{1}{2}$ $\left(\frac{12}{18}, \frac{4}{18}, \frac{9}{18}\right)$

2 Teach

Introduce the Lesson Have students tell about the picture. Ask a student to read the problem and identify the question being asked. Have students fill in the information sentences. Work through the model with students, emphasizing the renaming step. Have students complete the sentence to solve the problem.

Develop Skills and Concepts Explain that fractions and mixed numbers can only be added when the denominators are the same. Students must rename the fractions as equivalent fractions with the same denominator. Then the process becomes adding fractions with like denominators, as students learned in the previous lesson.

Name _____

Adding Fractions With Unlike Denominators

Allison is making fruit punch for the school party. After she mixes the orange juice and lime juice, how many liters of punch will she have?

We want to know the total number of liters of punch.

We know that Allison will mix $3\frac{1}{3}$ liters of orange juice and $4\frac{1}{2}$ liters of lime juice.

To find the total amount of punch, we add the amount of orange juice and lime juice.

We add $3\frac{1}{3}$ and $4\frac{1}{2}$.

Rename the fractions as equivalent fractions with the least common denominator.	Add the fractions.	Add the whole numbers. Simplify the fraction if necessary.
$3\frac{1}{3} = 3\frac{2}{6}$ $+ 4\frac{1}{2} = 4\frac{3}{6}$	$3\frac{1}{3} = 3\frac{2}{6}$ $+ 4\frac{1}{2} = 4\frac{3}{6}$ $\frac{5}{6}$	$3\frac{1}{3} = 3\frac{2}{6}$ $+ 4\frac{1}{2} = 4\frac{3}{6}$ $7\frac{5}{6}$

Allison will have $7\frac{5}{6}$ liters of punch.

Sometimes adding two fractions results in an improper fraction that can be simplified as a whole or mixed number.

$\frac{3}{8} + \frac{5}{8} = \frac{8}{8} = 1$ $\frac{2}{3} + \frac{5}{6} = \frac{4}{6} + \frac{5}{6} = \frac{9}{6} = 1\frac{3}{6} = 1\frac{1}{2}$

Getting Started

Add. Simplify answers if necessary.

1. $\frac{3}{5} + \frac{7}{10} = 1\frac{3}{10}$

2. $\frac{5}{6} + \frac{7}{8} = 1\frac{17}{24}$

3. $\frac{2}{3} + \frac{5}{6} + \frac{5}{9} = 2\frac{1}{18}$

4. $6\frac{1}{2}$ $+ 4\frac{1}{3}$ $10\frac{5}{6}$

5. $5\frac{1}{8}$ $+ 4\frac{1}{6}$ $9\frac{7}{24}$

6. $11\frac{5}{12}$ $+ 9\frac{1}{4}$ $20\frac{2}{3}$

7. $12\frac{2}{3}$ $+ 16\frac{1}{5}$ $28\frac{13}{15}$

Remind students that to add mixed numbers, they first add the fractional part and then add the whole number part and simplify the sum.

Point out that it is sometimes necessary for students to simplify the fraction in a sum when adding mixed numbers. Go over the steps for finding the sum for the two problems above Getting Started on page 117.

Then, call on volunteers to come to the board to show how to find the sum of each of the following problems:

$3\frac{1}{4} + 4\frac{7}{12}$ $\left(7\frac{5}{6}\right)$

$\frac{3}{4} + \frac{5}{8}$ $\left(1\frac{3}{8}\right)$

3 Practice

Have students complete all the exercises. Watch for students who add the numerators without renaming the fractions as equivalent fractions.

Practice

Add. Simplify answers if necessary.

1. $\frac{1}{3} + \frac{5}{6} = \underline{1\frac{1}{6}}$

2. $\frac{5}{8} + \frac{3}{4} = \underline{1\frac{3}{8}}$

3. $\frac{5}{10} + \frac{1}{2} = \underline{1}$

4. $\frac{5}{8} + \frac{7}{12} = \underline{1\frac{5}{24}}$

5. $\frac{5}{6} + \frac{5}{9} = \underline{1\frac{7}{18}}$

6. $\frac{3}{4} + \frac{5}{8} + \frac{9}{16} = \underline{1\frac{15}{16}}$

7. $\frac{3}{7} + \frac{2}{3} = \underline{1\frac{2}{21}}$

8. $\frac{3}{4} + \frac{11}{12} = \underline{1\frac{2}{3}}$

9. $\frac{11}{16} + \frac{2}{4} + \frac{7}{8} = \underline{2\frac{1}{16}}$

10. $5\frac{3}{5}$
$+ 3\frac{3}{15}$
$\overline{8\frac{4}{5}}$

11. $7\frac{3}{4}$
$+ 6\frac{1}{8}$
$\overline{13\frac{7}{8}}$

12. $9\frac{3}{8}$
$+ 9\frac{4}{16}$
$\overline{18\frac{5}{8}}$

13. $7\frac{3}{4}$
$+ 8\frac{1}{6}$
$\overline{15\frac{11}{12}}$

14. $12\frac{1}{4}$
$+ 7\frac{5}{12}$
$\overline{19\frac{2}{3}}$

15. $10\frac{2}{3}$
$+ 15\frac{1}{5}$
$\overline{25\frac{13}{15}}$

16. $9\frac{1}{3}$
$+ 16\frac{1}{2}$
$\overline{25\frac{5}{6}}$

17. $14\frac{2}{9}$
$+ 8\frac{1}{6}$
$\overline{22\frac{7}{18}}$

18. $16\frac{5}{12}$
$+ 10\frac{3}{8}$
$\overline{26\frac{19}{24}}$

19. $21\frac{3}{10}$
$+ 36\frac{1}{2}$
$\overline{57\frac{4}{5}}$

20. $47\frac{1}{3}$
$+ 39\frac{3}{8}$
$\overline{86\frac{17}{24}}$

21. $28\frac{7}{15}$
$+ 82\frac{3}{10}$
$\overline{110\frac{23}{30}}$

Problem Solving

Solve each problem.

22. Marti went cross-country skiing. She passed the first checkpoint at $\frac{7}{10}$ of a mile and the second checkpoint $\frac{9}{10}$ of a mile later. She stopped at the third checkpoint after $\frac{8}{10}$ of a mile more. How far had Marti skied?

$2\frac{2}{5}$ miles

23. To paint his room, Adam used $2\frac{1}{8}$ quarts of white paint, $1\frac{1}{6}$ quarts of blue paint, and $2\frac{1}{3}$ quarts of yellow paint. How much paint did he use?

$5\frac{5}{8}$ quarts

(Now Try This!)

Find the missing numbers. None of the fractions are improper. Possible answer:

$\frac{2}{3} + \frac{1}{2} + \frac{1}{6} + \frac{1}{4} + \frac{1}{3} + \frac{1}{3} + \frac{3}{4} = 3$

Lesson 6-3 • Adding Fractions With Unlike Denominators

Now Try This! Students practice estimating and adding as they follow the arrows along the fraction path, to fill in the missing numbers.

Possible answers include: 1, 1, 2, 3; 1, 3, 1, 3; 2, 4, 1, 1; 1, 4, 2, 1

4 Assess

Ask students to identify the four steps to adding mixed numbers with fractions that have unlike denominators. (Possible answer: 1. Rename the fractions. 2. Add the fractions. 3. Add the whole numbers. 4. Simplify the fraction in the sum if necessary.)

For Mixed Abilities

Common Errors • Intervention

Some students may write the fractions but forget to write the whole numbers when renaming and, therefore, forget to add the whole numbers for the answer. Encourage them, after they have obtained an answer, to go back and check the answer with the problem to determine if the answer is reasonable.

Enrichment • Number Sense

1. Have students copy and complete the following addition table. You might provide students with an answer key and tell them they score one point for every correct answer.

+	$2\frac{1}{3}$	$3\frac{1}{8}$	$8\frac{1}{5}$	$9\frac{1}{10}$
$4\frac{1}{6}$	$(6\frac{1}{2})$	$(7\frac{7}{24})$	$(12\frac{11}{30})$	$(13\frac{4}{15})$
$6\frac{1}{2}$	$(8\frac{5}{6})$	$(9\frac{5}{8})$	$(14\frac{7}{10})$	$(15\frac{3}{5})$
$7\frac{1}{4}$	$(9\frac{7}{12})$	$(10\frac{3}{8})$	$(15\frac{9}{20})$	$(16\frac{7}{20})$

2. Write the numbers 3, 4, 6, and 8 on the board. Challenge students to use two of these numbers as numerators and two as denominators to write two fractions, each less than 1, that have the greatest possible sum. $(\frac{3}{4} + \frac{6}{8} = 1\frac{1}{2})$

More to Explore • Measurement

Have students make a scale drawing of their "dream bedroom," using the scale of $\frac{1}{4}$ in. = 1 foot. Have them include furniture, decorations, and color schemes they would like if they could have any kind of bedroom they wanted. Encourage them to use materials other than pencil and paper. Display their drawings.

T118

pages 119–120

1 Getting Started

Objective
• To add mixed numbers with renaming

Warm Up • Mental Math
Have students rename each expression using exponents. Then, find the product.
1. $2 \times 2 \times 2$ (2^3; 8)
2. 4×4 (4^2; 16)
3. $10 \times 10 \times 10 \times 10$ (10^4; 10,000)
4. $3 \times 3 \times 3$ (3^3; 27)
5. 6×6 (6^2; 36)
6. $1 \times 1 \times 1 \times 1 \times 1 \times 1$ (1^6; 1)
7. $10 \times 10 \times 10$ (10^3; 1,000)
8. 8×8 (8^2; 64)

Warm Up • Pencil and Paper
Have students rename each fraction as a whole number or a mixed number.
1. $\frac{5}{5}$ (1)
2. $\frac{7}{3}$ ($2\frac{1}{3}$)
3. $\frac{15}{12}$ ($1\frac{1}{4}$)
4. $\frac{17}{10}$ ($1\frac{7}{10}$)
5. $\frac{11}{8}$ ($1\frac{3}{8}$)
6. $\frac{32}{16}$ (2)

Name _____

Renaming Mixed Numbers in Sums

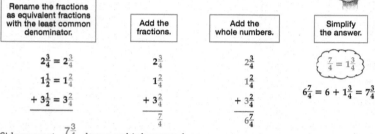

Sidney's goal is to improve his study habits. He kept track of the time he spent on homework last week. How many hours did he spend on his homework?

We want to know the number of hours Sidney spent on homework.

We know that on Monday he spent _$\frac{2\frac{3}{4}}{}$_ hours; on Tuesday, _$\frac{1\frac{1}{2}}{}$_ hours; and on Wednesday, _$\frac{3\frac{1}{2}}{}$_ hours.

To find the total time he spent on homework, we add the number of hours he did homework each day. We add _$2\frac{3}{4}$_, _$1\frac{1}{2}$_, and _$3\frac{1}{2}$_.

Rename the fractions as equivalent fractions with the least common denominator.	Add the fractions.	Add the whole numbers.	Simplify the answer.
$2\frac{3}{4} = 2\frac{3}{4}$	$2\frac{3}{4}$	$2\frac{3}{4}$	$\frac{7}{4} = 1\frac{3}{4}$
$1\frac{1}{2} = 1\frac{2}{4}$	$1\frac{2}{4}$	$1\frac{2}{4}$	
$+ 3\frac{1}{2} = 3\frac{2}{4}$	$+ 3\frac{2}{4}$	$+ 3\frac{2}{4}$	$6\frac{7}{4} = 6 + 1\frac{3}{4} = 7\frac{3}{4}$
	$\frac{7}{4}$	$6\frac{7}{4}$	

Sidney spent _$7\frac{3}{4}$_ hours on his homework.

Getting Started

Add. Simplify answers if necessary.

1. $3\frac{2}{3}$
   $+ 4\frac{1}{3}$
   8

2. $5\frac{3}{5}$
   $+ 6\frac{2}{3}$
   $12\frac{4}{15}$

3. $3\frac{1}{2}$
   $4\frac{2}{3}$
   $+ 5\frac{1}{4}$
   $13\frac{5}{12}$

4. $9\frac{2}{3}$
   $6\frac{3}{4}$
   $+ 5\frac{1}{6}$
   $21\frac{7}{12}$

Copy and add.

5. $3\frac{4}{5} + 8\frac{2}{3}$ $12\frac{7}{15}$

6. $7\frac{5}{6} + 15\frac{3}{8}$ $23\frac{5}{24}$

7. $25\frac{1}{2} + 29\frac{3}{5} + 52\frac{3}{4}$ $107\frac{17}{20}$

---

2 Teach

Introduce the Lesson Have a student identify the problem to be solved and what information is given. (the number of hours Sidney did homework; Monday, Sidney worked $2\frac{3}{4}$ hours, Tuesday he worked $1\frac{1}{2}$ hours, and Wednesday he worked $3\frac{1}{2}$ hours.) Tell students to fill in the information sentences. Work through the steps in the model with students and then have them write the answer to the solution sentence.

Develop Skills and Concepts Remind students that addition with mixed numbers involves two addition problems, one with fractions and one with whole numbers.

When the sum of the fractions is greater than or equal to 1, it is renamed as a mixed number. Then, the mixed number is added to the sum of the whole numbers. Point out that the answer $6\frac{7}{4}$ is not actually wrong. However, it

is customary to simplify any answer involving fractions and mixed numbers.

Two errors are particularly common. Students rename equivalent fractions incorrectly or simplify answers incorrectly. Provide practice in simplifying answers by having students work these exercises at the board.

$3\frac{1}{2}$ $9\frac{2}{5}$ $4\frac{5}{6}$
$+ 4\frac{3}{4}$ $+ 3\frac{2}{3}$ $+ 7\frac{4}{5}$
($8\frac{1}{4}$) ($13\frac{1}{15}$) ($12\frac{19}{30}$)

3 Practice

Have students complete all the exercises. Suggest students draw a diagram for troublesome problems.

T119

Practice

Add. Simplify answers if necessary.

1. $4\frac{1}{2}$
    $+ 5\frac{1}{3}$
    ___
    $9\frac{5}{6}$

2. $5\frac{3}{4}$
    $+ 7\frac{1}{4}$
    ___
    13

3. $6\frac{3}{4}$
    $+ 4\frac{5}{8}$
    ___
    $11\frac{3}{8}$

4. $8\frac{2}{3}$
    $+ 6\frac{5}{8}$
    ___
    $15\frac{7}{24}$

5. $17\frac{5}{8}$
    $15\frac{5}{6}$
    $+ 4\frac{3}{4}$
    ___
    $38\frac{5}{24}$

6. $21\frac{2}{3}$
    $8\frac{1}{2}$
    $+ 5\frac{5}{6}$
    ___
    36

7. $26\frac{3}{8}$
    $14\frac{5}{6}$
    $+ 23\frac{3}{4}$
    ___
    $64\frac{23}{24}$

8. $18\frac{7}{8}$
    $46\frac{5}{6}$
    $+ 38\frac{3}{4}$
    ___
    $104\frac{11}{24}$

Copy and add.

9. $7\frac{5}{8} + 9\frac{3}{4}$ $17\frac{3}{8}$

10. $6\frac{7}{12} + 9\frac{5}{6}$ $16\frac{5}{12}$

11. $8\frac{2}{3} + 9\frac{1}{2}$ $18\frac{1}{6}$

12. $9\frac{7}{12} + 8\frac{5}{8}$ $18\frac{5}{24}$

13. $8\frac{3}{8} + 5\frac{2}{3}$ $14\frac{1}{24}$

14. $11\frac{9}{16} + 12\frac{7}{8}$ $24\frac{7}{16}$

15. $25\frac{7}{9} + 18\frac{11}{12}$ $44\frac{25}{36}$

16. $27\frac{2}{5} + 14\frac{2}{3}$ $42\frac{1}{15}$

17. $21\frac{6}{7} + 83\frac{4}{5}$ $105\frac{23}{35}$

18. $9\frac{1}{6} + 5\frac{3}{4} + 8\frac{1}{2}$ $23\frac{5}{12}$

19. $12\frac{2}{3} + 8\frac{3}{4} + 9\frac{1}{6}$ $30\frac{7}{12}$

20. $42\frac{1}{2} + 29\frac{5}{8} + 47\frac{5}{6}$ $119\frac{23}{24}$

Problem Solving

Solve each problem.

21. The world's longest piece of spaghetti was $2\frac{3}{4}$ feet long. Mr. O'Malley made a piece of spaghetti $3\frac{1}{2}$ feet longer. How long was Mr. O'Malley's piece of spaghetti?
    $6\frac{1}{4}$ feet

22. A railroad crew repaired $3\frac{1}{4}$ miles of track on Monday, $5\frac{2}{3}$ miles on Wednesday, and $2\frac{5}{6}$ miles on Friday. How many miles of track did the crew repair during the week?
    $11\frac{3}{4}$ miles

23. A recipe for whole grain bread uses $3\frac{3}{5}$ cups of wheat flour and $2\frac{3}{4}$ cups of rye flour. How much flour is used in the recipe?
    $6\frac{7}{20}$ cups

24. Mr. Peterson fills his car with gasoline each Monday.

    Monday, December 6 $11\frac{3}{10}$ gallons

    Monday, December 13 $10\frac{2}{5}$ gallons

    Monday, December 20 $12\frac{7}{10}$ gallons

    Monday, December 27 $11\frac{3}{5}$ gallons

    How many gallons of gasoline did he buy in December?   46 gallons

120

Lesson 6-4 • Renaming Mixed Numbers in Sums

4 Assess

Have students simplify each sum.

$3\frac{5}{4}$ $(4\frac{1}{4})$ $9\frac{7}{5}$ $(10\frac{2}{5})$ $5\frac{8}{6}$ $(6\frac{1}{3})$

5 Mixed Review

1. $10\frac{3}{20} + 12\frac{11}{20} + 15\frac{5}{20}$ $(37\frac{19}{20})$

2. $103,208 - 98,170$ (5,038)

3. $3,284 \div 24$ (136 R20)

4. $\$500.20 - \$370.37$ ($129.83)

5. Simplify: $\frac{90}{15}$ (6)

6. $2,976 + 15,385 + 64$ (18,425)

7. $492,184 \div 7$ (70,312)

8. $6,000 \times 700$ (4,200,000)

9. $12\frac{1}{4} = \frac{?}{4}$ (49)

10. $\$92.16 \times 72$ ($6,635.52)

For Mixed Abilities

Common Errors • Intervention

Some students may forget to rename an answer such as $4\frac{6}{5}$ when they are adding. Have them work with partners to practice the renaming in the following manner.

$$4\frac{6}{5} = 4 + \frac{6}{5}$$
$$= 4 + \frac{5}{5} + \frac{1}{5}$$
$$= 4 + 1 + \frac{1}{5} = 5\frac{1}{5}$$

Enrichment • Number Sense

Prepare worksheets or write the following problems on the board. Have students match the sums of the addition problems in the left-hand column with the addition problems in the right-hand column that have the same sums.

1. $2\frac{3}{4} + 2\frac{1}{2}$ (c.)   a. $2\frac{2}{3} + 1\frac{7}{12}$ $(4\frac{1}{4})$

2. $\frac{5}{6} + 3\frac{5}{12}$ (a.)   b. $3\frac{27}{28} + 4\frac{1}{4}$ $(8\frac{3}{14})$

3. $1\frac{2}{3} + \frac{2}{3}$ (d.)   c. $1\frac{7}{8} + 3\frac{3}{8}$ $(5\frac{1}{4})$

4. $4\frac{5}{7} + 3\frac{1}{2}$ (b.)   d. $1\frac{5}{18} + 1\frac{1}{18}$ $(2\frac{1}{3})$

5. $3\frac{3}{4} + 2\frac{1}{2}$ (e.)   e. $\frac{5}{6} + 5\frac{5}{12}$ $(6\frac{1}{4})$

More to Explore • Graphing

Have each student take their pulse, counting the number of beats in 15 seconds. Multiply by four to find beats per minute. Record the results.

Pulse	
Beats per Minute	**Number of Students**
50–60	(tally)
61–70	
71–80	
81–90	
over 90	

Divide students into groups. Have each group do a different type of exercise. Have students walk up and down stairs, jog around the school, and so on. Have them take their pulses after the exercise and tally the new results on the board.

Ask each student to make a histogram. Suggest they use different colors for before and after exercise.

T120

6-5 Subtracting Fractions and Mixed Numbers

pages 121–122

1 Getting Started

Objective
• To subtract mixed numbers whose fractions have like denominators

Materials
index cards

Warm Up • Mental Math
Divide the class into four teams for an Equivalent Fraction Bee. Name a fraction. The first member of each team must name an equivalent fraction. Students who answer correctly may sit down. Students who do not answer correctly go to the end of the team line. The first team to have all members seated wins the game.

Warm Up • Pencil and Paper
Have students find each difference.
1. $15 - 8$ (7)
2. $18 - 6$ (12)
3. $17 - 9$ (8)
4. $12 - 5$ (7)
5. $38 - 17$ (21)
6. $62 - 43$ (19)
7. $564 - 324$ (240)
8. $301 - 259$ (42)
9. $872 - 486$ (386)

2 Teach

Introduce the Lesson Have a student read the problem and tell what is to be solved. (how many more pounds of pretzels Tashiki needs) Ask what information is given. (Tashiki needs $3\frac{3}{8}$ pounds of pretzels and he has only $1\frac{1}{8}$ pounds.) Remind students that subtraction is used to compare two quantities or to find how much is left.

• Have students fill in the information sentences.

• Work through each step of the model with students. Point out that since the fractions have the same denominators, students can subtract the numerators and write the difference over the denominator. They can then subtract the whole numbers and simplify the answer if needed.

• Have them complete the solution sentence.

Name _____

Subtracting Fractions and Mixed Numbers

Tashiki is preparing a snack for the all-day hike. He has only $1\frac{1}{8}$ pounds of pretzels. How many more pounds will he need?

Snacks for Hike	
Nuts	$2\frac{2}{8}$ pounds
Pretzels	$3\frac{3}{8}$ pounds
Raisins	$\frac{3}{8}$ pounds

We want to know how many more pounds of pretzels Tashiki needs.

We know that he needs $3\frac{3}{8}$ pounds for the trail mix, and that he has only $1\frac{1}{8}$ pounds.

To find the amount he still needs, we subtract the amount he has from the amount required by the recipe. We subtract $1\frac{1}{8}$ from $3\frac{3}{8}$.

Subtract the fractions. Write the difference over the denominator.	Subtract the whole numbers. Simplify the fraction if necessary.

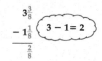

$$3\frac{3}{8}$$
$$-1\frac{1}{8}\qquad 3-1=2$$
$$\frac{2}{8}$$

$$3\frac{3}{8}$$
$$-1\frac{1}{8}$$
$$2\frac{2}{8}=2\frac{1}{4}$$

Tashiki needs $2\frac{1}{4}$ more pounds of pretzels.

Getting Started

Subtract. Simplify answers if necessary.
1. $\frac{5}{9} - \frac{3}{9} = \frac{2}{9}$
2. $\frac{7}{12} - \frac{3}{12} = \frac{1}{3}$
3. $\frac{3}{4} - \frac{1}{4} = \frac{1}{2}$

4. $6\frac{5}{6}$ $-4\frac{1}{6}$ $2\frac{2}{3}$
5. $59\frac{7}{10}$ $-26\frac{2}{10}$ $33\frac{1}{2}$
6. $17\frac{2}{3}$ $-8\frac{1}{3}$ $9\frac{1}{3}$
7. $527\frac{7}{16}$ $-214\frac{3}{16}$ $313\frac{1}{4}$

Lesson 6-5 • Subtracting Fractions and Mixed Numbers 121

Develop Skills and Concepts Emphasize that fractions and mixed numbers can only be subtracted when the denominators are the same. Remind students to examine their answers closely to be sure that they are in simplest form. Have students complete these problems to reinforce the skill.

$\frac{7}{8} - \frac{3}{8}$ $(\frac{1}{2})$
$3\frac{9}{10} - 1\frac{5}{10}$ $(2\frac{2}{5})$
$16\frac{7}{9} - \frac{4}{9}$ $(16\frac{1}{3})$

3 Practice

Have students complete all the exercises. Remind students to write all answers in simplest form.

T121

© Pearson Education, Inc./Dale Seymour Publications/Pearson Learning Group. All rights reserved. Copying strictly prohibited.

Practice

Subtract. Simplify answers if necessary.

1. $\frac{9}{16} - \frac{7}{16} = \frac{1}{8}$

2. $\frac{5}{7} - \frac{3}{7} = \frac{2}{7}$

3. $\frac{5}{9} - \frac{2}{9} = \frac{1}{3}$

4. $\frac{19}{24} - \frac{15}{24} = \frac{1}{6}$

5. $\frac{5}{6} - \frac{1}{6} = \frac{2}{3}$

6. $\frac{9}{10} - \frac{5}{10} = \frac{2}{5}$

7. $8\frac{3}{4}$
$- 3\frac{1}{4}$
$5\frac{1}{2}$

8. $12\frac{7}{8}$
$- 7\frac{3}{8}$
$5\frac{1}{2}$

9. $16\frac{7}{10}$
$- 9\frac{2}{10}$
$7\frac{1}{2}$

10. $13\frac{4}{5}$
$- 6\frac{3}{5}$
$7\frac{1}{5}$

11. $35\frac{11}{12}$
$- 12\frac{5}{12}$
$23\frac{1}{2}$

12. $42\frac{14}{16}$
$- 29\frac{4}{16}$
$13\frac{5}{8}$

13. $39\frac{17}{20}$
$- 26\frac{2}{20}$
$13\frac{3}{4}$

14. $71\frac{5}{8}$
$- 46\frac{3}{8}$
$25\frac{1}{4}$

15. $123\frac{2}{3}$
$- 97\frac{1}{3}$
$26\frac{1}{3}$

16. $206\frac{9}{10}$
$- 167\frac{3}{10}$
$39\frac{3}{5}$

17. $614\frac{17}{20}$
$- 246\frac{9}{20}$
$368\frac{2}{5}$

18. $531\frac{22}{24}$
$- 345\frac{5}{24}$
$186\frac{17}{24}$

Problem Solving

Solve each problem.

19. Dan spends $3\frac{1}{6}$ hours working in his garden every Saturday. During the week, he works $10\frac{5}{6}$ hours in his garden. How much longer does Dan work in his garden during the week than on Saturday? $7\frac{2}{3}$ hours

20. The largest pumpkin at the country fair weighs $56\frac{7}{8}$ pounds. The smallest weighs $7\frac{5}{8}$ pounds. How much heavier is the largest pumpkin than the smallest pumpkin? $49\frac{1}{4}$ pounds

Use the chart to solve Problems 21 and 22.

21. How much farther does Angie run than Margo? $7\frac{1}{5}$ miles

22. Lola runs $3\frac{1}{10}$ miles farther than Margo. How much less than Pat does Lola run? $4\frac{7}{10}$ miles

Running Distance	
Pat	$26\frac{3}{10}$ miles
Margo	$18\frac{5}{10}$ miles
Angie	$25\frac{7}{10}$ miles

Lesson 6-5 • Subtracting Fractions and Mixed Numbers

4 Assess

Have students show the steps for subtracting $5\frac{7}{8} - 2\frac{3}{8}$. $(3\frac{1}{2})$

5 Mixed Review

1. $\$4.09 \times 17$ ($\$69.53$)

2. $\frac{3}{5} + \frac{3}{10} + \frac{3}{4}$ $(1\frac{13}{20})$

3. $12\frac{5}{8} + 14\frac{5}{6}$ $(27\frac{11}{24})$

4. $11,113 \div 36$ (308 R25)

5. Simplify: $\frac{45}{72}$ $(\frac{5}{8})$

6. $16\frac{1}{3} = \frac{?}{3}$ (49)

7. 9×10^4 (90,000)

8. $15\frac{3}{4} = \frac{?}{4}$ (63)

9. $82,195 - 64,058$ (18,137)

10. $4,827 \div 4$ (1,206 R3)

For Mixed Abilities

Common Errors • Intervention

Some students may forget to subtract the whole-number part when they are subtracting mixed numbers. Have students work in pairs using manipulatives to model the problem.

Enrichment • Number Sense

1. Duplicate or write the following fractions on the board.

$\frac{1}{16}, \frac{3}{16}, \frac{5}{16}, \frac{7}{16}, \frac{9}{16}, \frac{11}{16}$

Direct students to solve the following problems.

(Answers will vary.)

Subtract two fractions to get $\frac{1}{2}$.

Subtract two fractions to get $\frac{1}{4}$.

Subtract two fractions to get $\frac{1}{8}$.

2. Write the following exercises on the board, or duplicate for students. Have students match the subtraction problems in the left-hand column that have the same difference as the problems in the right-hand column.

1. $\frac{7}{10} - \frac{3}{10}$ (b.)

a. $9\frac{5}{6} - 1\frac{1}{6}$ $(8\frac{2}{3})$

2. $6\frac{11}{12} - 5\frac{5}{12}$ (c.)

b. $\frac{3}{5} - \frac{1}{5}$ $(\frac{2}{5})$

3. $\frac{8}{9} - \frac{5}{9}$ (d.)

c. $2\frac{3}{4} - 1\frac{1}{4}$ $(1\frac{1}{2})$

4. $8\frac{11}{12} - \frac{3}{12}$ (a.)

d. $\frac{2}{3} - \frac{1}{3}$ $(\frac{1}{3})$

More to Explore • Logic

Duplicate the following for students.

Jack and Jill are two working parents who like to arrive home together each day. However, they work in opposite directions. Jill's company is 12 miles from home and the speed limit along her route is 30 miles per hour. Jack's road home has a speed limit of 45 miles per hour. If on Tuesday, they both left work at 4:00 and arrived home at the same time, what time did they get home? How far from home is Jack's office?

(4:24, 18 miles)

T122

6-6 Subtracting Fractions With Unlike Denominators

pages 123–124

1 Getting Started

Objective
- To subtract mixed numbers whose fractions have unlike denominators

Warm Up • Mental Math
Have students round each of the following to the nearest hundred.

1. 136 (100)
2. 478 (500)
3. 354 (400)
4. 529 (500)
5. 1,486 (1,500)
6. 2,816 (2,800)
7. 5,158 (5,200)
8. 7,271 (7,300)
9. 6,149 (6,100)
10. 9,907 (9,900)

Warm Up • Pencil and Paper
Have students write equivalent fractions, using the least common denominator, for each fraction pair.

1. $\frac{3}{8}, \frac{2}{3}$ $\left(\frac{9}{24}, \frac{16}{24}\right)$
2. $\frac{1}{4}, \frac{2}{5}$ $\left(\frac{5}{20}, \frac{8}{20}\right)$
3. $\frac{7}{10}, \frac{4}{5}$ $\left(\frac{7}{10}, \frac{8}{10}\right)$
4. $\frac{3}{4}, \frac{7}{12}$ $\left(\frac{9}{12}, \frac{7}{12}\right)$
5. $\frac{1}{3}, \frac{3}{5}$ $\left(\frac{5}{15}, \frac{9}{15}\right)$
6. $\frac{3}{10}, \frac{1}{6}$ $\left(\frac{9}{30}, \frac{5}{30}\right)$

2 Teach

Introduce the Lesson Have a student read the problem, tell what is to be solved, and tell what information is given. Ask students why subtraction is the appropriate operation for solving this problem. (You want to know how much is used.) Tell students to fill in the information sentences. Work through the three steps in the model with students and then have them complete the solution sentence.

Develop Skills and Concepts Point out that, just as in addition of fractions and mixed numbers with unlike denominators, students must rename the fractions as equivalent fractions before subtracting.

Emphasize that students are building on skills they already know:

- finding common denominators
- naming equivalent fractions
- subtracting fractions
- subtracting whole numbers
- renaming fractions in simplest form

Have students work at the board to solve each problem.

$\frac{9}{10} - \frac{3}{10}$ $\left(\frac{3}{5}\right)$; $\frac{6}{7} - \frac{1}{3}$ $\left(\frac{11}{21}\right)$

$7\frac{2}{3} - 1\frac{5}{9}$ $\left(6\frac{1}{9}\right)$; $4\frac{1}{2} - 2\frac{3}{10}$ $\left(2\frac{1}{5}\right)$

Remind students to subtract the numerators only.

3 Practice

Have students complete all the exercises. Ask students to suggest ways to check their answers. (addition, estimation)

Name _____

Subtracting Fractions With Unlike Denominators

Duncan is feeding the chickens on his uncle's farm. When he started there were $4\frac{1}{2}$ buckets of chicken feed. How much feed has he used?

We want to know how much chicken feed Duncan has used. We know that he started with $4\frac{1}{2}$ buckets of feed, and he has $1\frac{1}{4}$ buckets left.

To find the amount used, we subtract the amount left from the original amount. We subtract $1\frac{1}{4}$ from $4\frac{1}{2}$.

Rename the fractions as equivalent fractions with the least common denominator.	Subtract the fractions.	Subtract the whole numbers. Simplify the fraction if needed.
$4\frac{1}{2} = 4\frac{2}{4}$ $-1\frac{1}{4} = 1\frac{1}{4}$	$4\frac{1}{2} = 4\frac{2}{4}$ $-1\frac{1}{4} = 1\frac{1}{4}$ $\frac{1}{4}$	$4\frac{1}{2} = 4\frac{2}{4}$ $-1\frac{1}{4} = 1\frac{1}{4}$ $3\frac{1}{4}$

Duncan has used $3\frac{1}{4}$ buckets of feed.

Getting Started

Subtract.

1. $15\frac{5}{8}$ $-7\frac{1}{3}$ $8\frac{7}{24}$

2. $87\frac{2}{3}$ $-39\frac{1}{6}$ $48\frac{1}{2}$

3. $533\frac{3}{4}$ $-526\frac{3}{5}$ $7\frac{3}{20}$

4. $51\frac{5}{6}$ -17 $34\frac{5}{6}$

Copy and subtract.

5. $\frac{7}{8} - \frac{1}{4}$ $\frac{5}{8}$

6. $\frac{5}{6} - \frac{1}{2}$ $\frac{1}{3}$

7. $\frac{9}{10} - \frac{6}{15}$ $\frac{1}{2}$

Practice

Subtract.

1. $9\frac{4}{5}$
    $-\ 6\frac{3}{10}$
    ___
    $3\frac{1}{2}$

2. $11\frac{3}{4}$
    $-\ 5\frac{2}{3}$
    ___
    $6\frac{1}{12}$

3. $14\frac{7}{9}$
    $-\ 7\frac{1}{3}$
    ___
    $7\frac{4}{9}$

4. $17\frac{2}{3}$
    $-\ 9\frac{1}{5}$
    ___
    $8\frac{7}{15}$

5. $26\frac{5}{16}$
    $-\ 14$
    ___
    $12\frac{5}{16}$

6. $47\frac{9}{10}$
    $-\ 28\frac{5}{6}$
    ___
    $19\frac{1}{15}$

7. $36\frac{2}{3}$
    $-\ 18\frac{5}{12}$
    ___
    $18\frac{1}{4}$

8. $82\frac{5}{6}$
    $-\ 46\frac{7}{12}$
    ___
    $36\frac{1}{4}$

9. $112\frac{2}{3}$
    $-\ 66\frac{3}{5}$
    ___
    $46\frac{1}{15}$

10. $625\frac{5}{6}$
    $-\ 148\frac{5}{12}$
    ___
    $477\frac{5}{12}$

11. $907\frac{2}{3}$
    $-\ 319\frac{3}{8}$
    ___
    $588\frac{7}{24}$

12. $536\frac{45}{100}$
    $-\ 243\frac{3}{25}$
    ___
    $293\frac{33}{100}$

Copy and subtract.

13. $\frac{5}{8} - \frac{1}{2}$ $\frac{1}{8}$

14. $\frac{8}{9} - \frac{1}{6}$ $\frac{13}{18}$

15. $\frac{7}{8} - \frac{2}{3}$ $\frac{5}{24}$

16. $\frac{7}{10} - \frac{1}{2}$ $\frac{1}{5}$

17. $\frac{5}{6} - \frac{1}{4}$ $\frac{7}{12}$

18. $\frac{2}{3} - \frac{1}{5}$ $\frac{7}{15}$

Problem Solving

Solve each problem.

19. The drill bit is $1\frac{3}{4}$ inches long.
    How thick is the wood?
    $\frac{5}{8}$ of an inch

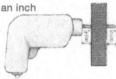

20. This year, workers in the town of
    Tidyville collected $425\frac{2}{3}$ pounds of
    newspapers. Last year, they collected
    $387\frac{1}{4}$ pounds. How much more did
    the workers collect this year?
    $38\frac{5}{12}$ pounds

(Now Try This!)

Look for a pattern to help you complete the three problems. Write the next problem in
the pattern. Write the answer.

1. $\frac{1}{2} + \frac{1}{4} = \frac{3}{4}$ 2. $\frac{1}{2} + \frac{1}{4} + \frac{1}{8} = \frac{7}{8}$ 3. $\frac{1}{2} + \frac{1}{4} + \frac{1}{8} + \frac{1}{16} = \frac{15}{16}$

   $\frac{1}{2} + \frac{1}{4} + \frac{1}{8} + \frac{1}{16} + \frac{1}{32} = \frac{31}{32}$

Lesson 6-6 • Subtracting Fractions With Unlike Denominators

Now Try This! Each fraction being added is a unit
fraction. Its numerator is 1. Discuss the relationship
between the denominators and have students write the
next problem in the pattern. Encourage students to
think about ways in which the solution is related to the
addends. There are several patterns here, including:

numerator of the sum = sum of denominators ÷ 2

numerator of the sum = 1 less than the denominator
of the last addend

4 Assess

Ask, *What is the first thing you need to do to find the
difference of two mixed numbers with fractions that have unlike
denominators?* (Rename the fractions as equivalent
fractions with the least common denominator.)

For Mixed Abilities

Common Errors • Intervention

Some students may find a
common denominator but forget
to change the numerators.

Incorrect	Correct
$3\frac{3}{4} = 3\frac{3}{12}$	$3\frac{3}{4} = 3\frac{9}{12}$
$-\ 1\frac{1}{3} = 1\frac{1}{12}$	$-\ 1\frac{1}{3} = 1\frac{4}{12}$
$2\frac{1}{6}$	$2\frac{5}{12}$

Have them make a drawing to
show the equivalent fraction.

Enrichment • Number Sense

1. Have students complete the
   table by following the rule.
   $\square + \triangle = 8\frac{5}{6}$

$\square$	2	$5\frac{3}{4}$	$7\frac{2}{3}$	$3\frac{1}{2}$	$4\frac{2}{9}$
$\triangle$	$(6\frac{5}{6})$	$(3\frac{1}{12})$	$(1\frac{1}{6})$	$(5\frac{1}{3})$	$(4\frac{11}{18})$

2. Have students identify the rule
   and complete the table.
   $(\square - 2\frac{1}{2} = \triangle)$

$\square$	$2\frac{1}{2}$	$5\frac{7}{8}$	$6\frac{8}{9}$	$3\frac{2}{3}$	$4\frac{3}{4}$	$10\frac{1}{2}$
$\triangle$	0	$3\frac{3}{8}$	$4\frac{7}{18}$	$(1\frac{1}{6})$	$(2\frac{1}{4})$	(8)

More to Explore • Application

Have students bring in sales
catalogs from variety stores. Tell
them to use the order form
provided in the back to order
whatever items they want, not to
exceed $100. Tell them to
complete their order, including
item, price, sales tax, delivery
charge, and so on, and stay within
their budget. Then, have students
compare orders to find similarities
and differences in what they
ordered.

T124

6-7 Subtracting Mixed Numbers With Regrouping

pages 125–126

1 Getting Started

Objective
• To subtract mixed numbers with regrouping

Materials
index cards

Warm Up • Mental Math
Have students rename each number as a whole number or mixed number in simplest form.

1. $2\frac{8}{7}$ $(3\frac{1}{7})$ 6. $1\frac{21}{19}$ $(2\frac{2}{19})$

2. $5\frac{5}{3}$ $(6\frac{2}{3})$ 7. $4\frac{10}{7}$ $(5\frac{3}{7})$

3. $8\frac{10}{9}$ $(9\frac{1}{9})$ 8. $7\frac{8}{5}$ $(8\frac{3}{5})$

4. $14\frac{8}{5}$ $(15\frac{3}{5})$ 9. $6\frac{9}{5}$ $(7\frac{4}{5})$

5. $11\frac{17}{11}$ $(12\frac{6}{11})$ 10. $3\frac{8}{2}$ (7)

Warm Up • Pencil and Paper
Have students find each difference and write it in simplest form.

1. $7\frac{1}{4} - 2\frac{3}{4}$ $(4\frac{1}{2})$

2. $5\frac{1}{8} - 1\frac{5}{8}$ $(3\frac{1}{2})$

3. $9\frac{2}{7} - 5\frac{6}{7}$ $(3\frac{3}{7})$

4. $11\frac{2}{5} - 8\frac{3}{5}$ $(2\frac{4}{5})$

5. $4\frac{1}{6} - 1\frac{5}{6}$ $(2\frac{1}{3})$

6. $10\frac{6}{11} - 4\frac{3}{11}$ $(6\frac{3}{11})$

2 Teach

Introduce the Lesson Have students read the problem and tell what is to be solved. Ask what information is given. Explain that to subtract $3\frac{3}{4}$ from $6\frac{1}{4}$, students need to regroup before subtracting. Work through the regrouping with students. Show the subtraction. Have a volunteer explain how to simplify the difference.

Develop Skills and Concepts Point out that this lesson ties together a series of skills students have already learned. Review the situations when regrouping and renaming is indicated: when the subtrahend must be regrouped for subtraction, and when the fractions are renamed as fractions with a common denominator and the difference is renamed to simplify. Have students complete the following problems at the board.

Name _____

Subtracting Mixed Numbers With Regrouping

Martha is making pottery in art class. She needs $6\frac{1}{4}$ pounds of clay to make six cups. How many more pounds of clay does Martha need?

We want to know how much more clay Martha needs.

We know that she needs _$6\frac{1}{4}$_ pounds of clay, and she has _$3\frac{3}{4}$_ pounds.

To find the difference, we subtract the amount she has from the total amount she needs for the cups.

We subtract _$3\frac{3}{4}$_ from _$6\frac{1}{4}$_.

Regroup the minuend.	Subtract and simplify.

$$6\frac{1}{4} = 5\frac{5}{4}$$
$$- 3\frac{3}{4} = 3\frac{3}{4}$$

$$6\frac{1}{4} = 5 + 1 + \frac{1}{4}$$
$$= 5 + \frac{4}{4} + \frac{1}{4}$$
$$= 5\frac{5}{4}$$

$$6\frac{1}{4} = 5\frac{5}{4}$$
$$- 3\frac{3}{4} = 3\frac{3}{4}$$
$$\overline{\quad 2\frac{2}{4} = 2\frac{1}{2}}$$

Martha needs _$2\frac{1}{2}$_ more pounds of clay.

Getting Started

Regroup each mixed number.

1. $7\frac{2}{3} = 6\frac{5}{3}$ 2. $4\frac{5}{6} = 3\frac{11}{6}$ 3. $8 = 7\frac{4}{4}$ 4. $2\frac{1}{2} = 1\frac{3}{2}$

Subtract.

5. $\begin{array}{r} 9\frac{3}{8} \\ - 6\frac{5}{8} \\ \hline 2\frac{3}{4} \end{array}$ 6. $\begin{array}{r} 7 \\ - 4\frac{1}{2} \\ \hline 2\frac{1}{2} \end{array}$ 7. $\begin{array}{r} 23\frac{1}{6} \\ - 17\frac{5}{6} \\ \hline 5\frac{1}{3} \end{array}$ 8. $\begin{array}{r} 431\frac{1}{7} \\ - 118\frac{5}{7} \\ \hline 312\frac{3}{7} \end{array}$

$\begin{array}{r} 7\frac{1}{2} \\ - 4\frac{9}{10} \\ \hline (2\frac{3}{5}) \end{array}$ $\begin{array}{r} 8\frac{3}{4} \\ - 4\frac{9}{10} \\ \hline (3\frac{17}{20}) \end{array}$ $\begin{array}{r} 9\frac{1}{4} \\ - 3\frac{3}{5} \\ \hline (5\frac{13}{20}) \end{array}$

3 Practice

Have students complete all the exercises. Tell them that each answer should be in simplest form.

Practice

Subtract.

1. $9\frac{1}{8}$
    $-6\frac{3}{4}$
    ___
    $2\frac{3}{8}$

2. $16\frac{3}{10}$
    $-7\frac{4}{5}$
    ___
    $8\frac{1}{2}$

3. $12\frac{1}{2}$
    $-6\frac{2}{3}$
    ___
    $5\frac{5}{6}$

4. $18\frac{3}{8}$
    $-15\frac{5}{12}$
    ___
    $2\frac{23}{24}$

5. $86\frac{2}{5}$
    $-29\frac{3}{4}$
    ___
    $56\frac{13}{20}$

6. $43\frac{5}{16}$
    $-26\frac{5}{8}$
    ___
    $16\frac{11}{16}$

Copy and subtract.

7. $7\frac{3}{8} - 2\frac{3}{4}$ $4\frac{5}{8}$

8. $24\frac{3}{4} - 9\frac{5}{6}$ $14\frac{11}{12}$

9. $28\frac{1}{2} - 19\frac{7}{8}$ $8\frac{5}{8}$

10. $1\frac{3}{5} - \frac{3}{10}$ $1\frac{3}{10}$

11. $38\frac{2}{7} - 19\frac{3}{5}$ $18\frac{24}{35}$

12. $82\frac{1}{3} - 37\frac{3}{4}$ $44\frac{7}{12}$

13. $61\frac{1}{2} - 36\frac{2}{3}$ $24\frac{5}{6}$

14. $58\frac{1}{9} - 16\frac{5}{12}$ $41\frac{25}{36}$

15. $75\frac{3}{8} - 66\frac{5}{6}$ $8\frac{13}{24}$

16. $374\frac{1}{5} - 196\frac{2}{3}$ $177\frac{8}{15}$

17. $621\frac{2}{3} - 438\frac{7}{8}$ $182\frac{19}{24}$

18. $901\frac{4}{15} - 728\frac{7}{10}$ $172\frac{17}{30}$

Problem Solving

Solve each problem.

19. Audrey worked $14\frac{1}{2}$ hours one week and $9\frac{3}{5}$ hours the next week. How many more hours did Audrey work the first week than the second?
    $4\frac{9}{10}$ hours

20. Miguel filled his gas tank with 15 gallons of gas. On Monday, he used $6\frac{2}{3}$ gallons and on Friday, he used another $3\frac{3}{4}$ gallons. How many gallons does Miguel have left?
    $4\frac{7}{12}$ gallons

[Now Try This!]

Choose a number from each row. Each number must be from a different column.

Find the sum. ___7___

Choose again. ___7___

What do you find? __They are the same.__

Try it again.

Columns

Rows			
$2\frac{3}{4}$	$2\frac{1}{4}$	2	$2\frac{1}{2}$
$2\frac{1}{4}$	$1\frac{3}{4}$	$1\frac{1}{2}$	2
2	$1\frac{1}{2}$	$1\frac{1}{4}$	$1\frac{3}{4}$
$1\frac{1}{2}$	1	$\frac{3}{4}$	$1\frac{1}{4}$

126

Lesson 6-7 • Subtracting Mixed Numbers With Regrouping

Now Try This! Make sure students choose fractions to add from different rows and different columns. Remind them to reduce answers to simplest form.

4 Assess

Ask students why they will need to regroup the minuend to find the difference $8\frac{1}{5} - 3\frac{4}{5}$. Then, have students show the regrouping. (Possible answer: There are not enough fifths to subtract. Regroup $8\frac{1}{5}$ as $7\frac{6}{5}$.)

For Mixed Abilities

Common Errors • Intervention

Some students might not make the whole number one less when they rename a mixed-number minuend.

Incorrect	Correct
$4 = 4\frac{5}{5}$	$4 = 3\frac{5}{5}$
$-1\frac{3}{5} = 1\frac{3}{5}$	$-1\frac{3}{5} = 1\frac{3}{5}$
$3\frac{2}{5}$	$2\frac{2}{5}$

Have these students work with partners and use manipulatives to model the problem.

Enrichment • Number Sense

Have students work in pairs. Have them write each number on an index card.

$1, 2, 3, 4, 5, 6, 7, 8, 9, \frac{1}{3}, \frac{1}{2}, \frac{1}{4}, \frac{3}{8},$
$\frac{1}{9}, 1\frac{1}{3}, 2\frac{1}{3}, 3\frac{3}{4}, 4\frac{2}{3}, 5\frac{5}{6}, 6\frac{1}{8}, 7\frac{4}{9}$

Direct students to shuffle the cards and place them face down in a stack. Then, have each student select two cards and subtract the lesser number from the greater number. The student whose answer is greater wins a point. Repeat until one player wins 5 points.

More to Explore • Probability

Have students look at the sports page and survey the batting averages of their favorite baseball team. A batting average is the number of times the player has hit the ball divided by the number of times at bat (minus any times the batter walks). Ask students which would be considered a better batting average: 0.300 or 0.200. (0.300)

ESL/ELL STRATEGIES

Clarify the difference between *renaming* and *regrouping*. Using examples on the board, show that renaming involves changing denominators so that two fractions have the same denominator. Regrouping involves changing a whole number and a fraction during the subtraction process.

6-8 Subtracting With Renaming and Regrouping

pages 127–128

1 Getting Started

Objective
• To subtract mixed numbers with renaming and regrouping

Warm Up • Mental Math
Have students name each mixed number as an improper fraction.

1. $3\frac{1}{2}$ $\left(\frac{7}{2}\right)$
2. $5\frac{5}{6}$ $\left(\frac{35}{6}\right)$
3. $7\frac{2}{5}$ $\left(\frac{37}{5}\right)$
4. $4\frac{7}{8}$ $\left(\frac{39}{8}\right)$
5. $6\frac{3}{11}$ $\left(\frac{69}{3}\right)$
6. $8\frac{9}{10}$ $\left(\frac{89}{10}\right)$
7. $2\frac{11}{12}$ $\left(\frac{35}{12}\right)$
8. $1\frac{15}{16}$ $\left(\frac{31}{16}\right)$
9. $10\frac{17}{20}$ $\left(\frac{217}{20}\right)$
10. $4\frac{3}{15}$ $\left(\frac{63}{15}\right)$

Warm Up • Pencil and Paper
Have students answer and simplify the following:

1. $7\frac{7}{8} - 3\frac{1}{5}$ $\left(4\frac{27}{40}\right)$
2. $18\frac{3}{8} + 5\frac{4}{5}$ $\left(24\frac{7}{40}\right)$
3. $7\frac{8}{9} - 6\frac{1}{2}$ $\left(1\frac{7}{18}\right)$
4. $4\frac{5}{6} - 1\frac{2}{7}$ $\left(3\frac{23}{42}\right)$
5. $9\frac{3}{4} - 6\frac{1}{3}$ $\left(3\frac{5}{12}\right)$

2 Teach

Introduce the Lesson Have students read the problem to identify the question and the given information. To solve this problem, $3\frac{5}{6}$ must be subtracted from $5\frac{1}{4}$. Point out that in order to subtract $3\frac{5}{6}$ from $5\frac{1}{4}$, students must rename and regroup the greater number. Work through the renaming, regrouping, and subtraction steps to complete the sentences, and solve the problem.

Develop Skills and Concepts Write this model on the board to show regrouping.

$$6 = 5 + 1$$
$$= 5 + \frac{3}{3}$$
$$= 5\frac{3}{3}$$

Name _____

Subtracting With Renaming and Regrouping

Chuck is cutting ribbons to make award badges. How much ribbon will be left after he cuts a $3\frac{5}{6}$-inch strip?

We want to know how much ribbon will be left. We know the ribbon is $5\frac{1}{4}$ inches long, and Chuck needs a strip $3\frac{5}{6}$ inches long.

To find the amount left, we subtract the length he cuts from the length he started with. We subtract $3\frac{5}{6}$ from $5\frac{1}{4}$.

Rename the fractions as equivalent fractions with the least common denominator.	Regroup the minuend.	Subtract.
$5\frac{1}{4} = 5\frac{3}{12}$	$5\frac{1}{4} = 5\frac{3}{12} = 4\frac{15}{12}$	$5\frac{1}{4} = 5\frac{3}{12} = 4\frac{15}{12}$
$-3\frac{5}{6} = 3\frac{10}{12}$	$-3\frac{5}{6} = 3\frac{10}{12} = 3\frac{10}{12}$	$-3\frac{5}{6} = 3\frac{10}{12} = 3\frac{10}{12}$
		$1\frac{5}{12}$

Chuck will have $1\frac{5}{12}$ inches of ribbon left.

Getting Started

Subtract.

1. $7\frac{1}{3}$
   $-4\frac{1}{2}$ $2\frac{5}{6}$

2. $8\frac{3}{8}$
   $-3\frac{5}{6}$ $4\frac{13}{24}$

3. $12\frac{1}{9}$
   $-6\frac{2}{3}$ $5\frac{4}{9}$

4. $26\frac{1}{8}$
   $-18\frac{5}{12}$ $7\frac{17}{24}$

5. $58\frac{3}{4}$
   $-23\frac{5}{6}$ $34\frac{11}{12}$

6. $436\frac{3}{5}$
   $-189\frac{7}{10}$ $246\frac{9}{10}$

Copy and subtract.

7. $9\frac{1}{5} - 2\frac{7}{10}$ $6\frac{1}{2}$

8. $60\frac{3}{8} - 27\frac{9}{10}$ $32\frac{19}{40}$

9. $703\frac{2}{9} - 677\frac{5}{6}$ $25\frac{7}{18}$

Have students regroup each of the mixed numbers using the same format.

$$8\frac{1}{4} = (7 + 1\frac{1}{4})$$
$$= (7 + 1 + \frac{1}{4})$$
$$= (7 + \frac{4}{4} + \frac{1}{4})$$
$$= (7\frac{5}{4})$$

$$4\frac{3}{5} = (3 + 1\frac{3}{5})$$
$$= (3 + 1 + \frac{3}{5})$$
$$= (3 + \frac{5}{5} + \frac{3}{5})$$
$$= (3\frac{8}{5})$$

Point out that to subtract, the fractions must have like denominators. Have students complete the following problems at the board showing the regrouping.

$7\frac{1}{3}$ $\left(6\frac{4}{3}\right)$	$2\frac{7}{16}$ $\left(1\frac{23}{16}\right)$	9 $\left(8\frac{5}{5}\right)$
$-3\frac{2}{3}$ $-3\frac{2}{3}$	$-1\frac{9}{16}$ $-1\frac{9}{16}$	$-7\frac{4}{5}$ $-7\frac{4}{5}$
$\left(3\frac{2}{3}\right)$	$\left(\frac{7}{8}\right)$	$\left(1\frac{1}{5}\right)$

3 Practice

Have students complete all the exercises.

Practice

Regroup each mixed number.

1. $6\frac{3}{4} = 5\frac{7}{4}$
2. $9\frac{2}{3} = 8\frac{5}{3}$
3. $5\frac{5}{9} = 4\frac{14}{9}$
4. $6 = 5\frac{6}{6}$

5. $7\frac{5}{8} = 6\frac{13}{8}$
6. $4\frac{5}{12} = 3\frac{17}{12}$
7. $12\frac{5}{7} = 11\frac{12}{7}$
8. $1\frac{3}{5} = \frac{8}{5}$

Subtract.

9. $9\frac{1}{4}$
   $- 7\frac{3}{4}$
   $\overline{1\frac{1}{2}}$

10. 8
    $- 5\frac{7}{10}$
    $\overline{2\frac{3}{10}}$

11. $16\frac{3}{8}$
    $- 9\frac{5}{8}$
    $\overline{6\frac{3}{4}}$

12. $25\frac{2}{5}$
    $- 18\frac{3}{5}$
    $\overline{6\frac{4}{5}}$

13. $23\frac{1}{3}$
    $- 16\frac{2}{3}$
    $\overline{6\frac{2}{3}}$

14. $86\frac{5}{11}$
    $- 49\frac{7}{11}$
    $\overline{36\frac{9}{11}}$

15. $128\frac{1}{16}$
    $- 92\frac{13}{16}$
    $\overline{35\frac{1}{4}}$

16. $309\frac{1}{12}$
    $- 153\frac{11}{12}$
    $\overline{155\frac{1}{6}}$

Problem Solving

Solve each problem.

17. At the track meet, Alexa jumped $14\frac{1}{4}$ feet. Marlene jumped $11\frac{3}{4}$ feet. How much farther than Marlene did Alexa jump?
    $2\frac{1}{2}$ feet

18. Dex wants to collect 100 pounds of aluminum cans. He has made two collection trips. On the first trip he collected $36\frac{3}{8}$ pounds and on the second, he collected $42\frac{5}{8}$ pounds. How many more pounds of aluminum cans does he need to reach his goal?
    21 pounds

(Now Try This!)

Study fraction addition on the clock. Complete the addition table using the clock to help find the sums.

$\frac{1}{5} + \frac{2}{5} = \frac{3}{5}$
$\frac{3}{5} + \frac{4}{5} = \frac{2}{5}$
$\frac{2}{5} + \frac{3}{5} = 0$

+	0	$\frac{1}{5}$	$\frac{2}{5}$	$\frac{3}{5}$	$\frac{4}{5}$
0	0	$\frac{1}{5}$	$\frac{2}{5}$	$\frac{3}{5}$	$\frac{4}{5}$
$\frac{1}{5}$	$\frac{1}{5}$	$\frac{2}{5}$	$\frac{3}{5}$	$\frac{4}{5}$	0
$\frac{2}{5}$	$\frac{2}{5}$	$\frac{3}{5}$	$\frac{4}{5}$	0	$\frac{1}{5}$
$\frac{3}{5}$	$\frac{3}{5}$	$\frac{4}{5}$	0	$\frac{1}{5}$	$\frac{2}{5}$
$\frac{4}{5}$	$\frac{4}{5}$	0	$\frac{1}{5}$	$\frac{2}{5}$	$\frac{3}{5}$

Lesson 6-8 • Subtracting With Renaming and Regrouping

Now Try This! Help students use the clock to fill in the table, using the first three examples given. Have them complete the table and explain any patterns they find.

4 Assess

To subtract mixed numbers ask students why they would rename the fractions and why they would regroup the fractions. (Possible answer: You rename the fractions when the fractions have unlike denominators. You regroup the fractions when there is not enough to subtract.)

For Mixed Abilities

Common Errors • Intervention

Once students have found a common denominator, some may subtract upside-down in order to avoid regrouping. Discuss how, when subtracting the fraction parts, the minuend must be larger than the subtrahend; when it is not, they must regroup.

Incorrect	Correct
$4\frac{1}{3} = 4\frac{4}{12}$	$4\frac{1}{3} = 4\frac{4}{12} = 3\frac{16}{12}$
$- 2\frac{3}{4} = 2\frac{9}{12}$	$- 2\frac{3}{4} = 2\frac{9}{12} = 2\frac{9}{12}$
$\overline{2\frac{5}{12}}$	$\overline{1\frac{7}{12}}$

Enrichment • Application

1. Have students solve the following problem: Lauren worked 6 hours on Wednesday. She worked $2\frac{1}{2}$ hours less than that on Thursday and $1\frac{2}{3}$ hours more on Friday than on Thursday. How many hours did she work all together in the three days? ($14\frac{2}{3}$ hours) How many more hours did Lauren work on Wednesday than on Friday? ($\frac{5}{6}$ hours)

2. Have students compare these numbers.

   $3\frac{6}{3} \,(>)\, 3\frac{1}{2}$ $2\frac{3}{4} \,(>)\, 1\frac{4}{3}$

   $4\frac{12}{3} \,(>)\, 6\frac{1}{4}$ $6\frac{9}{8} \,(<)\, 7\frac{8}{9}$

   $5\frac{2}{3} \,(=)\, 4\frac{5}{3}$

More to Explore • Biography

Albert Einstein was born in 1879 in Germany. His creative genius was sparked at age five when his father showed him a pocket compass. From then on, Einstein felt compelled to unravel the mysteries of science. His first job after college was in a Swiss patent office. During that time, he wrote three papers; each generated a new branch of physics. When the Nazis deprived Einstein of his property, position, and citizenship in Germany, he accepted a position at the new Institute for Advanced Study in Princeton, New Jersey. Einstein is best known for his theory of relativity and the now famous formula $E = mc^2$.

T128

6-9 Practice Adding and Subtracting Fractions

pages 129–130

1 Getting Started

Objective

- To solve mixed practice and application problems involving fractions

Warm Up • Mental Math

Have students simplify each expression.

1. $(4 \times 2) + 6$ (14)
2. $(27 \div 3) - 5$ (4)
3. $8 + 6 - 4$ (10)
4. $(6 \times 3) - 1$ (17)
5. $4 + (35 \div 7)$ (9)
6. $17 - 8 + 5$ (14)
7. $(56 \div 8) + 9$ (16)
8. $35 + 20 - 10$ (45)

Warm Up • Pencil and Paper

Have students estimate the sums to the nearest $\frac{1}{2}$ and explain.

1. $15\frac{2}{3} + 5\frac{5}{6}$ $(21\frac{1}{2})$
2. $40\frac{1}{9} + 10\frac{1}{25}$ (50)
3. $5\frac{3}{8} + 6\frac{1}{5}$ $(11\frac{1}{2})$
4. $19\frac{7}{8} + 5\frac{5}{6}$ (26)
5. $3\frac{3}{4} + 6\frac{2}{3}$ $(10\frac{1}{2})$

Name _____

Practice Adding and Subtracting Fractions

Add or subtract. Simplify answers if necessary.

1. $3\frac{1}{4}$ $+ 4\frac{1}{4}$ $\overline{7\frac{1}{2}}$

2. $8\frac{3}{7}$ $+ 9\frac{5}{7}$ $\overline{18\frac{1}{7}}$

3. $7\frac{7}{12}$ $- 5\frac{1}{12}$ $\overline{2\frac{1}{2}}$

4. 8 $- 2\frac{3}{5}$ $\overline{5\frac{2}{5}}$

5. $12\frac{2}{3}$ $+ 6\frac{1}{2}$ $\overline{19\frac{1}{6}}$

6. $18\frac{3}{5}$ $- 11\frac{1}{4}$ $\overline{7\frac{7}{20}}$

7. $62\frac{1}{3}$ $- 48\frac{2}{3}$ $\overline{13\frac{2}{3}}$

8. $79\frac{3}{4}$ $+ 26\frac{5}{6}$ $\overline{106\frac{7}{12}}$

9. $61\frac{1}{5}$ $- 14\frac{3}{10}$ $\overline{46\frac{9}{10}}$

10. $53\frac{2}{3}$ $- 18\frac{5}{7}$ $\overline{34\frac{20}{21}}$

11. $96\frac{1}{8}$ $- 48\frac{5}{6}$ $\overline{47\frac{7}{24}}$

12. $43\frac{5}{9}$ $- 16\frac{5}{6}$ $\overline{26\frac{13}{18}}$

13. $465\frac{1}{2}$ $- 183\frac{3}{5}$ $\overline{281\frac{9}{10}}$

14. $803\frac{1}{12}$ $- 675\frac{5}{8}$ $\overline{127\frac{11}{24}}$

15. $721\frac{4}{15}$ $- 239\frac{7}{10}$ $\overline{481\frac{17}{30}}$

16. $915\frac{2}{3}$ $- 625\frac{7}{15}$ $\overline{290\frac{1}{5}}$

17. $816\frac{3}{10}$ $- 177\frac{5}{8}$ $\overline{638\frac{27}{40}}$

18. $279\frac{7}{9}$ $- 158\frac{4}{5}$ $\overline{120\frac{44}{45}}$

19. $106\frac{3}{8}$ $- 97\frac{2}{3}$ $\overline{8\frac{17}{24}}$

20. $540\frac{2}{3}$ $- 215\frac{6}{7}$ $\overline{324\frac{17}{21}}$

Copy. Then, add or subtract.

21. $\frac{3}{4} + \frac{2}{3} + \frac{1}{6}$ $1\frac{7}{12}$

22. $\frac{1}{2} + \frac{5}{9} + \frac{2}{3}$ $1\frac{13}{18}$

23. $\frac{1}{5} + \frac{3}{10} + \frac{7}{15}$ $\frac{29}{30}$

24. $\frac{9}{16} - \frac{1}{6}$ $\frac{19}{48}$

25. $\frac{4}{7} - \frac{2}{5}$ $\frac{6}{35}$

26. $\frac{7}{12} - \frac{3}{16}$ $\frac{19}{48}$

27. $5\frac{2}{3} - 4\frac{3}{5}$ $1\frac{1}{15}$

28. $8\frac{7}{8} - 7\frac{1}{6}$ $1\frac{17}{24}$

29. $14\frac{1}{3} - 9\frac{4}{5}$ $4\frac{8}{15}$

30. $8\frac{1}{2} + 12\frac{5}{8}$ $21\frac{1}{8}$

31. $16\frac{2}{9} - \frac{5}{6}$ $15\frac{7}{18}$

32. $6\frac{9}{10} + 7\frac{3}{5}$ $14\frac{1}{2}$

33. $25\frac{1}{8} - 16\frac{1}{2}$ $8\frac{5}{8}$

34. $34\frac{2}{3} - 16\frac{1}{5}$ $18\frac{7}{15}$

35. $47\frac{5}{8} + 87\frac{9}{16}$ $135\frac{3}{16}$

36. $26\frac{3}{7} + 96\frac{6}{9}$ $123\frac{13}{63}$

37. $57\frac{7}{8} - 42\frac{3}{16}$ $15\frac{11}{16}$

38. $94 - 87\frac{7}{12}$ $6\frac{5}{12}$

39. $124\frac{2}{5} - 86\frac{7}{10}$ $37\frac{7}{10}$

40. $836\frac{5}{6} - 429\frac{2}{3}$ $407\frac{1}{6}$

41. $415\frac{7}{8} + 329\frac{5}{8}$ $745\frac{1}{2}$

42. $607\frac{1}{3} - 498\frac{1}{2}$ $108\frac{5}{6}$

43. $943\frac{5}{7} - 485\frac{5}{7}$ 458

44. $821\frac{3}{5} + 468\frac{9}{25}$ $1,289\frac{24}{25}$

Lesson 6-9 • Practice Adding and Subtracting Fractions

---

2 Teach

Introduce the Lesson Tell students that this practice page will help them review the material in Chapter 6 so they can identify areas that need additional study. Remind students that in order to add or subtract fractions or mixed numbers, they must first name the fractions with common denominators. Writing fractions with like denominators requires students to write equivalent fractions with the least common denominator.

Develop Skills and Concepts Adding mixed numbers may require students to rename the sum if the fraction is an improper fraction. If students do not remember how to rename a sum, refer them to Lesson 6-4.

Subtracting may also require regrouping. If students do not remember how to regroup a minuend, refer them to the previous lesson. Finally, remind students to write all answers in simplest form.

3 Practice

This lesson reviews the skills and concepts that were presented in this chapter. Have students complete all the exercises. Review answers with the students.

Problem Solving

Use information from the charts to solve each problem.

Gas Consumption	
Speed in Miles per Hour (MPH)	Miles per Gallon (MPG)
35	24
45	$20\frac{1}{4}$
55	$18\frac{1}{3}$
65	$15\frac{1}{5}$

1. How does the MPG change when the speed is reduced from 45 MPH to 35 MPH?
   $3\frac{3}{4}$ MPG more

2. How does the MPG change when the speed is reduced from 65 MPH to 55 MPH?
   $3\frac{2}{15}$ MPG more

3. How does the MPG change when the speed is increased from 35 MPH to 65 MPH?
   $8\frac{4}{5}$ MPG less

4. What is the change in ACT stock?
   $5\frac{7}{8}$

5. What is the yearly low for RPJ stock?
   $97\frac{3}{4}$

6. What is the yearly low for DCM stock?
   $66\frac{5}{8}$

7. What is the yearly high for STV stock?
   $93\frac{3}{8}$

8. How much more did RPJ stock change than STV stock?
   4

9. What is the combined yearly high for both DCM and RPJ stock?
   $183\frac{5}{8}$

Price of Stocks			
Name of Stock	Yearly High	Yearly Low	Change
ACT	$42\frac{1}{2}$	$36\frac{5}{8}$	?
RPJ	$106\frac{1}{4}$	?	$8\frac{1}{2}$
DCM	$77\frac{3}{8}$	?	$10\frac{3}{4}$
STV	?	$88\frac{7}{8}$	$4\frac{1}{2}$

Cases Packed	
Ron	$12\frac{1}{3}$ cases
Al	$9\frac{1}{5}$ cases
Rich	$15\frac{3}{4}$ cases

10. How many cases in all did the boys pack?
    $37\frac{17}{60}$ cases

11. How many more cases did Rich pack than Al?
    $6\frac{11}{20}$ cases

12. How many more cases did Al and Rich pack together than Ron?
    $12\frac{47}{60}$ cases

13. How many more cases did Rich pack than Ron?
    $3\frac{5}{12}$ cases

Lesson 6-9 • Practice Adding and Subtracting Fractions

---

4 Assess

Ask, *When would you regroup a mixed number sum?*
(Possible answer: when the fraction in the sum is an improper fraction)

5 Mixed Review

1. $25,113 \div 58$ (432 R57)
2. $6,280,192 + 4,372,959$ (10,653,151)
3. $\frac{6}{8} + \frac{3}{4} + \frac{3}{5}$ ($2\frac{1}{10}$)
4. 296×321 (95,016)
5. Simplify: $\frac{47}{7}$ ($6\frac{5}{7}$)
6. $24\frac{3}{8} - 15\frac{7}{8}$ ($8\frac{1}{2}$)
7. 7^3 (343)
8. $2,725 \times 14$ (38,150)
9. $28 - 6 \times 3$ (10)
10. $571,036 \div 8$ (71,379 R4)

For Mixed Abilities

Common Errors • Intervention

Watch for students who, when they must regroup for subtraction, regroup as they would for whole numbers.

Incorrect	Correct
$3\frac{1}{4} = 2\frac{11}{4}$	$3\frac{1}{4} = 2\frac{5}{4}$

Have these students work with partners and use fraction bars to practice regrouping.

Enrichment • Number Sense

1. Tell students to write word problems to fit the even numbered exercises, 1–20, on page 129. Have students exchange word problems and solve them.

2. Have students complete the fraction chains.

   $3\frac{1}{2} + \frac{2}{3} - 1\frac{4}{5} + 3\frac{5}{6} - 2 = $ ($4\frac{1}{5}$)

   $15 - 3\frac{3}{4} - 2\frac{1}{2} + 9\frac{5}{8} - 4\frac{5}{6} = $ ($13\frac{13}{24}$)

More to Explore • Logic

Have students solve the following problem.

Mike and Mac are twins who love to run races against each other. One day they decided to race from opposite points 2 miles apart, to a central spot. Each twin ran 4 miles an hour and they both reached the center at the same time.

Their dog Flash ran back and forth between them as they raced. Since Flash could run at 10 miles per hour, he started with Mike, ran to Mac, then back to Mike and so on, until all three met in the center. The twins knew they had each run 1 mile, but they wondered how far Flash had run. Can you help them figure out how far Flash ran? ($2\frac{1}{2}$ miles; The twins, running at 4 miles an hour, reached the center in 15 minutes. At 10 miles an hour and running for $\frac{1}{4}$ of an hour, Flash would have run $2\frac{1}{2}$ miles.)

1 Getting Started

Objective
- To solve problems by trying, checking, and revising

Warm Up • Mental Math
Ask students to tell the measurement of time represented by the following.

1. bimonthly (2 months)
2. fortnight (2 weeks)
3. centenarian (100 years)
4. biannual (6 months)
5. sesquicentennial (150 years)
6. diamond anniversary (75 years)
7. octogenarian (80 years)
8. generation (30 years)

Warm Up • Pencil and Paper
Have students write the following.

1. the sum of 503 and 505 (1,008)
2. how many nickels in $4.55 (91)
3. the sum of the first 5 even numbers (20)
4. 8,465 + 7,859 (16,324)
5. the sum of the odd numbers from 100 to 105 (309)
6. $\frac{1}{5}$ of 20 (4)
7. $\frac{1}{2}$ of 15 (7.5)

Name _____

Problem Solving: Try, Check, and Revise

Kerry weighs 15 pounds more than Holly. When they step on the scale together, the scale reads 181 pounds. How much does each weigh?

 SEE

We want to know how much Kerry and Holly each weigh.

We know that Kerry is __15__ pounds heavier than Holly.

We know that their total weight is __181__ pounds.

⭐ **PLAN**

Using the facts from the problem we can make a guess of Holly's weight. We add 15 to our guess to get Kerry's weight. We can add the weights of the children to check to see if the total is 181 pounds. Making a list will help us to keep a record and to adjust our guesses.

⭐ **DO**

Holly's Weight	Kerry's Weight	Total Weight
50	65	115
60	75	135
70	85	155
80	95	175
81	96	177
82	97	179
83	98	181

Holly weighs __83__ pounds. Kerry weighs __98__ pounds.

 CHECK

We can check our solution to see if it fits with the facts of the problem.

Is Holly's weight plus 15 equal to Kerry's weight? __83__ + 15 = __98__

Do the two weights add up to 181 pounds? __83__ + __98__ = __181__

---

2 Teach

Introduce the Lesson Tell students that this strategy allows them to guess a possible answer, try it, check to see if it makes sense, and then revise the guess. Ask students to share other times where they might guess an answer before trying to solve it. (estimating an amount or measurement; predicting the outcome of a trial or experiment; the number of jelly beans in a contest)

Develop Skills and Concepts Present the problem to the class. Explain that problems like this are best solved by a careful system of trying, checking, and revising. To prevent repetition and wild guessing, students should keep a list of guesses tried and rejected, and the reason for their rejection. Each guess that does not directly result in a solution can lead to a more accurate next guess. In this way, students will practice using critical thinking skills by analyzing the mathematical situation. Caution students that this kind of problem sometimes has more than one solution. Work through the stages of the solution and have students enter the missing data in the chart. Emphasize the importance of checking all problems, but especially when using the guessing strategy.

3 Apply

Have students complete the problems independently.

Solution Notes
Discuss the strategy of how to solve Problems 1, 2, and 3. Demonstrate by adding a series of numbers together and only give the students the sum.

1. Point out key words: *consecutive* and *even*.
2. Point out key words: *consecutive* and *odd*.
3. Suggest students work backward to solve.
4. Suggest students use the inverse operation, subtraction.

Apply

Solve each problem.

1. The numbers 2, 4, 6 are a set of 3 consecutive even numbers. Find a set of 3 consecutive even numbers whose sum is 300.
   98, 100, 102

2. Find a set of 4 consecutive odd numbers whose sum is 304.
   73, 75, 77, 79

3. Find a set of 3 consecutive whole numbers whose sum is 303.
   100, 101, 102

4. The sum of Mark's and Bob's ages is 11. Mark is 10 years older than Bob. How old is each boy?
   Bob is $\frac{1}{2}$ and Mark is $10\frac{1}{2}$.

5. In a collection of quarters and nickels there are three more nickels than quarters. How many nickels are there if the collection is worth $2.25?
   7 quarters, 10 nickels

6. Sarah is four years younger than Joe. In five years Joe will be twice Sarah's age now. How old are they now?
   Joe is 13, Sarah is 9.

7. Each letter represents a single digit and the same letter represents the same digit each time. Find the value of each digit.

SEND	9,567
+ MORE	+ 1,085
MONEY	10,652

8. Pete counted the chickens and pigs in his grandfather's barn. When he counted the heads, he got 20. When he counted the legs, he got 64. How many chickens were in the barn?
   8 chickens

Problem-Solving Strategy: Using the Four-Step Plan

★ **SEE** What do you need to find?

★ **PLAN** What do you need to do?

★ **DO** Follow the plan.

★ **CHECK** Does your answer make sense?

---

9. Write fractions to tell what part of a dollar is represented by a penny, a nickel, a dime, a quarter, and a half-dollar. Without adding the 5 fractions, tell why their sum would be less than 1.
   The fractions are $\frac{1}{100}, \frac{1}{20}, \frac{1}{10}, \frac{1}{4}$, and $\frac{1}{2}$. Answers may vary.

10. Heather has some horses. Take away one-half, one-fourth, and one-fifth of Heather's horses, and she would still have 2 horses left. Explain how you can find how many horses Heather actually has now.
    Answers will vary.

11. The sum of two fractions is $\frac{1}{2}$. What can you tell about the fractions?
    Each fraction must be less than $\frac{1}{2}$.

12. If each of 2 fractions is less than 1, what is the smallest whole number that is greater than their sum?
    2

For Mixed Abilities

More to Explore • Logic

Duplicate the following for students to solve.

Three space cadets were on an exploratory mission on the planet Zorg when they met three Zorgans. The two groups traveled together for a while until they came to a river of molten lava. They could cross the lava with their anti-gravity sled, but the cadets had a problem.

The sled could only carry 2 beings at a time, and only one cadet and one Zorgan knew how to operate the sled. The cadets knew that Zorgans remained peaceful only as long as they were with an equal number or more humans. They would attack if there were more of them than humans. How could the cadets get everyone across the lava?

(If the cadets are A, B, and C and the Zorgans are X, Y, and Z, the trips would be as follows: X and Y cross; X returns alone. X and Z cross; X returns. A and B cross; A and Y return. X and A cross; A and Z return. A and C cross; X returns alone. X and Y cross; X returns alone. X and Z cross.)

---

5. Remind students to use systematic, trial and error guessing.

6. Suggest students write the equation $(x + 4) + 5 = 2x$ and solve for x.

7. Help students see why it is likely that $m = 1$, and have them continue.

8. This is a good example of a problem in which random guessing and checking will take many steps and may not lead to a solution. Making a table and using methodical guesses will help students find the solution. Note that students must conclude that since there are 20 heads, there must be a total of 20 animals.

Higher-Order Thinking Skills

9. **Synthesis:** A possible answer is that since the coins do not add up to 1 whole dollar, the sum of the corresponding fractions would also be less than 1 whole.

10. **Analysis:** A sample explanation is the following. Add the fractions to find that you would be taking away $\frac{19}{20}$ of Heather's horses. If she had 20 horses,

she would have 1 left; but since the problem says she would have 2 left, she must have 40 horses.

11. **Analysis:** Students may be tempted to say that the fractions are $\frac{1}{4}$; however, although the sum of $\frac{1}{4}$ and $\frac{1}{4}$ does equal $\frac{1}{2}$, there are an infinite number of other pairs of fractions with sums equal to $\frac{1}{2}$.

12. **Synthesis:** The greatest that each fraction could be would be a number very close but not equal to 1. So, the sum would be very close but never equal to 2.

4 Assess

Ask, *When do you revise your guess?* (Possible answer: when the guess does not match the conditions in the original problem)

1 Getting Started

Objective
• To use a calculator to evaluate a formula

Vocabulary
formula

Materials
calculators

Warm Up • Mental Math
Have students name each in simplest form.

1. $\frac{3}{5} + \frac{1}{5} = (\frac{4}{5})$
2. $\frac{7}{8} - \frac{3}{8} = (\frac{1}{2})$
3. $\frac{2}{15} + \frac{4}{15} = (\frac{2}{5})$
4. $\frac{8}{10} + \frac{3}{10} = (1\frac{1}{10})$
5. $\frac{4}{9} + \frac{2}{9} = (\frac{2}{3})$
6. $\frac{9}{10} - \frac{3}{10} = (\frac{3}{5})$
7. $\frac{7}{12} - \frac{4}{12} = (\frac{1}{4})$
8. $\frac{7}{9} - \frac{4}{9} = (\frac{1}{3})$

Warm Up • Pencil and Paper
Have students complete the following codes using their calculators.

1. $12 \times 6 \div 3$ (24)
2. $341 \times 7 - 419$ (1,968)
3. $(746 - 569) \times 506$ (89,562)
4. $(48 \times 26) \div 52$ (24)
5. $38 \times 64 - 1,386$ (1,046)
6. $(7,055 \div 5) \times 349$ (492,439)
7. $(264 + 97) \times 35$ (12,635)

2 Teach

Introduce the Lesson Explain that the metric unit used to measure temperature is degrees *Celsius* and the customary unit used to measure temperature is degrees *Fahrenheit*. Often students will hear weather reports where the temperature is given in degrees Celsius. Tell students today they will learn to use a formula to rename a Celsius reading to a Fahrenheit reading. Remind them that a formula is a rule that uses symbols or letters to represent numbers.

Work through the formula on page 133 for changing dimes to pennies with students. Call on a student to read the formula for naming °C as °F. Point out that students should replace the C in the formula with the given

---

Calculator: Formulas

It's Algebra!

Temperature can be measured on the Celsius and Fahrenheit scales. On the Celsius scale, water boils at 100 degrees. On the Fahrenheit scale, water boils at 212 degrees. These two readings are equal to each other. What is the equivalent Fahrenheit reading for the temperature on the bank?

A **formula** is a statement that contains letters that represent numbers. For example, a formula that changes the value of dimes to pennies is written

$$10 \times d = p$$

To find out how many pennies there are in 12 dimes, we replace d by 12 and complete the calculator code.

$$10 \boxed{\times} 12 \boxed{=} \quad \overset{\text{pennies}}{\boxed{120}}$$

The formula for renaming a Celsius reading to a Fahrenheit reading is written

$$(9 \times C) \div 5 + 32 = F$$

To find out the Fahrenheit equivalent for 20 degrees Celsius, we replace C with 20 and complete the calculator code.

$$9 \boxed{\times} \underline{20} \boxed{\div} 5 \boxed{+} 32 \boxed{=} \quad \overset{\text{Fahrenheit}}{\boxed{68}}$$

The temperature at the bank is __68__ degrees Fahrenheit.

Getting Started

Use the formula for changing the value of dimes to pennies.

1. 346 dimes = __3,460__ pennies
2. 457 dimes = __4,570__ pennies
3. 9,658 dimes = __96,580__ pennies

Use the formula for changing a Celsius reading to a Fahrenheit reading.

4. 10°C = __50__ °F
5. 45°C = __113__ °F
6. 35°C = __95__ °F

---

temperature in degrees Celsius and then work through the steps shown.

Develop Skills and Concepts Discuss the models on page 133. Then, write the following formulas on the board and help students explain each one:

$d = r \times t$
(distance = rate × time) (multiply the rate (miles per hour) times the time (hours) to find the distance in miles)

$s \div 12 = D$
divide the number of single eggs by 12 to find the number of dozen

$[(F - 32) \times 5] \div 9 = C$
subtract 32 from the number of degrees Fahrenheit, multiply that difference by 5, and then divide that product by 9 to rename Fahrenheit temperature as Celsius temperature

It's Algebra! The concepts in this lesson prepare students for algebra.

Practice

Use the formula for changing the value of dimes to pennies.

1. 808 dimes = _8,080_ pennies

2. 1,246 dimes = _12,460_ pennies

3. 7,981 dimes = _79,810_ pennies

Use the formula for changing a Celsius reading to a Fahrenheit reading.

4. 5°C = _41_ °F

5. 30°C = _86_ °F

6. 90°C = _194_ °F

7. 50°C = _122_ °F

8. 75°C = _167_ °F

9. 30°C = _86_ °F

Problem Solving

Complete each table.

10. Find the distance traveled at 45 miles per hour using the formula rate times time equals distance.

$r \times t = d$ distance

$45 \boxed{\times} 2 \boxed{=} \boxed{90}$

t	Time in Hours	2	8	15	21	45
d	Distance in Miles	90	360	675	945	2,025

11. Find the number of dozens of eggs by dividing the number of single eggs by 12.

$s \div 12 = d$ dozens

$36 \boxed{\div} 12 \boxed{=} \boxed{3}$

s	Number of Single Eggs	36	144	180	288	768
d	Dozens	3	12	15	24	64

Now Try This!

A calculator can help find the sum of two fractions.

Find $\frac{3}{4} + \frac{5}{6}$.
Find the cross products. $3 \times 6 = 18$ $5 \times 4 = 20$
Add. $18 + 20 = 38$
Multiply denominators. $4 \times 6 = 24$
Put the sum over the denominator. $\frac{38}{24}$
Simplify. $1\frac{7}{12}$

Use a calculator to find these sums.

1. $\frac{1}{2} + \frac{1}{3} = $ $\frac{5}{6}$

2. $\frac{3}{8} + \frac{7}{15} = $ $\frac{101}{120}$

3. $\frac{5}{12} + \frac{5}{6} = $ $1\frac{1}{4}$

4. $\frac{9}{16} + \frac{3}{15} = $ $\frac{183}{240}$

Lesson 6-11 • Calculator: Formulas

For Mixed Abilities

Common Errors • Intervention

Watch for students who enter their numbers in an incorrect order when working with formulas and a calculator. Have them work with partners and take turns writing the calculator codes before they use the calculator to solve the problems.

Enrichment • Algebra

Explain that students can rewrite the formula $r \times t = d$ to find a formula for finding the time when the distance and rate are known. Work with students to derive the formula $t = d \div r$.

Then, ask students to write a formula for finding the rate when the distance and time are given. $r = d \div t$ Have students complete the following table using the formulas.

Rate	(37)	55	68	(78)
Time	5	(9)	4	23
Distance	185	495	(272)	1,794

More to Explore • Applications

Ask students to bring in packaging from breakfast, lunch, and dinner food items that lists how much of the recommended daily allowance of vitamins and minerals they provide. Let each student select one breakfast, lunch, and dinner item for a day's menu. Ask them to calculate the total amount of vitamins and minerals provided by their choices for the day.

ESL/ELL STRATEGIES

When introducing the two temperature scales, write the words on the board, capitalizing the stressed syllables. (FAHRenheit, CELsius) Next, say the words in isolation and ask students to repeat. Then, model how to say several different temperatures using both scales and have students repeat.

---

3 Practice

Have students use their calculators to complete all the exercises. Because this lesson involves many new concepts, you may wish to have students work in small groups to exchange ideas.

Now Try This! This activity requires students to read directions and follow an example. After students solve the problems using the method shown, have them check by subtraction.

4 Assess

Introduce the formula $P = 2l + 2w$ for finding the perimeter of a rectangle. Have students use the formula and a calculator to find the perimeter of a rectangle with a width of 108 inches and a length of 315 inches. (846 inches)

Chapter 6
Test

Name _____

Add. Simplify answers if necessary.

1. $\frac{3}{8} + \frac{3}{8} = \frac{3}{4}$

2. $\frac{5}{6} + \frac{4}{6} = 1\frac{1}{2}$

3. $3\frac{4}{15} + 5\frac{6}{15} = 8\frac{2}{3}$

4. $9\frac{1}{12} + 6\frac{5}{12} = 15\frac{1}{2}$

5. $\frac{7}{12} + \frac{3}{4} = 1\frac{1}{3}$

6. $\frac{2}{3} + \frac{1}{4} = \frac{11}{12}$

7. $8\frac{2}{3} + 7\frac{1}{5} = 15\frac{13}{15}$

8. $11\frac{2}{9} + 8\frac{1}{6} = 19\frac{7}{18}$

9. $7\frac{1}{4} + 8\frac{3}{4} = 16$

10. $16\frac{1}{2} + 12\frac{3}{4} = 29\frac{1}{4}$

11. $19\frac{2}{3} + 16\frac{5}{8} = 36\frac{7}{24}$

12. $36\frac{1}{2} + 18\frac{2}{3} + 55\frac{3}{4} = 110\frac{11}{12}$

Subtract. Simplify answers if necessary.

13. $\frac{5}{12} - \frac{1}{12} = \frac{1}{3}$

14. $\frac{9}{16} - \frac{7}{16} = \frac{1}{8}$

15. $8\frac{7}{8} - 2\frac{3}{8} = 6\frac{1}{2}$

16. $12\frac{7}{9} - 6\frac{1}{9} = 6\frac{2}{3}$

17. $\frac{5}{16} - \frac{1}{4} = \frac{1}{16}$

18. $\frac{5}{8} - \frac{1}{3} = \frac{7}{24}$

19. $39\frac{4}{5} - 18 = 21\frac{4}{5}$

20. $16\frac{2}{3} - 9\frac{5}{8} = 7\frac{1}{24}$

21. $23\frac{1}{5} - 9\frac{3}{5} = 13\frac{3}{5}$

22. $41 - 16\frac{2}{3} = 24\frac{1}{3}$

23. $79\frac{8}{15} - 48\frac{3}{10} = 31\frac{7}{30}$

24. $63 - 27\frac{6}{12} = 35\frac{1}{2}$

25. $43\frac{1}{4} - 28\frac{3}{8} = 14\frac{7}{8}$

26. $94\frac{5}{12} - 48\frac{7}{8} = 45\frac{13}{24}$

27. $653\frac{2}{5} - 286\frac{3}{4} = 366\frac{13}{20}$

28. $410\frac{1}{3} - 195\frac{1}{2} = 214\frac{5}{6}$

Alternate Chapter Test
You may wish to use the Alternate Chapter Test on page 380 of this book for further review and assessment.

Circle the letter of the correct answer.

1. $n + 9 = 16$
    $n = ?$
    a. 6
    b. 7
    c. 8
    d. NG

2. $n \times 4 = 12$
    $n = ?$
    a. 3
    b. 8
    c. 48
    d. NG

3. $12 \div 4 + 2 = n$
    $n = ?$
    a. 2
    b. 5
    c. 11
    d. NG

4. What is the place value of the 6 in 5,269,384?
    a. thousands
    b. ten thousands
    c. hundred thousands
    d. NG

5. 86,246
    + 64,695
    a. 140,941
    b. 150,941
    c. 151,934
    d. NG

6. $562.48
    − 296.78
    a. $265.70
    b. $267.70
    c. $334.30
    d. NG

7. 8.35×9
    a. $72.75
    b. $74.75
    c. $75.15
    d. NG

8. 427
    × 38
    a. 16,226
    b. 4,697
    c. 15,226
    d. NG

9. $440 \div 8$
    a. 5 R44
    b. 5 R5
    c. 50 R5
    d. NG

10. $28\overline{)7,290}$
    a. 26
    b. 26 R1
    c. 260 R10
    d. NG

11. $\frac{3}{n} = \frac{9}{12}$
    $n = ?$
    a. 3
    b. 4
    c. 12
    d. NG

12. Rename in simplest form. $\frac{24}{36}$
    a. $\frac{3}{4}$
    b. $\frac{6}{8}$
    c. $\frac{2}{3}$
    d. NG

score

STOP

Chapter 6 • Cumulative Assessment

page 136

Items	Objectives
1–2	To review basic facts (see pages 3–8)
3	To simplify an expression having two or more operations (see pages 11–12)
4	To identify place value through hundred thousands (see pages 19–20)
5–6	To add and subtract two 5-digit numbers (see pages 27–32)
7	To multiply money by 1-digit factors (see pages 45–46)
8	To multiply 3-digit numbers by 2-digit numbers (see pages 51–52)
9	To divide 3-digit numbers by 1-digit numbers (see pages 63–64)
10	To divide 4-digit numbers by 2-digit numbers (see pages 75–76)
11–12	To rename fractions (see pages 97–100)

Alternate Cumulative Assessment

Circle the letter of the correct answer.

1. $n + 5 = 13$
    $n = ?$
    a 7
    b 8
    c 9
    d NG

2. $n \times 6 = 30$
    $n = ?$
    a 5
    b 6
    c 7
    d NG

3. $21 \div 3 + 4 = n$
    $n = ?$
    a 3
    b 7
    c 11
    d NG

4. Give the place value of the 8 in 4,283,509.
    a thousands
    b ten thousands
    c hundred thousands
    d NG

5. 93,787
    + 86,269
    a 180,056
    b 180,956
    c 189,056
    d NG

6. $641.53
    − 497.58
    a $143.05
    b $143.95
    c $144.95
    d NG

7. $9.43
    × 8
    a $72.44
    b $74.44
    c $75.44
    d NG

8. 852
    × 67
    a 56,084
    b 57,084
    c 58,084
    d NG

9. $9\overline{)594}$
    a 6 R6
    b 666
    c 60 R6
    d NG

10. $35\overline{)6,311}$
    a 180
    b 180 R1
    c 180 R11
    d NG

11. $\frac{5}{n} = \frac{20}{36}$
    $n = ?$
    a 6
    b 8
    c 9
    d NG

12. Simplify: $\frac{18}{54}$
    a $\frac{3}{18}$
    b $\frac{2}{6}$
    c $\frac{1}{3}$
    d NG

7-1 Multiplying Fractions

pages 137–138

1 Getting Started

Objective
• To multiply fractions

Materials
index cards

Warm Up • Mental Math
Have students simplify each expression.

1. $7 \times 9 + 4$ (67)
2. $3 \times 8 + 9$ (33)
3. $6 \times 5 + 12$ (42)
4. $9 \times 8 + 6$ (78)
5. $7 \times 7 + 10$ (59)
6. $8 \times 6 - 3$ (45)
7. $4 \times 9 - 8$ (28)
8. $7 \times 8 - 11$ (45)
9. $3 \times 6 - 7$ (11)
10. $5 \times 8 - 6$ (34)

Warm Up • Pencil and Paper
Have students simplify each fraction or mixed number.

1. $\frac{18}{24}$ ($\frac{3}{4}$)   6. $2\frac{4}{14}$ ($2\frac{2}{7}$)
2. $\frac{27}{39}$ ($\frac{9}{13}$)   7. $\frac{17}{12}$ ($1\frac{5}{12}$)
3. $\frac{21}{56}$ ($\frac{3}{8}$)   8. $\frac{25}{4}$ ($6\frac{1}{4}$)
4. $1\frac{16}{40}$ ($1\frac{2}{5}$)   9. $\frac{40}{12}$ ($3\frac{1}{3}$)
5. $3\frac{17}{68}$ ($3\frac{1}{4}$)   10. $4\frac{10}{7}$ ($5\frac{3}{7}$)

Multiply and Divide Fractions

Lesson 7-1

Multiplying Fractions

Margaret is making several batches of biscuits. She will use $\frac{1}{2}$ of the flour in the measuring cup to make the first batch. The measuring cup is $\frac{3}{4}$ full. How much flour will Margaret use in the first batch?

We want to know how much flour Margaret will use.

We know that the measuring cup is $\frac{3}{4}$ full of flour, and Margaret needs $\frac{1}{2}$ of the flour.

To find what part something is of something else, we multiply. The word *of* means **times** or ×. We multiply to find one-half of the amount in the cup.

We multiply $\frac{3}{4}$ by $\frac{1}{2}$.

Multiply the numerators.	Multiply the denominators. Simplify the product if necessary.

$\frac{1}{2} \times \frac{3}{4} = \frac{1 \times 3}{} = \frac{3}{}$ $\frac{1}{2} \times \frac{3}{4} = \frac{1 \times 3}{2 \times 4} = \frac{3}{}$

Margaret will use $\frac{3}{8}$ of a cup of flour.

Getting Started

Multiply.

1. $\frac{2}{3} \times \frac{3}{4} = \frac{1}{2}$
2. $\frac{1}{5} \times \frac{3}{8} = \frac{3}{40}$
3. $\frac{5}{6} \times \frac{2}{3} = \frac{5}{9}$
4. $\frac{9}{10} \times \frac{5}{8} = \frac{9}{16}$
5. $\frac{3}{8} \times \frac{3}{5} = \frac{9}{40}$
6. $\frac{1}{2} \times \frac{4}{7} = \frac{2}{7}$
7. $\frac{3}{5} \times \frac{2}{5} \times \frac{1}{2} = \frac{3}{25}$
8. $\frac{1}{3} \times \frac{1}{3} \times \frac{1}{3} = \frac{1}{27}$

Lesson 7-1 • Multiplying Fractions

137

---

2 Teach

Introduce the Lesson Ask how many students cook at home. Have students identify the problem that is to be solved and the given information. Work through the multiplication steps with the class. Have students complete the sentences as they read aloud with you to solve the problem. Point out that the word *of* as in "$\frac{1}{2}$ of $\frac{3}{4}$" means $\frac{1}{2} \times \frac{3}{4}$. Remind students to simplify products.

Develop Skills and Concepts Multiplication of fractions may be easier than addition since the algorithm follows the student's natural inclination to multiply the numerators and then multiply the denominators.

Draw a diagram of a rectangle divided into eighths on the board. First, identify $\frac{3}{4}$ of the rectangle and then identify $\frac{1}{2}$ of the $\frac{3}{4}$. Make sure students understand that the regions identified are related to the whole rectangle.

Then, show that the product, $\frac{3}{8}$, results from multiplying the numerators and denominators of the two fractions.

Knowing when to multiply is often confusing for students. It is difficult to perceive multiplication with fractions as repeated addition. Lead students to compare multiplication of whole numbers to multiplication of fractions. Point out that the product of two whole numbers is always greater than either factor; the product of two fractions is always less than either factor.

3 Practice

Have students complete all the exercises. Remind them to simplify each product.

Practice

Multiply.

1. $\frac{2}{3} \times \frac{1}{2} = \underline{\frac{1}{3}}$ 2. $\frac{3}{5} \times \frac{4}{7} = \underline{\frac{12}{35}}$ 3. $\frac{3}{8} \times \frac{1}{5} = \underline{\frac{3}{40}}$ 4. $\frac{7}{8} \times \frac{5}{9} = \underline{\frac{35}{72}}$

5. $\frac{3}{4} \times \frac{3}{8} = \underline{\frac{9}{32}}$ 6. $\frac{3}{8} \times \frac{7}{9} = \underline{\frac{7}{24}}$ 7. $\frac{2}{11} \times \frac{5}{6} = \underline{\frac{5}{33}}$ 8. $\frac{4}{5} \times \frac{3}{10} = \underline{\frac{6}{25}}$

9. $\frac{1}{10} \times \frac{1}{10} = \underline{\frac{1}{100}}$ 10. $\frac{5}{6} \times \frac{2}{3} = \underline{\frac{5}{9}}$ 11. $\frac{9}{10} \times \frac{1}{3} = \underline{\frac{3}{10}}$ 12. $\frac{1}{4} \times \frac{4}{9} = \underline{\frac{1}{9}}$

13. $\frac{3}{7} \times \frac{7}{8} = \underline{\frac{3}{8}}$ 14. $\frac{5}{10} \times \frac{4}{5} = \underline{\frac{2}{5}}$ 15. $\frac{6}{7} \times \frac{7}{8} = \underline{\frac{3}{4}}$ 16. $\frac{3}{8} \times \frac{1}{6} = \underline{\frac{1}{16}}$

17. $\frac{3}{5} \times \frac{4}{7} = \underline{\frac{12}{35}}$ 18. $\frac{5}{8} \times \frac{4}{5} = \underline{\frac{1}{2}}$ 19. $\frac{3}{4} \times \frac{4}{9} = \underline{\frac{1}{3}}$ 20. $\frac{8}{9} \times \frac{5}{7} = \underline{\frac{40}{63}}$

21. $\frac{2}{3} \times \frac{5}{6} \times \frac{3}{5} = \underline{\frac{1}{3}}$ 22. $\frac{1}{5} \times \frac{2}{3} \times \frac{5}{8} = \underline{\frac{1}{12}}$ 23. $\frac{2}{3} \times \frac{2}{3} \times \frac{2}{3} = \underline{\frac{8}{27}}$ 24. $\frac{2}{5} \times \frac{3}{4} \times \frac{5}{6} = \underline{\frac{1}{4}}$

Problem Solving

Solve each problem.

25. Pat has a ribbon $\frac{3}{4}$ of a foot long. She needs $\frac{1}{3}$ of the ribbon for a decoration. How long a ribbon does she need?
$\frac{1}{4}$ of a foot

26. Mr. Ling bought $\frac{5}{6}$ of a pound of swiss cheese at the deli. He ate $\frac{3}{4}$ of it. How much was left?
$\frac{5}{24}$ of a pound

27. North America makes up about $\frac{1}{6}$ of the world's land surface. The United States is about $\frac{2}{5}$ of North America. What part of the world's surface is the United States?
$\frac{1}{15}$

28. A developer bought 20 acres of land. He built a shopping mall on $12\frac{2}{3}$ acres and a parking lot on $5\frac{1}{4}$ acres. How much land was left?
$2\frac{1}{12}$ acres

(Now Try This!)

Complete the first two magic squares. Remember, in a magic square, the sums of each row, column, and diagonal are the same. Subtract each fraction in square 2 from the corresponding fraction in square 1. Write the difference in the corresponding place in square 3. What do you notice?   **3 is also a magic square**

1.
$3\frac{3}{4}$	$\frac{1}{4}$	$2\frac{3}{4}$
$1\frac{1}{4}$	$2\frac{1}{4}$	$3\frac{1}{4}$
$1\frac{3}{4}$	$4\frac{1}{4}$	$\frac{3}{4}$

2.
2	$\frac{1}{4}$	$1\frac{1}{2}$
$\frac{3}{4}$	$1\frac{1}{4}$	$1\frac{3}{4}$
1	$2\frac{1}{4}$	$\frac{1}{2}$

3.
$1\frac{3}{4}$	0	$1\frac{1}{4}$
$\frac{1}{2}$	1	$1\frac{1}{2}$
$\frac{3}{4}$	2	$\frac{1}{4}$

Now Try This! Suggest that students begin with the first magic square and find the magic sum of the addends in the first row. Once students find the magic sum, they can find the missing addends in this square. They should begin by finding the missing addend in the third column. Once they have completed the first magic square, they should follow the directions for square 2 and square 3.

4 Assess

Ask students why $\frac{1}{2} \times \frac{3}{5}$ is less than $\frac{3}{5}$. (Possible answer: $\frac{1}{2} \times \frac{3}{5}$ is $\frac{1}{2}$ of $\frac{3}{5}$. This is a fractional part of $\frac{3}{5}$ and will be less than $\frac{3}{5}$.)

For Mixed Abilities

Common Errors • Intervention

Watch for students who multiply the numerators and write the product over one of the denominators.

Incorrect	Correct
$\frac{1}{2} \times \frac{3}{4} = \frac{3}{4}$	$\frac{1}{2} \times \frac{3}{4} = \frac{3}{8}$

Have these students work in pairs. Have them follow these steps to use rectangles to model multiplication of fractions.

1. Draw a rectangle that shows $\frac{3}{4}$.

2. Divide the rectangle in half to show $\frac{1}{2}$ of $\frac{3}{4}$.

$$\frac{1}{2} \times \frac{3}{4} = \frac{3}{8}$$

Enrichment • Number Sense

Have students solve the following:

Louis ate $\frac{1}{4}$ of $\frac{3}{4}$ of a pizza. Sally ate $\frac{1}{2}$ of $\frac{3}{4}$ of the pizza. Joe ate $\frac{3}{8}$ of $\frac{1}{2}$ of the pizza, and Erik ate $\frac{1}{2}$ of $\frac{1}{2}$ of the pizza. Who ate the most pizza? Which two ate the same amount? Was any of the pizza left over? (Sally ate the most. Louis and Joe ate the same amount. No pizza was left over.)

More to Explore • Application

Have students make a list of five items sold in the school supply store.

• Divide students into groups and have each group keep track of how many of each of these items is sold every day for a week.

• Ask the manager to explain the term *profit*, and give students matching wholesale and retail prices for each item.

• Ask students to find how much profit is made on each item, which day had the greatest profit, which item each day made the greatest profit, and how much the store's total profit was for the week.

1 Getting Started

Objective
- To use factoring to multiply fractions

Materials
index cards

Warm Up • Mental Math
Have students use the formula $(\square \times 7) - 4 = \triangle$ to name the $\triangle$.

$\square$	3	8	20	9
$\triangle$	(17)	(52)	(136)	(59)

Have students use the formula $(\square \div 3) + 8 = \triangle$ to name the $\triangle$.

$\square$	171	621	240	816
$\triangle$	(65)	(215)	(88)	(280)

Warm Up • Pencil and Paper
Have students find the GCF of the numerator and denominator of each fraction. Then, ask them to write the fraction in simplest form.

1. $\frac{8}{24}$ $(8; \frac{1}{3})$ 4. $\frac{9}{27}$ $(9; \frac{1}{3})$
2. $\frac{16}{36}$ $(4; \frac{4}{9})$ 5. $\frac{6}{15}$ $(3; \frac{2}{5})$
3. $\frac{9}{12}$ $(3; \frac{3}{4})$ 6. $\frac{4}{6}$ $(2; \frac{2}{3})$

Name _____

Factoring Before Multiplying

Edie was going to walk the $\frac{3}{10}$ of a mile from school to the dentist's office. She stopped at the park to feed the birds. If the park is $\frac{1}{3}$ of the total distance she has to walk, how far is the park from the school?

We want to find the distance from the school to the park. We know that it is $\frac{3}{10}$ of a mile from the school to the dentist, and it is $\frac{1}{3}$ of this distance from the school to the park. We multiply to find one third of the distance. We multiply $\frac{3}{10}$ by $\frac{1}{3}$.

We already know one way to multiply fractions. We can take a shortcut by dividing a numerator and a denominator by the same number before we multiply. This is called factoring. In this example, we can divide by 3.

$$\frac{\cancel{3}^{1}}{10} \times \frac{1}{\cancel{3}_{1}} = \frac{1 \times 1}{10 \times 1} = \frac{1}{10}$$

It is $\frac{1}{10}$ of a mile from the school to the park.

Sometimes we can use the shortcut more than once in a problem.

Factor by 10.	Factor by 3.	Factor by 3 again.
$\frac{10}{21} \times \frac{9}{30}$	$\frac{10}{21} \times \frac{9}{30}$	$\frac{10}{21} \times \frac{9}{30} = \frac{1}{7}$

Getting Started

Multiply. Factor first wherever possible.

1. $\frac{4}{5} \times \frac{3}{4} = \underline{\frac{3}{5}}$ 2. $\frac{5}{6} \times \frac{4}{5} = \underline{\frac{2}{3}}$ 3. $\frac{7}{8} \times \frac{4}{9} = \underline{\frac{7}{18}}$ 4. $\frac{2}{3} \times \frac{9}{10} = \underline{\frac{3}{5}}$

5. $\frac{7}{10} \times \frac{5}{8} = \underline{\frac{7}{16}}$ 6. $\frac{3}{8} \times \frac{7}{12} = \underline{\frac{7}{32}}$ 7. $\frac{9}{16} \times \frac{2}{3} = \underline{\frac{3}{8}}$ 8. $\frac{5}{14} \times \frac{21}{25} \times \frac{10}{15} = \underline{\frac{1}{5}}$

---

2 Teach

Introduce the Lesson Have a student read the problem and tell what is to be solved. Ask what information is given in the problem and the picture. Work through the factoring process with students to solve the problem.

Develop Skills and Concepts Explain that the model illustrates that if students divide any numerator and any denominator by common factors before multiplying, the product will be in simplest form. Write the following problem on the board: $\frac{3}{4} \times \frac{2}{3}$.

Have some students find the product by multiplying and then simplifying the product. $(\frac{1}{2})$ Have other students find the product by dividing the numerator and denominator by common factors and then multiplying. Have students compare answers. Point out that the shortcut is not a new skill, but is one that students

already know. It is helpful to write the problem with a single fraction bar:

$$\frac{3 \times 2}{4 \times 3}$$

Remind students that multiplication is commutative; they can rewrite this exercise as:

$$\frac{2 \times 3}{4 \times 3}$$

When the problem is written in this form, students can easily see that 3 can be canceled. Provide reinforcement by having students use the shortcut to find each product.

1. $\frac{3}{8} \times \frac{4}{9}$ $(\frac{1}{6})$ 2. $\frac{4}{5} \times \frac{3}{4}$ $(\frac{3}{5})$ 3. $\frac{5}{6} \times \frac{8}{25}$ $(\frac{4}{15})$

3 Practice

Have students complete all the exercises. Be sure students understand that there is more than one way to multiply fractions.

Practice

Multiply. Factor first wherever possible.

1. $\frac{4}{5} \times \frac{3}{4} = \underline{\frac{3}{5}}$ 2. $\frac{5}{6} \times \frac{7}{10} = \underline{\frac{7}{12}}$ 3. $\frac{2}{3} \times \frac{9}{10} = \underline{\frac{3}{5}}$ 4. $\frac{5}{8} \times \frac{6}{15} = \underline{\frac{1}{4}}$

5. $\frac{3}{7} \times \frac{5}{9} = \underline{\frac{5}{21}}$ 6. $\frac{4}{9} \times \frac{3}{10} = \underline{\frac{2}{15}}$ 7. $\frac{6}{15} \times \frac{20}{27} = \underline{\frac{8}{27}}$ 8. $\frac{3}{10} \times \frac{5}{7} = \underline{\frac{3}{14}}$

9. $\frac{8}{15} \times \frac{9}{10} = \underline{\frac{12}{25}}$ 10. $\frac{3}{4} \times \frac{5}{6} = \underline{\frac{5}{8}}$ 11. $\frac{7}{12} \times \frac{9}{14} = \underline{\frac{3}{8}}$ 12. $\frac{5}{16} \times \frac{8}{9} = \underline{\frac{5}{18}}$

13. $\frac{5}{8} \times \frac{1}{10} = \underline{\frac{1}{16}}$ 14. $\frac{15}{22} \times \frac{11}{20} = \underline{\frac{3}{8}}$ 15. $\frac{5}{12} \times \frac{8}{15} = \underline{\frac{2}{9}}$ 16. $\frac{10}{21} \times \frac{14}{15} = \underline{\frac{4}{9}}$

17. $\frac{9}{16} \times \frac{8}{15} = \underline{\frac{3}{10}}$ 18. $\frac{3}{20} \times \frac{15}{21} = \underline{\frac{3}{28}}$ 19. $\frac{5}{18} \times \frac{12}{25} = \underline{\frac{2}{15}}$ 20. $\frac{10}{11} \times \frac{11}{16} = \underline{\frac{5}{8}}$

21. $\frac{8}{15} \times \frac{7}{20} = \underline{\frac{14}{75}}$ 22. $\frac{1}{5} \times \frac{49}{50} \times \frac{15}{21} = \underline{\frac{7}{50}}$ 23. $\frac{5}{6} \times \frac{18}{75} \times \frac{10}{27} = \underline{\frac{2}{27}}$ 24. $\frac{7}{15} \times \frac{25}{28} \times \frac{49}{100} = \underline{\frac{49}{240}}$

Problem Solving

Solve each problem.

25. Kimo has planted vegetables in $\frac{2}{9}$ of his garden. Of the vegetables, $\frac{3}{4}$ are carrots. What part of the whole garden is planted with carrots?

    $\frac{1}{6}$

26. Tobor the Robot uses $\frac{9}{16}$ of a gallon of oil each day to remove its squeaks. So far today, it has used $\frac{1}{4}$ of a gallon. How much more oil will it use today?

    $\frac{5}{16}$ of a gallon

27. The milk container was $\frac{9}{10}$ full. Emmy drank $\frac{1}{3}$ of the milk in the container. What part of the milk was left?

    $\frac{3}{5}$

28. Maureen's gas tank was $\frac{7}{8}$ full. When she next looked at the gas gauge, it showed the tank was $\frac{1}{4}$ full. What part of the tank had she used?

    $\frac{5}{8}$

> ### [Now Try This!]
>
> Use these fractions to make each leg of the magic triangle add up to 1.
>
> $\frac{1}{20}, \frac{1}{10}, \frac{3}{20}, \frac{1}{5}, \frac{1}{4}, \frac{3}{10}, \frac{7}{20}, \frac{2}{5}, \frac{9}{20}$
>
>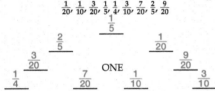

Lesson 7-2 • Factoring Before Multiplying

> **Now Try This!** Point out that these fractions are names for 1: $\frac{2}{2}, \frac{3}{3}, \frac{4}{4}, \frac{5}{5}, \frac{10}{10}$, and $\frac{20}{20}$. They can be used to rename fractions. Be sure students remember that they must find a common denominator before they can add fractions with unlike denominators.

4 Assess

Have students name the common factors that can be used to simplify this multiplication expression.

$\frac{10}{21} \times \frac{14}{15}$ (5 and 7)

For Mixed Abilities

Common Errors • Intervention

Watch for students whose answers may not be in simplest form because they have not factored completely. Have these students work with partners, each working the same problem. Have them compare their answers, helping each other find all the possible ways they can factor.

Enrichment • Number Sense

Have students match the multiplication exercises that have the same products.

1. $\frac{3}{8} \times \frac{5}{12}$ (c)   a. $\frac{1}{3} \times \frac{7}{8}$ ($\frac{7}{24}$)

2. $\frac{7}{8} \times \frac{2}{3} \times \frac{1}{2}$ (a)   b. $\frac{1}{6} \times \frac{1}{9}$ ($\frac{1}{54}$)

3. $\frac{2}{9} \times \frac{1}{10} \times \frac{5}{6}$ (b)   c. $\frac{5}{8} \times \frac{2}{5}$ ($\frac{1}{4}$)

4. $\frac{10}{11} \times \frac{2}{3} \times \frac{3}{5}$ (d)   d. $\frac{4}{5} \times \frac{5}{11} \times \frac{5}{5}$ ($\frac{4}{11}$)

Ask students to complete each multiplication.

5. $\frac{3}{4} \times \frac{1}{8} \times \frac{2}{3} = $ ($\frac{1}{16}$)

6. $\frac{7}{12} \times \frac{2}{3} \times \frac{6}{7} = $ ($\frac{1}{3}$)

7. $\frac{5}{9} \times \frac{3}{4} \times \frac{2}{15} = $ ($\frac{1}{18}$)

8. $\frac{2}{5} \times \frac{10}{11} \times \frac{11}{12} = $ ($\frac{1}{3}$)

9. $\frac{5}{8} \times \frac{2}{25} \times \frac{10}{25} = $ ($\frac{1}{50}$)

More to Explore • Measurement

Prepare a measurement scavenger hunt. Measure ten objects in the room using metric units. Record the length, width, and height of each on the board, without identifying the object. Tell students they are to find the objects with the matching measurements. This activity could be used over a day's class time, with students working individually when they have finished other work.

ESL/ELL STRATEGIES

> Explain that a *factor* is a number by which a larger number can be divided. Give an example such as: *Six can be divided by two or by three. Two and three are factors of six.* Then, use the example in the book to clarify how factoring works.

T140

1 Getting Started

Objective
- To multiply mixed numbers and fractions

Warm Up • Mental Math
Have students name each missing factor.

1. $6 \times (100) = 600$
2. $50 \times (7) = 350$
3. $90 \times (70) = 6,300$
4. $100 \times (900) = 90,000$
5. $80 \times (30) = 2,400$
6. $600 \times (60) = 36,000$
7. $2 \times (500) = 1,000$
8. $40 \times (8,000) = 320,000$

Warm Up • Pencil and Paper
Have students write each mixed number or whole number as a fraction.

1. $2\frac{1}{3}$ $\left(\frac{7}{3}\right)$
2. $4\frac{2}{3}$ $\left(\frac{14}{3}\right)$
3. $3\frac{3}{5}$ $\left(\frac{18}{5}\right)$
4. 5 $\left(\frac{5}{1}\right)$
5. 7 $\left(\frac{7}{1}\right)$
6. 15 $\left(\frac{15}{1}\right)$
7. $5\frac{1}{2}$ $\left(\frac{11}{2}\right)$
8. 72 $\left(\frac{72}{1}\right)$
9. $6\frac{5}{6}$ $\left(\frac{41}{6}\right)$

2 Teach

Introduce the Lesson Have a student read the problem and tell what is to be solved. (how wide the pathway is) Ask what information is given in the problem and the picture. Have students explain why the answer is found by multiplication. (need to find a total width)

Point out that the only new step in this multiplication is to rename the mixed numbers as improper fractions. Have students fill in the information sentences. Work through the model with students and then have them complete the solution sentence.

Develop Skills and Concepts Explain that when a mixed number is renamed as an improper fraction, students can use the procedure they already know for multiplying

fractions. Emphasize that if one factor is a whole number, it can be renamed as a fraction with a denominator of 1.

Provide reinforcement and practice by having students work these exercises at the board:

$3\frac{1}{2} \times 2\frac{2}{5}$ $\left(8\frac{2}{5}\right)$ $\frac{7}{8} \times 6$ $\left(5\frac{1}{4}\right)$ $32 \times \frac{3}{4}$ (24)

3 Practice

Have students complete all the exercises. Tell students to read the word problems carefully.

Multiplying Mixed Numbers

Mr. Jerome is putting a brick pathway $4\frac{1}{2}$ bricks wide in his yard. Each brick is $9\frac{1}{3}$ inches wide. How wide is the pathway?

We want to know the width of the pathway in inches.

We know that the pathway is $\underline{4\frac{1}{2}}$ bricks wide, and each brick is $\underline{9\frac{1}{3}}$ inches wide.

To find the total width, we multiply the number of bricks wide the pathway is, by the width of each brick. We multiply $\underline{4\frac{1}{2}}$ by $\underline{9\frac{1}{3}}$.

Rename the numbers as improper fractions.	Multiply the fractions. Factor wherever possible.
$4\frac{1}{2} \times 9\frac{1}{3}$ $\downarrow$ $\qquad$ $\downarrow$ $\frac{9}{2} \times \frac{28}{3}$	$\frac{\overset{3}{\cancel{9}}}{\cancel{2}} \times \frac{\overset{14}{\cancel{28}}}{\cancel{3}} = \frac{42}{1} = 42$

The pathway is $\underline{42}$ inches wide.

Getting Started

Multiply.

1. $\frac{3}{4} \times 3\frac{1}{5} = \underline{2\frac{2}{5}}$
2. $6\frac{1}{4} \times \frac{3}{5} = \underline{3\frac{3}{4}}$
3. $24 \times \frac{2}{3} = \underline{16}$
4. $2\frac{5}{8} \times 1\frac{1}{7} = \underline{3}$
5. $\frac{4}{5} \times 100 = \underline{80}$
6. $5\frac{1}{3} \times 5\frac{1}{4} = \underline{28}$
7. $\frac{3}{8} \times 2\frac{2}{3} \times 93 = \underline{93}$
8. $1\frac{1}{2} \times 5\frac{1}{3} \times 1\frac{1}{8} = \underline{9}$
9. $3\frac{1}{8} \times 6\frac{2}{3} \times 24 = \underline{500}$

Copy and multiply.

10. $\frac{2}{3} \times 150$ 100
11. $4 \times 5\frac{5}{6} \times 3$ 70
12. $2\frac{1}{2} \times 3\frac{1}{3} \times 4\frac{1}{8}$ $34\frac{3}{8}$

Practice

Multiply.

1. $\frac{5}{6} \times 1\frac{1}{10} = \underline{\frac{11}{12}}$

2. $4\frac{1}{5} \times \frac{3}{7} = \underline{1\frac{4}{5}}$

3. $\frac{5}{8} \times 64 = \underline{40}$

4. $3\frac{1}{3} \times 1\frac{1}{5} = \underline{4}$

5. $6\frac{1}{4} \times 2\frac{2}{5} = \underline{15}$

6. $9\frac{1}{3} \times 2\frac{1}{4} = \underline{21}$

7. $7\frac{1}{3} \times 2\frac{5}{6} = \underline{20\frac{7}{9}}$

8. $3\frac{1}{8} \times 1\frac{3}{10} = \underline{4\frac{1}{16}}$

9. $3\frac{7}{10} \times 4\frac{1}{6} = \underline{15\frac{5}{12}}$

10. $6\frac{2}{3} \times 4\frac{3}{8} = \underline{29\frac{1}{6}}$

11. $15\frac{3}{4} \times 2\frac{4}{7} = \underline{40\frac{1}{2}}$

12. $2\frac{7}{9} \times 2\frac{7}{10} = \underline{7\frac{1}{2}}$

13. $9\frac{3}{8} \times 5\frac{1}{5} = \underline{48\frac{3}{4}}$

14. $6\frac{3}{10} \times 2\frac{1}{9} = \underline{13\frac{3}{10}}$

15. $3\frac{2}{3} \times \frac{3}{22} = \underline{\frac{1}{2}}$

Copy and multiply.

16. $\frac{4}{9} \times 450$ 200

17. $5\frac{1}{4} \times 2\frac{2}{7}$ 12

18. $1\frac{3}{5} \times \frac{10}{11}$ $1\frac{5}{11}$

19. $2\frac{1}{4} \times 9\frac{1}{3}$ 21

20. $2\frac{2}{5} \times 2\frac{1}{2}$ 6

21. $6\frac{2}{3} \times 2\frac{5}{8}$ $17\frac{1}{2}$

22. $\frac{2}{3} \times 12 \times 2\frac{1}{4}$ 18

23. $3\frac{1}{3} \times 1\frac{1}{8} \times 2\frac{2}{5}$ 9

24. $1\frac{1}{5} \times \frac{7}{10} \times 2\frac{1}{12}$ $1\frac{3}{4}$

25. $1\frac{1}{14} \times 1\frac{1}{2} \times 2\frac{2}{3}$ $4\frac{2}{7}$

26. $\frac{3}{4} \times 1\frac{7}{9} \times \frac{3}{4}$ 1

27. $6 \times 2\frac{1}{12} \times 8$ 100

28. $1\frac{2}{5} \times 1\frac{15}{28} \times 3\frac{3}{4}$ $6\frac{3}{16}$

29. $7\frac{1}{9} \times 1\frac{15}{16} \times \frac{3}{7}$ $5\frac{19}{21}$

30. $3\frac{3}{5} \times 1\frac{1}{3} \times 1\frac{7}{8}$ 9

Problem Solving

Solve each problem. Use the circle graph to answer Problems 33 and 34.

31. The gravity on the moon is $\frac{1}{6}$ of the gravity on the earth. If a person weighs 180 pounds on the earth, how much will that person weigh on the moon?
    30 pounds

32. A recipe calls for $2\frac{3}{4}$ cups of white flour and $3\frac{1}{2}$ cups of whole wheat flour. How much flour will be needed to make the recipe $1\frac{3}{5}$ times larger?
    10 cups

Use the circle graph to solve Problems 33 and 34.

33. Mr. Williams earned $2,844 in November. How much did he spend on food and rent?
    $2,133

34. How much more did Mr. Williams spend for rent and other items than on food?
    $474

November Pay

Food $\frac{5}{12}$ · Rent $\frac{1}{3}$ · Other $\frac{1}{4}$

142

Lesson 7-3 • Multiplying Mixed Numbers

4 Assess

Write the three steps for multiplying mixed numbers each on a separate sheet of paper. Scramble the steps. Then, have the students rearrange the steps in the correct order. (Rename the mixed numbers as improper fractions. Factor wherever possible. Multiply the fractions.)

5 Mixed Review

1. $10^5 \times 7$ (700,000)

2. $376.25 + 295.48 + 651.95$ ($1,323.68)

3. $17\frac{1}{2} = \frac{?}{2}$ (35)

4. $\frac{7}{8} - \frac{2}{3}$ $\left(\frac{5}{24}\right)$

5. $16 + 4 \times 8$ (48)

6. $9,275 \div 37$ (250 R25)

7. $21,658 - 18,117$ (3,541)

8. $6\frac{4}{5} + 12\frac{5}{6}$ $\left(19\frac{19}{30}\right)$

9. $\frac{36}{42} = \frac{?}{7}$ (6)

10. 8.97×36 ($322.92)

For Mixed Abilities

Common Errors • Intervention

When multiplying mixed numbers, such as $2\frac{1}{5} \times 3\frac{1}{4}$, some students may multiply the whole numbers and the fractions separately.

Incorrect

$2\frac{1}{5} \times 3\frac{1}{4} = 6\frac{1}{20}$

Correct

$2\frac{1}{5} \times 3\frac{1}{4} = \frac{11}{5} \times \frac{13}{4}$

$= \frac{143}{20} = 7\frac{3}{20}$

Discuss with students how and why each mixed number should be written as a fraction before multiplying.

Enrichment • Application

Explain to students that they can use the Distributive Property to multiply a mixed number by a whole number. For example,

$14 \times 5\frac{1}{8} = (14 \times 5) + (14 \times \frac{1}{8})$

$\qquad = \quad 70 \quad + \quad \frac{7}{4}$

$\qquad = \quad 70 \quad + \quad 1\frac{3}{4} = 71\frac{3}{4}$

Have students use the Distributive Property to find each product.

1. $18 \times 4\frac{2}{3}$ (84)

2. $12 \times 7\frac{1}{10}$ $\left(85\frac{1}{5}\right)$

3. $10 \times 5\frac{3}{5}$ (56)

More to Explore • Biography

The man who had the greatest influence on the field of geometry since Euclid was David Hilbert, a German mathematician. Hilbert wrote extensively about geometry in his 1902 book, *Foundations of Geometry*. After receiving his Ph.D. from the University of Königsberg, Hilbert taught mathematics at the University of Göttingen, where he remained for the rest of his life.

Many mathematicians from the United States traveled to Göttingen to study with him. During his scholastic career, Hilbert listed twenty-three unsolved mathematical problems that he felt should be investigated further. Since his time, almost all of these problems have been solved.

T142

7-4 Reciprocals

pages 143–144

1 Getting Started

Objective
• To find reciprocals of fractions, mixed numbers, and whole numbers

Vocabulary
reciprocal

Warm Up • Mental Math
Have students tell

1. five fraction names for 1 $\left(\frac{2}{2}, \frac{3}{3}, \frac{4}{4}, \frac{5}{5}, \frac{6}{6}\right)$

2. five fraction names for $1\frac{1}{2}$ $\left(\frac{3}{2}, \frac{6}{4}, \frac{9}{6}, \frac{12}{8}, \frac{15}{10}\right)$

3. five fraction names for $2\frac{1}{4}$ $\left(\frac{9}{4}, \frac{18}{8}, \frac{27}{12}, \frac{36}{16}, \frac{225}{100}\right)$

4. five fraction names for 2 $\left(\frac{2}{1}, \frac{4}{2}, \frac{6}{3}, \frac{8}{4}, \frac{10}{5}\right)$

Warm Up • Pencil and Paper
Have students write each product in simplest form.

1. $\frac{3}{8} \times \frac{2}{15}$ $\left(\frac{1}{20}\right)$ 5. $17 \times \frac{5}{8}$ $\left(10\frac{5}{8}\right)$

2. $1\frac{1}{2} \times 10$ (15)   6. $2\frac{1}{4} \times 2\frac{1}{4}$ $\left(5\frac{1}{16}\right)$

3. $\frac{14}{15} \times \frac{5}{7}$ $\left(\frac{2}{3}\right)$ 7. $\frac{5}{9} \times 4\frac{1}{5}$ $\left(2\frac{1}{3}\right)$

4. $1\frac{9}{10} \times \frac{5}{19}$ $\left(\frac{1}{2}\right)$ 8. $3\frac{1}{8} \times 2\frac{2}{5}$ $\left(7\frac{1}{2}\right)$

Name _____

Reciprocals

When the product of two numbers is one, the numbers are called reciprocals of each other. What are the reciprocals of $\frac{2}{3}$, $2\frac{1}{4}$, and 5?

To find the reciprocal of a fraction, we exchange the positions of the numerator and the denominator.

Fraction	Reciprocal	Check
$\frac{2}{3}$ ⟶	$\frac{3}{2}$	$\frac{2}{3} \times \frac{3}{2} = \frac{6}{6} = 1$

To find the reciprocal of a mixed number or whole number, first name it as an improper fraction.

Mixed Number	Improper Fraction	Reciprocal	Check
$2\frac{1}{4}$ ⟶	$\frac{9}{4}$ ⟶	$\frac{4}{9}$	$\frac{9}{4} \times \frac{4}{9} = \frac{36}{36} = 1$

Whole Number	Improper Fraction	Reciprocal	Check
5 ⟶	$\frac{5}{1}$ ⟶	$\frac{1}{5}$	$\frac{5}{1} \times \frac{1}{5} = \frac{5}{5} = 1$

The reciprocals of $\frac{2}{3}$, $2\frac{1}{4}$, and 5 are $\underline{\frac{3}{2}}$, $\underline{\frac{4}{9}}$, and $\underline{\frac{1}{5}}$.

Getting Started

Write the reciprocal.

1. $\frac{5}{8}$ $\frac{8}{5}$ 2. $\frac{11}{3}$ $\frac{3}{11}$ 3. $2\frac{1}{5}$ $\frac{5}{11}$ 4. 9 $\frac{1}{9}$ 5. $3\frac{1}{3}$ $\frac{3}{10}$

Write the missing factors.

6. $\frac{3}{8} \times \underline{\frac{8}{3}} = 1$ 7. $4\frac{1}{2} \times \underline{\frac{2}{9}} = 1$ 8. $3 \times \underline{\frac{1}{15}} \times 5 = 1$

---

2 Teach

Introduce the Lesson Ask students to read the definition of *reciprocal*. Have a student explain how to find the reciprocal of a fraction. (interchange the numerator and the denominator) Point out the dual direction of the reciprocal relationship: $\frac{3}{2}$ is the reciprocal of $\frac{2}{3}$, and $\frac{2}{3}$ is the reciprocal of $\frac{3}{2}$. Explain that to find the reciprocal of a mixed number or whole number, students must rename the mixed number or whole number as an improper fraction.

Work through the model with students. Help students recognize that the products of reciprocals are fractional names for 1. Point out that 1 is its own reciprocal, since $1 \times 1 = 1$. Explain that zero has no reciprocal, since division by zero is impossible.

Develop Skills and Concepts Have students find each product.

$\frac{3}{4} \times \frac{4}{3}$ (1) $\frac{2}{3} \times \frac{3}{2}$ (1)

$\frac{1}{6} \times \frac{6}{1}$ (1) $\frac{5}{3} \times \frac{3}{5}$ (1)

Then, have students find the missing factor and write the product as a whole number.

$\frac{1}{3} \times \underline{\hspace{1cm}} = \frac{3}{3}$ $\left(\frac{3}{1}; 1\right)$

$\frac{6}{1} \times \underline{\hspace{1cm}} = \frac{6}{6}$ $\left(\frac{1}{6}; 1\right)$

$\frac{7}{3} \times \underline{\hspace{1cm}} = \frac{21}{21}$ $\left(\frac{3}{7}; 1\right)$

Ask students what they notice about the products. (Each product is 1.)

3 Practice

Have students complete all the exercises.

Practice

Write the reciprocal.

1. $\frac{7}{8}$ $\frac{8}{7}$
2. $2\frac{1}{3}$ $\frac{3}{7}$
3. 7 $\frac{1}{7}$
4. $4\frac{1}{8}$ $\frac{8}{33}$
5. $\frac{1}{9}$ 9

6. $\frac{5}{9}$ $\frac{9}{5}$
7. 12 $\frac{1}{12}$
8. $\frac{17}{100}$ $\frac{100}{17}$
9. 1 $\frac{1}{1}$
10. $6\frac{1}{4}$ $\frac{4}{25}$

11. $\frac{5}{3}$ $\frac{3}{5}$
12. $\frac{9}{10}$ $\frac{10}{9}$
13. 8 $\frac{1}{8}$
14. $9\frac{3}{8}$ $\frac{8}{75}$
15. 69 $\frac{1}{69}$

16. $\frac{3}{110}$ $\frac{110}{3}$
17. $4\frac{3}{4}$ $\frac{4}{19}$
18. $\frac{15}{4}$ $\frac{4}{15}$
19. 14 $\frac{1}{14}$
20. $\frac{3}{17}$ $\frac{17}{3}$

21. 75 $\frac{1}{75}$
22. $\frac{19}{2}$ $\frac{2}{19}$
23. $15\frac{2}{3}$ $\frac{3}{47}$
24. $\frac{15}{16}$ $\frac{16}{15}$
25. $6\frac{7}{9}$ $\frac{9}{61}$

Write the missing factors.

26. $\frac{2}{3} \times \frac{3}{2} = 1$
27. $\frac{5}{4} \times \frac{4}{5} = 1$
28. $2\frac{1}{2} \times \frac{2}{5} = 1$

29. $\frac{2}{15} \times 7\frac{1}{2} = 1$
30. $2\frac{1}{3} \times \frac{3}{7} = 1$
31. $4\frac{2}{3} \times \frac{3}{14} = 1$

32. $\frac{1}{6} \times 6 = 1$
33. $\frac{3}{8} \times \frac{8}{3} = 1$
34. $\frac{5}{2} \times \frac{2}{5} = 1$

35. $2\frac{9}{10} \times \frac{10}{29} = 1$
36. $\frac{1}{15} \times 15 = 1$
37. $\frac{3}{20} \times 6\frac{2}{3} = 1$

38. $\frac{3}{4} \times \frac{4}{3} = 1$
39. $8 \times 7 \times \frac{1}{56} = 1$
40. $2\frac{1}{3} \times \frac{4}{ } \times \frac{3}{7} = 4$

41. $7 \times 5 \times \frac{1}{35} = 1$
42. $3 \times \frac{1}{12} \times 4 = 1$
43. $\frac{1}{10} \times \frac{5}{3} \times 6 = 1$

(Now Try This!)

These cards are face up on a table.

1. Turn over two cards whose difference is $\frac{1}{2}$. $\frac{5}{6}, \frac{1}{3}$

2. Turn over two cards whose sum is $\frac{39}{40}$. $\frac{3}{8}, \frac{3}{5}$

3. Find the product of the two remaining cards. $1\frac{3}{5}$

4. Turn the cards face up again.

5. Turn over two cards whose product is $\frac{3}{7}$. $1\frac{1}{7}, \frac{3}{8}$

6. Turn over two cards whose product is less than $\frac{1}{4}$. $\frac{1}{3}, \frac{3}{5}$

7. Find the product of the two remaining cards. $1\frac{1}{6}$

Lesson 7-4 • Reciprocals

Now Try This! Encourage students to use trial and error to complete this activity. Remind students that to add or subtract fractions the fractions must have like denominators.

---

4) Assess

Ask, *Will the reciprocal of $3\frac{2}{5}$ be less than, equal to, or greater than 1?* (less than 1)

For Mixed Abilities

Common Errors • Intervention

Some students may try to write the reciprocal of a whole number or mixed number before writing the whole number or mixed number as a fraction. Discuss how, when you write the reciprocal, you exchange the numerator and denominator of the fraction and, therefore, you have to have a fraction first.

Enrichment • Number Sense

Have students find each product mentally. Have them write answers only.

1. $\frac{3}{8} \times \frac{11}{23} \times \frac{8}{3}$ $\left(\frac{11}{23}\right)$
2. $1\frac{1}{2} \times \frac{15}{16} \times \frac{2}{3}$ $\left(\frac{15}{16}\right)$
3. $\frac{91}{93} \times \frac{93}{91} \times \frac{7}{105}$ $\left(\frac{7}{105}\right)$
4. $\frac{17}{18} \times \frac{17}{18} \times \frac{18}{17}$ $\left(\frac{17}{18}\right)$
5. $\frac{57}{59} \times \frac{36}{43} \times 0 \times \frac{43}{36}$ (0)
6. $\frac{9}{10} \times \frac{9}{11} \times \frac{11}{9}$ $\left(\frac{9}{10}\right)$

More to Explore • Statistics

Show students how to use punched cards to tally data they have collected. Create a simple survey using 5 questions students can ask, such as, *Are you wearing red? Are you buying lunch today? Do you have blue eyes?*

Give students 3-by-5 index cards and ask them to mark three half-inch strips, labeling the strips 1 through 10. Tell them that for each yes, they will punch a hole in the top of the card in the appropriate strip. For each no, they will punch a hole and then open the top to make a slot.

When each student has surveyed ten others, show the class how to use the cards to answer questions. For example, to find out how many people answered yes to the first question, have them hold the cards in a stack and put a pencil through the first hole. The cards for people who answered no will fall away. When they count the number of cards remaining on the pencil they will know how many answered yes. Have them tally their results.

7-5 Dividing Fractions

1 Getting Started

Objective
• To divide fractions

Warm Up • Mental Math
Have students name

1. $\frac{1}{2}$ of 12 (6)
2. $\frac{1}{4}$ of 24 (6)
3. $\frac{1}{3}$ of 24 (8)
4. $\frac{1}{9}$ of 72 (8)
5. $\frac{1}{12}$ of 60 (5)
6. $\frac{3}{4}$ of 16 (12)
7. $\frac{2}{3}$ of 9 (6)
8. $\frac{7}{8}$ of 16 (14)
9. $\frac{5}{6}$ of 30 (25)
10. $\frac{3}{5}$ of 25 (15)

Warm Up • Pencil and Paper
Have students find each product.

1. $\frac{4}{5} \times \frac{3}{4}$ $\left(\frac{3}{5}\right)$
2. $\frac{2}{3} \times \frac{7}{8}$ $\left(\frac{7}{12}\right)$
3. $\frac{3}{10} \times 4$ $\left(1\frac{1}{5}\right)$
4. $\frac{1}{8} \times \frac{4}{5}$ $\left(\frac{1}{10}\right)$
5. $\frac{1}{3} \times \frac{5}{6}$ $\left(\frac{5}{18}\right)$
6. $\frac{2}{5} \times \frac{5}{8}$ $\left(\frac{1}{4}\right)$

Name _____

Dividing Fractions

Nick is cutting a board to make bookends. How many pieces $\frac{1}{3}$ of a foot long can he cut from the board?

We want to know how many equal-sized pieces Nick can cut. We know that the length of the board is __6__ feet, and each piece he cuts is __$\frac{1}{3}$__ of a foot long. To find the number of pieces, we divide the length of the board by the length of each piece. We divide __6__ by __$\frac{1}{3}$__.

To divide by a fraction, we multiply the dividend by the reciprocal of the divisor. Remember to factor where possible.

$$6 \div \frac{1}{3} = \frac{6}{1} \times \frac{3}{1} = 18$$

Nick can cut __18__ pieces of wood.

© Pearson Education, Inc./Dale Seymour Publications/Pearson Learning Group. All rights reserved. Copying strictly prohibited.

Getting Started
Divide.

1. $\frac{1}{5} \div \frac{2}{3} = \frac{3}{10}$
2. $9 \div \frac{1}{6} = 54$
3. $7 \div \frac{3}{4} = 9\frac{1}{3}$
4. $\frac{3}{8} \div \frac{1}{4} = 1\frac{1}{2}$
5. $\frac{5}{8} \div \frac{1}{2} = 1\frac{1}{4}$
6. $\frac{7}{12} \div \frac{5}{8} = \frac{14}{15}$
7. $\frac{3}{5} \div \frac{9}{10} = \frac{2}{3}$
8. $\frac{9}{10} \div \frac{3}{5} = 1\frac{1}{2}$
9. $\frac{7}{11} \div \frac{3}{7} = 1\frac{16}{33}$

Lesson 7-5 • Dividing Fractions

---

2 Teach

Introduce the Lesson Have a student read the problem and identify the question being asked. Discuss why division is used to solve this problem. (to find the number of equal-sized groups in a total)

Ask what information is given in the problem and the picture.

Have students complete the information sentences as they work through the model with you to solve the problem. Point out that students use what they already know about reciprocals and multiplication to divide fractions.

Develop Skills and Concepts Review the terms *dividend*, *divisor*, *quotient*, and *reciprocal*. Put $12 \div \frac{1}{4}$ on the board. As you work through the algorithm, stress the rule: to divide a fraction, multiply the dividend by the reciprocal of the divisor.

Remind students that in their work with fractions, they can use multiplication to check division. Review the method they used with whole numbers. (Quotient times the divisor equals the dividend.)

Have a student divide and check

$$18 \div \frac{1}{3} \quad (54)$$

Stress that division with fractions combines and extends skills students already know: writing reciprocals, multiplying fractions, and simplifying fractions.

3 Practice

Have students complete all the exercises. When students have solved the word problems, encourage rereading, estimating, and logical evaluation to test the reasonableness of each answer.

Practice

Divide.

1. $\frac{1}{8} \div \frac{2}{3} = \underline{\frac{3}{16}}$

2. $\frac{5}{8} \div \frac{5}{9} = \underline{1\frac{1}{8}}$

3. $\frac{5}{6} \div \frac{2}{9} = \underline{3\frac{3}{4}}$

4. $25 \div \frac{1}{3} = \underline{75}$

5. $\frac{5}{12} \div \frac{2}{3} = \underline{\frac{5}{8}}$

6. $\frac{3}{7} \div \frac{6}{7} = \underline{\frac{1}{2}}$

7. $8 \div \frac{4}{5} = \underline{10}$

8. $\frac{9}{20} \div \frac{6}{15} = \underline{1\frac{1}{8}}$

9. $\frac{9}{16} \div \frac{3}{4} = \underline{\frac{3}{4}}$

10. $12 \div \frac{8}{9} = \underline{13\frac{1}{2}}$

11. $\frac{5}{6} \div \frac{1}{2} = \underline{1\frac{2}{3}}$

12. $\frac{5}{6} \div \frac{1}{3} = \underline{2\frac{1}{2}}$

13. $\frac{3}{10} \div \frac{11}{15} = \underline{\frac{9}{22}}$

14. $\frac{5}{12} \div \frac{15}{16} = \underline{\frac{4}{9}}$

15. $6 \div \frac{2}{3} = \underline{9}$

16. $\frac{3}{8} \div \frac{5}{6} = \underline{\frac{9}{20}}$

17. $\frac{7}{16} \div \frac{7}{10} = \underline{\frac{5}{8}}$

18. $\frac{4}{5} \div \frac{8}{9} = \underline{\frac{9}{10}}$

19. $\frac{4}{15} \div \frac{3}{10} = \underline{\frac{8}{9}}$

20. $15 \div \frac{5}{8} = \underline{24}$

21. $\frac{4}{9} \div \frac{2}{3} = \underline{\frac{2}{3}}$

22. $\frac{7}{8} \div \frac{7}{8} = \underline{1}$

23. $\frac{5}{12} \div \frac{7}{16} = \underline{\frac{20}{21}}$

24. $\frac{5}{24} \div \frac{15}{16} = \underline{\frac{2}{9}}$

Problem Solving

Solve each problem.

25. Henry has 200 baseball cards. He gave $\frac{1}{5}$ of them to his brother. Then he gave $\frac{1}{4}$ of the remaining cards to his sister. How many cards does Henry have left?
120 cards

26. On a map, each $\frac{3}{8}$ inch represents one mile. If the distance between two towns on the map is $\frac{3}{4}$ inches, how far apart are the towns?
2 miles

27. Nina is putting peanuts in $\frac{3}{4}$-pound boxes. How many boxes will Nina need for 18 pounds of peanuts?
24 boxes

28. One glass of milk fills $\frac{3}{4}$ of a pitcher. How many glasses of milk can 6 pitchers hold?
8 glasses

⎰ Now Try This! ⎱

What fraction does each letter represent?

$A = \underline{\frac{1}{3}}$ $B = \underline{\frac{1}{6}}$ $C = \underline{\frac{1}{18}}$ $D = \underline{\frac{1}{27}}$

The rectangle is 1 unit. Find:

$A + B \ \underline{\frac{1}{2}}$ $A \times B \ \underline{\frac{1}{18}}$ $A \times C \div D \ \underline{\frac{1}{2}}$

$B - C \ \underline{\frac{1}{9}}$ $C \div B \ \underline{\frac{1}{3}}$

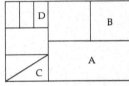

Lesson 7-5 • Dividing Fractions

For Mixed Abilities

Common Errors • Intervention

Some students may multiply using the reciprocal of the dividend instead of the divisor. Before students rewrite the problem as multiplication, have them circle the divisor to remind themselves that this is the number for which they find the reciprocal.

Enrichment • Number Sense

Challenge students to begin at Start of each flowchart and write the correct fraction in the empty box.

1.

$$\rightarrow \boxed{\frac{7}{3}} \times \boxed{\frac{1}{2}} \div \boxed{\frac{1}{2}} \times \boxed{\frac{3}{7}} \rightarrow \boxed{(1)}$$

2.

$$\rightarrow \boxed{\frac{4}{7}} \times \boxed{\frac{3}{4}} \div \boxed{\frac{3}{14}} \div \boxed{\frac{1}{2}} \times \boxed{\left(\frac{1}{4}\right)} \rightarrow \boxed{1}$$

Have students create their own flowcharts for others to solve.

More to Explore • Application

Have students research to find 10 interesting number facts, for example, the year the first automobile was built or the height of the Statue of Liberty. Have them translate these numbers into Roman numerals and list the fact description and number in two separate lists, mixing the order. Then, have them exchange lists with a partner to translate them into the Hindu-Arabic numeral and match the fact to the number. Have them exchange with several partners for more practice.

Now Try This! Remind students that they are comparing parts to a whole. Be sure they read the operation signs carefully.

---

4 Assess

Ask, *If you divide a whole number by a fraction, is the quotient less than, equal to, or greater than the whole number? Write an example.* (The quotient is greater than the whole number. Possible answer: $4 \div \frac{2}{5} = 10$)

T146

7-6 Dividing Mixed Numbers

pages 147-148

1 Getting Started

Objective
- To divide mixed numbers by whole and mixed numbers

Warm Up • Mental Math
Have students give the place value of the 4 in each number.

1. 72,487 (hundreds)
2. 346,802 (ten thousands)
3. 124,600 (thousands)
4. 42,615,708 (ten millions)
5. 136,714 (ones)
6. 39,649 (tens)

Warm Up • Pencil and Paper
Have students rename each mixed number as an improper fraction.

1. $3\frac{1}{4}$ $\left(\frac{13}{4}\right)$
2. $6\frac{3}{5}$ $\left(\frac{33}{5}\right)$
3. $1\frac{3}{5}$ $\left(\frac{8}{5}\right)$
4. $3\frac{2}{3}$ $\left(\frac{11}{3}\right)$
5. $7\frac{1}{2}$ $\left(\frac{15}{2}\right)$
6. $4\frac{1}{10}$ $\left(\frac{41}{10}\right)$

Have students find each quotient.

7. $\frac{5}{6} \div \frac{1}{3}$ $\left(2\frac{1}{2}\right)$
8. $\frac{4}{5} \div \frac{3}{10}$ $\left(2\frac{2}{3}\right)$
9. $\frac{5}{8} \div \frac{1}{6}$ $\left(3\frac{3}{4}\right)$

2 Teach

Introduce the Lesson Ask a student to read the problem and tell what is to be solved. Then, ask what information is given. Have students fill in the information sentences. Work through the model with students, emphasizing the renaming step. Have them complete the solution sentence to solve the problem.

Develop Skills and Concepts Point out that the only additional step in this lesson is to rename the mixed numbers and whole numbers as fractions. Then, the procedure is the same as in the previous division lesson. To provide practice and reinforcement, have students complete the following problems:

1. $3\frac{1}{2} \div \frac{7}{8}$ (4)
2. $1\frac{2}{3} \div 5$ $\left(\frac{1}{3}\right)$
3. $1\frac{1}{4} \div 1\frac{1}{2}$ $\left(\frac{5}{6}\right)$
4. $6\frac{3}{10} \div 9$ $\left(\frac{7}{10}\right)$
5. $2\frac{1}{2} \div 1\frac{2}{3}$ $\left(1\frac{1}{2}\right)$
6. $2\frac{2}{3} \div \frac{5}{9}$ $\left(4\frac{4}{5}\right)$

3 Practice

Have students complete all the exercises.

Name _____

Dividing Mixed Numbers

The Continental Railway crew must repair $13\frac{3}{4}$ miles of track this week. If they are to meet this goal, how many miles of track will the crew have to fix each day?

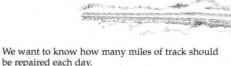

We want to know how many miles of track should be repaired each day.

We know that the Continental crew must repair __$13\frac{3}{4}$__ miles of track in __5__ days.

To find how many miles the crew will have to repair in one day, we divide the miles of track to be repaired by the number of days. We divide __$13\frac{3}{4}$__ by __5__.

Rename the numbers as improper fractions.	Multiply by the reciprocal of the divisor. Factor wherever possible.
$13\frac{3}{4} \div 5$ $\frac{55}{4} \div \frac{5}{1}$	$\frac{55}{4} \times \frac{1}{5} = \frac{11}{4} = 2\frac{3}{4}$

The crew must repair __$2\frac{3}{4}$__ miles of track each day.

Getting Started

Divide.

1. $2\frac{1}{4} \div 1\frac{1}{2} = \underline{1\frac{1}{2}}$
2. $5\frac{1}{3} \div 4 = \underline{1\frac{1}{3}}$
3. $6\frac{1}{4} \div 1\frac{2}{5} = \underline{4\frac{13}{28}}$
4. $6 \div 3\frac{1}{3} = \underline{1\frac{4}{5}}$
5. $\frac{5}{9} \div 1\frac{1}{2} = \underline{\frac{10}{27}}$
6. $1\frac{3}{4} \div 7 = \underline{\frac{1}{4}}$
7. $2\frac{1}{3} \div 2\frac{2}{3} = \underline{\frac{7}{8}}$
8. $1\frac{1}{2} \div 3\frac{2}{3} = \underline{\frac{9}{22}}$
9. $5\frac{3}{7} \div 1\frac{1}{3} = \underline{3\frac{9}{35}}$

Copy and divide.

10. $2\frac{1}{3} \div 1\frac{2}{5} = \underline{1\frac{2}{3}}$
11. $3\frac{3}{8} \div 1\frac{1}{2} = \underline{2\frac{1}{4}}$
12. $5 \div 1\frac{3}{7} = \underline{3\frac{1}{2}}$
13. $2\frac{1}{4} \div 3\frac{3}{8} = \underline{\frac{2}{3}}$

Lesson 7-6 • Dividing Mixed Numbers

147

Practice

Divide.

1. $1\frac{3}{8} \div 11 = \underline{\frac{1}{8}}$

2. $6 \div 3\frac{3}{5} = \underline{1\frac{2}{3}}$

3. $1\frac{1}{7} \div \frac{6}{7} = \underline{1\frac{1}{3}}$

4. $4\frac{1}{2} \div 1\frac{2}{3} = \underline{2\frac{7}{10}}$

5. $4\frac{1}{8} \div 1\frac{4}{7} = \underline{2\frac{5}{8}}$

6. $2\frac{3}{4} \div 2\frac{4}{9} = \underline{1\frac{1}{8}}$

7. $28 \div 3\frac{1}{2} = \underline{8}$

8. $5\frac{1}{3} \div 1\frac{7}{9} = \underline{3}$

9. $2\frac{2}{5} \div 1\frac{3}{5} = \underline{1\frac{11}{25}}$

10. $3\frac{1}{5} \div 3\frac{1}{10} = \underline{1\frac{1}{31}}$

11. $3\frac{2}{3} \div 2\frac{1}{6} = \underline{1\frac{9}{13}}$

12. $4\frac{1}{5} \div 1\frac{8}{13} = \underline{2\frac{3}{5}}$

Copy and divide.

13. $5\frac{7}{9} \div 1\frac{1}{3} \quad \underline{4\frac{1}{3}}$

14. $4\frac{2}{3} \div 1\frac{1}{15} \quad \underline{4\frac{3}{8}}$

15. $2 \div 1\frac{1}{7} \quad \underline{1\frac{3}{4}}$

16. $8\frac{1}{4} \div 3\frac{2}{3} \quad \underline{2\frac{1}{4}}$

17. $5\frac{3}{5} \div 14 \quad \underline{\frac{2}{5}}$

18. $1\frac{1}{8} \div 5\frac{1}{4} = \underline{\frac{3}{14}}$

19. $2\frac{7}{16} \div 1\frac{5}{8} \quad \underline{1\frac{1}{2}}$

20. $1\frac{1}{2} \div 2\frac{1}{3} \quad \underline{\frac{9}{14}}$

21. $7\frac{1}{2} \div 4\frac{1}{5} = \underline{1\frac{11}{14}}$

Problem Solving

Solve each problem.

22. Ben is filling a pail that holds $4\frac{1}{2}$ quarts. He is using a dipper that holds $\frac{3}{4}$ of a quart. How many times will Ben fill the dipper?
6 times

23. Doris ran $3\frac{1}{4}$ miles, $1\frac{3}{4}$ miles, and $2\frac{1}{2}$ miles on three different days. What is her average daily mileage?
$2\frac{1}{2}$ miles

24. An ant takes $1\frac{1}{4}$ hours to crawl one block. How many blocks can the ant crawl in 10 hours?
8 blocks

25. One lap of the Indianapolis Speedway Track is $2\frac{1}{2}$ miles. 125 miles is $\frac{1}{4}$ of the race. How many laps around the track is the race? 200 laps

[Now Try This!]

It's Algebra!

Circle the correct value for each letter to make the equation true.

$A = 1$

$B = \left(\frac{1}{3}\right) \text{ or } \frac{2}{3}$

$C = 2 \text{ or} \left(\frac{1}{2}\right)$

$D = \left(\frac{2}{3}\right) \text{ or } \frac{3}{2}$

$B + D - A = C \times 2 - A$

$E = 1$

$F = \left(\frac{3}{4}\right) \text{ or } \frac{4}{3}$

$G = 3 \text{ or} \left(2\right)$

$H = \left(\frac{1}{4}\right) \text{ or } \frac{1}{2}$

$(E + F) \times G \times \frac{1}{7} = F - H$

148

Lesson 7-6 • Dividing Mixed Numbers

For Mixed Abilities

Common Errors • Intervention

When the divisor is a mixed number, some students may write it as an improper fraction but then forget to write the reciprocal. Have these students work in cooperative-learning groups of three to solve problems, each member of the group having a different role. The first member writes the mixed number as an improper fraction, the second writes the reciprocal of the divisor, and the third finds the quotient. Members can change roles with each new problem.

Enrichment • Algebra

Challenge students to divide across the chart to name A and B, and to divide down to name C and D. (If they correctly name A, B, C, and D, then $A \div B = C \div D$.)

$3\frac{3}{4}$	$\frac{3}{4}$	$A = (5)$
$4\frac{1}{2}$	$2\frac{1}{4}$	$B = (2)$
$C = \left(\frac{5}{6}\right)$	$D = \left(\frac{1}{3}\right)$	$\left(2\frac{1}{2}\right)$

More to Explore • Probability

Explain to the class that for a genetic trait to be passed on, an organism must receive one gene for the trait from each of its parents. For example, the pea plant can carry one gene for tallness and one gene for shortness. One of the two genes will be passed on to each of its offspring.

Have students determine the chances of getting one gene or another by flipping a coin. Tell them to flip the coin 100 times. Let heads stand for the gene for shortness, and tails for the gene for tallness. Tell them to record the results on a chart. What are the chances of an offspring pea plant getting each of the genes?

Now Try This! Have students look at the first equation. Point out that $-A$ appears on both sides of the equal sign. We can rewrite the equation as $B + D = C \cdot 2$. Students can use guess and check to determine the values of B, C, and D that make the equation true. Students work with the second equation in a similar way.

4 Assess

Have students write a question that describes each division.

$\frac{3}{2} \div \frac{1}{2}$ (Possible answer: How many halves in $\frac{3}{2}$?)

$\frac{3}{2} \div 2$ (Possible answer: When you divide $\frac{3}{2}$ by 2, how large is each piece?)

Then, have students find each quotient. $(3; \frac{3}{4})$

T148

7-7 Estimating Products and Quotients

pages 149–150

1 Getting Started

Objective
- To use rounding or a compatible number to estimate products and quotients of mixed numbers

Vocabulary
compatible number

Materials
fraction number lines

Warm Up • Mental Math
Have students find each product or quotient.

1. 8×6 (48)
2. $72 \div 9$ (8)
3. $54 \div 6$ (9)
4. 7×8 (56)
5. $42 \div 6$ (7)
6. 6×5 (30)

Warm Up • Pencil and Paper
Have students find each product or quotient.

1. $\frac{1}{2} \times \frac{3}{5}$ $\left(\frac{3}{10}\right)$
2. $4\frac{1}{3} \times 3$ (13)
3. $1\frac{1}{2} \times 2\frac{3}{4}$ $\left(\frac{33}{8} \text{ or } 4\frac{1}{8}\right)$
4. $\frac{2}{3} \div \frac{1}{2}$ $\left(\frac{4}{3} \text{ or } 1\frac{1}{3}\right)$
5. $\frac{7}{8} \div \frac{3}{4}$ $\left(\frac{7}{6} \text{ or } 1\frac{1}{6}\right)$

Name _____

Estimating Products and Quotients

Some sixth-grade students are planning a pancake breakfast for the class. They plan to make $4\frac{1}{2}$ times more pancakes than the recipe yields. About how many cups of milk will they need?

We want to estimate the product $4\frac{1}{2} \times 1\frac{2}{3}$.

Use rounding.

Round each factor to the nearest whole number.

$4\frac{1}{2}$ rounds to __5__. $1\frac{2}{3}$ rounds to __2__.

Then, multiply. $5 \times \underline{2} = \underline{10}$

The sixth-grade students will need about __10__ cups of milk.

♥ PANCAKE BATTER ♥
$3\frac{1}{4}$ c pancake mix
$1\frac{2}{3}$ c milk
3^3 eggs
Makes 18 large pancakes

Use compatible numbers.

A **compatible number** is a number that is easy to use.

We will replace $1\frac{2}{3}$ with a number that makes multiplying by $4\frac{1}{2}$ easy. The number should be close to $1\frac{2}{3}$.

$4\frac{1}{2} \times 2$ is easy to multiply. __2__ is a compatible number.

We multiply $4\frac{1}{2}$ by __2__. $4\frac{1}{2} \times \underline{2} = \underline{9}$

The sixth-grade students will need about __9__ cups of milk.

We can also use a compatible number to estimate a quotient. Estimate $5 \div \frac{3}{4}$.

First, find the reciprocal of $\frac{3}{4}$. $\frac{4}{3}$

Look for a number that is close to 5 and makes multiplying by $\frac{4}{3}$ easy.

$6 \times \frac{4}{3}$ is easy to multiply. Replace 5 with 6.

$6 \div \frac{3}{4} = 6 \times \frac{4}{3} = \underline{8}$.

$5 \div \frac{3}{4}$ is close to __8__.

Getting Started

Estimate each product or quotient. Use rounding. Then use a compatible number.

1. $2\frac{1}{2} \times 3\frac{1}{3}$
   $3 \times 3 = 9$
   $\frac{3}{1} \times \frac{10}{3} = 10$

2. $4 \div 2\frac{1}{4}$
   $4 \div 2 = 2$
   $\frac{3}{1} \times \frac{4}{9} = 1\frac{1}{3}$

3. $4\frac{3}{8} \times 5\frac{7}{10}$
   $4 \times 6 = 24$
   $\frac{5}{1} \times \frac{57}{10} = 28\frac{1}{2}$

4. $8\frac{3}{4} \div 3\frac{1}{3}$
   $9 \div 3 = 3$
   $\frac{10}{1} \times \frac{3}{10} = 3$

Lesson 7-7 • Estimating Products and Quotients

---

2 Teach

Introduce the Lesson Read the problem aloud and ask a student to read and complete the information sentences. Explain that one way to estimate is to round each mixed number, and then compute. Work out the problem on the board while students follow in their texts.

Tell students that another way to estimate is to use a *compatible number*. This means changing one of the numbers to a number that is compatible with the denominator of the other fraction. Explain that if $1\frac{2}{3}$ is replaced by 2, they can multiply $4\frac{1}{2} \times 2$ easily. On the board, have a student show the steps to multiply $4\frac{1}{2} \times 2$.

Show students how to estimate the quotient $9 \div \frac{5}{2}$ using a compatible number. Remind students to change the quotient to a product before finding a compatible number. $\left(9 \div \frac{5}{2} = 9 \times \frac{2}{5}; 10 \div \frac{5}{2} = 4\right)$

Develop Skills and Concepts Have students round mixed numbers such as $1\frac{3}{5}$ (2), $12\frac{3}{10}$ (12), $4\frac{5}{8}$ (5), and $10\frac{1}{3}$ (10). They may use fraction number lines if they wish.

Provide several examples of estimating using compatible numbers. When one of the numbers being multiplied or divided is a whole number, this is the number that gets changed to a compatible number. For example, to estimate the product $15 \times \frac{3}{8}$, change 15 to a number that is compatible with 8 (16). So, $15 \times \frac{3}{8}$ is about $16 \times \frac{3}{8} = 6$.

When both numbers are mixed numbers, students need to decide which one to change to a compatible whole number. To multiply $5\frac{1}{4} \times 2\frac{1}{3}$, students will either change $5\frac{1}{4}$ to a number compatible with 3, or change $2\frac{1}{3}$ to a number compatible with 4. Show students how each estimate can be done.

T149

Practice

Estimate each product or quotient. Use rounding.

1. $5\frac{1}{2} \times 4\frac{2}{3}$
   $6 \times 5;\ 30$

2. $3\frac{4}{5} \times 1\frac{6}{7}$
   $4 \times 2;\ 8$

3. $8\frac{1}{2} \times 6\frac{4}{5}$
   $9 \times 7;\ 63$

4. $2\frac{2}{3} \times 5\frac{1}{6}$
   $3 \times 5;\ 15$

5. $14 \div 2\frac{1}{3}$
   $14 \div 2;\ 7$

6. $12 \div 3\frac{1}{6}$
   $12 \div 3;\ 4$

7. $8 \div 3\frac{4}{5}$
   $8 \div 4;\ 2$

8. $20 \div 4\frac{5}{6}$
   $20 \div 5;\ 4$

9. $9\frac{1}{2} \times 3\frac{2}{3}$
   $10 \times 4;\ 40$

10. $15 \div 2\frac{3}{5}$
    $15 \div 3;\ 5$

11. $6\frac{4}{5} \times 2\frac{3}{5}$
    $7 \times 3;\ 21$

12. $12 \div 6\frac{1}{5}$
    $12 \div 6;\ 2$

Estimate each product or quotient. Use a compatible number.

13. $8\frac{1}{4} \times 2\frac{2}{3}$
    $9 \times 2\frac{2}{3};\ 24$

14. $1\frac{2}{5} \times 9\frac{4}{5}$
    $1\frac{2}{5} \times 10;\ 14$

15. $2\frac{1}{2} \times 5\frac{1}{3}$
    $2\frac{1}{2} \times 6;\ 15$

16. $3\frac{1}{7} \times 6\frac{1}{3}$
    $3\frac{1}{7} \times 7;\ 22$

17. $14 \div \frac{5}{8}$
    $15 \div \frac{5}{8};\ 24$

18. $13 \div \frac{4}{7}$
    $12 \div \frac{4}{7};\ 21$

19. $8 \div \frac{3}{8}$
    $9 \div \frac{3}{8};\ 24$

20. $6 \div \frac{7}{11}$
    $7 \div \frac{7}{11};\ 11$

Copy. Then estimate each product or quotient. Show which method you used. Possible answers are given.

21. $2\frac{2}{3} \times 4\frac{1}{2}$
    Round $3 \times 5 = 15$

22. $6 \div 2\frac{5}{6}$
    Round $6 \div 3 = 2$

23. $3\frac{2}{5} \times 4\frac{3}{8}$
    Round $3 \times 4 = 12$

24. $15 \div 2\frac{5}{6}$
    Round $15 \div 3 = 5$

25. $13 \div \frac{3}{4}$
    Compatible numbers
    $12 \times \frac{4}{3} = 16$

26. $7\frac{1}{3} \times 2\frac{5}{6}$
    Compatible numbers
    $7\frac{1}{3} \times 3 = 22$

27. $3\frac{1}{4} \times 3\frac{4}{5}$
    Compatible numbers
    $3\frac{1}{4} \times 4 = 13$

28. $11 \div \frac{1}{4}$
    Compatible numbers
    $12 \div \frac{4}{1} = 48$

Problem Solving

Solve each problem.

29. Mr. Rose is installing a fence. His property is $12\frac{5}{6}$ feet by $20\frac{1}{6}$ feet. About how much area is his fenced-in property?

    about 260 square feet

30. Jessica is making a sugar cookie recipe for school. She needs to cut the recipe in half. The recipe calls for $4\frac{1}{4}$ cups of sugar, $1\frac{3}{4}$ cups of flour, and $\frac{3}{4}$ cup of butter. About how much sugar, flour, and butter will Jessica need?

    about 2 c sugar, 1 c flour, $\frac{1}{2}$ c butter

Lesson 7-7 • Estimating Products and Quotients

---

3 Practice

Have students complete all the exercises. Remind students of the two methods for estimating a product or quotient, rounding or using a compatible number.

4 Assess

Ask, *What compatible number would you use to estimate* $15 \div \frac{4}{5}$? (Possible answer: Use $16 \div \frac{4}{5}$ since $16 \div \frac{4}{5} = 16 \times \frac{5}{4}$, and 16 is compatible with 4.)

To estimate $9\frac{1}{4} \times \frac{2}{3}$, *would you use* $9 \times 2\frac{1}{3}$ *or* $9\frac{1}{4} \times 2$? (Possible answer: Use $9 \times 2\frac{1}{3}$ since 9 is compatible with 3.)

For Mixed Abilities

Common Errors • Intervention

Some students may forget to use a compatible number with the numerator of a mixed number or fraction when estimating a quotient. Have students rewrite the quotient as a product before finding a compatible number. For example, $17 \div 2\frac{1}{4} = 17 \div \frac{9}{4} = 17 \times \frac{4}{9}$. Now, have students highlight the denominator 9 and find a number close to 17 that is compatible to 9. (18)

Enrichment • Application

Have students make up scenarios about a pie bakery.

The bakery makes _____ (whole number) pies each day. Each pie requires _____ (mixed number) cups flour and _____ (mixed number) cups sugar.

About how much flour does the bakery use each week? About how much sugar does the bakery use each week? Have students exchange problems and solve.

More to Explore • Number Sense

Have students work in pairs. Each pair chooses 4 different digits from the list 0, 1, 2, ... 9. Have them write the largest number possible using the digits they chose. Next, have them write the smallest number possible. Then, have them subtract the smaller number from the larger.

Now, have them take the digits in the difference, write the largest number and smallest number possible, then subtract.

Have them continue with the digits of the difference until a "magic number" is discovered.

Remind them to check each subtraction by adding, and to make sure they are finding the largest and smallest numbers possible with each set of 4 digits.

T150

1 Getting Started

Objective
• To review addition, subtraction, multiplication, and division of fractions

Warm Up • Mental Math
Have students perform each operation from left to right.

1. $5 \times 4 + 12 \div 8$ (4)
2. $17 + 23 \div 5 \times 9$ (72)
3. $50 \div 10 \times 9 + 8$ (53)
4. $63 - 13 \div 5 \times 100$ (1,000)
5. $48 \div 6 + 4 \times 4$ (48)
6. $9 \times 6 - 14 \div 4$ (10)
7. $12 - 5 \times 7 + 6$ (55)
8. $88 - 8 \div 10 \times 100$ (800)

Warm Up • Pencil and Paper
Have students simplify each expression.

1. $4\frac{2}{5} + 3\frac{1}{3}$ $(7\frac{11}{15})$
2. $7\frac{3}{8} - 4\frac{2}{3}$ $(2\frac{17}{24})$
3. $3\frac{2}{3} \times 2\frac{1}{2}$ $(9\frac{1}{6})$
4. $6\frac{3}{4} \div 2\frac{1}{8}$ $(3\frac{3}{17})$
5. $(\frac{3}{4} + \frac{5}{6}) \times 2$ $(3\frac{1}{6})$
6. $(\frac{7}{8} - \frac{5}{6}) \div \frac{1}{12}$ $(\frac{1}{2})$

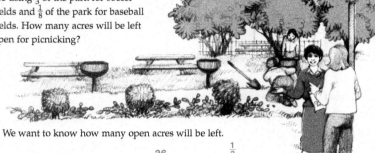

Name _____

Lesson 7-8

Practice With Fractions

The city is designing athletic fields for Highland Park. The city planners are using $\frac{1}{3}$ of the park for soccer fields and $\frac{1}{8}$ of the park for baseball fields. How many acres will be left open for picnicking?

Highland Park
36 Acres

We want to know how many open acres will be left.

We know Highland Park contains ___36___ acres, and ___$\frac{1}{3}$___ of it will be used for soccer and ___$\frac{1}{8}$___ of it for baseball.

To find how many acres will be left open, we must add to find what part will be used for the two sports and subtract that fraction from one which stands for the whole park. Then, we multiply to find the number of acres left for picnicking.

Find the total part used for fields.	Subtract from 1 which stands for the whole park.	Find $\frac{13}{24}$ of 36.
$\frac{1}{3} = \frac{8}{24}$ $+ \frac{1}{8} = \frac{3}{24}$ $\frac{11}{24}$	$1 = \frac{24}{24}$ $- \frac{11}{24} = \frac{11}{24}$ $\frac{13}{24}$	$\frac{13}{24} \times 36 = \underline{19\frac{1}{2}}$

There are ___$19\frac{1}{2}$___ acres left for picnicking.

Getting Started

Compute each problem. Simplify answers if necessary.

1. $5\frac{1}{3} + 4\frac{1}{2} = \underline{9\frac{5}{6}}$
2. $5\frac{1}{3} \times 4\frac{1}{2} = \underline{24}$
3. $5\frac{1}{3} - 4\frac{1}{2} = \underline{\frac{5}{6}}$
4. $5\frac{1}{3} \div 4\frac{1}{2} = \underline{1\frac{5}{27}}$

Copy and compute.

5. $43 - 6\frac{7}{9}$ $\underline{36\frac{2}{9}}$
6. $3\frac{1}{5} \times \frac{3}{8} \times \frac{13}{7}$ $\underline{2\frac{8}{35}}$
7. $3\frac{3}{4} \div 3\frac{1}{8}$ $\underline{1\frac{1}{5}}$
8. $5\frac{1}{2} + 6\frac{2}{3} + 7\frac{3}{4}$ $\underline{19\frac{11}{12}}$

Lesson 7-8 • Practice With Fractions

151

2 Teach

Introduce the Lesson Have a student read the problem and tell what is to be solved. Ask what information is given and have students fill in the information sentences. Work through the model with students and then have them complete the solution sentence. Point out that this is a multistep problem. Be sure students understand the computation for each step.

Develop Skills and Concepts Review the kinds of questions that can be answered by each of the operations.

• *addition*, how many in all when groups are unequal sizes

• *subtraction*, how much is left, or how much greater one value is compared with another

• *multiplication*, how many times greater one value is than another, or what part of a value another value is, or how many in all when groups are the same size

• *division*, how many in each equal-sized group or how many equal-sized groups

Provide practice in adding, subtracting, multiplying, and dividing fractions and mixed numbers by having students complete the following problems:

1. $\frac{1}{3} + \frac{2}{5}$ $(\frac{11}{15})$
2. $1\frac{3}{4} + 1\frac{1}{8}$ $(2\frac{7}{8})$
3. $\frac{7}{8} - \frac{2}{3}$ $(\frac{5}{24})$
4. $7\frac{9}{10} - 4\frac{2}{5}$ $(3\frac{1}{2})$
5. $\frac{7}{12} \times \frac{2}{3}$ $(\frac{7}{18})$
6. $5\frac{1}{2} \div 3\frac{2}{3}$ $(1\frac{1}{2})$

3 Practice

Have students complete all the exercises. Students should scan the page to see if they have any questions about the types of problems presented.

Practice

Compute each problem. Simplify answers if necessary.

1. $2\frac{3}{8} + 8\frac{1}{4} = \underline{10\frac{5}{8}}$

2. $5\frac{1}{3} + 2\frac{1}{4} = \underline{7\frac{7}{12}}$

3. $6\frac{1}{2} \times 5\frac{1}{2} = \underline{35\frac{3}{4}}$

4. $7\frac{1}{3} - 2\frac{1}{2} = \underline{4\frac{5}{6}}$

5. $9\frac{1}{4} + 7\frac{2}{3} = \underline{16\frac{11}{12}}$

6. $8\frac{1}{3} \times 4\frac{1}{5} = \underline{35}$

7. $6\frac{3}{8} - 4\frac{1}{7} = \underline{2\frac{13}{56}}$

8. $3\frac{1}{5} + 2\frac{1}{12} = \underline{5\frac{17}{60}}$

9. $4\frac{2}{5} \times \frac{10}{11} = \underline{4}$

10. $8\frac{1}{9} + 16\frac{3}{4} = \underline{24\frac{31}{36}}$

11. $8\frac{2}{3} \div 6 = \underline{1\frac{4}{9}}$

12. $14 - 3\frac{2}{7} = \underline{10\frac{5}{7}}$

13. $12\frac{1}{4} \times \frac{16}{21} = \underline{9\frac{1}{3}}$

14. $10\frac{1}{9} - 4\frac{3}{5} = \underline{5\frac{23}{45}}$

15. $8\frac{2}{5} \div 6 = \underline{1\frac{2}{5}}$

Copy and add, subtract, multiply, or divide.

16. $8\frac{1}{3} + 3\frac{2}{3}$ 12

17. $3\frac{1}{3} \div 6\frac{2}{9}$ $\frac{15}{28}$

18. $6\frac{1}{8} + 7\frac{5}{6}$ $13\frac{23}{24}$

19. $4\frac{2}{7} \times 1\frac{4}{10}$ 6

20. $1 \div \frac{2}{3}$ $1\frac{1}{2}$

21. $15\frac{1}{4} - 7\frac{5}{6}$ $7\frac{5}{12}$

22. $7\frac{3}{10} + 6\frac{9}{15}$ $13\frac{9}{10}$

23. $16 - 4\frac{1}{8}$ $11\frac{7}{8}$

24. $6\frac{2}{3} \div 1\frac{1}{9}$ 6

25. $\frac{4}{5} \times \frac{7}{8} \times \frac{6}{7}$ $\frac{3}{5}$

26. $2\frac{1}{2} + 5\frac{2}{3} + 6\frac{7}{12}$ $14\frac{3}{4}$

27. $2\frac{1}{2} \times 19\frac{5}{9} \times \frac{2}{5}$ $19\frac{5}{9}$

28. $3\frac{2}{3} + 5\frac{5}{8} + 2\frac{1}{2}$ $11\frac{19}{24}$

29. $7\frac{1}{2} \times 2\frac{2}{35} \times 1\frac{1}{2}$ $23\frac{1}{7}$

30. $15 \div 3\frac{4}{7}$ $4\frac{1}{5}$

Problem Solving

Solve each problem.

31. Mr. Alison is on a diet of 1,500 calories a day. For lunch, he has $\frac{2}{5}$ of his calories. How many calories can Mr. Alison have for his other meals?
900 calories

32. The library is $\frac{4}{5}$ of a mile from Rick's house. The library is $\frac{2}{3}$ of a mile from Mary's house. How much farther is it from Rick's to the library than from Mary's to the library?
$\frac{2}{15}$ of a mile

33. Hugh had $3\frac{1}{8}$ quarts of paint. He used $\frac{4}{5}$ of that paint to paint a bookcase. The amount he had left was $\frac{3}{4}$ of what he needed to paint some shelves. How much paint did Hugh need? $\frac{5}{6}$ quart

Lesson 7-8 • Practice With Fractions

4 Assess

Have students fill in the blanks to complete this statement.

Before you _____ or _____ mixed numbers, you need to change the mixed numbers to improper fractions.
(multiply; divide)

For Mixed Abilities

Common Errors • Intervention

When students work on a set of mixed problems, they may confuse the procedures for the four operations. Have them work with partners and discuss the proper procedure for each problem prior to solving the problem.

Enrichment • Number Sense

Direct students to write as many problems as possible resulting in the answer 3, using one of the basic operations and any two of the following numbers: $\frac{1}{2}$, $1\frac{1}{2}$, 2, $2\frac{1}{2}$, $4\frac{1}{2}$, 6

(Possible answers: $2\frac{1}{2} + \frac{1}{2}$; $4\frac{1}{2} - 1\frac{1}{2}$; $1\frac{1}{2} \div \frac{1}{2}$; $1\frac{1}{2} \times 2$)

More to Explore • Sets

The Venn diagram represents the number of nations participating in three economic organizations.

- Set U = All nations in the world

- Set A = nations in the International Financial Alliance (IFA)

- Set B = nations in the United Monetary Federation (UMF)

- Set C = nations in the Mutual Economic Compact (MEC)

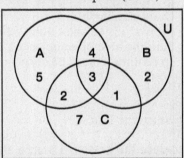

Use the numbers in the Venn diagram to answer each question.

1. How many nations participate in all 3 organizations? (3)

2. How many are in both the IFA and the MEC? (5)

3. How many are in both the IFA and the UMF? (7)

4. How many more nations are in the IFA than in the UMF? (4)

5. What is the total number of countries participating in one or more of these economic organizations? (24)

T152

pages 153–154

1 Getting Started

Objective
- To solve problems by solving a simpler but related problem

Warm Up • Mental Math
Have students perform each operation from left to right.

1. $\frac{1}{3}$ of $75 - 1 \div 12 \times 5 - 1 + 3$ (12)
2. $\frac{1}{2}$ of $56 \div 4 \times 4 - 3 \div 5$ (5)
3. $\frac{1}{5}$ of $20 + 6 - 2 \times$ itself $+ 6 \div 5$ (14)
4. $\frac{1}{4}$ of $40 \times 3 - 4 \div 2 + 3 \div 4$ (4)
5. $\frac{2}{3}$ of $9 \times$ itself $- 4 \div 8 + 2$ (6)
6. $\frac{3}{4}$ of $12 + 1 \times 3 + 3 \div 3 + 1 \div 12$ (1)
7. $\frac{3}{5}$ of $25 \div 3 \times 9 + 3 \div 6 - 8$ (0)

Warm Up • Pencil and Paper
Have students complete the following:

1. How many possible diagonals are there in a pentagon? (5)
2. How many 3s are used to count to 35? (11)
3. Draw a circle with 3 different chords. (Check students' work.)
4. Find the area of an object 62 by 46 centimeters. (2,852 sq cm)

2 Teach

Introduce the Lesson Direct a student to read the problem aloud. Remind students that the solution for this problem can be found by making an organized list or table. When the data require a large number of events, as in this sample, try to simplify the problem and look for a pattern to predict a reasonable solution. For example, in taking one additional flavor at a time, we see that the number of possible events increases as the square of the number of flavors.

Develop Skills and Concepts Once students see this pattern, they can stop listing events and predict that with 20 flavors there will be 20 times 20, or 400 events. One way to check this is by completing the entire list, but students will soon find this tedious and see the wisdom of working the simpler related problem.

T153

Problem Solving:
Solve a Simpler Problem

The Big Dipper Ice Cream Store sells 20 different flavors of ice cream. How many different two-scoop ice cream cones can be made? A cone with vanilla on the bottom and chocolate on the top is different from a cone with chocolate on the bottom and vanilla on the top.

⭐ **SEE**

We want to know how many different combinations of cones with two scoops can be made.

There are __20__ different flavors we can use.

⭐ **PLAN**

We can solve a simpler problem using fewer flavors. By recording our results in a table, we can find a pattern that will help us to solve the problem.

⭐ **DO**

Number of Flavors	Combinations	Number of Combinations
1 vanilla	$\frac{v}{v}$	1 (1 × 1)
2 vanilla chocolate	$\frac{v}{c}\ \frac{c}{v}\ \frac{c}{c}\ \frac{v}{v}$	4 (2 × 2)
3 vanilla chocolate strawberry	$\frac{v}{c}\ \frac{c}{v}\ \frac{v}{s}\ \frac{s}{v}\ \frac{c}{s}\ \frac{s}{c}\ \frac{v}{v}\ \frac{c}{c}\ \frac{s}{s}$	9 (3 × 3)
4 vanilla chocolate strawberry chocolate chip	$\frac{v}{c}\ \frac{c}{v}\ \frac{v}{s}\ \frac{s}{v}\ \frac{v}{cc}\ \frac{cc}{v}\ \frac{c}{s}\ \frac{s}{c}\ \frac{c}{cc}\ \frac{cc}{c}\ \frac{s}{cc}\ \frac{cc}{s}\ \frac{v}{v}\ \frac{c}{c}\ \frac{s}{s}\ \frac{cc}{cc}$	16 (_4_ × _4_)

Using 20 flavors, we can make __400__ different two-scoop ice cream cones.

⭐ **CHECK**

We can check by actually listing and counting the possible combinations of 20 different flavors of ice cream.

Lesson 7-9 • Problem Solving: Solve a Simpler Problem

153

---

3 Apply

Solution Notes

1. Have students count each successive rectangle, keeping a running total until they reach 45.
2. Begin with the greatest number of regions drawing one chord until the pattern is obvious.
3. Break the figure into smaller rectangles, find the area of each, and add.
4. Begin with the top layer, count the number of squares. Continue adding layers and counting until the pattern appears.
5. Count the number of diagonals in a triangle. Continue with each successive polygon and keep a table. Students can continue the pattern until reaching a dodecagon without drawing each figure.
6. A simpler problem would be to list the numbers that have a 5 in the ones place and in the tens place.

Apply

Solve a simpler but related problem to help solve each problem.

1. How many rectangles are found in this diagram?

   45 rectangles

   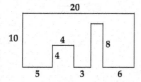

2. What is the greatest number of regions you can make by drawing 6 chords through a circle?

   22 regions

3. Find the area of this figure:

   168 square units

4. How many different squares are in this diagram?

   22 squares

5. How many diagonals can be drawn in a dodecagon?

   54 diagonals

6. Sally wrote the numbers from 1 to 100. How many times did she write the numeral 5?

   20 times

7. Everyone at the Student Council meeting shook hands with everyone else exactly once. If there were 105 handshakes exchanged, how many people were at the meeting?

   15 people

Problem-Solving Strategy: Using the Four-Step Plan

★ **SEE** What do you need to find?

★ **PLAN** What do you need to do?

★ **DO** Follow the plan.

★ **CHECK** Does your answer make sense?

---

8. All 48 students in the sixth-grade class will participate in an elimination checkers tournament. Each student will play a game. The loser is eliminated and the winner will continue to the next round. How many checker games will need to be played in order to determine a champion?

   47 games

9. What if one more rectangle were added to the diagram in Problem 1? How many rectangles would there be in the diagram then?

   55 rectangles

10. Look at the geometric figure shown in Problem 3. How could you change the figure so that the area is 184 square units?

    See Solution Notes.

11. Silly Willy said that the product of two numbers is always greater than either of the two numbers. Show at least one example that proves Silly Willy is incorrect.

    Answers will vary.

12. The sum of the two whole numbers 2 and 2 is the same as their product. Why is it not possible to find two fractions between 0 and 1 whose sum is the same as their product?

    See Solution Notes.

Lesson 7-9 • Problem Solving: Solve a Simpler Problem

For Mixed Abilities

More to Explore • Probability

Computers employ a *two-state system* because electrical signals are either on or off. Mazes are another example of a two-state system. At each point in the maze, a decision must be made about whether to go one way or the other. Over the course of a maze there are many possible routes. Ask each student to draw a maze, making it as complex as possible. Have them exchange mazes and try to work through the maze.

---

7. Begin with 2 people, one handshake, and continue until 105 handshakes are reached.

8. The simplest problem would be a tournament with 2 players, one game. Students will see multiples of four must be used. The pattern emerges that the number of games is one less than the number of players.

Higher-Order Thinking Skills

9. **Analysis:** Adding one rectangle to the diagram will add 10 more possible rectangles to the list.

10. **Analysis:** This calls for "adding" 16 more square units to the figure. Answers will vary, but a way would be to eliminate either the 4 by 4 opening or the 8 by 2 opening at the bottom of the figure.

11. **Analysis:** A sample answer is $\frac{1}{2} \times \frac{1}{4} = \frac{1}{8}$, where $\frac{1}{8}$ is less than both $\frac{1}{2}$ and $\frac{1}{4}$. (Any two proper fractions would be valid.)

12. **Analysis:** The product of any two fractions between 0 and 1 is always less than the smaller fraction, and the sum is always greater than the greater fraction.

4 Assess

Ask, *How many different two-scoop ice cream cones can be made using 25 flavors?* (625)

1 Getting Started

Objective
- To use the calculator to complete arrow diagrams of functions, inverse functions, and composite functions

Materials
calculators

Warm Up • Mental Math
Have students perform each operation from left to right.

1. $4 \times 2 + 6$ (14)
2. $12 \div 3 \times 2$ (8)
3. $12 \times 9 \div 4$ (27)
4. $34 + 10 + 10 - 4$ (50)
5. $7 \times 5 + 8$ (43)
6. $12 + 6 + 4$ (22)
7. $45 \div 15 + 20$ (23)
8. $54 \div 6 \times 9$ (81)
9. $8 \times 6 + 6$ (54)
10. $75 - 25 + 25$ (75)

Warm Up • Pencil and Paper
Have students use the numbers 1, 5, 6, and 12 with any operation signs to make a sentence whose answer is 1, 2, or 3. (Possible answers: $(1 + 5 + 6) \div 12 = 1$; $12 - 6 - 5 + 1 = 2$; $(6 + 12) \div (1 + 5) = 3$)

Name _____

Lesson 7-10

It's Algebra!

Calculator: Functions

A function can be used to find input or output numbers. A **function** is a rule. We write a function in a circle and use arrows to show the direction of the function.

Input	Function	Output		Input	Function	Output
6	+3	9		225	÷5	45

Sometimes the output and the function are known. We can use the inverse function to find the input. Notice the change in the arrow's direction.

Input	Function	Output		Output	Inverse Function	Input
?	×8	96		12	÷8	96

Multiplication and division are inverse operations.

Two functions that are used together are called **composite functions**.

Input 1	Function 1	Output 1 Input 2	Function 2	Output 2
16	×3	48	÷4	12

We can find a single function for these two functions.

16 — $\times\frac{3}{4}$ — 12

Use a calculator to complete each arrow diagram. Then, write a single function to combine the two functions.

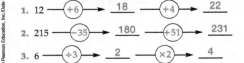

1. 12 —(+6)→ 18 —(+4)→ 22   12 —(+10)→ 22
2. 215 —(−35)→ 180 —(+51)→ 231   215 —(+16)→ 231
3. 6 —(÷3)→ 2 —(×2)→ 4   6 —($\times\frac{2}{3}$)→ 4

Lesson 7-10 • Calculator: Functions

155

---

2 Teach

Introduce the Lesson Draw a function machine on the board like the one below.

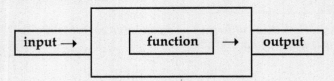

input → | function | → output

Explain that the input is the number put into the machine, the function is an operation (+, −, × or ÷) on the number, and the output is the result of the operation. Show how to find the output if the input is 6 and the function is +8. (14) Then, show how to find the input if the function is +8 and the output is 14.

Read through the models with students. Explain that a calculator can also perform composite functions such as + and +, × and +, or − and −.

Develop Skills and Concepts Discuss the models on page 155. Point out that inverse operations can be used to check operations. Have students provide examples to show how addition checks subtraction and multiplication checks division. Call on students to complete the following arrow diagrams at the board.

$13 \rightarrow - 6 \rightarrow$ (7)

(7) $\rightarrow + 6 \rightarrow 13$

$3 \rightarrow + 4 \rightarrow$ (7) $\rightarrow + 2 \rightarrow$ (9)

$3 \rightarrow + 6 \rightarrow$ (9)

It's Algebra! The concepts in this lesson help prepare students for algebra.

3 Practice

Have students complete the exercises using their calculators. Watch for students who combine functions inaccurately.

Practice

Use a calculator to complete each arrow diagram.

1. $10 \longrightarrow \times 5 \longrightarrow \underline{50}$
2. $16 \longrightarrow +25 \longrightarrow \underline{41}$
3. $\underline{8} \longrightarrow +9 \longrightarrow 17$
4. $28 \longrightarrow \div 7 \longrightarrow \underline{4}$
5. $\underline{15} \longrightarrow \times 8 \longrightarrow 120$
6. $\underline{8} \longrightarrow \times 9 \longrightarrow 72$
7. $64 \longrightarrow -18 \longrightarrow \underline{46}$
8. $\underline{14} \longrightarrow \times 7 \longrightarrow 98$
9. $\underline{38} \longrightarrow +38 \longrightarrow 76$
10. $89 \longrightarrow -27 \longrightarrow \underline{62}$
11. $17 \longrightarrow \times 12 \longrightarrow \underline{204}$
12. $300 \longrightarrow \div 15 \longrightarrow \underline{20}$
13. $53 \longrightarrow -15 \longrightarrow \underline{38}$
14. $\underline{14} \longrightarrow \times 25 \longrightarrow 350$
15. $\underline{56} \longrightarrow +53 \longrightarrow 109$
16. $\underline{400} \longrightarrow \div 16 \longrightarrow 25$
17. $9 \longrightarrow \times 7 \longrightarrow +7 \longrightarrow \underline{70}$
18. $48 \longrightarrow \times 5 \longrightarrow \div 12 \longrightarrow \underline{20}$
19. $34 \longrightarrow +9 \longrightarrow -9 \longrightarrow \underline{34}$
20. $56 \longrightarrow \div 7 \longrightarrow \times 7 \longrightarrow \underline{56}$

Use a calculator to complete each arrow diagram.
Then, write a single function to combine the two functions.

21. $36 \longrightarrow \times 7 \longrightarrow \div 2 \longrightarrow \underline{126}$ $36 \longrightarrow \frac{7}{2} \longrightarrow \underline{126}$
22. $306 \longrightarrow +5 \longrightarrow -3 \longrightarrow \underline{308}$ $306 \longrightarrow +2 \longrightarrow \underline{308}$
23. $73 \longrightarrow -16 \longrightarrow -10 \longrightarrow \underline{47}$ $73 \longrightarrow -26 \longrightarrow \underline{47}$

Use the function shown to complete each table.

24. Input $\longrightarrow \times 12 \longrightarrow$ Output

Input	1	4	6	7	12	16
Output	12	48	72	84	144	192

25. Input $\longrightarrow +15 \longrightarrow$ Output

Input	1	9	25	40	62	80
Output	16	24	40	55	77	95

26. Input $\longrightarrow \times \frac{2}{3} \longrightarrow$ Output

Input	6	12	24	30	48	42
Output	4	8	16	20	32	28

27. Input $\longrightarrow \times 5 \longrightarrow +3 \longrightarrow$ Output

Input	2	6	8	10	15	28
Output	13	33	43	53	78	143

156

Lesson 7-10 • Calculator: Functions

For Mixed Abilities

Common Errors • Intervention

Some students may have difficulty understanding how the two parts of the problem are related, such as in Exercises 21 to 23. Have the students describe the functions that were used in the first part of each problem and then describe what function would result if they were combined.

Enrichment • Application

Have students write a single function to combine the two functions in each problem.

1. $48 \rightarrow \times 4 \rightarrow \div 3 \rightarrow 64$ $(\times \frac{4}{3})$
2. $671 \rightarrow +8 \rightarrow -5 \rightarrow 674$ $(+3)$
3. $65 \rightarrow \div 5 \rightarrow \times 2 \rightarrow 26$ $(\times \frac{2}{5})$
4. $94 \rightarrow -17 \rightarrow -13 \rightarrow 64$ (-30)
5. $439 \rightarrow -18 \rightarrow +49 \rightarrow 470$ $(+31)$
6. $7 \rightarrow \times 3 \rightarrow \times 2 \rightarrow 42$ $(\times 6)$

More to Explore •
Number Sense

Have students assign the numbers 1–26 in order to the letters of the alphabet and find each of these:

1. the spelling word that has the highest value
2. the value of antidisestablishmentarianism
3. the name of 3 fruits with a value less than 50
4. the name of 2 sports with a value more than 100
5. a sentence with a value of between 90 and 100

(Answers will vary.)

ESL/ELL STRATEGIES

Clarify that an *operation* is a mathematical process like adding, subtracting, multiplying, or dividing. Then, walk students through some of the calculator exercises. For example: *The input is 6. The function is a combination of the operation of addition and the number 3. The output is 9.*

4 Assess

Have students create a composite function machine and determine the output for an input of 350. (Check students' work.)

5 Mixed Review

1. $\$168.79 \times 4$ ($675.16)
2. $6{,}003{,}001 - 2{,}571{,}385$ (3,431,616)
3. $41{,}270 \div 9$ (4,585 R5)
4. Simplify: $\frac{37}{4}$ ($9\frac{1}{4}$)
5. 407×318 (129,426)
6. $38 + 127 + 659 + 8$ (832)
7. $25 - 56 \div 7$ (17)
8. $43\frac{5}{8} - 27\frac{11}{12}$ ($15\frac{17}{24}$)
9. $\frac{2}{3} + \frac{4}{7}$ ($1\frac{5}{21}$)
10. $6{,}343 \div 208$ (30 R103)

T156

Chapter 7
Test

Name _____

Multiply. Factor wherever possible.

1. $\frac{2}{3} \times \frac{3}{8} = \frac{1}{4}$

2. $\frac{5}{9} \times \frac{3}{15} = \frac{1}{9}$

3. $\frac{7}{8} \times \frac{4}{5} = \frac{7}{10}$

4. $\frac{5}{6} \times \frac{9}{10} = \frac{3}{4}$

5. $16 \times \frac{3}{4} = 12$

6. $\frac{5}{8} \times 6 = 3\frac{3}{4}$

7. $15 \times \frac{3}{10} = 4\frac{1}{2}$

8. $24 \times \frac{9}{16} = 13\frac{1}{2}$

9. $6\frac{1}{2} \times 18 = 117$

10. $2\frac{1}{3} \times 5\frac{1}{7} = 12$

11. $3\frac{1}{3} \times 2\frac{1}{10} = 7$

12. $5\frac{1}{7} \times 3\frac{1}{6} = 16\frac{2}{7}$

13. $3\frac{4}{5} \times 6\frac{1}{4} = 23\frac{3}{4}$

14. $8\frac{2}{9} \times 5\frac{1}{2} = 45\frac{2}{9}$

Write the reciprocal.

15. $\frac{2}{3}$ $\frac{3}{2}$

16. $\frac{5}{8}$ $\frac{8}{5}$

17. $2\frac{1}{3}$ $\frac{3}{7}$

18. $4\frac{1}{5}$ $\frac{5}{21}$

19. 18 $\frac{1}{18}$

20. $\frac{17}{5}$ $\frac{5}{17}$

Divide.

21. $\frac{1}{2} \div \frac{3}{4} = \frac{2}{3}$

22. $\frac{5}{8} \div \frac{15}{16} = \frac{2}{3}$

23. $\frac{3}{7} \div \frac{6}{7} = \frac{1}{2}$

24. $\frac{3}{8} \div \frac{5}{6} = \frac{9}{20}$

25. $9 \div \frac{3}{4} = 12$

26. $\frac{5}{8} \div 10 = \frac{1}{16}$

27. $15 \div \frac{5}{7} = 21$

28. $\frac{2}{3} \div 18 = \frac{1}{27}$

29. $2\frac{1}{2} \div 1\frac{1}{5} = 2\frac{1}{12}$

30. $4\frac{2}{3} \div 1\frac{1}{9} = 4\frac{1}{5}$

31. $7\frac{1}{7} \div 3\frac{1}{3} = 2\frac{1}{7}$

32. $4\frac{1}{6} \div 6\frac{2}{3} = \frac{5}{8}$

33. $2\frac{5}{11} \div \frac{3}{7} = 5\frac{8}{11}$

34. $5\frac{5}{6} \div 2\frac{2}{3} = 2\frac{3}{16}$

Alternate Chapter Test

You may wish to use the Alternate Chapter Test on page 382 of this book for further review and assessment.

Circle the letter of the correct answer.

1 $n \times 3 = 15$
$n = ?$
   a. 5 ⃝
   b. 6
   c. 45
   d. NG

2 $12 - 6 \div 3 = n$
$n = ?$
   a. 2
   b. 6
   c. 10 ⃝
   d. NG

3 What is the place value of the 3 in 682,351?
   a. ones
   b. tens
   c. hundreds ⃝
   d. NG

4 $134.56 + 97.85$
   a. $231.41
   b. $232.31
   c. $232.41 ⃝
   d. NG

5 $31,052 - 15,785$
   a. 15,267 ⃝
   b. 15,367
   c. 24,733
   d. NG

6 39×5
   a. 155
   b. 195 ⃝
   c. 1,545
   d. NG

7 615×34
   a. 4,305
   b. 20,810
   c. 20,910 ⃝
   d. NG

8 $808 \div 7$
   a. 12 R4
   b. 115
   c. 115 R5
   d. NG ⃝

9 $32\overline{)8,156}$
   a. 25 R24
   b. 254
   c. 254 R28 ⃝
   d. NG

10 Complete the equivalent fraction.
$\frac{5}{9} = \frac{?}{18}$
   a. 5
   b. 9
   c. 10 ⃝
   d. NG

11 $3\frac{5}{6} + 2\frac{1}{4}$
   a. $5\frac{1}{12}$
   b. $5\frac{3}{5}$
   c. $6\frac{1}{12}$ ⃝
   d. NG

12 $6\frac{1}{5} - 3\frac{7}{10}$
   a. $2\frac{1}{2}$ ⃝
   b. $3\frac{3}{10}$
   c. $3\frac{7}{10}$
   d. NG

☐ score   **STOP**

page 158

Items	Objectives
1	To review basic facts of division (see pages 7–8)
2	To simplify an expression (see pages 11–12)
3	To identify place value (see pages 19–20)
4–5	To add and subtract two numbers up to 6 digits (see pages 27–28)
6–7	To multiply by 1-digit and 2-digit numbers (see pages 45–46, 51–52)
8–9	To divide by 1-digit and 2-digit numbers (see pages 69–70, 75-76)
10	To find equivalent fractions (see pages 97–98)
11	To add two mixed numbers with regrouping (see pages 119–120)
12	To subtract mixed numbers, rename the minuend (see pages 125–126)

Alternate Cumulative Assessment

Circle the letter of the correct answer.

1. $n \times 4 = 28$
$n = ?$
a 5
b 6
c 7 ⃝
d NG

2. $16 - 8 \div 2 = n$
$n = ?$
a 4
b 8
c 12 ⃝
d NG

3. Give the place value of the 4 in 75,483,902.
a ten thousands
b hundred thousands ⃝
c millions

4. $263.74 + 89.69$
a $352.43
b $353.33
c $353.43 ⃝
d NG

5. $42,075 - 18,397$
a 23,678 ⃝
b 23,778
c 24,678
d NG

6. 56×4
a 204
b 224 ⃝
c 2,024
d NG

7. 729×18
a 12,922
b 13,022
c 13,122 ⃝
d NG

8. $8\overline{)994}$
a 124
b 124 R1
c 124 R4
d NG ⃝

9. $49\overline{)9,961}$
a 23 R4
b 23 R14
c 203 R14 ⃝
d NG

10. $\frac{3}{7} = \frac{n}{35}$
$n = ?$
a 12
b 15 ⃝
c 18
d NG

11. $7\frac{4}{9} + 7\frac{11}{12}$
a $14\frac{13}{36}$
b $15\frac{12}{36}$
c $15\frac{13}{36}$ ⃝
d NG

12. $7\frac{1}{4} - 5\frac{5}{8}$
a $1\frac{5}{8}$ ⃝
b $1\frac{7}{8}$
c $2\frac{5}{8}$
d NG

8-1 Tenths and Hundredths

pages 159–160

1 Getting Started

Objective
• To read and write decimals to the hundredths place

Warm Up • Mental Math
Have students add to each amount to make $1.00.

1. $0.95 ($0.05)
2. $0.67 ($0.33)
3. $0.50 ($0.50)
4. $0.88 ($0.12)
5. $0.75 ($0.25)
6. $0.96 ($0.04)
7. $0.35 ($0.65)
8. $0.48 ($0.52)

Warm Up • Activity
Draw rectangles divided into tenths on the board to show $\frac{5}{10}$, $\frac{4}{10}$, and $\frac{9}{10}$. Have students write the fraction that tells what part of each region is shaded and what part of each region is not shaded. ($\frac{5}{10}$, $\frac{5}{10}$; $\frac{4}{10}$, $\frac{6}{10}$, $\frac{9}{10}$, $\frac{1}{10}$)

Name _____

Add and Subtract Decimals

Lesson 8-1

It's Algebra!

Tenths and Hundredths

Numbers to the left of the decimal point represent whole numbers. Numbers to the right of the decimal point represent a fractional or decimal part of one.

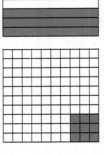

This one region is divided into __10__ equal parts.
Each part is $\frac{1}{10}$ of the region. The shaded part is $\frac{3}{10}$.
The unshaded part is __$\frac{7}{10}$__ of the region.
Every fraction has a decimal equivalent.

Fraction		Decimal
$\frac{1}{10}$	=	0.1

The zero in 0.1 says there are no ones in the decimal number.

This region is divided into __100__ equal parts.
Each part is __$\frac{1}{100}$__ of the region. The shaded part is __$\frac{9}{100}$__. The unshaded part is __$\frac{91}{100}$__ of the region.

Fraction		Decimal
$\frac{1}{100}$	=	0.01

We can use a place-value chart to understand and read decimals.

tens	ones	tenths	hundredths
1	5	2	7

We read 15.27 as **fifteen and twenty-seven hundredths.**

We say *and* in a number only to represent a decimal point.

Getting Started

Write the decimal.

1. $3\frac{25}{100}$ __3.25__   2. seven and four tenths __7.4__

Write the decimal number in words.

3. 6.3 __six and three tenths__

Lesson 8-1 • Tenths and Hundredths   159

---

2 Teach

Introduce the Lesson Read the description of *decimals* at the top of page 159. Explain that the word *decimal* means "based on 10." Have students read and complete the sentences about tenths. Point out that the fraction and decimal notation name the same number.

Then, have students read and complete the sentences about hundredths. Emphasize that because our numeration system is based on tens, it can be extended with decimals, which name fractions that have denominators of 10, 100, and other powers of 10.

Develop Skills and Concepts Point out that when students read decimals, they should read the decimal point as *and*. Write each decimal on the board and have students read them aloud:

7.2 (seven and two tenths)

12.8 (twelve and eight tenths)

3.41 (three and forty-one hundredths)

21.64 (twenty-one and sixty-four hundredths)

Explain to students that when they write a number less than 1, they should write a zero in the ones place. To reinforce the concepts in this lesson, have students write each fraction as a decimal.

1. $\frac{3}{10}$ (0.3)
2. $\frac{8}{10}$ (0.8)
3. $\frac{77}{100}$ (0.77)

3 Practice

Have students complete all the exercises. Remind them that one tenth is one part of 10, and one hundredth is one part of 100.

Practice

Write the decimal.

1. $7\frac{3}{100}$ _7.03_   2. $6\frac{1}{10}$ _6.1_   3. $28\frac{5}{10}$ _28.5_   4. $39\frac{5}{100}$ _39.05_

5. $4\frac{8}{10}$ _4.8_   6. $12\frac{50}{100}$ _12.50_   7. $36\frac{76}{100}$ _36.76_   8. $52\frac{1}{10}$ _52.1_

9. $116\frac{3}{100}$ _116.03_   10. $203\frac{4}{100}$ _203.04_   11. $91\frac{40}{100}$ _91.40_   12. $57\frac{9}{10}$ _57.9_

13. four and two tenths _4.2_

14. sixteen hundredths _0.16_

15. seventy hundredths _0.70_

16. thirty hundredths _0.30_

17. five tenths _0.5_

18. one tenth _0.1_

19. eighty-seven hundredths _0.87_

20. six and two hundredths _6.02_

Write the decimal numbers in words.

21. 8.43 _eight and forty-three hundredths_

22. 12.6 _twelve and six tenths_

23. 29.05 _twenty-nine and five hundredths_

24. 127.38 _one hundred twenty-seven and thirty-eight hundredths_

25. 214.3 _two hundred fourteen and three tenths_

26. 576.49 _five hundred seventy-six and forty-nine hundredths_

Problem Solving

Solve each problem.

27. Bill had $10. He spent $3. What decimal represents the part of the money Bill spent?
0.3

28. If you toss a coin 100 times and it turns up heads 55 of those times, what decimal represents the number of times it turns up tails?
0.45

29. Nancy has 100 coins. She has 46 nickels. What decimal represents the part of the coins that are not nickels?
0.54

30. Of the 10 people on the bus, 4 are adults and 7 are female. Write a decimal to represent each of the following groups on the bus: adults, children, females, males.
04, 0.6, 0.7, 0.3

160

Lesson 8-1 • Tenths and Hundredths

---

4 Assess

Have students write one-tenth and one-hundredth as a decimal. (0.1; 0.01)

5 Mixed Review

1. $7,295 \times 31$ (226,145)
2. $3\frac{1}{3} \times 4\frac{2}{5}$ $(14\frac{2}{3})$
3. $700 \times 6,000$ (4,200,000)
4. $62,176,490 + 490 + 322,158$ (62,499,138)
5. $\frac{12}{25} \times \frac{5}{6}$ $(\frac{2}{5})$
6. $17,903 - 14,627$ (3,276)
7. $7 - 3\frac{7}{10}$ $(3\frac{3}{10})$
8. $19,028 \div 36$ (528 R20)
9. 3^5 (243)
10. $21\frac{2}{3} + 12\frac{3}{4} + 8\frac{5}{6}$ $(43\frac{1}{4})$

For Mixed Abilities

Common Errors • Intervention

Some students will confuse the names for the decimal places with the names for the whole number places. Discuss how the *th* indicates a decimal place and how decimals take the name of the place in the decimal that is farthest to the right.

Enrichment • Number Sense

Have students perform each operation and write their answer as a decimal.

1. $\frac{7}{10} \times \frac{2}{10} =$ (0.14)

2. $5\frac{9}{10} + 3\frac{11}{100} =$ (9.01)

3. $8 - 2\frac{17}{100} =$ (5.83)

4. $\frac{10}{11} \div \frac{100}{11} =$ (0.10)

5. $3\frac{3}{10} \times 17 =$ (56.1)

6. $8\frac{7}{10} + \frac{3}{10} =$ (9.0)

More to Explore • Application

Have students bring to class a favorite family recipe having a minimum of 5 ingredients. Tell students to adjust the measurement of ingredients in their recipes enough so they could serve each person in the class, including the teacher. As an extension, have students group their recipes into interesting meals and calculate the total amount of each ingredient they would need per meal.

ESL/ELL STRATEGIES

Review the pronunciation and meaning of all key mathematical terms. Write them on the board, capitalizing the stressed syllables; for example, MINuend, SUBtrahend, DIFFerence. Help students explain each term in their own words. For example: *The minuend is the smaller number*.

T160

8-2 Thousandths

pages 161–162

1 Getting Started

Objective
- To read and write decimals through thousandths

Materials
40 index cards

Warm Up • Mental Math
Have students identify each number as prime or composite.

1. 14 (composite)
2. 2 (prime)
3. 31 (prime)
4. 57 (composite)
5. 11 (prime)
6. 29 (prime)
7. 27 (composite)
8. 35 (composite)
9. 63 (composite)
10. 23 (prime)

Warm Up • Pencil and Paper
Have students write >, <, or = to compare each pair of numbers.

1. $\frac{8}{10}$ (<) $\frac{9}{10}$ 5. $4\frac{9}{10}$ (>) $4\frac{9}{100}$
2. $\frac{3}{10}$ (>) $\frac{3}{100}$ 6. $5\frac{3}{10}$ (=) $5\frac{30}{100}$
3. $7\frac{1}{10}$ (=) $\frac{71}{10}$ 7. $\frac{91}{10}$ (>) $9\frac{1}{100}$
4. $\frac{35}{100}$ (<) $3\frac{5}{10}$ 8. $\frac{6}{10}$ (<) $\frac{60}{10}$

Name _____

Lesson 8-2

Thousandths

Beth put an odometer on her bike two years ago. The odometer shows the number of miles she has ridden since then. She took the bike to get a license, and had to read the odometer number to the inspector. How many miles had she ridden?

A place-value chart helps us read this number.

hundreds	tens	ones	tenths	hundredths	thousandths
4	3	1 .	0	4	6

4 hundreds
3 tens
1 one
0 tenths
4 hundredths
6 thousandths

The place value of the third digit to the right of the decimal point is __thousandths__. The decimal number on Beth's odometer is read __four hundred thirty-one and forty-six thousandths__

In reading a decimal number, say the number to the right of the decimal point as a whole number followed by the name of the last place value.

Getting Started

Write the decimal.

1. $\frac{86}{1000}$ __0.086__   2. $7\frac{236}{1000}$ __7.236__   3. $12\frac{5}{1000}$ __12.005__

4. nine and fifty-seven thousandths __9.057__

Write the decimal number in words.

5. 8.456 __eight and four hundred fifty-six thousandths__

---

2 Teach

Introduce the Lesson Have students study the place-value chart. The chart shows the value of the digit in each place so that the number can be read and understood. Emphasize that decimals are another way to represent fractions that have denominators of powers of 10.

Help students read the number aloud. Point out that they should read the whole number part, read the decimal point as *and*, and then read the decimal part. Stress that students read the decimal part just as they would a whole number. However, they should give the number the *last name* of the place farthest to the right that contains a digit.

For example, 0.072 is read 72 just as a whole number would be read. It is given the *last name* of thousandths because the digit farthest to the right is in the thousandths place.

Develop Skills and Concepts Draw a place-value chart from thousands to thousandths. Point out that each place is 10 times greater than the place to its right and $\frac{1}{10}$ of the value of the place to its left.

Write **4,236.57** on the chart and have a student read the number. Write **9** in the thousandths column of the chart and have students extend the pattern.

Have students practice reading the following numbers aloud, making sure they pronounce the *th* ending of *tenths*, *hundredths*, and *thousandths*:

1. 42.675 (forty-two and six hundred seventy-five thousandths)

2. 1.064 (one and sixty-four thousandths)

3. 0.003 (three thousandths)

Practice

Write the decimal.

1. $\frac{181}{1000}$ __0.181__   2. $\frac{316}{1000}$ __0.316__   3. $\frac{9}{10}$ __0.9__   4. $\frac{47}{1000}$ __0.047__

5. $\frac{14}{100}$ __0.14__   6. $\frac{5}{1000}$ __0.005__   7. $4\frac{216}{1000}$ __4.216__   8. $4\frac{30}{1000}$ __4.030__

9. $12\frac{75}{1000}$ __12.075__   10. $12\frac{75}{100}$ __12.75__   11. $16\frac{305}{1000}$ __16.305__   12. $15\frac{65}{1000}$ __15.065__

13. nine thousandths __0.009__

14. three and two hundred five thousandths __3.205__

15. four and twelve thousandths __4.012__

16. seventy-two hundredths __0.72__

17. eight and five thousandths __8.005__

18. six hundred thousandths __0.600__

19. two hundred ninety-three and one hundred forty-four thousandths __293.144__

20. one thousand twenty-five and forty-eight thousandths __1,025.048__

Write each decimal number in words.

21. 9.326 __nine and three hundred twenty-six thousandths__

22. 5.378 __five and three hundred seventy-eight thousandths__

23. 0.896 __eight hundred ninety-six thousandths__

24. 4.6 __four and six tenths__

25. 6.025 __six and twenty-five thousandths__

26. 12.34 __twelve and thirty-four hundredths__

Problem Solving

Solve each problem.

27. Mr. Ryan drove 1,000 miles on his vacation. On the first day, he drove 312 miles. What decimal represents the part of the whole trip Mr. Ryan drove on the first day?
0.312

28. In a poll of 1,000 people, 629 people chose dogs as their favorite pet. What decimal represents the part of all the people that did not choose dogs as their favorite pet?
0.371

Lesson 8-2 • Thousandths

162

For Mixed Abilities

Common Errors • Intervention

Some students may have difficulty writing the decimal for a decimal fraction. Have these students work with partners and two sets of index cards, one set containing decimal fraction names for numbers, one per card, and the other set containing the corresponding decimal names.

Mix up each set of cards. Place the fraction cards face down. Taking turns, one student turns over the top fraction card and the other then searches through the decimal cards to find the matching name.

Enrichment • Number Sense

Have students write the next two numbers in each sequence. Then, ask students to create their own decimal number sequences for friends to complete.

1. 4.611, 4.612, 4.613, (4.614, 4.615)

2. 0.425, 0.430, 0.435, (0.440, 0.445)

3. 4.042, 4.044, 4.046, (4.048, 4.050)

4. 729.573, 729.576, 729.579, (729.582, 729.585)

5. 5.007, 5.008, 5.009, (5.010, 5.011)

More to Explore • Application

Have students design their own color-by-number picture using fractions. First, have students design a picture to be colored. Have them choose a basic color key using simplified fractions; for example, green = $\frac{1}{2}$, red = $\frac{1}{3}$, blue = $\frac{1}{4}$. Then, have them designate the areas of the picture for each color by writing an addition or subtraction problem that will result in the correct fractional answer.

---

3 Practice

Have students complete all the exercises.

4 Assess

Have students write five tenths, five hundredths, and five thousandths as decimals. (0.5; 0.05; 0.005)

5 Mixed Review

1. $957 + 12,680 + 3,207$
(16,844)

2. $\$4.95 \times 36$ ($178.20)

3. $2\frac{1}{3} + 3\frac{5}{8} + 7\frac{3}{4}$ ($13\frac{17}{24}$)

4. $8 \div \frac{3}{4}$ ($10\frac{2}{3}$)

5. $1,976 \times 8$ (15,808)

6. $\$5,001.78 - \$3,270.95$
($1,730.83)

7. $\frac{3}{8} \times \frac{2}{9} \times \frac{6}{7}$ ($\frac{1}{14}$)

8. $16\frac{1}{3} - 7\frac{5}{8}$ ($8\frac{17}{24}$)

9. $8,094 \div 426$ (19)

10. 17×10^4 (170,000)

T162

pages 163–164

Objective
• To identify place value of a digit to millionths

Materials
*place-value chart from millions to millionths; index cards

Warm Up • Mental Math
Ask students if the fractions in each pair below are equivalent.

1. $\frac{3}{4}, \frac{6}{8}$ (yes)   6. $\frac{1}{2}, \frac{7}{14}$ (yes)
2. $\frac{2}{3}, \frac{6}{9}$ (yes)   7. $\frac{2}{3}, \frac{4}{6}$ (yes)
3. $\frac{6}{10}, \frac{3}{4}$ (no)   8. $\frac{9}{12}, \frac{2}{3}$ (no)
4. $\frac{3}{5}, \frac{6}{10}$ (yes) 9. $\frac{1}{4}, \frac{4}{16}$ (yes)
5. $\frac{2}{6}, \frac{1}{5}$ (no)   10. $\frac{3}{8}, \frac{7}{16}$ (no)

Warm Up • Activity
Draw a place-value chart labeled from thousands to thousandths on the board. Have students copy the chart, write each number on the chart, and then write the value of the 4 in each number.

1. 468.375 (hundreds)
2. 3,964.018 (ones)
3. 2,806.473 (tenths)
4. 1,965.014 (thousandths)
5. 3,917.047 (hundredths)

Name _____

Place Value to Millionths

In his annual physics lecture at the university, Dr. Johnson wrote this number on the chalkboard. Can you read it?

To read the number, we have to understand the decimal place-value system.

Each place is **10 times** greater than the place on its right.

Each place is $\frac{1}{10}$ of the value of the place on its left.

These principles are the same on both sides of the decimal point.

millions	hundred thousands	ten thousands	thousands	hundreds	tens	ones	tenths	hundredths	thousandths	ten thousandths	hundred thousandths	millionths
				9	0	3	2	8	5	6	1	4

The place value of the 1 is <u>hundred thousandths</u>.
The place value of the 4 is <u>millionths</u>.

We read Dr. Johnson's decimal number as <u>nine hundred three and two hundred eighty-five thousand, six hundred fourteen millionths</u>.

Decimals that name the same number are called **equivalent decimals**. For example:

0.5 = 0.50 = 0.500

Getting Started
Write the place value of 3 in each number.

1. 12.976483 <u>millionths</u>   2. 8.004531 <u>hundred thousandths</u>   3. 9.45632 <u>ten thousandths</u>

Write the decimal.

4. two hundred and fifty-six ten thousandths <u>200.0056</u>

Write the decimal number in words.

5. 7.2468 <u>seven and two thousand four hundred sixty-eight ten thousandths</u>

Lesson 8-3 • Place Value to Millionths

Introduce the Lesson Have a student read the introductory problem. Using the place-value chart, discuss the value of each digit in the number 903.285614.

Review the relationship between two adjoining places. Each place in a decimal number has a value that is 10 times the value of the place at its right and $\frac{1}{10}$ the value of the place at its left.

Have students read the number and complete the instructional sentences.

Then, help students use what they know about equivalent fractions to recognize equivalent decimals. Point out that $\frac{5}{10}$, $\frac{50}{100}$, and $\frac{500}{1,000}$ are equivalent fractions; the corresponding decimals, 0.5, 0.50, and 0.500 are also equivalent.

Develop Skills and Concepts Use the place-value chart to have students practice reading decimals. Have students write the following numbers in the chart and then read each number:

347.15674 (three hundred forty-seven and fifteen thousand six hundred seventy-four hundred thousandths)

29.304765 (twenty-nine and three hundred four thousand seven hundred sixty-five millionths)

When discussing equivalent decimals, point out that adding zeros to the right of a decimal does not change its value. Write the following examples on the board:

0.2 = 0.20 = 0.200 = 0.2000
4.07 = 4.070 = 4.0700 = 4.07000

Contrast writing zeros to the right of a whole number (250 < 2,500) with writing zeros to the right of decimals. (250 = 250.00)

Practice

Write the place value of 5 in each number.

1. 7.3259 ____thousandths____

2. 14.56832 ____tenths____

3. 136.1245 ____ten thousandths____

4. 96.25483 ____hundredths____

5. 37.291451 ____hundred thousandths____

6. 139.675148 ____thousandths____

7. 9.326015 ____millionths____

8. 14.63251 ____ten thousandths____

Write the decimal.

9. six and five hundred thirty-six hundred thousandths ____6.00536____

10. thirteen and ninety-eight ten thousandths ____13.0098____

11. fifty-six millionths ____0.000056____

12. twelve and one thousand six hundred ten thousandths ____12.1600____

Write the decimal numbers in words.

13. 6.2439 ____six and two thousand four hundred thirty-nine ten thousandths____

14. 4.000329 ____four and three hundred twenty-nine millionths____

15. 0.47385 ____forty-seven thousand three hundred eighty-five hundred thousandths____

16. 2.1806 ____two and one thousand eight hundred six ten thousandths____

[Now Try This!]

Label each point with a fraction. Look for a pattern to help you.

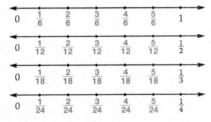

Lesson 8-3 • Place Value to Millionths

---

3 Practice

Have students complete all the exercises. Encourage them to use a place-value chart if needed.

Now Try This! Explain that these number lines do not imply that $\frac{1}{6} = \frac{1}{12} = \frac{1}{18} = \frac{1}{24}$. Each represents a different length, 1, $\frac{1}{2}$, $\frac{1}{3}$, and $\frac{1}{4}$. Have students focus on the number of points to be labeled between 0 and the given number on the right.

4 Assess

Have students write the number for sixty-five thousand (65,000), and sixty-five thousandths (0.065).

For Mixed Abilities

Common Errors • Intervention

Some students may have difficulty writing the decimal given its word name. Read the following decimals out loud as students write them on a place-value chart.

32.658	4.9756
297.0421	65.87291
3,964.1234	8.752461
3.5367	40.65867

Enrichment • Number Sense

Write the following decimals on index cards and place them face down on a table:

0.40, 42.2, 3.6, 74.700, 0.100, 33.3, 35.6, 17.4, 0.50, 0.30, 201.800, 0.600

Have students take turns choosing a card, reading the decimal on the card, and naming two decimals equivalent to the number they have chosen.

More to Explore • Number Sense

The writing called *cuneiform* is the system of numeration used by the Babylonians. The only symbol used was created in wet clay by a wedge-shaped piece of wood.

The symbol ▽ meant one. For ten, this symbol was placed on its side ◁. To write the number 14, the Babylonians wrote

◁▽▽▽▽ = 10 + 4 = 14

To write larger numbers, a space was left after a group of symbols to indicate that this value was to be multiplied by 60.

Write these cuneiform numerals on the board and show students how to determine their value.

◁◁◁◁▽▽ = 40 + 3 = 43

◁◁◁▽▽ ◁▽ = (32 × 60) + 11 = 1,920 + 11 = 1,931

Have students work in small groups to write their own cuneiform symbols and exchange with another group to convert them to Hindu-Arabic numerals.

T164

8-4 Comparing and Ordering Decimals

pages 165–166

1 Getting Started

Objective
• To compare and order decimals

Materials
index cards

Warm Up • Mental Math
Have students name the number that is

1. 9 more than 83. (92)
2. the difference between 67 and 49. (18)
3. the product of 5^2 and 2^2. (100)
4. 6 times the sum of 70 and 30. (600)
5. $\frac{1}{2}$ the difference between 17 and 9. (4)
6. the quotient when $\frac{3}{8}$ is divided by $\frac{3}{8}$. (1)
7. 10 times 3 nickels. (30 nickels or $1.50)

Warm Up • Pencil and Paper
Have students write >, <, or = to compare each pair of numbers.

1. 964 (<) 972
2. 72,601 (>) 72,106
3. 39,608 (<) 39,618
4. $\frac{4}{5}$ (>) $\frac{2}{5}$
5. $\frac{1}{2}$ (=) $\frac{5}{10}$
6. $2\frac{1}{3}$ (<) $3\frac{1}{2}$

2 Teach

Introduce the Lesson Have a student read the problem and tell what information is in the picture. (the sizes of the drill bits) Point out that the method for comparing decimals is the same as that for comparing whole numbers.

Review the meaning of the symbols > and <. As you discuss the sample problem, stress that when comparing decimals, they should begin with the digit in the place with the greatest place value and proceed from left to right, comparing each pair of digits up to the first pair that differ.

Develop Skills and Concepts Emphasize that before students compare numbers, they should align the numbers so that the decimal points are aligned. Then, have them write zeros so that all decimals have the same number of places.

T165

Comparing and Ordering Decimals

It's Algebra!

Abby needs to drill a hole in the birdhouse she is building. She needs to drill the largest hole that her drill can make. Which drill bit should Abby use?

To find the largest drill bit, we need to compare the three decimal sizes that she has. We compare

0.5625 , 0.687 , and 0.625 .

To compare decimals, write them in a column so that the decimal points are aligned. Write zeros to name equivalent decimals.

0.5625
0.6870   0.687 and 0.6870 are equivalent decimals.
0.6250   0.625 and 0.6250 are equivalent decimals.

Start at the left and compare digits. The ones are the same, but the tenths are not.

0.5625   Because 5 < 6, 0.5625 is the smallest drill bit.
0.6870
0.6250

Next compare hundredths.

0.6870   2 < 8   0.625 is the middle-sized drill bit.
0.6250   8 > 2   0.687 is the largest drill bit.

Abby should use the bit labeled 0.687 .

0.5625

0.687

0.625

Getting Started

Write >, <, or = in each circle.

1. 0.2431 (<) 0.2461
2. 7.19 (=) 7.190
3. 9.273 (<) 9.372

Order the numbers from least to greatest.

4. 2.59, 2.493, 2.571
   2.493, 2.571, 2.59
5. 6.1351, 6.13, 6.152, 6.143
   6.13, 6.1351, 6.143, 6.152

Lesson 8-4 • Comparing and Ordering Decimals

165

Work through the following comparison with students:

Compare: 4.632 and 4.6479

Align the decimal points.	4.632
	4.6479

Write zeros.	4.6320
	4.6479

Now, compare: 4.6320 and 4.6479.

Begin at the left and compare the digits in each place. Same number of ones, 4. Same number of tenths, 6. Compare hundredths.   3 < 4

4.6340 < 4.6479

Have students order these numbers from least to greatest.

7.06, 7.069, 7.013 (7.013, 7.06, 7.069)

It's Algebra! The concepts in this lesson prepare students for algebra.

Practice

Write >, <, or = in each circle.

1. 5.7 ⬤< 5.9
2. 3.26 ⬤= 3.260
3. 15.27 ⬤< 15.72
4. 0.029 ⬤< 0.039
5. 6.8325 ⬤> 6.8315
6. 10.03 ⬤= 10.030
7. 4.129 ⬤= 4.1290
8. 1.0025 ⬤< 1.025
9. 4.8317 ⬤< 4.9317
10. 29.02 ⬤< 29.20
11. 16.157 ⬤> 16.15
12. 0.2473 ⬤> 0.2437
13. 4.19 ⬤> 4.1823
14. 5.9620 ⬤< 5.9627
15. 3.2841 ⬤> 3.2814
16. 0.003 ⬤> 0.0003
17. 13.7 ⬤= 13.700
18. 9.2 ⬤> 9.199

Order the numbers from least to greatest.

19. 3.26, 3.45, 3.3

   3.26, 3.3, 3.45

20. 5.285, 5.825, 5.582

   5.285, 5.582, 5.825

21. 0.02, 0.002, 0.2

   0.002, 0.02, 0.2

22. 4.15, 4.29, 4.265

   4.15, 4.265, 4.29

23. 8.2416, 8.4261, 8.3416

   8.2416, 8.3416, 8.4261

24. 5.14, 5.1, 5.234, 5.2

   5.1, 5.14, 5.2, 5.234

Problem Solving

Use the chart to solve each problem.

25. Which was the fastest time?
    56.38 seconds

26. List the order of finish from first
    through fourth.
    Alan, Walt, Pete, Paul

The Big Bike Sprint	
Alan	56.38 seconds
Pete	56.5 seconds
Walt	56.48 seconds
Paul	56.54 seconds

Now Try This!

Use the digits 6, 7, and 8 to make six different decimal numbers.
Write a check next to the least.
Write an X next to the greatest.

✓ 0.678   0.687   0.768

  0.786   0.867   X 0.876

Lesson 8-4 • Comparing and Ordering Decimals

---

3 Practice

Have students complete all the exercises. Remind them to compare the digits in each place-value position from left to right.

Now Try This! You might have students work in pairs to complete this activity. Point out that any three digits can form six different numbers. Challenge them to use four digits and repeat this activity.

4 Assess

Name two decimals greater than 25.06. (Possible answer: 25.065 and 25.16)

For Mixed Abilities

Common Errors • Intervention

Some students may assume incorrectly that the decimal with more digits is the greater decimal. Have them write the two decimals one above the other with the decimal points aligned. Then, they can compare the numbers, one digit at a time, until they come to a place where the digits are different and can determine the greater number.

Enrichment • Number Sense

Have students work in pairs. Tell each student to write one decimal, up to thousandths, on each of 10 index cards. Tell partners to shuffle cards together and stack them face down. Have students take turns selecting two cards and writing a sentence comparing the numbers drawn. If the sentence is correct, the student wins a point. Extend the activity by having students compare 3 or 4 numbers.

More to Explore • Application

Have your students find examples of scale drawings in encyclopedias, newspapers, textbooks, or other reference materials. Have students tell how the actual size of an object can be determined. (First, measure the drawing, then multiply the measurements by the ratio of the scale given.) For example, if a drawing of a car is 6 in. in length and it is drawn on a scale of 1 in. = 16 in., the actual car length is 6 × 16 = 96 in.

Then, explain how to enlarge drawings. Show students how this can be done on a picture with a scale of 1 in. = 20 in. Measure the height and width of the picture and multiply by the number of times larger the picture will be. For example, to double the size of a drawing, multiply the length and height by 2. The scale would then be 1 in. = 10 in.

Have students make a simple scale drawing on graph paper and enlarge it. Display the finished product.

T166

8-5 Rounding Decimals

pages 167–168

1 Getting Started

Objective
- To round decimals to the nearest tenth, hundredth, or thousandth

Vocabulary
is approximately

Warm Up • Mental Math
Have students simplify each expression.

1. $(4 \times 2) + 7$ (15)
2. $7 + 5 + 2$ (14)
3. $6 + (54 \div 9)$ (12)
4. $5 + 0 + 6$ (11)
5. $3 + 3 + 3$ (9)
6. $(15 - 8) \times 3$ (21)
7. $6 \times 3 + 5$ (23)
8. $17 - 8 + 8$ (17)
9. $(5 \times 4) + 6$ (26)
10. $5 + 5 + 5$ (15)

Warm Up • Paper and Pencil
Have students round each number to the nearest ten:

1. 37 (40) 2. 52 (50) 3. 465 (470)

the nearest hundred:

4. 645 (600)   5. 329 (300)
6. 1,450 (1,500) 7. 5,096 (5,100)

the nearest thousand:

8. 2,476 (2,000) 9. 7,658 (8,000)
10. 8,529 (9,000) 11. 3,099 (3,000)

Rounding Decimals

Population density describes the number of people there are in each square mile of an area. Rounded to the nearest hundredth, what were the population densities of New York City in 1790 and 1980?

YEAR	POPULATION DENSITY
1790	7.179
1980	370.6032

NEW YORK CITY →

We want to know the population densities, rounded to the nearest hundredth.

We know the 1790 population density was __7.179__, and the 1980 density was __370.6032__.

To round a decimal number, circle the digit in the place you want to round.

    7.179   370.6032

Look at the digit to the right of the one circled.

If the digit is 5 or greater, add 1 to the circled number and drop all digits to the right.

    7.179   9 > 5 $7.179 \approx 7.18$

If the digit is less than 5, keep the circled number and drop all digits to the right.

    370.6032   3 < 5 $370.6032 \approx 370.60$

The sign $\approx$ is read "is approximately."

To the nearest hundredth, the population density of New York was __7.18__ in 1790, and __370.60__ in 1980.

Getting Started

Round to the nearest whole number.

1. 6.437 __6__
2. 94.86 __95__

Round to the nearest tenth.

3. 11.734 __11.7__
4. 215.963 __216.0__

Round to the nearest hundredth.

5. 7.999 __8.00__
6. 13.534 __13.53__

Round to the nearest thousandth.

7. 0.46581 __0.466__
8. 6.92952 __6.930__

---

2 Teach

Introduce the Lesson Have a student read the problem and tell what is to be done. Ask what information is given in the map and the chart. Explain that *population density* refers to the number of people who live in each square mile. Be sure students understand that the numbers do not actually mean fractional parts of people, but represent an average.

Explain that this information is easier to interpret and makes more sense if the numbers are expressed to the same place value. As students work through the model, point out that the steps for rounding a decimal are basically the same as the steps for rounding a whole number, with one difference. When a whole number is rounded, all digits to the right of the place being rounded to are replaced by zeros. When a decimal is rounded, all digits to the right of the place being rounded to are dropped or deleted.

Develop Skills and Concepts Emphasize that the key to rounding correctly is to identify those digits that are necessary to the rounding process. Some students may find this step confusing, especially when the digits involved are in the middle of the number.

Encourage students to circle the digit in the place to be rounded, in order to focus attention on that digit and the one to its right.

Help students round each decimal to the nearest tenth, hundredth, and thousandth.

1. 5.1457   (5.1; 5.15; 5.146)
2. 32.0643   (32.1; 32.06; 32.064)

3 Practice

Have students complete all the exercises.

Practice

Round to the nearest whole number.

1. 136.29 __136__
2. 14.765 __15__
3. 9.27 __9__
4. 12.565 __13__

5. 18.5 __19__
6. 37.032 __37__
7. 3.8091 __4__
8. 116.99 __117__

Round to the nearest tenth.

9. 7.39 __7.4__
10. 0.826 __0.8__
11. 25.45 __25.5__
12. 39.06 __39.1__

13. 115.26 __115.3__
14. 59.96 __60.0__
15. 210.53 __210.5__
16. 64.39 __64.4__

Round to the nearest hundredth.

17. 28.735 __28.74__
18. 19.158 __19.16__
19. 37.678 __37.68__
20. 9.241 __9.24__

21. 112.999 __113.00__
22. 7.046 __7.05__
23. 14.393 __14.39__
24. 416.595 __416.60__

Round to the nearest thousandth.

25. 8.53967 __8.540__
26. 0.76463 __0.765__
27. 12.98362 __12.984__
28. 4.58467 __4.585__

29. 16.37995 __16.380__
30. 42.64821 __42.648__
31. 3.54545 __3.545__
32. 7.91659 __7.917__

Problem Solving

Solve each problem.

33. The population density of Colorado is 27.89. What is the density of Colorado, rounded to the nearest tenth?
27.9

34. The population density of Kansas is 28.909. What is the density of Kansas rounded to the nearest hundredth?
28.91

[Now Try This!]

Complete the following. Use the pattern set up in the first sentence.

About 3.5 means "at least 3.45 and at most 3.54."

About 4.7 means "at least __4.65__ and at most __4.74__ ."

About 18.0 means "at least __17.95__ and at most __18.04__ ."

About 8.32 means "at least __8.315__ and at most __8.324__ ."

About 50¢ means "at least __45¢__ and at most __54¢__ ."

Lesson 8-5 • Rounding Decimals

Now Try This! Have students study the estimating pattern to observe how the "at least" and "at most" numbers are derived. Point out that 3.45 is the least number that will round to 3.5 since 3.44 rounds to 3.4. Point out that 3.54 is the greatest number that will round to 3.5 since 3.55 rounds to 3.6.

4 Assess

Have students round 9.6355 to the nearest tenth (9.6), hundredth (9.64), and thousandth (9.636).

For Mixed Abilities

Common Errors • Intervention

Some students may round progressively from left to right, digit by digit. Have these students place an arrow above the digit in the place to which the decimal is to be rounded and look at the digit in the place to the right to determine whether to make the digit under the arrow one greater or leave it alone. Remind students to drop all digits to the right of the digit under the arrow.

Enrichment • Application

Provide students with cash register receipts from a supermarket. Cover the totals. Ask students to round each amount on the receipt to the nearest dollar and estimate the sum. Then, have students compare their estimated sum to the actual sum given at the bottom of the receipt.

More to Explore • Number Sense

Duplicate the following for students:

The students in Mr. Hamilton's sixth-grade class had been pestering him to tell them his age. He finally consented to tell them, but only if they were willing to work for the answer.

Mr. Hamilton told his students, "If you multiply my age by any multiple of 3 under 30, your answer will be a 3-place number composed of only one digit."

His sixth graders never figured out his age. Can you? (37)

ESL/ELL STRATEGIES

Ask a student to contrast *rounding* and *estimating*, or supply the explanation yourself. *Rounding* is giving an approximate answer when you already know the exact number. *Estimating* is giving an approximate answer when you don't know the exact number.

T168

8-6 Estimating Sums and Differences

pages 169–170

1 Getting Started

Objective
• To use rounding to estimate sums and differences

Warm Up • Mental Math
Have students identify each missing number.
1. 5 times the number is 80. (16)
2. The number increased by 12 is 60. (48)
3. The number decreased by 11 is 19. (30)
4. The product of 3 numbers is 100, and two of the factors are 2 and 5. (10)

Warm Up • Pencil and Paper
Have students round each number to the nearest 10:
1. 36 (40) 2. 72 (70) 3. 58 (60)
4. 85 (90) 5. 143 (140) 6. 365 (370)

the nearest 100:
7. 365 (400)   8. 726 (700)
9. 451 (500)   10. 2,836 (2,800)
11. 204 (200)   12. 1,577 (1,600)

Estimating Sums and Differences

Stan has $920.75 in his savings account. He wants to buy the stereo and speakers while they are on sale. About how much of his savings will Stan have left after his purchase?

SALE
Stereo $374.99

Speakers $129.99

We want to know about how much money Stan will have left.

We need to estimate the total cost of the two items.

Then, we can find the difference between that total and the amount in Stan's account.

We know the stereo costs __$374.99__ and the speakers cost __$129.99__.

To estimate the total cost, we round the two amounts to the nearest $100 and add.

$$\begin{array}{r} \$374.99 \approx \$400 \\ 129.99 \approx + \ 100 \\ \hline \$500 \end{array}$$

To estimate how much will be left in his account, we subtract the estimated total cost from the estimate of his savings.

We subtract __$500__ from __$900__.

$$\begin{array}{r} \$920.75 \approx \$900 \\ - \ 500 \\ \hline \$400 \end{array}$$

Stan will have about __$400__ left in his savings account.

Getting Started

Round to the nearest dollar or whole number and estimate the answers.

1.	$50.68 + 37.26	$51 + 37 $88	2.	$69.48 − 14.85	$69 − 15 $54	3.	$72.50 − 51.96	$73 − 52 $21
4.	18.765 + 9.213	19 + 9 28	5.	32.139 + 16.73	32 + 17 49	6.	21.5 − 16.621	22 − 17 5

Copy and compute. Use estimation to check.

7. 43.5 + 36.73
   80.23

8. $24.18 − $13.65
   $10.53

9. $501.65 − $287.44
   $214.21

---

2 Teach

Introduce the Lesson Have a student read the problem and tell what is to be solved. Ask what information is given in the problem and the picture.

Point out that this is a two-step problem. First, students must estimate how much Stan will spend on the MP3 player and speakers. Then, they must subtract this estimated total from the estimated amount of money Stan has in the bank.

Explain that students should round each amount to the nearest $100 before computing. Then, they should use the rounded amounts to estimate answers.

Develop Skills and Concepts Emphasize that rounding and estimating are very important skills. We often need only an approximate sum or difference. Point out that one important reason for rounding is to produce numbers that can be computed mentally.

Note that in rounding to the nearest whole number or nearest dollar, students should examine the digit in the tenths place; they need not consider the digits in the hundredths place or in any other place farther to the right.

3 Practice

Have students complete all the exercises. Watch for students who compute first, then round the answer.

T169

Practice

Round to the nearest dollar or whole number and estimate the answers.

1.	17.64 − 9.85	18 − 10 8	2.	46.432 + 15.296	46 + 15 61	3.	47.896 + 15.5	48 + 16 64
4.	$38.16 − 19.39	$38 − 19 $19	5.	129.235 + 16.41	129 + 16 145	6.	$203.75 − 109.12	$204 − 109 $95
7.	83.705 + 28.9	84 + 29 113	8.	$17.50 + 12.75	$18 + 13 $31	9.	9.2875 − 6.834	9 − 7 2
10.	$16.42 + 35.89	$16 + 36 $52	11.	5.752 + 3.948	6 + 4 10	12.	76.894 − 62.387	77 − 62 15
13.	$116.99 − 20.49	$117 − 20 $97	14.	$49.65 + 34.81	$50 + 35 $85	15.	989 − 451	1,000 − 500 500

Copy and add or subtract.

16. 36.5 + 18.91
    55.41
17. $25.14 − $12.63
    $12.51
18. 4.964 + 8.175
    13.139
19. $48.36 + $42.15
    $90.51
20. 75.25 + 8.638
    83.888
21. 39.81 − 16.256
    23.554
22. 52.43 − 10.71
    41.72
23. $52.46 − $15.95
    $36.51
24. 109.25 − 78.96
    30.29
25. 427.81 − 118.04
    309.77
26. $56.94 + $40.01
    $96.95
27. 3.2985 + 7.216
    10.5145
28. 12.007 + 8.8
    20.807
29. 8.941 − 8.149
    0.792
30. $8,210 − $299
    $7,911
31. $87.86 − $36.12
    $51.74
32. $3.95 + $4.01
    $7.96
33. 52.5 + 93.77
    146.27

Apply

Solve these problems. Round answers to the nearest whole number.

34. Nan weighs 48.36 kilograms and Jan weighs 43.78 kilograms. About how much more than Jan does Nan weigh?
    4 kilograms

35. A sweater costs $46.87 and a shirt costs $18.95. About how much does it cost to buy both?
    $66

36. Mac ran 3.75 kilometers and Hugh ran 7.09 kilometers. About how much farther than Mac did Hugh run?
    3 kilometers

37. Mel bought a watch for $26.59 and a ring for $20.59. He gave the clerk $50. About how much change did he receive?
    $2

170

Lesson 8-6 • Estimating Sums and Differences

4 Assess

Ask students to describe an everyday situation in which they would estimate a sum or difference. (Possible answer: shopping)

5 Mixed Review

1. $15 \div \frac{5}{6}$ (18)
2. 5^2 (25)
3. $3\frac{3}{8} + 5\frac{3}{4} + 2\frac{5}{12}$ $(11\frac{13}{24})$
4. 176×84 (14,784)
5. $15 - 9\frac{7}{8}$ $(5\frac{1}{8})$
6. $4\frac{2}{5} \times 10$ (44)
7. $21,792 \div 3$ (7,264)
8. $807.38 − $576.21 ($231.17)
9. $\frac{15}{28} \div \frac{3}{4}$ $(\frac{5}{7})$
10. $9,541 + 16,026 + 14,300$ (39,867)

For Mixed Abilities

Common Errors • Intervention

Some students may estimate incorrectly because they mentally round correctly but then have difficulty remembering the rounded numbers. As students work each problem, have them write the rounded numbers next to the original problem and then add or subtract as appropriate.

Enrichment • Application

Give students pages from a mail-order catalog. Ask them to select 3 to 5 items they would like to buy. Have students round the price of each item and estimate the cost of all the items. After students have estimated and recorded the sum of their purchases, have them compute the actual amounts with calculators. Ask students to compare the estimated sum with the actual sum.

More to Explore • Logic

Duplicate the following for students to solve:

Four aliens from four different planets raced toward Earth in four different spacecraft, each having a different power source. Scientists on Earth predicted who would reach our planet first by knowing:

- The Astros came from Pluto.
- The red ship is solar powered.
- The Cosmos have an ion-powered ship.
- The yellow ship and the Betas come from neighboring planets.
- The Droids live on the planet closest to the sun.
- The nuclear-powered ship comes from Mars.
- The blue gravity-powered ship must come the farthest.
- The green ship will get to Earth first.

   What people from which planet will win the race? (the Betas, from Mars)

T170

8-7 Adding Decimals

pages 171–172

1 Getting Started

Objective
• To add decimals

Warm Up • Mental Math
Have students name the number of minutes in:

1. $\frac{1}{2}$ hour (30)
2. $\frac{1}{4}$ hour (15)
3. $\frac{1}{5}$ hour (12)
4. $\frac{1}{6}$ hour (10)
5. $\frac{1}{30}$ hour (2)
6. $\frac{1}{10}$ hour (6)
7. $\frac{1}{20}$ hour (3)
8. $\frac{1}{3}$ hour (20)
9. $\frac{1}{12}$ hour (5)
10. $\frac{1}{60}$ hour (1)

Warm Up • Pencil and Paper
Have students find each sum.

1.	326	2.	674	3.	407
	+ 471		+ 143		+ 566
	(797)		(817)		(973)

4. 283 + 517 (800)

Have students find each missing number.

5. (3,002) − 1,596 = 1,406
6. (7,105) − 3,968 = 3,137
7. (4,230) − 3,581 = 649
8. (6,341) − 4,876 = 1,465

2 Teach

Introduce the Lesson Have a student read the problem, identify the problem question, and describe the given information. Have students complete the sentences as they read aloud with you to solve the problem. As you work through the model, point out that the procedure for adding decimals is the same as that for adding whole numbers.

Develop Skills and Concepts Emphasize the following to students:

• Write the decimals so that the decimal points align vertically. Doing this will align the digits according to their place value.

Adding Decimals

Mrs. Garvin uses a pedometer to measure the distance she walks during the week. She was surprised by her weekly total, after she recorded that she had walked 2.57 miles on Friday. What was the total distance Mrs. Garvin walked during the week?

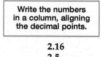

	Monday	Wednesday	Friday
Miles	2.16	2.5	2.57

We want to know how many miles Mrs. Garvin walked. We know she walked __2.16__ miles on Monday, __2.5__ miles on Wednesday, and __2.57__ miles on Friday.

To find the total mileage, we add the three distances together. We add __2.16__, __2.5__, and __2.57__.

Write the numbers in a column, aligning the decimal points.	Write equivalent decimals.	Add from right to left. Place the decimal point.
2.16	2.16	2.16
2.5	2.50	2.50
+ 2.57	+ 2.57	+ 2.57
		7.23

Mrs. Garvin walked a total of __7.23__ miles.

Getting Started _____

Add.

1.	5.769	2.	16.13	3.	9.275	4.	8.32
	+ 4.243		+ 29.75		+ 3.18		+ 5.7684
	10.012		45.88		12.455		14.0884

Copy and add.

5. 8.762 + 4.38 + 7.323 __20.465__

6. 7.294 + 8.6 + 9.5763 __25.4703__

• Write 0 to the left of the decimal point if there is no digit in the ones place.

• Add as though the numbers named whole numbers.

• Write the decimal for the sum directly below the decimal points in the addends.

3 Practice

Have students complete all the exercises.

Practice

Add.

1.
$$\begin{array}{r} 54.9 \\ + 16.7 \\ \hline 71.6 \end{array}$$

2.
$$\begin{array}{r} 35.16 \\ + 18.25 \\ \hline 53.41 \end{array}$$

3.
$$\begin{array}{r} 29.8 \\ + 15.76 \\ \hline 45.56 \end{array}$$

4.
$$\begin{array}{r} 28.57 \\ + 16.9 \\ \hline 45.47 \end{array}$$

5.
$$\begin{array}{r} 9.675 \\ + 3.429 \\ \hline 13.104 \end{array}$$

6.
$$\begin{array}{r} 25.3 \\ + 8.758 \\ \hline 34.058 \end{array}$$

7.
$$\begin{array}{r} 14.2965 \\ + 9.9813 \\ \hline 24.2778 \end{array}$$

8.
$$\begin{array}{r} 37.4815 \\ + 16.39 \\ \hline 53.8715 \end{array}$$

9.
$$\begin{array}{r} 29.61 \\ 13.75 \\ + 15.83 \\ \hline 59.19 \end{array}$$

10.
$$\begin{array}{r} 13.482 \\ 9.75 \\ + 6.824 \\ \hline 30.056 \end{array}$$

11.
$$\begin{array}{r} 6.7 \\ 8.18 \\ + 9.964 \\ \hline 24.844 \end{array}$$

12.
$$\begin{array}{r} 0.7384 \\ 0.9248 \\ + 0.3956 \\ \hline 2.0588 \end{array}$$

13.
$$\begin{array}{r} 80.431 \\ 72.1 \\ + 3.007 \\ \hline 155.538 \end{array}$$

14.
$$\begin{array}{r} 12.01 \\ 17.99 \\ + 200.472 \\ \hline 230.472 \end{array}$$

15.
$$\begin{array}{r} 86.68 \\ 24.87 \\ + 7.26 \\ \hline 118.81 \end{array}$$

16.
$$\begin{array}{r} 143.92 \\ 27.81 \\ + 101.79 \\ \hline 273.52 \end{array}$$

17.
$$\begin{array}{r} 28.942 \\ 7.801 \\ + 842.55 \\ \hline 879.293 \end{array}$$

18.
$$\begin{array}{r} 38.51 \\ 29.72 \\ + 99.004 \\ \hline 167.234 \end{array}$$

19.
$$\begin{array}{r} 501.34 \\ 20.75 \\ + 1.09 \\ \hline 523.18 \end{array}$$

20.
$$\begin{array}{r} 22.83 \\ 831.08 \\ + 75.29 \\ \hline 929.20 \end{array}$$

Copy and add.

21. 21.6 + 18.3 + 6.5
46.4

22. 12.16 + 8.05 + 7.2
27.41

23. 5.943 + 18.976 + 6.1
31.019

24. 125.382 + 186 + 59.2
370.582

25. 58.2463 + 16.9425
75.1888

26. 0.2965 + 0.954 + 0.36
1.6105

27. 47 + 1.96 + 3.3275
52.2875

28. 19.756 + 0.65 + 8.2903
28.6963

29. 17.896 + 21.07 + 12.81
51.776

30. 486.295 + 211.9094
698.2044

31. 81 + 201.75 + 821.403
1,104.153

32. 28.714 + 888.28 + 82
998.994

Problem Solving

Solve each problem.

33. Bill has three containers. One holds 11.25 liters of water, one holds 9.15 liters, and the third holds 8.105 liters. How much water does it take to fill the three containers? 28.505 liters

34. A male panda weighs 216.35 kilograms. A female panda weighs 36.5 kilograms more. What is the combined weight of the two pandas? 469.20 kilograms

Lesson 8-7 • Adding Decimals

---

4 Assess

Ask students what is wrong with this calculation.

$$\begin{array}{r} 7.25 \\ 10.5 \\ + 25.75 \\ \hline 108.75 \end{array}$$

(Possible answer: The decimal points are not lined up.)

Then, have students write the addition correctly and find the sum. (43.5)

For Mixed Abilities

Common Errors • Intervention

When students are adding decimals, some may add the decimal parts to the right of the decimal point and the whole-number parts to the left of the decimal point separately. Have students work the problems on a place-value chart, renaming from right to left, just as they do when adding whole numbers.

Enrichment • Number Sense

Have students work in groups. Ask each student to draw three rows of answer blanks as shown below:

—— —— —— ——

—— —— —— ——

+ —— —— —— ——

Have students take turns rolling two number cubes six times. After each roll, have them write those two digits in any answer blank. When all the blanks are filled, have students add the numbers. The student with the greatest sum wins one point. The first student to win 5 points wins the game.

More to Explore • Sets

Duplicate this for students.

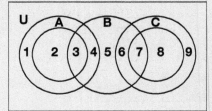

U = hungry people

A = people who like hot dogs

B = people who like hamburgers

C = people who like pizza

D = people who like mustard

E = people who like cheese.

Have students use the Venn diagram to create questions for other students to answer.

T172

8-8 Subtracting Decimals

pages 173–174

1 Getting Started

Objective
• To subtract decimals

Warm Up • Mental Math
Have students name the next three numbers in each sequence.

1. $\frac{1}{2}, \frac{1}{4}, \frac{1}{8}, \frac{1}{16}$, ($\frac{1}{32}, \frac{1}{64}, \frac{1}{128}$)
2. $\frac{1}{5}, \frac{1}{10}, \frac{1}{20}, \frac{1}{40}$, ($\frac{1}{80}, \frac{1}{160}, \frac{1}{320}$)
3. 1.97, 1.98, 1.99, (2, 2.01, 2.02)
4. 1, 2, 6, 24, (120, 720, 5,040)
5. 1, 4, 9, 16, (25, 36, 49)

Warm Up • Pencil and Paper
Have students find each difference.

1. 681 − 426 (255)
2. 304 − 97 (207)
3. 632 − 538 (94)

Have students find each missing number.

4. 5,149 − (3,685) = 1,464
5. 8,021 − (3,994) = 4,027
6. 6,302 − (267) = 6,035
7. 9,001 − (7,426) = 1,575

Name _____

Subtracting Decimals

The girls' 200-meter medley relay team set a new record at the swim meet. How much faster did Lauren swim than Doris?

We want to know the difference in times for the two swimmers.

200-meter Medley Relay		
Amy	50-m Backstroke	35.26 seconds
Doris	50-m Butterfly	33.7 seconds
Nancy	50-m Breaststroke	38.24 seconds
Lauren	50-m Freestyle	29.85 seconds

We know Doris swam the 50-meter butterfly relay in __33.7__ seconds, and Lauren swam the 50-meter freestyle relay in __29.85__ seconds.

To find how much faster Lauren swam than Doris, we subtract Lauren's time from Doris's time. We subtract __29.85__ from __33.7__.

Write the numbers in a column, aligning the decimal points.	Write equivalent decimals.	Subtract from right to left. Place the decimal point.
33.7 − 29.85	33.70 − 29.85	33.70 − 29.85 3.85

Lauren swam the 50-meter freestyle relay __3.85__ seconds faster than Doris swam the 50-meter butterfly relay.

Getting Started

Subtract.

1. 6.35 − 4.91 = 1.44
2. 5.958 − 2.68 = 3.278
3. 18.21 − 3.485 = 14.725
4. 7.3408 − 1.9659 = 5.3749

Copy and subtract.

5. 18.762 − 9.853 = 8.909
6. 36 − 7.032 = 28.968
7. 49.385 − 0.9786 = 48.4064

---

2 Teach

Introduce the Lesson Have students read the problem. Ask, *What do you need to find out?* (how much faster Lauren swam than Doris) Have students look at the chart of race results. Ask what information is given. Have students complete the information sentences and work through the subtraction algorithm. Point out that the procedure for subtracting decimals is the same as that for subtracting whole numbers.

Develop Skills and Concepts Remind students how to write equivalent decimals by filling all necessary places. Stress the importance of keeping corresponding places and decimal points aligned. Have students do the following subtraction exercises at the board:

18.43	26.97	3.0471
− 9.5	− 12.864	− 2.9765
(8.93)	(14.106)	(0.0706)

3 Practice

Have students complete all the exercises. Students need to be careful to align place values and decimal points. Remind students to annex zeros to the right of the decimals as needed so that both decimals have the same number of decimal places.

Practice

Subtract.

1.
$$\begin{array}{r} 9.61 \\ -\ 2.35 \\ \hline 7.26 \end{array}$$

2.
$$\begin{array}{r} 6.95 \\ -\ 2.43 \\ \hline 4.52 \end{array}$$

3.
$$\begin{array}{r} 9.671 \\ -\ 4.385 \\ \hline 5.286 \end{array}$$

4.
$$\begin{array}{r} 1.916 \\ -\ 0.187 \\ \hline 1.729 \end{array}$$

5.
$$\begin{array}{r} 13.824 \\ -\ 2.89 \\ \hline 10.934 \end{array}$$

6.
$$\begin{array}{r} 73.508 \\ -\ 15.323 \\ \hline 58.185 \end{array}$$

7.
$$\begin{array}{r} 18.7659 \\ -\ 7.3246 \\ \hline 11.4413 \end{array}$$

8.
$$\begin{array}{r} 14.683 \\ -\ 6.9285 \\ \hline 7.7545 \end{array}$$

9.
$$\begin{array}{r} 39.392 \\ -\ 16.596 \\ \hline 22.796 \end{array}$$

10.
$$\begin{array}{r} 128.16 \\ -\ 79.3852 \\ \hline 48.7748 \end{array}$$

11.
$$\begin{array}{r} 13.004 \\ -\ 9.7765 \\ \hline 3.2275 \end{array}$$

12.
$$\begin{array}{r} 25.3 \\ -\ 24.965 \\ \hline 0.335 \end{array}$$

13.
$$\begin{array}{r} 59.781 \\ -\ 21.947 \\ \hline 37.834 \end{array}$$

14.
$$\begin{array}{r} 883.01 \\ -\ 29.75 \\ \hline 853.26 \end{array}$$

15.
$$\begin{array}{r} 175.9291 \\ -\ 83.2107 \\ \hline 92.7184 \end{array}$$

16.
$$\begin{array}{r} 286.1 \\ -\ 24.314 \\ \hline 261.786 \end{array}$$

Copy and subtract.

17. 52.7 − 13.8
38.9

18. 24.37 − 18.26
6.11

19. 51.2 − 13.51
37.69

20. 0.679 − 0.296
0.383

21. 15.96 − 9.376
6.584

22. 48.321 − 19.9681
28.3529

23. 49 − 3.276
45.724

24. 76.423 − 38.9
37.523

25. 76.745 − 19.968
56.777

26. 15.003 − 14.386
0.617

27. 83.76516 − 14.83768
68.92748

28. 15.3802 − 6.94376
8.43644

29. 824.39 − 261.22
563.17

30. 282.07 − 41.703
240.367

31. 12.785 − 4.039
8.746

Problem Solving

Solve each problem. Use the chart for Problems 34 and 35.

32. From a roll of paper 3.67 meters long, Irv cuts a piece 1.48 meters long. How long is the piece that is left?
2.19 meters

33. For a cookout, Mary bought 4.59 kilograms of Swiss cheese and 2.65 kilograms of American cheese. How much cheese did Mary buy?
7.24 kilograms

34. How much longer is the sixth graders' record jump than that of the fifth graders?
1.84 meters

35. Gloria is in the sixth grade. Her long jump measured 6.955 meters. How far short of the record is Gloria's jump?
0.645 meters

Long Jump Records	
4th Graders	5.3 meters
5th Graders	5.76 meters
6th Graders	7.6 meters

174 Lesson 8-8 • Subtracting Decimals

---

4 Assess

On the board, write **56.2 − 9.0681**

Ask students whether or not they will need to write an equivalent decimal for 56.2, and explain why or why not. (Yes.; Possible answer: Both numbers should have the same number of decimal places to subtract. 56.2 should be written as 56.2000.)

5 Mixed Review

1. 21,256 ÷ 8 (2,657)
2. $5\frac{3}{8} + 4\frac{6}{7}$ ($10\frac{13}{56}$)
3. $\frac{11}{12} - \frac{3}{10}$ ($\frac{37}{60}$)
4. 181.3 − 107.276 (74.024)
5. 2,176 × 59 (128,384)
6. $3 \times \frac{5}{9} \times 4\frac{1}{3}$ ($7\frac{2}{9}$)
7. 40,201 − 38,376 (1,825)
8. 323.3 + 426 + 18.612 (767.912)
9. 275,620 + 308,176 (583,796)
10. $\frac{5}{6} \div \frac{2}{3}$ ($1\frac{1}{4}$)

For Mixed Abilities

Common Errors • Intervention

When copying decimals to subtract, some students may not line up the decimal points correctly. Have them work the problems on a place-value chart, which will help them to align the digits in their proper places.

Enrichment • Number Sense

Ask students to find the missing digits in each of the following problems:

1.
$$\begin{array}{r} 7.4\,(8) \\ -\ (3).(5)\,7 \\ \hline 3.9\ 1 \end{array}$$

2.
$$\begin{array}{r} 7\,4.1\ 5 \\ -\quad (8).4\,(0) \\ \hline (6)\ 5.7\ 5 \end{array}$$

3.
$$\begin{array}{r} 8\,3.(2)0(4) \\ -\ 3\,8.1\,4\,7 \\ \hline (4)5.0\,(5)\,7 \end{array}$$

4.
$$\begin{array}{r} 6.(0)6\ 5 \\ -\ (2).7\,3\,(6) \\ \hline (3).\,3(2)\ 9 \end{array}$$

5.
$$\begin{array}{r} 4.0\,7\,5(6) \\ -\ 1.9\,7(4)8 \\ \hline (2).1(0)0\ 8 \end{array}$$

6.
$$\begin{array}{r} (6).\,4(3)0 \\ -\ 2.0\,2\,8 \\ \hline 4.(4)0\ 2 \end{array}$$

More to Explore • Biography

A precocious youngster, Gottfried Wilhelm von Leibniz taught himself Latin and Greek by age 8, and entered the University of Leipzig at 15. This German mathematician and philosopher, who lived from 1646 to 1716, invented calculus independently from Isaac Newton. The son of a philosophy professor, Leibniz became a lawyer and worked for legal reform. While practicing law, he studied math and physics. Besides inventing a calculating machine, Leibniz wrote many works on his philosophy, named *monadology* or *leibnizianism*. Here Leibniz stated that each person and thing is a monad, or a completely separate being whose existence is in harmony with God and is separate from outer experience. Leibniz believed that humans should accept their lot in life and not try to change it because it was "the best of all possible worlds." Unfortunately, Leibniz's ideas were not widely accepted, and he died alone and unnoticed.

8-9 Practice Adding and Subtracting Decimals

pages 175–176

1 Getting Started

Objective
• To practice adding and subtracting decimals

Materials
newspaper advertisements or catalogs

Warm Up • Mental Math
Have students name the number that is 0.1 more than:

1. 5.8 (5.9)
2. 6.0 (6.1)
3. 2.9 (3.0)
4. 3.7 (3.8)
5. 3.42 (3.52)
6. 2.91 (3.01)
7. 7.06 (7.16)
8. 5.29 (5.39)

Warm Up • Pencil and Paper
Have students round each number to nearest whole number to estimate the sum or difference. Then, have them compute the exact sum or difference, and calculate the difference between the estimate and exact answer.

1. 4.684 + 2.795 = (8; 7.479; 0.521)
2. 3.76 + 4.7 + 2.946 = (12; 11.406; 0.594)
3. 3.065 − 1.459 = (2; 1.606; 0.394)
4. 6.5 − 1.946 = (5; 4.554; 0.446)

2 Teach

Introduce the Problem Have a student read the problem, tell what is to be solved, and tell what information is given. Have students complete the information sentences as they work through the model to solve the problem.

Develop Skills and Concepts Explain that in this lesson, students will practice addition and subtraction of decimals in both exercises and word problems. Encourage them to estimate before they compute as a way of evaluating the reasonableness of their answers. Remind students of the inverse nature of addition and subtraction; have students demonstrate how one operation can be used to check the other.

Name _____

Practice Adding and Subtracting Decimals

Randy made a chart to show the average yearly rainfall in different U.S. cities. On the average, how many more inches of rain fall in New York City than in Chicago?

Average Yearly Rainfall
Chicago   39.57 inches
New York City 48.15 inches
Los Angeles   12.23 inches

We want to know the difference in average yearly rainfall in New York City and Chicago. We know New York averages __48.15__ inches and Chicago averages __39.57__ inches.

To compare the rainfall in the two cities, we subtract the average Chicago rainfall from the average New York City rainfall.

We subtract __39.57__ from __48.15__.

$$\begin{array}{r} 48.15 \\ -\ 39.57 \\ \hline 8.58 \end{array}$$

New York City averages __8.58__ more inches of rain than Chicago per year.

Getting Started

Compute.

1.	2.	3.	4.
97.3	32.9	112.476	4.093
− 16.45	− 16.63	− 96.789	+ 8.97
80.85	16.27	15.687	13.063

Copy and compute.

5. 39.7 + 18.9 − 15.6 43.0
6. 5.89 − 3.341 + 14 16.549
7. 6.2 + 3.96 + 8.54 18.70
8. 24.61 + 18.21 − 3.75 39.07
9. 19.61 − 8.75 + 2.95 13.81
10. 3.01 + 120.72 − 8.08 115.65

Lesson 8-9 • Practice Adding and Subtracting Decimals

175

3 Practice

Have students complete all the exercises. They should skim the page to see if they spot difficulties with the types of problems presented. Have students work with peers on those problems they feel will be difficult.

Practice

Compute.

1.
$$\begin{array}{r} 16.91 \\ + 15.86 \\ \hline 32.77 \end{array}$$

2.
$$\begin{array}{r} 37.21 \\ - 19.58 \\ \hline 17.63 \end{array}$$

3.
$$\begin{array}{r} 32.5 \\ + 18.63 \\ \hline 51.13 \end{array}$$

4.
$$\begin{array}{r} 75.2 \\ - 17.58 \\ \hline 57.62 \end{array}$$

5.
$$\begin{array}{r} 126.2 \\ - 97.851 \\ \hline 28.349 \end{array}$$

6.
$$\begin{array}{r} 52.483 \\ + 78.925 \\ \hline 131.408 \end{array}$$

7.
$$\begin{array}{r} 89.58 \\ - 23.965 \\ \hline 65.615 \end{array}$$

8.
$$\begin{array}{r} 49.008 \\ - 15.779 \\ \hline 33.229 \end{array}$$

9.
$$\begin{array}{r} 156.17 \\ + 28.3954 \\ \hline 184.5654 \end{array}$$

10.
$$\begin{array}{r} 175.43 \\ - 98.7658 \\ \hline 76.6642 \end{array}$$

11.
$$\begin{array}{r} 415.673 \\ - 295.873 \\ \hline 119.800 \end{array}$$

12.
$$\begin{array}{r} 515.2843 \\ + 329.7689 \\ \hline 845.0532 \end{array}$$

Copy, and add or subtract.

13. $75.2 + 18.6 - 65.3$
28.5

14. $82.16 - 13.29 + 16.5$
85.37

15. $4.962 + 6.83 - 4.759$
7.033

16. $25.371 + 13.7 + 19.651$
58.722

17. $39 - 16.28 + 15.75$
38.47

18. $96.136 + 48.792 - 63.4248$
81.5032

Problem Solving

Solve each problem. Use the chart to answer Problems 21 and 22.

19. The maximum weight of a full container is 19.325 kilograms. Into the container, Larry put one object that weighs 9.5 kilograms and another that weighs 7.75 kilograms. How much more weight can Larry put into the container?
2.075 kilograms

20. In one year, 42.8 million passengers passed through O'Hare Airport in Chicago, and 37.9 million passengers went through the Atlanta Airport. The Los Angeles Airport had 4.5 million fewer passengers than did the Atlanta Airport. In all, how many passengers passed through these three airports?
114.1 million

21. Mr. Martinez kept track of the miles he drove in a car he rented at the airport. How far did Mr. Martinez drive?
1,025.41 miles

22. How much farther did Mr. Martinez drive on Monday and Tuesday than Wednesday and Thursday?
4.69 miles

Driving Record	
Monday	365.7
Tuesday	149.35
Wednesday	296.5
Thursday	213.86

176

Lesson 8-9 • Practice Adding and Subtracting Decimals

Common Errors • Intervention

Some students may forget to annex zeros when required, particularly in subtraction problems. Have them align the decimal points and then annex zeros so that there are the same number of places to the right of the decimal point in both numbers. Grid paper can be used with those students who also have difficulty keeping the digits aligned after aligning the decimal points.

Enrichment • Number Sense

Instruct students to complete the magic square. Explain that the magic number, or the sum for each row, column, and diagonal, should be 1.47.

0.76	(0.13)	0.58
0.31	(0.49)	(0.67)
0.4	(0.85)	(0.22)

More to Explore • Applications

Have students collect copies of schedules from a commercial, long-distance bus line. Have them study the schedules and practice reading them. Help them identify route numbers, departures, arrivals, and fares.

Let students pair off and take turns pretending to be a reservationist for the bus company and a customer. The reservationist's job is to answer questions for the customer about the schedule, such as *What is the cheapest fare to Chicago? What is the earliest bus to Dallas on Tuesday?* Have students change roles.

You may also have pairs of students research and report on what other types of transportation use schedules.

4 Assess

On the board, write **3.08 + 20.72 − 8.08**

Have students explain how they would calculate the answer. (Possible answer: First, write the addition in vertical form, aligning the decimal points, and find the sum. Next, subtract 8.08. Before you subtract, you will need to rewrite the sum 123.8 as 123.80.)

5 Mixed Review

1. $5,000.00 − $372.29
($4,627.71)

2. $170.3 − 123.6$
(46.7)

3. $6\frac{1}{4} + 5\frac{2}{3} + 3\frac{5}{8}$ $(15\frac{13}{24})$

4. 608×37 (22,496)

5. $5\frac{2}{3} \times 2$ $(11\frac{1}{3})$

6. 0.036×10^4
(360)

7. $6.07 + 15.3 + 21.795$
(43.165)

8. 4^4 (256)

9. $\frac{7}{10} - \frac{2}{5}$ $(\frac{3}{10})$

10. $62,186 \div 4$ (15,546 R2)

T176

8-10 Problem Solving: Draw a Diagram

pages 177–178

1 Getting Started

Objective
• To solve problems by using appropriate notation

Warm Up • Mental Math
Ask students what number is represented by each word.

1. trio (3)
2. ream (500)
3. decathlon (10)
4. octave (8)
5. couplet (2)
6. Pentateuch (5)
7. duet (2)
8. quintet (5)

Warm Up • Activity
Draw this Venn diagram on the board and discuss the information that is illustrated.

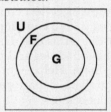

Ask students what conclusions they can reach about sets F and G.
(Possible answer: G is in F. Some of F is in G. Some of F is not in G.)

Name _____

Problem Solving: Draw a Diagram

Fred has a blue shirt, a white shirt, a yellow shirt, a pair of jeans, and a pair of black corduroy slacks. He also has a white jacket and a navy blue jacket. How many different three-piece outfits can he make?

★ SEE
We want to know how many different three-piece outfits Fred can make.

It takes a shirt, a pair of slacks, and a __jacket__ to make one outfit.

He has __3__ different shirts. He has __2__ different kinds of slacks. He has __2__ different jackets.

★ PLAN
Since we have many possible combinations we can use a **tree diagram** to show how many outfits Fred can make.

★ DO

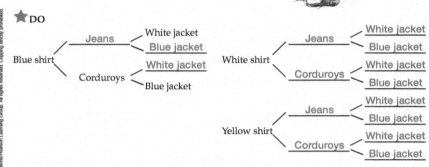

Fred can make __4__ different outfits that will use the blue shirt.

He can make __4__ outfits with the white shirt and __4__ outfits with the yellow shirt. Fred can make __12__ different three-piece outfits.

★ CHECK
We can check our work by being sure we recorded each type of slacks and each type of jacket with each kind of shirt.

---

2 Teach

Introduce the Lesson Have students take time to familiarize themselves with the problem. They must have a clear idea of the kind and number of each piece of clothing. They should clarify this in the SEE stage.

Develop Skills and Concepts Tell students that notation does not always mean numbers or equations. Notation can be any device that will enable them to clearly see and operate on data. An unusual example of this is the *tree diagram* used in the DO stage. All the possible combinations can be easily seen in a tree diagram. Be sure students summarize the diagram orally before they complete the solution statements independently.

Discuss with students other unusual notations they have used in solving problems. These might include Venn diagrams, making a list or table, or any of the problem-solving strategies.

3 Apply

Solution Notes
1. Use a Venn diagram.

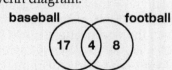

2. Use a tree diagram such as the following for the burger.

burger
- french fries
  - vanilla
  - chocolate
  - strawberry
  - banana
- home fries
  - vanilla
  - chocolate
  - strawberry
  - banana
- onion rings
  - vanilla
  - chocolate
  - strawberry
  - banana

Apply

Solve each problem.

1. There are 29 sixth graders in Miss Smith's class. Twenty-one students are on the baseball team. Twelve students are on the football team. How many students are on both the baseball and football teams?
   4 students

2. At Quick Burger you can order a regular burger, a double burger, or a giant burger. Side dishes are cole slaw, home fries, or salad. Milkshake flavors include vanilla, chocolate, strawberry, and banana. How many different meals can you buy that include a sandwich, a side order, and a milkshake?
   36 meals

3. Margaret, Fred, and Susan all teach at the same school. They teach math, science, and social studies. Margaret, who rides a bus to school, teaches in the room next to the math teacher. The social studies teacher gives Susan a ride to school each day. Who teaches what subject?
   Margaret science, Fred social studies, Susan math

4. Of the twenty-nine students who take music lessons,
   13 play violin,
   16 play piano,
   8 play trumpet,
   4 play piano and violin,
   1 plays trumpet and violin, and
   1 plays all three instruments.
   How many students play only piano? How many play only trumpet?
   11 students play only piano.
   6 students play only trumpet.

5. Francis, Linda, and Peggy love animals. Their pets are a dog, a cat, and a hamster. The pet's names are Daisy, Sasha, and Lucky. Peggy feeds Sasha each day before leaving for work. Francis, who owns Lucky, is allergic to cats. Linda walks her dog twice a day. Find the name of each pet and its owner.
   Peggy owns Sasha, the cat. Linda owns Daisy, the dog. Francis owns Lucky, the hamster.

Problem-Solving Strategy: Using the Four-Step Plan

★ SEE What do you need to find?

★ PLAN What do you need to do?

★ DO Follow the plan.

★ CHECK Does your answer make sense?

6. A scientist has four identical gold nuggets. He knows one is counterfeit. It has a slightly different weight. How can the scientist find the counterfeit nugget using only a balance scale?
   See Solution Notes.

7. Malcolm won a prize. He had a choice of receiving $150 right away or a daily payment beginning with $1.50 and doubling each day for 7 days. Which should Malcolm choose and why?
   7-day payment; he will receive $40.50 more.

8. The Math Club offers a prize to any person who writes a problem that can be solved using a tree diagram and the answer is 48 choices. What problem would you write?
   Answers will vary.

9. Maria added 3 decimals to get a sum of 5.27. If the sum of the first 2 addends is 3.79, and the sum of the last 2 addends is 3.63, what are the 3 addends added by Maria?
   1.64, 2.15, and 1.48

10. You have four addends. Each is a number of dollars and cents. If you round each amount to the nearest whole dollar and add, how far from the actual sum can you be?
    $2.00

Lesson 8-10 • Problem Solving: Draw a Diagram

More to Explore •
Number Sense

Tell students the following steps will enable them to guess another person's age:

1. Tell a person to multiply their age by three, add six to the answer, and then divide the sum by three.

2. Ask for the result.

3. Subtract two from that number and the student will get the person's age.

Have the students try this with each other and then mystify their family with their mind-reading ability.

---

3. Make a chart.

	Science	Math	Social Studies
Margaret	✓	✗	✗
Fred	✗	✗	✓
Susan	✗	✓	✗

4. Use a Venn diagram. Be sure that students understand that some students play more than one instrument.

5. Suggest students set up a chart as in problem 3.

6. Have students label the nuggets and cover all possible outcomes. The least number of weighings is 2.

Higher-Order Thinking Skills

7. **Evaluation:** The 7-day payments total $190.50, or $40.50 more than the one-time prize of $150.

8. **Synthesis:** Any set of factors of 48 will do as the basis: number pairs, such as 2 × 24 or 3 × 16, or number triples, such as 2 × 4 × 6, or number quadruples, such as 2 × 4 × 3 × 2, etc.

9. **Analysis:** Subtract 3.79 from 5.27 to get the third addend, 3.63 from 5.27 to get the first addend, and the sum of the first and third addends from 5.27 to get the second addend.

10. **Analysis:** The answer comes from rounding up 4 midway numbers at 50¢ each. Any other rounding, up or down, would result in less than 50¢.

4 Assess

Have students look back at the problem on page 177. Pose this problem, *Suppose Fred chooses the jacket first, then the shirt, and then the pants. How many different three-piece outfits can he make? Explain.* (12; Possible answer: The order the parts of the outfits are chosen does not affect the results.)

Chapter 8
Test

Alternate Chapter Test

You may wish to use the Alternate Chapter Test on page 384 of this book for further review and assessment.

What is the place value of the 3 in each number?

1. 6.02031 — ten thousandths

2. 5.13961 — hundredths

3. 9.24063 — hundred thousandths

4. 3.21 — ones

5. 15.3 — tenths

6. 231.5 — tens

Write <, =, or > in each circle.

7. 2.689 $<$ 2.698

8. 4.3216 $>$ 4.321

9. 7.685 $=$ 7.6850

10. 2.613 $<$ 2.631

11. 95.6231 $>$ 95.6223

12. 81.652 $>$ 81.562

Round to the nearest hundredth.

13. 7.9895   7.99

14. 4.3861   4.39

15. 15.673   15.67

Round to the nearest thousandth.

16. 15.6785   15.679

17. 14.21849   14.218

18. 4.00653   4.007

19. 256.2316   256.232

20. 38.3672   38.367

21. 9.6289   9.629

Estimate each sum and difference by first rounding the numbers to the nearest whole number.

22.	42.78 + 36.12	43 + 36 79	23.	$18.39 − 9.50	$18 − 10 $8	24.	15.372 − 12.986	15 − 13 2	25.	12.28 − 6.428	12 − 6 6

Add.

26. 14.64 + 58.97 = 73.61

27. 57.391 + 16.83 = 74.221

28. 12.241 + 16.9586 = 29.1996

29. 47.3829 + 25.7883 = 73.1712

Subtract.

30. 46.34 − 18.96 = 27.38

31. 52.385 − 16.79 = 35.595

32. 24.376 − 8.9378 = 15.4382

33. 75.2418 − 39.7519 = 35.4899

Circle the letter of the correct answer.

1 $12 - 6 \times 2 = n$
$n = ?$
a. 0 ⃝
b. 4
c. 12
d. NG

2 What is the place value of the 7 in 1,796,851?
a. ten thousands
b. hundred thousands ⃝
c. millions
d. NG

3 $\begin{array}{r} 62,412 \\ + 15,496 \end{array}$
a. 77,808
b. 77,908 ⃝
c. 78,908
d. NG

4 $\begin{array}{r} 46,250 \\ - 38,659 \end{array}$
a. 7,592
b. 12,408
c. 17,592
d. NG ⃝

5 546×7
a. 3,522
b. 3,582
c. 3,822 ⃝
d. NG

6 $\begin{array}{r} 429 \\ \times\ 36 \end{array}$
a. 3,861
b. 15,444 ⃝
c. 16,444
d. NG

7 $476 \div 6$
a. 7 R56
b. 79
c. 79 R2 ⃝
d. NG

8 $46\overline{)9,453}$
a. 25 R23
b. 205
c. 205 R23 ⃝
d. NG

9 $\begin{array}{r} 3\frac{2}{3} \\ + 3\frac{1}{6} \end{array}$
a. $6\frac{1}{3}$
b. $7\frac{5}{6}$
c. $6\frac{5}{6}$ ⃝
d. NG

10 $\begin{array}{r} 5\frac{1}{3} \\ - 2\frac{3}{4} \end{array}$
a. $2\frac{7}{12}$ ⃝
b. $3\frac{5}{12}$
c. $3\frac{7}{12}$
d. NG

11 $1\frac{3}{4} \times 2\frac{2}{3}$
a. $\frac{21}{32}$
b. $4\frac{1}{2}$
c. $4\frac{2}{3}$ ⃝
d. NG

□ score

STOP

Chapter 8 • Cumulative Assessment

page 180

Items	Objectives
1	To simplify an expression having two or more operations (see pages 11–12)
2	To identify place value through trillions (see pages 25–26)
3	To add two 5-digit numbers (see pages 27–28)
4	To subtract, renaming over zeros (see pages 33–34)
5–6	To multiply 3-digit numbers by 1- or 2-digit numbers (see pages 45–46, 51–52)
7–8	To divide up to 5-digit numbers by 1- or 2-digit numbers (see pages 63–64 and 75–76)
9–10	To add or subtract mixed numbers that require regrouping (see pages 119–120 and 125–126)
11	To multiply two mixed numbers (see pages 141–142)

Alternate Cumulative Assessment

Circle the letter of the correct answer.

1. $14 - 9 \times 0 = n$
$n = ?$
a. 0 ⃝
b. 9
c. 14
d. NG

2. Give the place value of the 7 in 67,259,103.
a. hundred thousands
b. millions ⃝
c. ten millions
d. NG

3. $\begin{array}{r} 84,735 \\ + 26,289 \end{array}$
a. 101,024
b. 110,924
c. 111,024 ⃝
d. NG

4. $\begin{array}{r} 57,470 \\ - 29,977 \end{array}$
a. 27,493 ⃝
b. 27,503
c. 27,593
d. NG

5. $\begin{array}{r} 398 \\ \times\ 6 \end{array}$
a. 1,888
b. 2,348
c. 2,388 ⃝
d. NG

6. $\begin{array}{r} 752 \\ \times\ 48 \end{array}$
a. 34,096
b. 35,696
c. 36,096 ⃝
d. NG

7. $7\overline{)409}$
a. 58
b. 58 R3 ⃝
c. 59
d. NG

8. $35\overline{)14,338}$
a. 49 R23
b. 409 R3
c. 409 R23 ⃝
d. NG

9. $\begin{array}{r} 7\frac{4}{5} \\ + 1\frac{3}{4} \end{array}$
a. $8\frac{11}{20}$
b. $9\frac{10}{20}$
c. $9\frac{11}{20}$ ⃝
d. NG

10. $\begin{array}{r} 8\frac{2}{7} \\ - 3\frac{5}{6} \end{array}$
a. $4\frac{19}{42}$ ⃝
b. $5\frac{18}{42}$
c. $5\frac{19}{42}$
d. NG

11. $10\frac{2}{3} \times 4\frac{1}{2}$
a. 38
b. $47\frac{5}{6}$
c. 48 ⃝
d. NG

T180

9-1 Multiplying by Powers of 10

pages 181–182

Multiply and Divide Decimals

Lesson 9-1

1 Getting Started

Objective
• To multiply a decimal by a power of 10

Warm Up • Mental Math
Have students simplify each expression.
1. $2 \times 5 \times 14$ (140)
2. $\frac{1}{2}$ of (9×4) (18)
3. $8 \times 2\frac{1}{2}$ (20)
4. $10 \times 10 \times 23$ (2,300)
5. $\frac{4}{5}$ of 20 (16)
6. $\frac{3}{4}$ of 32 (24)
7. $1\frac{1}{2} \times 10$ (15)
8. $\frac{2}{3}$ of 12 (8)

Warm Up • Mental Math
Have students write each product.
1. 10×6 (60)
2. 100×12 (1,200)
3. $1,000 \times 8$ (8,000)
4. 10×12 (120)
5. 100×91 (9,100)
6. $1,000 \times 185$ (185,000)

Have students write each number in standard form.
1. 10^1 (10)   3. 10^4 (10,000)
2. 10^3 (1,000) 4. 10^2 (100)

2 Teach

Introduce the Lesson Have students complete the patterns for multiplying decimals by powers of 10 at the top of page 181. Stress that students can use these patterns to find products quickly. Have students read and complete each rule for multiplying by a power of 10.

Develop Skills and Concepts Be sure students understand the rationale behind the shortcut of moving the decimal point to the right as many places as there are zeros in the power of 10. Point out that moving the decimal point one place to the right also moves each digit one place to the left. This movement increases the value of each digit 10 times.

Extend this explanation to multiplying by 100. For example, to multiply 27×100, the decimal point in 27, which is understood to be to the right of the ones place

Multiplying by Powers of 10

Complete the patterns to find a shortcut for multiplying by powers of ten.

$5.2 \times 10 = 52$ $5.2 \times 10^1 = 52$
$5.2 \times 100 = 520$ $5.2 \times 10^2 = 520$
$5.2 \times 1,000 = 5,200$ $5.2 \times 10^3 = 5,200$
$5.2 \times 10,000 = \underline{52,000}$ $5.2 \times \underline{10^4} = \underline{52,000}$

$0.8341 \times 10 = 8.341$ $0.8341 \times 10^1 = 8.341$
$0.8341 \times 100 = 83.41$ $0.8341 \times 10^2 = 83.41$
$0.8341 \times 1,000 = 834.1$ $0.8341 \times 10^3 = 834.1$
$0.8341 \times 10,000 = \underline{8,341}$ $0.8341 \times \underline{10^4} = \underline{8,341}$

Powers of 10
$10^1 = 10$
$10^2 = 100$
$10^3 = 1,000$
$10^4 = 10,000$

Multiplying by 10 or 10^1 moves the decimal point $\underline{1}$ place to the right.

Multiplying by 100 or 10^2 moves the decimal point $\underline{2}$ places to the right.

Multiplying by 1,000 or 10^3 moves the decimal point $\underline{3}$ places to the right.

Multiplying by 10,000 or 10^4 moves the decimal point $\underline{4}$ places to the right.

When you multiply by a power of ten, move the decimal point to the right the same number of places as the exponent. Write extra zeros to the right of the number if needed.

Getting Started

Multiply. Use mental math.
1. $7.3 \times 1,000 = \underline{7,300}$
2. $8.26 \times 10^2 = \underline{826}$
3. $10^4 \times 5.68 = \underline{56,800}$
4. $10 \times 4.032 = \underline{40.32}$
5. $4.815 \times 10^4 = \underline{48,150}$
6. $100 \times 8.265 = \underline{826.5}$

Write the missing factor in exponential form.
7. $2.5974 \times \underline{10^2} = 259.74$
8. $\underline{10^3} \times 5.48 = 5,480$
9. $356.7 \times \underline{10^1} = 3,567$

Lesson 9-1 • Multiplying by Powers of 10

181

(27.), must shift two places to the right, requiring that zeros be written in the tens and ones places (2,700).

Remind students to use zero as a placeholder only. It is not necessary to write zeros to the right of the decimal point when writing whole numbers.

Have volunteers work at the board to show how to find each product. Have them explain each step.

6.73×10 (67.3) 0.1×10 (1) 0.004×10^2 (0.4)

1.28×100 (128) $10^3 \times 0.05$ (50) $3.6457 \times 10,000$ (36,457)

3 Practice

Have students complete all the exercises. Encourage them to do the multiplication mentally.

Practice

Multiply. Use mental math.

1. $6.451 \times 10^2 = $ __645.1__

2. $1,000 \times 4.275 = $ __4,275__

3. $10^3 \times 19.75 = $ __19,750__

4. $0.0249 \times 10,000 = $ __249__

5. $10^2 \times 15.974 = $ __1,597.4__

6. $5.9 \times 10^2 = $ __590__

7. $8.375 \times 10 = $ __83.75__

8. $0.021 \times 10^4 = $ __210__

9. $1,000 \times 0.2465 = $ __246.5__

10. $10^3 \times 5.7615 = $ __5,761.5__

11. $10 \times 0.5 = $ __5__

12. $11.751 \times 10^1 = $ __117.51__

Write the missing factor in exponential form.

13. $7.5 \times \underline{10^2} = 750$

14. $\underline{10^3} \times 0.651 = 651$

15. $3.271 \times \underline{10^2} = 327.1$

16. $\underline{10^3} \times 4.61 = 4,610$

17. $17.95 \times \underline{10^1} = 179.5$

18. $\underline{10^3} \times 8.15 = 8,150$

19. $0.4731 \times \underline{10^2} = 47.31$

20. $96.8 \times \underline{10^3} = 96,800$

Now Try This!

Try this unusual way of multiplying.

Row 1	0	1	2	3	4	5	6	7	8	9	10	11	12
Row 2	1	2	4	8	16	32	64	128	256	512	1,024	2,048	4,096

The Row 1 number for 16 is __4__.

The Row 1 number for 8 is __3__.

The Row 1 number for 128 is __7__.

The Row 1 number for 128 is __7__.

The Row 1 number for 32 is __5__.

The Row 1 number for 4,096 is __12__.

Multiply these numbers.   **Add these numbers.**

$$\begin{array}{r} 16 \\ \times\ 8 \\ \hline 128 \end{array} \qquad \begin{array}{r} 4 \\ +\ 3 \\ \hline 7 \end{array}$$

$$\begin{array}{r} 128 \\ \times\ 32 \\ \hline 4,096 \end{array} \qquad \begin{array}{r} 7 \\ +\ 5 \\ \hline 12 \end{array}$$

Use the Row 1 numbers to find these products.

1. $8 \times 64 = $ __512__

2. $512 \times 4 = $ __2,048__

3. $32 \times 32 = $ __1,024__

182   Lesson 9-1 • Multiplying by Powers of 10

Now Try This! Discuss the relationship between the Row 1 and Row 2 numbers. Explain to students that this problem uses a type of code. The Row 1 number is an exponent. The Row 2 number is the power when the base is 2. For example, the number 3 in Row 1 is matched to the number 8 in Row 2 because 2^3 is 8.

4 Assess

Ask, *How many places to the right do you move the decimal point when you multiply by 100,000?* (five)

For Mixed Abilities

Common Errors • Intervention

If students have difficulty remembering how many places to move the decimal point, have them complete the following table for practice.

×	10^1	10^2	10^3
7	(70)	(700)	(7,000)
4.2	(42)	(420)	(4,200)
0.641	(6.41)	(64.1)	(641)
27.306	(273.06)	(2,730.6)	(27,306)

Enrichment • Number Sense

Write the following on the board:

Before 4.27

After 42.7

Have students compare the two numbers. Ask what was done to the before number to get the after number. (multiplied by 10)

Continue with the following:

Before	**After**
1. 36.42	3,642 (multiplied by 100)
2. 0.0734	734 (multiplied by 10,000)
3. 5.41	54.1 (multiplied by 10)

More to Explore • Geometry

Write the following words on the board for students to research and define: *decahedron, heptagon, octahedron, dodecagon, icosahedron, pentahedron,* and *tetrahedron.* As an extension, have them draw an original geometric figure and name it. Display their creations.

ESL/ELL STRATEGIES

Ask students to point out the exponents in the examples on the first page of this chapter. Then, have them read some of the numbers aloud. For example, *Ten to the first power of 10 is 10. Ten squared is 100. Ten to the third power of 10 is 1,000.*

9-2 Estimating Decimal Products

pages 183–184

1 Getting Started

Objective
- To use rounding or a compatible number to estimate decimal products

Vocabulary
compatible number

Materials
fractional number lines

Warm Up • Mental Math
Have students find each product.
1. 20×40 (800)
2. $1.23 \times 1,000$ (1,230)
3. 15×20 (300)
4. 300×6 (1,800)
5. 60×80 (4,800)
6. 3.5×100 (350)

Warm Up • Pencil and Paper
Have students write the compatible number that could be used to estimate each product. Then, estimate the product.
1. $5\frac{1}{2} \times 1\frac{3}{4}$ (2; 11) 4. $3\frac{2}{3} \times 5\frac{1}{8}$ (6; 22)
2. $2\frac{4}{5} \times 9\frac{1}{6}$ (10; 28) 5. $3\frac{3}{8} \times 9\frac{1}{3}$ (8; 27)
3. $4\frac{1}{4} \times 7\frac{2}{5}$ (8; 34) 6. $6\frac{3}{7} \times 15\frac{1}{2}$ (1; 90)

Name _____

Lesson 9-2

Estimating Decimal Products

Sara buys 0.68 pound of turkey at the deli counter for $4.29 a pound. About how much will the turkey cost?

We can use rounding or compatible numbers to estimate products.

REMEMBER A compatible number is a number that is close to the original number but is easier to use.

Use rounding.	Use compatible numbers.

Round each factor to the nearest whole number. Then, multiply.

$$\begin{array}{r} \$4.29 \longrightarrow \quad \$4 \\ \times\ 0.68 \longrightarrow \times\ 1 \\ \hline \$4 \end{array}$$

The turkey will cost about ___$4___.

Substitute numbers that are easier to use. Then, multiply.

$$\begin{array}{r} \$4.29 \longrightarrow \quad \$4 \\ \times\ 0.68 \longrightarrow \times\ 0.70 \\ \hline \$2.80 \end{array}$$

Use the fact $4 \times 7 = 28$. So, $4 \times 0.70 = 2.80$

The turkey will cost about ___$2.80___.

Getting Started

Estimate by rounding to the nearest whole number.

1. 3.76×1.89
   about ___8___
2. 2.3×0.9
   about ___2___
3. 5.43×3.46
   about ___15___

Estimate by rounding to the nearest tenth.

4. 0.23×0.58
   about ___0.12___
5. 0.37×0.62
   about ___0.24___
6. 0.18×0.42
   about ___0.08___

Use compatible numbers to estimate. Write the numbers you used.

7. 4.03×0.34
   about ___1.2___
   4×0.3
8. $\$18.05 \times 0.39$
   about ___$8___
   $\$20 \times 0.4$
9. 9.52×0.48
   about ___5___
   10×0.5
10. 2.3×5.28
    about ___10___
    2×5

Lesson 9-2 • Estimating Decimal Products

183

---

2 Teach

Introduce the Lesson Have a student read the problem and tell what operation will be used. Point out to students that the word *about* tells them to estimate. Remind students of the two ways they have estimated: rounding and using compatible numbers.

Explain that to estimate decimal products by rounding, each factor is rounded to the nearest whole number. Draw a number line on the board and mark where 4.29 and 0.68 are located. Have a student explain how to round each number. Then, complete the problem by rounding.

To use compatible numbers, we substitute one or more factors for a number that is easier to use. Multiplying 4×0.70 is easier than multiplying 4.29×0.68. We can use the fact that $4 \times 7 = 28$ to find 4×0.70. Just move the decimal point one place to the left in 28 to get 2.8. So, $4 \times 0.70 = 2.8$.

Develop Skills and Concepts Give another example of estimating a product using both methods. Renee is buying 8.5 yards of fabric that costs $4.75 per yard. How much will the fabric cost? Write the following on the board:

Rounding	Compatible Numbers
$\begin{array}{r}\$4.75 \\ \times\ 8.5 \\ \hline\end{array}$	$\begin{array}{r}\$4.75 \\ \times\ 8.5 \\ \hline\end{array}$

Call on volunteers to demonstrate each method for estimating the product. For the rounding method, have a student explain how to round each factor. ($5, 9) Have the student write the rounded numbers to the right of the example and find the product. ($45)

To use compatible numbers, ask students to find a pair of numbers close to 4.75 and 8.5 that are easy to multiply. (5 and 10) Have them write these numbers to the right of the example and find the product. ($50)

Have students compare the estimates.

T183

Practice

Estimate by rounding to the nearest whole number.

1. 14.36
   $\times\ 1.89$
   about __28__

2. 24.63
   $\times\ 3.12$
   about __75__

3. 120.32
   $\times\ \ \ \ 5$
   about __600__

4. 124.86
   $\times\ \ \ 2.1$
   about __250__

5. 40.12
   $\times\ \ \ 9.8$
   about __400__

6. 300.42
   $\times\ 19.67$
   about __6,000__

7. 49.64
   $\times\ 30.4$
   about __1,500__

8. 24.69
   $\times\ \ 9.8$
   about __250__

Estimate by rounding to the nearest tenth.

9. 0.54
   $\times\ 0.86$
   about __0.45__

10. 0.29
    $\times\ 0.72$
    about __0.21__

11. 0.93
    $\times\ 0.66$
    about __0.63__

12. 0.18
    $\times\ 0.13$
    about __0.02__

Use compatible numbers to estimate. Write the numbers you used.

13. 6.07
    $\times\ 0.48$
    about __3__
    6×0.5

14. 23.2
    $\times\ 2.3$
    about __40__
    20×2

15. 0.48
    $\times\ 28.43$
    about __15__
    0.5×30

16. 38.65
    $\times\ \ \ 3.7$
    about __160__
    40×4

17. 97.6
    $\times\ 2.3$
    about __200__
    100×2

18. 190.6
    $\times\ \ \ 3.4$
    about __600__
    200×3

19. 27.6
    $\times\ 4.6$
    about __150__
    30×5

20. 5.8
    $\times\ 8$
    about __48__
    6×8

Problem Solving

Solve each problem. Tell which method you used to estimate.

21. About how much will 4 pairs of shoes cost if each pair costs $18.99?
    about $80 using rounding

22. Each shirt costs $28.95. About how much will 12 shirts cost?
    about $300 using rounding

23. Gasoline costs $1.68 per gallon. Steve buys 14.23 gallons to fill his tank. About how much does he pay?
    about $28 using rounding

24. If Steve buys $10.50 worth of gasoline at $1.68 per gallon, will he get more than 7 gallons? Explain.
    He will get less than 7 gallons because $1.68 × 7 is about $14.

184

Lesson 9-2 • Estimating Decimal Products

---

3 Practice

Have students complete all the exercises. Students may use number lines to help them round each decimal to the nearest whole number.

4 Assess

Have students estimate the product in Exercise 11 by rounding to the nearest whole number. Compare the answers gotten by rounding to the nearest tenth and the nearest whole number. Which method is better? Why?
(Possible answer: Rounding to the nearest tenth will give an answer closer to the actual product, but rounding to the nearest whole number is much easier to calculate.)

For Mixed Abilities

Common Errors • Intervention

Some students may have difficulty choosing compatible numbers. Explain that they should choose numbers that make it easy to multiply mentally. Have students work in pairs to discuss which pair of numbers would be easier to use.

1. 18.8×4.3
   19 and 4 or 20 and 5

2. 27.8×6.7
   30 and 5 or 28 and 7

3. 8.5×13.9
   9 and 14 or 10 and 14

Enrichment • Application

Paula and Doug are hosting a dinner party for 24 of their friends. They have made a list of the foods and the amounts they need to buy at the deli.

1.75 lb of Cheddar cheese at $7.29 per lb

1.75 lb of Pepperjack cheese at $6.79 per lb

3 lb of sliced turkey at $6.89 per lb

3 lb of sliced ham at $5.99 per lb

They will also buy fruit platters that serve 8–10 people. Each platter costs $15.25. Estimate how much they will spend. ($112)

More to Explore • Number Sense

Have students make a magic square using the numbers 1 through 16. Have them arrange the numbers in such a way so that the sum in any row—vertically, horizontally, or diagonally—will be the same.

(1)	(15)	(12)	(6)
(8)	(10)	(13)	(3)
(14)	(4)	(7)	(9)
(11)	(5)	(2)	(16)

T184

pages 185-186

1 Getting Started

Objective
• To multiply a whole number by a decimal

Warm Up • Mental Math
Have students find each product.

1. 50×10^2 (5,000)
2. 0.086×10^2 (8.6)
3. 3×10^2 (300)
4. 3.91×10^3 (3,910)
5. 0.029×10^2 (2.9)
6. 14.67×10^1 (146.7)
7. 46.7×10^1 (467)
8. 3.592×10^2 (359.2)

Warm Up • Pencil and Paper
Have students find each product.

1. 34×7 (238)
2. 16×52 (832)
3. 723×46 (33,258)
4. 493×67 (33,031)
5. $5,198 \times 476$ (2,474,248)
6. $7,894 \times 385$ (3,039,190)

Name _____

Multiplying Whole Numbers by Decimals

The shotput can be thrown about 6.1 times farther on the Moon than on Earth. What would the Earth's shotput record be on the Moon?

We want to know the number of meters the shotput could be thrown on the Moon.

We know the shotput record on Earth is __24__ meters, and the shotput can be thrown __6.1__ times farther on the Moon.

To find the distance on the Moon, we multiply the number of Earth meters by the number of times farther it can be thrown on the Moon.

We multiply __24__ by __6.1__.

Multiply the same as with whole numbers.	Put as many decimal places in the product as there are in the decimal factor.	Estimate to check the answer.
$\begin{array}{r} 2\,4 \\ \times\ 6.1 \\ \hline 2\,4 \\ 144 \\ \hline 146\,4 \end{array}$	$\begin{array}{r} 2\,4 \\ \times\ 6.1 \leftarrow 1\ \text{place} \\ \hline 2\,4 \\ 144 \\ \hline 146.4 \leftarrow 1\ \text{place} \end{array}$	$\begin{array}{r} 20 \\ \times\ 6 \\ \hline 120 \end{array}$

The shotput record on the Moon would be __146.4__ meters.

Getting Started

Place the decimal point in the product.

1. $\begin{array}{r} 79 \\ \times\ 2.3 \\ \hline 181.7 \end{array}$
2. $\begin{array}{r} 3.9 \\ \times\ 48 \\ \hline 187.2 \end{array}$
3. $\begin{array}{r} 2.765 \\ \times\ 87 \\ \hline 240.555 \end{array}$
4. $\begin{array}{r} 839 \\ \times\ 0.44 \\ \hline 369.16 \end{array}$

Multiply.

5. $\begin{array}{r} 4.7 \\ \times\ 9 \\ \hline 42.3 \end{array}$
6. $\begin{array}{r} 27 \\ \times\ 0.17 \\ \hline 4.59 \end{array}$
7. $\begin{array}{r} 9.653 \\ \times\ 298 \\ \hline 2,876.594 \end{array}$
8. $\begin{array}{r} 187 \\ \times\ 3.25 \\ \hline 607.75 \end{array}$

Copy and multiply.

9. 4.503×47
211.641
10. 324×8.52
2,760.48
11. 5.39×51
274.89
12. 26×4.98
129.48

---

2 Teach

Introduce the Lesson Have a student read the problem and state the question being asked. Ask what information is given in the problem and the picture. Have students fill in the information sentences. Point out that to solve this problem, students multiply 24 by 6.1.

Work through the model with the class. Remind students that multiplying decimals is like multiplying whole numbers. The only new step is placing the decimal point correctly in the product.

Have students complete the solution sentence.

Develop Skills and Concepts To help students understand the rationale for placing the decimal point, you may wish to have students name the factors 24 and 6.1 as fractions and then compute the answer.
$(24 \times 6.1 = \frac{24}{1} \times \frac{61}{10} = 14\frac{64}{10} = 146.4)$

Point out that when students multiply as if the factors are whole numbers, they are multiplying the numerators of fractions. When they place the decimal point in the product, they are, in effect, multiplying the denominators.

Encourage students to estimate products as a method of checking the reasonableness of their answers.

3 Practice

Have students complete all the exercises.

Practice

Place the decimal point in the product.

1. 95
   $\times$ 5.7
   541.5

2. 485
   $\times$ 3.9
   1,891.5

3. 479
   $\times$ 4.5
   2,155.5

4. 396
   $\times$ 0.123
   48.708

5. 2.385
   $\times$ 12
   28.620

6. 968
   $\times$ 1.38
   1,335.84

7. 12.94
   $\times$ 168
   2,173.92

8. 0.9651
   $\times$ 18
   17.3718

Multiply.

9. 2.8
   $\times$ 7
   19.6

10. 39
   $\times$ 0.8
   31.2

11. 115
   $\times$ 2.3
   264.5

12. 7.38
   $\times$ 12
   88.56

13. 9.625
   $\times$ 26
   250.250

14. 592
   $\times$ 1.58
   935.36

15. 42.6
   $\times$ 495
   21,087.0

16. 8,246
   $\times$ 2.13
   17,563.98

Copy and multiply.

17. 16 × 4.15
   66.4

18. 3.02 × 46
   138.92

19. 5.198 × 25
   129.95

20. 43 × 1.65
   70.95

21. 4.3921 × 6
   26.3526

22. 29 × 7.246
   210.134

23. 176 × 1.09
   191.84

24. 3.761 × 24
   90.264

25. 684.12 × 28
   19,155.36

26. 45 × 7.207
   324.315

27. 86 × 65.826
   5,661.036

28. 76 × 982.1
   74,639.6

29. 58.32 × 95
   5,540.4

30. 26.82 × 19
   509.58

31. 52 × 9.65
   501.8

32. 245 × 0.5
   122.5

33. 3.201 × 77
   246.477

34. 0.002 × 69
   0.138

35. 1.961 × 84
   164.724

36. 22.2 × 881
   19,558.2

Problem Solving

Solve each problem.

37. Rhonda weighs 46 kilograms. Earline weighs 1.06 times as much as Rhonda. How much does Earline weigh?
   48.76 kilograms

38. Chris jumped 245 centimeters. His cousin Katy jumped 0.45 that distance. Chris's friend, Charlie, jumped 3.7 centimeters less than Katy. How far did Charlie jump?
   106.55 centimeters

186

Lesson 9-3 • Multiplying Whole Numbers by Decimals

4 Assess

Write the following on the board:

0.12
$\times$ 0.9
108

Ask students to insert the decimal point in the product.
(0.108)

5 Mixed Review

1. $\frac{14}{25} \times \frac{15}{28} \times \frac{5}{12}$ ($\frac{1}{8}$)

2. 627 × 3.8 (2,382.6)

3. 16 − 9$\frac{11}{12}$ (6$\frac{1}{12}$)

4. 143.278 − 96.32 (46.958)

5. 1,000 × 19.32 (19,320)

6. 6,050 − 2,143 (3,907)

7. 28,102 ÷ 18 (1,561 R4)

8. $\frac{5}{6} + \frac{1}{3} + \frac{7}{9}$ (1$\frac{17}{18}$)

9. 4,376 × 5 (21,880)

10. 12 ÷ 3$\frac{1}{2}$ (3$\frac{3}{7}$)

For Mixed Abilities

Common Errors • Intervention

In order to place the decimal point in the product, some students may count places from the left instead of from the right. Have them count the number of decimal places in the factors before they multiply and write this number beside the place for the product with an arrow to remind them of the direction to count.

346.58
$\times$ 9
← 2

Enrichment • Estimation

Have students choose, without computing, the most reasonable answer for each multiplication exercise.

1. 24.2 × 37
   (a.) 895.4   b. 89.54   c. 8.954

2. 2.8 × 375
   a. 105.05   (b.) 1,050   c. 1.050

3. 65 × 7.864
   (a.) 511.16   b. 51.116   c. 5.1116

4. 381 × 4.247
   a. 161.8107   b. 16,181.07
   (c.) 1,618.107

5. 9.85 × 136
   a. 133.96   (b.) 1,339.6   c. 13.396

More to Explore • Measurement

Have students draw an outline of their right hand on a large sheet of paper. Then, allow them to choose a supply of dry beans, pennies, paper clips, or other small objects. Tell students that they are to find the measure of the perimeter of their hand, using their selected object as a nonstandard unit of measurement.

Before they measure, have students estimate and record how many units they think they will need. Then, have them actually measure the outline of their hand by placing the objects end to end. Have them record the measurement and compare this measurement to their estimate.

T186

9-4 Multiplying Decimals by Decimals

pages 187–188

1 Getting Started

Objective
- To multiply a decimal by a decimal

Warm Up • Mental Math
Have students compare each pair of number expressions.

1. 7.6×10^1 (>) 0.76×10^1
2. 5.23×10^2 (=) 52.3×10^1
3. 0.065×10^2 (<) 0.65×10^2
4. 29.4×10^4 (>) 0.294×10^5
5. 0.96×10^2 (<) 9.6×10^3
6. 5.214×10^4 (>) 521.4×10^1
7. 0.334×10^2 (=) 3.34×10^1
8. 98×10^2 (<) 9.8×10^4

Warm Up • Pencil and Paper
Have students find each product.

1. 36×42 (1,512)
2. 27×156 (4,212)
3. 186×364 (67,704)
4. 24×6.8 (163.2)
5. 3.67×438 (1,607.46)
6. 0.45×936 (421.2)

Name _____

Lesson 9-4

Multiplying Decimals by Decimals

Mr. Nikomoto and Mr. Allen carpool to work. Mr. Nikomoto's car gets 29.7 miles to a gallon of gasoline. Mr. Allen's car gets 25.3 miles per gallon. How far can Mr. Nikomoto drive on the gas he just purchased?

We want to know how many miles Mr. Nikomoto can drive on a tank of gas.

We know that his car gets __29.7__ miles to a gallon of gas, and he just bought __8.6__ gallons.

To find the distance he can drive, we multiply the number of miles per gallon, by the number of gallons. We multiply __29.7__ by __8.6__.

Multiply the same as whole numbers.	Place the decimal point in the product. It has the same number of decimal places as the sum of them in the decimal factors.	Round to whole numbers to check the answer.

```
    29.7              29.7  ← 1 place           30
  × 8.6             × 8.6  ← + 1 place        ×  9
   178 2             178 2                     270
   2376              2376
  25542            255.42  ← 2 places
```

Mr. Nikomoto can drive __255.42__ miles on a tank of gas.

Getting Started

Place the decimal point in the product.

```
1.    7.61        2.    9.32
    × 1.5             × 0.26
    11,415           2,4232
```

Multiply.

```
3.    67.3        4.    4.79
    × 2.7             × 1.05
   181.71           5.0295
```

Copy and multiply.

5. 5.243×6.76 6. 0.37×0.73 7. 3.125×0.67
   35.44268   0.2701   2.09375

Lesson 9-4 • Multiplying Decimals by Decimals

187

---

2 Teach

Introduce the Lesson Have a student read the problem and tell what is to be solved. (how far Mr. Nikomoto can drive on the gas he just purchased) Have students identify the needed information in the problem and the picture. (Mr. Nikomoto's car gets 29.7 miles per gallon, and he bought 8.6 gallons.) Ask, *What information is given that is not needed to solve the problem?* (Mr. Allen's car gets 25.3 miles per gallon.)

Work through the model problem. Point out that multiplying decimals is like multiplying whole numbers. The only difference is that the decimal point must be placed in the product.

Have students complete the solution sentence.

Develop Skills and Concepts Point out that in the model, students are multiplying tenths by tenths. The answer therefore is hundredths. Have students verify the answer by multiplying the fractions $\frac{297}{10}$ and $\frac{86}{10}$. $\left(\frac{297}{10} \times \frac{86}{10} = \frac{25,542}{100}\right)$

Explain that multiplying the factors as if they were whole numbers is like multiplying the numerators of fractions. When they place the decimal point in the product, they are, in effect, multiplying the denominators.

Have students work **0.31 × 42.2** (13.082) and **1.75 × 23.6** (41.3) on the board for additional practice.

3 Practice

Have students complete all the exercises. Point out that it is not necessary to align the decimal points in the factors when multiplying.

T187

Practice

Place the decimal point in the product.

1. $\begin{array}{r} 4.31 \\ \times\ 1.5 \\ \hline 6.465 \end{array}$

2. $\begin{array}{r} 70.3 \\ \times\ 2.4 \\ \hline 168.72 \end{array}$

3. $\begin{array}{r} 1.836 \\ \times\ 4.9 \\ \hline 8.9964 \end{array}$

4. $\begin{array}{r} 6.03 \\ \times\ 1.25 \\ \hline 7.5375 \end{array}$

5. $\begin{array}{r} 4.731 \\ \times\ 2.54 \\ \hline 12.01674 \end{array}$

6. $\begin{array}{r} 6.24 \\ \times\ 0.73 \\ \hline 4.5552 \end{array}$

7. $\begin{array}{r} 14.9 \\ \times\ 0.075 \\ \hline 1.1175 \end{array}$

8. $\begin{array}{r} 7.581 \\ \times\ 3.18 \\ \hline 24.10758 \end{array}$

Multiply.

9. $\begin{array}{r} 4.3 \\ \times\ 0.6 \\ \hline 2.58 \end{array}$

10. $\begin{array}{r} 2.56 \\ \times\ 1.8 \\ \hline 4.608 \end{array}$

11. $\begin{array}{r} 3.246 \\ \times\ 4.2 \\ \hline 13.6332 \end{array}$

12. $\begin{array}{r} 9.36 \\ \times\ 0.8 \\ \hline 7.488 \end{array}$

13. $\begin{array}{r} 26.25 \\ \times\ 7.3 \\ \hline 191.625 \end{array}$

14. $\begin{array}{r} 4.96 \\ \times\ 3.75 \\ \hline 18.6000 \end{array}$

15. $\begin{array}{r} 4.961 \\ \times\ 2.37 \\ \hline 11.75757 \end{array}$

16. $\begin{array}{r} 0.921 \\ \times\ 0.652 \\ \hline 0.600492 \end{array}$

Copy and multiply.

17. 4.761×3.21
    15.28281

18. 0.81×0.96
    0.7776

19. 5.134×0.25
    1.2835

20. 2.8×1.694
    4.7432

21. 2.24×2.24
    5.0176

22. 0.68×0.47
    0.3196

Problem Solving

Solve each problem.

23. Gasoline costs $0.91 a gallon. How much do 14.6 gallons cost?
    $13.29

24. Alan worked 6.5 hours. If he earns $5.50 an hour, how much did Alan earn?
    $35.75

25. It rained 1.75 centimeters in one hour. How much rain fell in 0.6 of an hour?
    1.05 centimeters

26. The length of a garden is 4.25 times as long as the width. If the width is 5.6 meters, how long is the garden?
    23.8 meters

188

Lesson 9-4 • Multiplying Decimals by Decimals

For Mixed Abilities

Common Errors • Intervention

Some students may confuse the form for addition with the form for multiplication and line up the decimal points in the factors and product.

Incorrect	Correct
3.9	3.9
× 6.4	× 6.4
249.6	24.96

Have them compare the number of decimal places in the product with the total number of decimal places in the factors.

Enrichment • Number Sense

Have students identify the place value of the product.

1. tenths × tenths = (hundredths)

2. hundredths × tenths = (thousandths)

3. hundredths × thousandths = (hundred thousandths)

4. tenths × thousandths = (ten thousandths)

5. hundredths × hundredths = (ten thousandths)

6. tenths × ten thousandths = (hundred thousandths)

Then, have them write a numerical problem to illustrate each of the exercises. Have students exchange papers and compute the products of the exercises.

More to Explore • Application

Divide students into small groups and assign each group a different foreign country. Have students find the current value of the American dollar in that foreign currency.

Then, using the foreign money, have them calculate what it would cost to purchase some of their favorite things, for example, a fast food item, toys, clothing, or school supplies.

Have each group share its price calculations in an oral report to the class.

---

4 Assess

Ask how many decimal places will be in the product of 0.145×0.597. (six)

5 Mixed Review

1. $6,348 \div 21$ (302 R6)

2. $\$150.30 - \$10.95$ ($139.35)

3. 394×427 (168,238)

4. 7.21×0.35 (2.5235)

5. $25\frac{1}{5} + 14\frac{3}{7}$ ($39\frac{22}{35}$)

6. $10^4 \times 32$ (320,000)

7. $\frac{3}{4} \times \frac{2}{9} \times \frac{5}{6}$ ($\frac{5}{36}$)

8. 6.28×58 (364.24)

9. 0.438×10^2 (43.8)

10. $\frac{7}{8} \div \frac{3}{4}$ ($1\frac{1}{6}$)

T188

Lesson 9-5

1 Getting Started

Objective
- To add zeros to the product to locate the decimal point

Warm Up • Mental Math
Have students simplify each expression.

1. 4×1 (4)
2. 6×0 (0)
3. 17×0 (0)
4. 1×9 (9)
5. $46 \times 0 + 1$ (1)
6. $1 \times (3 + 3)$ (6)
7. $0 \times 29 + 6$ (6)
8. $(7 + 3) \times 1$ (10)

Warm Up • Pencil and Paper
Ask students to rewrite the exercise using decimals instead of fractions.

1. $\frac{3}{10} \times \frac{1}{10} = \frac{3}{100}$
   ($0.3 \times 0.1 = 0.03$)
2. $\frac{4}{10} \times \frac{3}{100} = \frac{12}{1,000}$
   ($0.4 \times 0.03 = 0.012$)
3. $\frac{7}{100} \times \frac{4}{10} = \frac{28}{1,000}$
   ($0.07 \times 0.4 = 0.028$)
4. $\frac{9}{100} \times \frac{8}{100} = \frac{72}{10,000}$
   ($0.09 \times 0.08 = 0.0072$)
5. $\frac{11}{100} \times \frac{12}{1,000} = \frac{132}{100,000}$
   ($0.11 \times 0.012 = 0.00132$)

Name _____

Zeros in the Product

A laptop can complete a command in 0.5 of the time it takes the personal computer to do the same job. How long does it take this laptop to complete a command?

We want to know how many seconds it takes the laptop to complete a command.

We know that a personal computer takes __0.0003__ seconds to complete a command, and the laptop takes __0.5__ times as long.

Find out how long it takes the laptop to complete a command.

We multiply __0.0003__ by __0.5__.

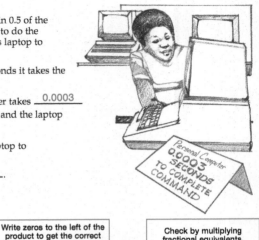
Personal Computer 0.0003 SECONDS TO COMPLETE COMMAND

Multiply.	Write zeros to the left of the product to get the correct number of places.	Check by multiplying fractional equivalents.
$\begin{array}{r} 0.0003 \\ \times\ \ 0.5 \\ \hline 15 \end{array}$	$\begin{array}{rl} 0.0003 & \leftarrow\ \ \ 4 \text{ places} \\ \times\ \ \ 0.5 & \leftarrow +1 \text{ place} \\ \hline 0.00015 & \leftarrow\ \ \ 5 \text{ places} \end{array}$	$\frac{3}{10,000} \times \frac{5}{10} = \frac{15}{100,000}$

It takes the laptop __0.00015__ seconds to complete a command.

Getting Started

Multiply.

1. $\begin{array}{r} 3.26 \\ \times\ 0.01 \\ \hline 0.0326 \end{array}$
2. $\begin{array}{r} 0.003 \\ \times\ \ \ 12 \\ \hline 0.036 \end{array}$
3. $\begin{array}{r} 0.03 \\ \times\ 0.06 \\ \hline 0.0018 \end{array}$
4. $\begin{array}{r} 4.24 \\ \times\ 0.005 \\ \hline 0.0212 \end{array}$
5. $\begin{array}{r} 0.02 \\ \times\ 0.03 \\ \hline 0.0006 \end{array}$
6. $\begin{array}{r} 9.21 \\ \times\ 0.005 \\ \hline 0.04605 \end{array}$
7. $\begin{array}{r} 0.008 \\ \times\ 0.005 \\ \hline 0.00004 \end{array}$
8. $\begin{array}{r} 6.01 \\ \times\ 0.006 \\ \hline 0.03606 \end{array}$

Copy and multiply.

9. 0.375×0.005
   0.001875
10. $0.07 \times 0.05 \times 0.02$
    0.00007
11. $0.06 \times 0.09 \times 5.8$
    0.03132

---

2 Teach

Introduce the Lesson Have a student read the problem and tell what is to be solved. Have students identify the needed information and fill in the information sentences. Work through the model, emphasizing the placement of extra zeros. Have students complete the solution sentence.

Develop Skills and Concepts Emphasize the following points:

- The number of decimal places in the product must equal the sum of the decimal places in the factors.

- You write zeros in the product when there are not enough decimal places. You write the zeros to the left of the digits in the product.

Encourage students to do the check step using one of these methods: estimating, reversing the factors, or renaming the factors as fractions.

Provide reinforcement by having students complete the following exercises on the board:

$\begin{array}{r} 0.04 \\ \times\ 0.06 \\ \hline (0.0024) \end{array}$
$\begin{array}{r} 3.16 \\ \times\ 0.007 \\ \hline (0.02212) \end{array}$
$\begin{array}{r} 7.04 \\ \times\ 0.008 \\ \hline (0.05632) \end{array}$

3 Practice

Have students complete all the exercises. Watch for students who fail to prefix the necessary zeros in the products.

Practice

Multiply.

1. 3.2 $\underline{\times\ 0.04}$ 0.128	2. 0.006 $\underline{\times\ \ \ \ 4}$ 0.024	3. 0.02 $\underline{\times\ 0.02}$ 0.0004	4. 4.1 $\underline{\times\ 0.005}$ 0.0205
5. 15.2 $\underline{\times\ 0.003}$ 0.0456	6. 0.006 $\underline{\times\ 0.008}$ 0.000048	7. 0.058 $\underline{\times\ \ 0.05}$ 0.00290	8. 0.009 $\underline{\times\ \ \ \ 16}$ 0.144

Copy and multiply.

9. 0.3×0.3
   0.09

10. 1.6×0.005
    0.008

11. 0.002×3.56
    0.00712

12. 0.007×2.3
    0.0161

13. 0.0006×2.21
    0.001326

14. 8.3×0.0041
    0.03403

15. 0.125×0.005
    0.000625

16. 4.04×0.004
    0.01616

17. 0.0047×0.59
    0.002773

18. $0.03 \times 0.04 \times 2.3$
    0.00276

19. $0.9 \times 4.7 \times 0.005$
    0.02115

20. $0.05 \times 0.02 \times 0.06$
    0.00006

Problem Solving

Solve each problem.

21. Each sheet of paper is 0.0025 inch thick. How thick is a stack of 50 sheets?
    0.125 inch

22. A piece of wire is 0.05 inch wide. How wide is a piece of wire 0.5 as wide?
    0.025 inch

Now Try This!

Write the missing numbers.

0.7	+	0.8	=	1.5	
+		+		+	
0.6	+	0.4	=	1.0	
=		=		=	
1.3	+	1.2	=	2.5	

2.25	−	0.45	=	1.8
−		−		−
1.65	−	0.15	=	1.5
=		=		=
0.6	−	0.3	=	0.3

Lesson 9-5 • Zeros in the Product

Now Try This! Have students study each puzzle before beginning to compute. The number in the lower right-hand corner of each puzzle provides a self-check.

4 Assess

Write the following on the board:

0.3
$\underline{\times\ 0.2}$
6

Have students insert the zeros to show the correct place value of the product. (0.06)

For Mixed Abilities

Common Errors • Intervention

When zeros must be annexed in the product, some students may place them to the right of the digits that already are there instead of between the decimal point and these digits. Have them rewrite a decimal problem, such as 0.2×0.04, in fraction form, $\frac{2}{10} \times \frac{4}{100} = \frac{8}{1,000}$, to show that the answer is smaller than hundredths or tenths.

Enrichment • Number Sense

Tell students to arrange the numbers within the group to form a correct multiplication exercise.

1. 0.0432
   0.0036
   12

   0.0036
   $\underline{\times\ \ \ \ \ 12}$
   0.0432

2. 0.07
   0.0245
   0.35

   0.07
   $\underline{\times\ \ 0.35}$
   0.0245

3. 0.009
   0.0522
   5.8

   0.009
   $\underline{\times\ \ \ \ 5.8}$
   0.0522

4. 0.00039
   28
   0.01092

   0.00039
   $\underline{\times\ \ \ \ \ \ \ 28}$
   0.01092

More to Explore • Biography

Analytic geometry and calculus are largely the invention of seventeenth-century French mathematician Pierre de Fermat. Fermat, the son of a leather merchant, was educated at home and worked as a lawyer and government official most of his life. Mathematics was only a hobby that Fermat developed at age 30.

Fermat was especially interested in the properties of prime numbers. He also worked with Blaise Pascal on the theory of probabilities, now widely used in insurance and statistics.

Fermat published almost nothing during his lifetime; he usually announced his discoveries in letters to his friends or marked his findings in the margins of the books he owned. His work was largely forgotten until it was rediscovered in the mid-1800s.

Dividing a Decimal by a Whole Number

pages 191–192

1 Getting Started

Objective
• To divide a decimal by a whole number

Warm Up • Mental Math
Have students tell the place value of the 3 in each number.

1. 74,316.2 (hundreds)
2. 452.03 (hundredths)
3. 39,408.167 (ten thousands)
4. 894.3067 (tenths)
5. 7,106.0473 (ten thousandths)

Warm Up • Pencil and Paper
Have students find each quotient.

1. $7\overline{)301}$ (43)
2. $4\overline{)864}$ (216)
3. $24\overline{)8,616}$ (359)
4. $61\overline{)10,858}$ (178)
5. $136\overline{)289,952}$ (2,132)
6. $645\overline{)506,970}$ (786)
7. $5\overline{)4,825}$ (965)
8. $35\overline{)9,240}$ (264)

Name _____

Dividing a Decimal by a Whole Number

Randi is going skiing this week. What was the average snowfall for the 3 preceding days?

We want to find the average snowfall for three days. To find an average, we find the sum of the numbers and divide by how many numbers there are.

Snowfall	
Friday	6.26 inches
Saturday	4.18 inches
Sunday	5.04 inches

We add _6.26_ , _4.18_ , and _5.04_ .

```
  6.26
  4.18
+ 5.04
 15.48
```

We divide _15.48_ by _3_ .

Divide the same as with whole numbers.	Put the decimal point in the quotient, directly above the decimal point in the dividend.	Check.

516
3)1548
−15
  4
− 3
  18
− 18
   0
	5.16
	3)15.48
	−15
	  4
	− 3
	  18
	− 18
	   0
		5.16
		× 3
		15.48

The average snowfall for the 3 days was _5.16_ inches.

Getting Started

Divide and check.

1. $6\overline{)140.4}$ → 23.4
2. $9\overline{)11.115}$ → 1.235
3. $23\overline{)143.75}$ → 6.25

Copy and divide.

4. 38.85 ÷ 5
   7.77
5. 501.6 ÷ 19
   26.4
6. 2,677.36 ÷ 683
   3.92

---

2 Teach

Introduce the Lesson Have a student read the problem, identify the question to be answered, and tell what information is given. Work through the division model with students. Remind them that to find an average, they should find the *sum* of the numbers and then divide by the *number* of numbers.

Develop Skills and Concepts Emphasize that dividing a decimal by a whole number is the same as dividing whole numbers. Students follow the same steps: divide, multiply, subtract, bring down, and compare. Encourage students to estimate the quotient *before* they compute to help in deciding where to place the first digit of the quotient.

Stress care in placing the decimal point in the quotient directly above the decimal point in the dividend. Explain the rationale for placing the decimal point in terms of the

inverse operation of multiplication. (The product will have as many decimal places as the multiplier and multiplicand combined.)

Students should use multiplication to check each division problem.

3 Practice

Have students complete all the exercises. You may want to have students use their calculators to check their answers. Then, have them find and correct their errors as needed.

Practice

Divide and check.

1. $7\overline{)88.2}$ → 12.6

2. $3\overline{)16.92}$ → 5.64

3. $8\overline{)27.616}$ → 3.452

4. $26\overline{)153.14}$ → 5.89

5. $65\overline{)1,553.5}$ → 23.9

6. $128\overline{)808.96}$ → 6.32

Copy and divide.

7. $3.52 \div 8$
   0.44

8. $121.23 \div 9$
   13.47

9. $14.668 \div 4$
   3.667

10. $37.24 \div 14$
    2.66

11. $135.2 \div 26$
    5.2

12. $313.11 \div 49$
    6.39

13. $1,515.8 \div 65$
    23.32

14. $7,516.8 \div 96$
    78.3

15. $3,685.92 \div 56$
    65.82

16. $1,350.5 \div 365$
    3.7

17. $1,842.54 \div 214$
    8.61

18. $19,495.25 \div 725$
    26.89

19. $351.4 \div 25$
    14.056

20. $72.4 \div 1,448$
    0.05

21. $5,007.6 \div 963$
    5.2

22. $60.25 \div 25$
    2.41

23. $24.8 \div 8$
    3.1

24. $296.5 \div 5$
    59.3

Problem Solving

Solve each problem.

25. A case of 12 bottles of juice contains 16.32 liters. How much does each bottle hold?
    1.36 liters

26. How far will a car travel in 8 hours if it averages 56.9 miles per hour?
    455.2 miles

Use the chart to answer Problems 27 and 28.

27. How much more did Morty earn on Friday than on Tuesday?
    $7.45

28. What is Morty's average daily wage?
    $46.91

Morty's Wages	
Monday	$47.23
Tuesday	$42.88
Wednesday	$45.67
Thursday	$48.44
Friday	$50.33

192

Lesson 9-6 • Dividing a Decimal by a Whole Number

For Mixed Abilities

Common Errors • Intervention

Some students may forget to place the decimal point in the answer. Have them place the decimal point in the quotient above the decimal point in the dividend first, before they divide. After they divide, have them check their answer by multiplying.

Enrichment • Number Sense

Challenge students to determine the number each letter represents in the division exercise.

1. $36\overline{)80.6B}$ → A.24
   (A = 2; B = 4)

2. $25\overline{)10.05E}$ → 0.4F2
   (E = 0; F = 0)

3. $49\overline{)2C7.63}$ → 5.8D
   (C = 8; D = 7)

4. $39\overline{)16.840G}$ → 0.H318
   (G = 2; H = 4)

More to Explore • Application

Have students calculate the average time for each of them to run the 50-yard dash. Measure a distance of 50 yards, marking the start and finish lines. Choose a record keeper, timer, and starter, and supply them with a stopwatch.

Divide the class into teams. Have each student run the dash, with the timer recording each time. Then, have students find the average time for each team, and the average for the girls, boys, and the entire class.

As an extension, point out that sports statistics are often expressed in averages. Ask students to bring in examples of these types of statistics for different sports. Discuss how they think these averages were calculated.

4 Assess

Ask, *Where do you place the decimal point in the quotient when you divide a decimal by a whole number?* (Possible answer: Put the decimal point in the quotient directly above the decimal point in the dividend.)

5 Mixed Review

1. 2^6 (64)

2. $\frac{2}{3} \div \frac{5}{6}$ ($\frac{4}{5}$)

3. $600 \times 3,000$ (1,800,000)

4. $\frac{3}{5} + \frac{1}{4} + \frac{1}{8}$ ($\frac{39}{40}$)

5. $62,010 - 37,370$ (24,640)

6. $27,489 \div 256$ (107 R97)

7. $18\frac{2}{3} - 15\frac{1}{6}$ ($3\frac{1}{2}$)

8. 0.002×0.015 (0.00003)

9. $4\frac{3}{8} \times 1\frac{5}{7}$ ($7\frac{1}{2}$)

10. $\$4.73 \times 24$ ($113.52)

9-7 Zeros in the Quotient

pages 193–194

1 Getting Started

Objective
• To divide a decimal by a whole number, quotients with zeros

Materials
graph paper

Warm Up • Mental Math
Have students round each number to the nearest tenth. Then, estimate the calculation.

1. $5.62 + 3.01 \quad (5.6 + 3 = 8.6)$
2. $0.86 - 0.49 \quad (0.9 - 0.5 = 0.4)$
3. $5.02 \times 12.03 \quad (5 \times 12 = 60)$
4. $4.78 \div 6 \quad (4.8 \div 6 = 0.8)$
5. $1.79 - 0.15 \quad (1.8 - 0.2 = 1.6)$
6. $6.07 + 4.49 \quad (6.1 + 4.5 = 10.6)$

Warm Up • Pencil and Paper
Have students find each quotient.

1. $9\overline{)3.825} \quad (0.425)$
4. $75\overline{)8.25} \quad (0.11)$

2. $5\overline{)10.95} \quad (2.19)$
5. $34\overline{)41.922} \quad (1.233)$

3. $7\overline{)215.6} \quad (30.8)$
6. $63\overline{)264.6} \quad (4.2)$

Name _____

Zeros in the Quotient

Marcie is setting up the experiment for today's science class. She has 0.336 liters of sugar water and is pouring equal amounts into the beakers. How much water will each beaker contain?

We want to know how much liquid Marcie will pour into each beaker.

We know she is pouring a total of __0.336__ liters into __6__ beakers.

To find how much will be poured into each beaker, we divide the total amount by the number of beakers.

We divide __0.336__ by __6__.

Place the decimal point in the quotient.	Write zeros in the quotient when needed.	Divide.
$6\overline{)0.336}$ with decimal point	$6\overline{)0.336}$ → 0.0	$\begin{array}{r} 0.056 \\ 6\overline{)0.336} \\ -30 \\ \hline 36 \\ -36 \\ \hline 0 \end{array}$

REMEMBER Every decimal place value in the dividend must be represented in the quotient.

Marcie will pour __0.056__ liters of water into each beaker.

Getting Started

Divide.

1. $8\overline{)0.184} \quad (0.023)$
2. $9\overline{)0.054} \quad (0.006)$
3. $6\overline{)0.354} \quad (0.059)$
4. $15\overline{)1.05} \quad (0.07)$
5. $38\overline{)0.304} \quad (0.008)$

Copy and divide.

6. $0.425 \div 5$
   0.085
7. $0.6499 \div 67$
   0.0097
8. $0.9345 \div 35$
   0.0267

---

2 Teach

Introduce the Lesson Have a student read the problem to tell what is to be solved. Ask what information is given in the problem and the picture. Point out that students will use the division skills they already know for dividing whole numbers and for dividing a decimal by a whole number.

Have students complete the information sentences. Then, work through the model and have a student complete the solution sentence.

Develop Skills and Concepts This lesson emphasizes the need for placing a zero in the tenths or hundredths place of the quotient. Remind students that zero serves as a placeholder. Emphasize the importance of careful alignment of decimal points in quotients and dividends.

Have students work the following division exercises on the board. Have other students check each quotient using multiplication.

$8\overline{)0.64} \quad (0.08)$
$23\overline{)2.047} \quad (0.089)$
$142\overline{)0.2982} \quad (0.0021)$

3 Practice

Have students complete all the exercises. Students may use graph paper to help them keep the digits aligned according to place value.

Practice

Divide.

1. $4\overline{)0.144}$ → 0.036

2. $9\overline{)0.576}$ → 0.064

3. $7\overline{)0.0441}$ → 0.0063

4. $7\overline{)0.0455}$ → 0.0065

5. $16\overline{)0.608}$ → 0.038

6. $29\overline{)0.2494}$ → 0.0086

Copy and divide.

7. $0.335 \div 5$
0.067

8. $0.0156 \div 4$
0.0039

9. $0.6904 \div 8$
0.0863

10. $0.888 \div 12$
0.074

11. $0.0448 \div 28$
0.0016

12. $0.0114 \div 19$
0.0006

13. $0.1323 \div 21$
0.0063

14. $0.2345 \div 35$
0.0067

15. $0.1372 \div 49$
0.0028

16. $3.276 \div 52$
0.063

17. $0.6432 \div 67$
0.0096

18. $6.975 \div 93$
0.075

19. $0.02744 \div 49$
0.00056

20. $20.224 \div 316$
0.064

21. $5.7246 \div 658$
0.0087

Now Try This!

Write the decimals that are

1. in A and B. 0.1, 0.2, 1.0

2. in A or B. 0.1, 0.2, 0.3, 0.6, 0.7, 0.8, 0.9, 1.0

3. not in A or B. 0.4, 0.5

4. in B and C. 0.1, 0.2, 0.7

5. in A or C. 0.1, 0.2, 0.3, 0.4, 0.5, 0.7, 0.8, 0.9, 1.0

6. in A and B and C. 0.1, 0.2

7. in A or B and C. 0.1, 0.2, 0.3, 0.7

8. not in B and not in C. 0.8, 0.9

194

Lesson 9-7 • Zeros in the Quotient

Now Try This! Explain that a Venn diagram is used to show groups or sets, and relationships between the groups or sets. Call attention to key words: *not*, *and*, and *or*.

4 Assess

Write the following on the board:

$9\overline{)0.0549}$

Have students find the quotient. (0.0061)

For Mixed Abilities

Common Errors • Intervention

Some students may forget to write zeros in the quotient when they are required and, therefore, write the quotient incorrectly.

Incorrect
.63
$3\overline{)0.189}$

Correct
.063
$3\overline{)0.189}$

Have these students do their work on grid paper, helping them to see that a digit must be in every square of the quotient between the decimal point and the last digit to the right in this dividend.

Enrichment • Number Sense

Challenge students to find the missing digits in this exercise.

```
           0. (0) 7  3
(4)3 )3. (1) (3) 9
      3  0  1
     (1) (2) (9)
     (1) (2) (9)
              0
```

Have students create their own missing number division exercises to exchange with their classmates.

More to Explore • Statistics

Have students work in groups. Give each group an outline map of the United States. Have groups take a survey of the class to find out the state in which each student was born. Have them make a tally of the information. You may wish to extend the survey by including another class.

Now, have students devise a color scheme for representing frequency on the map. For example, let yellow stand for 0 students born in that state, orange for 1 student, red for 2–6 students, and blue for 7 or more students. You may have to expand the system, depending on the distribution of births in your class.

When the survey is complete and the color scheme established, have groups color their maps to illustrate the data. Display their maps.

9-8 Dividing by Powers of 10

pages 195–196

1 Getting Started

Objective
- To divide a decimal by a power of 10

Warm Up • Mental Math
Have students name the next three numbers.

1. 0.4, 0.6, 0.8, (1.0, 1.2, 1.4)
2. 0.35, 0.4, 0.45, (0.5, 0.55, 0.6)
3. 4.1, 4.3, 4.5, (4.7, 4.9, 5.1)
4. 3.44, 4.44, 5.44, (6.44, 7.44, 8.44)
5. 0.493, 0.496, 0.499, (0.502, 0.505, 0.508)
6. 0.101, 0.211, 0.321, (0.431, 0.541, 0.651)
7. 5.7, 5.5, 5.3, (5.1, 4.9, 4.7)

Warm Up • Mental Math
Have students find the quotient.

1. $15,000 \div 10$ (1,500)
2. $67,000 \div 100$ (670)
3. $23,000 \div 1,000$ (23)
4. $8,000 \div 10$ (800)

Ask students to write each number in standard form.

1. 10^2 (100)
2. 10^4 (10,000)
3. 10^1 (10)
4. 10^3 (1,000)

Name _____

Dividing by Powers of 10

In an earlier lesson, we found a shortcut for multiplying by powers of 10. Complete the pattern to find a shortcut for dividing by powers of 10.

Powers of 10
$10^1 = 10$
$10^2 = 100$
$10^3 = 1,000$
$10^4 = 10,000$

$120 \div 10 = 12$ $120 \div 10^1 = 12$

$120 \div 100 = 1.2$ $120 \div 10^2 = 1.2$

$120 \div 1,000 = 0.12$ $120 \div 10^3 = 0.12$

$120 \div 10,000 = 0.012$ $120 \div \underline{10^4} = \underline{0.012}$

$1.6 \div 10 = 0.16$ $1.6 \div 10^1 = 0.16$

$1.6 \div 100 = 0.016$ $1.6 \div 10^2 = 0.016$

$1.6 \div 1,000 = \underline{0.0016}$ $1.6 \div 10^3 = \underline{0.0016}$

$1.6 \div 10,000 = \underline{0.00016}$ $1.6 \div \underline{10^4} = \underline{0.00016}$

Dividing by 10 or 10^1 moves the decimal point __1__ place to the left.

Dividing by 100 or 10^2 moves the decimal point __2__ places to the left.

Dividing by 1,000 or 10^3 moves the decimal point __3__ places to the left.

Dividing by 10,000 or 10^4 moves the decimal point __4__ places to the left.

When you divide by a power of 10, move the decimal point to the left the same number of places as the exponent in the divisor. Write any extra zeros needed to the left of the quotient digits.

Getting Started

Divide. Use mental math.

1. $154.89 \div 100 = \underline{1.5489}$

2. $0.063 \div 10^3 = \underline{0.000063}$

3. $14.296 \div 10^1 = \underline{1.4296}$

4. $127.9 \div 1,000 = \underline{0.1279}$

5. $39.58 \div 10^4 = \underline{0.003958}$

6. $4,629.2 \div 10^2 = \underline{46.292}$

Multiply or divide. Use mental math.

7. $1.5672 \times 10^4 = \underline{15,672}$

8. $4,893 \div 1,000 = \underline{4.893}$

9. $0.67 \div 10^2 = \underline{0.0067}$

10. $38.9 \times 10^2 = \underline{3,890}$

---

2 Teach

Introduce the Lesson Point out that the patterns at the top of page 195 demonstrate a shortcut for dividing decimals by powers of 10 quickly and accurately. Have students read each rule for dividing by a power of 10. Have them write an example of each rule on the board as you work through the lesson together.

Develop Skills and Concepts Be sure that students understand the shortcut of moving the decimal point to the left as many places as there are zeros in the power of 10. Remind them that when the decimal point is moved one place to the left, the place value of the digit is moved one place to the right.

Write **$4.2 \div 10 = 0.42$** on the board. The value of the 4 changed from (4×1) to $(4 \times \frac{1}{10})$; the value of the 2 changed from $(2 \times \frac{1}{10})$ to $(2 \times \frac{1}{100})$.

Point out that sometimes it is necessary to write extra zeros in the quotient. Put **$120 \div 10,000$** on the board. Point out that the decimal point must be moved four places to the left. In order to do this, it is necessary to write a zero to the left of the 1. ($120 \div 10,000 = 0.012$)

Have volunteers work these exercises on the board:

1. $4.82 \div 100 =$ (0.0482)
2. $68.4 \div 10 =$ (6.84)
3. $6.924 \div 10^3 =$ (0.006924)
4. $49.64 \div 10^2 =$ (0.4964)
5. $723.09 \div 10^4 =$ (0.072309)
6. $7.106 \div 1,000 =$ (0.007106)

3 Practice

Have students complete all the exercises, using the rule developed on page 195. Encourage them to compute mentally and write the answers only.

T195

Practice

Divide. Use mental math.

1. $46.9 \div 10 =$ _4.69_

2. $58.16 \div 10^2 =$ _0.5816_

3. $459.3 \div 100 =$ _4.593_

4. $5,963 \div 10^3 =$ _5.963_

5. $47.24 \div 1,000 =$ _0.04724_

6. $2,941.3 \div 10^2 =$ _29.413_

7. $46.58 \div 10^1 =$ _4.658_

8. $5.76 \div 100 =$ _0.0576_

9. $129.7 \div 10^3 =$ _0.1297_

10. $54.48 \div 100 =$ _0.5448_

Multiply or divide. Use mental math.

11. $81.9 \times 10^2 =$ _8,190_

12. $4.632 \times 1,000 =$ _4,632_

13. $129.6 \div 1,000 =$ _0.1296_

14. $85.26 \div 10^2 =$ _0.8526_

15. $19.24 \times 10 =$ _192.4_

16. $48.231 \div 10^2 =$ _0.48231_

17. $248.2 \times 1,000 =$ _248,200_

18. $3,279 \div 100 =$ _32.79_

Problem Solving

Solve each problem.

19. A plant cell measures 0.006 centimeters after it has been magnified 10 times. What was the original size of the cell?
0.0006 centimeters

20. A sheet of paper is 0.025 centimeters thick. How thick is a stack of 10,000 sheets of paper?
250 centimeters

Now Try This!

Use the powers of 2 to complete the statements.

2	4	8	16	32	64	128	256	512	1,024	2,048	4,096
2^1	2^2	2^3	2^4	2^5	2^6	2^7	2^8	2^9	2^{10}	2^{11}	2^{12}

1. $2^3 \times 2^2 = 8 \times 4 = 32 = 2^5$

2. $2^7 \div 2^3 = 128 \div 8 = 16 = 2^4$

3. $2^4 \times 2^3 =$ _$16 \times 8 = 128 = 2^7$_

4. $2^9 \div 2^4 =$ _$512 \div 16 = 32 = 2^5$_

5. $2^5 \times 2^2 =$ _$32 \times 4 = 128 = 2^7$_

6. $2^{11} \div 2^3 =$ _$2,048 \div 8 = 256 = 2^8$_

7. $2^6 \times 2^5 =$ _$64 \times 32 = 2,048 = 2^{11}$_

8. $2^8 \div 2^5 =$ _$256 \div 32 = 8 = 2^3$_

9. $2^4 \times 2^8 =$ _$16 \times 256 = 4,096 = 2^{12}$_

10. $2^{12} \div 2^7 =$ _$4,096 \div 128 = 32 = 2^5$_

Lesson 9-8 • Dividing by Powers of 10

Now Try This! Have students look for a pattern to follow as they complete these exercises. If you multiply two exponential expressions that have the same base, you can add the exponents to find the product. If you divide two exponential expressions that have the same base, you can subtract the exponents to find the quotient.

4 Assess

Write the following on the board:

$2,000 \div 10^6$

Have students show you how they move the decimal point from right to left to find the quotient. (0.002)

For Mixed Abilities

Common Errors • Intervention

Watch for students who move the decimal point to the right instead of to the left. Discuss that when you divide a decimal by 10, 100, or 1,000, the quotient always will be smaller than the other factor. To make a decimal smaller, you move the decimal point to the left.

Have students work with partners with exercises like the following:

$45.26 \div 10 = 4.526$

$45.26 \div 100 = 0.4526$

$45.26 \div 1,000 = 0.04526$

Enrichment • Number Sense

Write the following on the board:

Before	After
15.34	1.543 (divided by 10)

Have students compare the numbers below to determine what was done to the before number to get the after number.

	Before	After
1.	53.64	5.364 (divided by 10)
2.	7.36	0.00736 (divided by 1,000)
3.	394.756	3.94756 (divided by 100)
4.	69.3	0.00693 (divided by 10,000)
5.	0.02	0.0002 (divided by 100)

More to Explore • Sets

Have a team of students take a poll among the students in your class to find out how many of them have been to a professional football game, baseball game, or hockey game. Some will have attended more than one and some will never have attended any.

Have students write the results of the poll. For example, F = [all who have attended professional football]. Have them construct a Venn diagram using their information. Ask students what conclusions they can make by looking at the diagram.

T196

pages 197–198

Objective
• To divide a decimal by a decimal

Warm Up • Mental Math
Have students find each product.

1. 10^3 (1,000)
2. 2^4 (16)
3. 3^3 (27)
4. 2^6 (64)
5. 10^6 (1,000,000)
6. 5^3 (125)
7. 10^4 (10,000)
8. 9^3 (729)
9. 4^3 (64)
10. 7^3 (343)

Warm Up • Pencil and Paper
Have students find each quotient.

1. $5)\overline{73.15}$ (14.63)   5. $48)\overline{1.728}$ (0.036)

2. $4)\overline{927.2}$ (231.8)   4. $37)\overline{954.6}$ (25.8)

3. $3)\overline{7.356}$ (2.452)   6. $15)\overline{123.15}$ (8.21)

Name _____

Lesson 9-9

Dividing Decimals by Decimals

The Midtown Bakery is famous for its bran muffins, which the bakers make each morning. They use 3.6 pounds of flour every day. How long will their supply of flour last?

We want to know how many days the flour will last.

We know that the bakers have __75.6__ pounds of flour, and that they use __3.6__ pounds each day. To find the number of days the flour will last, we divide the total amount by the amount used each day. We divide __75.6__ by __3.6__.

Multiply the divisor by the power of 10 that makes it a whole number.	Multiply the dividend by the same power of 10, and divide.

$3.6_\wedge)\overline{75.6}$

(Multiply by 10.)

$$\begin{array}{r} 21 \\ 36)\overline{75.6_\wedge} \\ -72 \\ \hline 36 \\ -36 \\ \hline 0 \end{array}$$

(Multiply by 10.)

We can use a caret ($\wedge$) to show where the decimal point has been moved.

The flour will last __21__ days.

Getting Started _____

Put carets in the divisor and dividend to show where the decimal point moves.

1. $4.71)\overline{15.072}$ 2. $0.6)\overline{215.4}$ 3. $1.8)\overline{1.1718}$ 4. $0.025)\overline{0.975}$

Divide.

5. $2.6)\overline{10.14}$ (3.9)   6. $0.09)\overline{76.077}$ (845.3)   7. $0.008)\overline{5.6584}$ (707.3)

Copy and divide.

8. 9.074 ÷ 1.3   9. 1.332 ÷ 1.5   10. 0.1485 ÷ 16.5
   6.98   0.888   0.009

Lesson 9-9 • Dividing Decimals by Decimals   **197**

---

2 Teach

Introduce the Lesson Have a student read the problem, identify the question, and tell what information is given. Have students complete the sentences and work through the model division together. As you discuss the solution to the problem, emphasize the following rules about dividing by a decimal:

• The divisor must be renamed as a whole number.

• Whatever is done to make the divisor a whole number must be done to the dividend.

Develop Skills and Concepts To help students understand this process, have them complete these division exercises on the board:

$7)\overline{42}$ (6) $70)\overline{420}$ (6) $700)\overline{4,200}$ (6)

Point out that all three exercises have the same quotient. Ask how they differ. (Possible answer: In the second problem, the divisor and the dividend are 10 times greater than in the first problem. In the third problem, the divisor and the dividend are 100 times greater than in the first problem.)

Stress that multiplying both the divisor and the dividend by the same number does not change the quotient.

3 Practice

Have students complete all the exercises. Encourage students to question the reasonableness of their answers. Students can check their answers using multiplication. If their answer is correct, the quotient times the divisor will equal the dividend.

Practice

Put a caret in the divisor and dividend to show where the decimal moves.

1. $6.9\overline{)54.631}$ 2. $2.568\overline{)3.81467}$ 3. $0.06\overline{)424.20}$

Divide.

4. $0.4\overline{)27.24}$ **68.1**

5. $0.03\overline{)12.852}$ **428.4**

6. $2.7\overline{)0.54}$ **0.2**

7. $0.008\overline{)0.05056}$ **6.32**

8. $1.9\overline{)12.35}$ **6.5**

9. $0.038\overline{)0.23712}$ **6.24**

10. $0.15\overline{)0.00945}$ **0.063**

11. $7.3\overline{)332.88}$ **45.6**

12. $12.5\overline{)156.25}$ **12.5**

Copy and divide.

13. $1.52 \div 0.4$
   3.8

14. $0.01945 \div 0.005$
   3.89

15. $0.234 \div 0.06$
   3.9

16. $67.385 \div 0.01$
   6,738.5

17. $3.616 \div 0.08$
   45.2

18. $22.569 \div 0.3$
   75.23

19. $0.2576 \div 4.6$
   0.056

20. $1.8375 \div 0.25$
   7.35

21. $0.09312 \div 0.048$
   1.94

22. $3.5441 \div 0.083$
   42.7

23. $8.8038 \div 1.34$
   6.57

24. $1,855.45 \div 0.215$
   8,630

Problem Solving

Solve each problem. Round to the nearest hundredth.

25. A human can run 44.76 kilometers per hour. A human can run about 0.38 as fast as a cheetah. How fast can a cheetah run?
117.79 kilometers per hour

26. A box containing ball bearings weighs 635.2 grams. The box weighs 500 grams and each ball bearing weighs 2.6 grams. How many ball bearings are in the box?
52 ball bearings

198   Lesson 9-9 • Dividing Decimals by Decimals

For Mixed Abilities

Common Errors • Intervention

Some students may move the decimal point in the divisor but forget to move it in the dividend. Have them count the number of places they must move the decimal point in the divisor to make it a whole number, write that number, and then move the decimal point that number of places in the divisor and the dividend. Discuss how moving the decimal point the same number of places in both numbers is really multiplying both by the same number.

Enrichment • Number Sense

Challenge students to fill in the circles of this arrow puzzle.

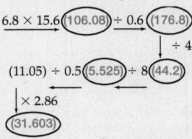

Ask students to create their own arrow puzzles to exchange with a friend.

More to Explore • Application

Have students research *adsorption*. Have them complete the following experiment to show how adsorption works:

- Take a few pieces of barbecue charcoal and crush them into smaller pieces.

- Mix a few drops of ink with water in a glass jar so that the water is dark.

- Now, sprinkle the crushed charcoal onto the surface of the water in the jar and shake.

- Pour the water through a sheet of coffee filter paper into a clean jar.

- What has happened? (The water is clear again.)

Have students explain why adsorbents are more efficient when they are divided into smaller bits.

---

4 Assess

Write the following on the board:

$0.08\overline{)3.216}$

Ask students to show how the decimal points move for the division. (0.08 ; 3.216) Then, have them divide to find the quotient. (40.2)

5 Mixed Review

1. $576.8 - 437.295$ (139.505)
2. $\$175.30 \div 5$ ($\$35.06$)
3. 0.052×0.06 (.00312)
4. $16.387 + 210.24$ (226.627)
5. $17\frac{2}{5} - 8\frac{2}{3}$ ($8\frac{11}{15}$)
6. $2.79 \div 9$ (0.31)
7. $25,275 \div 36$ (702 R3)
8. $6\frac{4}{5} \div 1\frac{2}{3}$ ($4\frac{2}{25}$)
9. 96.5×10^3 (96,500)
10. $278,302 - 149,656$ (128,646)

T198

pages 199–200

1 Getting Started

Objective
• To divide by a decimal, adding zeros to the dividend

Materials
calculators

Warm Up • Mental Math
Have students find each quotient or product.

1. $56 \div 10$ (5.6)
2. $23 \div 100$ (0.23)
3. $976 \div 10$ (97.6)
4. $485 \div 100$ (4.85)
5. $\frac{1}{10}$ of 56 (5.6)
6. $\frac{1}{100}$ of 23 (0.23)
7. $\frac{1}{10}$ of 976 (97.6)
8. $\frac{1}{100}$ of 485 (4.85)

Ask students what they noticed about these exercises. (Dividing by 10 is the same as multiplying by $\frac{1}{10}$. Dividing by 100 is the same as multiplying by $\frac{1}{100}$.)

Warm Up • Pencil and Paper
Ask students to write the missing factor in the equation.

1. $(10) \times 0.8 = 8$
2. $(10) \times 2.7 = 27$
3. $(100) \times 0.34 = 34$
4. $(1{,}000) \times 0.059 = 59$

2 Teach

Introduce the Lesson Have a student read the problem, tell what is to be solved, and what information is given. Point out that there are more decimal places in the divisor than in the dividend. Ask students what has to be done. (add zeros to the right of the decimal point in the dividend) Work through the division steps in the model with students to solve the problem.

Develop Skills and Concepts Recall that when the dividend is a whole number, any number of zeros may be written to the right of the decimal point.

Zeros in the Dividend

Shelby is working at a co-op produce store. She is putting 2.75 kilograms of rice into each sack. How many sacks of rice will Shelby fill from the basket of rice?

We want to know the number of sacks that can be filled with equal amounts of rice.

We know that the basket holds __33__ kilograms of rice, and each sack holds __2.75__ kilograms.

To find how many sacks Shelby can fill, we divide the total amount by the amount in each sack. We divide __33__ by __2.75__.

Write extra zeros after the decimal point in the dividend.	Multiply to make the divisor a whole number. Multiply the dividend by the same power of 10.	Divide.
$2.75\overline{)33.00}$	$2.75\overline{)33.00_\wedge}$	$2.75_\wedge\overline{)33.00_\wedge}$

$$2.75_\wedge\overline{)33.00_\wedge}$$
$$\begin{array}{r} 12 \\ \underline{27\ 5} \\ 5\ 50 \\ \underline{5\ 50} \\ 0 \end{array}$$

REMEMBER Any number of zeros can be written to the far right of a decimal number without changing its value.

Shelby will fill __12__ sacks of rice.

Getting Started

Divide.

1. $5.2\overline{)104}$ → 20
2. $0.6\overline{)0.3}$ → 0.5
3. $0.09\overline{)1.8}$ → 20
4. $1.2\overline{)360}$ → 300
5. $0.51\overline{)10.2}$ → 20
6. $0.032\overline{)40}$ → 1,250

Copy and divide.

7. $72 \div 1.6$
   45
8. $3.44 \div 0.08$
   43
9. $128 \div 0.8$
   160

Point out that zeros need to be annexed to the dividend only when the dividend contains fewer decimal places than the divisor. Have students work each division exercise on the board, explaining each step.

$0.7\overline{)42}$ (60) $7.2\overline{)216}$ (30) $0.22\overline{)31.90}$ (145) $0.03\overline{)82.80}$ (2,760)

3 Practice

Have students complete all the exercises. Encourage students to use multiplication to check each answer.

Practice

Divide.

1. $0.4\overline{)164}$ → 410

2. $0.07\overline{)16.1}$ → 230

3. $0.005\overline{)210}$ → 42,000

4. $1.3\overline{)312}$ → 240

5. $0.015\overline{)10.05}$ → 670

6. $0.048\overline{)3.36}$ → 70

7. $5.4\overline{)108}$ → 20

8. $0.67\overline{)455.6}$ → 680

9. $0.45\overline{)0.0162}$ → 0.036

Copy and divide.

10. $630 \div 0.3$
    2,100

11. $96 \div 0.6$
    160

12. $4.8 \div 0.08$
    60

13. $63 \div 1.4$
    45

14. $184.8 \div 0.33$
    560

15. $675 \div 0.9$
    750

16. $1,909 \div 0.83$
    2,300

17. $2.52 \div 0.07$
    36

18. $12.9 \div 2.15$
    6

Problem Solving

Solve each problem.

19. A rope is 17.4 meters long. How many 0.03-meter pieces can be cut from the rope?
    580 pieces

20. How long will it take to save $22.80 if you save $0.38 every other day?
    120 days

[Now Try This!]

A special operation is called **SquareAdd**. We square the first number and add the second. For example, $3 \text{ SquA } 2 = 3^2 + 2 = 11$

1. $5 \text{ SquA } 3 = \underline{28}$

2. $2 \text{ SquA } 7 = \underline{11}$

3. $6 \text{ SquA } 1 = \underline{37}$

Write yes or no.

4. Does $2 \text{ SquA } 5 = 5 \text{ SquA } 2$? $\underline{no}$

5. Does $4 \text{ SquA } 3 + 5 \text{ SquA } 3 = (4 + 5) \text{ SquA } 3$? $\underline{no}$

6. Does $5 \text{ SquA } 2 + 3 \text{ SquA } 2 = (5 \times 3) \text{ SquA } 2$? $\underline{no}$

Lesson 9-10 • Zeros in the Dividend

Now Try This! Have students read the directions and study the model. This activity focuses on the properties of an unfamiliar but sequential operation. Have students complete these exercises independently.

4 Assess

Write the following on the board:

$2.15\overline{)12.9}$ $0.005\overline{)2.5}$ $0.45\overline{)0.0162}$

Ask students to determine the number of zeros they would have to add to each dividend before they could divide. (1 zero; 2 zeros; no zeros)

For Mixed Abilities

Common Errors • Intervention

Some students may not recognize that they should place zeros in the dividend so they can move the decimal point the necessary number of places and simply place the decimal point at the end of the original dividend. Discuss how students must move the decimal point the same number of places in both the divisor and the dividend.

Enrichment • Number Sense

Have students complete the following exercises and then write a word problem for each:

1. $\$0.74\overline{)\$9.62}$ → (13)

3. $\$0.15\overline{)\$4.50}$ → (30)

2. $\$0.05\overline{)\$1.70}$ → (34)

4. $\$0.25\overline{)\$6.25}$ → (25)

More to Explore • Measurement

Have students conduct a survey of a plot of ground using metric measures following these steps:

1. Choose an interesting plot of ground.

2. Use a meterstick, string, and pegs to mark off a 4-meter-square plot.

3. Use the meterstick to further mark off the plot into four 1-square-meter sections.

4. On graph paper, make a scale drawing of the plot of land. Mark each section with an identifying letter.

5. List all the visible things found in the plot, both living and nonliving, by section and identify each by letter or number.

6. Mark the graph paper with the location of the things found in each section, using the identifying letter.

Encourage students to decorate plots as they wish. Display the students' plot diagrams.

T200

1 Getting Started

Objective
• To round the quotient

Warm Up • Mental Math
Ask students to round the decimal to the nearest tenth.

1. 4.63 (4.6)
2. 72.85 (72.9)
3. 615.38 (615.4)
4. 146.97 (147.0)

Ask students to round the decimal to the nearest hundredth.

1. 6.9379 (6.94)
2. 351.014 (351.01)
3. 17.985 (17.99)
4. 6.406 (6.41)

Warm Up • Paper and Pencil
Have students find each quotient.

1.
$$0.2\overline{)0.48} \quad (2.4)$$

2.
$$0.15\overline{)0.495} \quad (3.3)$$

3.
$$0.64\overline{)147.2} \quad (230)$$

Name _____

Rounding Quotients

Mrs. Heading bought supplies in bulk for the civic club's pancake breakfast. Besides flour, eggs, and butter, she bought maple syrup. How much did she pay per gallon of maple syrup?

MAPLE SYRUP
9.6 GALLONS
$12.35

We want to know the price of one gallon of syrup.

We know that Mrs. Heading bought ___9.6___ gallons of syrup, and she paid ___$12.35___.

To find the cost per gallon, we divide the total price by the number of gallons purchased. We divide ___$12.35___ by ___9.6___.

When a division continues to have a remainder, we can round the quotient. Sometimes it is necessary to write extra zeros to extend the dividend. In this problem, we carry the division to thousandths to find the cost to the nearest cent.

Multiply the divisor and dividend by 10.	Divide. Round the quotient to cents.

$$9.6_\wedge\overline{)\$12.3_\wedge 5}$$

$$\begin{array}{r} 1.286 \approx \$1.29 \\ 9.6_\wedge\overline{)\$12.3_\wedge 500} \\ -96 \\ \hline 27\ 5 \\ -19\ 2 \\ \hline 8\ 30 \\ -7\ 68 \\ \hline 6\ 20 \\ -5\ 76 \\ \hline 4\ 4 \end{array}$$

The symbol ≈ means is approximately.

REMEMBER Always carry the division one place value beyond the place to which you want to round.

Mrs. Heading paid about ___$1.29___ per gallon of syrup.

Getting Started _____

Divide. Round to the nearest tenth.

1.
$$5\overline{)9.21} \quad \begin{array}{c}1.84 \approx 1.8\end{array}$$

2.
$$4.6\overline{)8.7} \quad \begin{array}{c}1.89 \approx 1.9\end{array}$$

Divide. Round to the nearest hundredth.

3.
$$1.6\overline{)5.69} \quad \begin{array}{c}3.556 \approx 3.56\end{array}$$

4.
$$7.3\overline{)8.7} \quad \begin{array}{c}1.191 \approx 1.19\end{array}$$

---

2 Teach

Introduce the Lesson Have a student read the problem. Have students identify the question to be answered and tell what necessary information is given. Have them complete the information sentences. Discuss the division model used to solve the problem. Emphasize the rounding rule.

Develop Skills and Concepts Explain that when a division problem continues to have a remainder, students can round the quotient to any necessary place value, but the quotient will not be exact. The more decimal places in the quotient, the more exact it will be.

As you discuss the model division problem, point out that sometimes it is necessary to write extra zeros in the dividend. Ask students when they should annex zeros. (Possible answer: when they are to round to a given place and there are not enough decimal places in the dividend)

Explain that to round a quotient to the nearest hundredth, students must divide to one place beyond hundredths (thousandths) and then round.

Call attention to the symbol for *is approximately equal to*, ≈.

Write the following division exercises on the board. Have volunteers work out the division on the board. Have them round each quotient to the nearest tenth.

$$3.2\overline{)69.1} \quad (21.6)$$

$$0.25\overline{)6.41} \quad (25.6)$$

Have them round each quotient to the nearest hundredth.

$$1.6\overline{)3.24} \quad (2.03)$$

$$0.15\overline{)7.055} \quad (47.03)$$

Practice

Divide. Round each quotient to the nearest tenth.

1. $4\overline{)6.83}$ $\dfrac{1.70}{} \approx 1.7$

2. $8\overline{)9.44}$ $\dfrac{1.18}{} \approx 1.2$

3. $0.6\overline{)5.732}$ $\dfrac{9.55}{} \approx 9.6$

4. $0.3\overline{)8.175}$ $\dfrac{27.25}{} \approx 27.3$

5. $0.9\overline{)5.736}$ $\dfrac{6.37}{} \approx 6.4$

6. $1.3\overline{)6.845}$ $\dfrac{5.26}{} \approx 5.3$

7. $4.7\overline{)9.51}$ $\dfrac{2.02}{} \approx 2.0$

8. $3.6\overline{)5.3}$ $\dfrac{1.47}{} \approx 1.5$

Divide. Round each quotient to the nearest hundredth.

9. $7\overline{)3.964}$ $\dfrac{0.566}{} \approx 0.57$

10. $9\overline{)12.857}$ $\dfrac{1.428}{} \approx 1.43$

11. $0.2\overline{)5.615}$ $\dfrac{28.075}{} \approx 28.08$

12. $0.8\overline{)0.0651}$ $\dfrac{0.081}{} \approx 0.08$

13. $1.9\overline{)5.623}$ $\dfrac{2.959}{} \approx 2.96$

14. $7.6\overline{)0.178}$ $\dfrac{0.023}{} \approx 0.02$

15. $2.6\overline{)4.96}$ $\dfrac{1.907}{} \approx 1.91$

16. $3.8\overline{)9.7}$ $\dfrac{2.552}{} \approx 2.55$

Problem Solving

Solve each problem.

17. At the Snack Shoppe this week, you can buy 3 Super Submarine Sandwiches for $8.50. To the nearest cent, how much does 1 sandwich cost?
$2.83

18. Inga weighs 0.88 as much as Nick. If Nick weighs 146 pounds, how much does Inga weigh to the nearest pound?
128 pounds

19. Dinah's test scores this semester are 91.6, 89.25, 94.8, 83.35, and 97.4. The class average is 95.5. To the nearest hundredth, how much does Dinah's average differ from the class average?
4.22 lower

20. A lazy ant travels at a steady rate of 1.7 centimeters per minute. To the nearest tenth of a minute, how long does the ant take to travel 106.56 centimeters?
62.7 minutes

For Mixed Abilities

Common Errors • Intervention

Some students may stop dividing at the place to which they must round and not have the extra digit in the quotient essential for telling them how to round. Have them draw an arrow above the place to which they must round. Then, remind them that they need a digit to the right of that place.

Enrichment • Number Sense

Ask students to round the quotient to the nearest thousandth.

1. $4.2\overline{)95.86}$ (22.824)

2. $0.36\overline{)148.9}$ (413.611)

3. $6.2\overline{)573.2}$ (92.452)

4. $7.8\overline{)634.97}$ (81.406)

Challenge students to devise a flowchart showing how to round a quotient to the nearest hundredth when the divisor and the dividend are both decimals.

More to Explore • Application

Have students research how many miles away from the Sun each of the nine major planets in the solar system are. Have them then figure the travel time from each planet to the Sun in each of the following vehicles:

1. a car traveling 60 mph

2. a train traveling 110 mph

3. a passenger aircraft traveling 500 mph

3 Practice

Have students complete all the exercises. Remind them that if they round to the nearest tenth, they must divide to the hundredth place.

4 Assess

Tell students that they need to round to the nearest hundredth. Ask, *To what place should you divide?* (the thousandth place)

5 Mixed Review

1. $478.02 - 209.53 ($268.49)

2. $2.175 \div 25$ (.087)

3. $261\frac{9}{10} + 194\frac{3}{4}$ $(456\frac{13}{20})$

4. 5.23×0.08 (0.4184)

5. $5\frac{1}{2} \div \frac{3}{4}$ $(7\frac{1}{3})$

6. $128.73 - 79.8$ (48.93)

7. $\frac{8}{9} - \frac{5}{12}$ $(\frac{17}{36})$

8. $10.05 \div 10^3$ (0.01005)

9. $8 \times 3\frac{3}{4}$ (30)

10. $153 \div 0.3$ (510)

pages 203–204

1 Getting Started

Objective
• To rename a fraction as a decimal

Warm Up • Mental Math
Have students find each quotient or product.

1. $1.4 \div 2$ (0.7)
2. 0.7×0.9 (0.63)
3. $.04 \times 9$ (0.36)
4. $54 \div 0.6$ (90)
5. $0.064 \div 8$ (0.008)
6. $0.25 \div 5$ (0.05)
7. 8×0.9 (7.2)
8. $.042 \div 6$ (0.007)
9. 0.7×0.08 (0.056)
10. 0.4×8 (3.2)

Warm Up • Pencil and Paper
Have students rename each fraction as a whole or a mixed number.

1. $\frac{8}{4}$ (2)
2. $\frac{7}{3}$ ($2\frac{1}{3}$)
3. $\frac{9}{3}$ (3)
4. $\frac{47}{6}$ ($7\frac{5}{6}$)
5. $\frac{401}{35}$ ($11\frac{16}{35}$)
6. $\frac{293}{28}$ ($10\frac{13}{28}$)
7. $\frac{7}{5}$ ($1\frac{2}{5}$)
8. $\frac{17}{10}$ ($1\frac{7}{10}$)

Renaming Fractions as Decimals

Marie is the lead-off hitter on the softball team. What is Marie's batting average to the nearest thousandth?

Lincoln Batting Averages			
Girl	At Bat	Hits	Average
Marie	15	7	?
Rita	13	6	?
Joan	16	7	?

We want to know Marie's batting average.

We know that she has been at bat __15__ times, and she has had __7__ hits. Marie has hit the ball $\frac{7}{15}$ of the time.

To find her batting average, we name $\frac{7}{15}$ as a decimal.

Write the fraction as a division.

$$\frac{7}{15} = 15\overline{)7}$$

Divide the numerator by the denominator. Round the quotient to thousandths.

$$
\begin{array}{r}
0.4666 \approx 0.467 \\
15\overline{)7.0000} \\
-\underline{6\ 0} \\
1\ 00 \\
-\underline{90} \\
100 \\
-\underline{90} \\
100 \\
\end{array}
$$

Marie has a batting average of __0.467__.

Some fractions are renamed as decimals in tenths or hundredths. Others don't have an exact decimal equivalent. These we usually round to a particular place value.

Getting Started

Rename each fraction as a decimal. Round to the nearest hundredth.

1. $\frac{2}{3}$ 0.67
2. $\frac{5}{6}$ 0.83
3. $\frac{1}{16}$ 0.06
4. $\frac{1}{12}$ 0.08
5. $\frac{3}{8}$ 0.38

---

2 Teach

Introduce the Lesson Have a student read the problem, tell what is to be solved, and identify the information given in the chart. Be sure students understand that 7 out of 15 can be written as the fraction $\frac{7}{15}$. Point out that a fraction implies the division of the numerator by the denominator. In this case, $\frac{7}{15}$ can be interpreted as $7 \div 15$.

Continue by working through the model with students. Explain that because the division $15\overline{)7}$ does not result in a remainder of zero, the quotient is rounded to a convenient place value, the nearest thousandth.

Read the rule about decimals with and without remainders.

Develop Skills and Concepts Emphasize that students divide the numerator by the denominator when renaming a fraction as a decimal. Students often divide the denominator by the numerator because they are accustomed to dividing a large number by a smaller one. Encourage students to look at their answers; if the fraction represents a value less than 1, the decimal must also name a number less than 1.

Discuss the sentences at the end of the lesson. Have students write other fractions that can be renamed as decimals with no remainders, such as $\frac{1}{10}$, $\frac{1}{25}$, and $\frac{3}{4}$. Have students identify which of these fractions can be renamed as decimals with no remainder: $\frac{7}{8}$, $\frac{11}{15}$, and $\frac{9}{20}$. ($\frac{7}{8}$, $\frac{9}{20}$)

3 Practice

Have students complete all the exercises. Remind students that if the quotient is to be rounded to the nearest hundredth, they need to divide to the thousandth place.

T203

Practice

Rename each fraction as a decimal. Round to the nearest hundredth.

1. $\frac{1}{6}$ 0.17
2. $\frac{5}{12}$ 0.42
3. $\frac{1}{2}$ 0.50
4. $\frac{3}{16}$ 0.19
5. $\frac{5}{8}$ 0.63

6. $\frac{4}{7}$ 0.57
7. $\frac{3}{13}$ 0.23
8. $\frac{7}{11}$ 0.64
9. $\frac{3}{4}$ 0.75
10. $\frac{1}{8}$ 0.13

11. $\frac{7}{16}$ 0.44
12. $\frac{8}{9}$ 0.89
13. $\frac{1}{7}$ 0.14
14. $\frac{11}{15}$ 0.73
15. $\frac{17}{20}$ 0.85

16. $\frac{3}{11}$ 0.27
17. $\frac{17}{25}$ 0.68
18. $\frac{5}{9}$ 0.56
19. $\frac{7}{8}$ 0.88
20. $\frac{1}{12}$ 0.08

Problem Solving

Solve each problem. Round answers to thousandths.

21. The Eliot basketball team played 18 games and won 11 of them. What is the Eliot basketball team's winning average?
0.611

22. In the free throw contest, Ken made 9 shots and missed 5 shots. What is his shooting average?
0.643

23. Dennis was at bat 54 times and struck out only 14 times. What is his batting average?
0.741

24. Bill has 5 hits in 8 times at bat. How much will his batting average increase if he gets 2 hits in his next 2 turns at bat?
0.075

Now Try This!

When the only prime factors of the denominator are 2 or 5, the decimal will have no remainder.

$$\frac{1}{8} = 8\overline{)1.000}$$
$$\begin{array}{r} 0.125 \\ \underline{-8} \\ 20 \\ \underline{-16} \\ 40 \\ \underline{-40} \\ 0 \end{array}$$

When the prime factors of the denominator include factors other than 2 or 5, the decimal has a repeating remainder.

$$\frac{1}{3} = 3\overline{)1.000}$$
$$\begin{array}{r} 0.333 \\ \underline{-9} \\ 10 \\ \underline{-9} \\ 10 \end{array}$$

Circle the fractions that will have a repeating remainder.

$\frac{3}{4}$ $(\frac{1}{7})$ $(\frac{2}{3})$ $\frac{7}{8}$ $(\frac{1}{12})$ $(\frac{7}{9})$ $\frac{3}{5}$ $(\frac{5}{11})$ $(\frac{4}{15})$ $\frac{3}{10}$

Lesson 9-12 • Renaming Fractions as Decimals

Now Try This! After studying the examples, students should be able to pick out fractions that will have repeating remainders without doing any computations.

4 Assess

Write these fractions on the board:

$\frac{1}{4}, \frac{1}{3}, \frac{1}{2}, \frac{2}{5}, \frac{3}{8}$

Ask students to rename each fraction as a decimal, rounding to the nearest thousandth if necessary. (0.25, 0.333, 0.5, 0.4, 0.375)

For Mixed Abilities

Common Errors • Intervention

Watch for students who do not set up the problem correctly and use the numerator as the divisor. Have them work with partners with equivalents that are familiar to them, such as $\frac{1}{2} = 0.5$ and $\frac{3}{4} = 0.75$. Students then will see that to obtain 0.5 and 0.75, they must divide the numerator by the denominator. Have them continue practicing, using less familiar equivalents.

Enrichment • Number Sense

Introduce the notation for repeating decimals as shown.

$\frac{1}{3} \longrightarrow 3\overline{)1.000}$ 0.333 $\frac{1}{3} = 0.\overline{3}$

$\frac{2}{11} \longrightarrow 11\overline{)2.0000}$ 0.1818 $\frac{2}{11} = 0.\overline{18}$

$\frac{7}{15} \longrightarrow 15\overline{)7.000}$ 0.466 $\frac{7}{15} = 0.4\overline{6}$

Have students write repeating decimals for each fraction.

1. $\frac{2}{3} = (0.\overline{6})$
2. $\frac{4}{11} = (0.\overline{36})$
3. $\frac{11}{12} = (0.91\overline{6})$
4. $\frac{7}{11} = (0.\overline{63})$
5. $\frac{4}{9} = (0.\overline{4})$
6. $\frac{5}{6} = (0.8\overline{3})$

More to Explore • Application

Invite a representative from a local bank to discuss savings and checking accounts. Ask the representative to bring a sample savings account book to show how to keep track of dates, withdrawals, deposits, and balances. Have the speaker define these terms and demonstrate the math used to arrive at a balance.

Encourage students to ask questions. Supply students with a sample bank account sheet showing a list of deposits and withdrawals, and have them supply correct figures in the balance column.

pages 205–206

1 Getting Started

Objective

- To solve a problem by working backward

Warm Up • Mental Math

Have students name the next number.

1. 0, 3, 6, 9, 12, 15, 18, 21 (24)
2. 2, 5, 7, 10, 12, 15, 17, 20 (22)
3. 56, 52, 51, 47, 46, 42, 41 (37)
4. 2, 4, 8, 16, 32, 64, 128 (256)
5. 6, 12, 13, 26, 27, 54, 55, 110, 111 (222)
6. 72, 36, 40, 20, 24, 12, 16, 8 (12)
7. 3, 2, 5, 4, 7, 6, 9, 8, 11, 10 (13)
8. 1, 4, 9, 16, 25, 36, 49, 64 (81)

Warm Up • Pencil and Paper

Have students simplify the expression.

1. $\frac{1}{2}$ of 68 (34)
2. $\frac{1}{3}$ of 81 (27)
3. 565 − 392 (173)
4. 2 × 515 (1,030)
5. (6 × 5) + 3 − 2 (31)
6. double 49 (98)
7. a half of 52 (26)
8. a third of 102 (34)

Problem Solving: Work Backward

Marlene spent Saturday afternoon at the carnival. She spent one half of her money on tickets for the rides. She spent $3.00 on games. She then spent one half of what was left on food. When Marlene arrived home she had $2.00 left. How much money did Marlene take to the carnival?

⭐ SEE

We want to know how much Marlene took to the carnival.

She spent one half of her money on the rides. She then spent __$3.00__ on games. She spent one half of what was left for food. Marlene had __$2.00__ left when she arrived home.

⭐ PLAN

Since we know that Marlene had $2.00 left and we know how she spent her money, we can work backward to find out how much she had at the start.

⭐ DO

We double what she had at the end to find out what she had before eating.

$2.00 × 2 = __$4.00__

We add to what she had before games. **$3.00 +** __$4.00__ **=** __$7.00__

We can double the amount she had left to find out how much she had before buying tickets.

__$7.00__ **× 2 =** __$14.00__   Marlene took __$14.00__ to the carnival.

⭐ CHECK

We can check our solution by working through the problem in the correct order.

$\frac{1}{2}$ of $14.00 = __$7.00__ $14.00 − $7.00 = __$7.00__
amount spent on rides   amount left

$7.00 − $3.00 = __$4.00__ $\frac{1}{2}$ of $4.00 = __$2.00__
amount after games   amount spent on food

$4.00 − $2.00 = __$2.00__ amount left

Lesson 9-13 • Problem Solving: Work Backward   **205**

---

2 Teach

Introduce the Lesson Have students read the problem and then suggest that problems like this can be solved by working backward through the data.

Develop Skills and Concepts At this point, students should have a working knowledge of inverse operations: subtraction and addition, multiplication and division. Remind them that taking one-half of a number is the same as dividing it by two. Likewise, dividing a number by three is the same as multiplying it by one-third.

Tell students that it will be helpful to label the answers for each of the steps. The best way to check this problem is to start with the solution and work through each of the steps.

3 Apply

Solution Notes

1. $7.00 − 1.00 = $6.00
   $6.00 + 1.75 = $7.75
2. 283 − 188 = 95 lb, first boy off; 188 − 103 = 85 lb, second boy; 103 lb remaining
3. 8 + 4 + 3 + 2 = 17 strawberries
4. Start with Sandy having 4 cookies.
5. 40 − 16 = 24, liked ice cream
   18 + 12 = 30, liked both flavors
   30 − 24 = 6, liked both
6. Bob, 45
   Charles, 45 − 9 = 36
   Don, 36 + 7 = 43
   Adam, 43 + 10 = 53

Higher-Order Thinking Skills

7. **Analysis:** Since $10.00 is $2.25 more than what he was paid, he would have $2.25 more on Thursday ($7.00 + $2.25 = $9.25).

Apply

Solve each problem.

1. José was paid for mowing the lawn on Monday. On Tuesday he spent $1.75 and on Wednesday he earned $1.00 walking the neighbor's dog. On Thursday, José had $7.00 left. How much was José paid for mowing the lawn?
$7.75

2. Three boys got on a scale together and found their combined weight to be 283 pounds. One boy stepped off the scale and it showed 188 pounds. When a second boy jumped off the scale, it showed 103 pounds. What was the weight of each of the three boys?
95, 85, and 103 pounds

3. Marilyn bought some strawberries. She gave two to Todd, three to Mark, and four to Brian. She then had eight left. How many strawberries did Marilyn buy?
17 strawberries

4. Gary, Joe, and Sandy came home from school and raided the cookie jar. Gary had twice as many cookies as Joe, and Joe had three more than Sandy. Sandy had four cookies. How many cookies did Gary and Joe have?
Gary had 14 cookies. Joe had 7.

5. There were 40 boys on the football team. 18 liked chocolate ice cream and 12 liked vanilla ice cream. 16 liked neither. How many boys liked both chocolate and vanilla ice cream?
6 boys

6. Adam, Bob, Charles, and Don were having a contest to see who could score highest on a dart board. Adam scored 10 more than Don, and Don scored 7 more than Charles. Charles scored 9 less than Bob. Bob scored 45. What were the scores of the others?
Adam 53, Bob 45, Don 43, Charles 36

7. What if, in Problem 1, José was paid $10.00 to mow the lawn? How much would he have left on Thursday?
$9.25

8. Write a problem that can be solved by working backward where the answer is the number in a dozen.
Answers will vary.

9. If you round an amount to the nearest whole dollar and multiply by a one-digit number, what is the greatest amount you can be from the actual answer?
$4.50

10. Five students are in a race. Two of them have already crossed the finish line. How many ways can the others cross the line? Prove that you are correct.
6 ways

Lesson 9-13 • Problem Solving: Work Backward

For Mixed Abilities

More to Explore • Geometry

Use this activity to reintroduce the idea of volume. Bring ten different containers to class. Try to include pairs of containers that hold the same amount but are not the same shape. Arrange the containers on a table with a dishpan filled with water, sand, or dried beans. Place the containers in random order on a table.

Then, have students go to the table in small groups and work out a system for testing the capacity of the containers so that they can be placed in order from smallest to largest.

- One method would be to use one small container to fill the others.

- Another method would be to line up the containers by eye and then check each pair.

Students will realize this is unreliable. Have them discuss how to set up a worldwide unit for volume and a method of measuring volume. Have them discuss how large they would make the unit and what they would call it.

---

8. **Synthesis:** Students should recognize that the answer is 12, the number of anything in a dozen.

9. **Analysis:** The greatest "error" you can make when rounding is 50¢, when you round up a midway number. The greatest one-digit number is 9. So, the greatest you can be from the actual product is 9×50¢.

10. **Evaluation:** The proofs may vary, but one way is to use A, B, and C for the remaining students and show the six different ways the three remaining students can cross the finish line.

ABC	ACB
BAC	BCA
CBA	CAB

4 Assess

Ask students what information in the problem allows them to work backward to solve. (Possible answer: They know how much money Marlene has left and how she spent her money.)

9-14 Calculator: Repeating Decimals

pages 207–208

1 Getting Started

Objective
• To rename fractions as repeating decimals and repeating decimals as fractions

Vocabulary
repeating decimals

Materials
calculators

Warm Up • Mental Math
Have students name the power of 10 that when multiplied by the decimal would make a whole number.

1. 4.62 (10^2) 5. 8.91 (10^2)
2. 0.973 (10^3) 6. 0.4218 (10^4)
3. 72.8 (10^1) 7. 734.1 (10^1)
4. 64.426 (10^3) 8. 82.643 (10^3)

Warm Up • Calculator
Have students write the quotient rounded to the nearest hundredth.

1. (1.83) 5)9.1368
3. (6.87) 0.98)6.734
2. (0.83) 0.7)0.5831
4. (0.16) 4)0.643

Name _____

Calculator: Repeating Decimals

Fractions can be thought of as divisions in which the numerator is divided by the denominator. Enter these fractions into a calculator and write what appears on the screen.

$\frac{1}{11}$ = 0.090909 $\frac{5}{9}$ = 0.555555 $\frac{8}{15}$ = 0.533333

These decimals are called **repeating decimals**. One way to write repeating decimals is to place a bar over the digit or digits that repeat.

$\frac{1}{11} = 0.\overline{09}$ $\frac{5}{9} = 0.\overline{5}$ $\frac{8}{15} = 0.5\overline{3}$

Enter these fractions into a calculator and write them using bars.

$\frac{1}{9}$ = $0.\overline{1}$ $\frac{7}{30}$ = $0.2\overline{3}$ $\frac{7}{66}$ = $0.1\overline{06}$

Follow these steps to rename repeating decimals as fractions:

• The numerator is equal to the difference between the entire decimal number and the digits that don't repeat.

• The denominator will always start with 9 and there will be one 9 for every repeating digit. The 9s will be followed by one 0 for every nonrepeating digit.

$0.41\overline{6} = \frac{375}{900}$

Use a calculator to check this.
Rename $0.4\overline{5}$ as a fraction.

The numerator will be __41__.

The denominator will be __90__.

The fraction will be __$\frac{41}{90}$__.

We check by entering the fraction into a calculator.

$\frac{41}{90}$ = $0.4\overline{5}$

$0.41\overline{6} = \frac{[?]}{[?]}$

$416 - 41 = \underline{375}$

One repeating digit
↓
900
↑
two nonrepeating digits

---

2 Teach

Introduce the Lesson Review the steps for renaming a fraction as a decimal.

1. Write the fraction as a division problem.

$\frac{2}{15}$ = 15)2

2. Divide the numerator by the denominator. Round the quotient.

0.133 = $0.1\overline{3}$ $\frac{2}{15} = 0.1\overline{3}$
15)2.000
  15
  50
  45
  50

Have a student read the explanation of repeating decimals. Have students write $\frac{2}{15}$ as a repeating decimal. ($0.1\overline{3}$) Work through each step of the model with students.

Develop Skills and Concepts When students write repeating decimals, emphasize that they should write the bar only over the digits that repeat. Have them use calculators to name each fraction as a repeating decimal.

$\frac{7}{11}$ ($0.\overline{63}$) $\frac{4}{9}$ ($0.\overline{4}$) $\frac{2}{33}$ ($0.\overline{06}$)

Have students name each repeating decimal as a fraction. Discuss each step of the procedure.

$0.7\overline{6}$ ($\frac{69}{90}$) $0.21\overline{3}$ ($\frac{192}{900}$)

T207

Practice

Write each fraction as a repeating decimal. Place a bar over any repeating digits.

1. $\frac{2}{3} =$ __0.$\overline{6}$__ 2. $\frac{3}{8} =$ __0.375__ 3. $\frac{7}{12} =$ __0.58$\overline{3}$__ 4. $\frac{7}{15} =$ __0.4$\overline{6}$__

5. $\frac{13}{24} =$ __0.541$\overline{6}$__ 6. $\frac{8}{90} =$ __0.0$\overline{8}$__ 7. $\frac{4}{7} =$ __0.$\overline{571428}$__ 8. $\frac{5}{13} =$ __0.$\overline{384615}$__

9. $\frac{3}{11} =$ __0.$\overline{27}$__ 10. $\frac{12}{37} =$ __0.$\overline{324}$__ 11. $\frac{1}{9} =$ __0.$\overline{1}$__ 12. $\frac{15}{16} =$ __0.9375__

13. $\frac{7}{8} =$ __0.875__ 14. $\frac{1}{12} =$ __0.08$\overline{3}$__ 15. $\frac{3}{4} =$ __0.75__ 16. $\frac{6}{11} =$ __0.$\overline{54}$__

Write each repeating decimal as a fraction. Simplify the fraction.

17. $0.2\overline{3} = \frac{7}{30}$ 18. $0.\overline{45} = \frac{5}{11}$ 19. $0.08\overline{3} = \frac{1}{12}$ 20. $0.2\overline{6} = \frac{4}{15}$

21. $0.3\overline{6} = \frac{11}{30}$ 22. $0.7\overline{3} = \frac{11}{15}$ 23. $0.91\overline{6} = \frac{11}{12}$ 24. $0.\overline{39} = \frac{13}{33}$

25. $0.7\overline{5} = \frac{34}{45}$ 26. $0.2\overline{87} = \frac{19}{66}$ 27. $0.4\overline{12} = \frac{68}{165}$ 28. $0.\overline{142857} = \frac{1}{7}$

(Now Try This!)

What has a part always on the go?

You can solve this riddle by decoding the answers to these nine problems.

1. $1.8 \times 5.31 =$ __9.558__

2. $2.05 \times 1.3 =$ __2.665__

3. $10.48 - 1.982 =$ __8.498__

4. $3.4 \times 0.07 =$ __0.238__

5. $8.82 \div 3.6 =$ __2.45__

6. $11.904 \div 3.72 =$ __3.2__

7. $6.4 \times 1.21 =$ __7.744__

8. $0.598 + 1.852 =$ __2.45__

9. $61.295 \div 23 =$ __2.665__

Code Table	
A	0.238
D	9.558
E	2.665
L	1.67
M	0.035
P	8.498
R	2.45
S	4.061
T	3.2
U	7.744

D	E	P	A	R	T	U	R	E
1	2	3	4	5	6	7	8	9

Lesson 9-14 • Calculator: Repeating Decimals

Now Try This! Have students use their calculators to solve the exercises. Have them match each answer to a letter in the code table to solve the riddle.

3 Practice

Have students complete all the exercises. Be sure each student has a calculator.

4 Assess

Have students write the repeating decimal for $\frac{1}{3}$, $\frac{2}{9}$, and $\frac{5}{11}$. ($0.\overline{3}$, $0.\overline{2}$, $0.\overline{45}$)

For Mixed Abilities

Common Errors • Intervention

Watch for students who do not use the fraction bar correctly and write $0.3\overline{6}$ for $0.\overline{36}$. Stress that the bar shows the digit or set of digits that repeats. Have students work with partners and use the bar to rename each repeating decimal:

1. $0.8383\ldots$ ($0.\overline{83}$)

2. $0.9111\ldots$ ($0.9\overline{1}$)

3. $0.876876\ldots$ ($0.\overline{876}$)

4. $0.047333\ldots$ ($0.047\overline{3}$)

Enrichment • Number Sense

Have students write a repeating decimal for the fraction.

1. $\frac{1}{3}$ ($0.\overline{3}$)

2. $\frac{1}{33}$ ($0.\overline{03}$)

3. $\frac{1}{333}$ ($0.\overline{003}$)

4. $\frac{1}{3333}$ ($0.\overline{0003}$)

5. $\frac{2}{3}$ ($0.\overline{6}$)

6. $\frac{2}{33}$ ($0.\overline{06}$)

7. $\frac{2}{333}$ ($0.\overline{006}$)

8. $\frac{2}{3333}$ ($0.\overline{0006}$)

Encourage students to experiment to see what patterns they discover.

More to Explore • Probability

Draw this number line on the board for students to duplicate:

L 10 9 8 7 6 5 4 3 2 1 0 1 2 3 4 5 6 7 8 9 10 R

Tell them to pretend that a man starts walking at 0. They will flip a coin; if it lands on heads, the man will move one step to the right. If the coin lands on tails, the man will move one step to the left.

Ask students to guess where the man will be after 10, after 20, and after 40 flips. Have them actually flip the coin 10, 20, and 40 times and record where the man lands in the end. Suggest that they list positions to the right with a number followed by R and positions to the left with a number followed by L.

Have students compare their results. Ask if any ended on 0.

T208

Chapter 9
Test

page 209

Items	Objectives
1–4	To multiply whole numbers and decimals (see pages 185–186)
5–8	To multiply decimals (see pages 187–188)
9–12	To multiply decimals, write zeros in the product to locate the decimal point (see pages 189–190)
13–16	To divide decimals by whole numbers (see pages 191–192)
17–24	To divide decimals by decimals (see pages 197–198)
21, 23–24	To divide by decimals with fewer decimal places in the dividend than in the divisor, with zeros in the dividend (see pages 199–200)
25–28	To rename a fraction as a decimal, rounding to the nearest thousandth (see pages 203–204)

Alternate Chapter Test

You may wish to use the Alternate Chapter Test on page 386 of this book for further review and assessment.

Multiply.

1. $\begin{array}{r} 5.6 \\ \times\ \ 8 \\ \hline 44.8 \end{array}$
2. $\begin{array}{r} 4.21 \\ \times\ \ 6 \\ \hline 25.26 \end{array}$
3. $\begin{array}{r} 25 \\ \times 1.6 \\ \hline 40.0 \end{array}$
4. $\begin{array}{r} 3.05 \\ \times\ \ 28 \\ \hline 85.40 \end{array}$

5. $\begin{array}{r} 3.85 \\ \times\ 0.3 \\ \hline 1.155 \end{array}$
6. $\begin{array}{r} 4.7 \\ \times 0.13 \\ \hline 0.611 \end{array}$
7. $\begin{array}{r} 3.761 \\ \times\ \ 2.5 \\ \hline 9.4025 \end{array}$
8. $\begin{array}{r} 9.64 \\ \times\ 5.3 \\ \hline 51.092 \end{array}$

9. $\begin{array}{r} 0.004 \\ \times\ \ 15 \\ \hline 0.060 \end{array}$
10. $\begin{array}{r} 0.04 \\ \times 0.03 \\ \hline 0.0012 \end{array}$
11. $\begin{array}{r} 0.006 \\ \times\ 0.4 \\ \hline 0.0024 \end{array}$
12. $\begin{array}{r} 5.03 \\ \times 0.008 \\ \hline 0.04024 \end{array}$

Divide.

13. $4\overline{)14.4}$ → 3.6
14. $3\overline{)1.68}$ → 0.56
15. $8\overline{)2.848}$ → 0.356
16. $7\overline{)44.1}$ → 6.3

17. $0.05\overline{)3.261}$ → 65.22
18. $0.9\overline{)0.036}$ → 0.04
19. $1.6\overline{)6.08}$ → 3.8
20. $4.3\overline{)1.505}$ → 0.35

21. $2.6\overline{)208}$ → 80
22. $0.5\overline{)0.1}$ → 0.2
23. $0.016\overline{)3.2}$ → 200
24. $6.5\overline{)52}$ → 8

Write each fraction as a decimal rounded to the nearest thousandth.

25. $\frac{5}{7} \approx$ ___0.714___
26. $\frac{7}{16} \approx$ ___0.438___
27. $\frac{3}{7} \approx$ ___0.429___
28. $\frac{11}{15} \approx$ ___0.733___

Circle the letter of the correct answer.

1 $9 - 3 \div 3 = n$
$n = ?$
a. 2
b. 8
c. 9
d. NG

2 $34,358$
$+ 6,582$
a. 40,840
b. 40,940
c. 41,940
d. NG

3 $30,465$
$- 16,587$
a. 13,872
b. 23,872
c. 26,122
d. NG

4 $1,596$
$\times \quad 9$
a. 14,364
b. 14,366
c. 14,464
d. NG

5 $27\overline{)365}$
a. 13 R14
b. 14 R13
c. 131 R4
d. NG

6 $2\frac{1}{3}$
$+ 3\frac{2}{5}$
a. $5\frac{3}{8}$
b. $5\frac{11}{15}$
c. $6\frac{11}{15}$
d. NG

7 $7\frac{1}{3}$
$- 4\frac{3}{4}$
a. $2\frac{1}{2}$
b. $2\frac{7}{12}$
c. $3\frac{7}{12}$
d. NG

8 $4\frac{1}{2} \times 3\frac{1}{3}$
a. $\frac{1}{15}$
b. $2\frac{7}{10}$
c. 15
d. NG

9 $2\frac{1}{2} \div 3\frac{1}{4}$
a. $\frac{10}{13}$
b. $1\frac{3}{10}$
c. $8\frac{1}{8}$
d. NG

10 What is the place value of the 6 in 13.2684?
a. tens
b. ones
c. tenths
d. NG

11 7.83
$+ 2.648$
a. 10.47
b. 10.478
c. 10.578
d. NG

STOP

score

Chapter 9 • Cumulative Assessment

page 210

Items	Objectives
1	To simplify an expression having two or more operations (see pages 11–12)
2–3	To add or subtract two numbers up to 6-digits (see pages 27–28, 33–34)
4	To multiply 4-digit factors by 1-digit factors (see pages 45–46)
5	To divide 3-digit numbers by 2-digit numbers (see pages 75–76)
6–7	To add or subtract mixed numbers (see pages 115–116, 125–126)
8–9	To multiply or divide mixed numbers (see pages 141–142, 147–148)
10	To identify the place value of digits in decimals through millionths (see pages 163–164)
11	To add decimals (see pages 171–172)

Cumulative Assessment

Circle the letter of the correct answer.

1. $12 - 6 \div 3 = n$
$n = ?$
a 2
b 9
c 10
d NG

2. $57,046$
$+ 2,984$
a 59,930
b 60,030
c 60,020
d NG

3. $80,537$
$- 16,748$
a 63,889
b 64,789
c 73,789
d NG

4. $3,187$
$\times \quad 8$
a 25,496
b 24,896
c 24,696
d NG

5. $36\overline{)985}$
a 27 R1
b 27 R3
c 27 R13
d NG

6. $4\frac{3}{10}$
$+ 6\frac{4}{15}$
a $10\frac{16}{30}$
b $10\frac{17}{30}$
c $11\frac{17}{30}$
d NG

7. $9\frac{1}{6}$
$- 5\frac{7}{9}$
a $3\frac{7}{18}$
b $4\frac{7}{18}$
c $3\frac{2}{3}$
d NG

8. $6\frac{2}{9} \times 3\frac{3}{8}$
a 18
b 21
c 24
d NG

9. $9\frac{5}{7} \div 5\frac{2}{3}$
a $1\frac{5}{7}$
b $55\frac{1}{21}$
c $5\frac{1}{7}$
d NG

10. Give the place value of the 3 in 24.7385.
a tens
b hundreds
c thousandths
d NG

11. 5.74
$+ 6.397$
a 12.13
b 12.037
c 12.137
d NG

© Pearson Education, Inc./Dale Seymour Publications/Pearson Learning Group. All rights reserved. Copying strictly prohibited.

1 Getting Started

Objective
• To add, subtract, and find equivalent units of time

Warm Up • Mental Math
Have students name three equivalent fractions for the given fraction. (Answers may vary.)

1. $\frac{6}{15}$ $(\frac{2}{5}, \frac{8}{20}, \frac{60}{150})$
2. $\frac{3}{7}$ $(\frac{6}{14}, \frac{15}{35}, \frac{24}{56})$
3. $\frac{5}{6}$ $(\frac{15}{18}, \frac{25}{30}, \frac{40}{48})$
4. $\frac{11}{12}$ $(\frac{55}{60}, \frac{33}{36}, \frac{88}{96})$
5. $\frac{4}{17}$ $(\frac{8}{34}, \frac{40}{170}, \frac{20}{85})$

Warm Up • Pencil and Paper
Have students find each quotient or product.

1. $720 \div 60$ (12)
2. $216 \div 24$ (9)
3. $245 \div 7$ (35)
4. $5,304 \div 52$ (102)
5. $2,436 \div 12$ (203)
6. 14×52 (728)
7. 60×83 (4,980)
8. 7×699 (4,893)
9. 12×96 (1,152)
10. 24×76 (1,824)

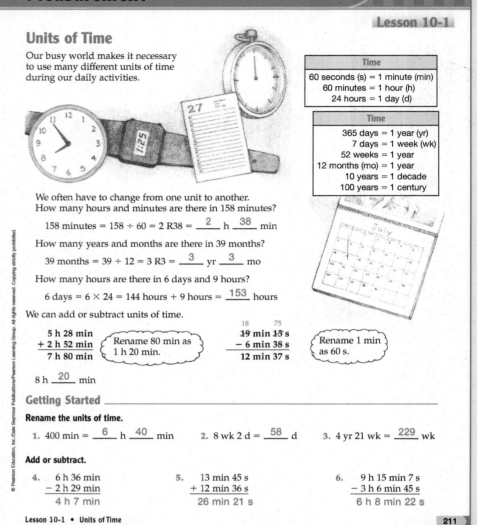

Name _____

Measurement

Units of Time

Our busy world makes it necessary to use many different units of time during our daily activities.

Time
60 seconds (s) = 1 minute (min)
60 minutes = 1 hour (h)
24 hours = 1 day (d)

Time
365 days = 1 year (yr)
7 days = 1 week (wk)
52 weeks = 1 year
12 months (mo) = 1 year
10 years = 1 decade
100 years = 1 century

We often have to change from one unit to another. How many hours and minutes are there in 158 minutes?

158 minutes = 158 ÷ 60 = 2 R38 = __2__ h __38__ min

How many years and months are there in 39 months?

39 months = 39 ÷ 12 = 3 R3 = __3__ yr __3__ mo

How many hours are there in 6 days and 9 hours?

6 days = 6 × 24 = 144 hours + 9 hours = __153__ hours

We can add or subtract units of time.

```
  5 h 28 min          Rename 80 min as
+ 2 h 52 min          1 h 20 min.
  7 h 80 min
```

8 h __20__ min

```
      18   75
   19 min 15 s        Rename 1 min
 −  6 min 38 s        as 60 s.
   12 min 37 s
```

Getting Started

Rename the units of time.

1. 400 min = __6__ h __40__ min
2. 8 wk 2 d = __58__ d
3. 4 yr 21 wk = __229__ wk

Add or subtract.

```
4.   6 h 36 min          5.   13 min 45 s          6.   9 h 15 min 7 s
   − 2 h 29 min             + 12 min 36 s             − 3 h 6 min 45 s
     4 h 7 min               26 min 21 s               6 h 8 min 22 s
```

Lesson 10-1 • Units of Time

2 Teach

Introduce the Lesson Explain to students that in this chapter they will work with various units of measure. Ask them to name various units for measuring time. Write their responses on the board.

Then, explain that the tables on page 211 are conversion tables that name various equivalent units for measuring time. Point out that students can use multiplication or division to convert or change one unit of time to another.

Work through the model with students, answering the questions, and finding the sum and difference.

Develop Skills and Concepts Explain that to change larger units into smaller units, students should multiply. To change smaller units into larger units, they should divide. Ask students to give an example of each kind of conversion.

When you discuss adding or subtracting units of time, emphasize that students must add or subtract like units, for example, minutes to minutes or hours to hours. Stress that when students add or subtract units of time, they will not rename by groups of 10. Encourage students to use the tables to help them rename.

Ask volunteers to complete each conversion equation on the board.

3 h 75 min = 4 h (15) min
1 h 83 min = 2 h (23) min
420 min = (7) h
9 wk 3 d = (66) d
3 yr 20 wk = (176) wk

3 Practice

Have students complete all the exercises. Encourage students to use the conversion tables if necessary.

Practice

Rename the units of time.

1. 6 wk 3 d = __45__ d
2. 27 mo = __2__ yr __3__ mo
3. 57 d = __8__ wk __1__ d
4. 146 min = __2__ h __26__ min
5. 63 h = __2__ d __15__ h
6. 265 min 14 s = __15,914__ s
7. 3 yr 12 wk = __168__ wk
8. 6 d = __144__ h
9. 3 wk 4 d = __25__ d
10. 8 h = __28,800__ s
11. 810 d = __2__ yr __80__ d
12. 75 wk = __1__ yr __161__ d
13. 487 min = __8__ h __7__ min
14. 3 h 4 min 59 s = __11,099__ s
15. 82 yr = __29,930__ d
16. 10 yr = __520__ wk
17. 107 mo = __8__ yr __11__ mo
18. 97 h = __349,200__ s

Add or subtract.

19.
```
  5 wk 4 d
+ 1 wk 8 d
  7 wk 5 d
```
20.
```
  3 d 14 h
+ 2 d 16 h
  6 d 6 h
```
21.
```
  14 h 12 min
-  9 h 36 min
   4 h 36 min
```
22.
```
  15 min 46 s
+ 17 min 18 s
  33 min 4 s
```
23.
```
  14 wk 3 d
-  6 wk 5 d
   7 wk 5 d
```
24.
```
  6 h 25 min
+ 2 h 39 min
  9 h 4 min
```
25.
```
  11 h  6 min
-  4 h 25 min
   6 h 41 min
```
26.
```
  2 yr 214 d
+ 3 yr 163 d
  6 yr 12 d
```
27.
```
  48 min 16 s
- 12 min 45 s
  35 min 31 s
```
28.
```
  38 wk 6 d
+  6 wk 6 d
  45 wk 5 d
```
29.
```
  14 min 29 s
+ 18 min 53 s
  33 min 22 s
```
30.
```
  10 h 15 min
-  6 h 35 min
   3 h 40 min
```

Problem Solving

Solve each problem.

31. Mary worked out in the gym 2 hours and 28 minutes on Monday. Her workout lasted 1 hour and 35 minutes on Tuesday. How long did Mary work out on Monday and Tuesday?
4 hours and 3 minutes

32. Don agreed to work for 8 hours. He worked 3 hours and 32 minutes on Saturday, and 2 hours and 25 minutes on Sunday. How much more time does Don need to work?
2 hours and 3 minutes

212   Lesson 10-1 • Units of Time

4 Assess

Ask students to identify the error in this calculation:

```
  2   1
  3 h 15 min
- 1 h 45 min
  1 h 70 min = 2 h 10 min
```
(1 hour was regrouped as 100 minutes.)

5 Mixed Review

1. $376.27 + $40.38 ($416.65)
2. 268 × 43 (11,524)
3. 215.36 − 113.857 (101.503)
4. $8\frac{3}{16} - 2\frac{5}{8}$ $(5\frac{9}{16})$
5. 3.28 ÷ 100 (0.0328)
6. $\frac{6}{7} \times \frac{5}{6} \times \frac{3}{10}$ $(\frac{3}{14})$
7. 80,010 − 43,175 (36,835)
8. 0.06 × 0.08 (0.0048)
9. $\frac{2}{3} + \frac{4}{7}$ $(1\frac{5}{21})$
10. $147.36 ÷ 3 ($49.12)

For Mixed Abilities

Common Errors • Intervention

Some students may convert units of time incorrectly because they multiply when they should divide or vice versa. For each exercise, have them write whether they are going from larger to smaller units, meaning there will be more and you multiply, or from smaller units to larger units, meaning there will be fewer and you divide.

Enrichment • Measurement

Duplicate these questions for students to answer.

1. If May 6 is a Tuesday, on what day will June 6 fall? (Friday)

2. What will be the date seven weeks from April 27? (June 15)

3. September 3 falls on Tuesday. On what day will it fall next year? (Wednesday) If the next year is a leap year, when will it fall? (Thursday)

4. New Year's Day is a Wednesday. On what day is Washington's Birthday, February 22? (Saturday)

To extend this lesson, ask students to write their own calendar problems. Have students exchange papers and solve each other's problems.

More to Explore • Application

Explain to students that Earth is very old and its age, in geological time, has been divided into eras. List the following on the board:

Paleozoic Era:
245,000,000–570,000,000 years ago

Mesozoic Era:
65,000,000–245,000,000 years ago

Cenozoic Era:
10,000–65,000,000 years ago

Ask students if they construct a timeline where 1 inch = 1,000 years, how many inches would be needed to represent each era. Make sure students first subtract to find the length of each era. (Paleozoic, 325,000 in.; Mesozoic, 180,000 in.; Cenozoic, 64,990 in.)

pages 213–214

1 Getting Started

Objective
• To read and use bus schedules

Materials
clock face with movable hands

Warm Up • Mental Math
Have students find each product or quotient.

1. $10^2 \times 3.2$ (320)
2. $0.48 \div 10^2$ (0.0048)
3. 0.3×0.9 (0.27)
4. $0.09 \div 3$ (0.03)
5. 6.1×0.2 (1.22)
6. $5.6 \div 0.8$ (7)
7. 0.6×0.04 (0.024)
8. $0.92 \div 0.2$ (4.6)
9. $5.921 \div 10^1$ (0.5921)
10. 0.675×10^1 (6.75)
11. 0.7×0.06 (0.042)
12. $8.64 \div .08$ (108)

Warm Up • Paper and Pencil
Have students complete the table.

Start	Work	Stop
9:00	1 h 15 min	10:15
12:05	55 min	(1:00)
2:45	40 min	(3:25)
7:30	1 h 45 min	(9:15)
11:20	2 h 10 min	(1:30)
5:25	45 min	(6:10)

2 Teach

Introduce the Lesson Have a student read the problem and tell what is to be solved. Ask what information is given in the problem and in the bus schedule. (Brigit takes the 10:15 bus from Milbrea, arriving at Sunnyvale at 12:00. She leaves Sunnyvale at 12:45 and arrives at Saratoga at 1:35.)

Have students complete the information sentences. Then, have them fill in the blanks in the addition model to find how long it will take Brigit to get to her grandmother's house. Complete the solution sentence.

Develop Skills and Concepts Explain that this problem involves finding elapsed time, which means actual traveling time plus the time spent waiting. Emphasize

that when students read a bus schedule, they must read the time the bus arrives and the time it leaves to find the total elapsed time.

Explain that the notation A.M. (ante meridian) is the time from midnight to noon and that P.M. (post meridian) is the time from noon to midnight. Twelve noon is written 12 P.M. and 12 midnight is written 12 A.M.

Have students give the elapsed time from the time they leave home in the morning for school to the time they return home in the afternoon.

3 Practice

Have students complete all the exercises. Remind students to rename minutes as hours and minutes when appropriate.

Name _____

Reading Schedules

Brigit is taking the bus from her home in Milbrea, to her grandmother's house in Saratoga. If Brigit leaves on the 10:15 bus, how long will it take her to get to her grandmother's?

Peninsula Express				Valley Lines		
City	Arrive	Leave		City	Arrive	Leave
Milbrea	10:00	10:15		Sunnyvale	12:30	12:45
Redwood City	10:40	11:00		Santa Clara	1:05	1:20
Palo Alto	11:20	11:40		Saratoga	1:35	1:55
Sunnyvale	12:00	12:15		Los Gatos	2:30	---
San Jose	1:00	---				

We want to know the total time needed for Brigit's trip.

We know she will take the 10:15 Peninsula Express bus from Milbrea to Sunnyvale, arriving at __12:00__.

She will leave Sunnyvale on the Valley Lines bus at __12:45__, and arrive at Saratoga at __1:35__.

To find the total time, we can add each of the times together.

Milbrea to Sunnyvale	10:15 to 12:00	__1__ h __45__ min
Waiting time	12:00 to 12:45	__0__ h __45__ min
Sunnyvale to Saratoga	12:45 to 1:35	+ __0__ h __50__ min
Total time		__1__ h __140__ min = __3__ h __20__ min

It will take Brigit __3__ hours __20__ minutes to make the trip.

Getting Started

Use the bus schedule to solve each problem.

1. How long does it take to travel from Redwood City to San Jose?
2 hours

2. How long does it take to travel from Palo Alto to Los Gatos?
2 hours and 50 minutes

3. It takes Robert 45 minutes to walk to the bus station. What is the latest time he can leave home and catch the bus leaving Santa Clara?
12:35

4. How much longer does it take to travel from Sunnyvale to Los Gatos than from Redwood City to Sunnyvale?
45 minutes

Practice

Use the bus schedules to solve each problem.

Illini Lines	Arrive	Leave
Aurora	1:05 P.M.	1:20 P.M.
Naperville	1:48 P.M.	2:05 P.M.
Wheaton	2:36 P.M.	2:45 P.M.
Chicago	4:00 P.M.	---

Plains Lines	Arrive	Leave
Peoria	---	8:00 A.M.
LaSalle	9:30 A.M.	9:45 A.M.
Ottawa	10:24 A.M.	10:50 A.M.
Aurora	12:00 P.M.	12:20 P.M.
Joliet	1:40 P.M.	---

Fox Lines	Arrive	Leave
Wheaton	2:38 P.M.	3:00 P.M.
Elgin	3:52 P.M.	4:10 P.M.
DeKalb	5:16 P.M.	5:35 P.M.
Rockford	6:31 P.M.	---

1. What time does the Plains Lines bus leave Ottawa?
   10:50 A.M.

2. How long does the Fox Lines bus stay in DeKalb?
   19 minutes

3. How long does the bus take to travel from LaSalle to Ottawa?
   39 minutes

4. How long does the Illini Lines bus stay in Naperville?
   17 minutes

5. How long does the bus take to travel from Ottawa to Joliet?
   2 hours and 50 minutes

6. How long does the bus take to travel from Naperville to Chicago?
   1 hour and 55 minutes

7. How long does the bus take to travel from Peoria to Joliet?
   5 hours and 40 minutes

8. How long does the bus take to travel from Wheaton to Rockford?
   3 hours and 31 minutes

[Now Try This!]

The clock at the corner bank says 7:30. Ted's watch is 25 minutes fast, and Ned's watch is 15 minutes slow. Ted and Ned agree to meet when Ted's watch says 8:15. What time will it be on Ned's watch? What time will it be on the clock at the bank?

Ned, 7:35; bank, 7:50

Now Try This! Have students read the problem carefully. Suggest that they list the information given to help them in working out a solution. Then, have students solve the problem independently.

4 Assess

Have students look at the bus schedule on page 213. Ask students to determine the time it takes to travel from Sunnyvale to Los Gatos on the Valley Lines bus. (1 h 45 min) Students may use various techniques to solve the problem.

For Mixed Abilities

Common Errors • Intervention

Some students may have difficulty working with schedules because they have difficulty finding elapsed time. Have them work with partners and a large clock face with movable hands to find the elapsed times between Time A and Time B.

Time A	Time B	
8:00	8:40	(40 min)
12:10	1:00	(50 min)
6:30	7:40	(1 h 10 min)
11:25	1:05	(1 h 40 min)

Enrichment • Application

1. Provide students with local bus schedules. Have students write time problems similar to the ones on page 214 using information on the bus schedules. After answering their own problems, have students exchange time problems and bus schedules, and solve.

2. Marilyn wakes up at 7:15. She has to be at work 1 hour and 20 minutes later. At what time does she have to be at work? (8:35) Have students write similar time problems for their classmates to solve.

More to Explore •
Number Sense

Divide students into groups to prepare an oral report on one of the following:

- How a calculator counts
- How an abacus computes
- Binary arithmetic
- Quinary arithmetic
- Duodecimal arithmetic

10-3 Adding and Subtracting Customary Units of Length

pages 215–216

1 Getting Started

Objectives
- To add and subtract customary units of length
- To find equivalent customary units of length

Warm Up • Mental Math
Have students round each money amount to the nearest dollar, ten dollars, or hundred dollars.
1. $4.72 ($5.00)
2. $10.16 ($10.00)
3. $9.87 ($10.00)
4. $631.40 ($600.00)
5. $0.74 ($1.00)
6. $47.38 ($50.00)

Warm Up • Paper and Pencil
Have students find the sum or difference.

1. 6 wk 4 d − 2 wk 6 d (3 wk 5 d)
2. 5 da 11 h + 2 da 19 h (8 d 6 h)
3. 7 min 46 s + 3 min 38 s (11 min 24 s)
4. 19 min 5 s − 8 min 28 s (10 min 37 s)
5. 10 h 15 min − 2 h 43 min (7 h 32 min)
6. 11 h 51 min + 3 h 48 min (15 h 39 min)

2 Teach

Introduce the Lesson Have a student read the problem and identify the problem question. (What is Robert's height in inches?) Ask what information is given in the problem and the picture. Have students complete the information sentences as they read aloud with you to solve the problem.

Call attention to the Remember statements at the end of the lesson. Point out to students that these are the same rules they used when they worked with units of time.

Develop Skills and Concepts Explain to students that renaming is often necessary in working with measures. Review inches, feet, and yards. Work through the following exercise with students, explaining how 1 foot

---

Name _____

Adding and Subtracting Customary Units of Length

It's Algebra!

Robert, Janis, and Jonathan are being measured for band uniforms. What is Robert's height in inches?

We want to rename Robert's height in inches.

We know that he is __6__ feet __4__ inches tall.

To rename feet and inches as inches, we multiply the number of feet by the number of inches in a foot, then add the extra inches.

We multiply __6__ by __12__ and add __4__.

$$\begin{array}{r} 12 \\ \times 6 \\ \hline 72 \\ + 4 \\ \hline 76 \end{array}$$

Robert is __76__ inches tall.

To rename a mixed number like $5\frac{1}{2}$ feet as inches, we multiply $5\frac{1}{2}$ by __12__.

$5\frac{1}{2} \times 12 = \frac{11}{2} \times 12 = $ __66__ inches

To rename smaller units as larger units like 48 inches as yards, we divide 48 by __36__.

$\frac{48}{36} = 1\frac{12}{36} = $ __$1\frac{1}{3}$__ yards

REMEMBER Multiply to rename larger units as smaller ones. Divide to rename smaller units as larger ones.

We can add or subtract measurements of length.

5 ft 6 in. + 7 ft 8 in. = 12 ft 14 in. = 13 ft 2 in.	(14 in. = 1 ft 2 in.)

3 yd 1 ft − 1 yd 2 ft = 1 yd 2 ft (1 yd = 3 ft; 3 ft + 1 ft = 4 ft)

Length
12 inches (in.) = 1 foot (ft)
3 feet = 1 yard (yd)
36 inches = 1 yard
5,280 feet = 1 mile (mi)
1,760 yards = 1 mile

Getting Started

Rename.
1. $7\frac{1}{2}$ yd = __$22\frac{1}{2}$__ ft
2. 5 ft 4 in. = __64__ in.
3. 68 in. = __$1\frac{8}{9}$__ yd

Add or subtract.
4. 6 ft 9 in. − 2 ft 11 in. = 3 ft 10 in.
5. 5 yd 2 ft + 3 yd 1 ft = 9 yd
6. 7 yd 2 ft 3 in. − 5 yd 1 ft 6 in. = 2 yd 9 in.

Lesson 10-3 • Adding and Subtracting Customary Units of Length

215

---

is renamed as 12 inches and the 12 inches is then added to the 5 inches to make 17 inches.

$$\begin{array}{rcl} 2 \text{ ft } 5 \text{ in.} & = & 1 \text{ ft } 17 \text{ in.} \\ - 1 \text{ ft } 9 \text{ in.} & & - 1 \text{ ft } 9 \text{ in.} \\ \hline & & 8 \text{ in.} \end{array}$$

Have students complete the following renaming exercises:

1. 3 ft = (36) in.
2. 1 yd = (3) ft
3. 12 in. = (1) ft
4. 3 ft = (1) yd
5. 5 ft = (60) in.
6. 21 ft = (7) yd
7. 72 in. = (2) yd
8. 108 in. = (9) ft
9. 6 ft = (2) yd
10. 39 in. = (1) yd (3) in.

Practice

Rename.

1. 5 ft 6 in. = __66__ in.
2. 96 in. = __8__ ft __0__ in.
3. 7 yd 2 ft = __23__ ft
4. 3 mi = __5,280__ yd
5. 6 yd 8 in. = __224__ in.
6. 5 mi = __26,400__ ft
7. 104 in. = __8__ ft __8__ in.
8. $6\frac{2}{3}$ ft = __80__ in.
9. 8 yd 2 ft = __26__ ft
10. 54 in. = __$4\frac{1}{2}$__ ft
11. 10,560 ft = __2__ mi
12. 2 yd 2 ft 6 in. = __102__ in.
13. 2 mi = __10,560__ ft
14. 18 yd 1 ft = __660__ in.
15. 784 ft = __261__ yd __1__ ft
16. 10,000 ft = __1__ mi __4,720__ ft
17. 880 ft = __293__ yd __1__ ft
18. 1 mi = __63,360__ in.

Add or subtract.

19.
```
  5 ft 5 in.
+ 2 ft 9 in.
  8 ft 2 in.
```
20.
```
  6 ft 3 in.
- 2 ft 7 in.
  3 ft 8 in.
```
21.
```
  4 ft 7 in.
+ 3 ft 5 in.
  8 ft
```
22.
```
  8 yd 1 ft
- 4 yd 2 ft
  3 yd 2 ft
```
23.
```
  6 yd 14 in.
+ 9 yd 21 in.
  15 yd 35 in.
```
24.
```
  7 mi 860 yd
+ 3 mi 925 yd
  11 mi 25 yd
```
25.
```
  17 ft 1 in.
-  9 ft 6 in.
   7 ft 7 in.
```
26.
```
  13 yd 3 ft
-  6 yd 2 ft
   7 yd 1 ft
```
27.
```
  12 yd 2 ft
+  6 yd 2 ft
  19 yd 1 ft
```
28.
```
  9 yd 14 in.
- 6 yd 18 in.
  2 yd 32 in.
```
29.
```
   6 ft 7 in.
+ 15 ft 9 in.
  22 ft 4 in.
```
30.
```
  3 yd 2 ft 9 in.
+ 5 yd 1 ft 7 in.
  9 yd 1 ft 4 in.
```

Problem Solving

Solve each problem.

31. Bart long jumped 15 feet 8 inches on his first try and 16 feet 5 inches on his second try. How much longer was Bart's second try?
9 inches

32. A fabric costs $5.40 per yard. A decorator bought $2\frac{1}{2}$ yards of blue fabric and $3\frac{1}{3}$ of yellow fabric. How much did the decorator pay for the fabric? $31.50

216   Lesson 10-3 • Adding and Subtracting Customary Units of Length

---

3 Practice

Have students complete all the exercises. Have them refer to the table of measures if they need help renaming.

4 Assess

Have students demonstrate the regrouping in both addition and subtraction by finding the sum and difference for the following examples:

```
  6 ft 3 in.
+ 2 ft 9 in.
  (8 ft 12 in. = 9 ft)
```

```
      5  15
  6 ft 3 in.
- 2 ft 9 in.
  (3 ft 6 in.)
```

For Mixed Abilities

Common Errors • Intervention

Some students may add and subtract customary units incorrectly because they rename the units as they would whole numbers; that is, they use tens instead of the proper conversion unit.

After these students have written the problem, have them write the conversion fact that will be needed. Then, they can refer to this fact as they rename and solve the problem.

Enrichment • Application

Challenge students to complete the following exercises:

1.
```
  3 yd 2 ft
×       4
  (14 yd 2 ft)
```
2.
```
  2 yd 1 ft 10 in.
×           9
  (23 yd 1 ft 6 in.)
```
3.
```
      (1 ft 4 in.)
  5)6 ft 8 in.
```
4.
```
      (3 yd 1 ft 6 in.)
  5)17 yd 1 ft 6 in.
```

More to Explore • Sets

Duplicate the following:

Mrs. Raymond, a fifth-grade teacher, was taking the members of two clubs from her school on a Saturday outing. The Shutterbugs, all photography enthusiasts, included Tim, Jane, Ron, Betty, Penny, Bob, and Phil. Dan, David, Sarah, Jane, Ted, Tim, and Penny called themselves the Pathfinders and loved hiking. Mrs. Raymond agreed to take both groups on a field trip to a nearby state park but was worried because her van only holds 12 people. Help Mrs. Raymond figure out how many students will be going on the outing by drawing a Venn diagram.

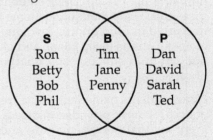

Ask, *Who would be in section B?* (members of both clubs) Ask, *Will Mrs. Raymond have room for everyone, including herself?* (Yes.)

T216

Customary Units of Capacity and Weight

pages 217–218

1 Getting Started

Objective
- To add and subtract customary units of capacity or weight
- To find equivalent customary units of capacity or weight

Materials
*cup, pint, quart, gallon containers; *sand; *10 pennies; *pan balance; *weights (oz and lb)

Warm Up • Mental Math
Have students give the elapsed time first in hours and minutes, and then in minutes.

1. 9:00 A.M. to 2:10 P.M.
   (5 h 10 min; 310 minutes)
2. 6:15 P.M. to 12:35 A.M.
   (6 h 20 min; 380 minutes)
3. 2:35 A.M. to 7:25 A.M.
   (4 h 50 min; 290 minutes)
4. 4:50 P.M. to 11:45 P.M.
   (6 h 55 min; 415 minutes)

Warm Up • Pencil and Paper
Have students find each product or quotient.

1. $432 \div 16$ (27)
2. $448 \div 8$ (56)
3. $32 \times 2 \times 2 \times 4$ (512)
4. $(4 \times 15) \div 2$ (30)
5. 14×16 (224)

2 Teach

Introduce the Lesson Have students examine the table of measures. Have them name items that are usually measured in fluid ounces (perfume), cups (liquids in recipes), quarts (milk, juice), and gallons (gasoline). Point out that fluid ounces, cups, quarts, and gallons are units of capacity.

Have students name items that are usually measured in ounces (spices), pounds (potatoes), and tons (trucks). Point out that ounces, pounds, and tons are units of weight.

Have students read the problem, identify the problem question, and determine what information is given. Then, have students complete the information sentences

as they read aloud with you to solve the two-step problem.

Develop Skills and Concepts Discuss the table of measures. Have cup, pint, quart, and gallon containers available as well as a supply of sand for pouring from one container to another. Have students demonstrate the relationships among the various measures. Ask questions such as *How many cups are in a gallon?* (16) *How many pints are in a gallon?* (8) *How many fluid ounces are in a pint?* (16)

Remind students to multiply when changing from a larger to a smaller unit and to divide when changing from a smaller to a larger unit. Encourage them to use the table of measures to help in renaming.

Write the following exercises on the board:

3 qt 2 pt	12 lb 5 oz
+ 1 qt 2 pt	− 3 lb 9 oz
(1 gal 2 qt)	(8 lb 12 oz)

Name _____

Customary Units of Capacity and Weight

Isabelle is mixing salad dressing to be put on each of the 32 tables set up for the awards banquet. Each dressing bottle holds 6 fluid ounces. How many pints of salad dressing should Isabelle mix?

Capacity	Weight
8 fluid ounces (fl oz) = 1 cup (c)	16 ounces (oz) = 1 pound (lb)
2 cups = 1 pint (pt)	2,000 pounds = 1 ton (T)
2 pints = 1 quart (qt)	
4 quarts = 1 gallon (gal)	

We want to know how many pints of salad dressing Isabella should prepare. We know there are __32__ tables, and that each will contain a __6__-ounce bottle.

To find the total number of fluid ounces needed, we multiply the number of tables, by the number of fluid ounces each will be served.

To rename fluid ounces as pints, we divide the total fluid ounces by the number of ounces in one pint.

We multiply __32__ by __6__ and divide by __16__.

$$
\begin{array}{r}
32 \\
\times\ 6 \\
\hline
192
\end{array}
$$

32 ← number of tables
× 6 ← number of fluid ounces per table
192 ← total ounces needed

number of fluid → 16)192 ← number of fluid ounces
ounces in 1 pint

$$
\begin{array}{r}
12 \\
16\overline{)192} \\
16 \\
\hline
32 \\
32 \\
\hline
0
\end{array}
$$

12 ← number of pints

Isabelle needs to mix __12__ pints of salad dressing.

Getting Started

Rename.

1. $3\frac{1}{2}$ pt = __56__ fl oz
2. 11 qt = __2__ gal __3__ qt
3. 136 oz = __$8\frac{1}{2}$__ lb

Add or subtract.

4.
4 gal 3 qt
+ 2 gal 2 qt
7 gal 1 qt

5.
15 lb 8 oz
− 9 lb 12 oz
5 lb 12 oz

6. $3\frac{1}{2}$ lb $+ 2\frac{3}{4}$ lb = __$6\frac{1}{4}$__ lb

Practice

Rename.

1. 5 qt = __1__ gal __1__ qt

2. $5\frac{1}{2}$ gal = __44__ pt

3. 4 lb 6 oz = __70__ oz

4. 18 qt = __$4\frac{1}{2}$__ gal

5. 5,000 lb = __$2\frac{1}{2}$__ T

6. 52 fl oz = __3__ pt __4__ fl oz

7. $7\frac{1}{2}$ pt = __15__ c

8. 76 fl oz = __9__ c __4__ fl oz

9. $4\frac{3}{4}$ lb = __76__ oz

10. $5\frac{3}{5}$ T = __11,200__ lb

11. $3\frac{1}{4}$ pt = __$6\frac{1}{2}$__ c

12. 4 gal 8 pt = __40__ pt

Add or subtract.

13.
```
   2 c  4 fl oz
 + 3 c 15 fl oz
 ─────────────
   7 c  3 fl oz
```

14.
```
   9 lb  7 oz
 − 6 lb 10 oz
 ───────────
   2 lb 13 oz
```

15.
```
   5 qt 1 pt
 + 3 qt 5 pt
 ──────────
  11 qt
```

16.
```
   6 gal 3 qt
 − 2 gal 5 qt
 ───────────
   3 gal 2 qt
```

17. $5\frac{1}{4}$ gal + $3\frac{2}{3}$ gal = __$8\frac{11}{12}$ gal__

18. $9\frac{1}{5}$ T + $5\frac{3}{4}$ T = __$14\frac{19}{20}$ T__

Problem Solving

Solve each problem.

19. A recipe calls for $2\frac{1}{2}$ cups of milk and 6 fluid ounces cream. How many fluid ounces are needed for the recipe?

   26 fluid ounces

20. Robert has $5\frac{1}{2}$ pounds of cashews. He is putting them in 4-ounce bowls for a party. How many bowls of cashews can Robert fill?

   22 bowls

Now Try This!

There are ten pennies and a pan balance. One of the pennies is heavier than the others. How can you find the heavy penny in 3 weighings or less?

Weigh 2 sets of 5¢ each. Keep the heavier set. Weigh 2 sets of 2¢ each. If the coins balance, the odd penny is heavier. If they do not balance, keep the heavier set. Weigh 2 sets of 1¢ each. The heavier is the heaviest penny.

Lesson 10-4 • Customary Units of Capacity and Weight

Common Errors • Intervention

Some students may not simplify their answers when they are adding customary units of capacity and weight. Have them work with partners to write the conversion unit and then write each measure in simplest form:

1. 3 pt 3 c (2 c = 1 pt; 4 pt 1 c)

2. 5 lb 30 oz
   (16 oz = 1 lb; 6 lb 14 oz)

3. 4 gal 10 qt
   (4 qt = 1 gal; 6 gal 2 qt)

4. 30 c (2 c = 1 pt, 2 pt = 1 qt, 4 qt = 1 gal; 1 gal 3 qt 1 pt)

Enrichment • Estimation

Have students estimate how much water might be wasted by a leaky faucet. Have them collect the water that drips from a faucet in 1 hour and use this amount to estimate how much water would be wasted in 1 day, 3 days, 1 week, 1 month, and 1 year.

More to Explore • Application

Use the stock market section of the newspaper to give your students practice in multiplying decimals. Provide each student with a copy of the stock report. Explain where students can find price-per-share information. Have each student select three companies, and ask them to buy a set number of stocks in each company. Have students keep a record of the price of the stock over a 2-week period on a chart that shows date, price per unit, number of units, and total value for each stock. Have students calculate how much of a profit or loss they made on each stock.

ESL/ELL STRATEGIES

If students are unfamiliar with American measures of capacity, bring to class a measuring cup with both metric and U.S. customary unit markings. Pour different amounts of water into a glass bowl to illustrate cup, pint, and quart along with similar metric amounts.

---

3 Practice

Have students complete all the exercises. Suggest that students use the tables on page 217 if necessary.

Now Try This! Supply a pan balance scale and 10 pennies for students to use to help find the solution.

4 Assess

Have students explain how they could change 3 quarts to pints. (Possible answer: Because there are 2 pints in 1 quart, I can multiply 3 quarts by 2 to get 6 pints.)

pages 219–220

1 Getting Started

Objectives
- To rename metric units of length
- To compare metric units of length

Vocabulary
meter

Materials
*large metric place-value chart; metric rulers

Warm Up • Mental Math
Have students rename each measure.

1. $2\frac{1}{2}$ h = (150) min
2. 7 gal = (28) qt
3. 5 pt = $(2\frac{1}{2})$ qt
4. 3 lb = (48) oz
5. 5 T = (10,000) lb
6. 90 s = $(1\frac{1}{2})$ min

Warm Up • Mental Math
Have students find each product or quotient.

1. 4×0.01 (0.04)
2. 8.2×100 (820)
3. $340 \div 100$ (3.4)
4. $92 \div 0.1$ (920)
5. 5.6×10^3 (5,600)
6. $0.245 \div 10^2$ (0.00245)
7. $10^5 \times 29.53$ (2,953,000)
8. $8.1 \div 10^4$ (0.00081)

2 Teach

Introduce the Lesson Have a student read the first two paragraphs and study the table. Point out that the metric system of measurement has the same pattern as the decimal system of numeration. Each unit to the left of the meter is 10 times the unit to the right. Each unit to the right of the meter is $\frac{1}{10}$ the unit to the left.

Discuss the meaning of each prefix, stressing that the prefixes are used in naming units smaller and larger than a meter. Remind students to multiply to rename a larger unit as a smaller unit and to divide to rename a smaller unit as a larger one.

Read through the lesson with students, completing each renaming.

Develop Skills and Concepts Display a large metric place-value chart. Have students compare the values of adjoining places. Point out that to change from

- mm to cm, divide by 10
- mm to dm, divide by 100
- mm to m, divide by 1,000

Ask students to describe the pattern. Provide other examples, this time with multiplication.

Point out that these prefixes remain constant throughout the metric system.

Then, explain that to compare units of measurement, the first step is to make both units of measurement the same. Have students compare the following units:

1. 6.8 m (>) 69.2 cm
2. 13.32 m (>) 1,300 cm
3. 720 cm (=) 7.2 m
4. 4.28 m (<) 4,281 cm

Name _____

Metric Units of Length

The **meter** is the basic unit of length in the metric system.

The metric system is based on powers of 10. Each unit is 10 times greater than the next smaller unit and 0.1 the size of the next larger unit.

kilometer (km)	hectometer (hm)	decameter (dam)	basic unit (m)	decimeter (dm)	centimeter (cm)	millimeter (mm)
1,000 m	100 m	10 m	1 m	0.1 m	0.01 m	0.001 m

How many millimeters are in 1.5 meters?

> To rename a larger unit as a smaller unit, multiply by the corresponding power of 10.

To rename 1.5 meters as millimeters, multiply by 1,000.

$1.5 \times 1,000 = \underline{1,500}$

There are $\underline{1,500}$ millimeters in 1.5 meters.

How many kilometers are in 3,500 meters?

> To rename a smaller unit as a larger unit, divide by the corresponding power of 10.

To rename 3,500 meters as kilometers, divide by 1,000.

$3,500 \div 1,000 = \underline{3.5}$

There are $\underline{3.5}$ kilometers in 3,500 meters.

Getting Started

Rename.

1. 3 cm = $\underline{30}$ mm
2. 2 m = $\underline{200}$ cm
3. 25 mm = $\underline{0.025}$ m
4. 20 cm = $\underline{2}$ dm
5. 16.5 cm = $\underline{165}$ mm
6. 86,000 mm = $\underline{86}$ m

Write <, =, or >.

7. 43 m $\underline{<}$ 4.3 km
8. 26 cm $\underline{=}$ 260 mm
9. 350 dm $\underline{>}$ 3.5 m
10. 8.2 m $\underline{<}$ 82 km
11. 760 cm $\underline{>}$ 7.6 dm
12. 1,200 mm $\underline{=}$ 1.2 m

Practice

Find each missing number.

1. 1 cm = __10__ mm
2. 2.5 km = __2,500__ m
3. 15 dm = __1.5__ m

4. 6 m = __600__ cm
5. 0.025 m = __25__ mm
6. 525 m = __0.525__ km

7. 146 dam = __1.46__ km
8. 28.6 m = __28,600__ mm
9. 0.78 dm = __0.078__ m

10. 250 m = __0.25__ km
11. 326 hm = __32,600__ m
12. 4.95 m = __495__ cm

13. 12 mm = __1.2__ cm
14. 6 km = __6,000__ m
15. 75 cm = __0.75__ m

16. 0.38 dam = __3.8__ m
17. 25 dm = __250__ cm
18. 159.6 mm = __15.96__ cm

Write <, =, or >.

19. 45 cm __>__ 4.5 mm
20. 3 m __<__ 300 dm
21. 5 km __>__ 500 m

22. 1.5 m __=__ 150 cm
23. 560 m __=__ 5.6 hm
24. 180 cm __>__ 1.8 mm

25. 200 cm __>__ 0.2 m
26. 0.321 km __>__ 3,210 cm
27. 0.6 km __>__ 60 m

28. 500 mm __=__ 50 cm
29. 2.3 m __>__ 230 mm
30. 3.01 cm __<__ 301 mm

31. 586 mm __>__ 5.86 cm
32. 0.25 km __>__ 250 cm
33. 0.111 km __>__ 1,110 cm

34. 8 km __=__ 800 dam
35. 0.19 dm __<__ 0.19 m
36. 24 m __<__ 2,400 dm

Problem Solving

Solve each problem.

37. Jackie ran in a 10,000-meter race. How many kilometers did Jackie run?
    10 kilometers

38. Nat's kite has a tail 1.5 times longer than Bill's kite. If the tail on Bill's kite is 3.6 meters long, how many centimeters long is the one on Nat's kite? 540 centimeters

Add the length of each side to find the perimeter.

39.
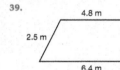
4.8 m
2.5 m   2.5 m
6.4 m
16.2 m

40.
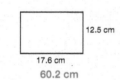
12.5 cm
17.6 cm
60.2 cm

Lesson 10-5 • Metric Units of Length

3 Practice

Have students complete all the exercises. Remind them to rename measures so that both measures are in the same unit before comparing.

4 Assess

Have students write the steps needed to compare 3 m to 30 cm. (Possible answer: 1. Change 3 m to 300 cm. 2. Compare 300 cm to 30 cm. 3. Write the correct symbol. Thus, 3 m > 30 cm.)

For Mixed Abilities

Common Errors • Intervention

Some students may compare metric measures of length incorrectly because they do not rename one of them so that both are expressed in the same metric unit. Have them work with partners, first choosing which of the two measures to rename, then using the unit of the other measure to rename, and finally comparing the two measures.

Enrichment • Measurement

Have students write >, <, or = to compare the measures.

1. 6.4 km (<) 6,482 m

2. 4,164 m (>) 41.64 cm

3. 3.136 km (>) 336 m

4. 0.7 cm (=) 7 mm

5. 6.4 m (=) 640 cm

More to Explore • Number Sense

Remind students that our place-value system is based on the number 10. Direct them to the base-five system. Explain to students that the place values become one, five, and twenty-five instead of one, ten, and hundred. Also, the only digits used are 0, 1, 2, 3, and 4. Write the following on the board:

31_{five} means 3 fives and 1 one
In base ten this is $15 + 1 = 16$.

24_{five} means 2 fives and 4 ones
In base ten this is $10 + 4 = 14$.

Give the students other base-five numerals and have them convert them to base-ten numerals.

ESL/ELL STRATEGIES

You may wish to help students compare U.S. customary measures of length with comparable metric measurements. For example, say, *1 inch is about 2.5 centimeters; 1 foot is about 30 centimeters; 1 yard is a little less than 1 meter.*

10-6 Metric Units of Capacity and Mass

pages 221–222

1 Getting Started

Objectives
- To rename metric units of capacity and mass
- To compare metric units of capacity and mass

Vocabulary
liter, gram

Materials
*large metric place-value chart

Warm Up • Mental Math
Have students simplify each expression.

1. $3 \times 10 \times 10 \div 100$ (3)
2. $50 \div 5 \times 6 + 3$ (63)
3. $7 \times 3 + 9 - 10$ (20)
4. $\frac{3}{4}$ of $12 - 2$ (7)
5. $8 \times 7 \div 2 \div 7$ (4)
6. $\frac{5}{6}$ of 30×2 (50)

Warm Up • Pencil and Paper
Have students rename each measure.

1. 3 mm = (0.003) m
2. 3.2 m = (320) cm
3. 6 km = (6,000) m
4. 1.5 dm = (15) cm
5. 0.026 m = (26) mm
6. 42 cm = (4.2) dm
7. 3,500 m = (3.5) km
8. 5,926 mm = (5.926) m
9. 56 cm = (0.56) m
10. 2.4 m = (2,400) mm

2 Teach

Introduce the Lesson Have a student read the first problem. Explain that *liter* is the basic metric unit of capacity. Write these equivalencies on the board:
  1 mL = 0.001 L and 1 kL = 1,000 L.

Have a student read the second problem. Explain that the *gram* is the basic metric unit of mass. Mass is the amount of matter in an object. Point out that weight takes into account the force of gravity. Mass does not. The mass of an object on Earth or on the Moon will be the same. The weight of an object on Earth or on the Moon will be different. At Earth's surface, the mass and the weight of an object are considered to be the same.

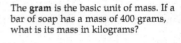

Metric Units of Capacity and Mass

The **liter** is the basic unit of capacity in the metric system. If a jar holds 2.35 liters of juice, how many milliliters does it hold?

The **gram** is the basic unit of mass. If a bar of soap has a mass of 400 grams, what is its mass in kilograms?

1 liter (L) = 1,000 milliliters
1 mL = 0.001 L

> To rename a larger unit as a smaller unit, multiply by the corresponding power of 10.

To rename 2.35 liters as milliliters, multiply by 1,000.

2.35 × 1,000 = __2,350__

The jar holds __2,350__ milliliters.

1 gram (g) = 1,000 milligrams (mg)
1,000 g = 1 kilogram (kg)
1,000 kg = 1 metric ton (t)

> To rename a smaller unit as a larger unit, divide by the corresponding power of 10.

To rename 400 grams as kilograms, divide by 1,000.

400 ÷ 1,000 = __0.4__

The soap has a mass of __0.4__ of a kilogram.

Getting Started

Rename.

1. 6.5 kg = __6,500__ g
2. 536 mL = __0.536__ L
3. 2.6 t = __2,600__ kg
4. 56 L = __56,000__ mL
5. 875 g = __0.875__ kg
6. 20 kg = __0.02__ t

Write <, =, or >.

7. 6.91 kg __>__ 691 g
8. 700 mL __=__ 0.7 L
9. 0.21 g __<__ 2,100 mg
10. 48 g __>__ 4.8 mg
11. 7,800 kg __<__ 78 t
12. 200 mL __=__ 0.2 L

Write these equivalencies on the board: **1,000 mg = 1 g; 1,000 g = 1 kg;** and **1,000 kg = 1 t.** Remind students that the prefixes *milli-* and *kilo-* are the same for units of length, capacity, and mass. *Milli-* means $\frac{1}{1,000}$ and *kilo-* means 1,000.

Develop Skills and Concepts Students most often have difficulty determining which power of 10 to multiply or divide by to name an equivalent unit. To clarify, use a large metric place-value chart to show that renaming milliliters to liters involves moving three places to the left, or dividing by 10^3.

Have students work with the place-value chart to show renaming from kilograms to milligrams, moving six places to right, or multiplying by 10^6, and renaming from liters to kiloliters, moving three places to left or dividing by 10^3.

Practice

Rename.

1. 648 g = __0.648__ kg

2. 5.1 kg = __5,100__ g

3. 5.268 mL = __0.005268__ L

4. 5,860 kg = __5.86__ t

5. 0.61 L = __610__ mL

6. 4.21 t = __4,210__ kg

7. 500 mL = __0.5__ L

8. 3.96 kg = __3,960,000__ mg

9. 5.9 kg = __5,900__ g

10. 9,256 mg = __9.256__ g

11. 859 kg = __0.859__ t

12. 2.961 L = __2,961__ mL

Write <, =, or >.

13. 5.6 L __=__ 5,600 mL

14. 0.215 t __>__ 21.5 kg

15. 5,000 mg __<__ 50 g

16. 6,249 mg __=__ 6.249 g

17. 0.5 kg __=__ 500 g

18. 0.82 L __>__ 82 mL

Problem Solving

Solve each problem.

19. A large truck has a mass of about 1.2 metric tons. What is the mass of the truck in kilograms?
1,200 kilograms

20. Tom has a mass of 82.6 kilograms, Ted 76.5 kilograms, and Rich 80.3 kilograms. What is the average mass of the boys?
79.8 kilograms

21. Ronnie is making hamburger patties that have a mass of 126 grams. How many kilograms of hamburger will Ronnie need to buy to serve 15 hamburgers?
1.89 kilograms

22. A pitcher holds 1.5 liters of juice. Marty has poured 2 glasses of juice from the pitcher. If each glass holds 480 milliliters of juice, how many milliliters of juice are left in the pitcher? 540 milliliters

---

(Now Try This!)

There are two buckets. One holds exactly 5 liters of water. The other holds exactly 7 liters of water. How can you use the two buckets to measure 3 liters of water?

Fill the 5-liter bucket. Empty it into the 7-liter bucket. Fill the 5-liter bucket again. Fill the 7-liter bucket from the 5-liter bucket. Three liters remain in the 5-liter bucket.

222   Lesson 10-6 • Metric Units of Capacity and Mass

---

3 Practice

Have students complete all the exercises.

Now Try This! Suggest that students make a table to help them solve the problem. Encourage try, check, and revise as a problem-solving strategy.

4 Assess

Ask what decimal part of a liter is a milliliter.
(one thousandth of a liter)

For Mixed Abilities

Common Errors • Intervention

Some students may rename incorrectly because they multiply when they should divide and vice versa. Have them first decide if they are renaming a larger unit as a smaller unit, in which case they multiply, or a smaller unit as a larger unit, in which case they divide.

Enrichment • Application

Have students collect labels that give weight in grams or kilograms. Have them display their labels and write word problems using the information on the labels. Have them exchange their word problems to solve.

More to Explore • Application

Duplicate the following crossword puzzle. Have students fill in the blanks using the given clues.

```
            5.        7.
1. D E N O M I N A T O R
   I     U           E
   V     M           X
   I     M     6.
2. I D E N T I T Y
   D     R           
   E     A     6.
3. E X A M P L E S
         T     E
         O     F
4. F R A C T I O N
```

Across

1. The bottom number of a fraction is called the _____.

2. To use 1 in a multiplication problem is the _____ property.

3. $\frac{1}{2}, \frac{3}{16}, \frac{7}{25}$ are _____ of fractions.

4. A rational number consisting of a numerator and a denominator is a _____.

Down

1. We invert fractions to _____.

5. The top number in a fraction is the _____.

6. $(\frac{8}{3}, \frac{3}{8})$ The fraction on the _____ is improper.

7. In the series $\frac{1}{2}, \frac{2}{4}$, and $\frac{4}{8}$, the _____ number would be $\frac{5}{10}$.

T222

1 Getting Started

Objectives
- To read Fahrenheit and Celsius temperatures
- To find a change in temperature

Vocabulary
Fahrenheit (°F), Celsius (°C)

Warm Up • Mental Math
Have students find each sum or difference.

1. 7 + 5 (12)
2. 8 + 5 (13)
3. 6 + 8 (14)
4. 9 + 7 (16)
5. 7 + 6 (13)
6. 13 − 6 (7)
7. 15 − 9 (6)
8. 12 − 7 (5)
9. 14 − 6 (8)
10. 15 − 8 (7)

Warm Up • Pencil and Paper
Have students perform each operation.

1. 12.5 + 32.7 (45.2)
2. 41.3 − 18.6 (22.7)
3. 2.8 × 5.2 (14.56)
4. 16.12 + 23.9 (40.02)
5. 34.44 ÷ 8.2 (4.2)
6. 86.25 − 39.7 (46.55)
7. 3.15 × 4.5 (14.175)
8. 2.8 + 9.32 + 12.6 (24.72)

Name _____

Fahrenheit and Celsius Temperatures

You can measure temperature in degrees **Fahrenheit (°F)** or in degrees **Celsius (°C)**. The Fahrenheit scale is used in the customary system of measurement. The Celsius scale is used by scientists to measure temperature.

At what temperature does water freeze on the Fahrenheit scale? __32°F__

On the Celsius scale? __0°C__

Riki measured the outside temperature at noon and then again at 8 P.M. The temperature at noon was 6°C. The temperature at 8 P.M. was ⁻2°C. What was the change in temperature?

You can use the thermometer like a number line.

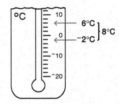

There was a __8°__ change in temperature.

Getting Started

Write each temperature in Fahrenheit or Celsius.

1. __74°F__   2. __23°F__   3. __−8°C__   4. __15°C__

Find each change in temperature.

5. 24°C to 15°C 9°
6. 68°F to 42°F 26°
7. −2°C to −6°C 4°
8. 5°C to −3°C 8°

---

2 Teach

Introduce the Lesson Explain that there are two scales used to read temperatures. In the United States, the Fahrenheit scale is used. The Celsius scale is used in almost all other countries and by scientists. Have a student read the opening paragraph and answer the questions.

Have another student read the problem about Riki. Work with students as they use the thermometer to determine the change in temperature. Point out that each tic mark on the thermometer is 2°. Have students count by 2s from 6° to ⁻2°, and complete the information sentence.

Develop Skills and Concepts Present this problem and work together with students to solve.

One day , the lowest temperature recorded in Maine was −12°F. The highest temperature in Maine that day was 25°F. What was the change in temperature?

Draw a vertical number line on the board from 25 to −15. Draw arrows toward the numbers −12, 0, and 25. Starting at −12, ask, *How many units is −12 from 0?* (12) Then ask, *How many more units do we go to get to 25?* (25) So, since 12 + 25 = 37, the difference in temperature is 37.

Point out that in this instance, the word *difference* does not mean "to subtract."

It's Algebra! The concepts in this lesson prepare students for algebra.

3 Practice

Have students complete all the exercises. Students can draw number lines to find the change in temperature.

Practice

Write each temperature in Fahrenheit or Celsius.

1.
   35°F

2.
   3°C

3.
   −7°C

4.
   101°F

Find each change in temperature.

5. 16°C to 23°C
   7°

6. 72°F to 64°F
   8°

7. 8°C to −1°C
   9°

8. −3°C to 3°C
   6°

9. 35°F to 48°F
   13°

10. −6°C to −2°C
    4°

11. 101°F to 95°F
    6°

12. 9°C to −2°C
    11°

Problem Solving

Solve each problem.

13. The temperature in Bismarck, North Dakota was −7°F one day in January. On that same day, the temperature in Miami, Florida was 67°F. What was the difference in temperatures?
    74°

14. The temperature is 36°. You turn on an air conditioner. Is the temperature reading in degrees Fahrenheit or in degrees Celsius?
    degrees Celsius

Now Try This!

It's Algebra!

The outside temperature is 30°C. Tanisha uses a formula to change this temperature reading to degrees Fahrenheit.

$$°F = \frac{9}{5} \times °C + 32 \longrightarrow °F = \frac{9}{5} \times 30 + 32$$
$$= 54 + 32$$
$$= 86°F$$

Use the formula to change each temperature to degrees Fahrenheit.

1. 5°C 41°F

2. 35°C 95°F

3. 20°C 68°F

4. 15°C 59°F

Now Try This! Be sure that students understand how to use the formula for changing Celsius temperature to Fahrenheit temperature. Emphasize that the temperature is the same.

4 Assess

Have students decide if the temperature in each situation is in degrees Fahrenheit or degrees Celsius.

1. turning on the heat when it is 30° outside (Fahrenheit)

2. wearing shorts when it is 22° outside (Celsius)

3. a body temperature of about 35° (Celsius)

4. snow falling when it is 10° outside (Fahrenheit)

For Mixed Abilities

Common Errors • Intervention

Some students may have difficulty finding the change in temperature if one or both temperatures are negative. Tell students that they can use a number line turned vertically to count the number of units between the temperatures.

Have students work in pairs to find the change in temperature.

1. ⁻1°F to 6°F (7°)

2. ⁻10°C to ⁻2°C (8°)

3. ⁻5°F to 23°F (28°)

Enrichment • Application

Challenge students to use the Internet to find the highest and lowest temperatures ever recorded. Then, have them find the difference in the temperatures. Students could also use a newspaper to find the lowest and highest readings for one day in the world, in the United States, or in their state.

More to Explore • Algebra

Did you know that crickets chirp faster when it is warm out and slower when it is cold? In fact, you can estimate the outside temperature by counting the number of times a cricket chirps! Here is the formula:

Fahrenheit temperature = number of chirps in 15 seconds + 40

Use this formula to find the temperature if a cricket chirps

1. twice every second (70°F)

2. eight times every 3 seconds (80°F)

To find the temperature on the Celsius scale, use this formula:

Celsius temperature = (number of chirps in 15 seconds + 13) ÷ 2

Find the temperature on the Celsius scale if a cricket chirps

1. 35 times in 15 seconds (24°C)

2. 16 times every 5 seconds (30.5°C)

1 Getting Started

Objective
• To solve problems by using formulas and writing open sentences

Warm Up • Mental Math
Have students perform the indicated operation and write the answer in simplest form.

1. 10 min × 8 (1 h 20 min)
2. 15 min × 4 (1 h)
3. 35 min × 5 (2 h 55 min)
4. 30 min + 75 min (1 h 45 min)
5. 65 min + 80 min (2 h 25 min)
6. 2 h − 35 min (1 h 25 min)

Warm Up • Pencil and Paper
Have students find the missing number.

1. $64 = (6 \times 7) + (\underline{(2)} \times 11)$
2. $74 = 81 - \underline{(7)}$
3. $100 = 325 - \underline{(225)}$
4. $\frac{1}{3}$ of $\underline{(126)} = 42$
5. $3.14 \times 6 \times 6 = \underline{(113.04)}$
6. $50 \div (2 \times \underline{(5)}) = 5$
7. $3.14 \times 0.3 \times 0.3 = \underline{(0.2826)}$
8. $\frac{1}{2}$ of $76 - \underline{(4)} = 34$

Name _____

Problem Solving: Use a Formula

A rectangular swimming pool has a 140-meter perimeter. What is the area of the pool if it is 50 meters long?

★ SEE
We want to know the area of the swimming pool.

The perimeter of the pool is __140__ meters.

The length of the pool is __50__ meters.

★ PLAN
We need to know the width to find the area. We use the formula for finding perimeter to help us write an open sentence to find the width of the pool. Once we know the width, we can then multiply it by the length to find the area of the pool.

★ DO

$$\text{Perimeter} = (\text{length} \times 2) + (\text{width} \times 2)$$
$$140 = (50 \times 2) + (\text{width} \times 2)$$
$$140 = (100) + (\text{width} \times 2)$$

Since we are missing an addend in this example, we will need to subtract to find the value of the width times 2.

$$\begin{array}{r} 140 \\ -\ 100 \\ \hline 40 \end{array}$$

The width doubled is __40__, so by dividing by 2 we find the measure of one width.

$$\text{Area} = \text{length} \times \text{width}$$
$$\underline{1,000} = 50 \times \underline{20}$$

The area of the pool is __1,000__ square meters.

★ CHECK
We can check our work by substituting the actual length and width in the perimeter and area formulas.

$$\text{Perimeter} = (2 \times \text{length}) + (2 \times \text{width})$$
$$? = (2 \times 50) + (2 \times \underline{20})$$
$$\underline{140} = \underline{100} + \underline{40}$$

$$\text{Area} = \text{length} \times \text{width}$$
$$\underline{1,000} = 50 \times \underline{20}$$

---

2 Teach

Introduce the Lesson Review the problem by reading it and working through the SEE step. Tell students that using what we know to arrive at what we do not know is one of the most important life skills that we can develop. That is what we do when we use a basic formula to arrive at a solution.

Develop Skills and Concepts Carefully lead students through the reasoning in the DO step. Point out that we place the equal signs under each other as we work through the formula. Call attention to the CHECK used in this problem. Using reversal formulas to check solutions reinforces the importance of the knowledge chain and working backward.

Its Algebra! The concepts in this lesson prepare students for algebra.

3 Apply

Solution Notes

1. Use the same formula as in the lesson example. The width is 40 meters.
2. Perimeter $= 3 \times$ side
   $= 3 \times 18$
   $= 54$ in.
3. Vol $=$ length $\times$ width $\times$ height
   $= 5 \times 3 \times 9$
   $= 135$ cu cm
4. See Exercise 3 to start.
   Vol $\div$ (length • width) $=$ height
   $1,260 \div (18 \times 7) = h$
   $1,260 \div 126 = h$
   $h = 10$ inches
5. area circle $= 3.14 \times 8 \times 8$
   $= 3.14 \times 64$
   $= 200.96$ sq in.

Apply

Solve each problem.

1. A rectangular playground has a perimeter of 200 meters. Its length is 60 meters. What is its width?

   40 meters

2. An equilateral triangle is a triangle having all three sides the same length. Write a formula to find the perimeter of an equilateral triangle. If one side of a equilateral triangle is 18 inches, what is the perimeter?

   Perimeter = 3 × side; 54 inches

3. The volume of a rectangular solid is found by multiplying the length, width, and height. Determine the volume of a rectangular solid having a length of 5 centimeters, a width of 3 centimeters, and a height of 9 centimeters.

   135 cubic centimeters

4. The volume of a rectangular solid is 1,260 cubic inches. The length of the solid is 18 inches and the width is 7 inches. Write a formula for finding the height and use it to find the missing measurement.

   height = Volume ÷ (length × width)
   10 inches

5. The area of a circle can be found by this formula:

   Area = 3.14 × radius × radius

   Determine the area of a circle whose radius is 8 inches.

   200.96 square inches

6. A computer floppy disk cover is a square 5.25 inches on a side. The hole in the center has a radius of approximately 0.5 of an inch. Determine the area of the disk cover.

   26.7775 square inches

7. Suppose the length of the rectangle in Problem 1 was 50 meters instead of 60 meters. How would this make the rectangle special?

   It would be a square.

8. A picometer is a metric unit of measure. A picometer is one trillionth of a meter. How does 1 picometer compare to 1 millimeter?

   It is one millionth of a millimeter.

9. The formula for the number of handshakes possible in a group of n people is $H = \frac{1}{2} \times n \times (n-1)$. How many handshakes are possible with 5 people? Explain the answer when you try the formula with 1 person.

   10 handshakes possible with 5 people

10. Lois has a garden shaped like a rectangle. The length is a whole number of meters as is the width. The perimeter is 20 meters. What are the possible dimensions of the rectangle?

    See Solution Notes.

For Mixed Abilities

More to Explore • Application

Duplicate the following problem for students to solve. Omit the numbers in parentheses in the table.

Miss Kenneth left her grade book on the picnic table, and it rained. Some of the digits in the spelling test grade report were blurred by getting wet. Help Miss Kenneth fill in the missing digits in the report.

Spelling Test Grades					
	Test 1	Test 2	Test 3	Test 4	Average
Martha	78	(80)	90	85	83
Bruce	75	78	92	83	(82)
Jennifer	(76)	85	95	84	85
Kathy	65	73	(8)2	88	(7)7
Ruth	9(2)	88	(93)	87	9(0)

6. Area = area of square − area of circle
   = (5.25 × 5.25) − (3.14 × 0.5 × 0.5)
   = 27.5625 − 0.785
   = 26.7775 sq in.

Higher-Order Thinking Skills

7. **Synthesis:** Each side of the rectangle would be the same—50 meters long.

8. **Synthesis:** Since a meter is 1,000 millimeters, then a picometer is 1,000 times $\frac{1}{1,000,000,000}$ or $\frac{1}{1,000,000}$ of a millimeter.

9. **Synthesis:** With 1 person, the following formula gives the result:

   $H = \frac{1}{2} \times 1 \times (1 - 1)$

   $= \frac{1}{2} \times (0)$

   $= 0$

   Since, other than our 1 person, there is nobody else in the group, no handshakes are possible.

10. **Synthesis:** Students can approach the problem by starting a table that soon reveals that the sum of the length and width must be 10 meters.

Width	Length
1	9
2	8
3	7
4	6
5	5

4 Assess

Write the formula for perimeter and for area on the board. Have students use the formulas to find the area of a pool that has a width of 20 meters and a perimeter of 120 meters. (800 square meters)

Chapter 10
Test

Items   Objectives

1–3,   To add and subtract units
7–8   of time (see pages
211–212)

4–6, 9 To add and subtract units
of customary length
(see pages 215–216)

10–15   To add and subtract units
of customary capacity and
weight (see pages
217–218)

16–30   To change metric units to
equivalent units (see pages
219–222)

31–36   To find the change in
temperature in Fahrenheit
and Celsius (see pages
223–224)

CHAPTER 10
TEST

Name _____

Add or subtract.

1. 5 h 18 min + 3 h 24 min **8 h 42 min**	2. 18 min 48 s + 9 min 18 s **28 min 6 s**	3. 6 h 4 min − 2 h 36 min **3 h 28 min**
4. 8 ft 3 in. − 2 ft 6 in. **5 ft 9 in.**	5. 8 yd 29 in. + 6 yd 16 in. **15 yd 9 in.**	6. 9 ft 8 in. + 4 ft 7 in. **14 ft 3 in.**
7. 12 min 48 s − 6 min 49 s **5 min 59 s**	8. 14 h 15 min − 12 h 50 min **1 h 25 min**	9. 8 yd 11 in. + 16 yd 48 in. **25 yd 23 in.**

Add or subtract.

10. 9 gal 2 qt + 6 gal 3 qt **16 gal 1 qt**	11. 3 pt 7 fl oz + 9 pt 5 fl oz **12 pt 12 fl oz**	12. 15 lb 9 oz − 6 lb 15 oz **8 lb 10 oz**
13. 9 lb 6 oz + 4 lb 15 oz **14 lb 5 oz**	14. 2 gal 1 qt − 1 gal 2 qt **3 qt**	15. 17 pt 1 fl oz − 5 pt 7 fl oz **11 pt 10 fl oz**

Find each missing number.

16. 25 cm = __250__ mm   17. 6,246 m = __6.246__ km   18. 515 mm = __0.515__ m

19. 840 mm = __84__ cm   20. 6.2 km = __6,200__ m   21. 48 mm = __0.048__ m

22. 720,000 mm = __0.72__ km   23. 572 t = __572,000__ kg   24. 0.02 g = __20__ mg

25. 920 kg = __920,000,000__ mg   26. 53 cm = __5.3__ dm   27. 747 t = __747,000__ kg

28. 2.92 km = __2,920__ m   29. 0.003 t = __3,000__ g   30. 0.0006 g = __0.6__ mg

Find each change in temperature.

31. 20°C to 31°C   32. 62°F to 48°F   33. 5°C to −3°C
    11°C   **14°C**   **8°C**

34. 35°F to 41°F   35. −9°C to 1°C   36. −2°C to −7°C
    6°C   **10°C**   **5°C**

© Pearson Education, Inc./Dale Seymour Publications/Pearson Learning Group. All rights reserved. Copying strictly prohibited.

Chapter 10 • Test

Alternate Chapter Test

You may wish to use the Alternate
Chapter Test on page 388 of this book
for further review and assessment.

Circle the letter of the correct answer.

① 53
× 36

a. 577
b. 1,898
c. 1,908 ⓒ
d. NG

② 9)348

a. 3 R86
b. 38
c. 38 R6 ⓒ
d. NG

③ 2 1/8
+ 5 1/3

a. 7 2/11
b. 7 11/24 ⓑ
c. 8 11/24
d. NG

④ 9 1/5
− 2 2/3

a. 6 8/15 ⓐ
b. 7 1/2
c. 7 8/15
d. NG

⑤ 2 1/4 × 5 1/3

a. 27/64
b. 1/12
c. 12 ⓒ
d. NG

⑥ 6 2/3 ÷ 1 1/4

a. 3/16
b. 5 1/3 ⓑ
c. 8 1/3
d. NG

⑦ What is the place value
of the 9 in 5.2974?

a. tenths
b. hundredths ⓑ
c. thousandths
d. NG

⑧ 2.3
4.68
+ 1.925

a. 7.805
b. 8.905 ⓑ
c. 8.005
d. NG

⑨ 25.06
− 9.875

a. 15.195
b. 15.285
c. 16.285
d. NG ⓓ

⑩ 3.4
× 0.08

a. 0.072
b. 0.272 ⓑ
c. 2.72
d. NG

⑪ 0.008)4.624

a. 5.78
b. 57.8
c. 578 ⓒ
d. NG

⑫ Write 1/8 as a decimal
rounded to tenths.

a. 0.1 ⓐ
b. 0.13
c. 0.125
d. NG

☐ score

🛑 STOP

page 228

Items	Objectives
1	To multiply two 2-digit factors (see pages 51–52)
2	To divide 3-digit numbers by 1-digit numbers (see pages 63–64)
3–4	To add and subtract mixed numbers (see pages 119–120, 127–128)
5–6	To multiply and divide mixed numbers (see pages 141–142, 147–148)
7	To identify the place value of a digit in decimals through millionths (see pages 163–164)
8–9	To add and subtract decimals (see pages 171–174)
10–11	To multiply and divide decimals (see pages 189–190, 197–198)
12	To rename fractions as decimals, rounding to the nearest tenth (see pages 203–204)

Alternate Cumulative Assessment

Circle the letter of the correct answer.

1. 47
× 68

a 3,196 ⓐ
b 2,796
c 2,196
d NG

2. 7)349

a 4 R96
b 49 R6 ⓑ
c 496
d NG

3. 4 1/2
+ 4 3/7

a 8 13/14 ⓐ
b 9
c 9 1/14
d NG

4. 8 1/4
− 5 1/3

a 2 3/4
b 2 11/12 ⓑ
c 3 11/12
d NG

5. 1 4/5 × 2 2/5

a 4 2/3
b 4 3/5
c 4 4/5
d NG ⓓ

6. 7 7/8 ÷ 3 3/11

a 2 12/32
b 2 13/32 ⓑ
c 25 68/88
d NG

7. Give the place value of the 2 in 47.5269.

a tenths
b hundredths ⓑ
c thousandths
d NG

8. 7.4
3.96
+ 5.872

a 16.132
b 17.232 ⓑ
c 17.332
d NG

9. 38.04
− 9.568

a 27.472
b 28.488
c 28.572
d NG ⓓ

10. 3.7
× 0.09

a 3.33
b 0.333 ⓑ
c 0.0333
d NG

11. 0.006)2.142

a 357 ⓐ
b 35.7
c 3.57
d NG

11-1 Basic Geometric Ideas

pages 229-230

1 Getting Started

Objective
- To identify points, lines, line segments, rays, intersecting lines, and parallel lines

Vocabulary
point, line segment, endpoint, congruent, line, ray, parallel

Materials
rulers or straightedges

Warm Up • Mental Math
Have students round to the nearest thousand.
1. 5,326 (5,000)
2. 41,642 (42,000)
3. 76,547 (77,000)
4. 8,964 (9,000)
5. 364,765 (365,000)
6. 7,501 (8,000)
7. 9,972 (10,000)
8. 39,486 (39,000)

Warm Up • Paper and Pencil
Have students compute the answer.
1. Divide the sum of $3\frac{5}{8} + 1\frac{1}{4}$ by $1\frac{1}{8}$. $\left(4\frac{1}{3}\right)$
2. Find the difference between the sum of $15\frac{2}{3} + 6\frac{3}{4}$ and the sum of $3\frac{1}{2} + 4\frac{5}{6}$. $\left(14\frac{1}{12}\right)$
3. What is the average of $5\frac{1}{2}$, $6\frac{3}{4}$, and $4\frac{7}{8}$? $\left(5\frac{17}{24}\right)$
4. Find the product of $\frac{2}{3}$ of 18 and $\frac{3}{4}$ of 16. (144)

2 Teach

Introduce the Lesson Have individual students read each geometric term and its definition aloud. Have them discuss the picture related to the term. Then, ask students to look around the room for objects that suggest points and line segments. For example, the corner where two sides of a table meet might suggest a point; the edge of the table might suggest a line segment.

Explain that the arrowhead at each end of a line shows that the line goes on forever in both directions. The arrowhead at one end of a ray shows that the ray goes on forever in that direction only.

T229

Geometry

Lesson 11-1

Basic Geometric Ideas

A **plane** is a flat surface that extends forever in all directions. A **point** is any position on the plane. A plane is named by any three of its points. Other basic geometric figures are also named by points they contain.

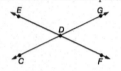

Plane *XYZ*

A **line segment** connects two points that are called **endpoints**. If two line segments are the same length, they are **congruent**.

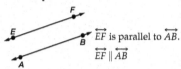

line segment *EF* or $\overline{EF}$
line segment *FE* or $\overline{FE}$

A **line** has no endpoints. It extends forever in opposite directions.

line *AB* or $\overleftrightarrow{AB}$
line *BA* or $\overleftrightarrow{BA}$

A **ray** is part of a line with only one endpoint.

ray *DF* or $\overrightarrow{DF}$

Some lines intersect or meet. The point of intersection for $\overleftrightarrow{EF}$ and $\overleftrightarrow{CG}$ is point *D*.

If there is no point of intersection, the lines are **parallel**.

$\overleftrightarrow{EF}$ is parallel to $\overleftrightarrow{AB}$.

$\overleftrightarrow{EF} \parallel \overleftrightarrow{AB}$

Getting Started

Name each figure.

1.
   ray *XY*

2.
   line *MN*

3.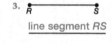
   line segment *RS*

Draw and label.

4. line segment $\overline{ST}$

5. $\overleftrightarrow{AB} \parallel \overleftrightarrow{VW}$

Lesson 11-1 • Basic Geometric Ideas

229

Develop Skills and Concepts Emphasize that a point is a specified location that has no length, no width, and no thickness. A point is represented by a small dot and is labeled by a capital letter.

Be sure students understand the difference between line segments and lines. A line continues without end in both directions and it cannot be measured. Explain that any two points on a line name that line. A line segment is named by its two endpoints and can be measured.

Explain that a ray has one endpoint and continues forever in one direction. Emphasize that a ray is named by its endpoint and another point through which it passes. Stress that the endpoint is always named first.

The term *congruent* is introduced when discussing line segments. Point out that when segments are congruent, they have the same measurement.

Have students identify objects in the room that suggest parallel lines and intersecting lines.

Name the figure.

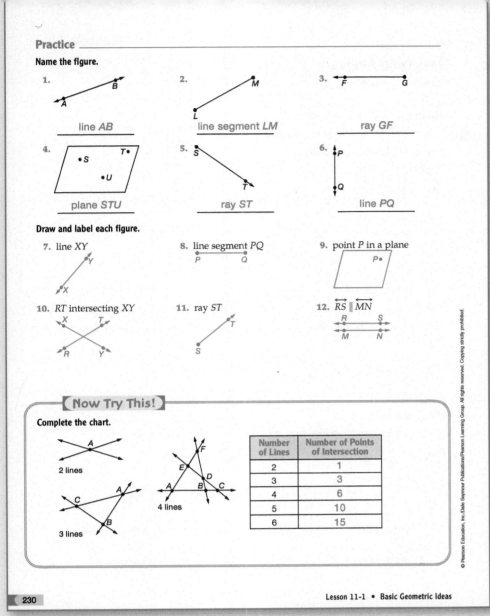

1. line *AB*

2. line segment *LM*

3. ray *GF*

4. plane *STU*

5. ray *ST*

6. line *PQ*

Draw and label each figure.

7. line *XY*

8. line segment *PQ*

9. point *P* in a plane

10. *RT* intersecting *XY*

11. ray *ST*

12. $\overleftrightarrow{RS} \parallel \overleftrightarrow{MN}$

Now Try This!

Complete the chart.

2 lines

3 lines

4 lines

Number of Lines	Number of Points of Intersection
2	1
3	3
4	6
5	10
6	15

For Mixed Abilities

Common Errors • Intervention

Some students may confuse the geometric symbols for naming a line and a line segment. Draw line *RS* on the board.

Have students go to the board, and discuss and point to the points that are on line *RS* (all points on the line), those that are on line segment *RS* (points *R* and *S*, and all of the points in between), and those that are on line *RS* but not on line segment *RS* (all points to the left of *R* and all points to the right of *S*). Lead them to realize that all the points on line segment *RS* are on line *RS*.

Enrichment • Geometry

Challenge students to name 11 line segments in the diagram.

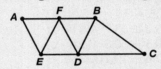

($\overline{AB}$, $\overline{AF}$, $\overline{FB}$, $\overline{AE}$, $\overline{EF}$, $\overline{FD}$, $\overline{BC}$, $\overline{DB}$, $\overline{EC}$, $\overline{ED}$, and $\overline{DC}$)

More to Explore • Application

Display a sales tax chart showing the local tax rate.

Discuss how to compute the amount of sales tax on an item and determine the total cost. For example, if an item costs $1.98, multiply it by the tax rate 6% in order to find the sales tax. Round to the nearest cent. Then, add the sales tax to the cost of the item.

Have students bring in pictures of five items with prices marked. Have them exchange pictures and calculate the total price of each item, including sales tax.

ESL/ELL STRATEGIES

Students should add the lesson vocabulary words to their notebooks. They should list key words, write simple definitions, and include one or more examples of each.

3 Practice

Have students complete all the exercises.

Now Try This! Suggest that students use a ruler or straightedge to draw lines to continue the pattern if needed. To extend the lesson, have students describe the pattern in the table and continue the table. (The number of points of intersection is increased by the previous number of line segments.)

4 Assess

Ask students to identify the number of endpoints in a line, line segment, and ray. (none; two; one)

pages 231–232

1 Getting Started

Objective
• To identify straight, right, obtuse, and acute angles

Vocabulary
vertex, right angle, acute angle, obtuse angle, straight angle

Materials
*demonstration clock face with movable hands; index cards

Warm Up • Mental Math
Dictate the following exercises:
1. $142 - 0.1$ (141.9)
2. $95 - 0.5$ (94.5)
3. $314 - 0.35$ (313.65)
4. $52 - 1.6$ (50.4)
5. $135 - 5.7$ (129.3)
6. $25.2 + 33.7$ (58.9)
7. $50.5 + 50.5$ (101)
8. $12.6 + 10.7$ (23.3)

Warm Up • Paper and Pencil
Have students write the fraction as a decimal. If necessary, round the decimal to the nearest hundredth.
1. $\frac{2}{3}$ (0.67)   5. $\frac{6}{7}$ (0.86)
2. $\frac{5}{9}$ (0.56)   6. $\frac{8}{11}$ (0.73)
3. $\frac{4}{5}$ (0.8)   7. $\frac{3}{8}$ (0.38)
4. $\frac{1}{6}$ (0.17)   8. $\frac{11}{12}$ (0.92)

Name _____

Classifying Angles

An **angle** is formed by two rays with a common endpoint. The rays intersect at the **vertex**. An angle is named with the vertex in the middle. What type of an angle is angle ABC?

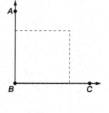

The angle formed can be named in 3 different ways.

angle ABC or ∠ABC
angle CBA or ∠CBA
angle B or ∠B

We can use the corner of this page to classify angles.

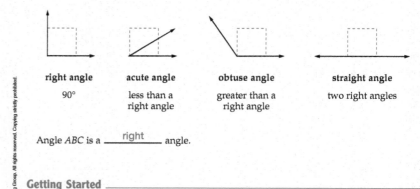

right angle	acute angle	obtuse angle	straight angle
90°	less than a right angle	greater than a right angle	two right angles

Angle ABC is a ____right____ angle.

Getting Started
Name each angle three different ways. Name the rays and vertex. Classify each angle.

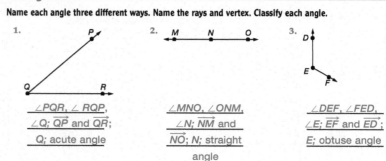

1. ∠PQR, ∠RQP, ∠Q; $\overrightarrow{QP}$ and $\overrightarrow{QR}$; Q; acute angle

2. ∠MNO, ∠ONM, ∠N; $\overrightarrow{NM}$ and $\overrightarrow{NO}$; N; straight angle

3. ∠DEF, ∠FED, ∠E; $\overrightarrow{EF}$ and $\overrightarrow{ED}$; E; obtuse angle

---

2 Teach

Introduce the Lesson Review the terms *endpoint*, *rays*, and *intersect*. Draw an acute angle, an obtuse angle, and a right angle on the board. Ask students to identify the angle with the greatest and smallest opening between the rays. (obtuse angle; acute angle) Ask students to locate various angles in the classroom.

Have a student read the paragraph on page 231. Point out the angle symbol and the three ways of naming an angle. Have students identify point B as the vertex. Then, give each student an index card and explain that a corner of the index card forms a right angle. Tell them to place a corner of the index card at the vertex of each angle in order to classify it. Work through the lesson with students and classify ∠ABC as a right angle.

Develop Skills and Concepts Emphasize the following:

• Two rays with a common endpoint form an angle.

• The point at which the rays meet is the vertex of the angle.

• An angle is usually named with three letters. The middle letter must name the vertex.

• Right angles are used as a reference for telling whether an angle is acute, straight, or obtuse.

Draw these angles on the board:

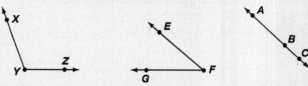

Have students use their index cards to classify each angle. Then, have them name each angle three different ways. (∠XYZ, ∠Y, ∠ZYX, obtuse; ∠EFG, ∠F, ∠GFE, acute; ∠ABC, ∠B, ∠CBA, straight)

Practice

Name each angle three different ways. Name the rays and vertex. Classify each angle.

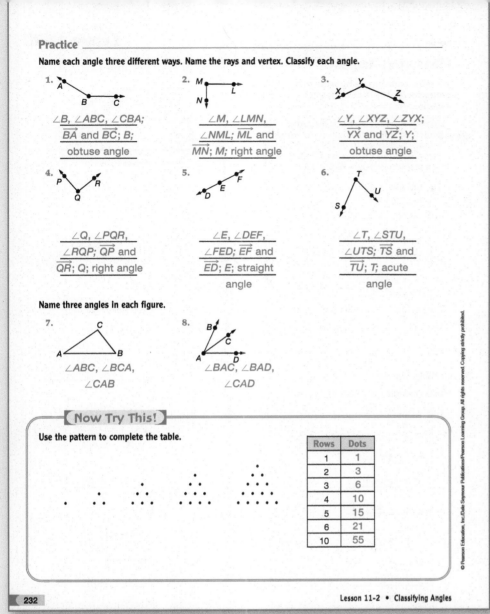

1.
∠B, ∠ABC, ∠CBA;
$\overrightarrow{BA}$ and $\overrightarrow{BC}$; B;
obtuse angle

2.
∠M, ∠LMN,
∠NML; $\overrightarrow{ML}$ and
$\overrightarrow{MN}$; M; right angle

3.
∠Y, ∠XYZ, ∠ZYX;
$\overrightarrow{YX}$ and $\overrightarrow{YZ}$; Y;
obtuse angle

4.
∠Q, ∠PQR,
∠RQP; $\overrightarrow{QP}$ and
$\overrightarrow{QR}$; Q; right angle

5.
∠E, ∠DEF,
∠FED; $\overrightarrow{EF}$ and
$\overrightarrow{ED}$; E; straight
angle

6.
∠T, ∠STU,
∠UTS; $\overrightarrow{TS}$ and
$\overrightarrow{TU}$; T; acute
angle

Name three angles in each figure.

7.
∠ABC, ∠BCA,
∠CAB

8.
∠BAC, ∠BAD,
∠CAD

Now Try This!

Use the pattern to complete the table.

Rows	Dots
1	1
2	3
3	6
4	10
5	15
6	21
10	55

Lesson 11-2 • Classifying Angles

For Mixed Abilities

Common Errors • Intervention

Some students may confuse straight, right, obtuse, and acute when classifying angles. Practice by displaying a clock face with movable hands. Call on students to move the clock's hands to show each of the following times. Then, have students name the kind of angle formed by each time shown.

3:00 (right) 4:00 (obtuse)

6:00 (straight) 1:00 (acute)

11:15 (obtuse) 9:00 (right)

5:00 (obtuse) 2:00 (acute)

10:00 (acute) 12:20 (obtuse)

Enrichment • Geometry

Draw the following polygon on the board:

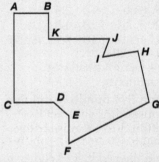

Ask students to find and classify the angles formed by the sides of the figure.

(right: ∠ABK, ∠BKJ, ∠ACD, ∠BAC, ∠IHC; acute: ∠EFG, ∠HIJ, ∠IJK; obtuse: ∠CDE, ∠DEF, ∠FGH)

More to Explore • Measurement

Use cardboard as woodwork to outline one or two windows and a door on a bulletin board. Invite a paperhanger to your class to demonstrate measuring, cutting, and hanging wallpaper to cover the board, explaining working around the door and windows. Have the guest take students step by step through the measurement and math processes needed for the job. Students can use pencil and paper or calculators to find the amount of wallpaper that is needed.

---

3 Practice

Have students complete all the exercises. Point out that they should use a right angle, such as the corner of an index card, as a reference for judging whether an angle is acute, straight, or obtuse.

Now Try This! Have students study the dot patterns to see a relationship between the dots and the number of rows. If necessary, have students draw patterns for 6 and 10.

4 Assess

Ask, *What kind of angle is greater than a right angle?* (obtuse angle)

1 Getting Started

Objective
• To measure and draw angles with a protractor

Vocabulary
protractor

Materials
demonstration protractor; protractors; rulers or straightedges; duplicated sheet of angles

Warm Up • Mental Math
Have students subtract 1.5 from each number.

1. 2.8 (1.3) 6. 10.4 (8.9)
2. 5.5 (4.0) 7. 3.3 (1.8)
3. 6.9 (5.4) 8. 2.6 (1.1)
4. 7.2 (5.7) 9. 4.0 (2.5)
5. 8.1 (6.6) 10. 7.12 (5.62)

Warm Up • Calculator
Have students:

1. From five hundred and five thousandths subtract five hundred five thousandths. (499.5)
2. Find the product of eighty-four hundredths and six tenths. (0.504)
3. Divide four hundred twelve by three and six tenths. Round the quotient to the nearest hundredth. (114.44)

2 Teach

Introduce the Lesson Remind students that two rays with a common endpoint form an angle. Draw and label three angles on the board: acute, obtuse, and right. Have students name each angle three ways and tell if it is acute, obtuse, right, or straight.

Have a student read the paragraph on page 233. Explain that angles are measured in degrees. Use a demonstration protractor at the board or on a transparency to show students how to measure an angle. Give each student a protractor and help students measure some angles you provide on worksheets.

Emphasize that angles do not have to be in the same position to be congruent, but their measures must be the same.

Name _____

Measuring Angles

The basic unit of angle measure is the **degree** (°). A **protractor** is used to measure angles. What is the measure of angle *ABC*?

To read a protractor, place the center of the protractor's straight edge on the vertex. One ray must pass through 0° on the protractor. The measure of this angle is read on the outside scale. The measure of angle *ABC* is 120°. We write:

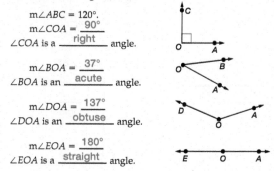

m∠ABC = 120°.
m∠COA = __90°__
∠COA is a __right__ angle.

m∠BOA = __37°__
∠BOA is an __acute__ angle.

m∠DOA = __137°__
∠DOA is an __obtuse__ angle.

m∠EOA = __180°__
∠EOA is a __straight__ angle.

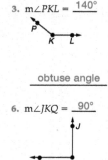

Angles with the same measure are called **congruent angles**.

Getting Started

Give the measure and classify each angle.

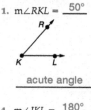
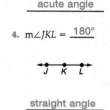

1. m∠RKL = __50°__
   acute angle

2. m∠QKL = __90°__
   right angle

3. m∠PKL = __140°__
   obtuse angle

4. m∠JKL = __180°__
   straight angle

5. m∠RKS = __20°__
   acute angle

6. m∠JKQ = __90°__
   right angle

Develop Skills and Concepts Discuss the steps needed to draw an angle congruent to a given angle.

• Measure the given angle.

• Draw one side (ray) of the angle with a ruler or straightedge.

• Place the protractor with its arrow at the endpoint of the ray and make a mark at the appropriate number of degrees.

• Draw the other ray so that its endpoint starts at the vertex and it passes through the mark.

Emphasize that angles may be congruent even when their sides are different lengths.

Practice

Give the measure and classify each angle.

1. m∠XOZ = __90°__
   right angle

2. m∠VOY = __90°__
   right angle

3. m∠VOZ = __150°__
   obtuse angle

4. m∠XOU = __90°__
   right angle

5. m∠UOZ = __180°__
   straight angle

6. m∠WOY = __60°__
   acute angle

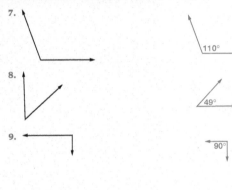

Problem Solving

Give the measure of each angle. Use your protractor to draw an angle congruent to each angle.

7.

8.

9.

10.

110°

49°

90°

140°

11. Draw $\overrightarrow{AB}$. Draw an angle of 45° with ray $\overrightarrow{AB}$ as one side.

45°

A   B

12. Draw $\overleftrightarrow{XY}$ with point T. Draw ∠XTA to measure 135°.

135°

X   T   Y

A

234

---

3 Practice

Have students complete all the exercises. Be sure students choose the correct scale for measuring and drawing angles. Remind students that one ray should be on the 0° mark.

4 Assess

Ask students to use their protractors to draw an acute angle and an obtuse angle. Have them give the measures of the angles they draw.

5 Mixed Review

1. $\frac{3}{8} \div \frac{1}{4}$ $(1\frac{1}{2})$

2. 0.46354 ÷ 0.086 (5.39)

3. 7 × 62,109 (434,763)

4. 514.923 − 206.05 (308.873)

5. $5 \times 3\frac{1}{3} \times 4\frac{1}{5}$ (70)

6. 700 × 600 (420,000)

7. $10\frac{5}{8} + 3\frac{3}{4} + 6\frac{1}{6}$ $(20\frac{13}{24})$

8. 12,176 ÷ 203 (59 R199)

9. 16.3 × 10⁴ (163,000)

10. 320,170 − 47,265 (272,905)

For Mixed Abilities

Common Errors • Intervention

Some students may read the incorrect scale on their protractor when they are measuring an angle. Before they begin to measure, have them identify the angle by writing *acute* (less than 90°), *right* (90°), or *obtuse* (greater than 90° but less than 180°). Then, have them use this label to help them choose the correct scale on the protractor.

Enrichment • Geometry

Provide a handout showing the figure below. Give each student a protractor.

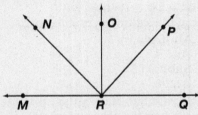

Have students name the following:

1. an angle congruent to ∠MRN (∠ORP)

2. an angle congruent to ∠PRQ (∠NRO)

3. two angles congruent to ∠NRP (∠MRO and ∠ORQ)

More to Explore • Measurement

Tell students that the weight of an object on the Moon is about 0.17 times its Earth weight. Have students gather ten common objects and find their Earth weight in metric units. Then, have students calculate the Moon weight of these same objects by multiplying the Earth weight by 0.17.

T234

11-4 Lines and Angles

pages 235–236

1 Getting Started

Objective
• To identify perpendicular and parallel lines

Vocabulary
vertical angles, perpendicular, transversal

Materials
*demonstration protractor; protractors; rulers or straightedges

Warm Up • Mental Math
Write the following on the board:

Multiply by 3.
Add 4.
Subtract 5.
Add 8.

Say a number. Have students name the answer to each step above. Repeat several times using larger numbers, fractions, or decimals.

Warm Up • Calculator
Write the formulas for the perimeter and area of a rectangle on the board. Have students compute the answer.

1. In a rectangle, the width is 20 in., the length is 25 in. What is the perimeter? (90 in.) The area? (500 sq. in.)
2. The length is 24 cm. The area is 288 cm². What is the width? (12 cm)
3. The width is 8 ft. The perimeter is 36 ft. What is the length? (10 ft)

2 Teach

Introduce the Lesson Have a student read the first paragraph and answer the question. (Four angles are formed.) Ask for examples from everyday life to illustrate the concept of intersecting lines. (roads on a map that cross each other, the grid on a checkerboard)

Work through the models with students, calling attention to the drawings showing *vertical angles, perpendicular lines,* and *parallel lines* with a *transversal.* Ask students how many angles are formed in the bottom drawing of the parallel lines with the transversal. (8) Have students identify the vertical angles. (∠7 and ∠10; ∠8 and ∠9; ∠11 and ∠14; ∠12 and ∠13)

Name _____   Lesson 11-4

Lines and Angles

In a plane, two lines either intersect or are **parallel.** What can we say about the angles formed by intersecting lines? We will use letters to represent lines and numbers to represent angles.

When two lines intersect, __4__ angles are formed. The angles opposite each other are called **vertical angles.**

Angles 1 and __3__ are vertical angles. Use your protractor to measure angles 2 and 4. What do you think is true of vertical angles?
They have equal measures.

If two lines intersect and form a right angle, the lines are **perpendicular.** We write: $c \perp d$ All the angles formed by perpendicular lines are right angles. If m∠5 = 90°, then m∠5 = m∠6 = m∠7 = m∠8.

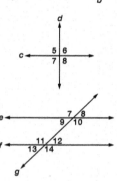

If two lines do not intersect, they are parallel. We write: $e \parallel f$. A line that intersects two other lines is called a **transversal.** Use your protractor to find each of the following.

m∠7 = __134°__   m∠11 = __134°__
m∠10 + m∠12 = __180°__

Getting Started

1. Name the perpendicular segments.

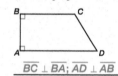

$\overline{BC} \perp \overline{BA}; \overline{AD} \perp \overline{AB}$

2. Name the parallel segments.

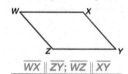

$\overline{WX} \parallel \overline{ZY}; \overline{WZ} \parallel \overline{XY}$

Develop Skills and Concepts Draw intersecting lines on the board. Label the lines using letters and the angles formed using numbers. Have students identify the vertical angles. Then, have students measure the angles formed with a demonstration protractor to verify that vertical angles have equal measures.

Note that the measure of an angle is written m∠ABC; the m stands for "measure." Repeat the activity with perpendicular lines and with parallel lines intersected by a transversal. You may also wish to identify the *corresponding angles* and have students verify that they also have equal measures.

3 Practice

Have students complete the page. Give each student a protractor to use. Remind students that right angles are indicated by the small square.

Practice

Name the perpendicular segments in each figure.

1.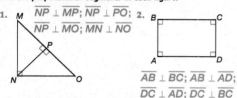
$\overline{NP} \perp \overline{MP}$; $\overline{NP} \perp \overline{PO}$;
$\overline{NP} \perp \overline{MO}$; $\overline{MN} \perp \overline{NO}$

2.
$\overline{AB} \perp \overline{BC}$; $\overline{AB} \perp \overline{AD}$;
$\overline{DC} \perp \overline{AD}$; $\overline{DC} \perp \overline{BC}$

3.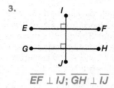
$\overline{EF} \parallel \overline{IJ}$; $\overline{GH} \perp \overline{IJ}$

Name the parallel segments in each figure.

4.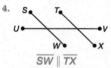
$\overline{SW} \parallel \overline{TX}$

5.
$\overline{BC} \parallel \overline{FE}$; $\overline{CD} \parallel \overline{AF}$; $\overline{AB} \parallel \overline{DE}$

6.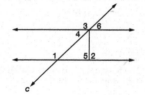
$\overline{LR} \parallel \overline{MN}$; $\overline{LR} \parallel \overline{MP}$; $\overline{LR} \parallel \overline{NP}$;
$\overline{RQ} \parallel \overline{MN}$; $\overline{RQ} \parallel \overline{MP}$; $\overline{RQ} \parallel \overline{NP}$;
$\overline{LQ} \parallel \overline{MN}$; $\overline{LQ} \parallel \overline{MP}$; $\overline{LQ} \parallel \overline{NP}$

Problem Solving

Use your protractor to measure each angle.

7. $m\angle 1 = \underline{120°}$

8. $m\angle 2 = \underline{90°}$

9. $m\angle 3 = \underline{120°}$

10. $m\angle 4 = \underline{60°}$

11. $m\angle 5 = \underline{90°}$

12. $m\angle 6 = \underline{60°}$

13. $m\angle 2 + m\angle 5 = \underline{180°}$

14. $m\angle 3 + m\angle 6 = \underline{180°}$

15. $m\angle 1 + m\angle 4 = \underline{180°}$

16. $m\angle 3 + m\angle 4 = \underline{180°}$

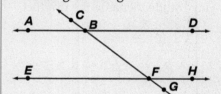

[Now Try This!]

The angle pairs formed when a transversal cuts two parallel lines have special names.

$\angle 3$ and $\angle 6$, and $\angle 4$ and $\angle 5$ are **alternate interior angles**. Pairs of alternate interior angles are congruent.

$\angle 3$ is congruent to $\angle 6$. $\angle 4$ is congruent to $\underline{\angle 5}$.

$\angle 3$ and $\angle 5$ are **same side interior angles**. The sum of the measures of same side interior angles is 180°.

$m\angle 3 + m\angle 5 = 180°$. $m\angle 4 + \underline{m\angle 6} = 180°$

236

Lesson 11-4 • Lines and Angles

Now Try This! Introduce students to the special relationship between the alternate interior angles formed when parallel lines are cut by a transversal. To help them to remember the angles' position, point out that the word *alternate* means "on either side of the transversal" and *interior* means "inside the parallel lines."

Then, introduce the same-side interior angles.

4 Assess

Have students draw and label intersecting lines, perpendicular lines, and parallel lines.

For Mixed Abilities

Common Errors • Intervention
Some students may confuse parallel and perpendicular. Practice by giving each student a worksheet showing various examples of perpendicular lines and parallel lines with a transversal. Have students trace the perpendicular lines with a blue crayon, the parallel lines with a red crayon, and transversals with a green crayon.

Then, have students number the angles and name pairs of vertical angles.

Enrichment • Geometry
Have students draw and label the following drawing:

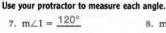

Tell them to measure all the angles. Have them list all the angles that are congruent to $\angle ABC$. ($\angle BFE$, $\angle DBF$, and $\angle HFG$)

Then, have them list all the angles that are congruent to $\angle ABF$. ($\angle CBD$, $\angle EFG$, and $\angle BFH$)

Finally, have them name the sum of the measures of the following angles:

$\angle ABF + \angle FBD$ (180°)

$\angle ABC + \angle CBD$ (180°)

$\angle EFB + \angle BFH$ (180°)

$\angle GFH + \angle HFB$ (180°)

More to Explore • Probability
Have students consider another random walk problem. Tell each student to let his or her imaginary pedestrian go for 5 walks of 25 steps each. For every walk the pedestrian takes, students are to record the number of times the pedestrian crosses 0. See if students can guess an average number of times a walker will cross 0 in a walk of 25 steps. (The theoretical average is about 3.)

T236

1 Getting Started

Objective
• To identify parts of a circle and the measure of an arc

Vocabulary
circle, center, circumference, radius, chord, diameter, central angle, arc

Materials
compasses; rulers or straightedges; *demonstration compass; *demonstration protractor

Warm Up • Mental Math
Have students name the product.
1. $4 \times 10 \times 10$ (400)
2. 3×10^3 (3,000)
3. 10^4 (10,000)
4. 4^3 (64)
5. 2^5 (32)
6. 5^2 (25)
7. 9^2 (81)
8. 10^5 (100,000)
9. 6×10^4 (60,000)
10. 3^2 (9)

Warm Up • Paper and Pencil
Have students find the perimeter of the following rectangles using the formula $P = 2l + 2w$.
1. length = 6 in., width = 3 in.
   (18 in.)
2. length = 12 ft, width = 10 ft
   (44 ft)
3. length = 9.3 yd, width = 8.4 yd
   (35.4 yd)

2 Teach

Introduce the Lesson Use the demonstration compass to draw a circle on the board. Then, have students try to draw circles accurately without using a tracing pattern or compass. Discuss their results and point out that it is very difficult to do. Explain that a compass is used to draw circles accurately. Use a circle on the board to introduce and discuss each of the concepts on page 237.

Develop Skills and Concepts Ask if a diameter is a chord. (Yes, it is a line segment with its endpoints on the circle.) Ask if a chord is a diameter. (not unless it passes through the center) Have students describe the central angle formed by any diameter. (straight angle) Discuss radius. Point out that the plural of *radius* is *radii*.

Name _____

Parts of a Circle

A **circle** is the set of all points the same distance from a point called the **center**. The circular line around a circle is called the **circumference**. We use a compass to construct the circumference of a circle.

A line segment from the center of the circle to its edge is called a **radius**. Line segment AO is a radius in circle O.

A line segment with its endpoints on the circle is called a **chord**. Line segment DE is a chord.

A **diameter** is a chord that passes through the center. Its length is twice the length of the radius. Line segment BC is a diameter.

An angle with its vertex at the center is called a **central angle**. Angle AOC is a central angle.

An **arc** is part of the circumference that is formed by a central angle. The curved line from A to C is an arc.

We write: $\overset{\frown}{AC}$.

The measure of an arc is the same as the measure of the central angle.

We write: $m\overset{\frown}{AC} = m\angle AOC$

A circle contains 360 degrees.

WEATHER STATION

Getting Started
Complete the following.

1. $\overline{MN}$ is a(n) __radius__.

2. $m\angle NMO =$ __60°__ ; $m\angle LMO =$ __120°__

3. $\overline{XY}$ is a(n) __chord__.

4. $\overset{\frown}{ON}$ is a(n) __arc__.

5. The length of the diameter is __6 cm__.

Continue with questions such as the following:

• How many radii may be drawn in a circle? (an infinite number)

• If one radius is x inches, what is true about the measure of another radius in the same circle? (It is also x inches.)

• If one radius is x inches, what is the measure of the diameter in the same circle? ($2x$ inches)

Have a student use a demonstration protractor to measure and draw a central angle of 45°. Have students determine the measure of the other central angle formed by this radius. ($360° - 45° = 315°$)

Practice

Complete the following.

1. $\overline{YO}$ is a(n) __radius__.

2. $\overset{\frown}{YZ}$ is a(n) __arc__.

3. $\overline{XZ}$ is a(n) __diameter__.

4. $m\overset{\frown}{YX}$ = __135°__.

5. The length of the diameter is __2 inches__.

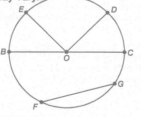

Draw circle O. Then, label these parts of the circle. Answers may vary.

6. arc BE

7. central angle EOD

8. diameter BC

9. chord FG

10. radius OD

Problem Solving

Complete the table. Remember $C = \pi \times d$.

11.

radius (r)	2 in.	5 in.	4 in.	6.5 in.	4.2 in.	4.38 in.	7.05 in.
diameter (d)	4 in.	10 in.	8 in.	13 in.	8.4 in.	8.76 in.	14.1 in.
circumference (c)	12.56	31.4	25.12	40.82	26.376	27.5064	44.274

【Now Try This!】

Use your compass and ruler to copy the figure. Then, draw other figures with your compass and ruler.

Lesson 11-5 • Parts of a Circle

---

3 Practice

Have students complete all the exercises. Review the notation for line segment, angle, and arc.

Now Try This! Provide each student with a compass and a ruler. Have students look for semicircles as they explore ways to copy the figure.

4 Assess

Have students use a compass to draw a circle. Then, have them label each part of the circle: radius, chord, diameter, and central angle. (Check students' work.)

For Mixed Abilities

Common Errors • Intervention

Some students may confuse radius and diameter. Have them work with partners and an 8-inch circle with the center labeled. Have them use a ruler to draw and measure a radius (4 in.) and a diameter (8 in.). Ask, *How does the length of a radius compare to the length of a diameter?* (It is half as long.) *How does the length of a diameter compare to the length of a radius?* (It is twice as long.)

Enrichment • Number Sense

Give students the following approximation for π, which is rounded to the nearest ten thousandth: 3.1416. Have them calculate the decimal value of $\frac{22}{7}$ to the nearest ten thousandth and compare the two approximations. ($\frac{22}{7} = 3.1429$ is greater) Then, have students draw two circles of different sizes, measure each diameter to the nearest inch, and find the circumference of each circle, using both 3.1416 and $\frac{22}{7}$ for π. Have them compare their results.

More to Explore • Sets

Duplicate the following for students to solve:

Draw a Venn diagram showing the following sets. Then, answer the questions.

The Universal set represents all pet owners: *A* is the set of dog owners, *B* is the set of cat owners, and *C* is the set of hamster owners.

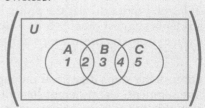

- What must Sections 2 and 4 represent? (2 = dog and cat owners; 4 = cat and hamster owners)

- What do Sections 1, 3, and 5 represent? (single pet owners)

- Where would your family be represented? (Answers will vary.)

T238

11-6 Bisecting Line Segments and Angles

pages 239–240

1 Getting Started

Objective
• To bisect a line segment and an angle with a compass

Vocabulary
bisect, midpoint

Materials
*demonstration compass; compasses; protractors; centimeter ruler; worksheets

Warm Up • Mental Math
Have students name the fraction in simplest form.

1. $\frac{9}{15}$ $(\frac{3}{5})$ 5. $\frac{14}{18}$ $(\frac{7}{9})$

2. $\frac{12}{16}$ $(\frac{3}{4})$ 6. $\frac{25}{100}$ $(\frac{1}{4})$

3. $\frac{24}{36}$ $(\frac{2}{3})$ 7. $\frac{85}{12}$ $(7\frac{1}{12})$

4. $\frac{50}{100}$ $(\frac{1}{2})$ 8. $\frac{29}{3}$ $(9\frac{2}{3})$

Warm Up • Paper and Pencil
Have students find the following:

1. $\frac{1}{2}$ of $6\frac{1}{2}$ $(3\frac{1}{4})$ 5. $\frac{1}{4}$ of $8\frac{4}{5}$ $(2\frac{1}{5})$

2. $\frac{1}{2}$ of $18\frac{2}{3}$ $(9\frac{1}{3})$ 6. $\frac{1}{4}$ of $36\frac{8}{9}$ $(9\frac{2}{9})$

3. $\frac{1}{3}$ of $12\frac{3}{4}$ $(4\frac{1}{4})$ 7. $\frac{1}{5}$ of $65\frac{5}{6}$ $(13\frac{1}{6})$

4. $\frac{1}{3}$ of $27\frac{6}{7}$ $(9\frac{2}{7})$ 8. $\frac{1}{5}$ of $30\frac{10}{11}$ $(6\frac{2}{11})$

Name _____

Bisecting Line Segments and Angles

To bisect means "to divide into two congruent parts." We bisect line segments and angles with a compass and straightedge.

To bisect a line segment, follow these steps:

• Draw a line segment and label the endpoints A and B. With your compass, select a radius that is larger than half the length of $\overline{AB}$. With point A as center, construct an arc above and an arc below $\overline{AB}$.

• Keep your compass setting unchanged. With point B as center, construct two arcs that intersect the arcs constructed in the first step. Label the intersections C and D.

• Draw $\overleftrightarrow{CD}$. $\overleftrightarrow{CD}$ bisects $\overline{AB}$. $\overline{AM}$ is congruent to $\overline{MB}$. M is the **midpoint** of $\overline{AB}$. $\overleftrightarrow{CD}$ is perpendicular to $\overline{AB}$.

We write: $\overleftrightarrow{CD} \perp \overline{AB}$.

To bisect an angle, follow these steps:

• Draw $\angle ABC$. With B as center, construct an arc intersecting $\angle ABC$ at points M and N.

• Keep your compass setting unchanged. With M and N as centers, construct two arcs intersecting in the interior of $\angle ABC$. Label the intersection D.

• Draw $\overrightarrow{BD}$. $\overrightarrow{BD}$ bisects $\angle ABC$. $\angle ABD$ is congruent to $\angle DBC$.

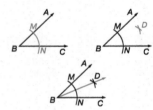

Getting Started

Bisect each segment or angle.

1.

2.

---

2 Teach

Introduce the Lesson Have a student read the definition of *bisect*. Point out that the prefix *bi-* means "two." Have students suggest other words with the same prefix and explain how the prefix helps define the word. (bicycle, biweekly, biped)

Discuss the steps described on page 239 for bisecting a line segment. Use a demonstration compass and a straightedge to demonstrate each step. Be sure students understand that the point at which the line bisector intersects the line is the midpoint of the line and that the line bisector is perpendicular to the line.

Then, discuss the steps described on page 239 for bisecting an angle. Again, use a demonstration compass and a straightedge to demonstrate each step.

Develop Skills and Concepts Ask students to name the measure of the two segments formed when a line segment 6 cm is bisected. (3 cm each) Have them measure and draw a line segment 6 cm long, bisect it, and measure the two resulting segments.

Then, ask students to name the measure of the two angles formed when a 60° angle is bisected. (30° each) Have them verify the answer by using a protractor to draw a 60° angle, bisect the angle with a compass, and measure the resulting angles.

3 Practice

Have students complete all the exercises. Be sure each student has a compass, protractor, and centimeter ruler.

T239

Practice

Bisect each segment or angle.

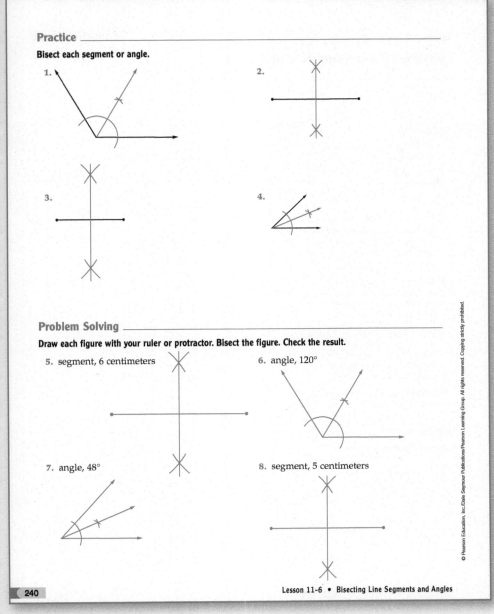

1.

2.

3.

4.

Problem Solving

Draw each figure with your ruler or protractor. Bisect the figure. Check the result.

5. segment, 6 centimeters

6. angle, 120°

7. angle, 48°

8. segment, 5 centimeters

Lesson 11-6 • Bisecting Line Segments and Angles

---

4 Assess

Ask students to describe a perpendicular bisector.
(Possible answer: A perpendicular bisector divides a line segment in half and is perpendicular to the line segment.)

5 Mixed Review

1. 78 × 321 (25,038)
2. 174.3 × 0.651 (113.4693)
3. 636,290 + 58,195 + 391,476 (1,085,961)
4. 2.7 ÷ 10³ (0.0027)
5. $2,011.90 ÷ 62 ($32.45)
6. 25 ÷ 3⅓ (7½)
7. 6⁴ (1,296)
8. 195.387 + 29.48 + 356.002 (580.869)
9. $1.85 × 16 ($29.60)
10. $\frac{11}{12} - \frac{5}{6}$ $(\frac{1}{12})$

11-7 Copying Line Segments and Angles

pages 241–242

1 Getting Started

Objective
• To duplicate a line segment and angle with a compass and straightedge

Materials
*demonstration compass; compasses; straightedges; protractors

Warm Up • Mental Math
Have students estimate each product.
1. 374 × 27 (12,000)
2. 631 × 92 (54,000)
3. 752 × 38 (32,000)
4. 4,926 × 55 (300,000)
5. 8,049 × 64 (480,000)

Warm Up • Paper and Pencil
Have students find the sum or difference.

1. 5 lb 11 oz
   + 9 lb 14 oz
   (15 lb 9 oz)

2. 9 wk 3 da
   − 2 wk 5 da
   (6 wk 5 da)

3. 5 yd 1 ft 8 in.
   − 1 yd 2 ft 11 in.
   (3 yd 1 ft 9 in.)

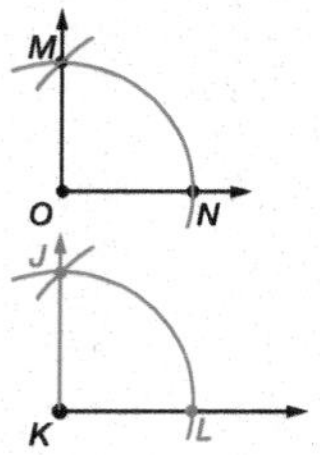

2 Teach

Introduce the Lesson Draw the following figures on the board:

vertical line segment CD, acute angle EFG, obtuse angle ABC, ray ST, horizontal line segment XY, and right angle PQO.

Have students name each line segment, angle, or ray. (CD or DC; ∠EFG, ∠GFE, or ∠F; ∠ABC, ∠CBA, or ∠B; ST; XY or YX; ∠PQO, ∠OQP, or ∠Q) Then, work through the steps on page 241 with students to show them how to copy each figure using only a compass and a straightedge.

Copying Line Segments and Angles

We use a compass and a straightedge to copy geometric figures.

To copy AB, follow these steps:

Draw a ray. Label the endpoint E.

Set the points of the compass on A and B. Mark an arc on the ray.

Label the intersection G. EG is congruent to AB.

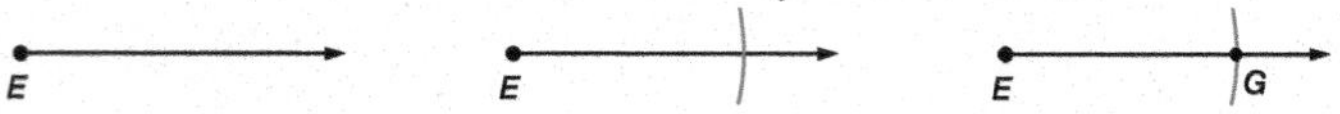

To copy ∠PQR, follow these steps:

• Set the compass point at Q and draw an arc through both rays. Label the intersections X and Y.

• Keep your compass setting unchanged. Draw the same arc at point S. Label the intersection U.

• Using your compass, measure the distance from Y to X. Measure the same distance on the new arc made through point U. Label this point V.

• Draw $\vec{SV}$. ∠VSU is congruent to ∠PQR.

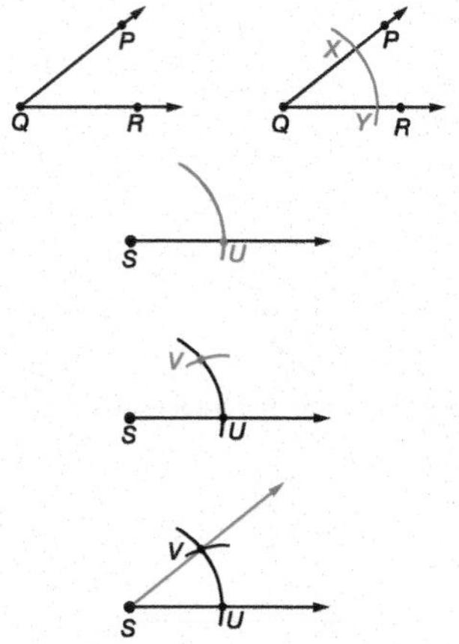

Getting Started

Copy each figure on the given ray.

1. $\overline{AB}$

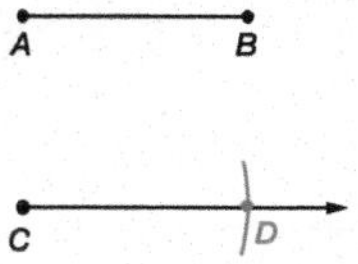

2. ∠MON

Develop Skills and Concepts Draw a line segment on the board and label it AB. Using a demonstration compass and a straightedge, follow the steps described on page 241 to copy AB. Draw several other line segments on the board and have students go to the board and copy each. Next, draw a 45° angle on the board and label it ∠ABC. Using a demonstration compass and a straightedge, follow the steps described on page 241 to copy ∠ABC. Then, have students draw angles on the board. Have other students copy them on the board.

3 Practice

Have students complete all the exercises. Be sure each student has a compass, protractor, and straightedge to do the construction.

Practice

Copy each figure on the given ray.

1. $\overline{AB}$

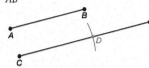

2. $\overline{LM}$

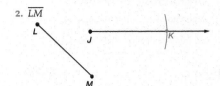

3. $\angle A$

4. $\angle J$

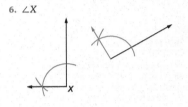

5. $\angle T$

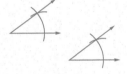

6. $\angle X$

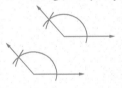

Problem Solving

7. Draw an acute angle with your protractor. Use a compass and straightedge to copy the angle. Check the new angle with your protractor.

8. Draw an obtuse angle with your protractor. Use a compass and straightedge to copy the angle. Check the new angle with your protractor.

Lesson 11-7 • Copying Line Segments and Angles

For Mixed Abilities

Common Errors • Intervention

Watch for students whose constructions are inaccurate because they are changing the setting of the compass when they should not be in order to draw different arcs. Carefully review the procedure, stressing at what steps the arcs must be the same size.

Enrichment • Geometry

1. Tell students to draw a line segment. Then, ask them to construct a triangle with sides the length of the segment. They should use only a compass and a straightedge to copy the segment.

2. Tell students to draw a triangle ABC. Then, challenge them to copy each of the angles of the triangle using a compass and a straightedge.

More to Explore • Application

Duplicate the following for students and have them use calculators to complete it:

What does it cost to borrow money? The class is going to borrow $600 for 6 months at 8% interest. The repayment schedule is set for them to pay back $100 each month plus interest on the balance due. The first payment is calculated as follows:

Balance Due:	$600.00
× 0.08 interest	$48.00
+ Payment	+ $100.00
Total Payment	$148.00

Have students calculate the payment schedule for the next 5 months. Then, find out how much the class paid back in total ($768) and how much interest the class paid. ($168)

4 Assess

Have students use a compass and a straightedge to copy a line segment and an angle. Then, have them use a ruler and a protractor to verify their construction. Accept slight variations.

5 Mixed Review

1. $18,070 - 12,654$ (5,416)
2. $3\frac{3}{5} - \frac{5}{6}$ ($2\frac{23}{30}$)
3. 0.056×10^4 (560)
4. 65×721 (46,865)
5. 0.05×0.4 (0.02)
6. 7^3 (343)
7. $6 \times 5\frac{1}{3} \times 2\frac{1}{2}$ (80)
8. $8.502 \div 100$ (0.08502)
9. $32,198 \div 46$ (699 R44)
10. $268 - 157.38$ (110.62)

1 Getting Started

Objective
- To identify regular and irregular polygons

Vocabulary
polygon, sides, vertex, diagonal, regular polygon

Materials
drinking straws or pipe cleaners; glue; graph paper

Warm Up • Mental Math
Have students estimate each quotient.

1. $3,424 \div 4$ (750)
2. $4,348 \div 7$ (600)
3. $1,607 \div 52$ (40)
4. $4,788 \div 38$ (120)
5. $41,856 \div 83$ (500)
6. $12,060 \div 61$ (200)

Warm Up • Paper and Pencil
Have students do the calculation.

1. $(19.5 - 3.06) + 245$ (261.44)
2. $(6.7 \times 2.05) - 0.26$ (13.475)
3. $(3.210 \div 0.3) \times 4.8$ (51.36)
4. $(95.2 + 6.79) - 83.001$ (18.989)
5. $(8.4 - 3.56) \times 7.9$ (38.236)
6. $(1.3608 \div 0.24) \times 9.26$ (52.5042)

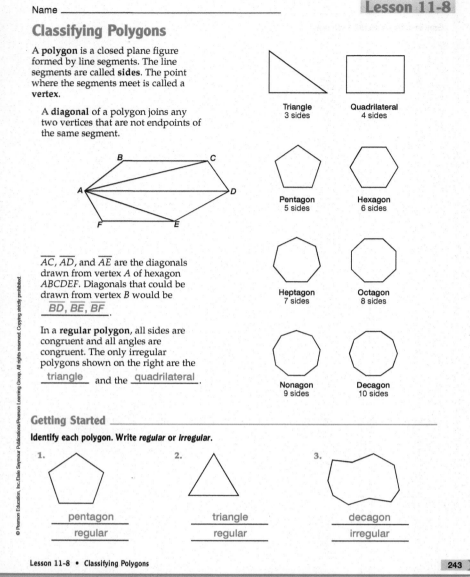

Name _____

Lesson 11-8

Classifying Polygons

A **polygon** is a closed plane figure formed by line segments. The line segments are called **sides**. The point where the segments meet is called a **vertex**.

A **diagonal** of a polygon joins any two vertices that are not endpoints of the same segment.

Triangle — 3 sides
Quadrilateral — 4 sides
Pentagon — 5 sides
Hexagon — 6 sides
Heptagon — 7 sides
Octagon — 8 sides
Nonagon — 9 sides
Decagon — 10 sides

$\overline{AC}$, $\overline{AD}$, and $\overline{AE}$ are the diagonals drawn from vertex A of hexagon $ABCDEF$. Diagonals that could be drawn from vertex B would be __$\overline{BD}$, $\overline{BE}$, $\overline{BF}$__.

In a **regular polygon**, all sides are congruent and all angles are congruent. The only irregular polygons shown on the right are the __triangle__ and the __quadrilateral__.

Getting Started _____

Identify each polygon. Write *regular* or *irregular*.

1. pentagon — regular
2. triangle — regular
3. decagon — irregular

Lesson 11-8 • Classifying Polygons

243

2 Teach

Introduce the Lesson Have a student read the definition of a polygon. Ask why a box is not a polygon. (Possible answer: A box is not a plane figure but a space or solid figure.) Then, discuss the polygons shown on page 243. Have students identify the number of sides and vertices in each. Point out that in every polygon the number of angles corresponds to the number of sides. Define *diagonals*. Ask how many diagonals can be drawn in figure *ABCDE*. (9)
Then, draw the following figures on the board:

A   B   C   D

Ask why each figure is not a polygon. (Figure A is not simple, which means the sides cross or intersect; Figure B is not closed; Figure C has one curved side; Figure D is not simple, not closed, and has curved sides.)

Have a student read the last paragraph, describing a *regular polygon*. Ask why the octagon is called regular. (All eight sides and all eight angles are congruent.) Point out that if a polygon is not regular, it is called irregular.

Develop Skills and Concepts Write the following prefixes and their meanings on the board: *tri-*, **three**; *quad-*, **four**; *penta-*, **five**; *hexa-*, **six**; *septa-*, **seven**; *octa-*, **eight**; *nona-*, **nine**; *deca-*, **ten**.

After students have had a chance to look at the list, write the following words on the board: **bicycle**, **tricycle**, **quadriphone**, **hexameter**, **octopus**, and **decimal**.

Help students understand how the prefix of each word helps define it. Ask a volunteer to read each definition from a dictionary.

Practice

Identify each polygon. Write *regular* or *irregular*.

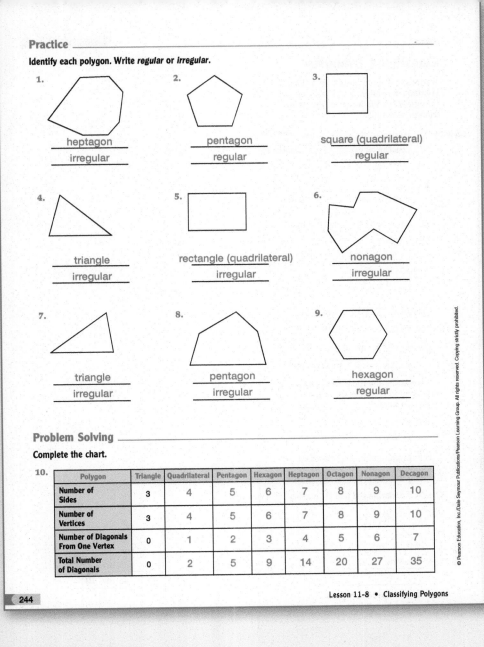

1. heptagon
   irregular

2. pentagon
   regular

3. square (quadrilateral)
   regular

4. triangle
   irregular

5. rectangle (quadrilateral)
   irregular

6. nonagon
   irregular

7. triangle
   irregular

8. pentagon
   irregular

9. hexagon
   regular

Problem Solving

Complete the chart.

10.

Polygon	Triangle	Quadrilateral	Pentagon	Hexagon	Heptagon	Octagon	Nonagon	Decagon
Number of Sides	3	4	5	6	7	8	9	10
Number of Vertices	3	4	5	6	7	8	9	10
Number of Diagonals From One Vertex	0	1	2	3	4	5	6	7
Total Number of Diagonals	0	2	5	9	14	20	27	35

Lesson 11-8 • Classifying Polygons

---

3 Practice

Have students complete all the exercises.

4 Assess

Ask, *How many sides are there to a pentagon, hexagon, and octagon?* (five, six, eight)

5 Mixed Review

1. 627,146 − 308,109 (319,037)
2. 33.670 ÷ 2.6 (12.95)
3. 472 + 308 + 1,275 + 691 (2,746)
4. 0.18 × 2.71 (0.4878)
5. 2,800 ÷ 40 (70)
6. 6,121 × 56 (342,776)
7. 195 − 58.37 (136.63)
8. 68,109 ÷ 3 (22,703)
9. 0.302 ÷ 10 (0.0302)
10. $\frac{2}{3} \times \frac{9}{14}$ $\left(\frac{3}{7}\right)$

For Mixed Abilities

Common Errors • Intervention

Some students may have difficulty identifying polygons. Have them work with partners, using drinking straws or pipe cleaners. Have them use these materials to construct models of polygons that can be glued onto a sheet of paper. Have students label each model by giving the name of each polygon, the number of sides, the number of angles, and whether it is regular or irregular.

Enrichment • Geometry

Draw the following figure on the board:

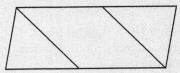

Ask students to tell how many triangles and how many quadrilaterals there are in the figure. (2 triangles, 4 quadrilaterals) Challenge students to draw other figures with overlapping polygons. Have them exchange papers and find the number of each type of polygon drawn. Students should outline each.

More to Explore • Statistics

Have students make a *contingency table* that shows the results of a survey in which the results are tabulated according to two classifications. Ask students to raise their hands if they have blond hair and blue eyes, blond hair and brown eyes, brown hair and blue eyes, then brown hair and brown eyes. Continue to include black and red hair, blue and brown eyes. Explain that their table should be set up as shown:

	Blond	Brown	Black	Red
Blue				
Brown				

After making all tally marks, ask students if people with blond hair are more likely to have blue or brown eyes. (blue) Ask if people with brown hair are more likely to have blue or brown eyes. (brown)

T244

11-9 Classifying Triangles

pages 245–246

1 Getting Started

Objectives

- To classify triangles by sides and angles
- To find the measure of a third angle when the other two are known

Vocabulary

triangle, equilateral, isosceles, scalene

Materials

scissors

Warm Up • Mental Math

Have students multiply each measure by 3.

1. 6 in. (18 in.)   4. 17 in. (51 in.)
2. 25 ft (75 ft)   5. 143 ft (429 ft)
3. 42 yd (126 yd) 6. 306 yd (918 yd)

Warm Up • Paper and Pencil

Have students rename the measure in the unit indicated.

1. 35 cm = (0.35) m
2. 0.032 g = (32) mg
3. 105 m = (105,000) mm
4. 5.2 kg = (5,200) g
5. 6.01 L = (6,010) mL
6. 7.5 cm = (75) mm
7. 93 m = (0.093) km
8. 5,120 cm = (51.2) m
9. 9.9 km = (9,900) m
10. 4.65 L = (4,650) mL

2 Teach

Introduce the Lesson Review the following concepts:

- An acute angle is less than 90°.
- An obtuse angle is greater than 90°.
- A right angle is equal to 90°.
- A straight angle is equal to 180°.

Read the first paragraph that defines a triangle. Work through the model with students. Write the following on the board: **equilateral**, **scalene**, and **isosceles**. Call on students to draw triangles that match each word. Then, ask students to classify each triangle by its largest angle. Finally, have students read to learn how to find the

measure of the third angle of a triangle if they know the measures of the other two angles.

Explain that m∠A is read *the measure of angle A.*

Develop Skills and Concepts Illustrate the concept that the sum of the measures of the angles of any triangle is 180° by having students draw and label any triangle *ABC*. Tell them to first cut out the triangle. Then, have them cut off the angles and place them next to each other. Point out that the three angles together have the same measure as a straight angle (180°); they lie on a straight line.

Ask the following questions: *Can a triangle have two right angles?* (No. A triangle must have three angles whose sum is 180°; two right angles equal 180°.) *Can a right triangle also be an isosceles triangle?* (Yes. One angle is 90°; the others are 45° each.) *Can an obtuse triangle contain a right angle?* (No. The sum of the measures of an obtuse angle and a right angle is greater than 180°.)

T245

Name _____

Classifying Triangles

A **triangle** is a polygon with 3 sides and 3 angles. We name a triangle by its vertices. We call this triangle *ABC*. We write: △*ABC*. What kind of triangle is △*ABC*?

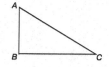

We classify a triangle by sides.

equilateral   isosceles   scalene

3 sides congruent   2 sides congruent   0 sides congruent

We also classify a triangle by its largest angle.

acute   obtuse   right

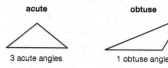

3 acute angles   1 obtuse angle   1 right angle

Classified by sides, △*ABC* is a ___scalene___ triangle.

Classified by angles, △*ABC* is a ___right___ triangle.

The sum of the angle measures of any triangle is 180°.

$$m\angle A + m\angle B + m\angle C = 180°$$

If we know the measures of two angles of a triangle, we can determine the measure of the third by adding the measures of the two known angles and subtracting the sum from 180°.

In triangle *ABC*, m∠*C* = 30° and m∠*B* = 90°.

30° + 90° = ___120___   180° − ___120___ = ___60___

The measure of angle *A* is ___60°___.

Getting Started

Classify the triangle and find the missing angle measure.

1.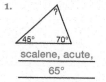
45° 70° ?

scalene, acute,
___65°___

2.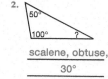
50°
100° ?

scalene, obtuse,
___30°___

3.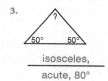
?
50° 50°

isosceles,
acute, 80°

Practice

Classify the triangle and find the missing angle measure.

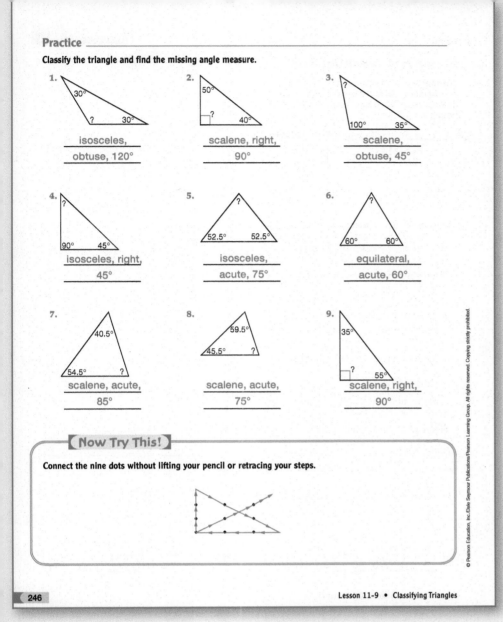

1. isosceles, obtuse, 120°

2. scalene, right, 90°

3. scalene, obtuse, 45°

4. isosceles, right, 45°

5. isosceles, acute, 75°

6. equilateral, acute, 60°

7. scalene, acute, 85°

8. scalene, acute, 75°

9. scalene, right, 90°

[Now Try This!]

Connect the nine dots without lifting your pencil or retracing your steps.

246

Lesson 11-9 • Classifying Triangles

---

3 Practice

Have students complete all the exercises.

Now Try This! Be sure students understand that they are not to lift their pencils or retrace any line. Encourage experimentation.

4 Assess

Draw a triangle and have students classify the triangle by sides and by angles.

For Mixed Abilities

Common Errors • Intervention

Some students may simply subtract the measure of one of the two given angles from 180° when they are trying to find the measure of the third angle of a triangle. Have them work with partners and a set of triangles that are labeled *ABC*, with one of the angle measures missing. Have them complete a chart to find the measure of the missing angle.

$\angle A$	$\angle B$	$\angle C$	$\angle A + \angle B + \angle C$

Enrichment • Geometry

Ask students how many triangles they can find in the drawing. (16)

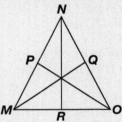

More to Explore • Measurement

Have students design a map for a road rally. Their maps must contain:

- Markings for direction

- A scale of 1 inch = 10 miles

- A list of at least eight different commands to follow.

Have them mark the beginning point *A* and give each command according to direction and distance. For example, point B is 20 miles NE of point *A*.

ESL/ELL STRATEGIES

Discuss the meaning of △ (*triangle*), m (which stands for *measure*) and ∠ (which stands for *angle*). Have students read a sentence, such as m∠*ABC* = 55°.

T246

11-10 Classifying Quadrilaterals

pages 247–248

1 Getting Started

Objective
- To identify parallelograms, rectangles, rhombuses, squares, and trapezoids

Vocabulary
quadrilateral, parallelogram, rectangle, rhombus, square, trapezoid

Warm Up • Mental Math
Have students rename the measure in the unit indicated.

1. 48 in. = (4) ft
2. 7 yd = (21) ft
3. 6 ft = (2) yd
4. 12 pt = (6) qt
5. 9 ft = (108) in.
6. 15 gal = (60) qt
7. 4 yd = (144) in.
8. 4 lb = (64) oz
9. 160 oz = (10) lb
10. 5 T = (10,000) lb

Warm Up • Mental Math
Ask students to solve the following:

1. A square has a perimeter of 144 in. What is the length of a side? (36 in.)

2. A rectangle is 12 ft long and has a perimeter of 64 ft. What is the width? (20 ft)

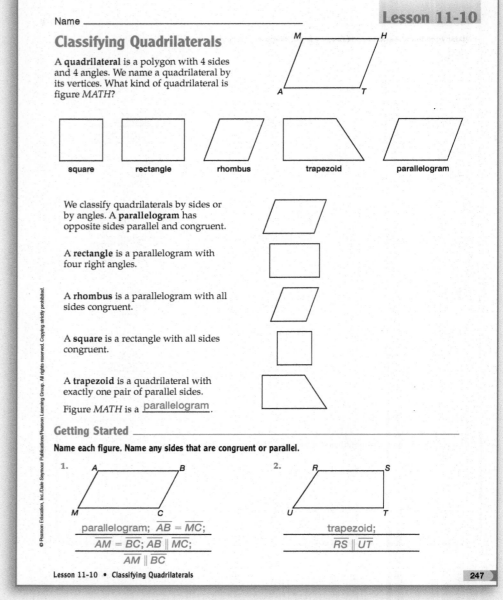

Classifying Quadrilaterals

A **quadrilateral** is a polygon with 4 sides and 4 angles. We name a quadrilateral by its vertices. What kind of quadrilateral is figure *MATH*?

square   rectangle   rhombus   trapezoid   parallelogram

We classify quadrilaterals by sides or by angles. A **parallelogram** has opposite sides parallel and congruent.

A **rectangle** is a parallelogram with four right angles.

A **rhombus** is a parallelogram with all sides congruent.

A **square** is a rectangle with all sides congruent.

A **trapezoid** is a quadrilateral with exactly one pair of parallel sides.

Figure *MATH* is a <u>parallelogram</u>.

Getting Started
Name each figure. Name any sides that are congruent or parallel.

1. parallelogram; $\overline{AB} = \overline{MC}$; $\overline{AM} = \overline{BC}$; $\overline{AB} \parallel \overline{MC}$; $\overline{AM} \parallel \overline{BC}$

2. trapezoid; $\overline{RS} \parallel \overline{UT}$

Lesson 11-10 • Classifying Quadrilaterals

247

2 Teach

Introduce the Lesson Review the meaning of *polygon*. Then, call on a student to read the definition of a *quadrilateral*. Point out that a quadrilateral is a special kind of polygon. Discuss the quadrilaterals shown on page 247. Review these concepts:

- Parallel lines go on forever and never intersect.
- *Congruent* means "equal to" or "having the same measure."
- A right angle is 90°.

Tell students to name quadrilaterals by their vertices.

Develop Skills and Concepts Ask the following questions and discuss the answers:

- *Are all rectangles parallelograms?* (Yes.)
- *Are all parallelograms rectangles?* (No.)
- *Are all squares rectangles?* (Yes.)
- *Are all rectangles squares?* (No.)
- *Are all rhombuses parallelograms?* (Yes.)
- *Are all parallelograms rhombuses?* (No.)

Develop the concept that the sum of the angle measures of a quadrilateral is 360°. Remind students that the sum of the angle measures of a triangle is 180°.

Tell students that a quadrilateral and one of its diagonals form two triangles. Draw a quadrilateral and one of its diagonals on the board to illustrate this point. Finally, show that the sum of the angle measures for the two triangles is the sum of the angle measures for the quadrilateral.

Practice

Name each figure. Name any sides that are congruent or parallel.

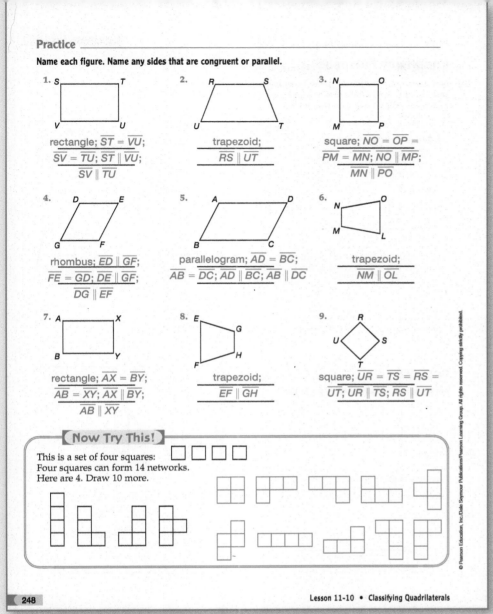

1. S   T / V   U

rectangle; $\overline{ST} = \overline{VU}$;
$\overline{SV} = \overline{TU}$; $\overline{ST} \parallel \overline{VU}$;
$\overline{SV} \parallel \overline{TU}$

2. R   S / U   T

trapezoid;
$\overline{RS} \parallel \overline{UT}$

3. N   O / M   P

square; $\overline{NO} = \overline{OP} = $
$\overline{PM} = \overline{MN}$; $\overline{NO} \parallel \overline{MP}$;
$\overline{MN} \parallel \overline{PO}$

4. D   E / G   F

rhombus; $\overline{ED} \parallel \overline{GF}$;
$\overline{FE} = \overline{GD}$; $\overline{DE} \parallel \overline{GF}$;
$\overline{DG} \parallel \overline{EF}$

5. A   D / B   C

parallelogram; $\overline{AD} = \overline{BC}$;
$\overline{AB} = \overline{DC}$; $\overline{AD} \parallel \overline{BC}$; $\overline{AB} \parallel \overline{DC}$

6. N   O / M   L

trapezoid;
$\overline{NM} \parallel \overline{OL}$

7. A   X / B   Y

rectangle; $\overline{AX} = \overline{BY}$;
$\overline{AB} = \overline{XY}$; $\overline{AX} \parallel \overline{BY}$;
$\overline{AB} \parallel \overline{XY}$

8. E   G / F   H

trapezoid;
$\overline{EF} \parallel \overline{GH}$

9. R U S T

square; $\overline{UR} = \overline{TS} = \overline{RS} = $
$\overline{UT}$; $\overline{UR} \parallel \overline{TS}$; $\overline{RS} \parallel \overline{UT}$

Now Try This!

This is a set of four squares:
Four squares can form 14 networks.
Here are 4. Draw 10 more.

Lesson 11-10 • Classifying Quadrilaterals

---

3 Practice

Have students complete all the exercises. Encourage them to use the models on page 247 to help name the figures.

Now Try This! Have students draw and cut out four squares of equal size to help them draw the ten networks.

4 Assess

Ask students to name the quadrilateral that has one pair of parallel sides. (trapezoid)

For Mixed Abilities

Common Errors • Intervention

Some students may confuse the different quadrilaterals. Draw the following polygons on the board: trapezoid, rhombus, parallelogram, rectangle, and square. Have students take turns to name each figure, describe its characteristics, and name a common object that has the same shape.

Enrichment • Number Sense

Ask students to find the number of diagonals in a triangle, a quadrilateral, a pentagon, a hexagon, a septagon, an octagon, a nonagon, and a decagon. Have them organize this data in a table. Challenge students to find a pattern that determines the number of diagonals. (number of diagonals = $\frac{1}{2}$ × number of vertices × number of vertices − 3)

Ask students to find the sum of the angle measures of a polygon. Have them draw:

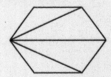

Have students find the sum of the measures of the hexagon. Then, write a formula. (720°; the number of triangles multiplied by 180° is the sum of the angle measures, $(n - 2) \times 180°$.)

More to Explore • Geometry

Duplicate the following toothpick puzzles to challenge students:

1. Here are 15 toothpicks forming 5 squares. Remove only 3 toothpicks so that there are 3 squares left.

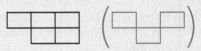

2. Move, but do not remove, two toothpicks to make this statement true.

$|\,|\,| = |$ $\left(\frac{|}{|} = |\right)$

T248

11-11 Congruent Polygons

pages 249–250

1 Getting Started

Objective
- To identify congruent polygons

Vocabulary
congruent figures, corresponding parts

Materials
tracing paper; graph paper; protractors; rulers; colored markers

Warm Up • Mental Math
Have students compute the sum.

1. $\frac{2}{7} + \frac{3}{7}$ $\left(\frac{5}{7}\right)$
2. $\frac{3}{8} + \frac{1}{4}$ $\left(\frac{5}{8}\right)$
3. $\frac{1}{3} + \frac{1}{6}$ $\left(\frac{1}{2}\right)$
4. $\frac{3}{5} + \frac{1}{15}$ $\left(\frac{2}{3}\right)$
5. $\frac{3}{10} + \frac{1}{5}$ $\left(\frac{1}{2}\right)$
6. $\frac{1}{2} + \frac{2}{5}$ $\left(\frac{9}{10}\right)$
7. $\frac{5}{12} + \frac{1}{3}$ $\left(\frac{3}{4}\right)$
8. $\frac{2}{5} + \frac{7}{15}$ $\left(\frac{13}{15}\right)$

Warm Up • Paper and Pencil
Have students simplify each expression.

1. $3 \times 2 + 9$ (15)
2. $(150 \div 10) \times 5$ (75)
3. $(\frac{2}{7}$ of $35) + 6$ (16)
4. $(0.5 \times 96) \div 8$ (6)
5. $(3 \times 16) + 3$ (51)

2 Teach

Introduce the Lesson Have a student read the first paragraph. Then, give students tracing paper and ask them to trace △ABC. Have them slide the tracing paper to match △PQR. Point out that since the tracings coincide, and the triangles have the same size and shape, they are *congruent*. Then, have students identify the corresponding sides and angles of △ABC and △PQR. Stress that congruent sides have the same length and congruent angles have the same measure.

Develop Skills and Concepts Emphasize that congruent figures are exactly the same. Point out that recognizing congruent figures is not always easy because they are not always in the same position. Tell students that position or motion does not change the size or the shape of a figure. Explain that tracings are very helpful in determining congruency. Trace a large trapezoid on the board several times, turning and flipping the pattern each time to show congruency.

3 Practice

Have students complete all the exercises. Provide them with tracing paper to help them identify congruent figures, congruent sides, and congruent angles.

Name _____ Lesson 11-11

Congruent Polygons

Figures that are the same size and shape are called **congruent figures**. The parts of each figure that match are called **corresponding parts**.

△ABC is congruent to △PQR.

We write: △ABC ≅ △PQR.

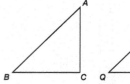

Corresponding Sides	Corresponding Angles
$\overline{AB} \leftrightarrow \overline{PQ}$	$\angle A \leftrightarrow \angle P$
$\overline{BC} \leftrightarrow \underline{QR}$	$\angle B \leftrightarrow \underline{\angle Q}$
$\overline{AC} \leftrightarrow \underline{PR}$	$\angle C \leftrightarrow \underline{\angle R}$

The corresponding parts of congruent figures are themselves congruent.

$\overline{AB} \cong \overline{PQ}$	$\angle A \cong \angle P$	We can conclude:
$\overline{BC} \cong \underline{QR}$	$\angle B \cong \underline{\angle Q}$	$\triangle ABC \cong \underline{\triangle PQR}$
$\overline{AC} \cong \underline{PR}$	$\angle C \cong \underline{\angle R}$.	

Getting Started

Are the figures congruent? Write yes or no.

1. 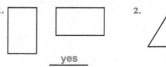 ___yes___

2.   ___no___

3.   ___yes___

These figures are congruent.

4. Name the congruent sides.

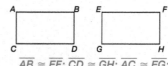

$\overline{AB} \cong \overline{EF}; \overline{CD} \cong \overline{GH}; \overline{AC} \cong \overline{EG};$
$\overline{BD} \cong \overline{FH}$

5. Name the congruent angles.

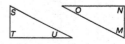

$\angle S \cong \angle M; \angle T \cong \angle N; \angle U \cong \angle O$

Lesson 11-11 • Congruent Polygons

Practice

Are the figures congruent? Write yes or no.

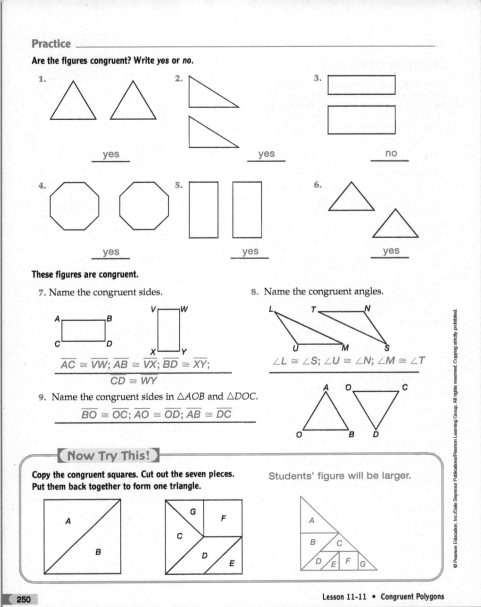

1. yes

2. yes

3. no

4. yes

5. yes

6. yes

These figures are congruent.

7. Name the congruent sides.

$\overline{AC} \cong \overline{VW}; \overline{AB} \cong \overline{VX}; \overline{BD} \cong \overline{XY};$

$\overline{CD} \cong \overline{WY}$

8. Name the congruent angles.

$\angle L \cong \angle S; \angle U \cong \angle N; \angle M \cong \angle T$

9. Name the congruent sides in $\triangle AOB$ and $\triangle DOC$.

$\overline{BO} \cong \overline{OC}; \overline{AO} \cong \overline{OD}; \overline{AB} \cong \overline{DC}$

Now Try This!

Copy the congruent squares. Cut out the seven pieces.
Put them back together to form one triangle.

Students' figure will be larger.

Lesson 11-11 • Congruent Polygons

Now Try This! Be sure students understand that they must use all seven pieces to form one triangle. Students may work in pairs to solve this puzzle.

4 Assess

Have students draw two congruent rectangles and explain why the rectangles are congruent. (Check students' drawings. Explanations should include congruent angles and congruent sides.)

For Mixed Abilities

Common Errors • Intervention

Some students may have difficulty naming the corresponding parts of congruent polygons. Have them work with partners, using pairs of labeled congruent polygons. Have students use different-colored markers to color and name each pair of corresponding parts.

Enrichment • Geometry

Draw the following pair of similar triangles:

Ask students to measure the angles of these triangles and tell what they notice. (m $\angle ACB$ = m$\angle EGF$; m$\angle ABC$ = $\angle EFG$: $\angle CAB$ = m$\angle GEF$) Ask how the lengths of the sides are related. (Each side of $\triangle EFG$ is half as long as its corresponding side in $\triangle ABC$.)

Challenge students to use protractors and rulers to draw other pairs of similar triangles.

More to Explore • Application

Have students bring a mail-order catalog to class. Discuss the prices listed on various items. Ask how many end in $0.98 or $0.99, and why they think merchants mark prices like this instead of in whole numbers. ($2.98 instead of $3.00) Have students choose ten items on the page and round the prices to the nearest dollar. Then, have them estimate the total. Remind them to include the local sales tax. (Tax rate can be expressed as a decimal.) After students have estimated the total price, have them use a calculator to compute the actual total. As an extension, have students create word problems for other students to estimate the answers by rounding the prices to the nearest dollar.

T250

11-12 Graphing Ordered Pairs

pages 251–252

1 Getting Started

Objective
• To name, locate, and graph ordered pairs

Vocabulary
graph

Materials
graph paper; rulers

Warm Up • Mental Math
Have students name the next three terms in the sequence.

1. $\frac{5}{6}, \frac{10}{12}, \frac{20}{24}$, $\left(\frac{40}{48}, \frac{80}{96}, \frac{160}{192}\right)$

2. 2, 13, 24, (35, 46, 57)

3. $3, 1\frac{1}{2}, \frac{3}{4}$, $\left(\frac{3}{8}, \frac{3}{16}, \frac{3}{32}\right)$

4. 0.7, 1.4, 2.1 (2.8, 3.5, 4.2)

5. 500, 100, 20 $\left(4, \frac{4}{5}, \frac{4}{25}\right)$

Warm Up • Pencil and Paper
Have students use the cross-products test to compare the following pairs of fractions:

1. $\frac{3}{8} \ (=) \ \frac{9}{24}$
2. $\frac{2}{9} \ (<) \ \frac{4}{16}$
3. $\frac{8}{10} \ (>) \ \frac{15}{19}$
4. $\frac{4}{5} \ (=) \ \frac{12}{15}$
5. $\frac{5}{7} \ (>) \ \frac{2}{5}$
6. $\frac{5}{6} \ (<) \ \frac{8}{9}$
7. $\frac{5}{12} \ (>) \ \frac{16}{39}$
8. $\frac{3}{4} \ (=) \ \frac{15}{20}$

Name _____

Lesson 11-12

It's Algebra!

Graphing Ordered Pairs

We can use an ordered pair of numbers to locate points on a plane. The ordered pair (2, 3) locates point P. The ordered pair (0, 0) is called the origin. Points P, A, and B are on the same straight line or **graph**. What ordered pairs locate points A and B?

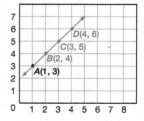

To locate points on a plane, follow these steps.

Find the horizontal distance from the origin.	Find the vertical distance from the origin.

Point A is 3 units from the origin.
Point A is 4 units from the origin.

Point A is located at ___(3, 4)___.
Point B is located at ___(5, 6)___.

We can graph a table of ordered pairs by joining points on the same line. Complete the table and the graph.

	A	B	C	D
n	1	2	3	4
$n + 2$	3	4	5	6

Getting Started

Name each point.

1. A (_2_ , _2_)   2. B (_3_ , _6_)

3. C (_6_ , _2_)   4. D (_7_ , _0_)

Locate and label each point.

5. W (5, 6)   6. X (0, 4)

7. Y (3, 3)   8. Z (5, 1)

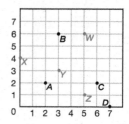

Lesson 11-12 • Graphing Ordered Pairs

251

---

2 Teach

Introduce the Lesson Direct attention to the two number lines drawn perpendicular to each other that form the graph at the top of page 251. Explain that points may be located anywhere on the plane surface by referring to the two number lines. Have a student read the first paragraph.

Point out that the numbers in parentheses form an ordered pair. Emphasize that the order is the key to locating points on a plane. Explain that the first number in an ordered pair shows the horizontal distance from the origin; the second number shows the vertical distance from the origin. The point is the intersection of the horizontal and vertical distances. Work through the steps for locating points with students.

Develop Skills and Concepts Have students move their pencil from the origin first horizontally, then vertically to locate and name each point on the plane. Refer to the table on page 251. Ask students what the rule is. ($n + 2$) Emphasize that the ordered pair is (n, $n + 2$). Have students note that the line joining these points forms a diagonal line.

Be sure students understand the importance of the order of the numbers in an ordered pair because (4, 2) is not the same as (2, 4). To demonstrate, have students locate (4, 2) and then (2, 4) on the graph next to the table on page 251.

It's Algebra! The concepts in this lesson prepare students for algebra.

3 Practice

Have students complete all the exercises. Remind them that to graph the points in the table, they draw a straight line.

Practice

Name each point.

1. A (_2_ , _3_) 2. B (_11_ , _9_)

3. C (_6_ , _0_) 4. D (_3_ , _8_)

5. E (_12_ , _4_) 6. F (_7_ , _5_)

7. G (_8_ , _9_) 8. H (_1_ , _10_)

9. I (_4_ , _2_) 10. J (_0_ , _5_)

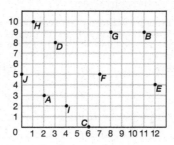

Locate and label each point.

11. P (3, 5) 12. Q (6, 2)

13. R (0, 4) 14. S (5, 5)

15. T (11, 2) 16. U (5, 0)

17. V (3, 10) 18. W (9, 6)

19. X (7, 1) 20. Y (12, 10)

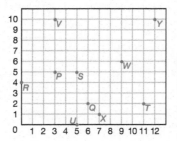

Problem Solving

Complete the table and graph each ordered pair.

21.

	A	B	C	D
n	2	4	6	8
1 × n	2	4	6	8

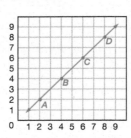

Lesson 11-12 • Graphing Ordered Pairs

4 Assess

Provide students with the coordinates of several points. Have students locate and label these points on a coordinate plane.

5 Mixed Review

1. 138 × 605 (83,490)
2. $3\frac{3}{8} \times 5\frac{1}{9}$ $(17\frac{1}{4})$
3. 38 + 17 (55)
4. 42,840 ÷ 71 (603 R27)
5. 18 − 9 × 2 (0)
6. $12 \div \frac{8}{15}$ $(22\frac{1}{2})$
7. 8 + 26 (34)
8. 20,100 − 12,694 (7,406)
9. 0.028 × 15 (0.42)
10. 98% of 50 (49)

For Mixed Abilities

Common Errors • Intervention

Some students may reverse the horizontal and vertical distances when graphing an ordered pair. Have them work with partners to graph ordered pairs that form simple figures. For example, the following points, when joined, will form a triangle with point A on the horizontal axis and point B on the vertical axis: A(4, 0), B(0, 10), C(8, 12).

Enrichment • Algebra

Provide graph paper and rulers. Have students draw perpendicular lines and label the origin. Then, have them label both lines from 1 through 10. Have students locate these points:

$A = (\frac{21}{7}, \frac{21}{3})$

$B = (\frac{1}{2}$ of 12, $\frac{1}{4}$ of 28)

$C = (\frac{72}{9}, \frac{125}{25})$

$D = (\frac{1}{7}$ of 49, $\frac{1}{3}$ of 9)

$E = (3 \times 1, 1 \times 3)$

$F = (\frac{68}{34}, \frac{55}{11})$

Have students connect the points in order, connecting F to A. Then, have them name the figure that is formed. (hexagon)

More to Explore • Statistics

Divide students into groups. Have groups research the frequency of election for various government officials, such as mayor, governor, senator, or president. Discuss the time differences and the reasons for each. Next, have students use the data gathered to determine the year of the first presidential election in which they will be able to vote. Have them then consider each of their parents' birth dates to determine the first year their parents voted in a presidential election and research who won.

T252

1 Getting Started

Objective
- To perform the transformations of translation, reflection, and rotation

Vocabulary
transformation, translation, reflection, rotation

Materials
graph paper

Warm Up • Mental Math
Have students name the next number in each sequence.
1. 95, 80, 65, 50, 35, 20 (5)
2. 72, 68, 65, 61, 58, 54 (51)
3. 3, 6, 8, 16, 18, 36, 38 (76)
4. 5, 4, 12, 11, 33, 32, 96 (95)
5. 49, 7, 36, 6 ,25, 5 (16)

Warm Up • Pencil and Paper
Have students simplify each fraction.
1. $\frac{15}{20}$ $(\frac{3}{4})$
2. $\frac{12}{18}$ $(\frac{2}{3})$
3. $\frac{25}{40}$ $(\frac{5}{8})$
4. $\frac{14}{35}$ $(\frac{2}{5})$
5. $\frac{18}{36}$ $(\frac{1}{2})$
6. $\frac{24}{60}$ $(\frac{2}{5})$
7. $\frac{27}{63}$ $(\frac{3}{7})$
8. $\frac{15}{50}$ $(\frac{3}{10})$

Name _____

Transformations

Byran traces a triangle on a coordinate plane. He labels the vertices A, B, and C. Then, he moves the triangle 4 units to the right. He traces a new triangle, and labels the vertices A', B', and C'.

The process by which a shape changes is called a **transformation**. $\triangle ABC$ becomes $\triangle A'B'C'$. The symbol A' is read, "A prime." This transformation is a slide or **translation**. Write the ordered pair for each vertex.

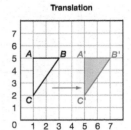
Translation

A (__1__ , __5__) A' (__5__ , __5__)
B (__3__ , __5__) B' (__7__ , __5__)
C (__1__ , __2__) C' (__5__ , __2__)

Here are two more transformations. The flip is called a **reflection**. The turn is called a **rotation**.

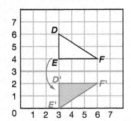
Reflection

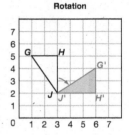
Rotation

Getting Started

Write the ordered pair for each vertex.

1. D (__3__ , __6__) D' (__3__ , __2__)
   E (__3__ , __4__) E' (__3__ , __0__)
   F (__6__ , __4__) F' (__6__ , __2__)

2. G (__1__ , __5__) G' (__6__ , __4__)
   H (__3__ , __5__) H' (__6__ , __2__)
   J (__3__ , __2__) J' (__3__ , __2__)

---

2 Teach

Introduce the Lesson Have a student read the opening paragraph. Read the second paragraph, explaining that the entire triangle was shifted and the three vertices are being used as guide points. Write the ordered pairs for the vertices of $\triangle ABC$ on the board. Have students find the ordered pairs for $\triangle A'B'C'$.

Draw the reflection and rotation examples on the board. Explain that the reflection is just a flip over the horizontal line. Since vertex E is one unit above the line, E' will be one unit below the line. For the rotation, tell students that $\triangle GHJ$ was turned one quarter (90°) in the clockwise direction. Students should see that the base of $\triangle GHJ$ forms a 90° angle with the base of $\triangle G'H'J'$.

Develop Skills and Concepts Draw a grid on the board and then draw $\triangle NOP$ on the grid, with vertices (1, 2), (3, 5), and (3, 2), respectively. Have students work in pairs to draw the following transformations:

1. slide $\triangle NOP$ left 3 units

2. flip $\triangle NOP$ over the x-axis

When students have completed the slide and flip, explain how to rotate $\triangle NOP$ a quarter turn clockwise. Draw $\triangle NOP$ on a grid on the board. The base of $\triangle NOP$ must form a 90° angle with the base of $\triangle N'O'P'$. Have students picture segment NP as the hand of a clock as you show them a quarter turn. Draw the base for $\triangle N'O'P'$. Vertex P' coincides with P. N' should be 2 units from P' as N is from P. N' is located at (3, 4). O' should be 3 units away from P', as O is from P. Vertex O' is located at (6, 2).

Practice

Do each transformation. Then, write the ordered pair for each vertex.

1. Slide △ABC down 3 units.

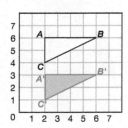

A (__2__ , __6__)   A' (__2__ , __3__)

B (__6__ , __6__)   B' (__6__ , __3__)

C (__2__ , __4__)   C' (__2__ , __1__)

2. Flip △DEF over the green line.

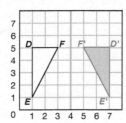

D (__1__ , __5__)   D' (__7__ , __5__)

E (__1__ , __1__)   E' (__7__ , __1__)

F (__3__ , __5__)   F' (__5__ , __5__)

3. Rotate △GHI a quarter turn counterclockwise.

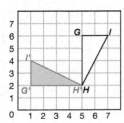

G (__5__ , __6__)   G' (__1__ , __2__)

H (__5__ , __2__)   H' (__5__ , __2__)

I (__7__ , __6__)   I' (__1__ , __4__)

4. Slide △KLM down 2 units and to the right 3 units.

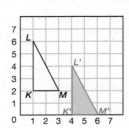

K (__1__ , __2__)   K' (__4__ , __0__)

L (__1__ , __6__)   L' (__4__ , __4__)

M (__3__ , __2__)   M' (__6__ , __0__)

Lesson 11-13 • Transformations

For Mixed Abilities

Common Errors • Intervention

Some student may try to draw the entire triangle in one step. Have these students draw △ABC in Exercise 1 on graph paper. Draw the horizontal line $y = 3$ for them, and ask them to flip △ABC over the line. Have them start with A and ask themselves, "Where does vertex A go when it is flipped across the line?" (2, 0) Have them label this point A'. Ask similar questions for vertices B and C. Then, have students connect the vertices to draw the triangle.

Enrichment • Geometry

Divide the class into groups of four. Each group is to make two transformation cards by following these steps:

1. Draw a red triangle on graph paper. Label each vertex with an ordered pair.

2. Using a pencil, perform a transformation on the red triangle.

3. Using a blue pen, perform another transformation. Label its vertices.

4. Erase the triangle drawn in pencil.

Groups should provide you with the transformations on their cards, such as "shift up 3, reflect over vertical line." Cards are exchanged until each group has done all other cards.

More to Explore • Biography

René Descartes was a French man, born in 1596. Descartes was often sick as a child and was allowed to sleep in every morning. According to legend, one day while lying in bed, Descartes watched a fly crawling on the ceiling. He tried to figure out how he could explain the fly's position. He realized that he could use the fly's distance from the walls to describe the position of the fly. The best way to do this was to draw a grid, place the fly on a point, and assign an ordered pair to its position. The coordinate plane was invented!

---

3 Practice

Have students complete all the exercises. Remind students to use the three vertices as guide points. Students should decide where the new vertices will be and then draw the triangle. When doing a rotation, students should first take the base of the triangle and decide what happens to it when it is rotated.

4 Assess

Ask students how they could check that they have done a flip correctly. (They can fold their paper along the line of reflection.) Ask students, *What would happen if you rotated a triangle a quarter turn clockwise four times?* (You would have the original triangle.) Ask students, *What would happen if you slid a triangle 2 units down, 3 units right, 2 units up, and then 3 units left?* (You would have the original triangle.)

T254

1 Getting Started

Objective
- To restate problems in more understandable terms

Warm Up • Mental Math
Have students work from left to right.

1. $\frac{1}{6}$ of $42 \times 8 - 1 \div 11 +$ itself $- 1$ (9)

2. $\frac{1}{8}$ of $40 \times 5 + 3 \div 4 + 1 \times$ itself (64)

3. $\frac{1}{10}$ of $19 + 0.1 \times$ itself $+ 1 - 5$ (0)

4. $\frac{3}{5}$ of $30 - 3 \div 5 \times$ itself $- 1$ (8)

5. $\frac{2}{3}$ of $60 - 4 \div 9 - 4 \times 56$ (0)

6. $\frac{3}{7}$ of $28 - 12 \times 45 \div 45$ (0)

7. $\frac{4}{9}$ of $72 + 3 \times 2 \div 7 - 1$ (9)

Warm Up • Paper and Pencil
Have students do the calculation.

1. 3×29 (87)

2. $\$3.65 + \$1.52$ (\$5.17)

3. 4×83 (332)

4. $\$4.46 - (\$1.83 + \$0.97)$ (\$1.66)

5. $5 + (\frac{1}{2}$ of $62)$ (36)

6. $\$3.00 - (\$0.54 + \$0.68)$ (\$1.78)

7. 9×56 (504)

8. $4 + (\frac{1}{2}$ of $48)$ (28)

Name _____

Problem Solving: Restate the Problem

Jonathan bought a ball, a book, and a drum for $26.25. The cost of the book was $\frac{1}{6}$ the cost of the drum; the cost of the ball was $\frac{1}{2}$ the cost of the book. What did each article cost?

⭐ SEE
We want to know the cost of each item.

He spent $\underline{\$26.25}$ altogether. The cost of the book was $\underline{\frac{1}{6}}$ the cost of the drum and the ball was $\underline{\frac{1}{2}}$ the cost of the book.

⭐ PLAN
We can restate the problem in our own words. The guess and check strategy and making a table will then be helpful in reaching a solution.

⭐ DO
The drum costs $\underline{6}$ times as much as the book. The book cost $\underline{2}$ times as much as the ball. We can use a table to help organize our guessing and checking.

Cost of Articles			
Cost of Ball	Cost of Book	Cost of Drum	Total Cost (26.25)
$1.00	$2.00	$12.00	$15.00
$1.50	$3.00	$18.00	$22.50
$2.00	$4.00	$24.00	$30.00
$1.75	$3.50	$21.00	$26.25

The ball cost $\underline{\$1.75}$. The book cost $\underline{\$3.50}$. The drum cost $\underline{\$21.00}$.

⭐ CHECK
We can check our solution by using the numbers in the original problem.

1. $\frac{1}{6} \times \underline{\$21.00} = \underline{\$3.50}$

2. $\frac{1}{2} \times \underline{\$3.50} = \underline{\$1.75}$

3. $\underline{\$21.00} + \underline{\$3.50} + \underline{\$1.75} = \underline{\$26.25}$

Lesson 11-14 • Problem Solving: Restate the Problem

255

---

2 Teach

Introduce the Lesson Problems like the one used to introduce this strategy are full of terms that are initially confusing. In this kind of problem, it is more important than ever that students restate the question and its facts in understandable terms. Encourage students to freely express what they know without structural and grammatical concerns. Emphasize that they should use understandable expressions and a correct interpretation of the facts.

Develop Skills and Concepts Work through each step to solve the problem with students. Explain that they may use additional tools, such as making a table. Point out that they may refer to the original language of the problem in the CHECK step.

3 Apply

Solution Notes

1. Students should notice that 3 apples cost the same as 18 strawberries. Therefore, 1 apple and 6 strawberries are equal in cost.

2. Students need to realize that Bob's age plus 10 must equal three times his age. By guessing and checking, they will arrive at the solution.

3. Emphasize that the glass costs $0.70 more than the soda to warn students who fall into the trap of subtracting $0.70 from $2.00. They can make a table and use guess and check.

4. Once students realize that 4 days after the day before yesterday is the same as 2 days after today, finding the solution will be much easier.

Apply

Solve each problem.

1. While at the supermarket, I noticed that three apples cost as much as one melon. A melon costs as much as 18 strawberries. How many strawberries cost as much as one apple?
   6 strawberries

2. In ten years Bob will be three times as old as he is now. How old is Bob now?
   5 years old

3. At the amusement park you can buy a cold drink in a souvenir glass for $2.00. The glass cost $0.70 more than the drink. How much does the drink cost?
   $0.65

4. What day is tomorrow if four days after the day before yesterday is Saturday?
   Friday

5. A large block weighs as much as 4 small blocks. Two large blocks and three small blocks weigh 88 pounds. How much does a small block weigh? How much does a large block weigh?
   A small block weighs 8 pounds. A large one weighs 32 pounds.

6. A ruler, a pencil, and a pen cost $2.70. The pencil was $1.00 more than the ruler. The pen was $0.30 more than the total cost of the pencil and ruler. How much did the pen cost?
   $1.50

7. What if, in Problem 3, the glass cost $\frac{1}{4}$ as much as the glass and drink combined? How much would the drink cost then?
   $1.50

8. Ann's Flower Shop is having a sale on mixed bouquets of fresh flowers. Small bouquets are $3.88 and large bouquets are $5.88. Use this information to write a problem where the answer is $0.24.
   See below.

9. Use mental computation to find the number that is twice the number that is one half of the number that is 10 less than 100.
   90

10. Use mental computation to find the number that is double one half of 99 added to the difference between 99 and itself.
    99

8. Possible answer: Ann's Flower Shop is having a sale on mixed bouquets of fresh flowers. Small bouquets cost $3.88; large bouquets cost $5.88. Dan bought one of each kind and paid with a $10 bill. How much change did he receive?

Lesson 11-14 • Problem Solving: Restate the Problem

For Mixed Abilities

More to Explore • Logic

This problem is one that has been discussed by professional mathematicians. Duplicate the following for students:

> There are three explorers: Matthew, Mark, and Luke. They know that two of them will be going on an expedition in the morning. Matthew goes to the leader of the expedition and asks for the name of one of the two who will be going. He explains that knowing one of the names will not give him any extra information because he already knows that at least one of the two will be going. When the leader tells him that Luke is one who will be going, Matthew leaves feeling very good. Why?

(Discuss with students the following explanation: Before Matthew talked to the leader, his chances of going were only 1 in 3, because 2 of the 3 were to be going. After talking to the leader, he figures his chances of going are 1 out of 2, because now he knows it is between Mark and himself.)

---

5. Since a large block weighs as much as 4 small blocks, students can substitute 8 small blocks for the 2 large ones on the scale. Therefore, 11 small blocks weigh 88 lb. One small block weighs 8 lb. Multiplication and subtraction will lead to the weight of 1 large block.

6. Once the prices are clearly related to each other, students can make a table and use the guess and check strategy.

Higher-Order Thinking Skills

7. **Analysis:** The glass costs $0.25, which means that the drink costs $1.75.

8. **Synthesis:** One possible problem is that a customer buys two bouquets, one small and one large, and gives Fernando a $10 bill; the change would be $0.24.

9. **Analysis:** Students should recognize that doubling and taking one-half undo each other. So, the number simply is 10 less than 100.

10. **Analysis:** Students should recognize that doubling and taking one-half negate each other, and the difference between any number and itself is 0; hence, 99 added to 0 is 99.

4 Assess

Have students look back at the problem and the restated problem. Ask, *Why did restating the problem make it easier to understand?* (Possible answer: Working with whole numbers is easier than working with fractions.)

Chapter 11
Test

page 257

Items	Objectives
1–3	To identify lines, segments, and rays (see pages 229–230)
4–6	To identify and name right, obtuse, and acute angles (see pages 231–232)
4–6, 12	To measure angles with a protractor (see pages 233–234)
7	To identify and name perpendicular lines (see pages 235–236)
8	To identify parts of a circle (see pages 237–238)
9	To copy an angle with a compass and a straightedge (see pages 241–242)
10	To bisect a line segment with a compass and a straightedge (see pages 239–240)
11	To identify a quadrilateral and draw its diagonals (see pages 247–248)
13	To identify congruent polygons and their parts (see pages 249–250)
14	To identify the diagonals of a polygon (see pages 243–244)

Alternate Chapter Test

You may wish to use the Alternate Chapter Test on page 390 of this book for further review and assessment.

T257

Name each figure.

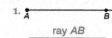

1. •⎯⎯⎯⎯→
   A   B
   _____ray AB_____

2. ←⎯•⎯⎯⎯•⎯→
   C   D
   _____line CD_____

3. •⎯⎯⎯⎯•
   E   F
   _____line segment EF_____

Use a protractor to measure each angle. Then, classify each angle.

4.
   _____60° acute_____

5. 
   _____90° right_____

6.
   _____115° obtuse_____

Complete each statement.

7.
   $\overline{AD}$ is perpendicular to _____$\overline{AB}$_____ .

8. 
   The radius is _____$\overline{OA}$_____ .

9. Copy angle P on $\overrightarrow{AB}$.

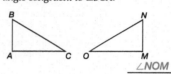

10. Bisect $\overline{MN}$.

11. Identify the polygon. Draw a diagonal.

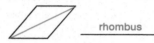 _____rhombus_____

12. Find m∠XZY.

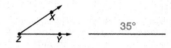 _____35°_____

13. The triangles are congruent. Name the angle congruent to ∠BCA.

∠NOM

14. Draw all possible diagonals.

Chapter 11 • Test

257

Circle the letter of the correct answer.

1.
$$4\tfrac{3}{5}$$
$$+\ 2\tfrac{2}{3}$$

a. $1\tfrac{1}{15}$
b. $4\tfrac{1}{15}$
c. $5\tfrac{1}{15}$
(d.) NG

2.
$$6\tfrac{1}{8}$$
$$-\ 2\tfrac{5}{6}$$

(a.) $3\tfrac{7}{24}$
b. $4\tfrac{7}{24}$
c. $8\tfrac{3}{7}$
d. NG

3. $1\tfrac{1}{2} \times 5\tfrac{1}{3}$

a. $5\tfrac{1}{6}$
(b.) 8
c. $8\tfrac{1}{6}$
d. NG

4. $2\tfrac{1}{3} \div 3\tfrac{1}{2}$

(a.) $\tfrac{2}{3}$
b. $\tfrac{3}{7}$
c. $8\tfrac{1}{6}$
d. NG

5. What is the place value of the 7 in 23.0178?

a. tenths
b. hundredths
(c.) thousandths
d. NG

6.
$$2.61$$
$$0.48$$
$$+\ 5.9$$

a. 8.89
(b.) 8.99
c. 9.00
d. NG

7. $16.5 - 9.36$

a. 6.71
(b.) 7.14
c. 7.26
d. NG

8.
$$3.61$$
$$\times\ 2.5$$

a. 9.05
(b.) 9.025
c. 90.25
d. NG

9. $0.03\overline{)0.762}$

(a.) 25.4
b. 254
c. 2,540
d. NG

10.
$$10\ \text{min}\ 38\ \text{s}$$
$$+\ 6\ \text{min}\ 25\ \text{s}$$

a. 15 min 3 s
b. 16 min 3 s
c. 21 min 3 s
(d.) NG

11.
$$10\ \text{lb}\ 4\ \text{oz}$$
$$-\ 3\ \text{lb}\ 8\ \text{oz}$$

(a.) 6 lb 12 oz
b. 7 lb 4 oz
c. 7 lb 12 oz
d. NG

12. 5 km = _?_ m

a. 50
b. 500
(c.) 5,000
d. NG

☐ score

STOP

page 258

Items	Objectives
1–2	To add and subtract mixed numbers (see pages 119–120, 127–128)
3–4	To multiply and divide mixed numbers (see pages 141–142, 147–148)
5	To identify the place value of a digit in decimals (see pages 165–166)
6–7	To add, subtract, multiply, and divide decimals (see pages 171–173)
8–9	To multiply and divide decimals (see pages 189–190, 197–198)
10	To add units of time (see pages 211–212)
11	To subtract units of weight (see pages 215–216)
12	To change metric units of length to equivalent units (see pages 219–220)

Alternate Cumulative Assessment

Circle the letter of the correct answer.

1.
$$5\tfrac{5}{8}$$
$$+\ \tfrac{7}{12}$$

a $1\tfrac{5}{24}$
b $5\tfrac{5}{24}$
(c) $6\tfrac{5}{24}$
d NG

2.
$$7\tfrac{3}{10}$$
$$-\ 4\tfrac{7}{15}$$

a $2\tfrac{4}{5}$
(b) $2\tfrac{5}{6}$
c $2\tfrac{13}{15}$
d NG

3. $9\tfrac{4}{5} \times 1\tfrac{3}{7}$

a $9\tfrac{12}{35}$
(b) 14
c $1\tfrac{4}{5}$
d NG

4. $8\tfrac{1}{10} \div 4\tfrac{1}{2}$

a $1\tfrac{2}{5}$
b $1\tfrac{3}{5}$
(c) $1\tfrac{4}{5}$
d NG

5. Give the place value of the 8 in 49.13789.

a thousandths
(b) ten thousandths
c hundred thousandths
d NG

6.
$$5.72$$
$$0.69$$
$$+\ 6.8$$

(a) 13.21
b 13.11
c 12.21
d NG

7.
$$14.6$$
$$-\ 8.93$$

a 5.57
(b) 5.67
c 5.73
d NG

8.
$$4.73$$
$$\times\ 5.6$$

a 2.6488
(b) 26.488
c 264.88
d NG

9. $0.04\overline{)0.2912}$

(a) 7.28
b 72.8
c 728
d NG

10.
$$22\ \text{min}\ 46\ \text{s}$$
$$+\ 7\ \text{min}\ 28\ \text{s}$$

a 29 min 14 s
(b) 30 min 14 s
c 30 min 15 s
d NG

11.
$$7\ \text{lb}\ 6\ \text{oz}$$
$$-\ 5\ \text{lb}\ 12\ \text{oz}$$

a 3 lb 6 oz
b 2 lb 6 oz
(c) 1 lb 10 oz
d NG

12-1 Perimeter

pages 259–260

1 Getting Started

Objectives
- To find the perimeter of a plane figure by adding sides
- To find the perimeter of a rectangle by using a formula

Vocabulary
perimeter

Materials
graph paper

Warm Up • Mental Math
Have students rename the measure in simplest form.
1. 71 s (1 min 11 s)
2. 31 in. (2 ft 7 in.)
3. 13 d (1 wk 6 d)
4. 1 yd 5 ft (2 yd 2 ft)
5. 95 in. (2 yd 1 ft 11 in.)
6. 125 min (2 h 5 min)

Warm Up • Pencil and Paper
Have students complete the following exercises:

1.		2.		3.	
	13		39		32
	12		40		28
	23		27		32
	+ 17		+ 25		+ 28
	(65)		(131)		(120)

4. $(2 \times 4\frac{1}{2}) + 3\frac{2}{5}$ $(12\frac{2}{5})$
5. $2\frac{1}{2} + (2 \times 1\frac{1}{3})$ $(5\frac{1}{6})$
6. $(2 \times 4.2) + 3.9$ (12.3)
7. $6.7 + (2.3 \times 2)$ (11.3)
8. $2 \times (\frac{4}{5} + \frac{3}{10})$ $(2\frac{1}{5})$

2 Teach

Introduce the Lesson Have a student read the problem and tell what is to be solved. Have another student identify the information given in the problem and in the picture. (The site is 40 ft long and 25 ft wide.) Ask students to complete the information sentences as they read aloud with you to solve the problem. Discuss the two methods shown to find the perimeter of the rectangle.

Develop Skills and Concepts Emphasize that perimeter is the sum of the measures of the sides of a plane, or two-dimensional (flat), figure. Ask students to name a plane figure that does not have a perimeter. (circle; its sides are

---

Perimeter

Lesson 12-1

It's Algebra!

Professor Landon is putting a fence around the archaeological dig he started by the river. The distance around a flat region is called the **perimeter**. What is the perimeter of Professor Landon's dig?

We want to know the perimeter, or the distance around the site.

We know that the length of the dig is __40__ feet and it is __25__ feet wide.

To find the perimeter of a rectangle, we find the **sum** of the lengths of the sides.

```
  40 ft
  25 ft
  40 ft
+ 25 ft
 130 ft
```

Or we can use the formula:

Perimeter = 2 × length + 2 × width or
$P = 2 \times (l + w)$
$P = 2 \times (40 + 25)$
$P = 2 \times 65$
$P = \underline{130}$

The perimeter of the dig is __130__ feet.

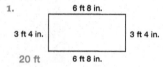

Getting Started

Find the perimeter of each figure.

1.
6 ft 8 in.
3 ft 4 in.   3 ft 4 in.
20 ft   6 ft 8 in.

2.
10 in.
12 in.
17 in.   9 in.
72 in.   5 in.
19 in.

Find the perimeter of each rectangle.

3. Length: 25 in.
   Width: 16 in.
   82 in.

4. Length: $3\frac{1}{2}$ ft
   Width: 7 ft
   21 ft

5. Length: 6 yd
   Width: 6 yd
   24 yd

Lesson 12-1 • Perimeter

259

---

not lines.) Note that a perimeter separates a plane into three sets of points: the points on the perimeter, the points inside the perimeter, and the points outside the perimeter.

Have students describe some situation in which they need to know the perimeter of a figure. (building a fence around a yard, making a picture frame) Have a student explain why the formula $2 \times (l + w)$ can be used to find the perimeter of any rectangular figure. (The two lengths are the same and the two widths are the same.) Have students derive a formula for finding the perimeter of a square (4s) and a regular octagon (8s).

It's Algebra! The concepts in this lesson prepare students for algebra.

3 Practice

Have students complete all the exercises. Remind students that they can add only similar units.

Practice

Find the perimeter of each figure.

1.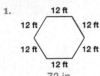
12 ft, 12 ft, 12 ft, 12 ft, 12 ft, 12 ft
72 in.

2.
6 ft 3 in., 12 ft 6 in., 12 ft 8 in.
31 ft 5 in.

3.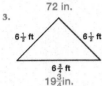
$6\frac{1}{2}$ ft, $6\frac{1}{2}$ ft, $6\frac{3}{4}$ ft
$19\frac{3}{4}$ in.

4.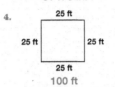
25 ft, 25 ft, 25 ft, 25 ft
100 ft

Find the perimeter of each rectangle.

5. Length: 21 ft
   Width: 7 ft
   56 ft

6. Length: $6\frac{3}{4}$ in.
   Width: $8\frac{1}{2}$ in.
   $30\frac{1}{2}$ in.

7. Length: 3.5 ft
   Width: 1.6 ft
   10.2 ft

8. Length: $12\frac{2}{3}$ ft
   Width: $8\frac{1}{2}$ ft
   $42\frac{1}{3}$ ft

9. Length: 16 in.
   Width: 24 in.
   80 in.

10. Length: 16 in.
    Width: $2\frac{1}{2}$ ft
    92 in. or $7\frac{2}{3}$ ft

11. Length: 27 ft
    Width: 18 in.
    684 in. or 57 ft

12. Length: $2\frac{1}{4}$ in.
    Width: $8\frac{1}{2}$ in.
    $21\frac{1}{2}$ in.

13. Length: $17\frac{3}{8}$ ft
    Width: $12\frac{2}{5}$ ft
    $59\frac{11}{20}$ ft

Problem Solving

Solve each problem. Draw a picture to help.

14. A lot in the shape of a triangle is 48 feet on each side. What is the perimeter of the lot?
    144 feet

15. The width of a rectangle is twice its length. The perimeter is 48 inches. How wide is the rectangle?
    16 inches

16. The Martins are fencing in a rectangular yard. The cost of the fence is $8.75 per foot. If the yard is 25 feet long and 30 feet wide, what will it cost to fence the yard?
    $962.50

17. Charmaine is buying fabric to make a dress. She needs a piece of silk $4\frac{3}{8}$ yards long by 36 inches wide. How many yards is the perimeter of Charmaine's fabric?
    $10\frac{3}{4}$ yd

18. A square has a perimeter of 64 inches. How long is each side?
    16 inches

19. Bela is putting a border of tile around her bathroom. Each tile is 6 inches wide. How many tiles will Bela need if the border of her bathroom is 6 feet long and 8 feet wide?   52 tiles

Lesson 12-1 • Perimeter

260

For Mixed Abilities

Common Errors • Intervention

When finding the perimeter of a rectangle, students may forget to add two lengths and two widths. Have them write the number of sides, write a measure for each side, and then add.

Enrichment • Number Sense

Have students solve these problems:

1. If the distance around a baseball diamond (a square) is 360 ft, how long is one side? (90 ft)

2. A regular hexagon has a perimeter of 96 inches. A regular pentagon has sides the same length as the hexagon. What is the perimeter of the pentagon? (80 in.)

3. Fencing costs $2.46 a foot. How much will it cost to enclose a rectangular garden $9\frac{1}{2}$ ft long and 42 in. wide? ($63.96)

4. Ask students to complete the chart.

l	w	P
3.4 ft	15 in.	(9.3 ft)
(3.5 ft)	1.5 ft	120 in.
9 ft	(7 yd)	20 yd

More to Explore • Geometry

In order to help your students understand the importance of rectangular solids, have them bring in examples of some found in everyday life. Some examples might be a puzzle cube, gift box, or cereal box. Display the examples and challenge your students to find the volume ($V = l \times w \times h$) and surface area of each.

---

4 Assess

Write the formula for the perimeter of a rectangle on the board.

$$P = 2 \times (l + w)$$

Have students use the formula to find the perimeter of a rectangle with a length of 20 ft and a width of 30 ft. (100 ft)

5 Mixed Review

1. $9 \times 2,765$ (24,885)
2. $28 + 37 + 19 + 48$ (132)
3. 0.095×10^4 (950)
4. $0.01035 \div 0.23$ (0.045)
5. $23 - 14\frac{7}{8}$ ($8\frac{1}{8}$)
6. $256.37 \div 10^2$ (2.5637)
7. $14 \div 1\frac{3}{4}$ (8)
8. 16.7×0.69 (11.523)
9. $\frac{7}{9} + \frac{7}{12}$ ($1\frac{13}{36}$)
10. $65,321 - 14,078$ (51,243)

12-2 Circumference

pages 261–262

1 Getting Started

Objective
• To find the circumference of a circle when the diameter is known

Vocabulary
circumference, diameter, pi

Materials
*string; *3 jar lids (same size); compasses and rulers; construction paper

Warm Up • Mental Math
Have students name the product.

1. $\frac{1}{2} \times \frac{2}{3}$ $(\frac{1}{3})$ 5. $\frac{3}{10} \times \frac{2}{3}$ $(\frac{1}{5})$

2. $\frac{3}{5} \times \frac{1}{3}$ $(\frac{1}{5})$ 6. $\frac{1}{8} \times \frac{2}{3}$ $(\frac{1}{12})$

3. $\frac{5}{6} \times \frac{1}{3}$ $(\frac{5}{18})$ 7. $\frac{2}{5} \times \frac{1}{2}$ $(\frac{1}{5})$

4. $\frac{3}{4} \times \frac{1}{2}$ $(\frac{3}{8})$ 8. $\frac{1}{12} \times \frac{1}{3}$ $(\frac{1}{36})$

Warm Up • Pencil and Paper
Have students write the product.

1. 4.6×7 (32.2)
2. 5.61×0.29 (1.6269)
3. 39.2×3.6 (141.12)
4. 7.12×8.1 (57.672)
5. 36.14×4.38 (158.2932)
6. 0.0673×0.98 (0.065954)

Name _____

Lesson 12-2

Circumference

It's Algebra!

The distance around a circle is called its **circumference**. The distance across the center of a circle is called its **diameter**. About how many times larger is the circumference of a circle than its diameter?

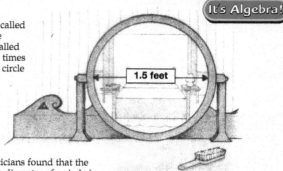

1.5 feet

The ancient Greek mathematicians found that the circumference divided by the diameter of a circle is always the same. They named this number with the Greek letter **pi**, π. We use 3.14 as the approximate value of π.

The circumference of a circle is ___3.14___ times larger than its diameter. This comparison can be shown by the **formula**:

Circumference = pi times diameter
$C = \pi \times d$
$C = 3.14 \times d$

To find the circumference of the bedroom mirror, we multiply pi by the diameter. We multiply ___3.14___ by ___1.5___.

```
    3.14
  × 1.5
    4.71
```

The circumference of the mirror is ___4.71 feet___.

Getting Started

Find the circumference of each circle.

1. $d = 4$ in.   12.56 in.

2. $d = 2.5$ yd   7.85 yd

3. $d = 2.25$ ft   7.065 ft

Complete the table.

4.

diameter	5.1 in.	7 ft	4.25 yd	8 ft
circumference	16.014 in.	21.98 ft	13.345 yd	25.12 ft

Lesson 12-2 • Circumference

261

---

2 Teach

Introduce the Lesson Have a student read the first paragraph defining the terms *circumference* and *diameter*. Cut a length of string that just encircles the lid of a jar. Put three lids in a row on the chalk tray. Stretch the string across the center of each lid, marking the length of the diameter each time. Show that the circumference of one lid is a little more than three diameters. Read the next paragraphs, which explain the pi (π), C, and d relationship. Point out that the circumference of any circle equals π times the diameter.

Develop Skills and Concepts Emphasize the following points:

• The circumference of a circle is the set of points that forms the circle.

• The circumference of a circle is always a little more than three times the length of its diameter.

• The Greek letter π is used to show the relationship between the circumference and the diameter.

• The decimal 3.14 is used as an approximation of π.

• To find the circumference of any circle, multiply the diameter by π, 3.14.

Explain that students can use what they know about the inverse relationship of multiplication and division. They can find the diameter of any circle when the circumference is known, by dividing the circumference by pi. Have a student express this as a formula. ($d = C \div \pi$) Then, ask students how they would find a radius if they knew the circumference of a circle. ($r = (\frac{C}{\pi}) \div 2$) Have students solve the following on the board: $d = ?$ if $C = 12.56$ in. (4 in.); and $r = ?$ if $C = 21.98$ yd (3.5 yd).

It's Algebra! The concepts in this lesson prepare students for algebra.

T261

Find the circumference of each circle.

1. $d = 3$ in. 9.42 in.

2. $d = 1.8$ ft 5.652 ft

3. $d = 9$ ft 28.26 ft

4. $d = 12$ ft 37.68 ft

5. $d = 7$ yd 21.98 yd

6. $d = 2\frac{1}{2}$ in. 7.85 in.

7. $d = 3.72$ in. 11.6808 in.

8. $d = 4.5$ yd 14.13 yd

9. $d = 7.05$ ft 22.137 ft

10. $d = 86$ yd 270.04 yd

11. $d = 29$ ft 91.06 ft

12. $d = .06$ yd 0.1884 yd

Complete the table.

13.

diameter	3.2 in.	4 ft	3.65 yd	5 ft	4.5 yd	6.65 ft	24 in.
circumference	10.048 in.	12.56 ft	11.461 yd	15.7 ft	14.13 yd	20.881 ft	75.36 in.

Problem Solving

Solve each problem.

14. A ball has a diameter of 9 inches. How far will the ball roll in one complete turn?

   28.26 inches

15. Martin is putting a low fence around a circular fishpond that has a diameter of 4 feet. The fence material costs $6.80 a foot. To the nearest cent, how much is the fence material?

   $85.41

16. A bicycle wheel has a diameter of 32 inches. How far will the bike travel after the wheel makes 6 full turns?

   602.88 inches

17. Ann is running a circular track that has a 133.33 yard diameter. To the nearest yard, how far is it around the track?

   419 yards

Now Try This!

How much smaller is the circumference of the small circle than the circumference of the large circle?

31.4 meters

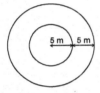

5 m   5 m

Lesson 12-2 • Circumference

For Mixed Abilities

Common Errors • Intervention

Some students may answer incorrectly because even though they are using the correct procedure, their computation is inaccurate. Have them work with partners to practice multiplying whole numbers and decimals by 3.14. Have them take turns using a calculator to check.

Enrichment • Measurement

Have each student use a piece of string to measure and record the circumference and the diameter of circular objects. Display a chart showing the results. Then, have students use a calculator to compute the circumference divided by the diameter of each circle (rounded to the nearest tenth). The average of these should be about 3.1.

More to Explore • Probability

Explain that the genes for one characteristic can dominate the genes for another. Give this example: Some pea plants have smooth round seeds and some have wrinkled seeds. The gene for round seeds (R) is dominant over the gene for wrinkled seeds (r) so that when they are combined in a new plant, Rr, the new plant will have round seeds. Ask students to use the following table to explain how many round and how many wrinkled seeds they would expect from these parent genes.

	Round (R)	Wrinkled (r)
Round (R)	(RR)	(Rr)
Wrinkled (r)	(rR)	(rr)

Ask students to express the probability that the offspring of these parents will have wrinkled seeds ($\frac{1}{4}$ or 25%) or round seeds. ($\frac{3}{4}$ or 75%)

---

3 Practice

Have students complete all the exercises. Remind them that d means "diameter" and that 3.14 is the approximate value of pi.

Now Try This! Be sure students read the activity carefully and identify what information is given before they start finding a circumference.

4 Assess

Ask, *How is the circumference of a circle similar to the perimeter of a polygon?* (Possible answer: Both circumference and perimeter measure the distance around a figure.)

12-3 Area of Rectangles and Squares

pages 263–264

1 Getting Started

Objective
• To find the area of rectangles or squares

Materials
1-cm graph paper; *square tiles

Warm Up • Mental Math
Have students perform the operations in the order given.
1. $26 + 10 \div 4 \times 9 + 4 \times 10 =$ (850)
2. $15 - 5 \times 5 \div 10 \times 6 + 8 =$ (38)
3. $58 + 50 \div 9 \times 12 \div 2 \div 9 =$ (8)
4. $40 \times 10 \div 50 \times 8 - 14 + 9 =$ (59)
5. $21 \times 4 \div 2 \div 7 \times 9 + 6 =$ (60)

Warm Up • Pencil and Paper
Have students find the product.

1.	27	2.	31	3.	312
	× 12		× 4.5		× 46
	(324)		(139.5)		(14,352)

4.	7.25	5.	8.9	6.	172
	× 122		× 7.2		× 13.6
	(884.5)		(64.08)		(2,339.2)

7.	0.75	8.	8.641	9.	3.05
	× 9.3		× 3.08		× 2.13
	(6.975)		(26.61428)		(6.4965)

Name _____

Area of Rectangles and Squares

It's Algebra!

The area of a region is the number of square units needed to cover the region. What is the area of Paul's patio in square meters?

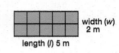

1m²

To find the area, we need to know how many square meters are needed to cover the region. To find the area, we use a formula for each shape.

To find the area of a rectangle, multiply the length by the width.	To find the area of a square, multiply the length of one side by itself.

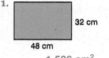

width (w) 2 m
length (l) 5 m

side (s) 3 m

$A = l \times w$
$A = 5 \times 2$
$A = \underline{10}$
Area = $\underline{10}$ m²

$A = s^2$
$A = 3 \times 3$
$A = \underline{9}$
Area = $\underline{9}$ m²

The area of the patio is $\underline{10}$ square meters added to $\underline{9}$ square meters. The total area is $\underline{19 \text{ square meters}}$.

Getting Started
Find the area of each figure.

1.

32 cm
48 cm
$A = \underline{1,536 \text{ cm}^2}$

2.

18 cm
$A = \underline{324 \text{ cm}^2}$

3.

2.5 mm
6.2 mm
$A = \underline{15.5 \text{ mm}^2}$

4. $l = 4.5$ m
$w = 9.8$ m
44.1 m²

5. $l = 123$ mm
$w = 214$ mm
26,322 mm²

6. $s = 8.25$ km
68.0625 km²

---

2 Teach

Introduce the Lesson Use square tiles or squares cut from construction paper to form the 5-by-2 rectangular region described in the exercise on page 263. Explain that there are 2 rows with 5 squares in each row. Have students count the squares to give the area in square units. Then, ask students how they could find the number of square units needed to cover the patio without counting. (multiply 5 by 2 to get 10)

Repeat this activity for a square region that is 4 squares long and 4 squares wide. Point out that s^2 is the same as $l \times w$ for a square because the length and width are the same. Work through the lesson with students determining the area of the patio.

Develop Skills and Concepts Compare the concept of area with that of perimeter. Stress that perimeter is a linear measure; it is the measure of the length of a line. Area is a square measure; it is a measure of the number of square units needed to cover a surface. Emphasize the importance of labeling answers as units or square units. Point out that the notation in the lesson, m², is the customary way of writing square meters.

Then, ask students how they would find the area of a rectangle that is 55 mm long and 26 cm wide. Point out that students must rename length as centimeters or rename width as millimeters so that length and width will be in similar units. Finally, have students find the area of the rectangle. (143 cm² or 14,300 mm²)

It's Algebra! The concepts in this lesson prepare students for algebra.

3 Practice

Have students find the areas of the figures in the exercises. Suggest that students draw a picture to help them visualize the figures that have not been drawn in the text.

T263

Practice

Find the area of each figure.

1.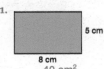
5 cm
8 cm
$A = \underline{40 \text{ cm}^2}$

2.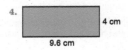
5 cm
12 cm
$A = \underline{60 \text{ cm}^2}$

3.
15 cm
$A = \underline{225 \text{ cm}^2}$

4.
4 cm
9.6 cm
$A = \underline{38.4 \text{ cm}^2}$

5.
6.5 cm
$A = \underline{42.25 \text{ cm}^2}$

6.
162 mm
17.5 cm
$A = \underline{283.50 \text{ cm}^2 \text{ or } 28,350 \text{ mm}^2}$

7. $l = 9$ m
$w = 6$ m
54 m^2

8. $l = 16$ km
$w = 5$ km
80 km^2

9. $l = 4.3$ cm
$w = 6$ cm
25.8 cm^2

10. $s = 12.1$ mm
146.41 mm^2

11. $l = 116$ mm
$w = 45$ mm
5,220 mm^2

12. $l = 8.6$ m
$w = 5.3$ m
45.58 m^2

13. $l = 14$ m
$w = 9.5$ m
133 m^2

14. $s = 45$ mm
2,025 mm^2

15. $l = 215$ mm
$w = 35$ mm
7.525 mm^2

16. $s = 9.3$ km
86.49 km^2

17. $l = 126$ cm
$w = 7.8$ cm
982.8 cm^2

18. $l = 4.25$ m
$w = 123$ cm
5.2275 m^2 or
52,275 cm^2

Problem Solving

Solve each problem. Draw a picture to help.

19. Joe is buying cork to cover a bulletin board. The bulletin board is 0.85 meters long and 53 centimeters wide. What is the area of the bulletin board?
4,505 cm^2 or 0.4505 m^2

20. Jeanette is painting a rectangular wall. The wall is 6 meters long and 3.2 meters high. A window in the wall is 1.5 meters long and 1 meter high. What is the area of the wall that must be painted?
17.7 m^2

Lesson 12-3 • Area of Rectangles and Squares

---

4 Assess

Have students find the area of a rectangle with a length of 4 ft and a width of 3 ft, and the area of a square with a side of 4 ft. (12 sq ft; 16 sq ft)

5 Mixed Review

1. $\$2.78 \times 29$ ($\$80.62$)
2. $800 \times 4,000$ (3,200,000)
3. $\frac{8}{17} = \frac{4}{?}$ (8.5)
4. 5.7×0.05 (0.285)
5. $27,680 \div 68$ (407 R4)
6. $265,107 + 84,193$ (349,300)
7. $6\frac{3}{8} + 2\frac{4}{9}$ ($8\frac{59}{72}$)
8. $0.01260 \div 0.35$ (0.036)
9. $4\frac{1}{8} \times 8$ (33)
10. $196.5 - 87.365$ (109.135)

For Mixed Abilities

Common Errors• Intervention

Some students may confuse perimeter with area. Have them work with partners and different rectangles drawn on graph paper. Have them count the number of units that a little bug would have to walk to go around the edge of one of the rectangles, the perimeter. Then, have them count the number of units the bug would have to paint to completely cover the rectangle, the area.

Enrichment • Number Sense

Have students use rulers, yardsticks, and metersticks to determine the following:
how many square feet equal 1 sq yd (9 sq ft)

how many square inches equal 1 sq ft (144 sq in.)

how many square inches equal 1 sq yd (1,296 sq in.)

how many square centimeters equal 1 sq m (10,000 sq cm)

More to Explore • Biography

George David Birkhoff (1884–1944) was one of the first great American-born mathematicians. He published his first mathematical work at age 18 and received a Ph.D. at age 23. Birkhoff taught at the University of Wisconsin, Princeton University, and Harvard University. He had significant influence on many of the most productive mathematicians of the mid-twentieth century. Birkhoff's mathematical contributions included work on differential equations and the theory of coloring maps.

ESL/ELL STRATEGIES

Remind students that areas are always expressed in square units and that the word *square* always comes before the unit of measurement. Model the following statements and ask students to repeat: 10 square meters, 15 square feet, 25 square inches.

12-4 Relating Area and Perimeter

1 Getting Started

Objective
• To find the measurements of squares and rectangles using area and perimeter formulas

Materials
graph paper

Warm Up • Mental Math
Have students evaluate each power.

1. 2^3 (8)
2. 10^2 (100)
3. 8^2 (64)
4. 10^1 (10)
5. 10^4 (10,000)
6. 4^2 (16)
7. 5^3 (125)
8. 10^3 (1,000)
9. 7^3 (343)
10. 3^4 (81)

Warm Up • Pencil and Paper
Have students perform the operation and write the answer as a mixed number when possible.

1. $\frac{9}{4} \times \frac{2}{3}$ $(1\frac{1}{2})$ 5. $3\frac{3}{10} - 1\frac{4}{5}$ $(1\frac{1}{2})$
2. $\frac{5}{6} - \frac{5}{12}$ $(\frac{5}{12})$ 6. $\frac{24}{5} \div \frac{14}{15}$ $(5\frac{1}{7})$
3. $2\frac{1}{3} + 1\frac{5}{6}$ $(4\frac{1}{6})$ 7. $6\frac{1}{4} \times \frac{8}{15}$ $(3\frac{1}{3})$
4. $10 \div \frac{5}{3}$ (6)   8. $\frac{5}{8} + \frac{7}{12}$ $(1\frac{5}{24})$

2 Teach

Introduce the Lesson Read the opening problem. Explain that the problem tells the area and one side of the rectangle. We are to find the width. Have a student read and complete the information sentences. Write the formulas for perimeter and area on the board. Ask students what each variable represents.

Develop Skills and Concepts Provide an additional example of finding missing measurements. Draw a rectangle on the board and label the length 10 ft. Write the following on the board:

$A = 60$ ft^2
$P =$ (32 ft)
$l = 10$ ft
$w =$ (6 ft)

Name _____

Relating Area and Perimeter

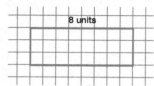 It's Algebra!

Maya plans to plant flowers along a wall that is 8 ft long. She has enough seeds to cover 24 square feet. How wide can the rectangular flower bed be?

We want to know the width of a rectangular flower bed.

We know that the area of the bed is ___24___ square feet.

We know that one side of the bed is ___8___ feet long.

Draw a rectangle on the grid paper below. The rectangle should be 8 units long and have an area of 24 square units.

8 units

Count the units. What is the width of the rectangle? ___3___ units

The flower bed can be ___3___ feet wide.

We can also use the area formula to find the width.

$A = l \times w$

$24 = 8 \times ?$

$24 = 8 \times 3$

Remember these formulas for finding perimeter and area of rectangles:

$P = 2l + 2w$
$A = l \times w$

Getting Started

Find each missing measurement.

1.

$A =$ 9 square units
$P =$ 12 units
$l =$ 3 units
$w =$ 3 units

2. 6 in.   ?

$A =$ 24 sq in.
$P =$ 20 in.
$l =$ 6 in.
$w =$ 4 in.

3. ?   ?

$A =$ 18 cm^2
$P =$ 18 cm
$l =$ 6 cm
$w =$ 3 cm

Lesson 12-4 • Relating Area and Perimeter

Explain that since the area and length are known, w can be found by using the formula

$A = l \times w$

$60 = 10 \times w$

$60 = 10 \times 6$

So, the width is 6 feet. Now, to find the perimeter, use the formula

$P = 2l + 2w$
$P = 2(10) + 2(6)$
$\quad = 20 + 12$
$\quad = 32$

It's Algebra! The concepts in this lesson prepare students for algebra.

Practice

Find each missing measurement.

1.
? | 2 in.

A = 12 sq in.
P = <u>16 in.</u>
l = <u>6 in.</u>
w = 2 in.

2.
9 m | ?

A = 18 m²
P = <u>22 m</u>
l = 9 m
w = <u>2 m</u>

3.
? | 3 ft

A = <u>27 sq ft</u>
P = 24 ft
l = <u>9 ft</u>
w = 3 ft

4.
? | 6 in.

A = <u>36 sq in.</u>
P = 24 in.
l = 6 in.
w = <u>6 in.</u>

5.
? | ?

A = 36 cm²
P = 30 cm
l = <u>12 cm</u>
w = <u>3 cm</u>

6.
? | ?

A = <u>25 cm²</u>
P = 20 cm
l = <u>5 cm</u>
w = <u>5 cm</u>

Problem Solving

Solve each problem.

7. Jan wants to frame a poster. One side measures 12 inches. The area of the poster is 108 square inches. How wide is the poster? How much frame will Jan need?
9 in.; 42 in.

8. Draw a rectangle that is 5 inches by 3 inches. What is the area? Double the dimensions and find the area. What happens to the area when the dimensions of a rectangle are doubled?
15 sq in.; 60 sq in.; area increases 4 times

9. Travis wants to construct a fence for his yard. The yard measures 120 feet long. The area of his yard is 7,200 square feet. How much fence will he need for his rectangular yard?
360 ft

10. John is carpeting his den. One side of the rectangular den is 22 feet long, and the perimeter of the room is 76 feet. What is the area of the room that he is carpeting?
352 sq ft

⟨ Now Try This! ⟩

It's Algebra!

Use grid paper to make as many different rectangles with an area of 16 square units as you can.

1. Which rectangle has the greatest perimeter? <u>16 × 1; 34 units</u>

2. Which rectangle has the least perimeter? <u>4 × 4; 16 units</u>

3. Given any area, predict the shape of the rectangle with the least perimeter. Explain your reasoning and show your work. a square

Lesson 12-4 • Relating Area and Perimeter

---

3 Practice

Have students complete all the exercises.

Now Try This! Use graph paper to draw as many rectangles with a perimeter of 24 units as you can.

1. Which rectangle has the greatest area? (6 × 6; 36 square units)

2. Which rectangle has the least area? (11 × 1; 11 square units)

4 Assess

Tell students that a rectangle has a perimeter of 14 m and a length of 5 m. Have students draw the rectangle and find the width and area. (w = 2 m; A = 10 m²)

For Mixed Abilities

Common Errors • Intervention

Sometimes students use an incorrect formula for perimeter. They may add one length and one width to get the perimeter instead of doubling each measurement. Have students work in pairs. Tell one partner to draw a rectangle and a square, and to label the width and length of each. Tell the second partner to write the formula for perimeter under each figure and then compute the perimeter. The students then switch roles.

Enrichment • Number Sense

Use graph paper to draw a rectangle of any size. Label the length and the width, and then find the area. Multiply the length and the width by 2. What is the new area? What is the new area if both length and width are multiplied by 3? What is the new area if both are multiplied by 4?

How does each new area compare to the original area? See if you can find the rule.

(If the length and the width are each multiplied by n, the new area is n^2 times the original area.)

More to Explore • Application

Have students participate in Motion Geometry. Divide students into teams. One member of each team leaves the room. You identify a starting point and an ending point in the classroom. The remaining team members have 4 minutes to write a series of commands for the teammate who is waiting outside the room to follow in order to get from start to finish. (No straight lines are allowed.) The commands must include direction and distance using any unit, and are limited to *forward*, *right turn*, and *left turn*.

Have the students who were waiting outside return and follow their group's commands, step by step. Measure the distance from where they end up, to the targeted endpoint. The group who comes closest wins. In case of a tie, the group using the fewest commands wins.

T266

pages 267–268

1 Getting Started

Objective
• To find the area of parallelograms

Vocabulary
height

Materials
parallelogram on heavy paper; construction paper; scissors; colored chalk; graph paper

Warm Up • Mental Math
Have students evaluate each power.

1. $(\frac{1}{2})^2$ $(\frac{1}{4})$

2. $(\frac{3}{4})^2$ $(\frac{9}{16})$

3. 0.5^2 (0.25)

4. $(\frac{2}{5})^2$ $(\frac{4}{25})$

5. 0.9^2 (0.81)

6. $(\frac{7}{8})^2$ $(\frac{49}{64})$

Warm Up • Pencil and Paper
Have students find the area of the rectangle.

1. $l = 9$ m, $w = 6.5$ m $(58.5$ m$^2)$
2. $l = 31$ km, $w = 9.8$ km $(303.8$ km$^2)$
3. $l = 12$ m, $w = 56$ cm $(6.72$ m^2; $67,200$ cm$^2)$
4. $l = 27$ cm, $w = 19$ cm $(513$ cm$^2)$
5. $l = 329$ m, $w = 87.6$ m $(28,820.4$ m$^2)$

2 Teach

Introduce the Lesson Discuss the problem posed on page 267. Display a large parallelogram drawn on a sheet of construction paper. Show students the *base* and the *height*. Recall that a parallelogram has opposite sides that are congruent and parallel. Demonstrate the steps for deriving the formula for finding the area of a parallelogram as described on page 267. Remind students to record area in square units.

Develop Skills and Concepts Draw several parallelograms on the board and mark the base of each with colored chalk. Have students use colored chalk to draw a line segment that represents the height of each parallelogram connected to the base. Emphasize that each line segment should be drawn from the side opposite to the indicated base and must form a right angle with that base. The line

---

Name _____

Area of Parallelograms

Justin is making a mosaic from tiles that are one square centimeter in area. How many tiles will he need for the parallelogram design at the right?

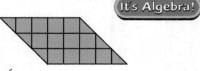

We can use what we know about finding the area of a rectangle to find the formula for the area of a parallelogram.

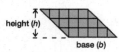

| Measure the base and the height of the parallelogram. The height is a **perpendicular line segment** between any pair of parallel sides. | Cut one end off the parallelogram and slide it to the other end. | You should have a rectangle with the same base and height as the parallelogram. |

To find the area of a parallelogram, we multiply the length of the base by the length of the height.

$A = b \times h$

$A = 5 \times 3$

$A = \underline{15}$

Area = $\underline{15}$ cm^2

Justin needs $\underline{15}$ tiles.

REMEMBER Always express area in square units.

Getting Started

Find the area of each parallelogram.

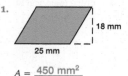

1. 18 mm, 25 mm
$A = \underline{450 \text{ mm}^2}$

2. 3 cm, 4.5 cm
$A = \underline{13.5 \text{ cm}^2}$

3. 5 cm, 5 cm
$A = \underline{25 \text{ cm}^2}$

4. $b = 5.2$ km
$h = 3.7$ km
19.24 km^2

5. $b = 85$ cm
$h = 24$ cm
2,040 cm^2

6. $b = 0.32$ m
$h = 2,415$ mm
772,800 mm^2
or 0.7728 m^2

---

segment showing the height should be drawn in various positions. For example,

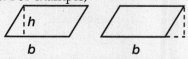

Point out that the base and the height of the parallelogram correspond to the length and the width of the rectangle.

It's Algebra! The concepts in this lesson prepare students for algebra.

3 Practice

Have students complete all the exercises. Remind students to be sure both base and height are expressed in the same unit before computing the area and to label their answers in square units.

Practice

Find the area of each parallelogram.

1.
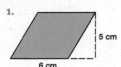
5 cm
6 cm

$A =$ ___30 cm²___

2.
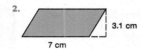
3.1 cm
7 cm

$A =$ ___21.7 cm²___

3.
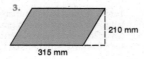
210 mm
315 mm

$A =$ ___66,150 mm²___

4. $b = 2$ km
$h = 5.6$ km
11.2 km²

5. $b = 112$ mm
$h = 36$ mm
4,032 mm²

6. $b = 4.3$ cm
$h = 5.8$ cm
24.94 cm²

7. $b = 22.4$ cm
$h = 8.7$ cm
194.88 cm²

8. $b = 6.7$ mm
$h = 0.09$ cm
6.03 mm² or
0.0603 cm²

9. $b = 3.02$ m
$h = 85$ cm
25,670 cm² or
2.567 m²

Problem Solving

Solve each problem.

10. How many square centimeters are in one square meter?
10,000 cm²

11. How many square millimeters are in one square meter?
1,000,000 mm²

12. The area of a parallelogram is 625 m². If the height is 25 m, how long is the base?
25 meters

13. If the base and height of a parallelogram are tripled, what happens to its area?
increases 9 times

> ### Now Try This!
> Given a string 16 centimeters long, make as many rectangles as possible whose length and width are whole numbers. Complete the table. Write the dimensions of the rectangle with the greatest area. length 4, width 4
>
> | l | 1 | 2 | 3 | 4 | 5 | 6 | 7 |
> |---|---|---|---|---|---|---|---|
> | w | 7 | 6 | 5 | 4 | 3 | 2 | 1 |
> | A | 7 | 12 | 15 | 16 | 15 | 12 | 7 |

Lesson 12-5 • Area of Parallelograms

Now Try This! You might give each student a sheet of graph paper to use as a grid for patterning rectangles. Have students work in pairs to complete this exercise.

4 Assess

Have students find the area of a parallelogram with a base of 6 ft and a height of 3 ft. (18 sq ft)

For Mixed Abilities

Common Errors • Intervention

Some students may incorrectly think the formula for the area of a parallelogram is $l \times w$. Have them work with partners, using graph paper to draw parallelograms. Have them use a crayon to color the base and the height along a grid line. Then, have them count the number of squares along the base and along the height and multiply to find the area.

Enrichment • Geometry

Ask students to find the area of the shaded region in the figure below. (24 cm²)

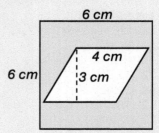
6 cm
6 cm
4 cm
3 cm

Have students create their own figures similar to the one above. After students solve their own problems and make answer keys, have them exchange papers and solve.

More to Explore • Application

Duplicate the following for students to solve:

Johnson City High School is holding a fund-raising dinner at a downtown hotel to pay for their new marching-band uniforms. The committee is paying $10 for each meal served. If they charge $20.95 for each ticket, how many people will have to attend the dinner for the school to raise $1,000? (at least 92 people)

12-6 Area of Triangles

pages 269–270

1 Getting Started

Objective
• To find the area of triangles

Materials
construction paper; colored chalk; graph paper

Warm Up • Mental Math
Have students name $\frac{1}{2}$ of each of the following:

1. 72 (36)
2. 770 (385)
3. 210 (105)
4. 424 (212)
5. 6,484 (3,242)
6. 37.8 (18.9)
7. 24.72 (12.36)
8. 38.64 (19.32)
9. 361.2 (180.6)
10. 340.2 (170.1)

Warm Up • Pencil and Paper
Each of the following exercises gives the measures of two angles of a triangle. Have students find the measure of the third angle and classify the triangle by angle.

1. 62°, 75° (43°; acute)
2. 90°, 47° (43°; right)
3. 38°, 51° (91°; obtuse)
4. 32°, 45° (103°; obtuse)
5. 28°, 90° (62°; right)

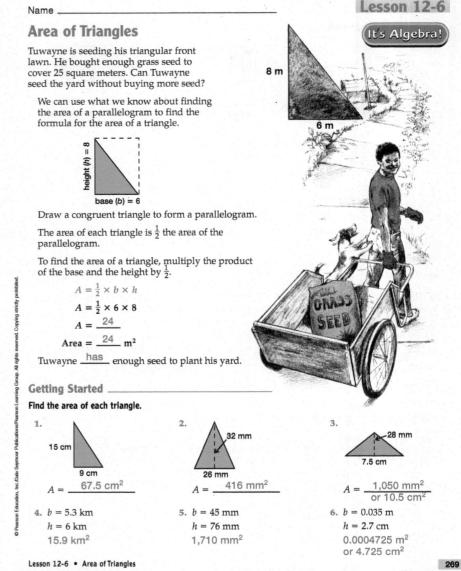

Name _____

Area of Triangles

Lesson 12-6

It's Algebra!

Tuwayne is seeding his triangular front lawn. He bought enough grass seed to cover 25 square meters. Can Tuwayne seed the yard without buying more seed?

We can use what we know about finding the area of a parallelogram to find the formula for the area of a triangle.

height (h) = 8
base (b) = 6

Draw a congruent triangle to form a parallelogram.

The area of each triangle is $\frac{1}{2}$ the area of the parallelogram.

To find the area of a triangle, multiply the product of the base and the height by $\frac{1}{2}$.

$A = \frac{1}{2} \times b \times h$

$A = \frac{1}{2} \times 6 \times 8$

$A = \underline{24}$

Area = $\underline{24}$ m²

Tuwayne __has__ enough seed to plant his yard.

8 m
6 m

Getting Started

Find the area of each triangle.

1.
15 cm
9 cm
$A = \underline{67.5 \text{ cm}^2}$

2.
32 mm
26 mm
$A = \underline{416 \text{ mm}^2}$

3.
28 mm
7.5 cm
$A = \underline{1,050 \text{ mm}^2}$ or 10.5 cm²

4. $b = 5.3$ km
$h = 6$ km
15.9 km²

5. $b = 45$ mm
$h = 76$ mm
1,710 mm²

6. $b = 0.035$ m
$h = 2.7$ cm
0.0004725 m² or 4.725 cm²

Lesson 12-6 • Area of Triangles

269

2 Teach

Introduce the Lesson Have a student read the problem, identify the question, and state the information given. Point out that to answer the question, students need to know the area of the lawn. Draw a triangle on the board and demonstrate the steps for deriving the formula for finding the area of a triangle as described on page 269. Have students complete the section and solve the problem.

Develop Skills and Concepts Draw several triangles on the board and mark the base of each with colored chalk. Have students use colored chalk to draw line segments representing the height of each triangle. Emphasize that the height is the length of a line segment from the vertex opposite the indicated base perpendicular to that base. The line segment showing the height should be drawn in various positions. For example,

h
b

h
b

3 Practice

Have students complete all the exercises. Remind students to label the areas in square units.

It's Algebra! The concepts in this lesson prepare students for algebra.

T269

Practice

Find the area of each triangle.

1.
8 cm
5 cm

$A =$ _____ 20 cm² _____

2.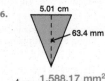
4 cm
3 mm

$A =$ _____ 0.6 cm² or 60 mm² _____

3.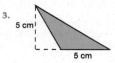
5 cm
5 cm

$A =$ _____ 12.5 cm² _____

4.
37 cm
68 cm

$A =$ _____ 1,258 cm² _____

5.
4.7 cm
76 mm

$A =$ _____ 1,786 mm² or 17.86 cm² _____

6.
5.01 cm
63.4 mm

$A =$ _____ 1,588.17 mm² or 15.8817 cm² _____

7. $b = 7$ m
$h = 8$ m
28 m²

8. $b = 15$ km
$h = 4$ km
30 km²

9. $b = 11$ mm
$h = 46$ mm
253 mm²

10. $b = 6.3$ km
$h = 8$ km
25.2 km²

11. $b = 8$ m
$h = 4.5$ m
18 m²

12. $b = 16$ cm
$h = 16$ cm
128 cm²

13. $b = 4.1$ mm
$h = 3.4$ mm
6.97 mm²

14. $b = 0.96$ m
$h = 0.08$ m
0.0384 m²

15. $b = 5.15$ km
$h = 2.6$ km
6.695 km²

16. $b = 24$ cm
$h = 4.5$ mm
540 mm² or
5.4 cm²

17. $b = 0.05$ m
$h = 21.6$ cm
0.0054 m² or
54 cm²

18. $b = 4.1$ m
$h = 117$ mm
239,850 mm²
or 0.23985 m²

Apply

Solve these problems. Draw a picture to help.

19. The 3 sides of a tent are congruent. Each base is 3 meters and each height is 3.6 meters. What is the total area of the three sides of the tent?
16.2 m²

20. An equilateral triangle has a base of 12 centimeters and a height of 10.3 centimeters. Find the perimeter and the area of the triangle.
$P = 36$ cm; $A = 61.8$ cm²

21. Find the area of the isosceles trapezoid on the right. (Hint: Find the area of 2 congruent triangles and a rectangle.)
$A = 36$ cm²

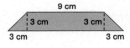
9 cm
3 cm   3 cm
3 cm   3 cm

270

Lesson 12-6 • Area of Triangles

For Mixed Abilities

Common Errors • Intervention

Some students may forget to multiply by $\frac{1}{2}$ when they are finding the area of a triangle. Have students work with partners and a sheet of graph paper. Have them draw a rectangle and multiply $l \times w$ to find the area. Next, have students connect two vertices of the rectangle to draw a diagonal and make two triangles. Ask, *How does the area of each triangle compare to the area of the rectangle?* (It is half.) *What formula can you use to find the area of one of the triangles?* ($\frac{1}{2} \times l \times w$)

Enrichment • Number Sense

Ask each student to draw a triangle and indicate the length of the base and the height. Have students find the area of the triangle. Then, ask them to find its area when only the height is doubled, when only the base is doubled, and then, when both the height and base are doubled. (When the height or the base is doubled, the area is doubled. When both the height and base are doubled, the area is four times as large.)

More to Explore • Biography

The last of his mother's children and the only one to live past infancy, Joseph Louis de Lagrange became a professor of mathematics at age 16. This French mathematician, who was born in Turin, Italy, taught at the Royal Artillery School in Turin and helped found the Royal Academy of Science there. Lagrange's most famous work, a book called *Analytical Mechanics*, studied the forces and motions involved in the orbits of the planets, the flow of liquids, and the vibration of strings. Lagrange also wrote about probability and helped to organize the French metric system. Because of overwork and low pay, Lagrange's health suffered, leaving him in a weakened state. Napoleon named him to the Legion of Honor and gave him the title of Count of the Empire 5 years before the mathematician's death.

---

4 Assess

Have students find the area of a triangle with a height of 10 in. and a base of 8 in. (40 sq in.)

5 Mixed Review

1. $9 \times 4,308$ (38,772)
2. $0.06 + 3.953 + 42.6$ (46.613)
3. $15 \div \frac{5}{8}$ (24)
4. $\frac{7}{8} - \frac{1}{6}$ ($\frac{17}{24}$)
5. $38,472 \div 4$ (9,618)
6. $25 - 72 \div 8$ (16)
7. $23\frac{4}{9} + 18\frac{5}{6}$ ($42\frac{5}{18}$)
8. 2.7×0.36 (0.972)
9. 9^3 (729)
10. $57.381 - 23.64$ (33.741)

12-7 Area of Circles

pages 271–272

© Pearson Education, Inc./Dale Seymour Publications/Pearson Learning Group. All rights reserved. Copying strictly prohibited.

1 Getting Started

Objective
• To find the area of circles

Materials
graph paper; compass

Warm Up • Mental Math
Ask students to name the most appropriate metric unit for measuring

1. their height
2. amount of water in a bathtub
3. the length of their thumb
4. mass of a vitamin pill
5. distance across the Atlantic Ocean
6. gas in a car's tank
7. mass of a watermelon
   (Answers will vary.)

Warm Up • Pencil and Paper
Have students find the circumference of the circle, using 3.14 for π.

1. $d = 8$ cm ($C = 25.12$ cm)
2. $d = 1.9$ m ($C = 5.966$ m)
3. $r = 2.5$ m ($C = 15.7$ m)
4. $d = 15$ cm ($C = 47.1$ cm)
5. $d = 0.96$ km ($C = 3.0144$ km)
6. $r = 3.14$ cm ($C = 19.7192$ cm)

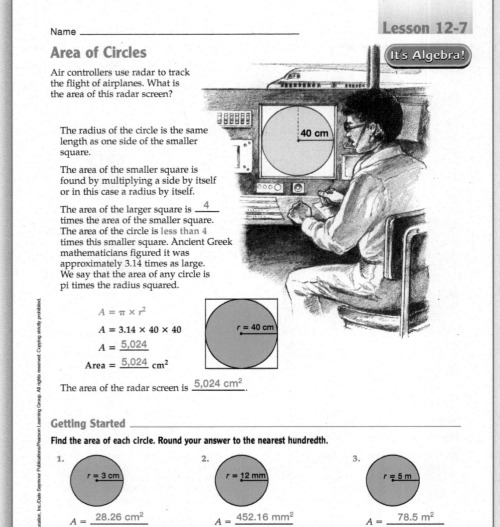

Name _____

Area of Circles

It's Algebra!

Air controllers use radar to track the flight of airplanes. What is the area of this radar screen?

The radius of the circle is the same length as one side of the smaller square.

The area of the smaller square is found by multiplying a side by itself or in this case a radius by itself.

The area of the larger square is ___4___ times the area of the smaller square. The area of the circle is less than 4 times this smaller square. Ancient Greek mathematicians figured it was approximately 3.14 times as large. We say that the area of any circle is pi times the radius squared.

$A = \pi \times r^2$
$A = 3.14 \times 40 \times 40$
$A = \underline{5,024}$
Area = $\underline{5,024}$ cm²

$r = 40$ cm

The area of the radar screen is ___5,024 cm²___.

Getting Started _____

Find the area of each circle. Round your answer to the nearest hundredth.

1. $r = 3$ cm
   $A = \underline{28.26 \text{ cm}^2}$

2. $r = 12$ mm
   $A = \underline{452.16 \text{ mm}^2}$

3. $r = 5$ m
   $A = \underline{78.5 \text{ m}^2}$

4. $r = 4$ cm
   50.24 cm²

5. $d = 12$ m
   113.04 m²

6. $r = 6.1$ km
   116.84 km²

Lesson 12-7 • Area of Circles

271

2 Teach

Introduce the Lesson Discuss the problem posed on page 271. Explain that a radar screen is a device for pinpointing the location of any object within its range. Have a student read the explanation of the relationship between π and r^2.

Demonstrate the development presented for deriving the formula for the area of a circle. Then, work through the problem with students, having them fill in the computed area of the radar screen.

Develop Skills and Concepts Help students use what they know to find the area of a circle when the measure of a diameter of 16 cm is given. Have students recall that the measure of a radius is $\frac{1}{2}$ the measure of a diameter. In this problem, the diameter is 16 cm, so the radius would be $\frac{1}{2}$ of 16 cm, or 8 cm. Then, use the formula $A = \pi \times r^2$. Write the following on the board:

$A = \pi \times r^2$
$A = 3.14 \times 8 \times 8$
Area = 200.96 cm²

Stress that r^2 means $r \times r$, not $2 \times r$. Emphasize that the area of a circle is always expressed in square units. Continue developing this skill by having students find the area of these circles: $r = 3$ cm ($A = 28.26$ cm²), $r = 9$ mm ($A = 254.34$ mm²).

It's Algebra! The concepts in this lesson prepare students for algebra.

3 Practice

Have students complete all the exercises. Watch for students who do not distinguish between the radius and the diameter.

T271

Practice

Find the area of each circle. Round your answer to the nearest hundredth.

1.

$r = 2$ cm

$A =$ _____12.56 cm²_____

2.
$r = 3$ cm

$A =$ _____28.26 cm²_____

3.
$d = 8$ cm

$A =$ _____50.24 cm²_____

4.
$r = 21$ mm

$A =$ _____1,384.74 mm²_____

5.
$d = 3.2$ cm

$A =$ _____8.04 cm²_____

6.
$r = 34$ m

$A =$ _____3,629.84 m²_____

7. $r = 4$ cm
50.24 cm²

8. $r = 1$ m
3.14 m²

9. $d = 10$ km
78.5 km²

10. $r = 9$ m
254.34 m²

11. $d = 14$ km
153.86 km²

12. $r = 1.8$ m
10.17 m²

13. $r = 36$ mm
4,069.44 mm²

14. $r = 9.5$ cm
283.39 cm²

15. $r = 0.1$ km
0.03 km²

16. $r = 2.5$ cm
19.63 cm²

17. $d = 100$ mm
7,850 mm²

18. $r = 10$ cm
314 cm²

Now Try This!

The area of a circular garden is 7,850 ft². About how much fencing is needed to enclose the garden? **314 ft**

Lesson 12-7 • Area of Circles

For Mixed Abilities

Common Errors • Intervention

When using the formula for the area of a circle, some students may multiply the radius times 2 and then square the resulting product. Discuss the fact that $A = \pi \times r^2$ can also be written as $A = \pi \times r \times r$. Encourage students to square the radius before they multiply by π.

Enrichment • Number Sense

Have students complete the following table to find the area of a circle as the radius increases:

Radius	Area
r	πr^2
$2 \times r$	$\pi \times (2 \times r)^2 = 4 \times \pi \times r^2$
$3 \times r$	$\pi \times (3 \times r)^2 = (9 \times \pi \times r^2)$
$4 \times r$	$\pi \times (4 \times r)^2 = (16 \times \pi \times r^2)$
$5 \times r$	$\pi \times (5 \times r)^2 = (25 \times \pi \times r^2)$

Students should conclude that if the radius of a circle is multiplied by a number, the area of the circle is multiplied by the square of that number.

More to Explore • Application

Tell students that they have been given their first part-time job. Ask them to set up a realistic time card to show a week's working schedule using the following data: work hours are after school or on Saturday, rate of pay is $3.35 per hour, $6\frac{1}{2}$ hours total work for the week. If necessary, put a model time card on the board. Have them set up the distribution of hours and find their total pay for the week. ($21.78) As an extension, have students research the terms *minimum wage* and *flex time*.

Now Try This! Point out that to solve this problem, students must find the circumference of the circular garden. They will need to find either the radius or the diameter of the garden. If necessary, suggest that students start by dividing 7,850 by 3.14. Then, they can determine what number squared equals that quotient.

4 Assess

Have students estimate the area of a circle with a diameter of 20 cm. Have them use 3 for π. (300 cm²)

© Pearson Education, Inc./Dale Seymour Publications/Pearson Learning Group. All rights reserved. Copying strictly prohibited.

1 Getting Started

Objective
• To identify prisms, pyramids, cones, cylinders, and spheres

Vocabulary
polyhedrons

Materials
models of square and rectangular prisms; triangular, square, and hexagonal pyramids; spheres; cylinders and cones; tracing paper; colored chalk; scissors; worksheets

Warm Up • Mental Math
Have students name the number.

1. 6^2 (36)
2. 10^2 (100)
3. 3^2 (9)
4. 7^2 (49)
5. 8^2 (64)
6. 4^2 (16)
7. $3^2 + 2^2$ (13)
8. $5^2 + 1^2$ (26)
9. $7^2 + 4^2$ (65)
10. $9^2 + 2^2$ (85)

Warm Up • Pencil and Paper
Have students find the area of the figure.

1. parallelogram, $b = 5.6$ cm, $h = 3.1$ cm (17.36 cm^2)
2. triangle, $b = 8\frac{1}{4}$ cm, $h = 10\frac{1}{2}$ cm ($43\frac{5}{16}$ cm^2)
3. square, $s = 10.01$ cm (100.2001 cm^2)
4. rectangle, $l = 3\frac{2}{3}$ cm, $w = 5\frac{2}{5}$ cm ($19\frac{4}{5}$ cm^2)
5. circle, $r = 7$ cm (153.86 cm^2)

2 Teach

Introduce the Lesson Draw a triangle, hexagon, pentagon, circle, rectangle, and square on the board. Ask students to identify and describe each figure. Then, display models of the solid figures described in the lesson. Explain that these solid figures are called polyhedrons. *Poly-* is the Greek word for "many"; *-hedron* is Greek for "faces" or "surfaces." Point out that the faces of many polyhedrons are polygons.

Use the models to discuss the properties of each solid figure as presented on page 273. Pay special attention to the number of faces, edges, and vertices each figure has.

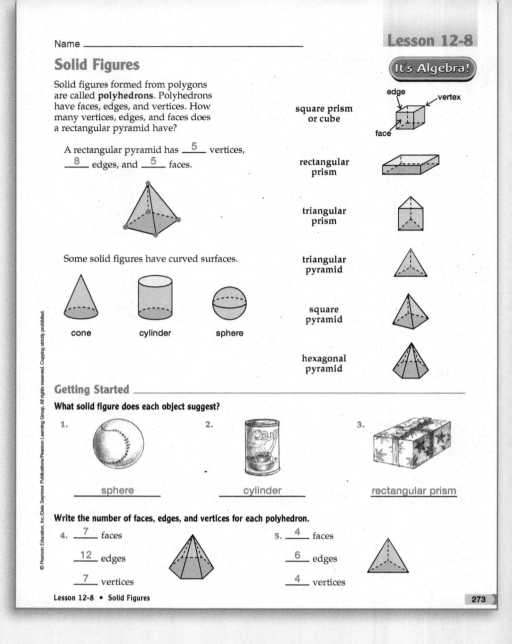

Name _____

Solid Figures

It's Algebra!

Solid figures formed from polygons are called **polyhedrons**. Polyhedrons have faces, edges, and vertices. How many vertices, edges, and faces does a rectangular pyramid have?

square prism or cube

edge   vertex   face

rectangular prism

triangular prism

triangular pyramid

square pyramid

hexagonal pyramid

A rectangular pyramid has __5__ vertices, __8__ edges, and __5__ faces.

Some solid figures have curved surfaces.

cone   cylinder   sphere

Getting Started

What solid figure does each object suggest?

1. _____ sphere
2. _____ cylinder
3. _____ rectangular prism

Write the number of faces, edges, and vertices for each polyhedron.

4. __7__ faces
__12__ edges
__7__ vertices

5. __4__ faces
__6__ edges
__4__ vertices

Lesson 12-8 • Solid Figures

273

Have students use colored chalk to mark and count the faces, edges, and vertices of the figures.

Develop Skills and Concepts Point out that a prism has two opposite parallel and congruent bases. Explain that a prism is named by the shape of these bases. For example, the bases of a square prism are squares.

Then, explain that a pyramid has only one base and it is named by the shape of the base. For example, a triangular pyramid has a triangular base. All other faces of a pyramid are triangular.

Finally, discuss the properties of cones, cylinders, and spheres. Point out that a cone is similar to a pyramid but with a circular base. Its only other surface is curved. A cylinder is similar to a prism but with two circular bases. Its only other surface is curved. If the curved surface could be peeled off, it would be rectangular in shape. The surface of a sphere is entirely curved.

Practice

What solid figure does each object suggest?

1.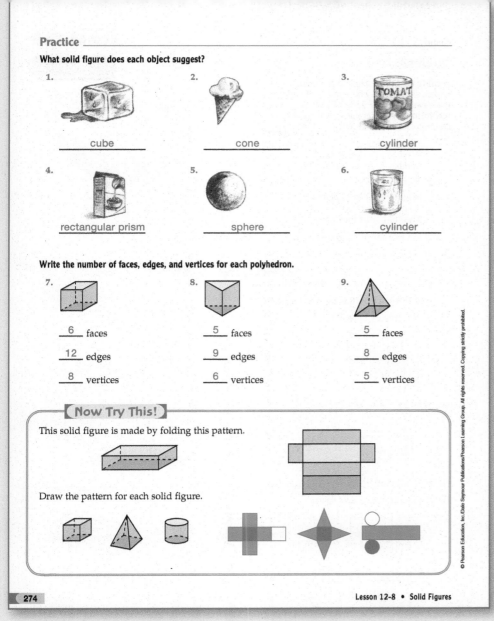

   cube

2. cone

3. cylinder

4. rectangular prism

5. sphere

6. cylinder

Write the number of faces, edges, and vertices for each polyhedron.

7.
   _6_ faces
   _12_ edges
   _8_ vertices

8.
   _5_ faces
   _9_ edges
   _6_ vertices

9.
   _5_ faces
   _8_ edges
   _5_ vertices

Now Try This!

This solid figure is made by folding this pattern.

Draw the pattern for each solid figure.

Lesson 12-8 • Solid Figures

For Mixed Abilities

Common Errors • Intervention

Some students may have difficulty identifying the faces, edges, and vertices of a solid figure. Have them work with partners with a pattern for a cube that they cut out and fold to make a model of a cube.

Have them touch the figure as they count the faces, edges, and vertices. Then, have them find other objects in the classroom that are shaped like a cube.

Enrichment • Geometry

Provide a worksheet with a large pattern for a triangular pyramid (regular tetrahedron).

Have students cut out and make the figure. Then, have them find and copy patterns for some other regular polyhedra. Students can make mobiles of the figures.

More to Explore • Statistics

Provide a list of the number of students in each grade in the school. Tell students to make a histogram to represent this information. Suggest that the grades be listed along the horizontal axis and the numbers in each class, in increments of 50, along the vertical axis. Display their completed graphs.

ESL/ELL STRATEGIES

As you introduce the names of the solid figures, write each one on the board, capitalizing the syllable that receives the strongest stress, for example, polyHEdron, hexAGonal PYRamid, and so on. Ask students to repeat the names.

---

3 Practice

Have students complete all the exercises. Encourage students to use models of the solid figures to count the number of faces, edges, and vertices.

Now Try This! Have students trace the unfolded pattern, cut it out, and fold it to form a rectangular prism. Then, ask students to draw a pattern for each figure shown. Have them cut and fold their patterns to construct the figures.

4 Assess

Ask students to name the figure that has two triangular bases and rectangles for the faces. (triangular prism)

T274

12-9 Volume

1 Getting Started

Objective
- To find the volume of rectangular prisms

Vocabulary
cubic unit, volume

Materials
inch or centimeter cubes; boxes of different sizes; centimeter rulers; heavy oak tag

Warm Up • Mental Math
Write the following on the board and have students determine if each number is divisible by 2, 3, 5, 9, or 10:

1. 80 (2, 5, 10)
2. 6,063 (3)
3. 405 (3, 5, 9)
4. 708 (2, 3)
5. 376 (2)
6. 2,376 (2, 3, 9)
7. 5,553 (3, 9)
8. 1,051 (none)

Warm Up • Pencil and Paper
Have students compute the area of the following:

1. rectangle, $l = 19$ in., $w = 15$ in. (285 in.^{2})
2. rectangle, $l = 7$ yd, $w = 3.5$ yd (24.5 yd^2)
3. rectangle, $l = 2$ ft, $w = 11$ in. (264 in.^{2} or $1\frac{5}{6}$ ft^2)
4. parallelogram, $b = 16.2$ m, $h = 14.7$ m (238.14 m^2)
5. parallelogram, $b = 42$ cm, $h = 150$ mm (63,000 mm^2 or 630 cm^2)

2 Teach

Introduce the Lesson Ask students how they would determine the number of cubes needed to fill a box that measures $3 \times 4 \times 3$. Make a 3×4 array with the inch or centimeter cubes. Explain that these 12 cubes cover the bottom layer of the box. Then, ask students how many layers are needed to fill the box. (3) Ask a student to construct the 3 layers. Have students observe that 36 cubes were used in all.

Then, discuss the example on page 275, paying particular attention to the terms *cubic unit* and *volume*. Direct attention to the illustration of the cube and work through the model with students. Then, have students complete the example to determine the volume of the prism.

Volume

It's Algebra!

A **cubic unit** has 3 dimensions: length, width, and height. The number of cubic units needed to fill a solid figure is called the **volume**. What is the volume of the rectangular prism in cubic centimeters?

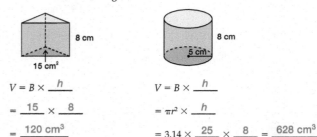

1 cm^3

We need to know the number of cubes in the whole figure. To find the total number of centimeter cubes, we multiply.

$$5 \times 4 \times 4 = \underline{80} \text{ centimeter cubes}$$
length width height

The volume of the box is $\underline{80}$ cm^3.

The formula for finding the volume of a rectangular prism is:

$V = l \times w \times h$ where $l \times w$ is the area of the base.

To find the volume of *any* prism or cylinder, multiply the area of the base times the height. The formula is:

$V = B \times h$ where B is the area of the base.

Find the volume of each figure.

8 cm   8 cm

15 cm^2 5 cm

$V = B \times \underline{h}$ $V = B \times \underline{h}$

$= \underline{15} \times \underline{8}$ $= \pi r^2 \times \underline{h}$

$= \underline{120} \text{ cm}^3$ $= 3.14 \times \underline{25} \times \underline{8} = \underline{628 \text{ cm}^3}$

Getting Started

Find the volume of each solid figure.

1.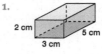
    2 cm   5 cm
    3 cm

    $V = \underline{30 \text{ cm}^3}$

2. 6 cm
    8 cm   12 cm

    $V = \underline{288 \text{ cm}^3}$

3. 14 mm
    5 mm

    $V = \underline{3,077.2 \text{ mm}^3}$

Next, discuss the general formula for finding the volume of a prism, $V = B \times h$. Point out that the capital B stands for the area of the base. This same formula can be used to find the area of a cylinder, where the area of the base is $\pi \times r^2$. Have students use the formula to find the area of the triangular prism and the cylinder.

Develop Skills and Concepts Tell students that volume is a measurement that applies only to three-dimensional figures. Point out that when students use the formula $V = l \times w \times h$, the measures of length, width, and height must be expressed in the same unit. Remind students that the answer will be in cubic units because there are three dimensions, length, width, and height.

Help students to derive a formula for finding the volume of a cube. ($V = s^3$) Recall that the measures of length, width, and height are all the same in a cube.

It's Algebra! The concepts in this lesson prepare students for algebra.

Find the volume of each solid figure.

1.

$V =$ ___83.2 mm³___

2.

$V =$ ___15.625 cm³___

3.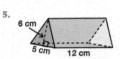

$V =$ ___635.85 m³___

4.

$V =$ ___864 mm³___

5.

$V =$ ___180 cm³___

6.

$V =$ ___753.6 cm³___

Find the volume of each rectangular prism.

7. $l = 4$ cm
   $w = 9$ cm
   $h = 12$ cm
   432 cm³

8. $l = 7$ m
   $w = 3.2$ m
   $h = 5$ m
   112 m³

9. $l = 6.3$ cm
   $w = 12.8$ cm
   $h = 9$ cm
   725.76 cm³

10. $l = 5.8$ m
    $w = 3.2$ m
    $h = 5.7$ m
    105.792 m³

Now Try This!

Each cube is 1 centimeter on each edge.

 = 1 cubic centimeter of water = 1 milliliter of water = 1 gram of water

How many liters will it take to fill an aquarium 50 by 30 by 20 centimeters?
30 liters

276

Lesson 12-9 • Volume

For Mixed Abilities

Common Errors • Intervention

Some students may have trouble using the formula to find the volume of a rectangular prism. Have them work with partners to make a rectangular prism using small wooden or plastic cubes. Have one partner count the number of cubes to find the length, width, and height and multiply to find the volume. Have the other partner verify the answer by counting all the cubes.

Enrichment • Number Sense

Have students find the volume of a rectangular prism that is 2 cm long, 3 cm wide, and 4 cm high. Then, have students answer the following questions:

- How many times greater is the volume if any one of the dimensions is doubled? (2)
- How many times greater is the volume if any two of the dimensions are doubled? (4)
- How many times greater is the volume if all three dimensions are doubled? (8)

More to Explore • Number Sense

Duplicate the following for students to solve:

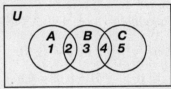

The universal set represents all people who like cookies:

- Set A includes those who like chocolate chipsters.
- Set B includes those who like peanut butter zipsters.
- Set C includes those who like fig fritzers.

What do sections 2 and 4 represent? (Section 2 represents people who like chipsters and zipsters; section 4 represents people who like zipsters and fritzers.)

---

3 Practice

Have students complete all the exercises.

Now Try This! Explain that the capacity of 1 cubic centimeter is 1 milliliter of water and that 1 milliliter of water weighs 1 gram. Remind students that 1 liter equals 1,000 milliliters. Have students solve the problem.

4 Assess

Have students find the volume of a rectangular prism with a length of 4 in., a width of 3 in., and a height of 5 in. (60 in.³)

T276

1 Getting Started

Objective
- To find the surface area of a rectangular prism and cube

Vocabulary
surface area

Materials
boxes of different sizes

Warm Up • Mental Math
Have students name the LCM of the group.
1. 2, 3, 9 (18)
2. 8, 9, 12 (72)
3. 4, 6, 16 (48)
4. 10, 20, 30 (60)

Have students name the GCF of the group.
5. 4, 10, 12 (2)
6. 36, 72, 27 (9)
7. 10, 85, 40 (5)
8. 7, 8, 20 (1)

Warm Up • Pencil and Paper
Have students find the volume of the rectangular prism.
1. $l = 5$ m, $w = 4$ m, $h = 8$ m ($V = 160$ m^3)
2. $l = 10$ cm, $w = 12.5$ cm, $h = 10$ cm ($V = 1,250$ cm^3)
3. $l = 9$ in., $w = 3$ in., $h = 15$ in. ($V = 405$ in.^{3})
4. $l = 13$ ft, $w = 11.5$ ft, $h = 9.8$ ft ($V = 1,465.1$ ft^3)

2 Teach

Introduce the Lesson Help students get a better understanding of the *surface area* of a rectangle by cutting through the edges of a box and laying it flat so students can see six faces. Call attention to the fact that opposite faces are congruent. Explain that students can find the surface area of the box or rectangular prism by finding the area of one front, one side, and one top, multiplying each area by 2, and then adding the products.

Emphasize that the surface area of any figure can be found by finding the area of each face and then adding all the areas. Work through the model with students to find the surface area of the rectangular prism.

Name _____

Surface Area

The total area on the outside of a solid figure is called the figure's **surface area**. Tina put a gift in a box the shape of a cube. How much paper is needed to cover the gift box?

We want to know the total area on the outside of the box.

We know that the shape of each face is a ___square___ and that each edge is __5__ inches long. To find the surface area, we find the area of one side and multiply by the number of congruent sides.

Area of 1 face = __5__ × __5__ = __25__ in.²

Area of 6 faces = 6 × __25__ = __150__ in.²

Tina will need __150__ square inches of paper.

A rectangular prism has 3 pairs of congruent faces: top and bottom; side and side; front and back. To find the surface area of a rectangular prism, follow these steps:

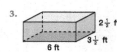

Multiply to find the area of the front. Double the product.	Multiply to find the area of one side. Double the product.	Multiply to find the area of the top. Double the product.	Add the products.
5 × 2 = __10__ ft²	4 × 2 = __8__ ft²	5 × 4 = __20__ ft²	
__10__ × 2 = __20__ ft²	__8__ × 2 = __16__ ft²	__20__ × 2 = __40__ ft²	__76__ ft²

The surface area of the rectangular prism is __76__ square feet.

Getting Started

Find the surface area of each prism.

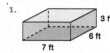

1. 7 ft, 6 ft, 3 ft
Surface Area = __162 ft²__

2. 3.5 yd
Surface Area = __73.5 yd²__

3. 6 ft, $3\frac{1}{2}$ ft, $2\frac{1}{2}$ ft
Surface Area = __$89\frac{1}{2}$ ft²__

Develop Skills and Concepts Be sure students understand that surface area is the sum of the areas of all the faces in a solid figure and is measured in square units.

Work with students to derive a formula for finding the surface area of a cube. ($6 \times s^2$)

It's Algebra! The concepts in this lesson prepare students for algebra.

Thinking Algebraically Extend the lesson by having students determine a method for finding the surface area of a square pyramid. ($4 \times \frac{1}{2}h \times b + s^2$) Point out to students that the height is the height of each triangular face, not the height of the pyramid.

3 Practice

Have students complete all the exercises. Remind students that a square foot is 12 in. long and 12 in. wide or 144 sq in. A cubic foot is a cube 12 in. long, 12 in. wide, and 12 in. high or 1,728 cubic inches.

Practice

Find the surface area of each prism.

1.
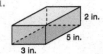
2 in.
5 in.
3 in.

Surface Area = __62 in.²__

2.

2 ft
2 ft
$1\frac{1}{2}$ ft

Surface Area = __20 ft²__

3.
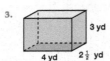
3 yd
4 yd
$2\frac{1}{2}$ yd

Surface Area = __59 yd²__

4.

12 in.

Surface Area = __864 in.²__

5.

6 yd
$1\frac{1}{2}$ yd
$2\frac{1}{3}$ yd

Surface Area = __53 yd²__

6.
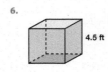
4.5 ft

Surface Area = __121.5 ft³__

Problem Solving

Solve each problem.

7. A box measures 6 inches wide, 9 inches long, and 5 inches deep. How many square inches of paper are needed to cover the box?

258 in.²

8. The length, width, and height of a rectangular prism are each doubled. What happens to the surface area?
increases 4 times

9. 1 square foot = __144__ square inches

1 square yard = __9__ square feet

1 square yard = __1,296__ square inches

10. 1 cubic foot = __1,728__ cubic inches

1 cubic yard = __27__ cubic feet

1 cubic yard = __46,656__ cubic inches

Now Try This!

It's Algebra!

The surface area of a sphere = $4 \times \pi \times r \times r$

The volume of a sphere = $\frac{4}{3} \times \pi \times r \times r \times r$ (use 1.3 for $\frac{4}{3}$)

Complete the table.

radius	1 in.	2 in.	3 in.	4 in.	5 in.
surface area	12.56 in.²	50.24 in.²	113.04 in.²	200.96 in.²	314 in.²
volume	4.082 in.³	32.656 in.³	110.214 in.³	261.248 in.³	510.25 in.³

278

Lesson 12-10 • Surface Area

Now Try This! Discuss the formulas for the surface area and volume of a sphere. Do one surface area and one volume exercise on the board before having students complete the table independently.

4 Assess

Have students determine the surface area and volume of a cube with a side of 3 cm. (54 sq in.; 27 cu in.)

For Mixed Abilities

Common Errors • Intervention

Watch for students who only find the area for the three faces that are visible in the drawing. Have them work with partners first to determine the number of faces on the figure in the drawing, next to list the dimensions of all six faces, and then to multiply and add to find the surface area.

Enrichment • Geometry

Provide heavy oak tag for students to make their own cubes from a pattern. (See page 274 for a pattern.)

1. Have students mark one corner of the cube A and the "opposite" corner Z. The shortest path from A to Z is a 3-edge route. Tell students to find as many 3-edge routes as possible. Have them repeat with 4-edge routes, 5-edge routes, and 6-edge routes.

2. Tell students to find the fewest number of colors needed to color a cube so that no two adjacent faces are the same color.

More to Explore • Probability

Explain that to understand the probability of an event occurring, it is useful to write a sample space. Tell students that if they have two different coins and they want to know if they will land on heads or tails, the sample space can be shown as (HH, HT, TH, TT). Another way to illustrate the sample space for two coins, showing the number of heads or tails in each flip is [(2, 0), (1, 1), (0, 2)]. This means that there could be (2 heads, no tails) or (1 head, 1 tail) or (no heads, 2 tails).

Have students make up a sample space for the distribution of two children, boys or girls, in a family of four. They can be the older or younger sibling. The first letter designates the older child. (BB, BG, GB, GG) Then, have them try the same problem for three children (GGG, GGB, GBB, GBG, BGB, BGG, BBG, BBB).

12-11 Problem Solving: Identify a Subgoal

pages 279–280

1 Getting Started

Objective
- To solve problems by identifying subgoals

Warm Up • Mental Math
Have students solve in simplest terms.

1. 4 in. × 6 (2 ft)
2. 10 in. × 8 (6 ft 8 in.)
3. 19 in. × 3 (4 ft 9 in.)
4. 13 in. × 5 (5 ft 5 in.)
5. 3 ft 2 in. × 6 (19 ft)
6. 5 ft 5 in. × 5 (27 ft 1 in.)
7. 10 ft 7 in. × 4 (42 ft 4 in.)
8. 15 ft 10 in. × 5 (79 ft 2 in.)

Warm Up • Pencil and Paper
Have students simplify each expression.

1. 15 ft 1 in. − 6 ft 5 in. (8 ft 8 in.)
2. 23,400 ÷ (65 × 72) (5)
3. $\frac{1}{2}$ of 439,406 (219,703)
4. 63,000 ÷ 9,000 (7)
5. 5 ft 6 in. − 3 ft 11 in. (1 ft 7 in.)
6. $3,038 ÷ 62 ($49)
7. 75,000,000 ÷ 25,000 (3,000)
8. $4,293 ÷ 81 ($53)

Name _____

Problem Solving: Identify a Subgoal

Four friends are planning a class party. Andy bought $5.00 worth of cold drinks. Shawn spent $2.50 on decorations. Beth bought a cake that cost $9.50. Rod spent $4.00 on snacks and bought paper products that cost $3.00. If they plan to share the cost of the party equally, who owes money to whom?

★ SEE

We want to share the cost equally among the four friends.

Andy spent $5.00 .

Shawn spent $2.50 .

Beth spent $9.50 .

Rod spent a total of $7.00 .

★ PLAN

Before we can determine which persons owe money and to whom, we need to find each one's equal share. We can do this by finding the average of all the expenses. Then we can determine how much above or below the average each person spent.

★ DO

```
  $5.00         $6.00
   2.50      4) $24.00
   9.50
 + 7.00       Each person's share is  $6.00 .
```

```
Andy    6.00     Shawn   $ 6.00
       − 5.00            −  2.50
       $1.00             $3.50

Beth   $9.50     Rod      $7.00
      − 6.00             − 6.00
      $3.50              $1.00
```

Andy _____ owes _____ Rod _____ $1.00 .

Shawn _____ owes _____ Beth _____ $3.50 .

★ CHECK

We can check by adding what one owes to what one spent, or subtracting what is owed from what was spent, whichever the case may be.

---

2 Teach

Introduce the Lesson This problem introduces a common situation that can be solved in a few short steps. Multistep problems like this one almost always require a certain sequence of operations. The PLAN step establishes the priority in this sequence. This process is called establishing subgoals.

Develop Skills and Concepts The first subgoal establishes the average amount each student needs to pay for the party. Once this is known, students use subtraction to produce the amount some of them owed or were owed. A careful listing of who owes whom and how much must be made. Point out that the CHECK step can be approached in different ways. With practice and experience, students will acquire a more arithmetic sense in planning the solution of multistep problems.

3 Practice

Solution Notes

1. Area of fence is
   4.5 × 300 = 1,350 sq ft
   $\frac{1,350}{270}$ = 5 gal @ $11.50 = $57.50

2. (400 × 600) = 240,000 sq ft
   $\frac{240,000}{43,560}$ = 5.5 acres

3. 80 in. wide = 10 bricks
   64 ft long = 192 bricks
   10 × 192 = 1,920 bricks

4. $\frac{93,000,000}{186,000}$ = 500 seconds
   $\frac{500}{60}$ = $8\frac{1}{3}$ minutes

5. 18 ft 2 in. =   17 ft 14 in.
                 − 7 ft 10 in.
                 ──────────────
                   10 ft 4 in.

Apply

Solve each problem.

1. A gallon of paint costing $11.50 covers 270 square feet with one coat. Find the cost of painting a fence with one coat on each side if the fence is 4 feet 6 inches high and encloses a rectangular lot 50 feet long and 25 feet wide.
$57.50

2. An acre contains 43,560 square feet. Approximately how many acres are in a lot 400 feet by 600 feet?
5.5 acres

3. How many bricks 8 inches long by 4 inches wide will be required to lay a walk 6 feet 8 inches wide by 64 feet long?
1,920 bricks

4. The sun is approximately 93 million miles from the Earth. The light from the sun travels at 186,000 miles per second. How long will it take the light to reach the Earth?
$8 \frac{1}{3}$ minutes

5. One board is 18 feet 2 inches long while the other is 7 feet 10 inches long. What is the difference in these two lengths?
10 feet 4 inches

6. A clock takes 5 seconds to chime 6 times. How long will it take to chime 12 times?
10 seconds

Problem-Solving Strategy: Using the Four-Step Plan

★ **SEE** What do you need to find?

★ **PLAN** What do you need to do?

★ **DO** Follow the plan.

★ **CHECK** Does your answer make sense?

---

7. Angelo is planting a rectangular garden 15 feet long by 8 feet wide. Angelica wants to plant a garden that has twice the area. Should she double both the length and width or just the length? Explain why.
Double only one dimension, the length.

8. Archie is making a circular garden. So is Luanne, but the radius of her garden is twice as long as the radius of Archie's garden. Tell how the area of Luanne's garden will compare to that of Archie's garden, and prove it.
See Solution Notes.

9. If the number of linear units in the circumference of a circle is the same as the number of square units in the area, how long is the radius?
2 units

10. Guess and then verify your guess. If all three dimensions of a shoe box are doubled, how does this affect the volume?
The volume is 8 times as great.

For Mixed Abilities

More to Explore •
Measurement

Tell students that it has been discovered that rectangles with the length-to-width ratio of 1.61803:1 are the most appealing to the human eye. These are called golden rectangles. Have students complete the following activities:

1. Using metric units, measure and record the lengths and widths of as many rectangles as you can find at home.

2. Using these measurements, form a proportion to test for each to see if they are golden rectangles.
$\frac{l}{w} = \frac{1.6}{1}$ (numbers rounded)

Make a list of the golden rectangles that students have found.

6.

          11 seconds

| | | | | | | | | | | | |

         5 seconds

The second ring ends 1 second.

Higher-Order Thinking Skills

7. **Analysis:** If she doubled both dimensions, the area would be 4 times as large.

8. **Evaluation:** Luanne's garden will be 4 times as large as Archie's garden. Proofs may vary but probably will be a variation of the following: If the radius of Archie's garden is r, then the radius of Luanne's is $2r$; thus, the area of Archie's garden is πr^2, and the area of Luanne's garden is $\pi(2r)^2$, or $4\pi r^2$.

9. **Synthesis:** Students can use guess and check to find the number for r that makes $2\pi r$ equal to πr^2. When 2 is substituted for r, both $2\pi r$ and πr^2 are equal to 4π.

10. **Analysis:** Doubling each dimension means doubling the volume three times, or volume $\times 2 \times 2 \times 2$, or $8 \times$ volume.

4 Assess

Have students write their own word problem for planning a class party. Have them write the solution on an index card. Then, exchange problems for others to solve. Have them compare these solutions with the solutions they wrote for their problem.

Chapter 12
Test

page 281

Items   **Objectives**

1–3   To find the perimeter or circumference of a figure (see pages 259–262)

4–6   To find the area of rectangles and squares (see pages 263–264)

7–9   To find the area of parallelograms (see pages 267–268)

10–12   To find the area of triangles (see pages 269–270)

13–15   To find the area of circles (See pages 271–272)

16–17   To use customary units when finding surface area (see pages 277–278)

18   To find the volume of rectangular prisms (see pages 275–276)

Name _____

Find the perimeter or circumference.

1.
3 in. 5 in. 5 in.
13 in.

2.
6 ft
8 ft   28 ft

3.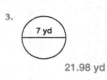
7 yd
21.98 yd

Find the area of each rectangle or square.

4.
2 cm
6 cm
$A =$ ___12 cm²___

5.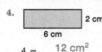
15 mm
$A =$ ___225 mm²___

6. $l = 0.09$ km
$w = 18$ cm
$A =$ ___162,000 cm²___

Find the area of these parallelograms.

7.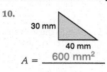
5 cm
6 cm
$A =$ ___30 cm²___

8.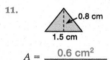
5.1 cm
5.1 cm
$A =$ ___26.01 cm²___

9. $b = 0.031$ km
$h = 45$ m
$A =$ ___1.395 m²___

Find the area of these triangles.

10.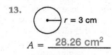
30 mm
40 mm
$A =$ ___600 mm²___

11.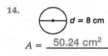
0.8 cm
1.5 cm
$A =$ ___0.6 cm²___

12. $b = 0.4$ m
$h = 25$ cm
$A =$ ___500 cm²___ or 0.05 m²

Find the area of these circles.

13.
$r = 3$ cm
$A =$ ___28.26 cm²___

14.
$d = 8$ cm
$A =$ ___50.24 cm²___

15. $r = 2.1$ m
$A =$ ___13.8474 m²___

Find the surface area of each box.   **Find the volume.**

16.
3 ft
2 ft
2 ft
Surface Area = ___32 ft²___

17.
$1\frac{1}{2}$ yd
$2\frac{4}{5}$ yd
$1\frac{1}{4}$ yd
Surface Area = ___$19\frac{3}{20}$ yd²___

18.
3 cm
6 cm
3 cm
$V =$ ___54 cm³___

Alternate Chapter Test

You may wish to use the Alternate Chapter Test on page 392 of this book for further review and assessment.

Circle the letter of the correct answer.

1 $3\frac{3}{4}$
$+ 2\frac{3}{8}$

a. $5\frac{1}{8}$
b. $5\frac{1}{2}$
c. $6\frac{1}{8}$
d. NG

2 $4\frac{1}{3}$
$- 1\frac{3}{4}$

a. $2\frac{7}{12}$
b. $3\frac{5}{12}$
c. $3\frac{7}{12}$
d. NG

3 $2\frac{1}{3} \times 1\frac{5}{7}$

a. $3\frac{3}{5}$
b. $5\frac{3}{5}$
c. 28
d. NG

4 $5\frac{1}{3} \div \frac{5}{6}$

a. $6\frac{2}{5}$
b. $3\frac{7}{12}$
c. $4\frac{4}{9}$
d. NG

5 What is the place value of the 0 in 9.0321?

a. tenths
b. hundredths
c. thousandths
d. NG

6 16.96
$+ 2.853$

a. 18.813
b. 19.113
c. 19.813
d. NG

7 $8.06 - 4.392$

a. 3.672
b. 3.678
c. 3.668
d. NG

8 6.3×0.09

a. 0.547
b. 5.47
c. 54.7
d. NG

9 $0.02\overline{)19.64}$

a. 9.82
b. 98.2
c. 982
d. NG

10 6 lb 9 oz
$+ 3$ lb 8 oz

a. 9 lb 1 oz
b. 10 lb 1 oz
c. 10 lb 7 oz
d. NG

11 Find the perimeter.

3 cm
5 cm

a. 15 cm
b. 15 cm²
c. 16 cm
d. NG

12 Find the missing angle measure.

35°
? 25°

a. 60°
b. 110°
c. 120°
d. NG

STOP

score

Cumulative Assessment

page 282

Items	Objectives
1–2	To add and subtract mixed numbers (see pages 119–120, 127–128)
3–4	To multiply and divide mixed numbers (see pages 141–142, 147–148)
5	To identify the place value of a digit in decimals through millionths (see pages 165–166)
6–7	To add and subtract decimals (see pages 171–172, 173–174)
8–9	To multiply and divide decimals (see pages 187–188, 197–198)
10	To add customary units of weight (see pages 215–216)
11	To find the perimeter of rectangles by using a formula (see pages 259–260)
12	To find the measure of a third angle of a triangle when the measures of the other two angles are known (see pages 245–246)

Alternate Cumulative Assessment

Circle the letter of the correct answer.

1. $6\frac{7}{8}$
$+ 3\frac{4}{9}$

a. $10\frac{23}{72}$
b. $10\frac{22}{72}$
c. $9\frac{23}{72}$
d. NG

2. $17\frac{4}{15}$
$- 9\frac{3}{5}$

a. $7\frac{2}{3}$
b. $7\frac{11}{15}$
c. $8\frac{11}{15}$
d. NG

3. $4\frac{2}{7} \times 5\frac{3}{5}$

a. $23\frac{34}{35}$
b. 24
c. $24\frac{1}{35}$
d. NG

4. $6\frac{2}{5} \div 5\frac{1}{3}$

a. $\frac{5}{6}$
b. $1\frac{1}{15}$
c. $1\frac{1}{5}$
d. NG

5. Give the place value of the 3 in 28.3597.

a. tenths
b. hundredths
c. thousandths
d. NG

6. 47.84
$+ 3.962$

a. 50.802
b. 51.702
c. 51.802
d. NG

7. 7.03
$- 5.684$

a. 1.346
b. 1.354
c. 1.446
d. NG

8. 0.07
$\times 9.4$

a. 0.648
b. 6.48
c. 64.8
d. NG

9. $0.04 \div 34.92$

a. 8.73
b. 87.3
c. 873
d. NG

10. 8 lb 5 oz
$+ 7$ lb 11 oz

a. 15 lb 15 oz
b. 16 lb
c. 16 lb 1 oz
d. NG

11. Find the perimeter of a rectangle 6 cm by 4 cm.

a. 20 cm
b. 24 cm
c. 24 cm²
d. NG

T282

13-1 Writing Ratios

pages 283–284

1 Getting Started

Objective
• To write a ratio to compare two numbers

Vocabulary
ratio

Warm Up • Mental Math
Have students estimate the sum.
1. $493 + 329$ (800)
2. $4,299 + 708$ (4,700)
3. $6,523 + 4,486$ (11,000)
4. $73,964 + 11,807$ (80,000)
5. $194,650 + 321,000$ (500,000)
6. $35 + 19$ (60)

Warm Up • Pencil and Paper
Have students find the sum or difference.
1. $\frac{3}{4} + \frac{1}{6}$ $\left(\frac{11}{12}\right)$
2. $\frac{2}{15} + \frac{1}{10}$ $\left(\frac{7}{30}\right)$
3. $\frac{1}{8} + \frac{5}{6}$ $\left(\frac{23}{24}\right)$
4. $6\frac{4}{9} + 5\frac{5}{12}$ $\left(11\frac{31}{36}\right)$
5. $\frac{1}{2} - \frac{1}{3}$ $\left(\frac{1}{6}\right)$
6. $\frac{11}{12} - \frac{9}{10}$ $\left(\frac{1}{60}\right)$
7. $\left(\frac{5}{6} + \frac{1}{4}\right) - \frac{2}{3}$ $\left(\frac{5}{12}\right)$
8. $\left(\frac{2}{3} - \frac{5}{9}\right) + \frac{11}{15}$ $\left(\frac{38}{45}\right)$

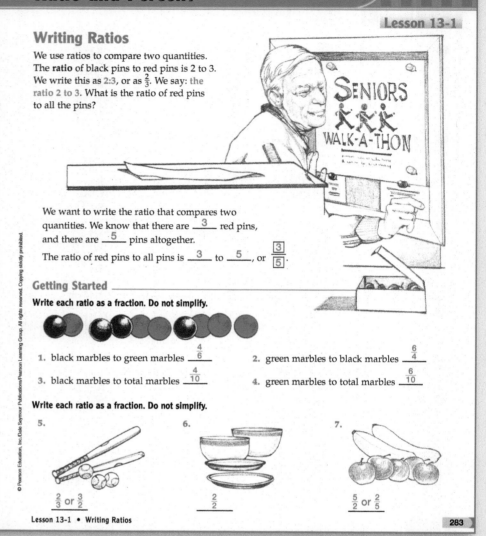

Name _____

Ratio and Percent

Lesson 13-1

Writing Ratios

We use ratios to compare two quantities. The **ratio** of black pins to red pins is 2 to 3. We write this as 2:3, or as $\frac{2}{3}$. We say: the ratio 2 to 3. What is the ratio of red pins to all the pins?

We want to write the ratio that compares two quantities. We know that there are __3__ red pins, and there are __5__ pins altogether.

The ratio of red pins to all pins is __3__ to __5__, or $\frac{3}{5}$.

Getting Started

Write each ratio as a fraction. Do not simplify.

1. black marbles to green marbles $\frac{4}{6}$
2. green marbles to black marbles $\frac{6}{4}$
3. black marbles to total marbles $\frac{4}{10}$
4. green marbles to total marbles $\frac{6}{10}$

Write each ratio as a fraction. Do not simplify.

5. $\frac{2}{3}$ or $\frac{3}{2}$
6. $\frac{2}{2}$
7. $\frac{5}{2}$ or $\frac{2}{5}$

Lesson 13-1 • Writing Ratios

283

2 Teach

Introduce the Lesson Have a student read the lesson aloud. Emphasize the definition and the way to use a ratio. Point out that order is important when writing a ratio. The ratio of black pins to red pins is 2 to 3, whereas the ratio of red pins to black pins is 3 to 2.

Have students give examples of other ratios, such as the number of chairs to the number of desks in the classroom or the number of windows to the number of doors.

Develop Skills and Concepts Ask students to give the ratio of the number of boys to the number of girls in the class. Write the ratio on the board. Then, ask students to give the ratio of the number of girls to the number of boys. Emphasize that the *order* of the numbers in a ratio is very important. A ratio of 5 to 2 is not the same as a ratio of 2 to 5, just as the fraction $\frac{5}{2}$ does not mean the same as $\frac{2}{5}$.

Point out that when students write a ratio as a fraction, the number of objects listed first in the comparison is the numerator, and the number of objects listed second is the denominator.

3 Practice

Have students complete all the exercises. Watch for students who reverse numerators and denominators.

Practice

Write each ratio as a fraction. Do not simplify.

1. pencils to rulers

$\frac{2}{2}$

2. length to width

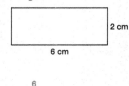

2 cm

6 cm

$\frac{6}{2}$

3. dimes to dollars

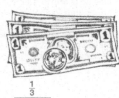

$\frac{1}{3}$

Problem Solving

Solve each problem.

4. Raisins cost 87¢ for 3 boxes. What is the ratio of boxes to cost?

$\frac{3}{87}$

5. There are 7 boys and 8 girls playing softball. What is the ratio of girls to players?

$\frac{8}{15}$

6. Jim ran 5 kilometers in 25 minutes. What is the ratio of time to distance?

$\frac{25}{5}$

7. Jeff has 18 tiger's eye marbles and Sharon has 15 agate marbles. What is the ratio of Sharon's marbles to Jeff's?

$\frac{15}{18}$

8. On Monday, Nancy got 2 hits in 4 tries. On Tuesday, she got 1 hit in 3 tries. What is the ratio of total hits to total tries?

$\frac{3}{7}$

9. Troubles, Silvia's dog, had 9 puppies. Seven were black and 2 were white. What is the ratio of white to black puppies?

$\frac{2}{7}$

[Now Try This!]

Some ratios are known as rates. A rate is a comparison of two quantities with two different units of measure. When we say that a car travels 100 miles in 2 hours, we are comparing miles to hours. A unit rate has a denominator of 1. The rate 50 miles per hour is a unit rate.

Is the ratio a rate? Write *yes* or *no*. If the ratio is a rate, change it to a unit rate.

1. 42 gal in 7 min
   yes; 6 gal per min

2. 165 miles in 3 hours
   yes; 55 miles per hour

3. 75 boys to 50 girls
   no

Now Try This! Discuss the similarities and differences between a ratio and a rate. Discuss some common rates, such as miles per hour and price per gallon.

4 Assess

Refer to Getting Started, Exercise 7 on page 283. Ask, *What is the ratio of bananas to fruit?* (2 to 7 or $\frac{2}{7}$)

5 Mixed Review

1. $27{,}225 \div 67$ (406 R23)
2. 0.06×3.58 (0.2148)
3. $17\frac{5}{8} - 12\frac{9}{10}$ ($4\frac{29}{40}$)
4. $\frac{1}{9} =$ _____% ($11\frac{1}{9}$)
5. $157 + 25 + 108$ (290)
6. $14{,}090 - 9{,}576$ (4,514)
7. $0.0276 \div 12$ (0.0023)
8. $\frac{16}{25} \times \frac{10}{21} \times \frac{15}{28}$ ($\frac{8}{49}$)
9. 10^4 (10,000)
10. $0.07 =$ _____% (7)

For Mixed Abilities

Common Errors • Intervention

Some students may write the numbers in reverse order. Have them work with partners, each drawing a mixed array of circles, triangles, squares, and pentagons, with no more than ten of each kind. Then, have them exchange papers and write the following ratios as fractions:

1. squares to triangles
2. triangles to pentagons
3. squares to pentagons
4. circles to squares
5. squares to circles
6. triangles to squares
7. circles to polygons

Have partners exchange papers and check each other's work.

Enrichment • Application

Ask students to check the team standings for a sport of their choice. Have them choose two teams and write the following ratios for each team:

1. games won to games lost
2. games lost to games won
3. games lost to games played
4. games won to games played

ESL/ELL STRATEGIES

Familiarize students with the language of ratios by suggesting things to compare and then eliciting questions and answers. For example, have students determine the number of girls and the number of boys in the class. Ask, *What is the ratio of girls to boys in this class? What is the ratio of boys to girls? What is the ratio of girls to students in this class?* Point out that the number of students in the class is the total number of students, boys and girls.

T284

pages 285–286

1 Getting Started

Objective
• To identify equal ratios

Vocabulary
equal ratio, cross-products test

Materials
*transparency showing squares, triangles, and hexagons

Warm Up • Mental Math
Have students name the product in simplest form.

1. $\frac{2}{3} \times \frac{1}{5}$ $(\frac{2}{15})$ 5. $\frac{1}{2} \times \frac{6}{7}$ $(\frac{3}{7})$

2. $\frac{3}{4} \times \frac{2}{3}$ $(\frac{1}{2})$ 6. $\frac{2}{3} \times \frac{3}{5}$ $(\frac{2}{5})$

3. $\frac{2}{3} \times \frac{1}{2}$ $(\frac{1}{3})$ 7. $\frac{2}{3} \times \frac{1}{4}$ $(\frac{1}{6})$

4. $\frac{1}{2} \times \frac{1}{2}$ $(\frac{1}{4})$ 8. $\frac{1}{5} \times \frac{2}{5}$ $(\frac{2}{25})$

Warm Up • Activity
Display a transparency that shows 6 squares, 7 triangles, and 10 hexagons. Have students write the ratio as a fraction.

1. squares to triangles $(\frac{6}{7})$

2. triangles to hexagons $(\frac{7}{10})$

3. hexagons to squares $(\frac{10}{6})$

4. squares to polygons $(\frac{6}{23})$

5. triangles to squares $(\frac{7}{6})$

6. hexagons to polygons $(\frac{10}{23})$

Equal Ratios

Donald is the quarterback for his junior high football team. He completed 10 out of 15 passes in Saturday's game. His cousin, Dick, played on the rival team. Compare the ratios of their completed passes.

We want to compare two ratios.

We know Donald completed __10__ out of __15__ passes, and Dick completed __12__ out of __16__ passes.

We write the ratio of Donald's completed passes to total passes as $\frac{10}{15}$. We write the ratio of Dick's completed passes as $\frac{12}{16}$.

One way to compare ratios is to rename them with a common denominator.

$\frac{10}{15} = \frac{2}{3} = \frac{8}{12}$

$\frac{8}{12} \neq \frac{9}{12}$

$\frac{12}{16} = \frac{3}{4} = \frac{9}{12}$ ↑ not equal

Another way to compare ratios is to use the cross-products test.

$\frac{10}{15} \diagdown\diagup \frac{12}{16}$ $10 \times 16 = 160$
$15 \times 12 = 180$
$160 \neq 180$

The boys' ratios of completed passes are __not equal__.

Getting Started

Use equivalent fractions. Write = or ≠.

1. $\frac{5}{8}$ __=__ $\frac{15}{24}$ 2. $\frac{3}{9}$ __≠__ $\frac{4}{15}$ 3. $\frac{6}{24}$ __=__ $\frac{4}{16}$

Use the cross-products test. Write = or ≠.

4. $\frac{9}{2}$ __≠__ $\frac{50}{12}$ 5. $\frac{8}{48}$ __=__ $\frac{20}{120}$ 6. $\frac{16}{40}$ __=__ $\frac{18}{45}$

---

2 Teach

Introduce the Lesson Display one math book and two reading books. Ask a student to write a fraction to show the ratio of math books to reading books. $(\frac{1}{2})$ Add one more math book and two more reading books and have another student write a fraction for the new ratio. $(\frac{2}{4})$ Add one more math book and two more reading books and repeat the procedure. $(\frac{3}{6})$ Explain that the ratio of math books to reading books has not changed because $\frac{1}{2}$, $\frac{2}{4}$, and $\frac{3}{6}$ are equal ratios.

Then, ask a student to read the problem and identify the given information. Work through the common denominator method and cross-products test in the model with the class.

Develop Skills and Concepts Explain that one procedure for finding equal ratios is the same as that for finding equivalent fractions: both terms of the original ratio are multiplied or divided by the same number. Similarly, students can compare two or more ratios when the ratios have a common denominator.

Point out that another way to compare ratios is to use the cross-products test as described on page 285. Ask why *cross products* is an appropriate name. (Possible answer: Multiplying the numerator of one fraction times the denominator of the other forms an X.)

Have students use the cross-products test to write = or ≠ for the following exercises:

1. $\frac{4}{5}$ (=) $\frac{144}{180}$ 2. $\frac{7}{2}$ (≠) $\frac{182}{54}$ 3. $\frac{13}{14}$ (=) $\frac{48.1}{51.8}$

Practice

Use equivalent fractions. Write = or ≠.

1. $\frac{5}{6}$ ≠ $\frac{40}{54}$
2. $\frac{3}{7}$ = $\frac{15}{35}$
3. $\frac{9}{24}$ —— $\frac{12}{32}$

4. $\frac{5}{6}$ ≠ $\frac{9}{21}$
5. $\frac{30}{36}$ = $\frac{5}{6}$
6. $\frac{8}{3}$ = $\frac{40}{15}$

7. $\frac{18}{27}$ ≠ $\frac{5}{45}$
8. $\frac{5}{7}$ ≠ $\frac{3}{5}$
9. $\frac{5}{16}$ ≠ $\frac{4}{9}$

10. $\frac{4}{17}$ ≠ $\frac{16}{54}$
11. $\frac{12}{15}$ = $\frac{48}{60}$
12. $\frac{14}{28}$ ≠ $\frac{16}{37}$

Use the cross products test. Write = or ≠.

13. $\frac{3}{4}$ ≠ $\frac{5}{6}$
14. $\frac{5}{6}$ ≠ $\frac{7}{9}$
15. $\frac{25}{35}$ ≠ $\frac{6}{14}$

16. $\frac{7}{12}$ ≠ $\frac{1}{2}$
17. $\frac{7}{24}$ ≠ $\frac{5}{9}$
18. $\frac{1}{8}$ = $\frac{5}{40}$

19. $\frac{14}{3}$ = $\frac{42}{9}$
20. $\frac{6}{9}$ = $\frac{8}{12}$
21. $\frac{5}{2}$ ≠ $\frac{21}{9}$

22. $\frac{8}{7}$ ≠ $\frac{12}{15}$
23. $\frac{2}{3}$ ≠ $\frac{7}{8}$
24. $\frac{7}{15}$ = $\frac{14}{30}$

Problem Solving

Solve each problem.

25. Pat drew a rectangle 6 inches wide and 5 inches long. Tom drew a rectangle 18 inches wide and 15 inches long. Was the ratio of width to length the same for both rectangles? Yes.

26. Sal mixed 4 cans of juice with 6 cans of water. Ben mixed 6 cans of juice with 8 cans of water. Was the concentration of juice to water the same? No.

〔 Now Try This! 〕

Use an arrow to show the direction the third wheel will move. Write the number of times it will turn.

The first wheel makes $\frac{1}{2}$ of a turn. ___1___

The first wheel makes 2 turns. ___$1\frac{3}{5}$___

Lesson 13-2 • Equal Ratios

For Mixed Abilities

Common Errors • Intervention

When students use the cross-products test, have them write the products above the fractions, not below them.

14   15

$\frac{2}{5}$ × $\frac{3}{7}$

Discuss how they really are comparing the numerators of fractions whose common denominator would be the product of the two denominators. Since $\frac{14}{35}$ ≠ $\frac{15}{35}$, then $\frac{2}{5}$ ≠ $\frac{3}{7}$.

Enrichment • Number Sense

Draw picture ratios such as the ones below on the board for students to solve. Have students determine if the ratios are equal.

$\frac{\bigcirc}{\triangle\triangle\triangle}$ $\frac{\bigcirc\bigcirc}{\triangle\triangle\triangle\triangle}$ ($\frac{1}{3}$ ≠ $\frac{2}{4}$)

$\frac{XX}{\square\square\square}$ $\frac{XXX\backslash}{\square\square\square\square\square}$ ($\frac{2}{3}$ = $\frac{3.5}{5.25}$)

More to Explore • Application

Duplicate the following for students to solve:

> The work schedule for the Seaway Canning factory was posted. Ralph discovered he was supposed to work Monday, Wednesday, and Friday from 6:30 A.M. to 11:30 A.M., and Saturday from 1:30 P.M. to 7:30 P.M. Ralph, who regularly is paid $4.75 an hour, was happy about this schedule because he is paid time and a half on Saturdays. What were his total earnings for the week? ($114)

---

3 Practice

Have students complete all the exercises. Watch for students who multiply the numerator and denominator of the same ratio instead of cross-multiplying.

Now Try This! Discuss the wheels with students and in which machine they might be found. Then, have students work in pairs or small groups to solve the problems.

4 Assess

Ask students which method they prefer to use to identify equal ratios. Then, have them use that method to decide if $\frac{5}{8}$ and $\frac{20}{32}$ are equal ratios. (Check students' work. $\frac{5}{8}$ and $\frac{20}{32}$ are equal ratios.)

T286

1 Getting Started

Objective
• To use cross products to find the missing term in a proportion

Vocabulary
proportion

Warm Up • Mental Math
Have students name the missing term in the equal ratio.

1. $\frac{2}{3} = \frac{(6)}{9}$

2. $\frac{6}{8} = \frac{(18)}{24}$

3. $\frac{5}{9} = \frac{(10)}{18}$

4. $\frac{49}{63} = \frac{7}{(9)}$

5. $\frac{8}{10} = \frac{4}{(5)}$

6. $\frac{12}{32} = \frac{(3)}{8}$

7. $\frac{3}{10} = \frac{(6)}{20}$

8. $\frac{16}{100} = \frac{4}{(25)}$

Warm Up • Pencil and Paper
Have students find the product or quotient.

1. $\begin{array}{r} 4.63 \\ \times\ \ 3.8 \\ \hline (17.594) \end{array}$ 4. $4\overline{)105.6}$ (26.4)

2. $\begin{array}{r} 7.34 \\ \times\ 0.91 \\ \hline (6.6794) \end{array}$ 5. $0.63\overline{)5.166}$ (8.2)

3. $3.2\overline{)20.16}$ (6.3)   6. $7.9\overline{)39.5}$ (5)

2 Teach

Introduce the Lesson Explain to students that solving proportions involves using skills and concepts they have already mastered. Stress that the first term in each ratio stands for cream paint and the second term stands for tan paint.

As you work through the model section with the class, point out that students will use cross products to find a missing term in the proportion. First, they write an equation that shows the cross products are equal. Then, they divide to find the missing term.

Have students fill in the solution sentence.

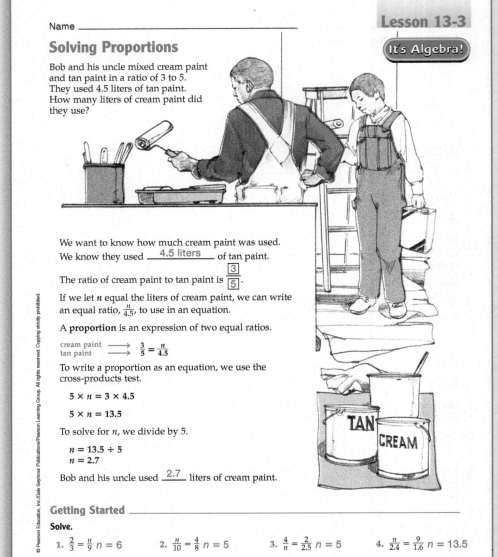

Name _____

Solving Proportions

It's Algebra!

Bob and his uncle mixed cream paint and tan paint in a ratio of 3 to 5. They used 4.5 liters of tan paint. How many liters of cream paint did they use?

We want to know how much cream paint was used. We know they used __4.5 liters__ of tan paint.

The ratio of cream paint to tan paint is $\frac{3}{5}$.

If we let n equal the liters of cream paint, we can write an equal ratio, $\frac{n}{4.5}$, to use in an equation.

A **proportion** is an expression of two equal ratios.

cream paint $\longrightarrow$ $\frac{3}{5} = \frac{n}{4.5}$
tan paint $\longrightarrow$

To write a proportion as an equation, we use the cross-products test.

$5 \times n = 3 \times 4.5$

$5 \times n = 13.5$

To solve for n, we divide by 5.

$n = 13.5 \div 5$
$n = 2.7$

Bob and his uncle used __2.7__ liters of cream paint.

Getting Started

Solve.

1. $\frac{2}{3} = \frac{n}{9}$ $n = 6$
2. $\frac{n}{10} = \frac{4}{8}$ $n = 5$
3. $\frac{4}{n} = \frac{2}{2.5}$ $n = 5$
4. $\frac{n}{2.4} = \frac{9}{1.6}$ $n = 13.5$

Lesson 13-3 • Solving Proportions

Develop Skills and Concepts Remind students that if two ratios are equal, their cross products are equal. Write $\frac{2}{4} = \frac{3}{6}$, then $2 \times 6 = 4 \times 3$ on the board. Point out that the two multiplications are written as an equation because we know that the cross products are equal.

Students can use this concept to solve a proportion where one term in the proportion is unknown. Review the idea that multiplication and division are inverse operations. Once students write the multiplication equation, they will divide to find the value of n.

Have students solve each equation.

1. $3 \times n = 18$ (6)   4. $n = 63 \div 7$ (9)
2. $n = 6 \times 8$ (48)   5. $n \times 3 = 4 \times 9$ (12)
3. $n \times 10 = 6 \times 5$ (3)   6. $n \div 9 = 8$ (72)

Practice

Solve.

1. $\frac{3}{4} = \frac{n}{16}$
   $n = 12$

2. $\frac{n}{8} = \frac{21}{24}$
   $n = 7$

3. $\frac{5}{6} = \frac{15}{n}$
   $n = 18$

4. $\frac{n}{9} = \frac{12}{27}$
   $n = 4$

5. $\frac{1}{n} = \frac{5}{25}$
   $n = 5$

6. $\frac{10}{7} = \frac{n}{70}$
   $n = 100$

7. $\frac{3}{n} = \frac{15}{40}$
   $n = 8$

8. $\frac{4}{6} = \frac{n}{12}$
   $n = 8$

9. $\frac{6}{n} = \frac{9}{12}$
   $n = 8$

10. $\frac{4}{16} = \frac{5}{n}$
    $n = 20$

11. $\frac{n}{2.5} = \frac{3}{5}$
    $n = 1.5$

12. $\frac{1.5}{6} = \frac{n}{12}$
    $n = 3$

13. $\frac{12}{n} = \frac{8}{3}$
    $n = 4.5$

14. $\frac{5.6}{2} = \frac{n}{4}$
    $n = 11.2$

15. $\frac{n}{3.5} = \frac{1}{7}$
    $n = 0.5$

16. $\frac{3.6}{15} = \frac{1.2}{n}$
    $n = 5$

17. $\frac{2.8}{7} = \frac{n}{4}$
    $n = 1.6$

18. $\frac{18}{25} = \frac{n}{7}$
    $n = 5.04$

19. $\frac{n}{3.7} = \frac{4}{29.6}$
    $n = 0.5$

20. $\frac{1.5}{n} = \frac{1.2}{7}$
    $n = 8.75$

Problem Solving

Solve each problem.

21. Mary ran 2 kilometers in 15 minutes. At the same rate, how far can she run in 25 minutes?
    $3\frac{1}{3}$ kilometers

22. Glasses are on sale at 6 for $6.90. How much will Mrs. Cavell pay for 4 glasses?
    $4.60

23. To make punch, Tom uses 3 cups of lemonade for every 2 cups of grape juice. Tom used 12 cups of lemonade. How many cups of grape juice did he use?
    8 cups

24. A tree grows 3 inches every 5 months. How many inches will the tree grow in 9 months?
    5.4 inches

(Now Try This!)

Complete the table to find the ratio of the area to the perimeter of the squares.

Length of Side of Square	Area of Square	Perimeter of Square	A/P
1	1	4	$\frac{1}{4}$
2	4	8	$\frac{1}{2}$
3	9	12	$\frac{3}{4}$
4	16	16	1
5	25	20	$\frac{5}{4}$

Lesson 13-3 • Solving Proportions

For Mixed Abilities

Common Errors • Intervention

Some students may have difficulty solving the equation that results from a proportion. Discuss with them the inverse relationship between multiplication and division, and how one undoes the other. Then, discuss how this idea is used as you solve various equations step by step as done with the one shown below.

$$5 \times n = 13.5$$
$$n \times 5 = 13.5$$
$$n \times 5 \div 5 = 13.5 \div 5$$
$$n \times 1 = 2.7$$
$$n = 2.7$$

Enrichment • Application

Show students how to find and take their pulse, using a watch with a second hand to count the beats. Have students count the number of times their heart beats in 1 minute. Ask them to write the ratio of minutes to heartbeats. Then, have students write proportions to find the number of times their heart would beat in 4 minutes, 6 minutes, and 10 minutes.

More to Explore • Biography

Ever since the beginning of time, people have been wondering how the universe came into existence. One scientist with a theory on the origin of the solar system was Pierre Simon de Laplace, a French astronomer and mathematician. Laplace's theory states that the universe began as a huge, lens-shaped cloud of gas that rotated, cooled, contracted, and threw off planets and satellites. What was left formed the Sun.

Laplace's nebular hypothesis, as it came to be called, was accepted for a long time but has now been replaced by other theories. Laplace was born in France in 1749, the son of a farmer. This remarkable young man became a professor of mathematics in Paris at age 19. He helped establish the metric system during the French Revolution. Politically active under Napoleon, Laplace was named a marquis in 1817.

---

3 Practice

Have students complete all the exercises. In the word problems, students must first set up the proportion and then solve it. Remind them that the first term in each ratio of a proportion must stand for the same thing.

Now Try This! Be sure students understand that they are finding the ratio of the area of a square to its perimeter. Put the formulas for each on the board.

4 Assess

Ask students to describe the steps for solving a proportion. (Possible answer: Write the proportion as an equation using the cross-products test. Then, divide by the number that is multiplied by *n*.)

13-4 Similar Polygons

pages 289–290

1 Getting Started

Objective
- To use proportions to find measures of similar polygons

Vocabulary
similar polygons, corresponding angles, corresponding sides

Materials
index cards; colored markers

Warm Up • Mental Math
Have students find the area of the square.

1. $s = 12$ in. (144 in.^{2})
2. $s = 9$ ft (81 ft^2)
3. $s = 15$ yd (225 yd^2)
4. $s = 30$ ft (900 ft^2)
5. $s = 20$ in. (400 in.^{2})
6. $s = 100$ yd ($10,000$ yd^2)
7. $s = 50$ ft ($2,500$ ft^2)

Warm Up • Pencil and Paper
Have students solve the proportion.

1. $\frac{6}{8} = \frac{9}{n}$ ($n = 12$)
2. $\frac{3}{8} = \frac{n}{21}$ ($n = 7.875$)
3. $\frac{9}{15} = \frac{6}{n}$ ($n = 10$)
4. $\frac{12}{9} = \frac{n}{12}$ ($n = 16$)
5. $\frac{2.5}{4} = \frac{n}{6}$ ($n = 3.75$)
6. $\frac{0.99}{5} = \frac{5.94}{n}$ ($n = 30$)

2 Teach

Introduce the Lesson Work through the model section with students, showing how the lengths of corresponding sides of similar polygons can be expressed as ratios. Have students solve the proportion to answer the question. Use the model to emphasize the following:

- Polygons that have the same shape but different sizes are called *similar polygons*.

- The *corresponding angles* of similar polygons are congruent.

- The *corresponding sides* of similar polygons are proportional; their lengths can be expressed as ratios.

- To find the missing measure of one side of a polygon that is similar to another polygon, solve the proportion formed by the ratios of corresponding sides.

Name _____

Similar Polygons

Similar polygons are the same shape, but not always the same size. The angles or sides that are in the same relative positions are called **corresponding**. The corresponding angles of similar polygons are congruent. The corresponding sides of similar polygons are proportional. If triangle ABC is similar to triangle PQR, what is the measurement of line segment QR?

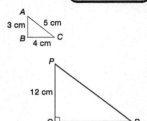

We need to know the length of one side of triangle PQR. We know triangles PQR and ABC are similar.

In the triangles, side AB corresponds to side __PQ__, and the ratio of $\overline{AB}$ to $\overline{PQ}$ is $\boxed{\frac{1}{4}}$.

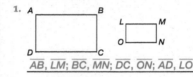

Side BC corresponds to side __QR__, and the ratio of $\overline{BC}$ to $\overline{QR}$ is $\boxed{\frac{4}{n}}$.

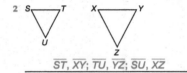

To find the length of $\overline{QR}$, write and solve a proportion.

$$\frac{m(AB)}{m(PQ)} = \frac{m(BC)}{m(QR)} \qquad \frac{3}{12} = \frac{4}{n}$$

$$3 \times n = 12 \times 4$$
$$3 \times n = 48$$
$$n = 48 \div 3$$
$$n = 16$$

The length of $\overline{QR}$ is __16__ centimeters.

Getting Started

The figures are similar. Name the corresponding sides.

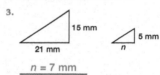

1. $\overline{AB}, \overline{LM}; \overline{BC}, \overline{MN}; \overline{DC}, \overline{ON}; \overline{AD}, \overline{LO}$

2. $\overline{ST}, \overline{XY}; \overline{TU}, \overline{YZ}; \overline{SU}, \overline{XZ}$

The figures are similar. Solve for n.

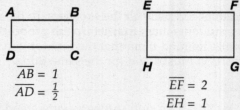

3. $n = 7$ mm

4. $n = 2$ m

Develop Skills and Concepts Draw and label two rectangles as shown below.

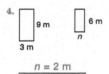

$\overline{AB} = 1$
$\overline{AD} = \frac{1}{2}$

$\overline{EF} = 2$
$\overline{EH} = 1$

Point out that side AB of the smaller rectangle corresponds to side EF of the larger rectangle. Ask students to name other pairs of corresponding sides. Then, explain that similar polygons are polygons that have the same shape but are not always the same size.

Practice

The figures are similar. Name the corresponding sides.

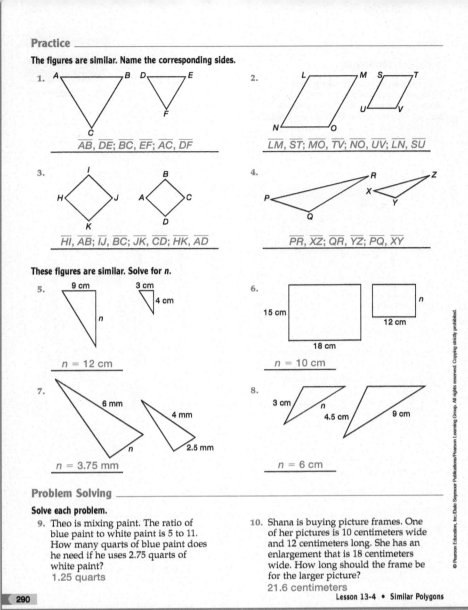

1. $\overline{AB}, \overline{DE}; \overline{BC}, \overline{EF}; \overline{AC}, \overline{DF}$

2. $\overline{LM}, \overline{ST}; \overline{MO}, \overline{TV}; \overline{NO}, \overline{UV}; \overline{LN}, \overline{SU}$

3. $\overline{HI}, \overline{AB}; \overline{IJ}, \overline{BC}; \overline{JK}, \overline{CD}; \overline{HK}, \overline{AD}$

4. $\overline{PR}, \overline{XZ}; \overline{QR}, \overline{YZ}; \overline{PQ}, \overline{XY}$

These figures are similar. Solve for n.

5. 9 cm, 3 cm, 4 cm, n

$n = 12$ cm

6. 15 cm, 18 cm, 12 cm, n

$n = 10$ cm

7. 6 mm, 4 mm, n, 2.5 mm

$n = 3.75$ mm

8. 3 cm, n, 4.5 cm, 9 cm

$n = 6$ cm

Problem Solving

Solve each problem.

9. Theo is mixing paint. The ratio of blue paint to white paint is 5 to 11. How many quarts of blue paint does he need if he uses 2.75 quarts of white paint?
   1.25 quarts

10. Shana is buying picture frames. One of her pictures is 10 centimeters wide and 12 centimeters long. She has an enlargement that is 18 centimeters wide. How long should the frame be for the larger picture?
    21.6 centimeters

---

3 Practice

Have students complete all the exercises. Remind them to write the ratios of corresponding sides and solve for the missing term.

4 Assess

Ask students to explain how they would set up the proportion for the similar rectangles in Exercise 6. (Possible answer: $\frac{15}{n} = \frac{18}{12}$)

1 Getting Started

Objectives

- To use a scale drawing to find actual length
- To use actual length to determine scale

Vocabulary

scale drawing

Materials

rulers; maps; yardsticks

Warm Up • Mental Math

Have students find the volume of the box.

1. $l = 6$ cm
   $w = 3$ cm
   $h = 2$ cm
   ($V = 36$ cm^3)
2. $l = 10$ m
   $w = 8$ m
   $h = 4$ m
   ($V = 320$ m^3)
3. $l = 9$ m
   $w = 6$ m
   $h = 3$ m
   ($V = 162$ m^3)

Warm Up • Pencil and Paper

Have students solve the proportion.

1. $\frac{4}{5} = \frac{12}{n}$ ($n = 15$)
2. $\frac{3}{8} = \frac{n}{10}$ ($n = 3.75$)
3. $\frac{5}{6} = \frac{6.5}{n}$ ($n = 7.8$)
4. $\frac{0.3}{12} = \frac{0.5}{n}$ ($n = 20$)
5. $\frac{0.6}{15} = \frac{n}{22.5}$ ($n = 0.9$)
6. $\frac{10}{2.5} = \frac{14}{n}$ ($n = 3.5$)

2 Teach

Introduce the Lesson Have students name some uses for scale drawings. (Possible answers: maps, blueprints, and models) Explain that a scale can be used to determine actual distances.

Ask a student to read the problem and identify the question and the given information. Note that the scale can be written as a ratio. Work through the model, having students measure the map distance from Lakeville to East City with centimeter rulers. Have students fill in the blanks and solve the problem.

Remind students to use cross products to solve proportions.

Scale Drawings

It's Algebra!

The scale on the map shows what every centimeter on the map represents in actual kilometers. Using a proportion, we can find the actual distance between any two points on the map. How far is it from Lakeville to East City?

We want to know the actual distance between two places on the map.

We know that the ratio in the scale drawing is $\frac{\boxed{2}}{\boxed{9}}$.

By measuring, we find that the map distance from Lakeville to East City is ___7___ centimeters. To find the actual distance from Lakeville to East City, we write a proportion and solve for n.

$\frac{2}{9} = \frac{7}{n}$ ⟵ map distance in centimeters
⟵ actual distance in kilometers

$2 \times n = 9 \times 7$
$2 \times n = 63$
$n = 63 \div \underline{2}$
$n = \underline{31.5}$

It is ___31.5___ kilometers from Lakeville to East City.

Getting Started

Measure the map distance and find the actual distance.

1. From South City to Centerville
   11.25 kilometers
2. From Pine to South City
   13.5 kilometers
3. From Centerville to East City
   27 kilometers
4. From Lakeville to Centerville
   13.5 kilometers

Develop Skills and Concepts Explain that in each scale ratio, the first number represents units used in the drawing and the second number represents units used to measure the actual distance or object. Encourage students to follow this procedure to find actual distances:

- Write the map scale as a ratio, comparing the map distance (in. or cm) to the actual distance.
- Measure the distance on the scale drawing or map; write a ratio comparing this distance to n.
- Write a proportion.
- Solve the proportion to find the missing distance.

If students need concrete reinforcement, set up a ratio table that shows actual distance per centimeter according to scale. (2 cm = 9 km, 3 cm = 13.5 km, and so on)

Practice

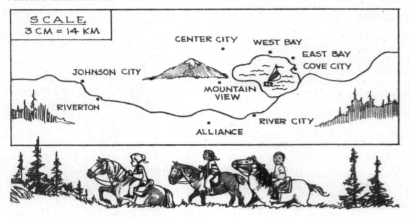

Measure the map distance and find the actual distance.

1. From Riverton to Center City
   28 kilometers

2. From Johnson City to River City
   35.9 kilometers

3. From Center City to West Bay
   9.8 kilometers

4. From East Bay to West Bay
   4.7 kilometers

5. From Mountain View to Center City
   7 kilometers

6. From River City to Riverton
   32.7 kilometers

7. From Alliance to Riverton
   23.8 kilometers

8. From East Bay to Cove City
   2.8 kilometers

Problem Solving

Solve each problem.

9. On a scale drawing of a kitchen,
   3 cm = 4 meters. If one side of the
   kitchen wall is 5 cm long in the drawing, what is its actual length?
   6.7 meters

10. On a scale drawing of a house,
    2 cm = 7 meters. The outside
    dimensions of the house are 10 meters
    by 15 meters. What will be the
    dimensions on the scale drawing?
    2.9 centimeters by 4.3 centimeters

11. The dinosaur model in the museum is
    built to a scale of 5 cm = 8 m. If the
    dinosaur was actually 42 meters in
    length, what is the length of the
    model?
    26.25 centimeters

12. The statue of George Washington
    in the park is built to a scale of
    9 in. = 1.75 ft. If the statue is 14 feet
    tall, what was George Washington's
    height in feet?
    6 feet

Lesson 13-5 • Scale Drawings

For Mixed Abilities

Common Errors • Intervention

Some students may have
difficulty organizing their work
when they are using a scale to
find actual distance. Help them
create four-column tables using
the headings shown below.

Map scale	Actual scale	Map distance	Actual distance

Enrichment • Application

1. Have students work in small
   groups with a map and rulers.
   Have students choose four
   places on their map to visit.
   Tell them to use the map scale
   to determine the actual
   distance of each leg of their
   trip. Have them add to find the
   total distance of their trip.

2. Have students measure the
   length and width of the
   classroom with yardsticks or
   metersticks. Then, have them
   decide on an appropriate scale
   and make a scale drawing of
   the room.

More to Explore • Sets

Duplicate the following Venn
diagram for students:

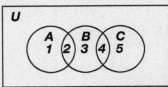

Challenge students to create their
own set information to coordinate
with the diagram. Have them
write a series of questions for
another student to answer about
their diagram. Have them
exchange their completed
diagram and questions for
classmates to solve.

---

3 Practice

Have students complete all the exercises. Watch for
students who do not set up parallel ratios.

4 Assess

Have students write the proportion needed to find the
actual distance from Riverton to Center City. $(\frac{3}{14} = \frac{6}{n})$

5 Mixed Review

1. $3\frac{3}{8} \times 5\frac{5}{9}$ $(18\frac{3}{4})$

2. 16.8×14.4 (241.92)

3. $9,376 - 5,757$ (3,619)

4. $326.43 \div 0.03$ (10,881)

5. 70×800 (56,000)

6. 32% of 226 (72.32)

7. $153.892 - 89.7$ (64.192)

8. $\frac{2}{7} + \frac{4}{5}$ $(1\frac{3}{35})$

9. $7,520 \times 43$ (323,360)

10. $18 \div \frac{6}{7}$ (21)

13-6 Percents

pages 293–294

1 Getting Started

Objective
• To write decimals as percents and percents as decimals

Vocabulary
percent

Materials
index cards

Warm Up • Mental Math
Have students solve for n.

1. $5 \times n = 100$ (20)
2. $8 \times n = 800$ (100)
3. $n \times 3 = 96$ (32)
4. $n \times 2 = 506$ (253)
5. $7 \times n = 707$ (101)
6. $9 \times n = 7,200$ (800)
7. $n \times 4 = 280$ (70)
8. $n \times 6 = 1,212$ (202)

Warm Up • Pencil and Paper
Have students find the product or quotient.

1. $36 \div 100$ (0.36)
2. 0.07×100 (7)
3. 5.29×100 (529)
4. $425 \div 100$ (4.25)
5. $9 \div 100$ (0.09)
6. 15×100 (1,500)
7. 0.67×100 (67)
8. $0.8 \div 100$ (0.008)

Name _____

Percents

There are 16 trumpet players and 25 drummers in the Allentown marching band. There are 100 members in the band altogether. We can write the ratio $\frac{16}{100}$ to compare the number of trumpet players to all the band members.

Another way to express the same comparison is to use a percent. **Percent** means that amount per 100 and is shown by the symbol %. We say 16% of the players are trumpeters. What percent of the band members are drummers?

We want to use a percent to compare the number of drummers to the total band.

We know that there are __25__ drummers and that there are __100__ musicians in all.

The ratio of drummers to total membership is $\boxed{\dfrac{25}{100}}$.

We say __25__ % of the musicians are drummers. We can write a percent as a decimal by dividing by 100.

$$16\% = 0.16$$
$$25\% = \underline{0.25}$$

We can write a decimal as a percent by multiplying by 100.

$$0.09 = 9\%$$
$$0.45 = \underline{45\%}$$

Getting Started

Write the percent.

1. shaded squares __18%__
2. unshaded squares __82%__

Write as a percent.

3. $0.65 = $ __65%__
4. $0.01 = $ __1%__
5. $0.25 = $ __25%__

Write as a decimal.

6. $16\% = $ __0.16__
7. $89\% = $ __0.89__
8. $75\% = $ __0.75__

Lesson 13-6 • Percents

293

---

2 Teach

Introduce the Lesson Explain that a percent is a special kind of ratio that always has a second term of 100 because *percent* means "per hundred." Ask students to name some common uses of percents. (sales tax, discounts, interest rates)

Work through the model with students, having them fill in the blanks. Remind students that a fraction is an implied division; $\frac{25}{100}$ means $25 \div 100$.

Develop Skills and Concepts Emphasize that a percent can be written as a decimal because it is a number divided by 100. Have students write each of the following percents as decimals: 23% (0.23), 61% (0.61), and 97% (0.97).

Emphasize also that inversely, a decimal can be written as a percent by multiplying by 100. Have students write each of the following decimals as a percent: 0.32 (32%), 0.77 (77%), and 0.02 (2%).

3 Practice

Have students complete all the exercises. In the Problem Solving section, encourage students to express the numbers in ratio form first, then as percents.

Practice

Write the percent.

1. shaded squares ___23%___

2. unshaded squares ___77%___

Write as a percent.

3. 0.39	4. 0.17	5. 0.02	6. 0.98	7. 0.37	8. 0.16
39%	17%	2%	98%	37%	16%

9. 0.05	10. 0.40	11. 0.51	12. 0.65	13. 0.39	14. 0.87
5%	40%	51%	65%	39%	87%

15. 1.87	16. 0.01	17. 0.57	18. 0.19	19. 2.87	20. 0.99
187%	1%	57%	19%	287%	99%

Write as a decimal.

21. 50%	22. 36%	23. 12%	24. 85%	25. 7%	26. 19%
0.50	0.36	0.12	0.85	0.07	0.19

27. 16%	28. 1%	29. 70%	30. 10%	31. 99%	32. 12%
0.16	0.01	0.70	0.10	0.99	0.12

33. 75%	34. 200%	35. 4%	36. 53%	37. 107%	38. 22%
0.75	2.00	0.04	0.53	1.07	0.22

Problem Solving

Solve each problem.

39. In a telephone poll, 4 out of 100 people had pets. What percent of the people called had pets?
4%

40. Robert bought a suit for $100. He paid $7 in sales tax. What percent was the sales tax?
7%

41. Student Council members want to raise $100 to buy new books for the library. So far, they have raised $83. What percent of their goal must they still raise?
17%

42. Leon wants to jog 100 miles this month. So far, he has jogged 45 miles. What percent of his goal does Leon have left?
55%

294

Lesson 13-6 • Percents

---

4 Assess

Have students write 5, 0.5, and 0.05 as a percent. (500%, 50%, 5%) Have students write 2%, 22%, and 220% as a decimal. (0.02, 0.22, 2.2)

5 Mixed Review

1. 327×406 (132,762)
2. $5\frac{1}{3} \times 1\frac{1}{8}$ (6)
3. $17 + 9 \times 5$ (62)
4. $168 - 127.382$ (40.618)
5. 40% of 25 (10)
6. 0.03×0.18 (0.0054)
7. $\$19.63 + \$107.52 + \$8.98$ ($136.13)
8. $27\frac{3}{5} - 19\frac{5}{8}$ ($7\frac{39}{40}$)
9. $\frac{5}{6} + \frac{7}{9} + \frac{1}{4}$ ($1\frac{31}{36}$)
10. $64,028 \div 8$ (8,003 R4)

For Mixed Abilities

Common Errors • Intervention

If students have difficulty relating percents and decimals, have them practice with partners. Have each pair of students write each of the following numbers on a separate index card.

0.44	0.35	0.5	0.46	0.99
0.02	0.07	0.27	0.90	0.08
35%	50%	46%	2%	7%
27%	90%	99%	8%	44%

Separate the percents and decimals into two piles. Shuffle each pile and place facedown. Students take turns picking the top card from each pile. If the cards name the same value, the student keeps the cards; otherwise the cards are discarded. The student with the most matched cards is the winner.

Enrichment • Number Sense

Have students draw squares, 10 units on a side, on graph paper. Have them color the interior squares according to these instructions: 30% red, 16% blue, 10% orange, 14% yellow, 5% green, and 25% purple.

More to Explore • Probability

Explain that flower color is determined by the genes of the parent flower plants. The diagram shows the possible combinations of genes for color in snapdragons.

	R	W
R	RR	RW
W	RW	WW

Each seed will have two genes, one from each parent. If each parent plant gives the seed an R gene, the plant will have red flowers. If both parents pass on a W gene, the plant will have white flowers. If one parent gives an R gene and the other gives a W gene, the flowers will be pink. This is called incomplete dominance.

Ask students to express the probability that a snapdragon will be pink ($\frac{1}{2}$ or 50%), red ($\frac{1}{4}$ or 25%), or white. ($\frac{1}{4}$ or 25%).

T294

13-7 Renaming Fractions, Decimals, and Percents

pages 295–296

1 Getting Started

Objectives

- To rename fractions and decimals as percents

- To rename percents as fractions and decimals

Warm Up • Mental Math

Have students rename the fraction in simplest form.

1. $\frac{6}{100}$ $(\frac{3}{50})$ 4. $\frac{66}{100}$ $(\frac{33}{50})$

2. $\frac{24}{100}$ $(\frac{6}{25})$ 5. $\frac{46}{100}$ $(\frac{23}{50})$

3. $\frac{85}{100}$ $(\frac{17}{20})$ 6. $\frac{52}{100}$ $(\frac{13}{25})$

Warm Up • Pencil and Paper

Have students round the quotient to the nearest hundredth and multiply the rounded quotient by 100.

1. $3\overline{)6.45}$ 5. $7\overline{)0.948}$
   (2.15; 215)   (0.14; 14)

2. $9\overline{)34.9}$ 6. $4\overline{)76.1}$
   (3.88; 388)   (19.03; 1,903)

3. $6\overline{)4.86}$ 7. $2\overline{)9.40}$
   (0.81; 81)   (4.70; 470)

4. $5\overline{)7.23}$ 8. $8\overline{)69.84}$
   (1.45; 145)   (8.73; 873)

Name _____

Renaming Fractions, Decimals, and Percents

We can use what we have learned about ratios to rename equivalent fractions, decimals, and percents. What percent of the geometric figures are triangles?

We want to name a ratio as a percent.

We know that the ratio of triangles to geometric figures is $\frac{2}{3}$.

To rename $\frac{2}{3}$ as a decimal and a percent, follow these steps:

Divide the numerator by the denominator. Carry the division to hundredths. Write the remainder as a fraction.	Write the decimal as a percent by multiplying by 100.

$$0.66\frac{2}{3}$$
$$3\overline{)2.00}$$
$$\underline{18}$$
$$20$$
$$\underline{18}$$
$$2$$

$0.66\frac{2}{3} = \underline{66\frac{2}{3}\%}$

$\underline{66\frac{2}{3}}$ % of the geometric figures are triangles.

To rename a percent as a decimal and a fraction, follow these steps:

Write the percent as a decimal by dividing by 100.	Write the decimal as a fraction by writing the value over 100.	Rename the fraction in simplest form.
$75\% = 0.75$	$0.75 = \frac{75}{100}$	$\frac{75}{100} = \frac{3}{4}$

Getting Started

Write as a percent.

1. $\frac{3}{5} = \underline{60\%}$ 2. $\frac{7}{25} = \underline{28\%}$ 3. $\frac{1}{2} = \underline{50\%}$ 4. $\frac{1}{3} = \underline{33\frac{1}{3}\%}$ 5. $\frac{1}{8} = \underline{12\frac{1}{2}\%}$

Write as a fraction.

6. $82\% = \underline{\frac{41}{50}}$ 7. $16\% = \underline{\frac{4}{25}}$ 8. $25\% = \underline{\frac{1}{4}}$ 9. $48\% = \underline{\frac{12}{25}}$ 10. $15\% = \underline{\frac{3}{20}}$

---

2 Teach

Introduce the Lesson Have students read the problem, identify the question, and state the given information. Explain that if the denominator of a fraction is not 100, students must follow the given steps to rename the fraction as a percent. Work through the model with students.

Point out that to rename percents as fractions, the procedure is reversed.

Develop Skills and Concepts Emphasize these points:

- A percent represents a number divided by 100 ($\frac{n}{100}$).

- To rename a percent as a fraction, write the percent as a fraction with 100 as the second term, simplifying if necessary.

- Point out that to rename a percent as a decimal, divide by 100, as implied in the fraction.

- To rename a fraction as a decimal, divide the numerator by the denominator.

- To rename a decimal as a percent, multiply by 100.

Have students work at the board to complete the following chart and explain their work:

Decimal	Fraction	Percent
(0.14)	$(\frac{7}{50})$	14%
0.58	$(\frac{29}{50})$	(58%)
(0.8)	$\frac{4}{5}$	(80%)
(0.875)	$\frac{7}{8}$	(87.5%)
0.3	$(\frac{3}{10})$	(30%)

3 Practice

Have students complete all the exercises. Remind them to write zeros in the dividend, to the hundredths place, to facilitate any division.

T295

Practice

Write as a percent.

1. $\frac{3}{5} = $ __60%__

2. $\frac{1}{8} = $ __$12\frac{1}{2}$%__

3. $\frac{1}{6} = $ __$16\frac{2}{3}$%__

4. $\frac{3}{7} = $ __$42\frac{6}{7}$%__

5. $\frac{5}{8} = $ __$62\frac{1}{2}$%__

6. $\frac{3}{4} = $ __75%__

7. $\frac{1}{3} = $ __$33\frac{1}{3}$%__

8. $\frac{7}{8} = $ __$87\frac{1}{2}$%__

9. $\frac{9}{20} = $ __45%__

10. $\frac{16}{25} = $ __64%__

11. $\frac{17}{50} = $ __34%__

12. $\frac{5}{9} = $ __$55\frac{5}{9}$%__

13. $\frac{3}{11} = $ __$27\frac{3}{11}$%__

14. $\frac{17}{20} = $ __85%__

15. $\frac{1}{9} = $ __$11\frac{1}{9}$%__

Write as a fraction.

16. $35\% = $ __$\frac{7}{20}$__

17. $40\% = $ __$\frac{2}{5}$__

18. $75\% = $ __$\frac{3}{4}$__

19. $16\% = $ __$\frac{4}{25}$__

20. $6\% = $ __$\frac{3}{50}$__

21. $10\% = $ __$\frac{1}{10}$__

22. $21\% = $ __$\frac{21}{100}$__

23. $58\% = $ __$\frac{29}{50}$__

24. $65\% = $ __$\frac{13}{20}$__

25. $70\% = $ __$\frac{7}{10}$__

26. $35\% = $ __$\frac{7}{20}$__

27. $88\% = $ __$\frac{22}{25}$__

28. $90\% = $ __$\frac{9}{10}$__

29. $79\% = $ __$\frac{79}{100}$__

30. $8\% = $ __$\frac{2}{25}$__

Problem Solving

Solve each problem.

31. Kuni tried to kick 10 goals. He scored 6 times. What is Kuni's scoring percent?
60%

32. Todd scored 90% on a test. Peggy received 8 out of 9 correct. Who has the higher score?
Todd

33. A wizzlemaker makes 250 wizzles per minute. Of these, 200 wizzles wizzle well. What percent of the wizzles wizzle well?
80%

34. Audrey planted $\frac{3}{25}$ of her garden with carrots, $\frac{1}{10}$ with beans, and $\frac{7}{20}$ with tomatoes. What percent of her garden is still unplanted?
43%

Now Try This!

Complete the pattern.

$0.99 = \frac{99}{100} = $ __99__ %

$1.25 = $ __125__ %

$1.06 = $ __106__ %

$1.00 = \frac{100}{100} = $ __100__ %

$1.09 = $ __109__ %

$150 = $ __15,000__ %

$1.01 = \frac{101}{100} = $ __101__ %

$1.8 = $ __180__ %

$2.00 = $ __200__ %

Lesson 13-7 • Renaming Fractions, Decimals, and Percents

For Mixed Abilities

Common Errors • Intervention

Some students may fail to write the fraction in simplest form. Stress that they should always check the numerator and denominator for common factors and divide both by the greatest common factor. The simplest form is no more correct than other equivalent factors, but it is desired because it is easier to work with.

Enrichment • Number Sense

Show students how to use the cross-products test to name a fraction or decimal as a percent.

• Name the decimal as a fraction.

• Write an equivalent ratio with n as the first term and 100 as the second term.

• Cross multiply to solve for n.

$0.75 = \frac{3}{4}$

$\frac{3}{4} = \frac{n}{100}$

$4 \times n = 300$

$n = 300 \div 4$

$n = 75\%$

Have students use this method to write the fraction as a percent.

1. $\frac{7}{8}$ ($87\frac{1}{2}$%)

4. 0.916 (91.6%)

2. $\frac{11}{12}$ ($91\frac{2}{3}$%)

5. 5.7 (570%)

3. $\frac{1}{16}$ ($6\frac{1}{4}$%)

6. 0.094 (9.4%)

More to Explore • Geometry

Tell students that they will be making a goniometer. Have each student cut two strips of stiff cardboard, 5 inches by 1 inch. Tell them to lay one strip directly over the other and fasten them together. This will form a movable V.

Now, have students bring in a variety of angular, not rounded, rocks. To measure the angles of the rocks with the goniometer, have students first "pinch" the rock securely with the cardboard strips, hold the brace firmly, move the goniometer to a piece of paper, and trace the angle formed by the inside of the strips. Have students then measure each angle with a protractor.

Now Try This! Ask students when it would be possible to have more than 100% of something. Explain that the method for renaming fractions and decimals as percents remains the same for percents greater than 100%.

4 Assess

Have students match the fraction in Column A to the correct percent in Column B.

Column A	Column B
$\frac{1}{2}$	30%
$\frac{3}{4}$	$33\frac{1}{3}$%
$\frac{1}{5}$	50%
$\frac{3}{10}$	75%
$\frac{1}{3}$	20%

($\frac{1}{2}$, 50%; $\frac{3}{4}$, 75%; $\frac{1}{5}$, 20%; $\frac{3}{10}$, 30%; $\frac{1}{3}$, $33\frac{1}{3}$%)

13-8 Finding a Percent of a Number

pages 297–298

1 Getting Started

Objective
• To find the percent of a number

Materials
small paper bags; red and white chips

Warm Up • Mental Math
Ask students if the number is divisible by 3.

1. 125 (No.)
2. 536 (No.)
3. 729 (Yes.)
4. 6,963 (Yes.)

Ask students if the number is divisible by 9.

5. 123 (No.)
6. 681 (No.)
7. 657 (Yes.)
8. 5,283 (Yes.)

Warm Up • Pencil and Paper
Have students find the product. Have them round money amounts to the nearest cent.

1. $\frac{18}{100} \times 300$ (54)

4. $\begin{array}{r} 462 \\ \times\ \ 0.09 \\ \hline (41.58) \end{array}$

2. $\frac{41}{100} \times 200$ (82)

5. $\begin{array}{r} \$16.59 \\ \times\ \ 0.48 \\ \hline (\$7.96) \end{array}$

3. $\begin{array}{r} 65 \\ \times\ 0.36 \\ \hline (23.4) \end{array}$

6. $\begin{array}{r} \$9.25 \\ \times\ 0.37 \\ \hline (\$3.42) \end{array}$

2 Teach

Introduce the Lesson Have a student read the problem and identify what is to be solved and what is known. As you work through the model, point out that the first step in finding the percent of a number is to write the percent as a decimal. The second step is to multiply the number by the decimal.

Have students think the following:

What number is 6% of $18.50?

$n = 0.06 \times \$18.50$

T297

Name _____

Lesson 13-8

Finding a Percent of a Number

Dawn bought her father a shirt for Father's Day. The sales tax on clothing is 6%. How much tax did Dawn pay? What was the total cost of the shirt?

Shirts $18.50
Sweaters $42.00
Jackets $65.00

We need to know how much Dawn paid for her father's gift.

We know that the shirt cost $18.50 and that the sales tax rate is 6 %.

To find the amount of tax, we need to find 6% of $18.50.

To find the percent of a number, follow these steps:

Rename the percent as a decimal.	Multiply the cost of the shirt by the rate.
6% = 0.06	$\begin{array}{r} \$18.50 \\ \times\ \ 0.06 \\ \hline \$1.1100 \end{array}$

The amount of sales tax was $1.11.

To find the total cost of the shirt, we add $18.50 and $1.11.

$\begin{array}{r} \$18.50 \\ +\ \ 1.11 \\ \hline \$19.61 \end{array}$

Dawn paid a total of $19.61 for her father's gift.

Getting Started

Write the percent of each number.

1. 25% of 28 = 7
2. 30% of 30 = 9
3. 75% of 48 = 36

Write the percent of each number. Use a decimal.

4. 12% of 70 = 8.4
5. 15% of 116 = 17.4
6. 8% of 36 = 2.88

Lesson 13-8 • Finding a Percent of a Number

297

Work through the rest of the information sentences with students, having them find the total cost by adding the price of the shirt to the sales tax.

Develop Skills and Concepts Point out that to find the percent of a number, students should refer to multiplying decimals. Point out that to find 70% of 165, they could think of 70% as a decimal (0.70) or a fraction $\frac{70}{100}$. Write the following on the board:

$$0.70 \times 165 = 115.5 \qquad \frac{70}{100} \times 165 = 115\frac{1}{2}$$

Explain that in this case, using decimals is easier.

Explain that it may be easier to use fractions for percents such as 20%, 25%, 50%, and 75% because they are easily remembered. Write the following on the board:

$$25\% \text{ of } 24 = \frac{1}{4} \times 24 = 6 \qquad 50\% \text{ of } 130 = \frac{1}{2} \times 130 = 65$$

Practice

Write the percent of each number.

1. 20% of 35 = __7__

2. 50% of 94 = __47__

3. 40% of 18 = __$7\frac{1}{5}$__

4. 75% of 64 = __48__

5. 30% of 16 = __$4\frac{4}{5}$__

6. 10% of 120 = __12__

7. 60% of 420 = __252__

8. 25% of 248 = __62__

9. 90% of 1,400 = __1,260__

10. 86 % of 230 = __$197\frac{4}{5}$__

11. 25% of 32 = __8__

12. 6% of 30 = __$1\frac{4}{5}$__

13. 21% of 70 = __14.7__

14. 37% of 19 = __7.03__

15. 68% of 215 = __146.2__

16. 6% of 315 = __18.9__

17. 29% of 116 = __33.64__

18. 48% of 39 = __18.72__

19. 67% of 87 = __58.29__

20. 1% of 750 = __7.5__

21. 96% of 600 = __576__

22. 200% of 10 = __20__

23. 28% of 82 = __22.96__

24. 32% of 98 = __31.36__

Problem Solving

Solve each problem.

25. The Lincoln School has 460 students. 45% of the students are boys. How many boys are in Lincoln School?
207 boys

26. Chris and Terry own and operate the Myers Inn. Chris owns 56% of the hotel. What percent of the Myers Inn belongs to Terry?
44%

27. The Swim Club needs to raise $800. So far, they have raised 70% of what they need. How much more money does the Swim Club need?
$240

28. At the restaurant, Mr. Wilson's chicken dinner cost $7.25. Mrs. Wilson's dinner cost $6.95. The check included a 5% tax. How much did the Wilsons pay for dinner?
$14.91

29. Jerome bought a $25 item and paid a 4% sales tax. His friend bought the same item in another state for $25, but paid a 5% sales tax. How much less tax than his friend did Jerome pay?
25 cents

Lesson 13-8 • Finding a Percent of a Number

---

3 Practice

Have students complete all the exercises. Encourage them to write equations for word problems, especially the ones that are multistep.

4 Assess

Ask students which number would be greater, 20% of 300 or 2% of 300? (20% of 300)

For Mixed Abilities

Common Errors • Intervention

When finding the percent of a number, some students may forget to write the percent as a decimal and simply multiply by the number without the percent sign. Have students organize their work using the following table to help them remember each step.

Problem	Percent as a decimal	Multiply
20% of 85	0.2	$0.2 \times 85 = 17$

Enrichment • Number Sense

Have students compare the pair of values by writing >, <, or =.

1. 35% of 90 (<) 75% of 50

2. 15% of 300 (=) 25% of 180

3. 50% of 36 (<) 60% of 45

4. 85% of 150 (>) 70% of 160

5. 70% of 40 (=) 40% of 70

6. 12% of 148 (>) 10% of 170

More to Explore • Probability

Remind students that probabilities are often expressed as fractions. Ask them to describe the probability of drawing a red card out of a deck of cards.
(1 out of 2, or $\frac{1}{2}$)

Explain that the probability can also be expressed as a decimal or as a percent. Have students give the decimal and percent equivalent for $\frac{1}{2}$. (0.50 or 50%) Remind them that any fraction can be changed to a decimal by dividing the numerator by the denominator.

Ask students to give the decimal and percent probabilities for the event.

1. The probability of tossing a coin and getting heads (0.50 or 50%)

2. The probability of drawing a club from a deck of cards ($\frac{13}{52} = \frac{1}{4}$ = 0.25 or 25%)

3. The probability of getting a six on the toss of one die ($\frac{1}{6}$ = 0.17 = 17%)

13-9 Percent Increase and Decrease

pages 299–300

1 Getting Started

Objective
- To find the percent increase or decrease

Vocabulary
percent increase, percent decrease

Warm Up • Mental Math
Have students name the number using exponents.

1. 81 (3^4)
2. 16 (2^4)
3. 27 (3^3)
4. 49 (7^2)
5. 8 (2^3)
6. 25 (5^2)
7. 9 (3^2)
8. 36 (6^2)

Warm Up • Pencil and Paper
Have students factor the number into primes. Remind them to use exponents when possible.

1. 18 (2×3^2)
2. 50 (2×5^2)
3. 42 ($2 \times 3 \times 7$)
4. 60 ($2^2 \times 3 \times 5$)
5. 28 ($2^2 \times 7$)
6. 72 ($2^3 \times 3^2$)
7. 45 ($3^2 \times 5$)
8. 100 ($2^2 \times 5^2$)

Name _____

Percent Increase and Decrease

Laura is shopping for hiking boots. The boots she wants increased in price from $80 last year to $100 this year. What is the percent increase in the price of the boots? Laura decides to buy a different model that originally cost $90 but now is on sale for $72. What is the percent decrease on the price of these boots?

The examples above show a change in price. We can find the percent of change by finding percent increase or percent decrease.

$100 $90 $72

Price increase from $80 to $100

We subtract to find the change from the original price to the new price.

$100 − $80 = <u>$20</u>

We use the difference to write a ratio.

amount of increase ⟶ 20
original amount ⟶ 80

Then, we write the ratio as a percent.

20 ÷ 80 = 0.25 = 25%

The percent increase in the price of hiking boots from last year to this year is <u>25</u> %.

Price decrease from $90 to $72

$90 − $72 = <u>$18</u>

amount of decrease ⟶ 18
original amount ⟶ <u>90</u>

18 ÷ 90 = 0.20 = <u>20</u> %

The sale price shows a percent decrease of <u>20</u> %.

Getting Started

Find the percent increase or decrease to the nearest percent.

1. salary increase from $7 per hour to $9 per hour
   29%

2. a loss in weight from 130 lb to 115 lb
   12%

3. an increase in time spent in school each day from 6 hours to 7 hours
   17%

4. temperature decrease from 40°C to 25°C
   38%

---

2 Teach

Introduce the Lesson Read the problem aloud. Explain that whenever there is a change—in price, time, temperature, weight—there is a percent increase or a percent decrease. Tell students that because the boots went from $85 to $100, we will find the percent increase. For the boots that were $90 but are on sale for $72, we will find the percent decrease.

Have students work through the model to find the percent increase for those boots that had a price increase from $85 to $100. Then, have students work through the model to find the percent decrease for the boots that went on sale.

Explain that whether we are finding a percent increase or a percent decrease, we first find the difference in price by subtracting the smaller amount from the larger. Then, we divide the difference by the original amount.

Develop Skills and Concepts Give an additional example of finding a percent increase.

A 6-month-old baby weighed 16 pounds. By his first birthday, he weighed 24 pounds. What is the percent increase?

- First, find the difference in the two weights.

   24 lb − 16 lb = 8 lb

- Then, divide the difference by the original weight, 16 lb.

   8 lb ÷ 16 lb = 0.5 = 50%

The percent increase is 50%.

Write the amount of change.

1. rise in temperature from 50°F to 82°F

   change: _32°_

2. weight gain from 90 lb to 105 lb

   change: _15 lb_

3. price drop from $24 to $18

   change: _$6_

4. decrease in class size from 32 students to 25 students

   change: _7 students_

Find the percent increase or decrease to the nearest percent.
Write *increase* or *decrease*.

5. rise in temperature from 50°F to 82°F
   64%; increase

6. weight gain from 90 lb to 105 lb
   17%; increase

7. price drop from $24 to $18
   25%; decrease

8. decrease in class size from 32 students to 25 students
   22%; decrease

9. a stock's price rises from $2 to $3
   50%; increase

10. water in a pool falls from 550 gallons to 400 gallons
    27%; decrease

Problem Solving _____

Use the price list to solve each problem.

11. Darla is the buyer for Thrift Mart. She buys items at the wholesale cost and sells them at retail price. What is the percent increase of gloves and belts?

    gloves: _80%_
    belts: _300%_

12. When scarves go on sale, what is the percent decrease from the retail price?
    11%

13. Dwayne has a sale coupon for $\frac{1}{3}$ off the retail price of hats. How much money will he save and what is the percent decrease?
    $5; 33%

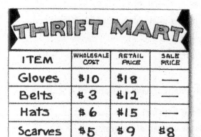

ITEM	WHOLESALE COST	RETAIL PRICE	SALE PRICE
Gloves	$10	$18	—
Belts	$3	$12	—
Hats	$6	$15	—
Scarves	$5	$9	$8

3 Practice

Have students complete all the exercises. Remind students of the steps needed to find a percent increase or decrease: find the difference, divide the difference by the original amount, and then write the answer as a percent.

4 Assess

Have students review their work for Exercise 5. Suppose the temperature dropped from 50°F to 18°F. Would the percent increase or decrease be the same as the percent increase in Exercise 5? (Yes.) Why? (The differences are equal, $82 - 50 = 50 - 18 = 32$, and then so are the percentages, $\frac{32}{50} = 64\%$.)

For Mixed Abilities

Common Errors • Intervention

Some students may be confused on the process of finding a percent increase or decrease. For each exercise below, have students decide if the situation represents a percent increase or a percent decrease. Then, have them find the percent increase or decrease by following the steps.

1. (increase) price from $12 to $15
   difference: ($3)
   original amount: ($12)
   difference ÷ original amount: (0.25)
   percent increase: (25%)

2. (decrease) attendance from 30 to 24
   difference: (6)
   original amount: (30)
   difference ÷ original amount: (0.20)
   percent decrease: (20%)

Enrichment • Application

It is not uncommon for the value of a house to increase by 100% or more over a period of several years. Suppose that in 1980, the value of a house was $60,000, and in 1990, its value was $120,000. What was the percent increase? (100%) What if, by 2000, the value of the house was $180,000? What would the percent increase from 1980 to 2000 be? (200%)

More to Explore • Number Sense

Tell students that the following steps will enable them to guess another person's age:

1. Tell a person to multiply his or her age by 3, add 6 to the answer, and then divide the sum by 3.

2. Ask for the result.

3. Subtract 2 from that number and you will get the person's age.

Have students try this with each other and then mystify their family with their mind-reading ability!

T300

1 Getting Started

Objective
• To find simple interest

Vocabulary
simple interest

Materials
worksheets

Warm Up • Mental Math
Have students estimate the product.

1.	426 × 38 (16,000)	4.	650 × 25 (21,000)
2.	761 × 52 (40,000)	5.	7,099 × 248 (1,400,000)
3.	3,811 × 63 (240,000)	6.	6,591 × 829 (5,600,000)

Warm Up • Pencil and Paper
Have students copy and complete the table.

Fraction	Decimal	Percent
$\frac{1}{25}$	(0.04)	(4%)
$\frac{7}{50}$	(0.14)	(14%)
($\frac{46}{100}$)	0.46	(46%)
($\frac{4}{5}$)	0.8	(80%)
($\frac{24}{25}$)	(0.96)	96%
($\frac{59}{100}$)	(0.59)	59%

2 Teach

Introduce the Lesson Explain that banks pay interest to use the money that customers deposit in an account. Point out that banks not only pay interest but also charge interest on money a customer borrows from the bank. Tell students that interest rates are expressed as percents and are a percent of the amount borrowed or deposited.

Have a student read the problem. Work through the steps for computing the interest for 1 year and the balance at the end of the year. Point out that the amount of interest is expressed as a decimal. The amount of money in the bank is multiplied by this decimal.

Simple Interest

Interest Rates	
Money Market	8%
Pass Book	5%
Certificates	$12\frac{1}{2}$%

Miss King deposits $850 in a money market account. How much interest will her account earn in one year?

We want to know the amount of interest Miss King's deposit will earn in one year.

We know that she deposits __$850__ in a money market account.

The rate of interest is __8%__.

To find the **interest** for one year, we multiply the amount of the deposit by the rate of interest.

We multiply __$850__ by __8%__.

To compute interest for one year, follow these steps:

Name the rate of interest as a decimal.	Multiply the amount of the deposit by the rate of interest.
8% = __0.08__	$850 × 0.08 $68.00

Miss King will earn __$68__ in interest for one year.

To find the total balance in Miss King's account, we add the deposit and the interest.

$850 ← amount deposited
+ 68 ← interest
$918

Miss King will have a balance of __$918__ in her account after one year.

Getting Started

Write the interest for one year.

1. 3% on $250 = __$7.50__   2. 8% on $980 = __$78.40__   3. 12% on $35,600 = __$4,272__

Write the total amount after one year.

4. $1,256 at 5% = __$1,318.80__   5. $5,000 at 6% = __$5,300__   6. $725 at 7% = __$775.75__

Develop Skills and Concepts Write the formula for finding simple interest on the board:

interest (*i*) = principal (*p*) × rate of interest (*r*) × time (*t*)

Explain that principal is the amount of money in an account or the amount of money borrowed. Demonstrate how to find the simple interest on $775 at a 7% rate of interest for 1 year.

$$i = p \times r \times t$$
$$i = \$775 \times 7\% \times 1$$
$$i = \$775 \times 0.07 \times 1$$
$$i = \$54.25$$

Have students calculate the simple interest for 1 year for 6% on $400 ($24) and 5% on $875 ($43.75).

Practice

Write the interest for one year.

1. 5% on $360 = __$18__
2. 6% on $580 = __$34.80__
3. 4% on $175 = __$7__
4. 10% on $750 = __$75__
5. 7% on $700 = __$49__
6. 7% on $800 = __$56__
7. 8% on $6,000 = __$480__
8. 12% on $1,960 = __$235.20__
9. 9% on $800 = __$72__
10. 16% on $2,000 = __$320__
11. 3% on $786 = __$23.58__
12. 11% on $72 = __$7.92__

Write the total amount after one year.

13. $825 at 8% = __$891__
14. $1,200 at 9% = __$1,308__
15. $2,500 at 7% = __$2,675__
16. $1,860 at 6% = __$1,971.60__
17. $1,325 at 5% = __$1,391.25__
18. $2,290 at 7% = __$2,450.30__
19. $1,450 at 5% = __$1,522.50__
20. $3,050 at 11% = __$3,385.50__
21. $975 at 12% = __$1,092__
22. $3,000 at 12% = __$3,360__
23. $629 at 8% = __$679.32__
24. $330 at 4% = __$343.20__

Problem Solving

Solve each problem.

25. Scott put $450 in a bank for one year. The bank paid 6% interest. Margie put $440 in a different bank, also for one year. This bank paid 9% interest. At the end of the year, who will have the greater amount? How much greater?
Margie; $2.60 more

26. The Morgans bought a new car for $9,500. They paid $\frac{2}{5}$ of the price in cash, and borrowed the rest at 8% for one year. If they repaid their loan at the end of the year, how much did they have to pay for the car?
$9,956

27. Mr. Chen plans to borrow $2,500 for one year. One bank charges 7% interest and another bank charges 8% interest. How much can Mr. Chen save if he borrows the money from the bank that charges the lower interest rate? $25

28. Nancy put $500 in a one-year certificate at 8% interest. After the first year, she put the total back in the same account. How much will Nancy have after the second year?
$583.20

For Mixed Abilities

Common Errors • Intervention

Some students may forget to add the interest to the original amount when finding the total amount. Have them write the following steps on an index card and use it as a guide when they are solving such problems.

1. Name the rate of interest as a decimal.
2. Multiply the deposited amount by the decimal to find the interest.
3. Write the interest as a dollar amount.
4. Add the interest to the deposited amount.

Enrichment • Algebra

Show students a shortcut for finding the total amount $(i + p)$. Point out that the total amount = 100% + the rate of interest. Therefore, to find the total amount after 1 year on $925 at 6%, multiply $925 by 106% or $925 \times 1.06 = 980.50. Have students use this shortcut to find the total amount after 1 year.

1. $1,350 at 9% ($1,471.50)
2. $675 at 13.5% ($766.13)

More to Explore • Applications

Have students estimate the cost of owning a car for 1 year. Have them research the original costs, including a down payment and other fees; insurance costs; maintenance costs, including tires, gas, oil, license, and tune-ups; parking fees and toll charges.

ESL/ELL STRATEGIES

Preteach the meaning of the terms *money market account* (a special kind of bank account), *deposit* (money you put in the bank), *interest* (money the bank pays you for use of your money), and *total balance* (the full amount of money in the account, including deposits and interest).

---

3 Practice

Have students complete all the exercises. Remind them that the total is the principal plus interest.

4 Assess

Ask students to determine the interest on $1,000 at 6% for 1 year. ($60)

5 Mixed Review

1. 54,000 ÷ 600 (90)
2. 75% of 440 (330)
3. 3 × 7 + 8 (29)
4. 678,309 + 427,981 (1,106,290)
5. $\frac{2}{3} \times \frac{5}{6} \times \frac{3}{10}$ ($\frac{1}{6}$)
6. 1.28×10^3 (1,280)
7. $\frac{7}{12} - \frac{3}{8}$ ($\frac{5}{24}$)
8. 73 × 52 (3,796)
9. $\frac{8}{9} \div \frac{2}{3}$ ($1\frac{1}{3}$)
10. 83.22 ÷ 3.8 (21.9)

pages 303–304

1 Getting Started

Objective
- To solve problems by defining and collecting necessary data

Warm Up • Mental Math
Dictate the following slowly and have students calculate the answer.
1. (25% of 16) + 6 − 2 × 2 (6)
2. (20% of 25) × 8 + 2 × 2 (44)
3. (10% of 500) − 7 ÷ 1 + 3 (46)
4. (50% of 72) ÷ 12 + 2 (5)
5. (10% of 1,000) − 2 ÷ 2 (99)
6. $(33\frac{1}{3}\%$ of 27) + 1 + 20 − 2 (28)
7. (10% of 800) + 9 ÷ 1 − 9 (80)
8. $(14\frac{2}{7}\%$ of 70) − 2 × 2 + 2 (8)

Warm Up • Pencil and Paper
Have students solve the following:
1. 1,484,966 ◯ 1,478,899 (>)
2. 4,389 ◯ 12,201 (<)
3. 14 × 16 (224)
4. Find the average: 8, 5, 7, 9, 6. (7)
5. 72 × 15 (1,080)
6. Find the average: 46, 52, 85, 97. (70)
7. (2 × 4.2) + (2 × 6.5) (21.4)
8. $(2 \times 1\frac{1}{2}) + (2 \times 2\frac{1}{3})$ $(7\frac{2}{3})$

Problem Solving: Collect Data

Elaine is writing reports on each planet and the sun for her science class. When she finishes the reports, she would like to put them in order beginning with the sun and then arranging the other planets in order according to their distance from the sun. The planet farthest from the sun will be the last chapter. In what order should Elaine arrange the chapters?

⭐ SEE

We want to arrange the planets in order according to their distance from the sun. We need to know the name of each planet and its distance from the sun.

⭐ PLAN

Since the data we need to solve the problem is not given, we will need to look in a reference book to find the name of each planet and its distance from the sun. Once we have obtained this data, we can put the planets in order.

⭐ DO

Planet	Earth	Jupiter	Mars	Mercury	Neptune	Pluto	Saturn	Uranus	Venus
Distance from the sun*	93	484	142	36	2,792	3,664	887	1,782	67

* in millions of miles   Distances may vary by year. Order of planets should not change.

Elaine should put the reports in the following order:

Mercury , Venus , Earth , Mars , Jupiter ,
Saturn , Uranus , Neptune , Pluto .

Reference books are just one of the many types of resources from which we can collect data.

⭐ CHECK

We can check our work to be sure we have correctly copied the information and that we have put the planets in the correct order. When we are collecting data, we can use more than one reference book to verify the correctness of the data.

---

2 Teach

Introduce the Lesson Realistically, the solutions of many problems depend on facts that we do not immediately have. Discuss with students the various sources of facts available to them: encyclopedias, almanacs, atlases, dictionaries, and books of tables and lists.

Develop Skills and Concepts Extend the discussion by asking such questions as *Where would you look for baseball statistics? Current populations of cities? Lengths of bridges? Calories used in various exercises? Weights and measures? Dates of historic events?* and so on.

Ask students why it is good practice to get data from more than one source. (for verification) Be sure to discuss the possibility of finding contradictory data and how to handle this. (evaluate the worth of the source; go to a third source)

In working through the sample problem on this page, have students suggest possible sources for the needed data about planets. Have one or more students do the research and share the facts with the rest of the class.

3 Apply

Solution Notes
1. Review finding an average. It may be helpful to let students use calculators to do computations. Discuss how to interpret the decimal part of an average.
2. Students will have to make a survey and tally to complete the problem. Review the validity of their sampling and discuss problems that can arise when making a survey and tabulating results.
3. Necessary data can be found in an encyclopedia or almanac.
4. Again students will need to refer to an almanac or encyclopedia. It may be beneficial to have them compare the same data presented in two different sources to note any disparity.

T303

Apply

Solve each problem. You will need to collect data from outside sources for Exercises 1–9.

1. What is the average height of the sixth graders in your class?
   Answers will vary.

2. What is the favorite pet among the second graders in your school?
   Answers will vary.

3. List the five Great Lakes in order from largest to smallest.
   Lake Superior, Lake Huron, Lake Michigan, Lake Erie, Lake Ontario

4. What are the five longest rivers in the United States?
   Mississippi River, Missouri River, Yukon River, Arkansas River, Colorado River

5. What is the perimeter of the cover of your math book to the nearest centimeter?
   98 centimeters

6. List the seven continents in order from the most populated to the least populated.
   Asia, Africa, Europe, North America, South America, Australia, Antarctica

7. Eight states have an area of over 100,000 square miles. In order from largest to smallest, they are Alaska, Texas, California, Montana, New Mexico, Arizona, Nevada, and Colorado. List these eight states in order beginning with the state that has the greatest population and ending with the state that has the least population.
   California, Texas, Arizona, Colorado, New Mexico, Nevada, Montana, Alaska

8. The school wants to buy carpeting for your classroom. Find the area of the room to the nearest square meter.
   Answers will vary.

Problem-Solving Strategy: Using the Four-Step Plan

★ SEE What do you need to find?
★ PLAN What do you need to do?
★ DO Follow the plan.
★ CHECK Does your answer make sense?

9. How many letters are in the first name of each student in your class? What is the most and least common number of letters?
   Answers will vary.

10. There were 800 students in school in September. By January, enrollment increased 10%. Then, by June, enrollment decreased 10%. Were there more, fewer, or the same number of students in June as in September? Explain how you know.
    fewer students

11. A suit in a clothing store is on sale at 20% off. An employee of the store gets a discount of 10% on any item for sale in the store. If the employee buys the suit, is the price less than, equal to, or more than the original price less 30%? Verify your answer.
    more than

12. Tyrone and Luis were hiking across Horace Heights to the camp store. Tyrone walked 80% of the distance and then stopped to rest. Luis walked 50% of the distance, stopped to rest, and then walked another 30% of the distance before stopping to rest again. At this point along the trail, who is ahead?
    neither

For Mixed Abilities

More to Explore • Number Sense

Remind students that they have learned that golden rectangles are eye-appealing and have the length to width ratio of 1.61803:1.

Have them divide into groups to conduct a survey. Have the groups ask 20 people each to draw a rectangle. When they have all samples, have students measure each rectangle and test it according to the golden rectangle ratio of $\frac{l}{w} = \frac{1.6}{1}$. Round the result.

Have them chart their data, label the golden rectangles found, and display their charts to compare findings.

---

5. Review finding perimeter.
6. Direct students to use an outside source.
7. Students will find an almanac to be the best source. Tell them to find the most current data available.
8. Review how to find area. Discuss that when laying carpet, which is measured in square yards, it is also necessary to allow for waste.
9. Again, students will have to make a tally to collect data.

Higher-Order Thinking Skills

10. **Synthesis:** 800 plus 80 (10% of 800), or 880 students in January; 880 less 88 (10% of 880), or 792 students in June.

11. **Evaluation:** Students can verify the answer by using some "made-up" prices. For example, if the suit was originally $100, then the sale price is $100 − (20% of $100), or $80. The additional discount of 10% is on the sale price; thus, the employee's cost is $80 − (10% of $80), or $72. In comparison, a discount of 30% on the original price gives a final price of $70.

12. **Analysis:** 80% of any number is the same as 50% of the number plus 30% of the same number. This can be verified with the Distributive Property.
    $$0.8n = 0.5n + 0.3n$$
    $$= (0.5 + 0.3)n$$
    $$= 0.8n$$

13. **Analysis:** It does not matter since the amounts of money that Alex would get are equal. 60% of 80 is $48, as is 80% of $60.

4 Assess

Ask students where they might find sports statistics.
(world almanac, Internet)

pages 305–306

1 Getting Started

Objective
- To use the calculator to find original and sale prices

Vocabulary
rate of discount, sale price

Materials
calculators; newspaper advertisements; catalogs

Warm Up • Mental Math
Have students tell how much to add to the number to get 100.

1. 40 (60)   5. 48 (52)
2. 50 (50)   6. 36 (64)
3. 20 (80)   7. 95 (5)
4. 65 (35)   8. 10 (90)

Warm Up • Pencil and Paper
Have students find the amount.

1. 75% of $40 ($30)
2. 165% of $90 ($148.50)
3. 32% of $65 ($20.80)
4. 112% of $49 ($54.88)
5. 15% of $72 ($10.80)
6. 125% of $150 ($187.50)

Name _____

Lesson 13-12

Calculator: Discount and Sale Price

When merchants have sales on their merchandise, they sometimes advertise a certain percent off the original price. This percent is called the rate of discount. When the original price is multiplied by the rate of discount, we can find the amount of discount or discount. What is the amount of discount on the radio that is on sale?

What is the **sale price**?

RADIOS regularly $48⁰⁰ SALE 25% OFF

We want to know the amount we save by buying the radio on sale.

The original price is __$48__ and the rate of discount is __25%__. To find the discount, we find __25%__ of __$48__.

48 ⊠ 25 [%] [12]

The discount on the radio is __$12__.

To find the sale price, or the cost after discount, we subtract the discount from the original price.

48 ⊟ 12 [=] [36]

The sale price is __$36__.

There is another method for finding the sale price.

We know that the sale price is "25% off" so the sale price must be 75% of the original price.

48 ⊠ 75 [%] [36]

Originally the radio cost 100%. If I don't have to pay 25% of this cost, I will have to pay 75% of it.

---

2 Teach

Introduce the Lesson Display several newspaper advertisements containing phrases such as 25% off and 10% discount. Discuss the meaning of *original price* (price of an item when it is not on sale), *rate of discount* (percent used to compute the amount of discount), *amount of discount* (amount of money subtracted from the original price to get the sale price; product of original price times rate of discount), and *sale price* (price of item when it is on sale).

Work through the model with students. Have them compute the amount of discount and the sale price. Then, complete the solution sentence.

Develop Skills and Concepts Discuss the two methods presented for finding the sale price of an item. Point out that the first method requires two steps: finding the amount of discount and subtracting that amount from the original price. The second method requires only one

step, but before students multiply to find the sale price, they must subtract the rate of discount from 100.

Continue developing this skill by having students use their calculators to solve the following problems:

1. original price = $460, rate of discount = 40%, sale price = ($276)

2. original price = $250, rate of discount = 30%, sale price = ($175)

3 Practice

Have students complete all the exercises. Watch for students who confuse rate of discount with amount of discount.

Practice

Complete each code.

1. $65 \boxed{\times} 12 \boxed{\%}$ $\boxed{7.8}$

2. $12 \boxed{\times} 65 \boxed{\%}$ $\boxed{7.8}$

3. $16 \boxed{\times} 225 \boxed{\%}$ $\boxed{36}$

4. $9{,}000 \boxed{\times} 0.5 \boxed{\%}$ $\boxed{45}$

5. $800 \boxed{\times} 6.25 \boxed{\%}$ $\boxed{50}$

6. $265 \boxed{\times} 9.4 \boxed{\%}$ $\boxed{24.91}$

Use a calculator to complete each equation.

7. 80% of $25 = $20

8. 6.25% of $80 = $5

9. 6.4% of $950 = $60.80

10. 236% of $75 = $177

11. 62.5% of $16 = $10

12. 500% of $6 = $30

Problem Solving

Use a calculator to find each discount.

13.

Original Price	$75	$68	$49.50	$648
Rate of Discount	5%	4%	10%	3%
Discount	$3.75	$2.72	$4.95	$19.44

14.

Original Price	$240	$125	$488	$960
Rate of Discount	15%	40%	12.5%	62.5%
Sale Price	$204	$75	$427	$360

Now Try This!

When Pat saved $5,000 for one year at 8% interest, she ended with $5,400 in her account.

Shelby saved her money at 8% compound interest. Each quarter (3 months), the bank figured Shelby's interest on the money in her account at the beginning of the quarter. Complete the following to see how much Pat would have ended with at 8% compounded interest. 8% compounded quarterly is 2% each 3 months.

Quarter 1	Quarter 2	Quarter 3	Quarter 4
$5,000	$5,100	$5,202	$5,306.04
× 2%	× 2%	× 2%	× 2%
$ 100	$ 102	104.04	106.12
+ 5,000	+ 5,100	+ 5,202.00	+ 5,306.04
$5,100	$5,202	$5,306.04	$5,412.16

Pat would have had $5,412.16.

Lesson 13-12 • Calculator: Discount and Sale Price

Now Try This! This activity introduces students to compound interest problems. Point out that 8% represents the rate of interest for 1 full year. Compounded quarterly (four times a year), the rate is 8% ÷ 4, or 2% per quarter. Have students note also that the principal changes as the amount of interest from the previous quarter is added to it.

4 Assess

Have students find the sale price of an item that costs $100 and is on sale for 25% off. ($75)

For Mixed Abilities

Common Errors • Intervention

Some students may forget to subtract the amount of discount from the original price when they are multiplying by the rate of discount to find the sale price. As they work each problem, have them fill in the information in each of the following categories.

Original price	Rate of discount	Amount of discount	Sale price

Enrichment • Application

Have students determine which is the better buy.

1. a $350 television set at 15% off or at a sale price of $300 (15% off)

2. a $95 portable radio at 25% off or at a sale price of $75 (25% off)

3. a $79 jacket at 12% off or reduced by $9.50 (reduced by $9.50)

More to Explore • Logic

Review syllogisms with students. Give the following example:

• All trees on Franklin Street are elms.

• Jeff lives on Franklin Street.

• The trees in Jeff's yard are elms.

Remind students that it is possible to write weak syllogisms, or even false syllogisms. Point out that in the example, the fact that Jeff might live on a corner, with trees from another street in his yard, has not been accounted for. This fact makes the syllogism weak. If one of the statements is false, the whole syllogism is false.

Have students work in groups to write three true and two false syllogisms, leaving out one of the statements in each. Have the groups exchange and complete the missing statements and identify the syllogisms as true or false.

T306

Chapter 13
Test

page 307

Items	Objectives
1–4	To write the ratio of two numbers (see pages 283–284)
5–8	To solve a proportion (see pages 287–288)
9	To find missing measures of two similar polygons (see pages 289–290)
10	To use a scale drawing to find actual length (see pages 291–292)
11–12	To change fractions to percents (see pages 295–296)
13–14	To change percents to fractions (see pages 295–296)
15–18	To find percent of a number (see pages 297–298)
19–20	To find percent increase or decrease (see pages 299–300)
21–22	To find the total amount after one year at a given rate of simple interest (see pages 301–302)

Alternate Chapter Test

You may wish to use the Alternate Chapter Test on page 394 of this book for further review and assessment.

T307

Write each ratio as a fraction.

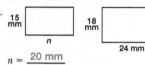

1. forks to spoons $\frac{2}{2}$

3. spoons to silverware $\frac{2}{4}$

2 cm · 4 cm (rectangle)

2. width to length $\frac{2}{4}$

4. length to perimeter $\frac{4}{12}$

Solve each proportion.

5. $\frac{3}{4} = \frac{n}{16}$ $n = \underline{12}$

6. $\frac{n}{9} = \frac{10}{30}$ $n = \underline{3}$

7. $\frac{7}{14} = \frac{5}{n}$ $n = \underline{10}$

8. $\frac{8}{n} = \frac{5}{9}$ $n = \underline{14\frac{2}{5}}$

The figures are similar. Solve for n.

9. 15 mm, n ; 18 mm, 24 mm

$n = \underline{20\ mm}$

Use the scale to find the actual distance.

10. 4 cm, A ———— B

Scale: 1 cm = 8 m $\underline{32\ m}$

Write each fraction as a percent.

11. $\frac{4}{5} = \underline{80\%}$

12. $\frac{7}{20} = \underline{35\%}$

Write each percent as a fraction.

13. $30\% = \underline{\frac{3}{10}}$

14. $48\% = \underline{\frac{12}{25}}$

Find the percent of each number.

15. 70% of 43 = $\underline{30.1}$

16. 15% of 84 = $\underline{12.6}$

17. $5\frac{1}{2}\%$ of $96 = $\underline{\$5.28}$

18. 9.6% of $650 = $\underline{\$62.40}$

Find the percent increase or decrease.

19. price increase from $25 to $30
20%

20. fall in temperature from 45° to 27°
40%

Write the total amount after one year.

21. $3,000 at 2% $3,060

22. $1,500 at 3.5% $1,552.50

Chapter 13 • Test

Circle the letter of the correct answer.

1. $2\frac{4}{5}$ $+ 3\frac{2}{3}$
a. $1\frac{7}{15}$
b. $5\frac{7}{15}$
c. $6\frac{7}{15}$ ✓
d. NG

2. $5\frac{1}{8}$ $- 3\frac{2}{3}$
a. $1\frac{11}{24}$ ✓
b. $2\frac{11}{24}$
c. $3\frac{11}{24}$
d. NG

3. $3\frac{1}{3} \times 1\frac{1}{5}$
a. $\frac{1}{4}$
b. 2
c. 4 ✓
d. NG

4. $2\frac{1}{3} \div 1\frac{1}{6}$
a. $\frac{1}{2}$
b. 2 ✓
c. $2\frac{13}{18}$
d. NG

5. What is the place value of the 7 in 2.1076?
a. tenths
b. hundredths
c. thousandths ✓
d. NG

6. $3.46 + 2.7$
a. 5.16
b. 6.10
c. 6.16 ✓
d. NG

7. $16.1 - 8.304$
a. 7.804
b. 7.806
c. 12.204
d. NG ✓

8. 4.1×0.03
a. 0.123 ✓
b. 1.23
c. 12.3
d. NG

9. $16 \div 0.01$
a. 0.16
b. 1.6
c. 16
d. NG ✓

10. 16 ft 6 in. $-$ 6 ft 9 in.
a. 9 ft 7 in.
b. 9 ft 9 in. ✓
c. 10 ft 3 in.
d. NG

11. Find the area.
a. 24 cm² ✓
b. 48 cm²
c. 56 cm²
d. NG

12. Find the volume.
a. 9 m³
b. 18 m³
c. 27 m³ ✓
d. NG

score

STOP

308

Chapter 13 • Cumulative Assessment

page 308

Items	Objectives
1–2	To add or subtract two mixed numbers (see pages 119–120, 125–126)
3–4	To multiply or divide mixed numbers (see pages 141–142, 147–148)
5	To identify the place value of a digit in decimals through millionths (see pages 163–164)
6–7	To add or subtract decimals (see pages 171–172, 173–174)
8	To multiply decimals, write zeros in the product to locate a decimal point (see pages 189–190)
9	To divide by decimals with fewer decimal places in the dividend than in the divisor (see pages 199–200)
10	To subtract customary units of length (see pages 215–216)
11	To find the area of triangles (see pages 269–270)
12	To find the volume of cubes (see pages 275–276)

Alternate Cumulative Assessment

Circle the letter of the correct answer.

1. $3\frac{6}{7}$ $+ 4\frac{2}{3}$
a $7\frac{11}{21}$
b $8\frac{10}{21}$
c $8\frac{11}{21}$ ✓
d NG

2. $9\frac{5}{12}$ $- 3\frac{9}{16}$
a $5\frac{40}{48}$
b $5\frac{41}{48}$ ✓
c $5\frac{42}{48}$
d NG

3. $3\frac{3}{4} \times 6\frac{2}{5}$
a $18\frac{6}{20}$
b 21
c 24 ✓
d NG

4. $2\frac{5}{8} \times 1\frac{5}{9}$
a $1\frac{11}{16}$
b $1\frac{12}{16}$
c $2\frac{1}{8}$
d NG ✓

5. Give the place value of the 9 in 4.08379.
a thousandths
b hundred thousandths ✓
c ten thousandths

6. 5.83 $+ 2.9$
a 7.73
b 8.70
c 8.73 ✓
d NG

7. $37.3 - 9.806$
a 27.404
b 27.506
c 27.594
d NG ✓

8. 0.07×5.2
a 0.364 ✓
b 3.64
c 36.4
d NG

9. $14 \div 0.04$
a 0.35
b 3.5
c 35
d NG ✓

10. 12 ft 3 in. $-$ 7 ft 11 in.
a 4 ft 2 in.
b 4 ft 4 in. ✓
c 5 ft 4 in.
d NG

11. Find the volume of a 2-cm cube.
a 4 cm³
b 6 cm³
c 8 cm³ ✓
d NG

T308

14-1 Stem-and-Leaf Plots

pages 309–310

1 Getting Started

Objective
- To use a stem-and-leaf plot to answer questions
- To construct a stem-and-leaf plot from a list of data

Vocabulary
stem-and-leaf plot, outlier

Materials
calculators; highlighters

Warm Up • Mental Math
Have students give the answer.

1. 4×7 (28)   5. $27 \div 3$ (9)
2. $16 \div 4$ (4)   6. 6×6 (36)
3. $8 + 5$ (13)   7. 7×9 (63)
4. $15 - 7$ (8)   8. $54 \div 9$ (6)

Warm Up • Pencil and Paper
Have students find the average for each set of data.

1. prices of 5 books: $12, $17, $9, $10, $14 ($12.40)
2. ages of 6 students: 7, 7, 10, 9, 7, 14 (9 years)
3. hours worked in 4 weeks: 25, 31, 28, 29 (28.25 hours)
4. prices of 5 pairs of shoes: $35, $52, $41, $64, $39 ($46.20)

Graphs and Probability

Lesson 14-1

Stem-and-Leaf Plots

The table shows how long Ari practices the trumpet each day. How much time does Ari most often spend practicing?

We know that Ari practiced for __10__ days.

The table shows the amount of time he spent each day. We need to arrange the data to see the amount of time Ari most often spends practicing.

We can organize data in a **stem-and-leaf plot**. A stem-and-leaf plot arranges the data according to place value.

First we write the data from least time to greatest time.

25, 27, 30, 35, __35__, __35__, __38__, __45__, __48__, __70__

Then, we make a stem-and-leaf plot.

- The stems are the tens digits.
- The leaves are the ones digits listed in order from least to greatest.

What are the leaves for stem 4? __5, 8__

What are the leaves for stem 7? __0__

What amount of time does Ari most often spend practicing? __35 min__

Notice how the numbers cluster between the 20s and 40s. The number 70 is far from the cluster. It is called an **outlier**, because it is not typical of most of the data.

Time Spent Practicing For Ten Days

30 min	25 min	35 min
45 min	38 min	35 min
48 min	70 min	27 min
35 min		

Minutes Practicing

Stem	Leaf
2	5 7
3	0 5 5 5 8
4	5 8
5	
6	
7	0

Getting Started

Use the stem-and-leaf plot above to answer each question.

1. What was the least amount of time Ari practiced?
   25 min

2. What was the greatest amount of time Ari practiced?
   70 min

3. What is the range of the data?
   45 min

4. Ari says that half the time he practices at least 35 minutes. Is he correct? Explain.
   Yes; more than half the data are 35 min or greater.

Lesson 14-1 • Stem-and-Leaf Plots

309

---

2 Teach

Introduce the Lesson Read the problem aloud. Explain that a stem-and-leaf plot makes reading a list of numbers easier. Give students time to write the data in order from least time to greatest time. Have a student read the list while you write it on the board.

Copy the stem-and-leaf plot from the student edition on the board. Explain that since the stems are the tens digits, we write 2 through 7 in the stem column because the numbers in the list go from 25 to 70.

For each number 20 through 38, point to the number in the list and show students how it is represented in the stem-and-leaf plot. The stem for 2 has two leaves, 5 and 7. This represents the numbers 25 and 27. There are three 35s in the list, so we have one stem of 3, and three leaves of 5.

Have a student read the questions and provide answers.

Develop Skills and Concepts Guide students in making a stem-and-leaf plot for the following situation.

Middlebury Junior High has played 12 basketball games this year. The numbers of points scored were 58, 62, 54, 53, 60, 62, 63, 73, 59, 61, 54, and 62.

Make a stem-and-leaf plot on the board as students make the plot at their desks. Write the scores on the board and ask a student to order the list from least to greatest. (53, 54, 54, 58, 59, 60, 61, 62, 62, 62, 63, 73) Write **Points Scored** on the board, and draw the frame for a stem-and-leaf plot. Have a student write the stems on the board. Have another student write the leaves. Ask, *How many leaves will there be?* (12) *What is the most common number of points scored?* (62)

Explain the concepts of *outlier* and *range*. An outlier is a number that is not typical of most of the data. Ask students what the outlier is in this example. (73) The range is the highest number minus the lowest number. Ask students to find the range for this example. (20)

Use the stem-and-leaf plot on the right to answer Exercises 1–4.

Rick's Science Test Scores

Stem	Leaf
7	0
8	8 9 9
9	2 2 2 5 6 7
10	0

1. How many tests did Rick take?
   11 tests

2. Which test score is the outlier?
   70

3. Rick calls himself an A student. Is this a true statement? Explain.
   True; most of Rick's scores are A's.

4. Which test grade did Rick receive most often?
   92

Use the stem-and-leaf plot on the right to answer Exercises 5–8.

Hours of TV Sixth Graders Watch in a Week

Stem	Leaf
0	0 1 5 8
1	0 1 4 5 6 8 9
2	2 2 2 2 3 5 7
3	0 5

5. How many students were involved in this survey?
   20

6. What was the least number of hours spent watching TV? What was the greatest number of hours?
   0 hours; 35 hours

7. Parents say that many sixth graders watch about 22 hours of TV a week. Are they correct? Explain.
   Correct; 22 hours is the most frequent number of hours reported.

8. What is the range of the data?
   35 hours

Use the data to construct a stem-and-leaf plot. Then, answer Exercises 9–12.

9. The light bulb company made the following claim: Our light bulbs last up to 80 hours. Is this a true statement?
   Yes; 80 hours is in the range.

Lifetime of Light Bulbs to the Nearest Hour

66	52	57	60	41
63	72	46	66	55
54	66	73	66	80

Lifetime of Light Bulbs

Stem	Leaf
4	1 6
5	2 4 5 7
6	0 3 6 6 6 6
7	2 3
8	0

10. Can the company make the claim: Our light bulbs last 80 hours? Explain why this is misleading.
    It leads consumers to believe that most light bulbs last 80 hours.

11. How many hours did most light bulbs last?
    66 hours

12. What is the range of the data?
    39 hours

For Mixed Abilities

Common Errors • Intervention

Some students may write only one leaf for a repeated data number. Have students make a stem-and-leaf plot for the following set of data.

The ages of 20 people entering an amusement park were recorded: 12, 25, 25, 35, 36, 55, 52, 12, 14, 18, 43, 44, 25, 26, 17, 18, 12, 13, 52, 53.

Have students highlight all tens digits with one color and all the ones digits with another color. One color represents the stems, and the other represents the leaves. There should be 5 stems and 20 leaves in the stem-and-leaf plot. Have students count the leaves; if there are not 20, have them find their error.

Stem	Leaf
1	2 2 2 3 4 7 8 8
2	5 5 5 6
3	5 6
4	3 4
5	2 2 3 5

Enrichment • Application

Have students work in groups of four. Have each group choose a favorite professional basketball or football team. Challenge students to use the Internet to find the number of points scored by the team for ten games during the same season. Each group is to make a stem-and-leaf plot for its data. Encourage students to be creative, as their work will be displayed in the classroom. Provide markers, posterboard, rulers, and other art supplies.

3 Practice

Have students complete all the exercises. Remind students that if a number is repeated in their list of data, the number of leaves must match the number of times the number appears. Also, remind students to count their leaves and make sure they match the number of data in their list.

4 Assess

Have students review their stem-and-leaf plot for Exercises 9–12. Ask them what would happen if their leaf list for the stem 6 looked like 0 3 6. (Possible answer: Only one 6 is listed, so you would think that there was only one 66 in the list of data. When making a stem-and-leaf plot, we have to account for repeated numbers. Otherwise, data is left out.)

ESL/ELL STRATEGIES

Before introducing the information on graphs, preteach these expressions: *to poll people, to collect data, to plot a line,* and *to make a graph.* Use each term in a sentence and help students explain it in their own words.

14-2 Double-Bar Graphs

pages 311–312

1 Getting Started

Objective
• To interpret and construct double-bar graphs

Vocabulary
double-bar graph

Materials
science or social studies textbooks; bar graphs from magazines or newspapers

Warm Up • Mental Math
Have students solve for n.
1. $(9 + 6) \div n = 3$ (5)
2. $8 \times 5 \div n = 4$ (10)
3. $8 \div 2 + n = 12$ (8)
4. $12 \div 4 \times n = 21$ (7)
5. $4 \times n - 5 = 31$ (9)

Warm Up • Pencil and Paper
Have students find the percent of the number.
1. 20% of 240 (48)
2. 15% of 600 (90)
3. 75% of 350 (262.5)
4. 30% of 700 (210)
5. 5% of 160 (8)
6. 8% of 90 (7.2)
7. 65% of 95 (61.75)
8. 40% of 200 (80)

Name _____

Double-Bar Graphs

A bar graph is another way to report numerical information. For example, this **double-bar graph** shows how a group of TV viewers in two different cities rate a certain television program. Use the data in the chart to complete the bar graph. Use two different bars to represent the different cities. The key tells you what each bar represents.

Television Ratings		
	City A	City B
Very Good	60%	55%
Good	20%	20%
Fair	15%	10%
Poor	5%	15%

Getting Started

For this bar graph, 800 viewers in City A and 600 viewers in City B were polled. Use the graph to solve the problems.

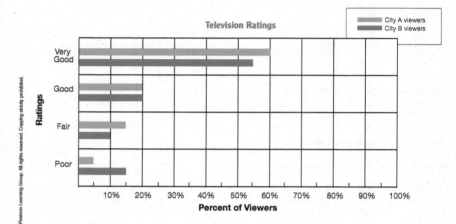

1. What percent of City A viewers rated the program fair?
   15%

2. What percent of City B viewers rated the program poor?
   15%

3. How many City A viewers rated the program very good?
   480

4. How many City B viewers rated the program good?
   120

5. How many viewers rated the program fair?
   180

6. How many viewers did not rate the program very good?
   590

Lesson 14-2 • Double-Bar Graphs

311

---

2 Teach

Introduce the Lesson Explain that bar graphs allow the reader to see differences easily and make comparisons quickly. Have students find examples of horizontal and vertical bar graphs in their social studies or science textbook.

Explain that to read a horizontal bar graph, students should find the appropriate bar, move across left to right, and then down to the scale. Point out that the model on page 311 is a double-bar graph because there are two sets of data for each category.

Develop Skills and Concepts Discuss the graph on page 311 by asking, *What is the title of the graph?* (Television Ratings) *What do the two different bars represent?* (City A viewers and City B viewers) *What information is given on the vertical scale?* (ratings) *On the horizontal scale?* (percent of viewers) *What is the interval between each percent on the horizontal scale?* (10%)

Tell students that when they make a bar graph, their choice of style, the available space, and the type of data determine the width of the bars, the space between the bars, and the unit size on the scale. Emphasize, however, that each of these factors must be consistent on the graph.

3 Practice

Have students complete all the exercises. Some students may align a ruler from the end of the bar to the scale axis. Have students work in pairs to construct the graph in Problem Solving.

T311

Practice

For this bar graph, 180 boys and 160 girls were polled. Use the graph to answer each question.

Types of Teenagers' Part-time Work

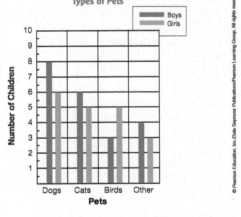

1. What percent of boys work at restaurants?
   45%

2. What percent of the girls work in supermarkets?
   10%

3. How many teenagers work at odd jobs?
   76

4. How many boys work in restaurants?
   81

5. How many of the boys work in department stores?
   27

6. How many boys do not work in department stores?
   153

7. What percent of girls do not work in department stores?
   75%

8. What kind of part-time work do most girls do?
   restaurant

9. How many girls work in restaurants or at odd jobs?
   104

10. What percent of boys do not work in supermarkets or at odd jobs?
   60%

Problem Solving

Use the data to make a double-bar graph.

Types of Pets	Boys	Girls
Dogs	8	6
Cats	6	5
Birds	3	5
Other	4	3

Types of Pets

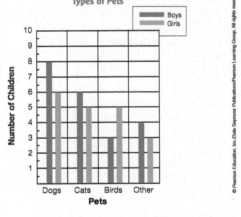

312

Lesson 14-2 • Double-Bar Graphs

4 Assess

Have students refer to the double-bar graph for Exercises 1–10. Ask, *What kind of work do most girls do?* (restaurant)

For Mixed Abilities

Common Errors • Intervention

Some students will confuse the data for the two categories when constructing a double-bar graph. Have them look at the construction as two separate tasks. Discuss how first they can use the data for Category A and draw all the bars for that category. Then, they can go back and use the data for Category B and draw all the bars for that category.

Enrichment • Application

Have students make a survey of the number of hours all sixth graders watch television each day of the week. Instruct students to break the data down to show the number of hours television is watched by boys and the number of hours television is watched by girls. Have students construct a data table and a double-bar graph. Finally, ask students to write and answer five questions about the information on their graphs.

More to Explore • Application

Remind students about the body units used for measurement they previously studied. Tell them that this week they will be testing a new measure, the bite. (This would be a good activity to coordinate with a nutrition unit.)

Each day, during an appropriate snack time, have students eat a different fruit or vegetable. Have them keep a chart of exactly how many bites it took them to eat each.

Have students graph their results and compare. Let them discuss whether or not this is an accurate measurement unit and what some obvious variables are. (bite size, fruit size, and so on)

14-3 Double-Line Graphs

pages 313–314

1 Getting Started

Objective
• To interpret and construct double-line graphs

Vocabulary
double-line graph

Warm Up • Mental Math
Have students round the number to the nearest ten, hundred, and thousand.

1. 43,674 (43,670; 43,700; 44,000)
2. 306,544 (306,540; 306,500; 307,000)
3. 2,011,469 (2,011,470; 2,011,500; 2,011,000)
4. 600,926 (600,930; 600,900; 601,000)

Warm Up • Pencil and Paper
Have students estimate the answer and then compute the answer.

1. 346 + 807 (1,100; 1,153)
2. 7,811 + 9,643 (18,000; 17,454)
3. 60,908 + 3,520 (64,000; 64,428)
4. 794 − 456 (300; 338)
5. 9,126 − 594 (8,400; 8,532)
6. 38,564 − 19,601 (20,000; 18,963)
7. 365 + 241 + 96 (700; 702)

Name _____

Double-Line Graphs

This double-line graph shows the high values of 2 stocks for one week.

The high value for stock Y on Monday was $50.

The weekly low for stock X was on Friday.

Its value on that day was $70.

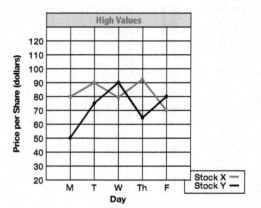

Getting Started

Use the graph above to solve each problem.

1. What is the difference between the high values of stocks X and Y on Thursday?
   about $30

2. On what day of the week did Stock Y reach its highest value?
   Wednesday

Use the data to make a double-line graph. Be sure to fill in the key.

Time Interval	Number of Cars W-E	Number of Cars N-S
8–9 A.M.	44	35
9–10 A.M.	20	25
10–11 A.M.	29	20
11–12 A.M.	14	24
12–1 P.M.	24	15
1–2 P.M.	40	38

Number of Cars at Intersection

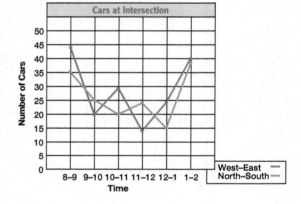

---

2 Teach

Introduce the Lesson Tell students that line graphs are used to show data changes over a period of time. For example, yearly temperatures and rainfall are often shown on line graphs. Discuss the double-line graph on page 313. After students look at the table, key, and labels on each axis, ask them to explain what the graph shows. (the high prices of two stocks during the course of a week) Explain to students that to read a line graph, they should move across the horizontal scale, left to right, and then go up to the point. Work with students as they fill in the information sentences.

Develop Skills and Concepts To help students interpret a line graph, ask questions such as *What was the value of stock X on Tuesday?* ($90) *On what day did stock Y have a value of $50 per share?* (Monday) *What is the interval on the vertical scale?* ($10) *What is the interval on the horizontal scale?* (1 day)

To help students construct a double-line graph from a table, have them follow these steps:

1. Choose and label the scales.

2. Locate the points on the grid.

3. Draw lines to connect the points.

4. Write a title.

5. Make a key to show what each type of line represents.

3 Practice

Have students complete all the exercises. Remind students to carefully notice the intervals on each scale.

Practice

Use the graph to solve Exercises 1–6.

1. What is Tom's best test score?
   97

2. What is Mary's best test score?
   94

3. How many more points did Tom receive than Mary on Test 3?
   9

4. How many more points did Mary receive than Tom on Test 2?
   6

5. What is Tom's average test score?
   86

6. What is Mary's average test score?
   87

7. Use the data to make a double-line graph of the temperatures.

Dan's and Ann's Temperatures		
Time	**Ann**	**Dan**
8 A.M.	101°	98°
10 A.M.	100°	98°
12 NOON	101°	99°
2 P.M.	102°	100°
4 P.M.	101°	99°
6 P.M.	99°	98°
8 P.M.	99°	98°

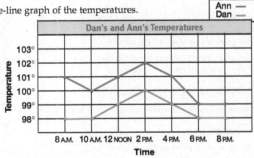

8. Use the chart to make a double-line graph of the rainfall.

Monthly Rainfall		
Month	**Iona**	**Racine**
March	20 cm	25 cm
April	30 cm	35 cm
May	35 cm	30 cm
June	30 cm	40 cm
July	25 cm	25 cm
August	15 cm	20 cm

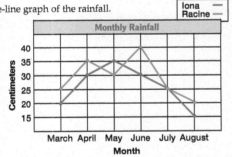

314

Lesson 14-3 • Double-Line Graphs

For Mixed Abilities

Common Errors • Intervention

Some students may read incorrectly the vertical scale for values that fall between units. Have them use a straightedge or a sheet of paper to help them align the point on the line graph with the value on the vertical scale.

Enrichment • Application

Have students locate double-line graphs in periodicals, corporate reports, and scientific journals. Ask them to explain the graphs to the class, pointing out the significant data and trends, and telling how the two lines are useful in making comparisons.

More to Explore • Statistics

Have students graph these three sets of imaginary test scores using a line graph:

Score	Test 1	Test 2	Test 3
0–20	2	4	0
21–40	8	15	5
41–60	15	9	9
61–80	9	6	16
81–100	2	2	6

Ask students to compare and explain the differences between the three graphs. Point out that the graph for Test 1 shows what we call normal distribution. It is an even distribution around a central point. Explain that the graph of Test 2 is skewed to the right because the tail on the right side is longer than the tail on the left. The third graph is skewed to the left. Display the graphs.

---

4 Assess

Have students look at the graph of test scores for Exercises 1–6. Ask students what the difference is from Tom's best score to his lowest. (22 points)

5 Mixed Review

1. 3,481 × 48 (167,088)
2. 48% of 675 (324)
3. 63.5 − 39.603 (23.897)
4. $200 - 156\frac{3}{5}$ $(43\frac{2}{5})$
5. 8^3 (512)
6. ⁻21 + ⁻35 (⁻56)
7. 3.58 × 6.5 (23.27)
8. $\frac{7}{10} \div \frac{3}{4}$ $(\frac{14}{15})$
9. 23,100 ÷ 57 (405 R15)
10. $6\frac{5}{7} - 3\frac{2}{3}$ $(3\frac{1}{21})$

14-4 Circle Graphs

pages 315–316

1 Getting Started

Objective
• To interpret circle graphs

Vocabulary
circle graph

Materials
*demonstration compass and protractor; protractors and rulers; compasses

Warm Up • Mental Math
Have students name the missing angle measure of the triangle and classify the triangle by angle.

1. 65°, 70° (45°; acute)
2. 50°, 40° (90°; right)
3. 25°, 35° (120°; obtuse)
4. 83°, 27° (70°; acute)
5. 60°, 40° (80°; acute)

Warm Up • Pencil and Paper
Have students write the fraction as an equivalent decimal and as a percent.

1. $\frac{5}{8}$ (0.625; 62.5%)
2. $\frac{7}{25}$ (0.28; 28%)
3. $\frac{9}{20}$ (0.45; 45%)
4. $\frac{9}{25}$ (0.36; 36%)
5. $\frac{5}{4}$ (1.25; 125%)
6. $\frac{2}{3}$ (0.667; 66.7%)

2 Teach

Introduce the Lesson Have a student read the first paragraph and tell what is to be solved. As you talk about the circle graph, explain that the inside, or interior, of the circle represents an entire quantity, a whole unit or 100%. Ask what 100% is in this problem. (24 hours) Explain that a fractional part of the whole is represented by an equivalent fractional part of the circle and is usually named as a percent. For example, in the model on page 315, 25% is $\frac{1}{4}$ of 100%, so 25% is represented by $\frac{1}{4}$ of the circle. Work through the rest of the lesson with students. Point out how the circle graph makes it easy to compare parts to each other or to compare a part to the whole quantity.

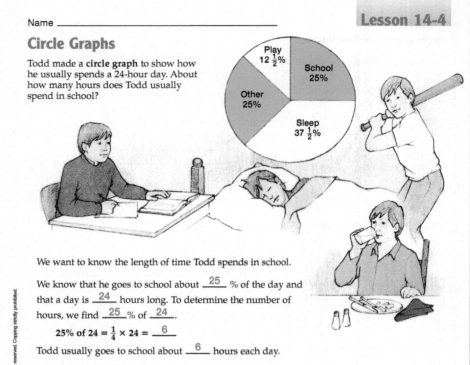

Name _____

Lesson 14-4

Circle Graphs

Todd made a **circle graph** to show how he usually spends a 24-hour day. About how many hours does Todd usually spend in school?

Play 12½%
School 25%
Other 25%
Sleep 37½%

We want to know the length of time Todd spends in school.

We know that he goes to school about __25__ % of the day and that a day is __24__ hours long. To determine the number of hours, we find __25__% of __24__.

25% of 24 = $\frac{1}{4}$ × 24 = __6__

Todd usually goes to school about __6__ hours each day.

Getting Started

Use the circle graph above to solve each problem.

1. How many hours did Todd spend sleeping?
   9
2. How many hours does Todd spend playing?
   3
3. Todd spends 10% of his time eating. How many hours is this?
   2.4
4. How many hours is Todd either in school or sleeping?
   15
5. How many hours is Todd not sleeping?
   15
6. What is the sum of the percents in the graph?
   100%

Lesson 14-4 • Circle Graphs

315

Develop Skills and Concepts Note that the greater the percent, the greater the section on the graph. Emphasize that the sum of the percents shown on a circle graph must equal 100%. Draw a circle on the board and show students that the sum of the measures of all the central angles in a circle is 360°. Tell students that to show that 26% of a worker's paycheck is spent for food, they figure 26% of the whole circle is 0.26 × 360° = 93.6°, or about 94°. Therefore, the central angle that will be labeled Food (26%) will measure 94°. Use a demonstration protractor to draw this angle.

3 Practice

Have students complete all the exercises. Watch for students who confuse percent of people with number of people.

Practice

Use the circle graph to solve Exercises 1–4.

1. How many people took the survey?
1,200

2. How many people rated the show good?
312

3. How many people rated the show fair or poor?
288

4. How many people rated the show very good?
240

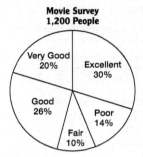

**Movie Survey
1,200 People**

Very Good 20%
Excellent 30%
Good 26%
Poor 14%
Fair 10%

Use the circle graph to solve Exercises 5–8.

5. How much is spent for housing?
$972

6. How much is spent for food?
$720

7. How much is spent for the car and utilities?
$828

8. How much more is spent for utilities than entertainment?
$108

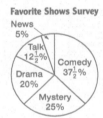

$3,600 Monthly Budget

Utilities 8%
Car 15%
Housing 27%
Other 15%
Food 20%
10%
5%
Savings
Entertainment

Now Try This!

Use your protractor to construct a circle graph with the following data:

Favorite Shows Survey	
Show	**Percent**
Comedy	$37\frac{1}{2}$
Mystery	25
Drama	20
Talk	$12\frac{1}{2}$
News	5

Favorite Shows Survey

News 5%
Talk $12\frac{1}{2}$%
Comedy $37\frac{1}{2}$%
Drama 20%
Mystery 25%

Lesson 14-4 • Circle Graphs

Now Try This! Remind students that the sum of the angles in a circle equals 360°. Tell them to complete the measure of each angle before they begin, making sure their total is 360.

4 Assess

Have students use the graph for Exercises 5–8 on page 316 to find both the total percent and the total dollar amount spent for car, utilities, and housing. (50%; $1,800)

For Mixed Abilities

Common Errors • Intervention

Some students may answer questions concerning a circle graph incorrectly because they do not compute with percents correctly. Review with them the procedure for finding a percent of a number before they answer questions concerning a circle graph.

Enrichment • Application

Explain that a budget is a plan for using money. Have students list categories for a budget on the board. Have students work in pairs to construct a data table and a circle graph showing their budgets for 1 week.

More to Explore • Number Sense

This is an exercise in mental multiplication using the Distributive Property.

Tell students that we can use this property when multiplying two 2-digit numbers. Put the example **15 × 12 = 15 × (10 + 2)** on the board and tell students that this is the same as 15 × 10 + 15 × 2 or 150 + 30 or 180.

Divide students into teams and have them take turns completing the following mentally:

1. $7 \times 99 = 7 \times (100 - 1)$
$= 7 \times 100 - 7 \times 1 = 700 - 7$
$= (693)$

2. $25 \times 22 = 25 \times (20 + 2)$
$= 25 \times 20 + 25 \times 2 = 500 + 50$
$= (550)$

3. $15 \times 103 = 15 \times (100 + 3)$
$= 15 \times 100 + 15 \times 3$
$= 1,500 + 45$
$= (1,545)$

ESL/ELL STRATEGIES

Review with students the definitions of *percent*, *interior*, and *fractional part*. Be sure that students understand the concept of a circle's central angle measures totaling 360°.

T316

1 Getting Started

Objective
- To learn mean, median, and mode and how to read and make a histogram

Vocabulary
mean, median, range, histogram

Materials
graph paper

Warm Up • Mental Math
Have students estimate the sum.

1. 32 + 73 + 29 (130)
2. 16 + 94 + 65 (180)
3. 145 + 270 + 384 (800)
4. 609 + 352 + 465 (1,500)
5. 1,406 + 3,921 + 4,596 (10,000)
6. 75 + 186 + 32 (310)

Warm Up • Pencil and Paper
Have students find the average and round quotients to the nearest hundredth.

1. 56, 69, 70, 41, 38, 93 (61.17)
2. 10.6, 11.4, 12.67, 9.1 (10.94)
3. $2\frac{5}{6}$, $3\frac{1}{3}$, $4\frac{3}{4}$, $2\frac{1}{6}$, $1\frac{11}{12}$ (3)
4. 9.463, 10.7, 8.71, 9.93 (9.70)
5. 253, 169, 250, 254 (231.5)

Name _____

Histograms

The table of data represents the temperatures of 20 cities taken at the same time of the day.

One way to describe the temperatures is to look at the center or middle readings. The **mean** temperature is the average temperature. To find the mean temperature, we add all the temperatures and divide by the number of readings.

Temperature Readings °F									
30	48	46	25	20	25	50	30	48	34
22	35	25	40	49	26	50	24	28	35

We divide __690__ by __20__.

The mean temperature is __34.5°__.

The **median** is the midway temperature. We list the temperatures in order.

20, 22, 24, 25, 25, 25, 26, 28, 30, 30, 34, 35, 35, 40, 46, 48, 48, 49, 50, 50

There is an even number of scores. So the median is the mean of the two midway scores. The median temperature is __32°__.

Another way to describe the temperature is to look at the way the data is spread out.

The **range** is the difference between the largest and smallest numbers. The range of temperatures is __30°__.

A **histogram** is a bar graph that shows the number of times data occurs.

How many temperature readings were in the 30s?

We can make a table and draw the histogram.

Temperature Readings	Number of Readings
20–29	8
30–39	5
40–49	5
50–59	2

Histogram of Temperatures

There were __5__ temperature readings in the 30s.

---

2 Teach

Introduce the Lesson Explain that an average, or mean, gives general information about a collection of data. Have five students volunteer their heights in inches or centimeters and show students how to compute the average. Add the heights. Then, divide the sum by the number of heights. Use the five heights to explain how to find the median of a set of numbers. Have a student list the heights in order from smallest to largest and find the middle height. Use the same set of data to find the range. Subtract the smallest height from the largest. Point out that *mean*, *median*, and *range* are all statistical concepts. Mean and median are measures of central tendency; they show an area around which other data collects.

Develop Skills and Concepts Discuss the models on page 317. Point out that data must be organized or arranged before being interpreted. In the example, the frequency of temperature intervals is recorded in the table. Then, the data are reassembled in the histogram. The histogram actually shows the number of times the temperature is in the 50s, 40s, 30s, and 20s.

3 Practice

Have students complete all the exercises. Have them use graph paper to draw each histogram.

Find the mean, the median, and the range for each set of data.
Complete the table and histogram.

1. In eleven games, Marcia scored the following number of points:
14, 14, 15, 17, 17, 14, 16, 14, 16, 17, 11

Mean __15__   Median __15__   Range __6__

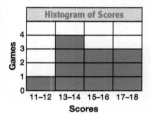

Marcia	
Scores	Games
11–12	1
13–14	4
15–16	3
17–18	3

Histogram of Scores

How many scores were over 14? __6__

2. The heights of the sixth-grade children are given in centimeters.

138	135	143	137	145
147	149	143	139	149
158	149	147	152	150
163	161	158	153	151
164	162	158	160	164

Mean __151__   Median __150__   Range __29__

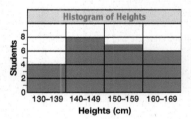

Sixth Grade	
Heights (cm)	Number of Students
130–139	4
140–149	8
150–159	7
160–169	6

Histogram of Heights

How many students were in the 140 to 159 centimeter range? __15__

318

4 Assess

Have students use the histogram for Exercise 1 on
page 318 as you ask, *In which interval did Marcia score the
most points?* (13–14)

For Mixed Abilities

Common Errors • Intervention

Some students may confuse mean
and median. Draw a number line
on the board from 30 to 60. Have
different volunteers go to the
board and mark points on the line
for the following scores: 48, 55, 42,
54, and 51. Ask students which is
the middle score. (51) Mark it with
an arrow. Discuss how *middle score*
and *median* mean the same thing:
there are the same number of
scores to the right of the median
as there are to the left. Then, have
students find the average. (50)
Discuss how the average, or
mean, is the number that if all the
numbers were this number, the
sum would be the same.

Enrichment • Statistics

Present the following set of data
to students:

In 15 games, the home team
scored the following number
of points: 85, 82, 115, 79, 88, 96,
85, 91, 90, 98, 78, 104, 99, 110,
and 95. Ask students to find
the mean, median, and range
of the set of data. Have them
make a histogram to show the
spread of scores. (mean: 93;
median: 91; range: 37)

More to Explore • Statistics

In addition to the normal and
skewed distributions, there are
three other possible line
distributions in a graph: J-shaped,
reverse J, and U-shaped. Ask
students to think of a distribution
that would give a graph in each of
these shapes. Have them draw a J,
a backward J, and a U on their
papers to help them visualize.
(Possible answers: J-shaped: the
number of games in a world series
where the fewest number is 4 but
the most is 7; reverse J: the
number of times a member of the
class has been to the hospital that
year, where the majority answer
would be zero, with a few at 1 or
2; U-shaped: the number of
deaths per 100 people at different
ages, which tends to be large in
the first year, tapers off in youth,
and rises again after age 25 or so.)

T318

14-6 Probability

pages 319–320

1 Getting Started

Objective
• To find the probability of an event

Vocabulary
probability, event

Materials
*crayons; pennies

Warm Up • Mental Math
Have students select the better estimate.

1. 50% of 500   250, 125 (250)
2. 25% of 40   20, 10 (10)
3. 75% of 60   45, 25 (45)
4. 110% of 100   110, 10 (110)
5. $33\frac{1}{3}$% of 12   9, 4 (4)
6. 10% of 60   20, 6 (6)

Warm Up • Pencil and Paper
On the chalk tray, place these crayons: 3 red, 4 green, 2 yellow, 1 black, 1 white, and 1 blue. Ask students to write a fraction to answer these and similar questions:

1. What part of the group is red? ($\frac{3}{12}$ or $\frac{1}{4}$)

2. What part is white? ($\frac{1}{12}$)

3. What part is green? ($\frac{4}{12}$ or $\frac{1}{3}$)

4. What part is yellow? ($\frac{2}{12}$ or $\frac{1}{6}$)

Name _____

Probability

The spinner has 9 equal parts. On each spin, it is equally possible that the pointer will stop on any of the 9 parts. Because 5 parts are labeled *A*, there are 5 possible outcomes where the pointer will land on *A*. The chance of stopping on *A* for any spin is 5 out of 9. The **probability** of stopping on *A* on any spin is $\frac{5}{9}$. What is the probability of the pointer stopping on a *B*?

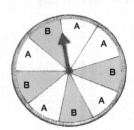

The probability of an **event** occurring is expressed as a ratio.

$$\frac{\text{number of chances to get an outcome}}{\text{number of possible outcomes}} = \frac{\text{number of parts marked } B}{\text{number of possible stops}} = \frac{4}{9}$$

The probability of stopping on *B* is $\frac{4}{9}$.

Probability is measured on a scale from 0 (impossible) to 1 (certain).

When a weather forecaster says the probability for rain is 70%, there is a pretty good chance (7 out of 10) that it will rain. What does it mean when a friend says, "The probability I will get a phone call today is only about 20%"?
There is little chance (2 out of 10) that I will get a phone call.

Getting Started

Write the probability of each event.

1. You will spin a C on the spinner above.
   0

2. The sun will set tonight.
   1

Find the probability if you draw one card without looking.

| 2 | 2 | 2 | 1 | 3 |

3. It will be a 3.
   $\frac{1}{5}$

4. It will be a 2.
   $\frac{3}{5}$

5. It will be an even number.
   $\frac{3}{5}$

6. It will be greater than 1.
   $\frac{4}{5}$

---

2 Teach

Introduce the Lesson Have a student read the problem aloud. Emphasize the terms *possible outcome*, *probability*, and *event*. As you discuss the model, explain that the probability of the pointer stopping on A or B is $\frac{5}{9} + \frac{4}{9}$ or $\frac{9}{9}$, which is equal to 1. A probability of 1 means there is a 100% chance of the pointer stopping on A or B.

Point out that the probability of the pointer stopping on C is $\frac{0}{9}$ or 0. A probability of 0 means there is 0% or no chance of the pointer stopping on C.

Explain that a probability can be written as a ratio ($\frac{1}{4}$) or as a percent (25%).

Develop Skills and Concepts Draw and label a circle spinner on the board that is divided into fourths, with two parts marked red, one green, and one yellow.

Tell students that the sections are of equal size. Ask students to identify the possible outcomes. (red, yellow, or green) Ask if the possible outcomes are equally likely. (No.)

Ask students to name the probability of the pointer stopping on yellow ($\frac{1}{4}$ or 25%), on green ($\frac{1}{4}$ or 25%), and on red ($\frac{2}{4}$ or 50%). Ask a student what the probability is of the spinner stopping on orange. ($\frac{0}{4}$ or 0%)

3 Practice

Have students complete all the exercises. Remind students that they must determine the possible outcomes first and then write the probability of each given event.

T319

Practice

Write the probability of each event.

1. You will get wet when you jump in the pool.

   1

2. Your best friend has a March birthday.

   $\frac{1}{12}$

3. The sun will rise in the west.

   0

4. Your birthday is on Friday in any year.

   $\frac{1}{7}$

5. You will catch the next ball that is thrown to you.

   $\frac{1}{2}$

6. You will go skating this weekend.

   Answers will vary.

7. A newborn baby will be a boy.

   $\frac{1}{2}$

8. You will go right home after school.

   Answers will vary.

Find the probability if you draw one marble without looking.

9. It will be black.

   $\frac{4}{11}$

10. It will be white.

    $\frac{5}{11}$

11. It will be green.

    $\frac{2}{11}$

12. It will not be green.

    $\frac{9}{11}$

Find the probability if you draw one card without looking.

13. It will be a 5.

    $\frac{1}{12}$

14. It will be a 10.

    $\frac{1}{12}$

15. It will be greater than 3.

    $\frac{9}{12}$

16. It will be less than 5.

    $\frac{4}{12}$

17. It is divisible by 2.

    $\frac{6}{12}$

18. It is prime.

    $\frac{5}{12}$

19. It has 3 for a factor.

    $\frac{4}{12}$

20. It is a multiple of 4.

    $\frac{3}{12}$

21. It is greater than 12.

    0

22. It is less than 0.

    0

23. It is an odd number.

    $\frac{6}{12}$

24. It is a factor of 30.

    $\frac{6}{12}$

320 Lesson 14-6 • Probability

4 Assess

Ask students what the probability is that you will get tails when a coin is flipped. ($\frac{1}{2}$)

For Mixed Abilities

Common Errors • Intervention

When finding probability, some students may correctly use the number of chances to get an outcome for the numerator, but instead of the number of possible outcomes, they incorrectly use the number of chances not to get an outcome for the denominator. Have them count the number of possible outcomes first and write it as the denominator before they count the number of ways they can get the outcome that they want.

Enrichment • Probability

Ask students to name possible outcomes when a paper cup is dropped. (up, down, on its side) Ask students which outcomes they think are most likely. Then, have students drop a paper cup 30 times from the same height and record the result of each drop. Have students combine their results and determine whether each outcome is equally likely. Have them discuss whether their data would change if they dropped the cup 30 more times.

More to Explore • Statistics

Have students research the meaning *actuarial table* used by insurance companies. Have students think of other industries that could make use of such a table. Students should note the current life expectancy for the average man and woman living in the United States. Then, have students calculate the number of presidential elections each man and woman could expect to vote in, in their lifetime.

1 Getting Started

Objective
- To use a tree diagram to list possible outcomes

Vocabulary
sample space

Warm Up • Mental Math
Have students find the area of the parallelogram.
1. $b = 6$ cm, $h = 5$ cm ($A = 30$ cm^2)
2. $b = 14$ m, $h = 4$ m ($A = 56$ m^2)
3. $b = 3.5$ m, $h = 3$ m ($A = 10.5$ m^2)
4. $b = 130$ cm, $h = 20$ cm ($A = 2,600$ cm^2)
5. $b = 10$ m, $h = 0.91$ m ($A = 9.1$ m^2)
6. $b = 1.96$ m, $h = 10$ m ($A = 19.6$ m^2)

Warm Up • Pencil and Paper
Have students use factor trees to write the prime factorization of the following numbers:
1. 28 (7×2^2)
2. 66 ($2 \times 3 \times 11$)
3. 125 (5^3)
4. 100 ($2^2 \times 5^2$)
5. 45 ($3^2 \times 5$)
6. 84 ($2^2 \times 3 \times 7$)
7. 50 (2×5^2)
8. 64 (2^6)
9. 225 ($3^2 \times 5^2$)

2 Teach

Introduce the Lesson Have a student read the problem and identify the question to be answered and the given information. Explain that the tree diagram shows all possible combinations or outcomes. Discuss the meaning of *sample space*.

Develop Skills and Concepts Remind students that when one penny is tossed, there are two possible outcomes, heads or tails. Explain that when two pennies are tossed, the possible outcomes are combinations, or pairs of heads and tails. Point out that tree diagrams provide a method for visualizing and naming these pairs. Draw the following tree diagram on the board. Explain that the diagram shows all the possible ways the two pennies can land. Tell students that H means heads and that T means tails.

Sample Space

Martina is going to fix a double dip frozen yogurt cone. How many possible outcomes can Martina choose?

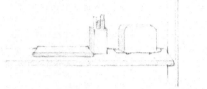

We can use a tree diagram to list the possible outcomes.

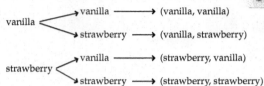

vanilla
→ vanilla ——→ (vanilla, vanilla)
→ strawberry ——→ (vanilla, strawberry)

strawberry
→ vanilla ——→ (strawberry, vanilla)
→ strawberry ——→ (strawberry, strawberry)

Martina can fix ___4___ different cones.

The possible outcomes can be written as a set of ordered pairs. This set is sometimes called a **sample space**. The possible outcomes listed as a sample space are (V, V), (V, S), (S, V), (S, S), where V stands for vanilla and S stands for strawberry.

Getting Started

Use a tree diagram to list the possible outcomes.
Write the possible outcomes as a sample space.

1. Nat is spinning two spinners. The first has the letters A, B, and C. The second has the letters Y and Z.

   (AY), (AZ), (BY), (BZ), (CY), (CZ)

   $A <^{Y \to (A, Y)}_{Z \to (A, Z)}$
   $B <^{Y \to (B, Y)}_{Z \to (B, Z)}$
   $C <^{Y \to (C, Y)}_{Z \to (C, Z)}$

2. Tad, John, Martha, Alice, and Patti are forming a committee of two. The committee will have one boy and one girl.

   (TM), (TA), (TP), (JM), (JA), (JP)

   Tad → Martha → (Tad, Martha)
   → Alice → (Tad, Alice)
   → Patti → (Tad, Patti)
   John → Martha → (John, Martha)
   → Alice → (John, Alice)
   → Patti → (John, Patti)

Penny 1	Penny 2	
H	H	HH
	T	HT
T	H	TH
	T	TT

Have students list the possible outcomes as a sample space. (HH, HT, TH, TT) Tell students that any one of the elements in a sample space is called a sample point. For example, TT is a sample point. You may wish to extend the lesson to three coins. Have students complete the diagram and list all possible outcomes.

3 Practice

Have students complete all the exercises. Watch for students who combine elements of the same item.

Use a tree diagram to list the possible outcomes.
Write the possible outcomes as a sample space.

1. Emile is spinning two spinners. The first has the numbers 1, 2, and 3. The second has the numbers 1 and 2.

   (1,1), (1,2), (2,1), (2,2), (3,1), (3,2)

   1 → 1 → (1, 1)
   1 → 2 → (1, 2)
   2 → 1 → (2, 1)
   2 → 2 → (2, 2)
   3 → 1 → (3, 1)
   3 → 2 → (3, 2)

2. Amy has a spinner with the numbers 1, 2, 3, 4, and a nickel. She spins the spinner and tosses the nickel.

   (1H), (1T), (2H), (2T), (3H), (3T), (4H), (4T)

   1 → heads → (1, heads)
   1 → tails → (1, tails)
   2 → heads → (2, heads)
   2 → tails → (2, tails)
   3 → heads → (3, heads)
   3 → tails → (3, tails)
   4 → heads → (4, heads)
   4 → tails → (4, tails)

3. Ben has a blue coat, a green coat, and a brown coat. He can wear his coat with brown slacks or blue slacks.

   (BLBR), (BLBL), (GRBR), (GRBL), (BRBR), (BRBL)

   blue coat → brown slacks → (blue coat, brown slacks)
   blue coat → blue slacks → (blue coat, blue slacks)
   green coat → brown slacks → (green coat, brown slacks)
   green coat → blue slacks → (green coat, blue slacks)
   brown coat → brown slacks → (brown coat, brown slacks)
   brown coat → blue slacks → (brown coat, blue slacks)

4. Juan is spinning two spinners. The first has the letters M, N, and O. The second has the numbers 1, 2, and 3.

   (M1), (M2), (M3), (N1), (N2), (N3), (O1), (O2), (O3)

   M → 1 → (M, 1)
   M → 2 → (M, 2)
   M → 3 → (M, 3)
   N → 1 → (N, 1)
   N → 2 → (N, 2)
   N → 3 → (N, 3)
   O → 1 → (O, 1)
   O → 2 → (O, 2)
   O → 3 → (O, 3)

5. Michelle is tossing a penny and a dime. She is recording the heads and tails.

   (HH), (HT), (TT), (TH)

   heads → heads → (heads, heads)
   heads → tails → (heads, tails)
   tails → heads → (tails, heads)
   tails → tails → (tails, tails)

6. Willie is making a sandwich. He has wheat bread and rye bread. He can use ham or beef.

   (WH), (WB), (RH), (RB)

   wheat → ham → (wheat, ham)
   wheat → beef → (wheat, beef)
   rye → ham → (rye, ham)
   rye → beef → (rye, beef)

For Mixed Abilities

Common Errors • Intervention

When drawing tree diagrams, some students might show all the possible outcomes for the first choice but then forget to write all the possible outcomes for subsequent choices. To correct these students, have them draw the tree diagram with all the blanks for all the possible combinations before filling them in.

Enrichment • Probability

Have students make a tree diagram to show the possible answers they could choose in a multiple-choice test having three questions. The first and second questions each have three choices: a, b, and c. The third question has two choices: a and b. Have students list the sample space.

(a, a, a), (a, a, b), (a, b, a), (a, b, b), (a, c, a), (a, c, b), (b, a, a), (b, a, b), (b, b, a), (b, b, b), (b, c, a), (b, c, b), (c, a, a), (c, a, b), (c, b, a), (c, b, b), (c, c, a), (c, c, b)

More to Explore • Biography

The American mathematician Norbert Wiener established the science of cybernetics, a branch of science that deals with sending information. Wiener likened the nervous systems in animals to the workings of machines. He proposed that the fields of biology and engineering be studied together as cybernetics. One part of cybernetics, called feedback, is very important. When, in the normal workings of a system, something goes wrong, information concerning the error is fed back to the controlling device to correct the error and to set the system back on track again.

Wiener's ideas led him to develop high-speed electronic computers during World War II. Wiener graduated from Tufts University when he was 14 and received his Ph.D. from Harvard University at age 18. He then taught at the Massachusetts Institute of Technology (MIT) from 1919 to 1960.

4 Assess

Create a tree diagram containing two jacket choices and three slacks choices. Have students list the sample space.
[(jacket 1, slack 1), (jacket 1, slack 2), (jacket 1, slack 3), (jacket 2, slack 1), (jacket 2, slack 2), (jacket 2, slack 3)]

5 Mixed Review

1. $3\frac{1}{3} \times 3\frac{1}{2} \times 4\frac{1}{5}$ (49)
2. -6×-5 (30)
3. 30% of 75 (22.5)
4. $50{,}308 - 38{,}726$ (11,582)
5. $15\frac{5}{9} + 8\frac{7}{12}$ ($24\frac{5}{36}$)
6. 6.03×0.17 (1.0251)
7. $^-7 - {}^-3$ ($^-4$)
8. $2\frac{4}{5} \div 2\frac{1}{3}$ ($1\frac{1}{5}$)
9. $63{,}000 \div 70$ (900)
10. 526×79 (41,554)

14-8 Compound Probability

pages 323–324

1 Getting Started

Objectives
• To find the probability when *and* or *or* is used

Materials
spinners; index cards

Warm Up • Mental Math
Have students name the decimal as a percent.

1. 0.41 (41%)   6. 0.99 (99%)
2. 0.65 (65%)   7. 0.11 (11%)
3. 0.04 (4%)   8. 1.73 (173%)
4. 0.2 (20%)   9. 0.57 (57%)
5. 0.4 (40%)   10. 0.855 (85.5%)

Warm Up • Pencil and Paper
Have students write the percent as a decimal.

1. 13% of 64 = (8.32)
2. 25% of 150 = (37.5)
3. 52% of 100 = (52)
4. 32% of 60 = (19.2)
5. 6% of 90 = (5.4)
6. 125% of 125 = (156.25)
7. 9% of 401 = (36.09)
8. 78% of 100 = (78)
9. 85% of 116 = (98.6)
10. 64% of 88 = (56.32)

Name _____

Compound Probability

The sixth-grade class is conducting a probability experiment with spinners. What is the probability that in one spin of both spinners,

• the first spinner is even and the second spinner is odd?
  P(even, odd)

• the sum of the digits is 2 or 5?
  P(sum is 2 or 5)

To answer these questions, follow these steps:

1. List the possible outcomes.

First Spinner	Second Spinner	Sample Space
1	1	(1,1), (1,2), (1,3)
1	2	(2,1), (2,2), (2,3)
1	3	
2	1	
2	2	
2	3	

2. List the ordered pairs from the sample space that have the first number even and the second number odd: (2,1), (2,3)

3. List the ordered pairs that have a sum of 2 or 5: (1,1), (2,3)

The probability that the first spinner is even and the second is odd is $\frac{2}{6}$.

The probability that the sum of the digits is 2 or 5 is $\frac{2}{6}$.

Getting Started

Use the spinners above to find each probability.

1. P(both digits are the same)
   $\frac{2}{6}$

2. P(sum is 4)
   $\frac{2}{6}$

3. P(sum is greater than 1)
   $\frac{6}{6}$

4. P(odd, odd)
   $\frac{2}{6}$

5. P(quotient is 1)
   $\frac{2}{6}$

6. P(product is 5)
   0

Lesson 14-8 • Compound Probability

323

---

2 Teach

Introduce the Lesson Prepare spinners similar to the ones on page 323. Use them to develop the concepts presented in the model problems. Point out the notation used for asking each probability. Be sure that students list the sample space (the set of all possible outcomes) correctly. Remind students that probability is written as a fraction.

Develop Skills and Concepts Explain that probability problems with the word *and* indicate that both events must occur at the same time and those with the word *or* indicate that the outcome contains at least one of the events.

Label a group of index cards with the digits 3, 5, and 7 and a second group with 2, 4, and 6. Place the cards in the chalk tray so that the numbers do not show. Have a student draw a tree diagram on the board using the even numbers as one set and the odd numbers as the other set.

Have students find the following:

P(5 and 4) $(\frac{1}{9})$; P(5 or 4) $(\frac{5}{9})$; P(sum < 10) $(\frac{6}{9})$;

P(product is a multiple of 6) $(\frac{5}{9})$

Have students randomly select a card from each set nine times to test the probabilities.

3 Practice

Have students complete all the exercises. Remind them that the notation P(. . .) is asking for the probability of an event.

Practice

Use the spinners to find each probability.

Sample Space
(2, 3), (2, 5),
(4, 3), (4, 5),
(6, 3), (6, 5)

1. P(product is 10)
$\frac{1}{6}$

2. P(sum is odd)
$\frac{6}{6}$

3. P(4 and 3)
$\frac{1}{6}$

4. P(4 and 5)
$\frac{1}{6}$

5. P(quotient is 2)
$\frac{1}{6}$

6. P(even, odd)
$\frac{6}{6}$

7. P(sum is less than 8)
$\frac{3}{6}$

8. P(product is a multiple of 3)
$\frac{4}{6}$

Use the spinners to find each probability.

Sample Space
(1, 2), (1, 3), (1, 1),
(2, 2), (2, 3), (2, 1),
(3, 2), (3, 3), (3, 1)

9. P(2 or 3)
$\frac{8}{9}$

10. P(2 and 3)
$\frac{2}{9}$

11. P(even, even)
$\frac{1}{9}$

12. P(prime, odd)
$\frac{6}{9}$

13. P(sum is 6)
$\frac{1}{9}$

14. P(sum is less than 6)
$\frac{8}{9}$

15. P(product is 6)
$\frac{2}{9}$

16. P(quotient is 2)
$\frac{1}{9}$

Three coins are tossed. Find each probability.

Nickel	Penny	Dime

17. All coins are heads.
$\frac{1}{8}$

18. All coins are the same.
$\frac{2}{8}$

19. There is exactly one head.
$\frac{3}{8}$

20. There are exactly two tails.
$\frac{3}{8}$

21. There are at least two tails.
$\frac{4}{8}$

22. There is at least one head and one tail.
$\frac{6}{8}$

Lesson 14-8 • Compound Probability

4 Assess

Tell students that two coins are tossed. Ask for the probability of both coins landing heads up. ($\frac{1}{4}$)

For Mixed Abilities

Common Errors • Intervention

Some students may try to find compound probability without first showing the sample space. Have them work with partners to practice with the following problem: Suppose you select a letter at random from the word *tea* and one from the word *sea*. Find the following:

P(both letters the same) ($\frac{2}{9}$)

P(two vowels) ($\frac{4}{9}$)

P(two consonants) ($\frac{1}{9}$)

P(one vowel, one consonant) ($\frac{4}{9}$)

Enrichment • Probability

Divide students into small groups. Provide each group of students with two spinners, each labeled with the numbers 1 through 6. Have students turn the spinner to determine the following:

P(4 or 1) ($\frac{20}{36}$ or $\frac{5}{9}$)

P(4 and 1) ($\frac{2}{36}$ or $\frac{1}{18}$)

P(even, even) ($\frac{9}{36}$ or $\frac{1}{4}$)

P(even, odd) ($\frac{9}{36}$ or $\frac{1}{4}$)

P(quotient = 2) ($\frac{5}{36}$)

P(product is multiple of 4)
($\frac{15}{36}$ or $\frac{5}{12}$)

P(sum < 9) ($\frac{26}{36}$ or $\frac{13}{18}$)

P(difference = 0) ($\frac{6}{36}$ or $\frac{1}{6}$)

More to Explore • Measurement

Draw and number 15 various abstract, geometric shapes, each on a separate sheet of paper. Make sure each pair of students has a metric ruler and paper to make calculations. Give each pair one of the drawings and tell partners that they have 4 minutes to determine the perimeter and area of each shape, showing all their work. Have them pass the drawing to the right when time is up, until students have completed measuring all 15 shapes. As an extension, include drawings of three-dimensional objects.

14-9 Multiplying Probabilities

pages 325–326

Objectives

- To distinguish between independent and dependent events
- To find probabilities of independent and dependent events by multiplying

Vocabulary

independent events, dependent events

Warm Up • Mental Math

Have students find the product.

1. $\frac{1}{3} \times \frac{2}{3}$ $\left(\frac{2}{9}\right)$
2. $\frac{1}{6} \times \frac{1}{3}$ $\left(\frac{1}{18}\right)$
3. $\frac{2}{5} \times \frac{1}{6}$ $\left(\frac{1}{15}\right)$
4. $\frac{1}{2} \times \frac{1}{2}$ $\left(\frac{1}{4}\right)$
5. $\frac{1}{4} \times \frac{2}{3}$ $\left(\frac{1}{6}\right)$
6. $\frac{3}{4} \times \frac{1}{2}$ $\left(\frac{3}{8}\right)$
7. $\frac{1}{3} \times \frac{3}{4}$ $\left(\frac{1}{4}\right)$
8. $\frac{5}{6} \times \frac{1}{6}$ $\left(\frac{5}{36}\right)$

Warm Up • Pencil and Paper

Have students find the sum.

1. $\frac{1}{2} + \frac{3}{4}$ $\left(1\frac{1}{4}\right)$
2. $\frac{3}{5} + \frac{3}{10}$ $\left(\frac{9}{10}\right)$
3. $\frac{2}{3} + \frac{1}{6}$ $\left(\frac{5}{6}\right)$
4. $\frac{1}{3} + \frac{3}{4}$ $\left(1\frac{1}{12}\right)$
5. $\frac{1}{4} + \frac{3}{8}$ $\left(\frac{5}{8}\right)$
6. $\frac{2}{3} + \frac{4}{9}$ $\left(1\frac{1}{9}\right)$

Name _____

Multiplying Probabilities

The names of six students are placed in a sack. One name will be drawn to win first prize. Another name will be drawn to win second prize. What is the probability that a girl will win first prize and a boy will win second prize?

If the first name is replaced, then the two events have no effect on each other. They are **independent events**.

- There will be 6 names to choose from for each event.

If the first name is *not* replaced, then the second drawing is affected by the first drawing. When one event is affected by another event, they are **dependent events**.

- There will be 6 names to choose from for the first prize, but only 5 names to choose from for the second prize.

We can find the probability of independent and dependent events by multiplying.

$$P(\text{girl, boy}) = P(\text{girl}) \cdot P(\text{boy})$$

With replacement

The probability of choosing a girl's name is $\frac{2}{3}$.

The probability of choosing a boy's name is $\frac{1}{3}$.

Multiply to find the probability that both events will occur.

$$\frac{2}{3} \times \frac{1}{3} = \frac{2}{9}$$

The probability that a girl will win first prize and a boy will win second prize is $\underline{2}$ out of 9.

Without replacement

The probability of choosing a girl's name is $\frac{2}{3}$.

The probability of choosing a boy's name is $\frac{2}{5}$.

Multiply to find the probability that both events will occur.

$$\frac{2}{3} \times \frac{2}{5} = \frac{4}{15}$$

The probability that a girl will win first prize and a boy will win second prize is $\underline{4}$ out of 15.

Getting Started

Find the probability. Write independent or dependent to describe the events.

1. A green marble is picked and replaced. Then a white marble is picked.

   $\frac{6}{25}$; independent

2. A white marble is picked and not replaced. Then a green marble is picked.

   $\frac{6}{20}$; dependent

Introduce the Lesson Read the opening paragraphs aloud. Explain independent and dependent events with another example.

- If Amy's name is drawn first, but we put her name back before drawing a second name, the set of names has not been changed. The two draws are independent events.

- If Amy's name is drawn first and we keep her name out during the second drawing, we are changing the original set of names. The two draws are dependent events.

Whether we have independent or dependent events, we can multiply probabilities to find a combined probability of two events occurring. Have a student read the "With replacement" column and complete the information

sentences. Have another student do the same for the "Without replacement" column.

Develop Skills and Concepts Use the same scenario as the opening problem to find the probability of picking two girls' names with replacement.

Since the first name drawn will be replaced before the second drawing, we have independent events. We want to find $P(\text{girl, girl})$. We can multiply two probabilities $P(\text{girl}) \cdot P(\text{girl})$.

Think Algebraically Have students decide what number must go in the blank so that the equation is true.

1. $3 \times \underline{(5)} = 15$
2. $16 \div \underline{(8)} = 2$
3. $\underline{(6)} \times 5 = 30$
4. $10 \times \underline{(6)} = 60$
5. $\underline{(24)} \div 8 = 3$
6. $54 \div \underline{(6)} = 9$

Practice

Write independent or dependent to describe the events.

1. Tuli draws 3 raffle tickets from the basket and then announces the winners.
   dependent

2. Pedro uses two spinners to spin a 5 followed by a 2.
   independent

3. Shua picks the first card for a magic trick. He returns the card to the deck and Elise picks the second card.
   independent

4. Without looking, Avi picks two marbles from a bag, first blue and then red.
   dependent

Find the probability. Two cards are picked without looking and not replaced.

5. picking a 5 followed by picking a 7
   $\frac{1}{56}$

6. picking two even numbers
   $\frac{3}{14}$

7. picking an even number followed by an odd number
   $\frac{2}{7}$

8. picking an odd number followed by a 4
   $\frac{1}{14}$

Two spinners are spun.

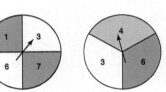

9. $P(3 \text{ and } 6) = \frac{1}{12}$

10. $P(\text{sum equal to } 9) = \frac{1}{6}$

11. $P(\text{odd, even}) = \frac{1}{2}$

12. $P(\text{prime, composite}) = \frac{1}{3}$

Problem Solving

Solve each problem.

13. Alex claims that given a deck of 10 cards numbered 1–10, he can pick an even-numbered card followed by an odd-numbered card more than half the time if he replaces the card before the second pick. Is he correct? Explain.
   No; the probability of picking an even number or an odd number is $\frac{1}{2}$. The probability of picking an even number followed by an odd number is $\frac{1}{4}$.

14. Julie wants to win the first prize and have her friend win the second prize. She thinks they have a 50-50 chance because two names will be drawn from 4 entries. What is the probability that she will win the first prize and then her friend the second prize?
   $\frac{1}{12}$

Lesson 14-9 • Multiplying Probabilities

326

---

3 Practice

Have students complete all the exercises. Remind students that they must decide whether the events are independent or dependent before determining the probability.

4 Assess

Ask students to determine if each of the following events is independent or dependent:

1. Two coins are tossed. (independent)

2. One card is drawn and not replaced. Another card is drawn. (dependent)

For Mixed Abilities

Common Errors • Intervention

Some student may have difficulty deciding if two events are independent or dependent. Have students work in pairs to act out events involving spinners, cards, and number cubes. Have them design probability experiments and observe whether the total number of possible outcomes changes with the second event. Stress that if they do not replace a card that is drawn, the events are dependent.

Enrichment • Application

We can multiply to find the probability for three or more events. A number cube has dots for the numbers 1 through 6. Suppose you roll each number cube once. What is the probability that you will get a 1, then a 2, and then a 3?

$$P(1, 2, 3) = P(1) \bullet P(2) \bullet P(3)$$
$$= \frac{1}{6} \bullet \frac{1}{6} \bullet \frac{1}{6} = \frac{1}{216}$$

Find the following probabilities:

1. $P(\text{all rolls are even})$ $(\frac{1}{27})$

2. $P(\text{all rolls are prime})$ $(\frac{1}{8})$

3. $P(\text{first roll is less than 3, second roll is a 4, and third roll is 2 or greater})$ $(\frac{5}{108})$

More to Explore • Biography

Sonya Kovalevskaya was born in Russia in 1850, a time when Russian universities were closed to women. Sonya developed an interest in mathematics because of a domestic error. Not enough wallpaper had been ordered to decorate her entire family home, so for several years Sonya's room was papered with notes her father had taken at calculus lectures. Studying her father's notes eventually helped her to understand differential calculus.

Higher education for women was only available abroad to married women. Determined to attend a university, Sonya eloped with Vladimir Kovalevsky. In Germany, she became a successful mathematics doctorate.

1 Getting Started

Objective
- To use given information to make predictions

Vocabulary
prediction

Warm Up • Mental Math
Draw this graph on the board:

Books Read

20			
15			
10			
5			
	Jill	Bill	Will

Ask these questions:
1. *Who read the most books?* (Will)
2. *How many books did Bill read?* (5)
3. *How many fewer books did Jill read than did Will?* (5)

Warm Up • Pencil and Paper
Have students draw
1. an acute angle
2. perpendicular lines
3. an isosceles triangle
4. a quadrilateral
5. parallel lines
6. an obtuse angle

2 Teach

Introduce the Lesson Read the opening paragraph. Tell students that they can use given information to predict what would happen in a related situation. Have a student read and complete the information sentences. Explain to students that we first found a probability with the given information. Then, we multiplied that probability by the number in question.

Develop Skills and Concepts Give students the following scenario and draw the table on the board.

At Fillmore Junior High, there are 320 students. Leon did an experiment in which he asked 40 of the students how many siblings they have. His results are in the table. Use Leon's results to predict how many siblings students have.

Name _____

Making Predictions

As part of quality control, a toy company tested a sample of jack-in-the-boxes and found 2 that were defective. Predict how many defective boxes there would be in a batch of 800.

Quality Control
2 out of 100 are defective

We want to know how many defective boxes we could expect in a batch of 800.

We know that 2 out of _100_ are defective.

We need to find how many out of _800_ are defective.

We can use probability to make predictions.

The probability of defective boxes is $\frac{2}{100}$ or $\frac{1}{50}$

Multiply the probability by 800 to find how many will be defective in a batch of 800.

$\frac{1}{50} \times 800 = \frac{1}{50} \times \frac{800}{1} = \frac{800}{50} = \underline{16}$

We can predict that _16_ jack-in-the-boxes will be defective in a batch of 800.

REMEMBER Simplify fractions before multiplying.

Getting Started

Make a prediction for each.

1. Bill is stocking oranges at the grocery store. In a box of 40 oranges he finds 3 that are bad. Predict how many bad oranges there will be in a case of 240 oranges.
   18

2. There are 120 red and black checkers in a bag. Marla did an experiment. Without looking she picked 8 red checkers and 16 black checkers from the bag and put them back. Predict about how many of the checkers in the bag are red. How many are black? (Hint: think about the ratio of red checkers picked to total checkers picked.)
   $\frac{1}{3}$ or about 40 are red; $\frac{2}{3}$ or about 80 are black.

Number of Siblings	Number of Students
0	JHT II
1	JHT JHT I
2	JHT JHT IIII
3	JHT I
4	II

Have students find

1. *P*(no siblings) (56)
2. *P*(2 siblings) (112)

Practice

Use the tally for Exercises 1–5.

1. Lauren picked a card from a bag without looking. She replaced the card each time and recorded the results of each trial. How many times did she pick a card?

   33

2. Using her results, predict how many times a star would be picked in 150 trials.

   73

3. Using her results, predict how many times a heart would be picked in 150 trials.

   50

4. Using her results, predict how many times a sun would be picked in 150 trials.

   27

5. If 240 cards were in the bag, how many cards with each picture would you predict there were?

   116 stars; 80 hearts; 44 suns

Card Experiment

CARD	TIMES PICKED
☀	⦀⦀⦀ I
♡	⦀⦀⦀ ⦀⦀⦀ I
☆	⦀⦀⦀ ⦀⦀⦀ ⦀⦀⦀ I

Use the graph for Exercises 6–8.

6. The bar graph shows the results of an experiment. A marble was picked from a bag without looking and replaced each time. How many times was a yellow marble picked?

   4

7. How many times was a blue marble picked? A white marble? A green marble?

   10 blue; 8 white; 2 green

8. If 50 marbles were in the bag, how many of each color would you predict there were?

   8 yellow; 21 blue; 17 white; 4 green

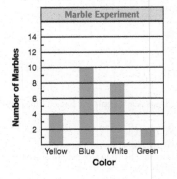

Marble Experiment

Problem Solving

Solve each problem.

9. Park rangers captured and tagged 25 raccoons and then released them back into the park. A month later they captured 100 raccoons. 20 had tags. How many raccoons are in the park?

   125

10. An inspector finds 3 damaged cans of paint in a quality check of 500. In a shipment of 1,500 cans, how many would you expect to be damaged?

    9

328

Lesson 14-10 • Making Predictions

For Mixed Abilities

Common Errors • Intervention

Some students may not remember how to write a ratio and then reduce the fraction. Have these students look back at Lesson 5-7 on simplifying fractions. Explain that the words *out of* tell us we can write a ratio, or fraction. The number before *out of* is the numerator, and the number after is the denominator. For each situation below, have them write the ratio and then reduce it.

1. 2 out of 10 apples were bad. What is the probability of a bad apple? ($\frac{2}{10}$ or $\frac{1}{5}$)

2. 24 out of 40 cars passed the safety test. What is the probability of passing the test? ($\frac{24}{40}$ or $\frac{3}{5}$)

3. 6 out of 54 students had the flu. What is the probability of having the flu? ($\frac{6}{54}$ or $\frac{1}{9}$)

4. 14 out of 21 people have brown hair. What is the probability of having brown hair? ($\frac{14}{21}$ or $\frac{2}{3}$)

Enrichment • Application

The table below shows the number of free throws attempted and made by four players during a basketball game. Estimate how many free throws each player would make if he had shot 100 free throws. Use compatible numbers to make the fractions easier to work with.

Player	Free Throws		
	Attempted	Made	Made in 100 Attempts
A	11	4	(40)
B	4	2	(50)
C	9	7	(70)
D	3	2	(66)

More to Explore • Number Sense

Check that each equation is correct. (Yes.) Use the pattern to write down the next three equations.

$1 + 2 + 1 = 2^2$

$1 + 2 + 3 + 2 + 1 = 3^2$

$1 + 2 + 3 + 4 + 3 + 2 + 1 = 4^2$

---

3 Practice

Have students complete all the exercises. Remind students that the first step is to find a probability using the known information. Then, multiply the probability by the number in question.

4 Assess

Have students review their work and answers for Exercises 2–4. Ask students what the answers to these exercises would be if Lauren did 300 trials instead of 150. (The answers would double: 150, 100, and 50.)

T328

pages 329–330

1 Getting Started

Objectives
- To identify why a graph is misleading
- To redraw a misleading graph so it is more accurate

Warm Up • Mental Math
Have students name the next number in the sequence.

1. 95, 80, 65, 50, 35, 20 (5)
2. 72, 68, 65, 61, 58, 54 (51)
3. 3, 6, 8, 16, 18, 36, 38 (76)
4. 5, 4, 12, 11, 33, 32, 96 (95)
5. 49, 7, 36, 6, 25, 5 (16)
6. 0, 5, 7, 14, 19, 21, 28 (33)
7. 39, 37, 35, 40, 38, 36, 41 (39)
8. 65, 60, 62, 57, 59, 54, 56 (51)

Warm Up • Pencil and Paper
Have students estimate the product.

1. 38 × 215 (8,000)
2. 172 × 45 (10,000)
3. 831 × 9 (8,000)
4. 575 × 330 (180,000)
5. 996 × 16 (20,000)
6. 64 × 668 (42,000)

Name

Problem Solving: Misleading Graphs

Sometimes graphs can be misleading. Companies will use misleading graphs to influence people. We need to learn to interpret such graphs to better understand the situation.

Bee's Bakery claims that their apple pie has less than 3 times the number of grams of fat than an apple pie from Abe's Bakery. Is this true?

⭐ **SEE**
We want to know if Abe's apple pie has ___3___ times the grams of fat as Bee's apple pie.

⭐ **PLAN**
We can look at the graph to determine the number of grams of fat in each apple pie. Next, multiply the number of grams of fat in Bee's apple pie by 3. Then, compare this number with the number of grams of fat in Abe's apple pie.

⭐ **DO**
Bee's apple pie has ___15___ grams of fat.
Abe's apple pie has ___25___ grams of fat.

___15___ × 3 = ___45___

Does Abe's apple pie have 3 times the fat in Bee's apple pie? ___No___

⭐ **CHECK**
We can check the accuracy of the statement by drawing a more accurate graph. We can begin the scale at 0.

Does Abe's apple pie have 3 times the fat as Bee's apple pie? ___No___

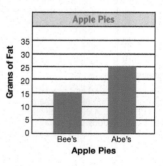

Lesson 14-11 • Problem Solving: Misleading Graphs

329

---

2 Teach

Introduce the Lesson Read the opening problem aloud. Refer students to the graph. Explain that the squiggly line on the bottom left corner of the graph means that numbers are missing. Ask, *Does it appear from the graph that Abe's apple pie has 3 times the fat as does Bee's pie?* (Yes, the bar is 3 times as high.) Tell students that we need to look further to see if this is true.

Have four different students read the SEE, PLAN, DO, and CHECK steps and complete the information sentences. Refer students to the accurate graph. Here, the distance between each number is 5; this distance was not used in the original graph. Explain that once the graph is redrawn with a scale starting at 0, we can see the claim is not true.

3 Apply

Solution Notes
1. The realtor should use the first line graph. In this line graph, the winter temperature in Fairbanks looks much colder than the temperature in Anchorage. The intervals on the vertical scales on the graphs are different, which makes the differences in temperature look greater on the first graph. A possible title for each graph is "Winter in Alaska."

2. On the graph, the bar representing San Diego is twice as long as the bar representing San Francisco. It appears as if the population of San Diego is twice the population of San Francisco because the scale does not begin at 0.

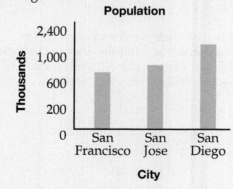

Apply

Use the graphs to answer each question.

1. A realtor wants to show that the winters in Anchorage, Alaska, are milder than the winters in Fairbanks. Which line graph should the realtor use? Why? What makes the two graphs look different? What title would you write for the graph?
See Solution Notes.

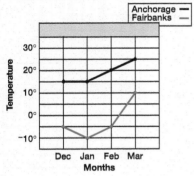

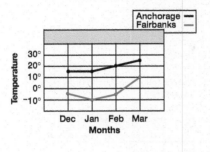

2. The mayor of San Diego claims that his city has twice the population of San Francisco. Is this an accurate statement? Explain. Draw an accurate graph of the information shown in the graph on the right.

See Solution Notes.

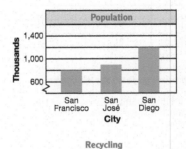

3. The recycling commission claims that the amount of plastic recycled in 2003 was 4 times the amount recycled in 1995. Is this an accurate statement? Explain. Draw a graph of the information shown in the graph on the right.

See Solution Notes.

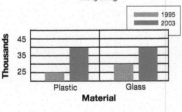

330

Lesson 14-11 • Problem Solving: Misleading Graphs

For Mixed Abilities

More to Explore• Statistics

Have students collect ads from magazines and newspapers that show the use of statistics. Post the ads in the classroom and ask students to look at each carefully. Have them determine whether the claim made by the advertiser is one that could actually be measured or if the statistic used represents an opinion. Encourage class discussion. Have students rearrange the ads, putting those that are measurable on one side of the board and those that seem to be opinion on the other.

---

3. On the graph, the bar representing the amount of plastic recycled in 2003 is 4 times as long as the bar representing the amount of plastic recycled in 1995 because the scale does not begin at 0.

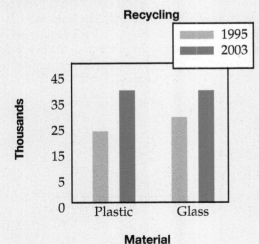

4 Assess

Draw the following graph on the board:

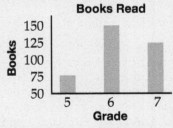

Ask students the following questions:

1. How many books did fifth graders read? (75)

2. How many books did sixth graders read? (150)

3. Sixth graders read how many times the books fifth graders did? (2 times)

4. In the graph, does it look as if sixth graders read twice as many? (No, it looks like 4 times as many.)

T330

Chapter 14
Test

page 331

Items Objectives

1–2 To interpret a double-bar graph (see pages 311–312)

3 To construct a line graph (see pages 313–314)

4 To interpret a circle graph (see pages 315–316)

5–8 To find the probability of single events and of compound events (see pages 319–326)

Use the double-bar graph to solve Problems 1–2.

1. In which months is there more than average rainfall?
   March, July

2. How much over the average did it rain in March?
   5 cm

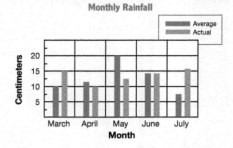

Monthly Rainfall

3. Use the following data to draw a line graph.

Students		
Grade	Lincoln School	Franklin School
1	26	32
2	29	24
3	21	26
4	25	30
5	23	28
6	31	29

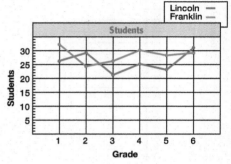

Use the circle graph to solve this problem.

4. If Bill's monthly expenses are $350, how much is spent on entertainment?
   $35

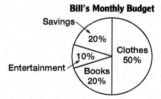

Bill's Monthly Budget

Use the spinners to find each probability.

5. $P(4)$
   $\frac{1}{4}$

6. $P(5)$
   $\frac{1}{2}$

7. $P(4, 5)$
   $\frac{1}{8}$

8. $P(\text{even, even})$
   $\frac{1}{4}$

Alternate Chapter Test

You may wish to use the Alternate Chapter Test on page 396 of this book for further review and assessment.

Circle the letter of the correct answer.

1
$$5\frac{2}{3}$$
$$+ 6\frac{5}{8}$$

a. $11\frac{7}{24}$
b. $11\frac{7}{11}$
c. $12\frac{7}{24}$
d. NG

2
$$9\frac{1}{4}$$
$$- 4\frac{2}{3}$$

a. $3\frac{1}{2}$
b. $4\frac{1}{2}$
c. $4\frac{7}{2}$
d. NG

3 $6\frac{1}{4} \times 3\frac{1}{5}$

a. $\frac{1}{20}$
b. $1\frac{25}{64}$
c. 20
d. NG

4 $4\frac{1}{2} \div 2\frac{1}{4}$

a. $\frac{1}{2}$
b. 2
c. $2\frac{1}{4}$
d. NG

5 $6.96 + 4.3$

a. 7.39
b. 11.2
c. 11.26
d. NG

6 $12.5 - 7.82$

a. 4.68
b. 5.38
c. 5.68
d. NG

7
$$5.6$$
$$\times 0.004$$

a. 0.0224
b. 0.224
c. 2.24
d. NG

8 $12 \div 0.04$

a. 3
b. 30
c. 300
d. NG

9 Find the area.

5 cm   3 cm
6 cm

a. 15 cm²
b. 18 cm²
c. 30 cm²
d. NG

10 Find the volume.

4 in.
4 in.
4 in.

a. 16 in.³
b. 24 in.³
c. 64 in.³
d. NG

11 $\frac{7}{20} = ?$

a. 7%
b. 35%
c. 70%
d. NG

score

STOP

page 332

Items	Objectives
1–2	To add or subtract mixed numbers (see pages 119–120, 127–128)
3–4	To multiply or divide mixed numbers (see pages 141–142, 147–148)
5–6	To add or subtract decimals (see pages 171–172, 173–174)
7	To multiply decimals, write zeros in the product to locate the decimal point (see pages 189–190)
8	To divide by decimals, with fewer places in the dividend than in the divisor and zeros in the dividend (see pages 199–200)
9	To find the area of parallelograms (see pages 267–268)
10	To find the volume of cubes (see pages 275–276)
11	To write a fraction as a percent (see pages 295–296)

Alternate Cumulative Assessment

Circle the letter of the correct answer.

1.
$$3\frac{1}{3}$$
$$+ 5\frac{4}{5}$$

a $8\frac{1}{15}$
b $8\frac{2}{15}$
c $9\frac{2}{15}$
d NG

2.
$$6\frac{2}{7}$$
$$- 2\frac{3}{8}$$

a $3\frac{17}{18}$
b $3\frac{52}{56}$
c $4\frac{51}{56}$
d NG

3. $6\frac{3}{4} \times 8\frac{2}{3}$

a $58\frac{5}{12}$
b $58\frac{1}{2}$
c $58\frac{5}{6}$
d NG

4. $5\frac{4}{9} \div 8\frac{1}{6}$

a $\frac{1}{3}$
b $\frac{2}{3}$
c 1
d NG

5.
$$5.83$$
$$+ 7.9$$

a 6.62
b 13.7
c 13.73
d NG

6.
$$14.6$$
$$- 8.94$$

a 5.66
b 5.74
c 5.76
d NG

7. 7.4×0.007

a 0.0518
b 0.518
c 5.18
d NG

8. $27 \div 0.03$

a 9
b 90
c 900
d NG

9. Find the area of a parallelogram 7 cm by 5 cm.

a 35 cm²
b 12 cm²
c 30 cm²
d NG

10. Find the volume of a 5-in. cube.

a 25 in.³
b 75 in.³
c 125 in.³
d NG

11. Write $\frac{3}{11}$ as a percent.

a 27.2%
b 27.3%
c 27.7%
d NG

T332

15-1 Integers

1 Getting Started

Objective
• To locate integers on a number line

Vocabulary
integers, opposite

Materials
40 index cards per group

Warm Up • Mental Math
Have students state the opposite of the situation.

1. 10 feet higher (10 feet lower)
2. $25 deposit ($25 withdrawal)
3. $20 increase ($20 decrease)
4. gain of 5 lb (loss of 5 lb)
5. 15 minutes late (15 minutes early)
6. 3 steps down (3 steps up)
7. 12° warmer (12° cooler)
8. 2 in. to the right (2 in. to the left)
9. multiply by 4 (divide by 4)
10. subtract 13 (add 13)

Warm Up • Pencil and Paper
Have students find the area of the triangle.

1. $b = 4$ m, $h = 7$ m ($A = 14$ m^2)
2. $b = 3$ m, $h = 10$ m ($A = 15$ m^2)
3. $b = 28$ cm, $h = 20$ cm ($A = 280$ cm^2)
4. $b = 15$ cm, $h = 5$ cm ($A = 37.5$ cm^2)

2 Teach

Introduce the Lesson Draw a number line on the board and label a center point 0. Have students label points 1 through 5. Remind students that these are whole numbers. Ask students to write + in front of each number to emphasize that these whole numbers are positive.

Point out that the opposites of these numbers can also be shown on the number line and that they are negative. Write the negative integers −1 through −5 on the number line.

Direct attention to the arrows at the ends of the number line. Explain that they indicate that the set of integers on the number line continues indefinitely in both directions. Work through the model with students to learn what the numbers labeled on the number line are called.

Name _____

Integers

Integers

Lesson 15-1

It's Algebra!

The electrical power was interrupted 5 seconds before radio station WXYZ was to announce the winner of the Fantasy Vacation contest. What number describes the time the electricity went off?

Before is the opposite of after, just as a negative number is the opposite of a positive number. If we say that positive 5 (⁺5) is 5 seconds after the announcement, and 0 is the exact time of the announcement, then negative 5 (⁻5) is 5 seconds before the station lost power.

We use ___−5___ to describe the time the electricity went out.

Positive numbers, negative numbers, and zero make up the set of integers. We can graph integers on a number line. Positive integers are to the right of zero. Negative integers are to the left of zero.

Locate the opposite of ⁻3 on the number line. The **opposite** of ⁻3 is ___+3___.

Opposites

```
  <———+———+———+———+———+———+———+———+———+———+———+———>
    ⁻5   ⁻4   ⁻3   ⁻2   ⁻1    0   ⁺1   ⁺2   ⁺3   ⁺4   ⁺5
              ↑                              ↑
          negative 3                     positive 3
```

Getting Started

Write the opposite.

1. up ___down___   2. left ___right___   3. increase ___decrease___

4. win 5 ___lose 5___   5. 4 above zero ___4 below zero___   6. find a dollar ___lose a dollar___

7. ⁺5 ___−5___   8. −3 ___+3___   9. ⁺15 ___−15___   10. ⁺26 ___−26___   11. 0 ___0___   12. −16 ___+16___

Locate each point on the number line. Write the corresponding letter.

13. A ⁺2   14. B ⁻3   15. C opposite of ⁻4

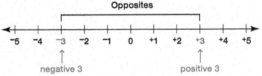

```
  <———+———+———+———+———+———+———+———+———+———>
    ⁻4   ⁻3   ⁻2   ⁻1    0   ⁺1   ⁺2   ⁺3   ⁺4
              B                        A    C
```

Develop Skills and Concepts As students study the lesson on page 333, point out that a set of *integers* consists of positive numbers, negative numbers, and zero. Note that fractions and decimals are not included in this set. Use the number line to emphasize these points:

• Whole numbers to the right of 0 are positive integers.

• Whole numbers to the left of 0 are negative integers. 0 is neither positive nor negative.

• Each integer has an opposite. The opposite of ⁺1 is ⁻1, the opposite of ⁺3 is ⁻3, and the opposite of ⁻4 is ⁺4.

• The opposite of 0 is 0.

It's Algebra! The concepts in this lesson prepare students for algebra.

Practice

Write the opposite.

1. east ___west___
2. profit ___loss___
3. right ___left___
4. saving $5 ___spending $5___
5. losing 8 yards ___gaining 8 yards___
6. plus 7 seconds ___minus 7 seconds___
7. 5 meters right ___5 meters left___
8. left 15 miles ___right 15 miles___
9. 10 kilometers south ___10 kilometers north___
10. 25 minutes before ___25 minutes after___
11. climbing up 500 feet ___climbing down 500 feet___
12. 20 degrees below zero ___20 degrees above zero___

13. $+6$ ___-6___
14. -17 ___$+17$___
15. $+15$ ___-15___
16. $+19$ ___-19___
17. -3 ___$+3$___
18. 0 ___0___
19. $+8$ ___-8___
20. $+25$ ___-25___
21. -1 ___$+1$___
22. -50 ___$+50$___
23. $+75$ ___-75___
24. -6 ___$+6$___

25. Locate each point on the number line. Write the corresponding letter.

A. $+2$ B. -2 C. 0   D. -6
E. -3 F. $+7$ G. opposite of $+3$ H. -7
I. opposite of -9

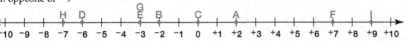

```
  H   D       G B     C     A         F     I
◄──┼──┼──┼──┼──┼┼─┼──┼──┼──┼──┼──┼──┼──┼──┼──┼──┼──┼──┼──►
 -10 -9 -8 -7 -6 -5 -4 -3 -2 -1  0  +1 +2 +3 +4 +5 +6 +7 +8 +9 +10
```

Problem Solving

Write an integer to describe these situations.

26. Burt got to the station 15 minutes before the train left.
    ___-15___

27. Angie earned $15.
    ___$+$15___

28. Tonya grew 6 centimeters in height.
    ___$+6$___

29. Dennis gained 17 yards on the third down.
    ___$+17$___

Solve these problems.

30. An elevator stopped at the 85th floor. Going 85 floors up, the elevator is $+85$. What integer shows going down 85 floors?
    ___-85___

31. The temperature dropped to 6 degrees below zero. Below zero is shown by a negative integer. What integer is used to show the temperature rising back to zero? ___$+6$___

334

Lesson 15-1 • Integers

---

3 Practice

Have students complete all the exercises. Emphasize the correct position of the + and − signs.

4 Assess

Ask students to tell the opposite of 25 feet above sea level. (25 ft below sea level)

For Mixed Abilities

Common Errors • Intervention

Some students may have difficulty understanding opposites when they are associated with numbers. Think of one number undoing the other, and think in terms of everyday experiences. For example, to undo walking 2 miles west, you would walk 2 miles east. To undo the spending of $2, you would save $2. Thus, $+2$ and -2 are opposite numbers.

Enrichment • Number Sense

Have pairs of students write the integers -20 through $+20$ on index cards, shuffle the cards, and deal 20 cards facedown to each of the two players. Have each player turn over the top card on their stack. The player whose card has the greater value keeps both of the cards. When all of the cards have been played, the player with the most cards wins.

More to Explore • Graphs

Have students conduct a survey of how fellow students spend their time during an average week when they are not in school. Be sure they include data for sleeping, eating, and other major time users. When students have completed their tally, have them find an average time spent for each major category.

Then, have students construct a circle graph for the weekly time spent on each activity by the *average* student. Remind them that they already know how much time is spent in school and should show that activity as part of the week. Display their graphs and compare.

ESL/ELL STRATEGIES

Explain how integers are different from and similar to the numbers students have been working with thus far. Such numbers can be fractions or decimals, but integers consist of only whole numbers, including zero, and their opposites.

T334

15-2 Comparing and Ordering Integers

pages 335–336

1 Getting Started

Objectives
- To compare integers
- To order integers

Warm Up • Mental Math
Round the decimal to the nearest tenth.

1. 1.35 (1.4)
2. 15.823 (15.8)
3. 8.96 (9.0)
4. 23.14 (23.1)

Round the decimal to the nearest hundredth.

1. 0.576 (0.58)
2. 9.442 (9.44)
3. 38.708 (38.71)
4. 0.345 (0.35)

Warm Up • Pencil and Paper
Have students write the correct sign for the equation.

1. 980 (−) 145 = 835
2. $25.98 (+) $132.63 = $158.61
3. 6 (×) 30 = 180
4. 610 (−) 39 = 571
5. 2,052 (÷) 54 = 38
6. 17 (×) 17 = 289

© Pearson Education, Inc./Dale Seymour Publications/Pearson Learning Group. All rights reserved. Copying strictly prohibited.

Name _____

Comparing and Ordering Integers

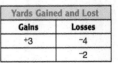

Dan records the gains and losses of his football team using integers. Which integer is greatest? Which integer is least?

We need to determine which of the integers listed is greatest.

We can use a number line to compare integers.

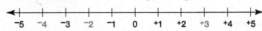

Given any two integers, the integer farther to the right on the number line is the greater integer.

Which integer is greatest? $^+3$

We can list the integers from greatest to least. $+3$, $^-2$, $^-4$

Which integer is least? $^-4$

We can list the integers from least to greatest. $^-4$, $^-2$, $^+3$

Yards Gained and Lost	
Gains	Losses
+3	⁻4
	⁻2

Getting Started

Use the number line below to compare the integers. Write >, <, or = in each circle.

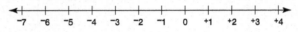

1. $^-2$ ⓧ $^+1$
2. $^+4$ ⓧ $^-5$
3. $^+3$ ⓧ $^+1$
4. $^-3$ ⓧ $^+3$
5. $^-5$ ⓧ 0
6. $^-6$ ⓧ $^+3$

Write the integers in order from greatest to least.

7. $^+1, ^-3, ^+2$
   $^+2, ^+1, ^-3$
8. $^+2, ^-4, ^+4$
   $^+4, ^+2, ^-4$
9. $0, ^-5, ^+1, ^-6$
   $^+1, 0, ^-5, ^-6$

---

2 Teach

Introduce the Lesson Read the opening problem. Draw the number line on the board. Remind students that the word *integers* means the counting numbers, their opposites, and zero. We place the positive integers to the right of zero and the negative integers to the left of zero on a number line.

Have a student read and complete the information sentences. Emphasize the fact that the numbers are greater as you move right on the number line.

Develop Skills and Concepts Draw a number line on the board and label the integers ⁻10 through ⁺10. Describe the following situation:

> During a 4-day period, the price of a certain stock went up 5, down 2, up 3, and down 8.

Ask students to write the integers that represent these increases and decreases. ($^+5, ^-2, ^+3, ^-8$) Then, ask four students to go to the board and mark each integer on the number line.

Ask the following questions:

1. *Which integer is greatest?* ($^+5$)

2. *Which integer is least?* ($^-8$)

Ask a volunteer to list the four integers in order from least to greatest. ($^-8, ^-2, ^+3, ^+5$)

Have students use the number line to compare the following pairs of integers:

1. $^-2$ (>) $^-8$ 2. $^+3$ (>) $^-2$ 3. $^-8$ (<) $^+5$

It's Algebra! The concepts in this lesson prepare students for algebra.

T335

Practice

Use the number line below to compare the integers. Write >, <, or = in each circle.

```
◄─┼──┼──┼──┼──┼──┼──┼──┼──┼──┼──┼──┼──┼──┼──┼──┼──┼──┼──┼──┼──┼─►
 ⁻10 ⁻9 ⁻8 ⁻7 ⁻6 ⁻5 ⁻4 ⁻3 ⁻2 ⁻1  0 +1 +2 +3 +4 +5 +6 +7 +8 +9 +10
```

1. 0 $>$ ⁻5

2. ⁻5 $>$ ⁻10

3. ⁻3 $<$ +3

4. ⁻8 $<$ +7

5. +2 $>$ ⁻4

6. +3 $<$ +8

7. +10 $=$ +10

8. ⁻6 $<$ +9

9. ⁻10 $<$ ⁻9

10. ⁻7 $=$ ⁻7

11. ⁻1 $>$ ⁻2

12. ⁻10 $<$ 0

Write the integers in order from greatest to least.

13. +3, ⁻9, +4, 0
    +4, +3, 0, ⁻9

14. ⁻4, +5, ⁻1, +2
    +5, +2, ⁻1, ⁻4

15. ⁻1, 0, ⁻6, +4
    +4, 0, ⁻1, ⁻6

16. ⁻10, 0, +5, ⁻3
    +5, 0, ⁻3, ⁻10

17. ⁻3, 0, ⁻9, ⁻6
    0, ⁻3, ⁻6, ⁻9

18. ⁻1, ⁻10, 0, +6
    +6, 0, ⁻1, ⁻10

Write the integers in order from least to greatest.

19. 0, ⁻4, +3, ⁻6
    ⁻6, ⁻4, 0, +3

20. +2, ⁻9, +1, ⁻8
    ⁻9, ⁻8, +1, +2

21. ⁻8, +8, 0, +2
    ⁻8, 0, +2, +8

22. +5, ⁻2, +1, ⁻3
    ⁻3, ⁻2, +1, +5

23. +9, ⁻1, ⁻5, +4
    ⁻5, ⁻1, +4, +9

24. ⁻3, 0, +3, ⁻4
    ⁻4, ⁻3, 0, +3

25. +10, ⁻6, 0, ⁻2
    ⁻6, ⁻2, 0, +10

26. 0, +1, ⁻2, ⁻1
    ⁻2, ⁻1, 0, +1

Problem Solving

Use the thermometer on the right to solve each problem.

27. On January 16 the low temperature in Buffalo, NY, was ⁻16°F. On the same day in Harrisburg, PA, the low temperature was ⁻10°F. Which city was colder?
    Buffalo

28. Yesterday the high temperature in Chicago and New York were opposite. The temperature in Chicago was ⁻19°F. What was the temperature in New York?
    19°F

29. These temperatures were recorded in Boston in the month of February: ⁻20°F, 23°F, ⁻14°F, and 11°F. List the temperatures from warmest to coldest.
    23°F, 11°F, ⁻14°F, ⁻20°F

30. These temperatures were recorded in March: Los Angeles, CA, 69°F; Madison, WI, ⁻25°F; Las Vegas, NV, 75°F; Vail, CO, ⁻8°F. Which city was the warmest? Which city was the coldest?
    Las Vegas, NV; Madison, WI

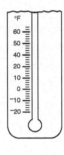

336   Lesson 15-2 • Comparing and Ordering Integers

Common Errors • Intervention

Some students may have difficulty comparing integers. Pair students and give each pair a set of 21 index cards, labeled ⁻10 through +10. Have students quiz each other. One student picks two cards, and the partner tells which integer is greater. Draw a large number line on the board for students to refer to.

Enrichment • Application

Have students work in pairs to use newspapers, magazines, or other textbooks to find examples of positive and negative integers. Examples could be temperatures or stock market prices. Once students have found eight integers, have them make a number line and mark each integer on the number line. Then, have them order the integers from least to greatest. Students may need to round decimal integers to the nearest tenth.

More to Explore • Number Sense

A magic square is one where the numbers add to the same sum whether you add across, down, or diagonally. Here is a 3 × 3 magic square.

8	1	6
3	5	7
4	9	2

You can form other magic squares using the given 3 × 3 square. First, memorize the square above. Then, ask someone to give you a number. Add this number to every number in the square above to come up with another magic square!

---

3 Practice

Have students complete all the exercises. Remind students that integers are greater as you move to the right on the number line. Students may want to draw a number line from ⁻10 through +10 to help them with the exercises.

4 Assess

Ask students the following: Suppose someone wrote 0, ⁻1, +4, ⁻6 for their answer to Exercise 15. What mistake did this person make? (The person looked only at the number and not the sign. The correct order is +4, 0, ⁻1, ⁻6.)

5 Mixed Review

1. 25% of 20 is _____. (5)
2. 10% of 35 is _____. (3.5)
3. 50% of 28 is _____. (14)
4. 20% of 50 is _____. (10)
5. 40% of 25 is _____. (10)
6. 35% of 100 is _____. (35)

15-3 Adding Integers With Like Signs

pages 337–338

1 Getting Started

Objective
• To add integers with like signs

Warm Up • Mental Math
Have students identify the rule and name the next numbers in the table.

n	0	1	2	5	9	16
$(n \times 4)$	0	4	8	(20)	(36)	(64)

The rule is ($n \times 4$).

Warm Up • Pencil and Paper
Ask students to find the sum.

1. $\frac{1}{2}$ of 440 + $\frac{1}{4}$ of 28 (227)
2. $\frac{3}{5}$ of 265 + $\frac{5}{6}$ of 366 (464)
3. $\frac{1}{10}$ of 60 + $\frac{2}{3}$ of 75 (56)
4. $\frac{3}{4}$ of 164 + $\frac{1}{8}$ of 32 (127)
5. $\frac{7}{9}$ of 117 + $\frac{4}{5}$ of 250 (291)

2 Teach

Introduce the Lesson Work through the model with students. Point out that a deposit is represented by a positive integer, and its opposite, a withdrawal, by a negative integer.

Emphasize that the signs of the integers tell which direction to move on a number line. Demonstrate using the number line to find the sums.

Emphasize that students move to the right on the number line when they add two or more positive numbers. They move to the left on the number line when they add two or more negative numbers.

Develop Skills and Concepts Ask students to compare the signs of the addends with the sign of the sum in their completed problems. (They are the same.) Emphasize the following:

• When all addends are positive, the sum is positive.

• When all addends are negative, the sum is negative.

• The sum of an integer and its opposite is 0.

Provide examples to demonstrate these generalizations. You may wish to discuss the relative values of the sum to its addends. The sum of two negative integers has a lesser value than either addend because it is farther left of zero than the addends.

It's Algebra! The concepts in this lesson prepare students for algebra.

3 Practice

Have students complete all the exercises. Be sure students read the exercises and label their answers appropriately.

Name _____

Adding Integers With Like Signs

Lesson 15-3

It's Algebra!

Sherry withdrew $9 from her account on May 1. She withdrew some more money on May 6. How much has Sherry withdrawn from her account so far in May?

We need to know the total amount Sherry has withdrawn so far in May.

We know she withdrew __$9__ on May 1 and __$6__ on May 6. We can represent the amounts Sherry withdrew with negative numbers. The $9 withdrawal is ⁻9.

The $6 withdrawal is __⁻6__. To find the total, we add the negative integers on the number line. Start at zero. Move 9 units to the left. Then move 6 more units to to the left.

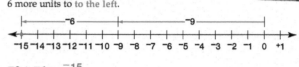

⁻9 + ⁻6 = __⁻15__

Sherry withdrew __$15__.

We can also add positive integers on the number line.

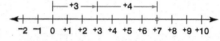

The sum of two positive integers is a __positive__ integer.
The sum of two negative integers is a __negative__ integer.

Getting Started

Show the addition on the number line. Complete the equation.

1.
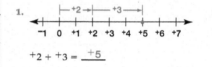

+2 + +3 = __+5__

2.

⁻3 + ⁻4 = __⁻7__

Add.

3. +8 + +9 = __+17__   4. ⁻6 + ⁻8 = __⁻14__   5. ⁻9 + ⁻3 = __⁻12__   6. +8 + 0 = __+8__

Lesson 15-3 • Adding Integers With Like Signs

337

Account Holder: Sherry Smith
Account Number: 555-00-555

Date: May 6, 2004

Amount Withdrawn: $6.00

Practice

Show the addition on the number line. Complete the equation.

1.
   $^+2 + {}^+5 = \underline{^+7}$

2.
   $^-4 + {}^-3 = \underline{^-7}$

3.
   $^-3 + {}^-5 = \underline{^-8}$

4.
   $^+1 + {}^+5 = \underline{^+6}$

Add.

5. $^-7 + {}^-8 = \underline{^-15}$ 6. $^+6 + {}^+4 = \underline{^+10}$ 7. $^+8 + {}^+5 = \underline{^+13}$

8. $^-3 + {}^-4 = \underline{^-7}$ 9. $^+7 + {}^+3 = \underline{^+10}$ 10. $^+3 + {}^+8 = \underline{^+11}$

11. $0 + {}^-5 = \underline{^-5}$ 12. $^-8 + {}^-4 = \underline{^-12}$ 13. $^+6 + 0 = \underline{^+6}$

14. $^-4 + {}^-6 = \underline{^-10}$ 15. $^-2 + {}^-10 = \underline{^-12}$ 16. $^+14 + {}^+8 = \underline{^+22}$

17. $^+3 + {}^+5 + {}^+4 = \underline{^+12}$ 18. $^-2 + {}^-5 + {}^-8 = \underline{^-15}$

19. $^+8 + 0 + {}^+6 = \underline{^+14}$ 20. $^-6 + {}^-1 + {}^-3 = \underline{^-10}$

Problem Solving

Solve each problem. Write an equation to show the addition.

21. On the first play, Bill lost 9 yards. On the second play, he lost 7 more yards. How many yards did Bill lose?

    16 yards; $^-9 + {}^-7 = {}^-16$

22. The temperature dropped 12 degrees the first hour and 6 more degrees the second hour. How many degrees did the temperature drop?

    18 degrees; $^-12 + {}^-6 = {}^-18$

[Now Try This!]

Use the integers $^-2, {}^-4, {}^-6, {}^-8, {}^-10, {}^-12, {}^-14, {}^-16,$ and $^-18$ to build a magic square with the magic number $^-30$.

$^-16$	$^-2$	$^-12$
$^-6$	$^-10$	$^-14$
$^-8$	$^-18$	$^-4$

Lesson 15-3 • Adding Integers With Like Signs

For Mixed Abilities

Common Errors • Intervention

Watch for students who think that because they are adding, the answer will always be positive. Help students understand that the sum of two negatives is negative by relating it to everyday life.

For example, if east is positive, then west would be negative. So, when you walk 4 blocks west ($^-4$) and 6 more blocks west ($^-6$), you are 10 blocks west ($^-10$) of where you were when you started.

Enrichment • Number Sense

Have students find the missing addend.

1. $^+4 + \underline{(^+8)} = {}^+12$

2. $^-6 + \underline{(^-5)} = {}^-11$

3. $^+30 + \underline{(^+15)} = {}^+45$

4. $^-15 + \underline{(0)} = {}^-15$

5. $^-7 + \underline{(^-9)} = {}^-16$

6. $^+13 + \underline{(^+13)} = {}^+26$

7. $^-16 + \underline{(^-29)} = {}^-45$

8. $\underline{(0)} + {}^+48 = {}^+48$

9. $\underline{(^-22)} + {}^-17 = {}^-39$

More to Explore • Number Sense

Tell students that in modular arithmetic, first introduced by Carl Friedrich Gauss in 1801, a number a is said to be congruent $\equiv$ to b in a particular modulo, m, if b is the remainder when divided by m. This is written $a \equiv b(\bmod\ m)$. For example, $8 \equiv 3(\bmod\ 5)$ reads 8 is congruent to 3 mod 5 because $8 \div 5 = 1$ remainder 3. Also, $13 \equiv 3(\bmod\ 5)$ because $13 \div 5 = 2$ remainder 3.

Have students list four other congruences for $3(\bmod\ 5)$.
[Possible answers: $18 \equiv 3(\bmod\ 5)$, $23 \equiv 3(\bmod\ 5)$, and $28 \equiv 3(\bmod\ 5)$]

Now Try This! One approach to completing the magic square is to identify how much more must be added to the known addends to obtain a sum of $^-30$ and then to try different combinations of addends to name that sum.

---

4 Assess

Ask, *If you add two negative integers, what is the sign of the sum?* (negative) *Will the sum be greater than or less than either addend?* (less than)

pages 339–340

1 Getting Started

Objective
• To add integers with unlike signs

Materials
masking tape; index cards; markers

Warm Up • Mental Math
Tell students to name the distance using a map scale of 2 cm = 5 km.
1. 4 cm (10 km)
2. 6 cm (15 km)
3. 10 cm (25 km)
4. 100 cm (250 km)
5. 50 cm (125 km)
6. 8 cm (20 km)

Warm Up • Mental Math
Have students find the sum.
1. $^+2 + {}^+7$ $(^+9)$
2. $^-1 + {}^-4$ $(^-5)$
3. $^-9 + {}^-8$ $(^-17)$
4. $^+9 + {}^+4$ $(^+13)$
5. $^+6 + {}^+8$ $(^+14)$
6. $^-6 + 0$ $(^-6)$
7. $^-4 + {}^-7$ $(^-11)$
8. $^+8 + {}^+7$ $(^+15)$
9. $^+5 + 0$ $(^+5)$
10. $^-4 + {}^-4$ $(^-8)$

Name _____

It's Algebra!

Mr. Allmon kept track of the price of United Computer stock for several months. This record shows how the price changed during one week. By how many points had the stock price changed after the market closed on Tuesday?

United Computer Stock	
Day	Point Change
Monday	⁻2
Tuesday	+5
Wednesday	⁻1
Thursday	⁻2
Friday	+3

We want to find one number that describes the point change after Monday's and Tuesday's trading. We know that on Monday the stock price went down __2__ points (⁻2), and on Tuesday the price went up __5__ points (⁺5). To find the total point change, we add these two integers.

We add __⁻2__ and __⁺5__.

We start at zero on the number line. We move 2 units to the left and then 5 units to the right. $^-2 + {}^+5 =$ __⁺3__

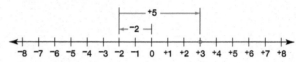

The stock had changed __⁺3__ points.

We can add any positive and negative integers on the number line.

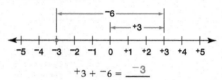

$^+3 + {}^-6 =$ __⁻3__

To add a positive and a negative integer, find the *difference* between the two numbers and keep the sign of the larger number.

Getting Started
Show the addition on the number line. Complete the equation.

1.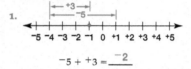

$^-5 + {}^+3 =$ __⁻2__

Add.

2. $^+2 + {}^-5 =$ __⁻3__   3. $^-4 + {}^+3 =$ __⁻1__

4. $^-2 + {}^+7 =$ __⁺5__   5. $^+6 + {}^-6 =$ __0__

---

2 Teach

Introduce the Lesson Make a number line on the floor with masking tape. Have students solve these problems by walking on the number line. Have them stand at 0 to start. Tell students to do the following: Take 3 steps to the right. Then, take 5 steps to the left. Ask how far they are from where they started? (2 steps) In which direction? (left)

Now, have students take 2 steps to the left. Then, have them take 4 steps to the right. Ask how far they are from where they started? (2 steps) In which direction? (right)

Call on students to read the problem on page 339 and to work through the model. Point out that an *increase* in the stock price is a move to the right. A *decrease* in the price is a move to the left. Emphasize that each addition begins at 0.

Develop Skills and Concepts Students may notice that they move twice in the same direction when they add two positive integers or two negative integers. Therefore, they actually add the two distances and write the sign that shows the direction of the moves.

They may also notice that when the addends have different signs, the move in one direction cancels or partially cancels the move in the other direction. Therefore, they actually subtract the two distances and label the sum with the sign of the addend farthest from 0.

Emphasize that the sum of an integer and its opposite is zero. Do not attempt to teach formal rules for adding integers at this time. It is important for students to learn to use the number line and to develop an understanding of distances and directions of moves on the number line.

It's Algebra! The concepts in this lesson prepare students for algebra.

Practice

Show the addition on the number line. Complete the equation.

1.

$$^+4 + ^-9 = \underline{^-5}$$

2.

$$^-9 + ^+3 = \underline{^-6}$$

Add.

3. $^+3 + ^-5 = \underline{^-2}$

4. $^-9 + ^+5 = \underline{^-4}$

5. $^-8 + ^+3 = \underline{^-5}$

6. $^+5 + ^-3 = \underline{^+2}$

7. $^-4 + ^+5 = \underline{^+1}$

8. $^+9 + ^-9 = \underline{0}$

9. $^-3 + ^+8 = \underline{^+5}$

10. $^-1 + ^+6 = \underline{^+5}$

11. $^-6 + 0 = \underline{^-6}$

12. $^-5 + ^+5 = \underline{0}$

13. $^+4 + ^-7 = \underline{^-3}$

14. $^-6 + ^+2 = \underline{^-4}$

15. $^+8 + ^-10 = \underline{^-2}$

16. $^-10 + ^+10 = \underline{0}$

17. $^+9 + ^-15 = \underline{^-6}$

Problem Solving

Solve each problem. Write each as an integer.

18. By Tuesday afternoon, the temperature had dropped 3 degrees. Wednesday, the sun came out, and the temperature rose 7 degrees. How much did the temperature change?

    4 degrees; $^+4$

19. An ant was trying to climb a wall. It moved 3 centimeters up the wall, then fell 8 centimeters down. How far did the ant get in these two moves?

    5 centimeters down; $^-5$

⸨ Now Try This! ⸩

Find the total at the end of the path.

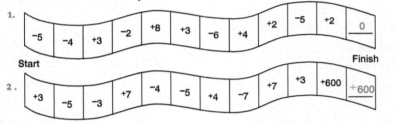

1. Start ... Finish

2.

340 Lesson 15-4 • Adding Integers With Unlike Signs

3 Practice

Have students complete all the exercises. Encourage them to draw number lines if necessary.

Now Try This! Some students may want to write each sum as they move along the path toward Finish. Encourage students to simplify their work by reminding them that an integer and its opposite equal 0.

4 Assess

Have students find the sum of $^-2 + ^+8$ and $^-8 + ^+2$.
($^+6$; $^-6$)

For Mixed Abilities

Common Errors • Intervention

Some students may continue to write the incorrect sign for the answer when they are adding integers with unlike signs. Have them work with partners and a set of index cards labeled $^-10$ through $^+10$ and a number line labeled $^-20$ through $^+20$. Have students place a marker at 0. They then take turns choosing two cards at random and moving the marker along the number line as the cards indicate. Have them write a number sentence to show the addition that they have modeled.

Enrichment • Number Sense

Have students solve the pairs of exercises and compare the sums.

1. $^+5 + ^-6 = (^-1)$;
   $^-6 + ^+5 = (^-1)$

2. $(^-4 + ^-2) + ^+3 = (^-3)$;
   $^-4 + (^-2 + ^+3) = (^-3)$

3. $^-7 + 0 = (^-7)$;
   $^+7 + 0 = (^+7)$

Have students identify each addition property represented. (Commutative Property, Associative Property, and Zero Property) Then, ask students to write other examples of each property.

More to Explore • Number Sense

Continue working with modular arithmetic. Refer to the previous More to Explore if necessary. Have students work in groups to list four congruences for 1(mod 5). [Possible answers: 6 ≡ 1(mod 5), 11 ≡ 1(mod 5), 16 ≡ 1(mod 5), and so on] Then, have them complete the following congruences from different modules:

1. 9 ≡ _____ (mod 4) (1)
2. 10 ≡ _____ (mod 7) (3)
3. 52 ≡ _____ (mod 10) (2)
4. 37 ≡ _____ (mod 6) (1)
5. 28 ≡ _____ (mod 3) (1)

T340

15-5 Subtracting Integers

pages 341–342

pages 341–342

1 Getting Started

Objective
• To subtract positive and negative integers

Materials
masking tape; scissors

Warm Up • Mental Math
Have students name the opposite of the number.

1. $^+3$ ($^-3$) 6. $^-5$ ($^+5$)
2. $^-3$ ($^+3$) 7. $^+16$ ($^-16$)
3. $^-8$ ($^+8$) 8. $^+7$ ($^-7$)
4. $^-41$ ($^+41$) 9. $^-32$ ($^+32$)
5. $^-11$ ($^+11$) 10. $^+6$ ($^-6$)

Warm Up • Pencil and Paper
Have students find the sum.

1. $^+8 + {}^+11$ ($^+19$)
2. $^+18 + {}^-26$ ($^-8$)
3. $^-14 + {}^-25$ ($^-39$)
4. $^+35 + {}^-48$ ($^-13$)
5. $^-6 + {}^+12$ ($^+6$)
6. $^+67 + {}^-83$ ($^-16$)
7. $^+37 + {}^-9$ ($^+28$)
8. $^-11 + {}^-79$ ($^-90$)

Name _____

Lesson 15-5

Subtracting Integers

It's Algebra!

The Weather Bureau predicts a cold snap in the next 24 hours. The temperature at 6 P.M. was $^+8°$. At 9 P.M., it is $^-4°$. What is the change in the temperature since 6 P.M.?

We want to know the change in temperature between 9 P.M. and 6 P.M.

We know that the temperature at 6 P.M. was $\underline{{}^+8}$, and it is $\underline{{}^-4}$ at 9 P.M.

To find the change or difference in the temperature, we subtract the 6 P.M. temperature from the 9 P.M. temperature.

We subtract $\underline{{}^+8}$ from $\underline{{}^-4}$.

To subtract, think of the subtraction equation $^-4 - {}^+8 = n$, as the addition equation $n + {}^+8 = {}^-4$. The only number added to $^+8$ that would equal $^-4$ is $^-12$.

We can also find the missing addend on the number line.

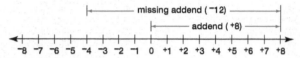

The temperature changed $\underline{{}^-12}$ degrees since 6 P.M.

To subtract an integer, rewrite the subtraction problem as an addition problem by adding the opposite of the subtrahend.

$^-4 - {}^+8 = {}^-12$

$^-4 + {}^-8 = {}^-12$

$^+8$ and $^-8$ are opposites.

Getting Started

Write the missing addend.

1. $\underline{{}^-9} + {}^+6 = {}^-3$
2. $^-2 + \underline{{}^-4} = {}^-6$
3. $\underline{{}^+10} + {}^-7 = {}^+3$

Subtract.

4. $^+5 - {}^-6 = \underline{{}^+11}$
5. $^-7 - {}^+3 = \underline{{}^-10}$
6. $^-4 - {}^-2 = \underline{{}^-2}$

Lesson 15-5 • Subtracting Integers

341

---

2 Teach

Introduce the Lesson Make a floor number line with masking tape. Tell students to begin each problem at 0.

1. They want to be 7 steps to the right of 0. They take 3 steps to the right. Ask how many more steps must they take to reach 7? (4) In which direction? (right)

2. They want to be 3 steps to the right of 0. They take 4 steps to the right. How many more steps must they take? (1) In which direction? (left)

Relate these sentences to the problems above:

1. $^+3 + n = {}^+7$ 2. $^+4 + n = {}^+3$
   $^+7 - {}^+3 = {}^+4$ $^+3 - {}^+4 = {}^-1$

Work through the model on page 341 with students.

Develop Skills and Concepts This lesson provides a rationale for subtracting integers by finding a missing addend. Students can use this method for all combinations of positive and negative integers. Have students examine the related addition sentence for each subtraction problem. Point out that the direction and the distance of the greater move on the number line give the sign and the number of the subtraction answer.

Students may be surprised to find that sometimes the difference is greater than the subtrahend. Write these sentences on the board: $^+2 - {}^-6 = {}^+8$ and $^+2 + {}^+6 = {}^+8$. Discuss how the sentences are alike and different. Allow students to pace them off on the floor number line.

Provide similar examples that lead students to conclude that subtracting a number is the same as adding its opposite.

It's Algebra! The concepts in this lesson prepare students for algebra.

Practice

Write the missing addend.

1. $^+5 + \underline{} = ^-3$ $\underline{^-8}$

2. $^-4 + \underline{} = ^-5$ $\underline{^-1}$

3. $\underline{} + ^-2 = ^-6$ $\underline{^-4}$

4. $\underline{} + ^+4 = ^+2$ $\underline{^-2}$

5. $^+7 + \underline{} = ^+9$ $\underline{^+2}$

6. $^-3 + \underline{} = ^-5$ $\underline{^-2}$

7. $\underline{} + ^-5 = 0$ $\underline{^+5}$

8. $^-8 + \underline{} = ^+7$ $\underline{^+15}$

9. $\underline{} + ^-1 = ^-4$ $\underline{^-3}$

Subtract.

10. $^+4 - ^-3 = \underline{^+7}$

11. $^-6 - ^+4 = \underline{^-10}$

12. $^-7 - ^-3 = \underline{^-4}$

13. $^+6 - ^+5 = \underline{^+1}$

14. $^-2 - ^-8 = \underline{^+6}$

15. $^-9 - ^+5 = \underline{^-14}$

16. $^-1 - ^-6 = \underline{^+5}$

17. $^+3 - ^-9 = \underline{^+12}$

18. $^-7 - 0 = \underline{^-7}$

Compute.

19. $^+7 + ^-4 + ^+5 = \underline{^+8}$

20. $^-6 - ^-3 - ^-1 = \underline{^-2}$

21. $^-4 + ^-6 - ^+3 = \underline{^-13}$

22. $^-9 - ^+4 - ^-3 = \underline{^-10}$

23. $^+6 - ^+3 + ^-2 = \underline{^+1}$

24. $^+3 - ^-3 + ^-2 = \underline{^+4}$

Problem Solving

Solve each problem. Write an equation to show the operation.

25. Sonia dove from the 10-meter board. She dove 6 meters under the water. How far did Sonia dive?
$^-10 + ^-6 = ^-16$; 16 meters

26. The average January temperature was $^+2°$. The average February temperature was $^-3°$. What was the difference between the two average temperatures? $^+2 - ^-3 = ^+5$; 5°

Now Try This!

Find the number at the end of the path.

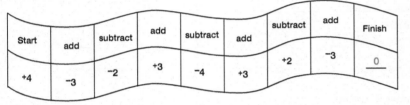

Start	add	subtract	add	subtract	add	subtract	add	Finish
+4	⁻3	⁻2	+3	⁻4	+3	+2	⁻3	0

For Mixed Abilities

Common Errors • Intervention

Some students may change the subtraction sign to addition but forget to change the sign of the number being subtracted. Discuss how subtracting a number is the same as adding its opposite. Use everyday experiences.

For example, if you earn $4 and then you spend (take away) $3, you have $1 left. If you earn $4 and you add to this a debt of $3, you really have only $1. These examples can be modeled by the following sentences:
$^+4 - ^+3 = ^+1$ and $^+4 + ^-3 = ^+1$.

Enrichment • Number Sense

Ask students to write $>$, $<$, or $=$ to complete the problem.

1. $^-7 - ^-7 \ (=) \ ^-7 + ^+7$
2. $^-8 - ^+3 \ (<) \ ^-5 - ^+5$
3. $^+4 - ^-6 \ (=) \ ^+2 - ^-8$
4. $^-2 - ^+10 \ (<) \ ^-2 - ^+8$
5. $^-7 - ^+10 \ (>) \ ^-15 - ^+6$
6. $^+6 - ^-13 \ (>) \ ^+11 - ^-4$
7. $^+3 - ^-3 \ (=) \ ^+3 + ^-3$
8. $^+1 - ^-5 \ (<) \ ^-2 - ^-10$

More to Explore • Measurement

Have students measure their height in centimeters. After compiling all the heights of the students in the class and recording them, have students compute what the total class height is. Then, have them do research to see what building or other landmark in their town they would be as tall as, collectively. Have them list as many possibilities as they can.

ESL/ELL STRATEGIES

Before introducing subtraction of integers, review the words *difference*, what you have when you subtract one number from another; *addend*, one of the numbers in a list of numbers to be added up; and *subtrahend*, a number to be subtracted from another number.

---

3 Practice

Have students complete all the exercises. Encourage them to use a number line if needed.

Now Try This! Have students suggest various ways to begin, such as combining all positive numbers and all negative numbers, or looking for opposites. Then, have students solve the problem.

4 Assess

Have students rewrite the subtraction problem as an addition problem and find the sum.

1. $^-5 - ^+2 \ (^-5 + ^-2; \ ^-7)$

2. $^+5 - ^-2 \ (^+5 + ^+2; \ ^+7)$

T342

1 Getting Started

Objective
- To multiply positive and negative integers

Warm Up • Mental Math
Have students name the sum or difference.

1. $^+3 + {^-5}$ ($^-2$)
2. $^+2 - {^-3}$ ($^+5$)
3. $^-5 + {^-7}$ ($^-12$)
4. $^-2 - 0$ ($^-2$)
5. $^-4 + {^+9}$ ($^+5$)
6. $^-5 - {^-8}$ ($^-13$)
7. $^-9 + {^+7}$ ($^-2$)
8. $^+3 - {^-2}$ ($^+1$)
9. $^+9 + {^+1}$ ($^+10$)
10. $^+9 - {^-9}$ ($^+18$)

Warm Up • Pencil and Paper
Have students find the sum.

1. $^+5 + {^+5} + {^+5}$ ($^+15$)
2. $^-12 + {^-12} + {^-12}$ ($^-36$)
3. $^+7 + {^+7} + {^+7} + {^+7}$ ($^+28$)
4. $^-16 + {^-16} + {^-16} + {^-16}$ ($^-64$)
5. $^+85 + {^+85} + {^+85} + {^+85}$ ($^+340$)
6. $^-94 + {^-94} + {^-94} + {^-94} + {^-94}$
   ($^-470$)

Name _____

Multiplying Integers

It's Algebra!

Alex and Sidra are making a poster for math class. They want to show how to multiply with integers. What integers are Alex and Sidra missing as products on their poster?

You can show how to find the products for the two examples on the number line. Remember, multiplication is repeated addition.

$^+3 \times {^+4} = {^+12}$

$^+3 \times {^-4} = {^-12}$

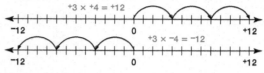

You can use the Commutative Property to show that:

$^-4 \times {^+3} = {^-12}$

Just as two negatives in a sentence mean positive, so a negative times a negative equals a positive.

$^-3 \times {^-4} = {^+12}$

For the poster, the complete equations are:

$^+3 \times {^+4} = \underline{^+12}$ $^+3 \times {^-4} = \underline{^-12}$

$^-3 \times {^+4} = \underline{^-12}$ $^-3 \times {^-4} = \underline{^+12}$

The product of two positive integers is a ___positive___ integer.

The product of two negative integers is a ___positive___ integer.

The product of a positive and a negative integer is a ___negative___ integer.

Getting Started _____

Multiply.

1. $^+5 \times {^-3} = \underline{^-15}$ 2. $^-2 \times {^-5} = \underline{^+10}$ 3. $^+6 \times {^+6} = \underline{^+36}$

4. $^-8 \times {^+2} = \underline{^-16}$ 5. $^-7 \times {^-7} = \underline{^+49}$ 6. $^+7 \times {^+7} = \underline{^+49}$

Complete each pattern.

7. $^+7 \times {^+2} = \underline{^+14}$ $^+7 \times {^+1} = \underline{^+7}$ $^+7 \times 0 = \underline{0}$ $^+7 \times {^-1} = \underline{^-7}$

---

2 Teach

Introduce the Lesson Remind students that multiplication can be thought of as repeated addition: $^+3 \times {^-4}$ means $^-4 + {^-4} + {^-4}$.

Direct attention to the signs of the two factors and the sign of the product in each problem. Recall that multiplication is commutative; that is, the factors can be reversed without changing the product. This property explains how $^-3 \times {^+4} = {^-12}$.

Work through the model on page 343 with students. Have a student read each completed multiplication rule aloud.

Develop Skills and Concepts Have students study the following patterns:

$^+6 \times {^+3} = {^+18}$ $^-6 \times {^+3} = {^-18}$

$^+6 \times {^+2} = {^+12}$ $^-6 \times {^+2} = {^-12}$

$^+6 \times {^+1} = {^+6}$ $^-6 \times {^+1} = {^-6}$

$^+6 \times 0 = 0$ $^-6 \times 0 = 0$

$^+6 \times {^-1} = {^-6}$ $^-6 \times {^-1} = {^+6}$

$^+6 \times {^-2} = {^-12}$ $^-6 \times {^-2} = {^+12}$

$^+6 \times {^-3} = {^-18}$ $^-6 \times {^-3} = {^+18}$

Help students match each of the following statements to the patterns.

- positive integer $\times$ positive integer = positive integer
- positive integer $\times$ negative integer = negative integer
- negative integer $\times$ positive integer = negative integer
- negative integer $\times$ negative integer = positive integer

It's Algebra! The concepts in this lesson prepare students for algebra.

Practice

Multiply.

1. $-3 \times +4 = \underline{-12}$
2. $-7 \times -8 = \underline{+56}$
3. $+4 \times -6 = \underline{-24}$
4. $+7 \times +6 = \underline{+42}$
5. $-2 \times 0 = \underline{0}$
6. $-9 \times +3 = \underline{-27}$
7. $-1 \times -1 = \underline{+1}$
8. $+4 \times -5 = \underline{-20}$
9. $-4 \times -6 = \underline{+24}$
10. $+8 \times -2 = \underline{-16}$
11. $+7 \times +4 = \underline{+28}$
12. $-9 \times -9 = \underline{+81}$

Complete each pattern.

13. $+4 \times +3 = +12$
    $+4 \times +2 = +8$
    $+4 \times +1 = \underline{+4}$
    $+4 \times 0 = \underline{0}$
    $+4 \times -1 = \underline{-4}$

14. $+3 \times +2 = +6$
    $+2 \times +2 = +4$
    $+1 \times +2 = \underline{+2}$
    $0 \times +2 = \underline{0}$
    $-1 \times +2 = \underline{-2}$

15. $-5 \times +3 = -15$
    $-5 \times +2 = -10$
    $-5 \times +1 = \underline{-5}$
    $-5 \times 0 = \underline{0}$
    $-5 \times -1 = \underline{+5}$

16. $-3 \times +3 = \underline{-9}$
    $-3 \times +2 = \underline{-6}$
    $-3 \times +1 = \underline{-3}$
    $-3 \times 0 = \underline{0}$
    $-3 \times -1 = \underline{+3}$

17. $+5 \times +2 = \underline{+10}$
    $+5 \times +1 = \underline{+5}$
    $+5 \times 0 = \underline{0}$
    $+5 \times -1 = \underline{-5}$
    $+5 \times -2 = \underline{-10}$

18. $-3 \times -6 = \underline{+18}$
    $-2 \times -6 = \underline{+12}$
    $-1 \times -6 = \underline{+6}$
    $0 \times -6 = \underline{0}$
    $+1 \times -6 = \underline{-6}$

Now Try This!

It's Algebra!

Write each missing factor.

1. $+3 \times n = +18$
   $n = \underline{+6}$
2. $+2 \times n = -8$
   $n = \underline{-4}$
3. $n \times -5 = -10$
   $n = \underline{+2}$
4. $n \times +6 = -30$
   $n = \underline{-5}$
5. $-4 \times n = +16$
   $n = \underline{-4}$
6. $-7 \times n = -49$
   $n = \underline{+7}$

Find each quotient.

7. $-54 \div -9 = \underline{+6}$
8. $+24 \div -6 = \underline{-4}$
9. $+40 \div +5 = \underline{+8}$
10. $-36 \times +6 = \underline{-6}$
11. $-56 \div -8 = \underline{+7}$
12. $0 \div -6 = \underline{0}$

Lesson 15-6 • Multiplying Integers

For Mixed Abilities

Common Errors • Intervention

If students have difficulty multiplying integers, have them work with partners to complete the following exercises and then describe the patterns

1. $+5 \times +4 = (+20)$
   $+5 \times +3 = (+15)$
   $+5 \times +2 = (+10)$
   $+5 \times +1 = (+5)$

2. $-5 \times +4 = (-20)$
   $-5 \times +3 = (-15)$
   $-5 \times +2 = (-10)$
   $-5 \times +1 = (-5)$

3. $+5 \times -4 = (-20)$
   $+5 \times -3 = (-15)$
   $+5 \times -2 = (-10)$
   $+5 \times -1 = (-5)$

Enrichment • Number Sense

Have students complete each puzzle by multiplying across and then down. A × B and C × D should each equal E.

-8	-6	A. $(+48)$
-7	$+4$	B. (-28)
C. $(+56)$	D. (-24)	E. $(-1,344)$

$+9$	-5	A. (-45)
-3	$+12$	B. (-36)
C. (-27)	D. (-60)	E. $(+1,620)$

$+6$	$+11$	A. $(+66)$
-9	-15	B. $(+135)$
C. (-54)	D. (-165)	E. $(+8,910)$

More to Explore • Application

Ask your students to bring in empty cereal boxes. Discuss such terms as *recommended daily allowance*, *nutrient*, *serving size*, *vitamin*, and *mineral*. This activity could be incorporated into a science nutrition unit.

Divide students into groups. Have them create problems involving ratio and percentage by using the charts given on the cereal boxes.

---

3 Practice

Have students complete all the exercises.

Now Try This! Remind students that multiplication and division are inverse operations. Have students check their quotients by multiplying. Encourage students to generalize rules that determine the sign in division exercises.

4 Assess

Ask students to determine the sign of each product.

1. $+5 \times +4$ (positive)
2. $-5 \times +4$ (negative)
3. $+5 \times -4$ (negative)
4. -5×-4 (positive)

T344

1 Getting Started

Objective
• To divide integers

Warm Up • Mental Math
Have students use the scale
1 cm = 6 m to find the actual map
distance.

1. 3 cm   (18 m)
2. 6 cm   (36 m)
3. 12 cm   (72 m)
4. 20 cm   (120 m)
5. 7 cm   (42 m)
6. 15 cm   (90 m)
7. 10 cm   (60 m)
8. 50 cm   (300 m)
9. 100 cm   (600 m)

Warm Up • Pencil and Paper
Have students find the product.

1. $^-7 \times {}^+8$ ($^-56$)
2. $^+3 \times {}^-12$ ($^-36$)
3. $^-5 \times {}^-20$ ($^+100$)
4. $^-6 \times {}^+12$ ($^-72$)
5. $^-3 \times {}^-15$ ($^+45$)

Name _____

Dividing Integers

Lesson 15-7

It's Algebra!

A scuba diver descended 56 ft from the surface of the ocean to the reef in 8 minutes. What was the diver's average rate of descent?

We need to find how many feet per minute the diver descended.

We know that the reef is __56__ feet below the surface.

We can write that as the integer __$^-56$__.

We know that it takes the diver __8__ minutes to descend.

We can divide integers to find the answer.

We know that multiplication and division are inverse operations. We can use what we know about multiplying integers to divide integers. Notice the sign for each number and the sign of the quotient.

$^+5 \times {}^+3 = {}^+15 \longrightarrow {}^+15 \div {}^+3 = {}^+5$

The quotient of two positive integers is a __positive__ integer.

$6 \times {}^-7 = {}^-42 \longrightarrow {}^-42 \div {}^-7 = {}^+6$

The quotient of two __negative__ integers is a __positive__ integer.

$^-7 \times {}^+8 = {}^-56 \longrightarrow {}^-56 \div {}^+8 = {}^-7$

The quotient of a __negative__ and a __positive__ integer is a __negative__ integer.

The quotient of $^-56 \div {}^+8$ is __$^-7$__.

The diver's average rate of descent is __$^-7$__ feet per minute.

Getting Started

Divide.

1. $^+25 \div {}^-5 =$ __$^-5$__
2. $^-12 \div {}^-4 =$ __$^+3$__
3. $^+40 \div {}^+8 =$ __$^+5$__
4. $^-27 \div {}^+3 =$ __$^-9$__
5. $^+48 \div {}^-6 =$ __$^-8$__
6. $^-28 \div {}^-7 =$ __$^+4$__

---

2 Teach

Introduce the Lesson Read the opening problem aloud. Have a student read and complete the information sentences. Explain that for every multiplication fact, we can use the same numbers to write division facts.

Remind students that the answer to a division problem is called a quotient. As you read through the multiplication and division facts, emphasize the positive and negative signs. Write each rule on the board.

Develop Skills and Concepts Give another example of dividing integers.

During the Jones's vacation, they drove to the lowest point of Death Valley, California, 280 feet below sea level. They started at sea level, and the drive took 14 minutes. What was their average rate of descent?

Ask a student to tell how to find the answer. ($^-280 \div {}^+14$) Explain that we use a negative sign in front of 280 because the car is descending. The time it took, 14 minutes, is positive. Now, have students find the quotient.

$$^-280 \div {}^+14 = {}^-20$$

The Jones's average rate of descent was $^-20$ feet per minute.

It's Algebra! The concepts in this lesson prepare students for algebra.

3 Practice

Have students complete all the exercises. Students may want to write the three rules for dividing integers before they begin and refer to them as they work.

T345

Complete each pattern.

1. $^+24 \div {}^+3 = \underline{{}^+8}$
   $^+24 \div {}^+2 = \underline{{}^+12}$
   $^+24 \div {}^+1 = \underline{{}^+24}$
   $^+24 \div {}^-1 = \underline{{}^-24}$
   $^+24 \div {}^-2 = \underline{{}^-12}$

2. $^+30 \div {}^+3 = \underline{{}^+10}$
   $^+30 \div {}^+2 = \underline{{}^+15}$
   $^+30 \div {}^+1 = \underline{{}^+30}$
   $^+30 \div {}^-1 = \underline{{}^-30}$
   $^+30 \div {}^-2 = \underline{{}^-15}$

3. $^-36 \div {}^+3 = \underline{{}^-12}$
   $^-36 \div {}^+2 = \underline{{}^-18}$
   $^-36 \div {}^+1 = \underline{{}^-36}$
   $^-36 \div {}^-1 = \underline{{}^+36}$
   $^-36 \div {}^-2 = \underline{{}^+18}$

4. $^-18 \div {}^+3 = \underline{{}^-6}$
   $^-18 \div {}^+2 = \underline{{}^-9}$
   $^-18 \div {}^+1 = \underline{{}^-18}$
   $^-18 \div {}^-1 = \underline{{}^+18}$
   $^-18 \div {}^-2 = \underline{{}^+9}$

5. $^+12 \div {}^+2 = \underline{{}^+6}$
   $^+12 \div {}^+1 = \underline{{}^+12}$
   $^+12 \div {}^-1 = \underline{{}^-12}$
   $^+12 \div {}^-2 = \underline{{}^-6}$
   $^+12 \div {}^-3 = \underline{{}^-4}$

6. $^-42 \div {}^+2 = \underline{{}^-21}$
   $^-42 \div {}^+1 = \underline{{}^-42}$
   $^-42 \div {}^-1 = \underline{{}^+42}$
   $^-42 \div {}^-2 = \underline{{}^+21}$
   $^-42 \div {}^-3 = \underline{{}^+14}$

Divide. Check your answer by multiplying.

7. $^+24 \div {}^-8 = \underline{{}^-3}$

8. $^-56 \div {}^+8 = \underline{{}^-7}$

9. $^-21 \div {}^-7 = \underline{{}^+3}$

10. $^+16 \div {}^+4 = \underline{{}^+4}$

11. $^+32 \div {}^-4 = \underline{{}^-8}$

12. $^-18 \div {}^+2 = \underline{{}^-9}$

13. $^-45 \div {}^-9 = \underline{{}^+5}$

14. $^+64 \div {}^+8 = \underline{{}^+8}$

15. $^+24 \div {}^-6 = \underline{{}^-4}$

16. $^-49 \div {}^+7 = \underline{{}^-7}$

17. $^-63 \div {}^-7 = \underline{{}^+9}$

18. $^+30 \div {}^+6 = \underline{{}^+5}$

Problem Solving

Solve each problem. Write each answer as an integer.

19. The temperature dropped from 0°F to $^-21$°F in 3 hours. The temperature fell an equal number of degrees each hour. What was the change in temperature in one hour?

    7 degrees; $^-7$

20. A submarine was at sea level and dove to $^-72$ feet in 9 seconds. What was the submarine's average rate of descent?

    8 feet per second; $^-8$

21. A turtle traveled 63 feet in 7 minutes. What was the turtle's average rate of speed per minute?

    9 feet per minute; $^+9$

22. A hot air balloon ascended 64 feet in 8 seconds. What was the hot air balloon's average rate of ascent?

    8 feet per second; $^+8$

For Mixed Abilities

Common Errors • Intervention

Some students may get confused with the sign rules for division. Tell students that they can always tell if their answer is correct by using multiplication to check.

Have students check each answer below by multiplying. They should explain why the answer is correct or not. If the answer is not correct, have students find the correct answer.

1. $^-35 \div {}^+7 = {}^+5$ (incorrect because $^+5 \times {}^+7 = {}^+35$; $^-5$ is the correct answer.)

2. $^+50 \div {}^+5 = {}^-10$ (incorrect because $^-10 \times {}^+5 = {}^-50$; $^+10$ is the correct answer.)

3. $^-8 \div {}^-4 = {}^+2$ (correct because $^+2 \times {}^-4 = {}^-8$)

4. $^-64 \div {}^-8 = {}^-8$ (incorrect because $^-8 \times {}^-8 = {}^+64$; $^+8$ is the correct answer.)

5. $^+9 \div {}^-3 = {}^-3$ (correct because $^-3 \times {}^-3 = {}^+9$)

6. $^+42 \div {}^-6 = {}^+7$ (incorrect because $^+7 \times {}^-6 = {}^-42$; $^-7$ is the correct answer.)

Enrichment • Application

A typical flying altitude for a passenger jet is about $5\frac{1}{2}$ miles. If it takes a jet 25 minutes to reach this height, what is the average rate of ascent in feet per minute? What is the average rate of ascent in feet per second? (1,161.6 feet per minute; 19.36 feet per second)

More to Explore •
Number Sense

Choose a 3-digit number whose ones and hundreds digits are different, for example, 168. Reverse the digits and find the number (861), then find the difference (861 − 168 = 693). Reverse the digits of the difference (396) and then add (693 + 369 = 1089). Do this for three different 3-digit numbers. What do you find? You have found a magic number!

4 Assess

For each fact, have students place a positive or negative sign in front of each number on the left side of the equation to make a true statement. There are two possibilities for each fact.

1. $25 \div 5 = {}^-5$ ($^-25 \div {}^+5$; $^+25 \div {}^-5$)

2. $48 \div 6 = {}^+8$ ($^+48 \div {}^+6$; $^-48 \div {}^-6$)

3. $18 \div 3 = {}^-6$ ($^+18 \div {}^-3$; $^-18 \div {}^+3$)

4. $28 \div 4 = {}^+7$ ($^+28 \div {}^+4$; $^-28 \div {}^-4$)

pages 347–348

Objective

- To review the strategies learned for solving problems

Warm Up • Mental Math

Have students compute.

1. $^-1 + {}^-4$ ($^-5$)
2. $^-5 + {}^+3$ ($^-2$)
3. $^+8 + {}^+5$ ($^+13$)
4. $^-7 - {}^+7$ (0)
5. $^-2 \times {}^-1$ ($^+2$)
6. $^-1 \times {}^-1 \times {}^-1$ ($^-1$)
7. $^-3 \times {}^+1 \times {}^-1$ ($^+3$)
8. $^-10 \div {}^-5$ ($^+2$)

Warm Up • Pencil and Paper

Have students find the following:

1. $\frac{1}{2}$ of \$7,000 (\$3,500)
2. $\frac{1}{5} = \frac{N}{30}$ ($N = 6$)
3. $\frac{1}{2}$ of \$6,500 (\$3,250)
4. $\frac{7}{4} = \frac{N}{32}$ ($N = 56$)

Name _____

Problem Solving: Logical Reasoning

Four friends each have one of the pets shown at the right. Sam does not have a snake. Ann has a dog. Pam's pet has 4 legs. Which pet does Dan have?

⭐ **SEE**

We want to know which pet belongs to Dan. There are __4__ people. There are __4__ pets.

⭐ **PLAN**

We need to think logically about the data in the problem. Making a table can help organize the information.

⭐ **DO**

Use the information given in the problem to complete the table. Write an **X** if the person does not have the pet and ✓ if they do.

	Snake	Bird	Dog	Cat
Sam	X		X	X
Ann	X	X	✓	X
Pam	X	X	X	✓
Dan			X	X

1. Sam does not have a snake, so that box gets an X.

2. Ann has a dog. Put a ✓ in Ann's dog box. Put an X to show that she does not have a snake, bird, or cat. If Ann has a dog, then her 3 friends do not. Put an X in the dog box for Sam, Pam, and Dan.

3. Pam's pet has 4 legs. Of the four pets only the dog and cat have 4 legs. Ann has the dog, so Pam must have the cat. Put a ✓ in Pam's cat box. Put an X to show that she does not have a snake or bird. If Pam has a cat, then her 3 friends do not. Put an X in the cat box for Sam, Ann, and Dan.

4. Look at the table. Which pet must Sam have? ____bird____

   Therefore, which pet must Dan have? ____snake____

⭐ **CHECK**

We can check by matching the solution with the information in the problem.

Introduce the Lesson This lesson reviews all the problem-solving strategies used this year. Students should be aware that not every strategy fits a situation and that the strategies often work most effectively paired together.

Develop Skills and Concepts Now that students have been exposed to these strategies and have had some success in using them, they should be able to transfer these skills to similar circumstances in future problems. In the practice accompanying this review, students get a chance to choose the strategy or combination of strategies to help them find solutions. Allow for individual approaches and originality in solutions.

Solution Notes

1. Begin by placing Mia behind Sara. Then, place Ben last. Phil will be either first or third. Since Phil is not first, Phil must be third.

2. Begin by looking for a 5 point difference between two scores to find the scores for Justin and Ken. Justin's score is 23 and Ken's score is 18. The two scores left are 15 and 22. Since Roy has fewer points than Jon, Roy's score is 15 and Jon's is 22.

3. Since the doctor is Smith's daughter and Smith lives next door to the lawyer, Smith must be the dentist. Since Smith and Alster are twins, Jani must be Smith's daughter. Jani is the doctor. Alster is the lawyer.

Apply

Use logical reasoning to solve each problem. Make a table to help you.

1. Phil, Mia, Ben, and Sara are standing on line. Mia is behind Sara. Phil is not first. Ben is last. In what order are they standing?

   Sara, Mia, Phil, Ben

2. Jon, Ken, Justin, and Roy are shooting baskets. Their scores so far are 15, 18, 22, and 23. Justin has 5 more points than Ken. Roy has fewer points than Jon. How many points does each boy have?

   Justin, 23; Ken, 18; Roy, 15; Jon, 22

3. The doctor, lawyer, and dentist in town are named Smith, Alster, and Jani. Smith lives next door to the lawyer. The doctor is Smith's daughter. Smith and Alster are twins. Which job does each person have?

   Smith, dentist; Alster, lawyer; Jani, doctor

4. Soo, Reggie, Michael, Ella, and Tati ran a 5 km race. Ella ran faster than Soo. Michael came in before Reggie. Soo and Tati were tied for second place. In what order did they finish the race?

   Ella, Soo, Tati, Michael, Reggie

5. Cars A, B, C, and D each take either exit 11, 25, 32, or 33 off the turnpike. Car A does not exit at an odd number. Car B sees Car D exit before Car A. Car C exits before Car D. Which car takes each exit?

   Car A, 32; Car B, 33; Car C, 11; Car D, 25

Problem-Solving Strategy: Using the Four-Step Plan

★ **SEE** What do you need to find?

★ **PLAN** What do you need to do?

★ **DO** Follow the plan.

★ **CHECK** Does your answer make sense?

---

6. Five hats are in the lost and found: blue, green, yellow, black, and orange. Amanda and Joan never wear yellow or green. Dwayne only wears black or green. Anita saw Jake wearing a blue hat. Joan's hat is not orange. Which hat belongs to each person?

   Amanda, orange; Joan, black; Dwayne, green; Jake, blue; Anita, yellow

7. Grades on the math test were 75, 82, 85, and 93. Lee's grade was 3 points higher than Kristin's. There was a 10-point difference between Stacey's grade and Lee's grade. What grade did Rose get?

   93

8. Four students are asked to stand in order of height from shortest to tallest. Joe is 5'2" inches tall. Deepak is shorter than Chris, but taller than Joe. Ali is 2 inches shorter than Chris but taller than Deepak. List the students in order of height from shortest to tallest.

   Joe, Deepak, Ali, Chris

For Mixed Abilities

More to Explore • Application

Put this diagram on the board and have students roughly copy it.

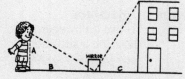

Pair students, give each pair a meterstick and mirror, and take students outside. Have one student of each pair put the mirror on the ground in front of the school and stand on the opposite side of it, facing the building. Tell the student to move around the mirror until he or she can see the reflection of the top edge of the building at the top edge of the mirror.

Point out that two similar triangles have been formed. Refer students to their diagram. Have the second student of the pair measure the length of *a*, *b*, and *c*. Take students back to the classroom and have them calculate the height of the building by using the proportion $\frac{a}{b} = \frac{?}{c}$. Check students' work.

---

4. Soo and Tati tied for second place. Since Ella ran faster than Soo, Ella is in first place. Since Michael came in before Reggie, Reggie must be last.

5. Car A exits at 32. (It is the only even number.) Since Car B sees Car D exit before Car A, Car B exits at 33. There are only two exits left, 11 and 25. Since Car C exits before Car D, Car C exits at 11 and Car D exits at 25.

6. Check a blue hat for Jake. Cross out all of the other boxes in that row and column. Cross out yellow and green for Amanda and Joan, and orange for Joan. Joan must be wearing a black hat. Check this box and cross out the other boxes in that column. Amanda wears orange. Dwayne must be wearing green and Anita a yellow hat.

7. Have students look for a 3-point difference in the grades. Since Lee's grade is 3 points higher than Kristin's, Lee got an 85 and Kristin got an 82. Stacey got a 75. So, Rose got a 93.

8. List the students from left to right in order from shortest to tallest. Deepak is to the right of Joe but to the left of Chris. Since Chris is 2 inches taller than Ali, Chris is the tallest and Ali is to the left of Chris.

4 Assess

Have students write their own logic problem and provide the solution.

Chapter 15
Test

page 349

Items	Objectives
1–3	To use an integer to describe a situation (see pages 333–334)
4–9	To compare integers (see pages 335–336)
10–13, 15, 17, 21	To add integers with like signs (see pages 337–338)
14, 16, 18–20, 22–25	To add integers with unlike signs (see pages 339–340)
26–37	To subtract integers (see pages 341–342)
38–45	To multiply integers (see pages 343–344)
46–53	To divide integers (see pages 345–346)

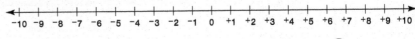

Name _____

Write an integer to describe each situation.

1. loss of 3 yards -3 2. savings of \$10 $+10$ 3. 12° below zero -12

Use the number line below to compare the integers. Write >, <, or = in each circle.

⟵ | | | | | | | | | | | | | | | | | | | | | ⟶
-10 -9 -8 -7 -6 -5 -4 -3 -2 -1 0 $+1$ $+2$ $+3$ $+4$ $+5$ $+6$ $+7$ $+8$ $+9$ $+10$

4. $2 \;(>)\; -3$ 5. $0 \;(<)\; 4$ 6. $-6 \;(<)\; 3$

7. $-7 \;(<)\; 0$ 8. $-1 \;(>)\; -5$ 9. $-2 \;(=)\; -5 + {}^+3$

Add.

10. $+3 + {}^+9 = +12$ 11. $-6 + {}^-5 = -11$ 12. $-3 + {}^-4 = -7$ 13. $+7 + {}^+2 = +9$

14. $+5 + 0 = +5$ 15. $-7 + {}^-3 = -10$ 16. $0 + {}^-3 = -3$ 17. $-2 + {}^-3 = -5$

18. $-3 + {}^+5 = +2$ 19. $+7 + {}^-8 = -1$ 20. $+6 + {}^-3 = +3$ 21. $-8 + {}^-9 = -17$

22. $-4 + {}^+3 = -1$ 23. $-6 + {}^+5 = -1$ 24. $-9 + {}^+3 = -6$ 25. $+4 + {}^-8 = -4$

Subtract.

26. $+9 - {}^+8 = +1$ 27. $-7 - {}^-3 = -4$ 28. $-6 - {}^+4 = -10$ 29. $-6 - {}^-8 = +2$

30. $+4 - {}^+6 = -2$ 31. $-5 - {}^+7 = -12$ 32. $0 - {}^-3 = +3$ 33. $+5 - {}^-3 = +8$

34. $-8 - {}^+2 = -10$ 35. $-3 - {}^-8 = +5$ 36. $+2 - {}^-4 = +6$ 37. $+8 - {}^-7 = +15$

Multiply.

38. $-3 \times {}^+8 = -24$ 39. $-7 \times {}^-7 = +49$ 40. $+5 \times {}^-4 = -20$ 41. $-1 \times {}^-9 = +9$

42. $+4 \times {}^+6 = +24$ 43. $-2 \times {}^+3 = -6$ 44. $-8 \times {}^-9 = +72$ 45. $+5 \times {}^+6 = +30$

Divide.

46. $-10 \div 2 = -5$ 47. $-20 \div {}^-5 = +4$ 48. $24 \div 6 = +4$ 49. $-6 \div {}^-3 = +2$

50. $40 \div 5 = +8$ 51. $-14 \div 7 = -2$ 52. $-28 \div {}^-4 = +7$ 53. $30 \div {}^-5 = -6$

Chapter 15 • Test

349

Alternate Chapter Test

You may wish to use the Alternate Chapter Test on page 398 of this book for further review and assessment.

Circle the letter of the correct answer.

① $4\frac{7}{8}$
  $+ 3\frac{1}{3}$

   a. $7\frac{5}{24}$
   b. $7\frac{8}{11}$
   c. $8\frac{5}{24}$
   d. NG

② $8\frac{2}{3} - 1\frac{4}{5}$

   a. $6\frac{13}{15}$
   b. $7\frac{2}{15}$
   c. $7\frac{13}{15}$
   d. NG

③ $3\frac{1}{5}$
  $\times 2\frac{1}{2}$

   a. $\frac{1}{8}$
   b. $6\frac{1}{10}$
   c. 8
   d. NG

④ $4\frac{1}{2} \div 3\frac{3}{4}$

   a. $\frac{5}{6}$
   b. $1\frac{1}{5}$
   c. $16\frac{3}{4}$
   d. NG

⑤ $12.2 + 4.83$

   a. 16.03
   b. 17.03
   c. 60.5
   d. NG

⑥ $12 - 3.08$

   a. 8.92
   b. 9.02
   c. 9.08
   d. NG

⑦ 0.09×0.06

   a. 0.54
   b. 0.054
   c. 0.0054
   d. NG

⑧ $6.08 \div 0.4$

   a. 1.52
   b. 15.2
   c. 152
   d. NG

⑨ Find the area.

   a. 8.75 cm²
   b. 12.5 cm²
   c. 10 cm²
   d. NG

(triangle: 4 cm, 5 cm, 3.5 cm, 5 cm)

⑩ 35% of 180

   a. 63
   b. 630
   c. 6,300
   d. NG

⑪ Both spinners are spun at the same time. What is P (b, odd number)?

   a. $\frac{2}{12}$
   b. $\frac{1}{12}$
   c. $\frac{3}{12}$
   d. NG

(spinners: a b / c d and 1 2 / 3)

STOP

☐   score

page 350

Items	Objectives
1–2	To add or subtract mixed numbers (see pages 119–120, 127–128)
3–4	To multiply or divide mixed numbers (see pages 141–142, 147–148)
5–6	To add or subtract decimals (see pages 171–172, 173–174)
7	To multiply decimals, write zeros in the product to locate the decimal point (see pages 189–190)
8	To divide decimals by decimals (see pages 197–198)
9	To find the area of triangles (see pages 269–270)
10	To find the percent of a number (see pages 297–298)
11	To find compound probability (see pages 323–324)

Alternate Cumulative Assessment

Circle the letter of the correct answer.

1. $7\frac{1}{6}$
   $+ 4\frac{8}{9}$

   a $11\frac{17}{18}$
   b 12
   c $12\frac{1}{18}$
   d NG

2. $9\frac{1}{6}$
   $- 3\frac{5}{12}$

   a $5\frac{2}{3}$
   b $5\frac{3}{4}$
   c $6\frac{3}{4}$
   d NG

3. $6\frac{4}{9} \times 3\frac{6}{7}$

   a 21
   b $23\frac{26}{27}$
   c $24\frac{6}{7}$
   d NG

4. $3\frac{6}{10} \div 2\frac{4}{5}$

   a $1\frac{2}{7}$
   b $1\frac{3}{7}$
   c $1\frac{4}{7}$
   d NG

5. $16.7 + 6.34$

   a 22.04
   b 23.04
   c 80.1
   d NG

6. $13 - 6.02$

   a 6.98
   b 7.02
   c 7.08
   d NG

7. 0.03
   $\times 0.08$

   a 0.24
   b 0.024
   c 0.0024
   d NG

8. $7.05 \div 0.3$

   a 2.35
   b 23.5
   c 235
   d NG

9. Find the area.

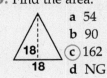

   a 54
   b 90
   c 162
   d NG

   (triangle: 18, 18)

10. Find 85% of 420.

   a 35.7
   b 357
   c 3,570
   d NG

16-1 Algebraic Expressions

1 Getting Started

Objectives
- To write an algebraic expression
- To evaluate an algebraic expression

Vocabulary
variable, algebraic expression

Warm Up • Mental Math
Have students simplify the expression.

1. $7 \times 2 - 5$ (9)
2. $8 + 5 - 6$ (7)
3. $42 \div 7 + 9$ (15)
4. $16 - 8 + 8$ (16)
5. $8 \times 5 - 10$ (30)
6. $81 \div 9 + 9$ (18)

Warm Up • Pencil and Paper
Have students use the Order of Operations to simplify the expression.

1. $(40 + 5) \times 6$ (270)
2. $80 \times 8 + 4$ (644)
3. $328 - (6 \times 30)$ (148)
4. $60 \times 70 - 160$ (4,040)
5. $42 + 5 \times 6$ (72)

Algebraic Expressions

Lesson 16-1

It's Algebra!

Lisa has a job after school as a dog walker. The number of hours she works each week varies. How much does Lisa earn each week?

LISA'S
Dog Walking
$7⁰⁰ an hour

What information is missing?

the number of __hours__ Lisa works each week

We can use a **variable** to represent the unknown number of hours. A variable is usually a letter used to stand for an unknown number or quantity.

We will let h represent the number of __hours__ Lisa works.

Now we can write an **algebraic expression** to represent this situation. An algebraic expression contains numbers, symbols, and at least one variable.

The algebraic expression __$7 \times h$__ represents how much Lisa earns each week.

Last week Lisa worked 12 hours. How much did she earn?

To find how much Lisa earned last week, we evaluate the algebraic expression.

We substitute __12__ for h, then find the value of the expression.

$7 \times h = 7 \times \underline{12} = \underline{84}$

Lisa earned __$84__ last week.

Here are some other ways to write "seven times h."

$7h \quad 7 \bullet h \quad (7)(h) \quad$ mean $7 \times h$

Getting Started

Write an algebraic expression for each situation. Then, substitute the given number for the variable. Evaluate the numerical expression to answer each question.

1. Al is 5 years younger than Sarah. __$s - 5$__
   Suppose Sarah is 16 years old. __$16 - 5$__
   How old is Al? __11__

2. Bob earns twice as much as Tim. __$2t$__
   Suppose Tim earns $3.75 an hour. __$2(3.75)$__
   How much does Bob earn? __$7.50__

Evaluate each algebraic expression for $x = 5$.

3. $x + 9.3$ 14.3
4. $15x$ 75
5. $x - 3\frac{1}{4}$ $1\frac{3}{4}$
6. $x \div 4$ 1.25
7. $2x + 12$ 22
8. $^-3 - x$ $^-8$
9. $2 \div x$ 0.4
10. ^-7x $^-35$

Lesson 16-1 • Algebraic Expressions

351

---

2 Teach

Introduce the Lesson Have students read the problem and discuss what information is needed to determine how much Lisa earns each week as a dog walker. (how much Lisa charges per hour; the number of hours she spends as a dog walker) Introduce the concept of a variable to represent a number that is not known.

Point out that the situation in the word problem can be represented by the algebraic expression $7 \times h$, where h stands for the number of hours Lisa walks dogs. Have students read the information sentences to determine how much Lisa earned last week. Have students substitute 12 for h and evaluate the numerical expression.

Develop Skills and Concepts Have students write an algebraic expression for the following situation:

Elan earns $3 more an hour than Chaya does at the supermarket. ($c + 3$)

Have students determine how much Elan earns an hour if Chaya earns $5 an hour ($8), $9 an hour ($12).

Write the following on the board:

1. twice as old as Julia (C)   A. $b - 6$
2. $8 more than the shirt (D)   B. $\frac{1}{3}t$
3. 6 fewer books (A)   C. $2j$
4. a third as many teachers (B)   D. $s + 8$

Have a volunteer match the situation on the left with the algebraic expression on the right.

It's Algebra! The concepts in this lesson prepare students for algebra.

Practice

Match each situation to an algebraic expression.

1. two years older than Aaron d
2. half as tall as Meg a
3. $10 less than the jeans e
4. 3 times as old as Paul b
5. 15° drop in temperature c

a. $\frac{1}{2}m$
b. $3p$
c. $t - 15$
d. $a + 2$
e. $j - 10$

Evaluate each algebraic expression for $b = {}^-8$.

6. $b + {}^-3$
$^-11$

7. b^2
64

8. $b - 4$
$^-12$

9. $b \div 2$
$^-4$

10. $3b + 4$
$^-20$

11. $b^2 \div 8$
8

12. $b^2 \div b$
$^-8$

13. $5b - 10$
$^-50$

Evaluate each algebraic expression for $x = 4$ and $y = {}^-12$.

14. $5(x + y)$
$^-40$

15. $2y - 2x$
$^-32$

16. $x - y$
16

17. $\frac{1}{4}x + y - 6$
$^-17$

18. $\frac{1}{2}x + \frac{1}{3}y$
$^-2$

19. $^-3 + y \div x + 12$
6

20. $x^2 - 12$
4

21. $x^2 - y$
28

22. $xy + x^2$
$^-32$

Problem Solving

Write an algebraic expression for each situation. Then, substitute the given number for the variable. Evaluate the numerical expression to answer each question.

23. Amy sells 8 more books than Kate. $k + 8$
Suppose Kate sells 25 books. $25 + 8$
How many books does Amy sell? 33 books

24. Mr. Re works 15 fewer hours than Mr. Lee. $l - 15$
Suppose Mr. Lee works 40 hours. $40 - 15$
How many hours does Mr. Re work? 25 hours

25. Jo made half as many signs as Beth. $\frac{1}{2}b$
Suppose Beth made 36 signs. $\frac{1}{2}(36)$
How many signs did Jo make? 18 signs

26. Don has twice as many CDs as Tom. $2t$
Suppose Tom has 9 CDs. $2(9)$
How many CDs does Don have? 18 CDs

27. Large drinks cost $1.50 more than small drinks. $s + \$1.50$
Suppose small drinks cost $3. $\$3 + \$1.50$
How much do large drinks cost? $4.50

28. There are 12 more girls than boys. $b + 12$
Suppose there are 19 boys. $19 + 12$
How many girls are there? 31 girls

352

Lesson 16-1 • Algebraic Expressions

3 Practice

Have students complete all the exercises.

4 Assess

Ask students to write an algebraic expression for *half as old as Dad*. Then, determine the age if Dad is 32. ($\frac{1}{2}d$; 16)

5 Mixed Review

1. $32 \times 147 = (4{,}704)$
2. $3{,}952 \div 26 = (152)$
3. $2^3 + 5 - 20 = (^-7)$
4. $(12 - 5) + (3 - 6) = (4)$
5. $2 \times 3 \times 4 \times 1 = (24)$
6. $(20 \div 5) - (3 - 10) = (11)$

For Mixed Abilities

Common Errors • Intervention

Some students may have difficulty writing algebraic expressions for situations. Have students match an operation to each of the following clue words:

more than ($+$)
less than ($-$)
twice as much ($\times$)
fewer ($-$)

Enrichment • Number Sense

Have students work in pairs to solve these crypt-arithmetic problems by finding what numbers each of the letters represents. You may want to warn them the problems become progressively harder.

1.
$$\begin{array}{r} 3A7 \\ -\ 1BA \\ \hline 212 \end{array}$$
(A = 5; B = 4)

2.
$$\begin{array}{r} 6C5 \\ +\ D2D \\ \hline 111C \end{array}$$
(C = 9; D = 4)

3.
$$\begin{array}{r} 3E6 \\ \times\ 2F \\ \hline 9968 \end{array}$$
(E = 5; F = 8)

4.
$$\begin{array}{r} G5G \\ \times\ 3H \\ \hline 908 \\ 136H \\ \hline 1G5H8 \end{array}$$
(G = 4; H = 2)

5.
$$\begin{array}{r} 70J4 \\ -\ 835 \\ \hline 6KK9 \end{array}$$
(J = 6; K = 2)

6.
$$\begin{array}{r} L9L9 \\ +\ N0N \\ \hline NLNL \end{array}$$
(L is any digit.; N is one more than L.)

Have students make up problems for each other, making sure that they give enough information so that the problems can be solved.

ESL/ELL STRATEGIES

Explain that an algebraic expression is like a sentence that uses numbers and symbols instead of words. Put several examples of algebraic expressions on the board and point out the numbers, symbols, and variables in each expression.

T352

1 Getting Started

Objective
• To solve addition and subtraction equations by using inverse operations

Vocabulary
equation, solution, inverse operations

Materials
calculators

Warm Up • Mental Math
Have students find the sum or difference.

1. $^+9 + {}^+6$ $(^+15)$ 6. $^+16 - {}^+7$ $(^+9)$
2. $^-5 + {}^-7$ $(^-12)$ 7. $^-9 - {}^-5$ $(^-4)$
3. $^-8 + {}^+6$ $(^-2)$ 8. $^+12 - {}^-3$ $(^+15)$
4. $^+7 + {}^-5$ $(^+2)$ 9. $^-6 - 2$ $(^-8)$
5. $^-6 + {}^-3$ $(^-9)$ 10. $^-14 - {}^-8$ $(^-6)$

Warm Up • Pencil and Paper
Have students find the product.

1. 2.5×0.6 (1.5)
4. $3,506 \times 72$ (252,432)
2. 116×84 (9,744)
5. 345×1.25 (431.25)
3. 6.33×8.5 (53.805)
6. 912×246 (224,352)

Addition and Subtraction Equations

An **equation** is a mathematical sentence that contains an equal sign, =. The equal sign tells us that the values of the expressions on both sides of the symbol are equal. Sometimes an equation contains a variable.

$$a + 19 = 26 \qquad b - 32 = 8 \qquad 15 + c = 10$$

To **solve** an equation, we find the value of the variable that makes the equation true. This value is a **solution** to the equation. Find the solutions to the equations above.

$$a = \underline{7} \qquad b = \underline{40} \qquad c = \underline{{}^-5}$$

We can write an equation to represent a situation in a word problem. Then, solve the equation to solve the problem.

Ellen has 28 pages to read in her biology book. She read 12 pages on Saturday. How many pages does Ellen have left to read?

We can write an **equation** to solve this problem.

$$12 + p = \underline{28}$$

The variable, p, represents the unknown number of pages Ellen has left to read. To solve the equation, we want to get the variable alone on one side of the equation. We can do this by using **inverse operations**.

$$12 + p = 28$$
$$12 - \underline{12} + p = 28 - \underline{12} \qquad \text{To undo adding 12, subtract } \underline{12}$$
$$0 + p = 16 \qquad\qquad \text{from both sides of the equation.}$$
$$p = 16$$

Ellen needs to read $\underline{16}$ more pages.

We can also solve a subtraction equation, such as $m - 30 = 50$.

To undo subtracting 30, $\underline{\text{add}}$ 30 to both sides of the equation. $m = \underline{80}$

Getting Started

Write what you would do to solve each equation. Then, solve.

1. $x + 8 = 15$
   $x = \underline{7}$
   Subtract 8.

2. $n - 5 = {}^-3$
   $n = \underline{2}$
   Add 5.

3. $d - {}^-3 = 5$
   $d = \underline{2}$
   Add $^-3$.

4. $y + {}^-3 = {}^-8$
   $y = \underline{{}^-5}$
   Subtract $^-3$.

Lesson 16-2 • Addition and Subtraction Equations

353

---

2 Teach

Introduce the Lesson Have a student read the first two paragraphs, including the equations. Remind students that variables stand for unknown numbers. To solve the first equation, ask, *What number when added to 19 gives 26?* (7) Point out that a must equal 7. Repeat this question for the second and third equations.

Read the problem aloud, having students follow in their books. Ask students to fill in the blanks.

Develop Skills and Concepts Remind students that inverse operations are those that undo each other. Ask students what the inverse operation for subtraction is (addition) and what operation undoes division (multiplication). Tell students that when we solve an equation, we want to get the variable alone on one side of the equation.

For each of the following, have students say what needs to be done to solve the equation.

1. $x - 3 = 12$ 2. $m + 15 = 20$ 3. $9 + z = {}^-5$
   (add 3)   (subtract 15)   (subtract 9)

Then, show students how to solve each equation. Write the steps as shown on the opening page. Underline what is done to each side of the equation to emphasize that the same operation must be done to both sides.

It's Algebra! The concepts in this lesson prepare students for algebra.

Think Algebraically Have students decide what number must go in the blank so that the equation is true.

1. $5 + \underline{(2)} = 7$ 4. $\underline{(15)} - 15 = 0$

2. $14 - \underline{(5)} = 9$ 5. $22 + \underline{(9)} = 31$

3. $\underline{(10)} + 10 = 20$ 2. $12 - \underline{(5)} = 7$

Practice

Write what you would do to solve each equation. Then, solve.

1. $m + 5 = 18$
$m = \underline{13}$
Subtract 5.

2. $t + 9 = 8$
$t = \underline{^-1}$
Subtract 9.

3. $x - 13 = 20$
$x = \underline{33}$
Add 13.

4. $a - 6 = {}^-10$
$a = \underline{^-4}$
Add 6.

5. $18 + y = 12$
$y = \underline{^-6}$
Subtract 18.

6. $b - {}^-1 = 2$
$b = \underline{1}$
Add $^-1$.

7. $c + 9 = {}^-1$
$c = \underline{^-10}$
Subtract 9.

8. $n + {}^-1 = {}^-3$
$n = \underline{^-2}$
Subtract $^-1$.

9. $d - 20 = 5$
$d = \underline{25}$
Add 20.

10. $k - 5 = {}^-8$
$k = \underline{^-3}$
Add 5.

11. $15 + p = {}^-15$
$p = \underline{^-30}$
Subtract 15.

12. $z - {}^-5 = {}^-5$
$z = \underline{^-10}$
Add $^-5$.

Solve each equation. Check the solution by substituting the value you found for the variable in the original equation.

13. $25 + y = 150$
$y = \underline{125}$

14. $w - 12 = 18$
$w = \underline{30}$

15. $h - {}^-3 = 5$
$h = \underline{2}$

16. $q + {}^-2 = 10$
$q = \underline{12}$

17. $n + 8 = {}^-12$
$n = \underline{^-20}$

18. $13 + a = 17$
$a = \underline{4}$

19. $c - 3 = {}^-5$
$c = \underline{^-2}$

20. $x + 120 = 100$
$x = \underline{^-20}$

21. $m - 16 = {}^-5$
$m = \underline{11}$

Problem Solving

Write an equation for each problem. Then, solve.

22. Nico spent $57 total on a baseball bat and glove. The glove cost $32. How much did the bat cost?
$32 + b = 57$; $25

23. Cheyenne spent $12 of her savings on a CD. She has $15 left. How much were her savings?
$s - 12 = 15$; $27

Lesson 16-2 • Addition and Subtraction Equations

For Mixed Abilities

Common Errors • Intervention

Some students may have difficulty identifying what to do to solve an equation. Encourage them to first find the variable in the equation and ask themselves, "What is being added to or subtracted from the variable?" and "What operation undoes this?"

Have students work in pairs. For each equation, have them write what operation is done to the variable and what the inverse operation is. Then, have them solve the equation.

1. $b - 8 = {}^-2$
(8 is subtracted; add 8; $b = 6$)

2. $5 + m = {}^-35$
(5 is added; subtract 5; $m = {}^-40$)

3. $a - {}^-3 = 10$
($^-3$ is subtracted; add $^-3$; $a = 7$)

Enrichment • Algebra

Students can solve equations containing fractions and decimals the same way they solve equations containing integers. Have them use inverse operations to solve each equation. Be sure they check their solutions.

1. $x + \frac{1}{2} = \frac{3}{4}$ ($x = \frac{1}{4}$)
2. $m - \frac{5}{6} = \frac{7}{12}$ ($m = 1\frac{5}{12}$)
3. $\frac{3}{8} + z = \frac{15}{8}$ ($z = 1\frac{1}{2}$)
4. $2.35 + y = 10.88$ ($y = 8.53$)
5. $25.6 = c + 18.9$ ($c = 6.7$)
6. $a - 7.135 = 15.44$ ($a = 22.575$)

More to Explore • Application

Have students find their approximate age in seconds. There are 365.25 days each year. Point out that we add 1 day every fourth year to account for the 0.25 part of a day. This is a leap year.

Have students fill in the blanks with the right numbers that 12-year-olds would use to find their age in seconds:

$12 \times 365.25 \times \underline{(24)} \times \underline{(60)}$
$\times \underline{(60)} =$ age in seconds

Now, have students replace the 12 with their own age. They can get an even better approximation by using half or quarter years (12.5 or 12.25) if appropriate.

T354

---

3 Practice

Have students complete all the exercises. Remind students that whatever is done to one side of the equation must be done to the other side. Students may want to underline the inverse operation on both sides to ensure they remember this.

4 Assess

Have students review their work for Exercise 5. Ask students what the result would be if 18 were added to both sides. ($36 + y = 30$) Ask students if they have solved the equation. (No.) Have them explain why or why not. (Possible answer: The variable is not alone on the left side of the equation.)

16-3 Multiplication and Division Equations

pages 355–356

1 Getting Started

Objective
- To solve multiplication and division equations by using inverse operations

Vocabulary
inverse operations

Warm Up • Mental Math
Have students simplify the expression.

1. $4 \times 0 \times 6$ (0)
2. $(8 + 2) \times 10$ (100)
3. $(5 - 3) + (15 + 12)$ (29)
4. $1 \times 2 \times 3 \times 4$ (24)
5. $9 + (3 \times 6)$ (27)
6. $2 \times (5 + 1) \times 1$ (12)
7. $16 + (8 \times 0) + 9$ (25)
8. $(5 \times 6) - (2 \times 8)$ (14)

Warm Up • Pencil and Paper
Have students write the next three numbers in the sequence.

1. 1, 2, 4, 8 (16, 32, 64)
2. 1, 3, 9, 27 (81, 243, 729)
3. 1, 4, 16, 64 (256; 1,024; 4,096)
4. 1, 5, 25, 125 (625; 3,125; 15,625)

Name _____

Multiplication and Division Equations

Seth makes $7 an hour at his job. He is saving money to buy a camera. How many hours does Seth need to work to have enough money for the camera?

We want to find how many __hours__ Seth needs to work.

We can write an equation to solve this problem.

$7h =$ __84__

The variable, h, represents the unknown number of hours Seth needs to work. To solve the equation, we want to get the variable alone on one side of the equation. We do this by using **inverse operations**.

$7h = 84$

$\frac{7h}{7} = \frac{84}{7}$ To undo multiplying by 7, divide both sides of the equation by __7__.

$h = 12$

Seth needs to work __12__ hours.

We can also use inverse operations to solve a division equation, such as $\frac{x}{8} = 3$.

To undo dividing by 8, __multiply__ both sides of the equation by 8. $x =$ __24__

REMEMBER The solution to an equation is the value of the variable that will make the equation true.

Getting Started

Write what you would do to solve each equation. Then, solve.

1. $5c = 40$
   $c =$ __8__
   Divide by 5.

2. $\frac{x}{5} = ^-6$
   $x =$ __$^-30$__
   Multiply by 5.

3. $^-2d = 8$
   $d =$ __$^-4$__
   Divide by $^-2$.

4. $\frac{m}{^-4} = 9$
   $m =$ __$^-36$__
   Multiply by $^-4$.

---

2 Teach

Introduce the Lesson Read the problem aloud, and ask a student to read and complete the information sentences. Remind students that to solve an equation, we want to get the variable alone on one side of the equation. Ask students how they would check the solution $h = 12$. (substitute 12 for *h* in the original equation) Have students check the solution. Discuss how to solve the division equation. Write 8 as $\frac{8}{1}$ and show students how the 8s *cancel* each other, or become ones, because $\frac{8}{8} = \frac{1}{1}$, on the left side of the equation.

Develop Skills and Concepts Remind students that inverse operations are those that undo each other. Ask students what operation undoes division (multiplication) and what operation undoes multiplication (division).

For each of the following, have students say what needs to be done to solve the equation:

1. $4c = ^-24$ 2. $\frac{n}{5} = ^-12$ 3. $^-9c = 36$
   (divide by 4)   (multiply by $^-5$)   (divide by $^-9$)

Show students how to solve each equation. Write the steps as shown on the opening page. Have students check each solution.

It's Algebra! The concepts in this lesson prepare students for algebra.

Think Algebraically Have students decide what number must go in the blank so that the equation is true.

1. $3 \times$ __(5)__ $= 15$
2. $16 \div$ __(8)__ $= 2$
3. __($^-6$)__ $\times ^-5 = 30$
4. $10 \times$ __(6)__ $= 60$
5. __(24)__ $\div 8 = 3$

Write what you would do to solve each equation. Then, solve.

1. $4x = 20$

   $x = \underline{5}$
   Divide by 4.

2. $9y = 27$

   $y = \underline{3}$
   Divide by 9.

3. $\frac{m}{6} = 8$

   $m = \underline{48}$
   Multiply by 6.

4. $\frac{a}{3} = 10$

   $a = \underline{30}$
   Multiply by 3.

5. $^-10n = 100$

   $n = \underline{^-10}$
   Divide by $^-10$.

6. $5c = ^-25$

   $c = \underline{^-5}$
   Divide by 5.

7. $^-2t = ^-12$

   $t = \underline{6}$
   Divide by $^-2$.

8. $\frac{k}{4} = 3$

   $k = \underline{^-12}$
   Multiply by $^-4$.

9. $10p = ^-50$

   $p = \underline{^-5}$
   Divide by 10.

10. $\frac{d}{7} = ^-9$

    $d = \underline{63}$
    Multiply by $^-7$.

11. $12h = 36$

    $h = \underline{3}$
    Divide by 12.

12. $\frac{z}{9} = 9$

    $z = \underline{81}$
    Multiply by 9.

Solve each equation. Check the solution by substituting the value you found for the variable in the original equation.

13. $\frac{n}{2} = 15$

    $n = \underline{^-30}$

14. $6t = 72$

    $t = \underline{12}$

15. $8h = ^-40$

    $h = \underline{^-5}$

16. $\frac{x}{10} = ^-3$

    $x = \underline{30}$

17. $^-1b = 15$

    $b = \underline{^-15}$

18. $\frac{c}{7} = ^-8$

    $c = \underline{^-56}$

19. $20c = 20$

    $c = \underline{1}$

20. $\frac{p}{9} = 12$

    $p = \underline{108}$

21. $^-7m = ^-49$

    $m = \underline{7}$

Problem Solving

Write an equation for each problem. Then, solve.

22. Claudia and Janel play on the basketball team. During one game, Janel scored twice as many points as Claudia. If Janel scored 34 points, how many did Claudia score?

    $2c = 34; \; 17$

23. When Frank's age is divided by 3, he gets his son Max's age. If Max is 12 years old, how old is Frank?

    $\frac{f}{3} = 12; \; 36$

356

Lesson 16-3 • Multiplication and Division Equations

---

3 Practice

Have students complete all the exercises. Remind students that whatever is done to one side of the equation must be done to the other side. Students may want to underline the inverse operation on both sides to ensure they remember this.

4 Assess

Have students review their work for Exercise 15. Ask students what the result would be if each side of the equation was divided by $^-40$. ($^-\frac{1}{5}h = 1$) Ask students if they have solved the equation. (No.) Have them explain why or why not. (Possible answer: The variable is not alone on the left side of the equation.)

For Mixed Abilities

Common Errors • Intervention

Some students may have difficulty identifying what to do to solve an equation. Encourage them to first find the variable in the equation and ask themselves, "What is being multiplied by the variable or what is the variable being divided by?" and "What operation undoes this?"

Have students work in pairs. For each equation, have them write what operation is done to the variable and what the inverse operation is. Then, have them solve the equation.

1. $4t = ^-60$ (t is multiplied by 4; divide by 4; $t = ^-15$)

2. $\frac{m}{3} = 10$ (m is divided by 3; multiply by 3; $m = 30$)

3. $\frac{y}{^-6} = ^-2$ (y is divided by $^-6$; multiply by $^-6$; $y = 12$)

Enrichment • Algebra

Students can solve equations containing fractions and decimals the same way they solve equations containing integers. Have them use inverse operations to solve each equation. Be sure they check their solutions.

1. $\frac{2}{3}x = 6$ ($x = 9$)

2. $0.25d = 4$ ($d = 16$)

3. $\frac{4}{5}m = ^-12$ ($m = ^-15$)

4. $1.5k = 12$ ($k = 8$)

5. $\frac{3}{8}c = 3$ ($c = 8$)

6. $2.4z = 8.4$ ($z = 3.5$)

More to Explore • Number Sense

The greatest number that can be shown on a calculator is 99,999,999. This may vary with some calculators. Have students list the ways they can obtain this number using as many operations as they can, addition, subtraction, multiplication, and division. Allow students to use a calculator. Display student responses on the bulletin board.

1 Getting Started

Objective
• To solve two-step equations

Vocabulary
two-step equations

Warm Up • Mental Math
Have students find the sum, difference, product, or quotient.

1. $^+28 \div {}^-7$ ($^-4$)
2. $^+5 - {}^-8$ ($^+13$)
3. $^-7 \times {}^-3$ ($^+21$)
4. $^-6 + {}^-8$ ($^-14$)
5. $^-54 \div {}^-9$ ($^+6$)
6. $^-8 \times {}^+6$ ($^-48$)
7. $^+9 + {}^-7$ ($^+2$)
8. $^-14 - {}^-6$ ($^-8$)
9. $^-15 + {}^+9$ ($^-6$)
10. $^+72 \div {}^-8$ ($^-9$)

Warm Up • Pencil and Paper
Have students write the number backward. Then, compare the two numbers. Write >, <, or =.

1. 4,362 (2,634; 4,362 > 2,634)
2. 17,682 (28,671; 17,682 < 28,671)
3. 39,333 (33,393; 39,333 > 33,393)
4. 33,343 (34,333; 33,343 < 34,333)
5. 91,119 (91,119; 91,119 = 91,119)

2 Teach

Introduce the Lesson Call on a student to read the problem aloud and tell what is to be found. (the cost of each notebook) Have students identify the facts that are given. (Two notebooks and a backpack cost $19. The backpack costs $15.) Have students complete the information sentences.

Ask, *What information is unknown?* (the cost of a notebook) Point out that we can use a variable to represent the cost of a notebook and write an equation to represent the situation. Write the following equation on the board:

2n + 15 = 19

Call on a student to tell what each piece of the equation represents. (2n is the cost of two notebooks; 15 is the cost of the backpack; 19 is the total cost of the notebooks and the backpack.) On the board, show the steps needed to solve the equation as students follow in their texts.

Develop Skills and Concepts Write the following equations on the board:

$$a - 9 = 8 \qquad b + 7 = 13 \qquad 8c = 56 \qquad \frac{d}{6} = 7$$

Have volunteers go to the board to solve the equations. ($a = 17$; $b = 6$; $c = 7$; $d = 42$) Stress the importance of using the inverse operation to solve equations. Point out that the inverse operation undoes the operation on the variable. The result of this action is that the variable stands alone on the left side of the equations, allowing us to easily identify the value of the variable.

Write the following equations on the board:

$$6m - 12 = 30 \qquad \frac{n}{5} + 17 = 32$$

Have volunteers go to the board to solve the equations. Have them explain the operations they will use to solve the equation and why. Then, have them solve the equation. (Possible answers: Add 12 to both sides of the equation to undo subtracting 12. Then, divide both sides of the equation by 6 to undo multiplying by 6. $m = 7$; subtract 17 from both sides of the equation to undo adding 17. Then, multiply both sides of the equation by 5 to undo dividing by 5. $n = 75$)

T357

Solving Two-Step Equations

Josh bought two notebooks and a backpack for a total of $19. How much did each notebook cost?

We need to find the cost of each notebook.

How much does the backpack cost? **$15**

What was the total cost of the items? **$19**

We can write an **equation** to solve this problem. Let the variable n represent the cost of each notebook.

Then, **$2n$** represents the cost of two notebooks.

$$\underline{2n} + 15 = 19$$

This is a **two-step equation**. A two-step equation has two operations. To solve the equation, we need to undo the two operations.

First, we will undo the addition. Then we will undo the multiplication.

$$2n + 15 - \underline{15} = 19 - \underline{15} \qquad \text{To undo adding 15, } \underline{\text{subtract}} \text{ 15 from both sides.}$$

$$2n = \underline{4}$$

$$\frac{2n}{2} = \frac{4}{2} \qquad \text{To undo multiplying by 2, } \underline{\text{divide}} \text{ both sides by 2.}$$

$$n = \underline{2}$$

To check our solution, substitute the 2 for n into the original equation. Is the equation true?

$$2(2) + 15 \; ? \; 19$$

$$4 + 15 \; ? \; 19$$

$$19 = 19 \; ✔$$

Each notebook costs **$2**.

Getting Started

Write what you would do to solve each equation. Then, solve and check.

1. $3m - 1 = 8$

   $m = \underline{3}$

   Add 1. Then, divide by 3.

2. $\frac{x}{2} + 3 = 10$

   $x = \underline{14}$

   Subtract 3. Then, multiply by 2.

Practice

1. Is 3 the solution to the equation $5x - 30 = {}^-15$? **Yes.**
   Explain. **Possible answer: If you substitute 3 for x in the equation, it is true.**

2. Is $^-8$ the solution to the equation $6 + 3m = 18$? **No.**
   Explain. **Possible answer: If you substitute $^-8$ for m in the equation, it is not true.**

Copy and solve each equation. Then, check your solution.

3. $4n + 3 = 11$
   $n = \underline{2}$

4. $^-2y + 6 = 14$
   $y = \underline{{}^-4}$

5. $10c - 3 = {}^-33$
   $c = \underline{{}^-3}$

6. $\frac{k}{5} + 8 = {}^-12$
   $k = \underline{{}^-100}$

7. $\frac{t}{3} - 9 = {}^-1$
   $t = \underline{24}$

8. $12x - 25 = 35$
   $x = \underline{5}$

9. $^-8m + 1 = {}^-47$
   $m = \underline{6}$

10. $4z - 22 = 10$
    $z = \underline{8}$

11. $^-4p + 7 = {}^-1$
    $p = \underline{2}$

12. $70 + 6w = 82$
    $w = \underline{2}$

13. $3b - 102 = 0$
    $b = \underline{34}$

14. $0 = 5r - {}^-15$
    $r = \underline{{}^-3}$

15. $5a + 12 = {}^-13$
    $a = \underline{{}^-5}$

16. $^-6q + 9 = 39$
    $q = \underline{{}^-5}$

17. $12h - 32 = {}^-8$
    $h = \underline{2}$

Problem Solving

Write an equation for each problem. Then, solve.

18. Coach Rivera bought 9 basketballs and 1 practice hoop for a total cost of $137. The basketball hoop cost $65. How much did each basketball cost?

    $9b + 65 = 137$; $8

19. An unknown number is divided by $^-3$. The result is added to 15 to yield a final answer of 27. What is the unknown number?

    $\frac{n}{^-3} + 15 = 27$; $^-36$

It's Algebra! The concepts in this lesson prepare students for algebra.

3 Practice

Have students complete all the exercises. Remind students that to solve a two-step equation, first you add or subtract and then you multiply or divide.

4 Assess

Have students tell how they would solve the equation $29x - 48 = 10$. (Possible answer: First, add 48 to both sides of the equation. Then, divide both sides of the equation by 29.)

For Mixed Abilities

Common Errors • Intervention

Some students may multiply or divide before they add or subtract when they try to solve a two-step equation. Have students carefully write the original equation on a sheet of paper. Next, have them circle the plus or minus sign in green, and the multiplication or division sign in red. Then, have them solve the equation one step at a time.

Enrichment • Algebra

Divide the class into two teams. Write the following equations on the board:

1. $3m - 2.1 = 4.5$
2. $\frac{n}{2} - 1.002 = 0.93$
3. $2a - 6\frac{2}{3} = 4\frac{1}{3}$
4. $5b - 3.6 = 1.2$
5. $3j + 3 = 3.75$
6. $2k + 2\frac{3}{4} = 4$
7. $\frac{5}{7}x + \frac{5}{7} = 5$
8. $\frac{1}{3}y - \frac{2}{7} = 5\frac{5}{7}$
9. $\frac{2}{5}c - \frac{1}{5} = 3$
10. $2d + 0.6 = 9$

Assign equations 1–5 to one team and equations 6–10 to the other team. Call on different team members to solve the equations. The student who has been asked to solve the equation may consult his or her teammates. Each team receives 1 point for each correct solution. ($m = 2.2$; $n = 3.864$; $a = 5.5$; $b = 0.96$; $j = 0.25$; $k = \frac{5}{8}$; $x = 6$; $y = 18$; $c = 8$; $d = 4.2$)

More to Explore • Probability

Ask students to guess the probability that two students have the same birthday. Have students survey the class and record their birthdays. Then, have them determine the probability of two students with the same birthday.

The probability of two birthdays being the same is 50% for a group of 23 people, 70% for a group of 30 people, 81% for a group of 35 people, and 97% for a group of 50 people.

Have students work in small groups to survey the birthdays of other classes in the school.

T358

16-5 Graphing Ordered Pairs of Integers

pages 359–360

1 Getting Started

Objective
• To locate and graph ordered pairs of integers

Vocabulary
coordinates

Materials
graph paper; number cubes; grid transparency

Warm Up • Mental Math
Divide the class into two teams, X and O, to play Five in a Row. Prepare a 5-by-5 grid transparency for the overhead projector. Label only the origin (0). Have team members call out ordered pairs. The first pair that is mentioned is the point you locate on the grid by writing the team name, either X or O. The first team to have 5 marks in a row, either vertically, horizontally, or diagonally, wins the game.

Warm Up • Pencil and Paper
Have students draw two perpendicular lines on graph paper, labeling the origin and points through 8 on each axis. Then, have students graph and label the following ordered pairs. (Check students' work.)

$A = (4, 3)$	$B = (3, 4)$	$C = (6, 2)$
$D = (1, 4)$	$E = (5, 5)$	$F = (7, 6)$
$G = (3, 1)$	$H = (2, 0)$	$I = (0, 5)$

2 Teach

Introduce the Lesson Have a student read the first paragraph on page 359. Have students locate point A by moving their finger 3 units to the right and 2 units down. Repeat the procedure for point B. Emphasize that the graph of each point is located by its coordinates.

Have students locate the following coordinates: $(^-2, ^-3)$ and $(^-4, ^+2)$. Have them tell which direction to move, right or left, and how many units up or down. Emphasize that the first coordinate indicates the number of units left ($-$) or right ($+$) of the origin. The second coordinate indicates the number of units up ($+$) or down ($-$).

Name _____

Graphing Ordered Pairs of Integers

It's Algebra!

We use ordered pairs of integers to locate points on a number grid. Point A is located by the **coordinates** of the ordered pair ($^+3$, $^-2$). The graph of point A is 3 units to the right and 2 units below the origin. Where is the graph of point B located?

The first coordinate of an ordered pair tells how far to move left or right of the origin. The second coordinate tells how far to move up or down.

Point B is __3__ units to the left of the origin and __4__ units above the line. The graph of point B is located by the ordered pair ($\underline{^-3}$, $\underline{^+4}$). We say: the coordinates of B are the ordered pair ($^-3$, $^+4$).

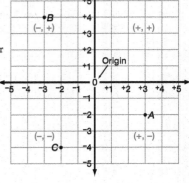

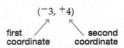

($^-3$, $^+4$)

first coordinate   second coordinate

What are the coordinates of point C?

Point C is __2__ units to the left of the origin and __4__ units below the origin.

The coordinates of point C are the ordered pair ($\underline{^-2}$, $\underline{^-4}$).

Getting Started

Write the coordinates of each point.

1. $A(\underline{^-4}, \underline{^+2})$

2. $B(\underline{^+3}, \underline{0})$

3. $C(\underline{0}, \underline{^-3})$

Graph each point.

4. $P(^-3, 0)$

5. $Q(^+2, ^-5)$

6. $R(^-2, ^-3)$

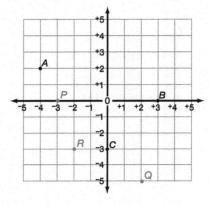

Develop Skills and Concepts Explain that the two axes of the graph are number lines, one horizontal and the other vertical. When the number lines (axes) intersect, they form four sections, or quadrants. Point out that students previously located points only in the upper-right quadrant of the graph because they worked with whole numbers exclusively.

Draw a pair of coordinate axes on the board and label them $^-5$ through $^+5$. Tell students to locate and graph the following points:

$A (^+4, ^-2)$	$B (^-1, ^-5)$	$C (^+5, ^+3)$
$D (^-6, 0)$	$E (0, ^+4)$	$F (^-2, ^+3)$

Have students note that when 0 is a coordinate, as in $(0, ^+4)$, no movement is indicated.

Then, mark and label one point in each quadrant. Ask students to name the coordinates of each point.

It's Algebra! The concepts in this lesson prepare students for algebra.

Practice

Write the coordinates of each point.

1. A(<u>−4</u>, <u>+3</u>) 2. B(<u>+2</u>, <u>+2</u>)

3. C(<u>0</u>, <u>−3</u>) 4. D(<u>−2</u>, <u>−3</u>)

5. E(<u>+3</u>, <u>−1</u>) 6. F(<u>−1</u>, <u>+3</u>)

7. G(<u>−3</u>, <u>0</u>) 8. H(<u>−4</u>, <u>−4</u>)

9. I(<u>+5</u>, <u>−3</u>)

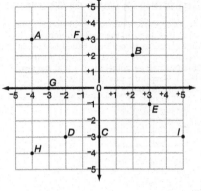

Graph each point.

10. R(⁺3, ⁺4) 11. S(⁻3, ⁻4)

12. T(⁺3, ⁻4) 13. U(⁻3, ⁺4)

14. V(⁻4, ⁻3) 15. W(⁻2, ⁺2)

16. X(⁺3, ⁻2) 17. Y(⁻5, ⁻5)

18. Z(⁻4, 0)

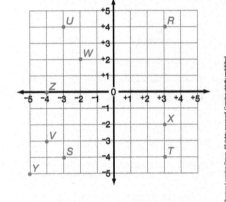

Problem Solving

For each problem, make a grid, graph the points, and connect them in order with line segments. Name the figure.

19. A(⁻3, ⁺3), B(⁺2, ⁺2), C(⁺2, ⁻2),
    D(⁻3, ⁻3), A(⁻3, ⁺3)
    trapezoid

20. L(⁻3, 0), M(⁺4, ⁺3), N(⁺4, ⁻3),
    L(⁻3, 0)
    triangle

21. P(⁺4, ⁺5), Q(⁺4, ⁺1), R(⁻4, ⁻5),
    S(⁻4, ⁺7), P(⁺4, ⁺5)
    trapezoid

22. W(⁻2, ⁻1), X(⁺2, ⁻1), Y(⁺2, ⁻4),
    Z(⁻2, ⁻4), W(⁻2, ⁻1)
    quadrilateral

Lesson 16-5 • Graphing Ordered Pairs of Integers

Common Errors • Intervention

Some students may not pay attention to the positive and negative signs, and graph ordered pairs as if the numbers were both positive. Have students practice with partners by graphing sets of points such as the following: (⁺4, ⁺3), (⁺4, ⁻3), (⁻4, ⁺3), and (⁻4, ⁻3). Students should recognize that the point for each ordered pair is located in a different section, or quadrant, of the number grid.

Enrichment • Geometry

Provide graph paper for this activity. Have each student draw vertical and horizontal axes, labeling them from ⁻15 to ⁺15. Ask students to draw a polygon on the grid, placing the vertices at the points of intersection of the grid lines. Have students letter consecutive points and list the coordinates for each point.

Direct students to exchange lists, graph the points on another grid, join the points in order, and then compare the new drawing with the original drawing.

More to Explore • Application

As an extended class project, give students the opportunity to create their own math filmstrip. Have them make a film about a mathematics application in their daily life. Provide students with a filmstrip kit, a blank cassette tape, and a videorecorder.

When the project is complete, present the filmstrip to another class.

3 Practice

Have students complete all the exercises. Remind students that when they graph coordinates, they need to look at the signs of the coordinates so that they place the coordinates in the correct quadrant.

4 Assess

Have students graph the following coordinates on a number grid. (Check students' work.)

A(3, 5) B (⁻3, 5) C(3, ⁻5) D(⁻3, ⁻5) E(5, ⁻3)

ESL/ELL STRATEGIES

When introducing the graphing of ordered pairs of integers, review the meaning of the terms *ordered*, *integers*, *coordinates*, *x-axis*, and *y-axis*. Ask students to explain each term in their own words, using board drawings to illustrate if they wish.

1 Getting Started

Objective
- To graph a table of values on a coordinate plane

Vocabulary
coordinate plane, *x*-axis, *y*-axis

Materials
graph paper

Warm Up • Mental Math
Have students find the sum or difference.
1. $^-3 + ^+2$ ($^-1$)
2. $^+4 + ^-4$ (0)
3. $^-3 - ^+3$ ($^-6$)
4. $^+5 + ^-3$ ($^+2$)

Warm Up • Pencil and Paper
Have students complete each table.

n	$^-1$	0	1	2
$3 \times n$	($^-3$)	(0)	(3)	(6)

n	$^-6$	$^-3$	1	4
$n + {}^+2$	($^-4$)	($^-1$)	(3)	(6)

n	6	4	3	1
$n - {}^-1$	(7)	(5)	(4)	(2)

Name _____

Graphing a Table of Values

It's Algebra!

Two perpendicular number lines that intersect at the origin form the **coordinate plane**. The horizontal number line is called the *x*-axis. The vertical number line is called the *y*-axis. The *x*-coordinate of point *A* is $^-2$, and the *y*-coordinate is $^-1$. Graph the points in this table on the coordinate plane.

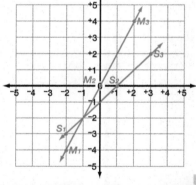

	B	C	D
x	$^-1$	$^+2$	$^+4$
y	0	$^+3$	$^+5$

We graph the coordinates

$B(^-1, 0)$, $C(^+2, {}^+3)$, $D(^+4, {}^+5)$.

Draw a line connecting the coordinates. Complete this table by adding $^+2$ to the *x*-coordinate to identify the *y*-coordinate.

	E	F	G
x	$^-3$	0	$^+3$
y	$^-1$	$^+2$	$^+5$

Since $0 + {}^+2 = {}^+2$, the coordinates of the second point are $(0, \underline{{}^+2})$. The coordinates of the third point are $(^+3, \underline{{}^+5})$.

Getting Started

Complete the table and graph the ordered pairs on the coordinate plane. Draw a line through the points.

1.

Multiply by $^+2$.			
x	$^-2$	0	$^+2$
y	$^-4$	0	$^+4$

2.

Subtract $^+1$.			
x	$^-2$	$^+1$	$^+3$
y	$^-3$	0	$^+2$

---

2 Teach

Introduce the Lesson Discuss the concepts presented in the models on page 361. Stress the vocabulary words, particularly *x-axis* and *y-axis*. Have students determine the rule for the first table. ($y = x + 1$) Explain that all points on the graph at the top of page 361 correspond to ordered pairs that fit the equation. Have students find three other ordered pairs of numbers that lie on the line. You may wish to explain that an equation that can be graphed as a line is called a linear equation.

Point out that the rule for the second table is $y = x + 2$. Have students name other ordered pairs that fit this equation.

Develop Skills and Concepts Write a table such as the following on the board. Ask students to name pairs of integers that have a sum of $^+6$. (Answers will vary.)

First Number $^+4$	Second Number $^+2$	Ordered Pair ($^+4$, $^+2$)
(0)	($^+6$)	(0, $^+6$)
($^+5$)	($^+1$)	($^+5$, $^+1$)
($^-1$)	($^+7$)	($^-1$, $^+7$)
($^-3$)	($^+9$)	($^-3$, $^+9$)

Then, draw a coordinate plane on the board. Have students locate the points that correspond to each ordered pair. Ask students what they observe about the points. (Possible answer: All the points lie in a straight line.)

It's Algebra! The concepts in this lesson prepare students for algebra.

Practice

Complete the table and graph the ordered pairs.
Draw a line through the points.

1.

Add −2.			
x	−3	0	+3
y	−5	−2	+1

2.

Subtract +3.			
x	−2	0	+2
y	−5	−3	−1

Complete the table and graph the ordered pairs.
Draw a line through the points.

3.

Multiply by +3.			
x	−2	0	+2
y	−6	0	+6

4.

Multiply by −2.			
x	−2	0	+2
y	+4	0	−4

Complete the table and graph the ordered pairs.
Draw a line through the points.

5.

Add −3.			
x	−1	0	+2
y	−4	−3	−1

6.

Subtract −3.			
x	−5	0	−1
y	−2	−3	+2

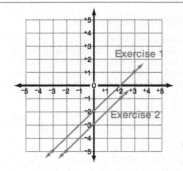

Exercise 1

Exercise 2

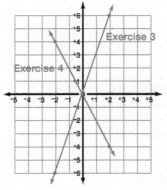

Exercise 3

Exercise 4

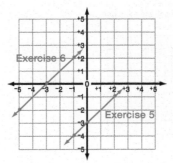

Exercise 6

Exercise 5

362

Lesson 16-6 • Graphing a Table of Values

---

3 Practice

Have students complete all the exercises. Watch for students who reverse the x-coordinate and the y-coordinate in ordered pairs.

4 Assess

Have students complete the table and graph the ordered pairs. Then, have them connect the points. (Check students' graphs.)

Add 3.			
x	−3	0	+3
y	(0)	(+3)	(+6)

For Mixed Abilities

Common Errors • Intervention

If students have difficulty graphing the ordered pairs in a table, have them first use the rule to complete the table; then, have them write the ordered pairs from the table in a more familiar form, that is, in the form of (x, y). In this form, they can locate more easily the point for each ordered pair on the coordinate plane.

Enrichment • Algebra

1. Ask students to complete the table. Then, have them write the rule above the table and graph the equation. (Check students' graphs.)

   $(2x + 3 = y)$

x	0	1	2	7	10
y	3	5	7	(17)	(23)

2. Have students complete the table for the equation $y = x^2$. Have them graph the equation and describe the graph.

x	0	1	−1	2	−2	3
y	(0)	(1)	(1)	(4)	(4)	(9)

   (Graph resembles a capital U.)

Enrichment • Number Sense

Divide students into two teams to play decimal baseball. Follow the basic rules of baseball. The first player up to "bat" has to mentally solve an addition or subtraction decimal problem written by the other team. If a correct answer is given, the player moves to first base, and so on. A wrong answer is a strikeout. The teacher acts as umpire and sets the point limit the game will be played to. This activity can be extended over a week's time in order to create a tournament within the class or against another class.

T362

16-7 Graphing Equations

pages 363–364

1 Getting Started

Objectives

- To graph an equation from a table of values

- To write equations

Materials

paper; rulers; colored pencils

Warm Up • Mental Math

Have students solve each equation.

1. $y + 3 = 12$ ($y = 9$)
2. $n - 5 = 6$ ($n = 11$)
3. $8b = 56$ ($b = 7$)
4. $24 \div x = 8$ ($x = 3$)
5. $4m = {}^-20$ ($m = {}^-5$)

Warm Up • Pencil and Paper

Have students solve each equation and then check their solution.

1. $3x - 5 = 19$ ($x = 8$)
2. $\frac{y}{3} + 2 = 7$ ($y = 15$)
3. ${}^-5c + 1 = 11$ ($c = {}^-2$)
4. $6m - 7 = 23$ ($m = 5$)
5. $\frac{a}{4} - 12 = 3$ ($a = 60$)
6. $2t + 8 = {}^-6$ ($t = {}^-7$)

Name _____

Lesson 16-7

Graphing Equations

It's Algebra!

Jeff is saving to buy new basketball equipment. At the end of each week he puts $3 in the bank. The table on the right shows how his savings increase each week. If Jeff has $6 in savings, how much will he have after he makes his deposit at the end of the week?

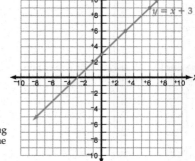

Add 3.

x	0	3	6	9
y	3	6	9	12

Look at the table. x represents Jeff's savings. ___y___ represents how much he has after his deposit.

Complete the table. Then, write the rule in the graph.

The rule for this table is add __3__.

The equation that describes this rule is $y = x + $ __3__

We can graph the equation $y = x + 3$ by graphing the coordinates in the table. Then, we connect the coordinates with a straight line.

We can use the table, the equation, or the graph to find y when x is 6.

If Jeff has $6, he will have __$9__ after he makes his deposit.

Getting Started

Use the equation to complete the table. Then, graph the equation.

1.

$y = x - 5$		
x	5	3	1
y	0	$^-2$	$^-4$

2.

$y = x + 4$		
x	$^-2$	0	4
y	2	4	8

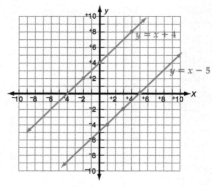

Lesson 16-7 • Graphing Equations

363

---

2 Teach

Introduce the Lesson Read the opening problem aloud. Have students follow in their books as you read the information sentences. As you read, ask students to fill in the blanks. Emphasize the fact that the table contains solutions to the equation $y = x + 3$.

Write the ordered pairs **(0, 3)**, **(3, 6)**, **(6, 9)**, and **(9, 12)** on the board and have students find these points on the line. Explain that the equation, table, and graph all give us the same information. For each format, explain to students how to find y when x is 6.

Develop Skills and Concepts Write the following on the board: the equation $y = x - 6$, a coordinate plane, and the table.

x	$^-2$	0	8
y	$^-8$	($^-6$)	(2)

Tell students that we will complete the table and then use it to graph the equation. Ask students how to find y when $x = {}^-2$. (substitute $^-2$ for x) Write $y = {}^-2 - 6 = {}^-8$ on the board. Do the same for $x = 0$ and $x = 8$.

Complete the table. Tell students that we have just found three solutions to the equation. Have three students go to the board and ask each student to write one of the ordered pairs in the table and graph it. Explain that we connect the three points with a straight line and draw arrow tips on the ends. We now have a graph of the equation.

It's Algebra! The concepts in this lesson prepare students for algebra.

T363

Practice

Complete each table. Then, write the equation.

1.

x	3	4	5	6	7
y	5	6	7	8	9

equation: y = __x + 2__

2.

x	-3	0	3	6	9
y	6	9	12	15	18

equation: y = __x + 9__

3.

x	8	6	4	0	-4
y	4	2	0	-4	-8

equation: y = __x - 4__

Use the equation to complete each table. Then, graph the equation.

4. y = x + 6

x	4	0	-2	-6
y	10	6	4	0

5. y = 8 - x

x	8	5	2	-1
y	0	3	6	9

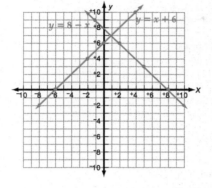

Problem Solving

Make a table of values for the problem. Then, graph the values.

6. Amy saves the $10 she earns as an allowance each week. Make a graph to show her savings. How much will she have in savings after 6 weeks?

x	0	1	2	3	4	5	6
y	0	10	20	30	40	50	60

$60.00

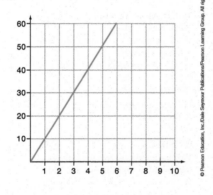

Lesson 16-7 • Graphing Equations

---

3 Practice

Have students complete all the exercises. Remind students to draw arrow tips on the ends of their lines. These arrow tips tell us that the line continues indefinitely. Also, remind students that when they complete the table, they have found solutions to the equation.

4 Assess

Have students look at their work for Exercise 4. Tell students to make a table with three new x values of their choice. Have them find the y values and then graph the ordered pairs on the same coordinate axes as Exercise 4. Ask them what they notice. (Possible answer: These three ordered pairs lie on the line; it doesn't matter which ordered pairs are used to graph the equation.)

For Mixed Abilities

Common Errors • Intervention

Some students may be confused about the connection between a table of values and a graph. Ask students to copy the equation y = x − 3 and the following table:

x	-1	0				
y	(-4)	(-3)				

Have students find the y values for x = -1 and x = 0. Have students then choose four additional x values and find the corresponding y values. Now, have students write the six ordered pairs they found. Next, have them graph their ordered pairs and connect them with a straight line.

Students exchange papers to see that the graph is the same regardless of the values in the table.

Enrichment • Application

Melissa is doing a science project on recycling. She found out that in 1997, 120 households in her community recycled. By 2000, that number had increased to 200, and by 2003, 320 households were recycling. She wants to predict about how many households will be recycling in the year 2007.

Draw coordinate axes with the y-axis representing the number of households that recycle and the x-axis representing the year (count the year 1997 as 0). Graph the three data points. The points will not lie on a straight line, but you can draw a straight line that you feel best fits the points. Use your line to predict how many households will be recycling in 2007.

More to Explore • Geometry

You can make curved lines by drawing several straight lines. Using graph paper, make coordinate axes and label the numbers 1 through 6 on the positive x- and y-axes, and -1 through -6 on the negative x- and y-axes. Use your ruler to connect all the 1s and -1s, all the 2s and -2s, and so on. Make your design more exciting by using colored pencils.

T364

pages 365-366

1 Getting Started

Objective
- To review the strategies learned for solving problems

Warm Up • Mental Math
Have students simplify the expression.

1. $^-1 + ^-4 + ^+6$ ($^+1$)
2. $^-5 + ^+3 + ^-8 + ^+3$ ($^-7$)
3. $^+8 + ^+5 + ^-7 - ^-5$ ($^+11$)
4. $^-7 - ^+7 + ^-5 + ^+7$ ($^-12$)
5. $^-2 \times ^-1 \times ^+2 \times ^-1$ ($^-4$)
6. $^-1 \times ^-1 \times ^-1 \times ^-1 \times ^-1$ ($^-1$)
7. $^-3 \times ^+1 \times ^-1 \times ^+1$ ($^+3$)
8. $^-1 \times ^-1 \times ^-1 \times ^-2$ ($^+2$)

Warm Up • Pencil and Paper
Have students find the following:

1. $\frac{1}{2}$ of $7,000 ($3,500)
2. $\frac{1}{5}$ of $10,000 ($2,000)
3. $\frac{1}{2}$ of $6,500 ($3,250)
4. $\frac{1}{3}$ of $3,600 ($1,200)

2 Teach

Introduce the Lesson This lesson reviews all the problem-solving strategies used this year. Students should be aware that not every strategy fits a situation and that the strategies often work most effectively paired together. Now that students have been exposed to these strategies, and have had some success in using them, they should be able to transfer these skills to similar circumstances in future problems.

Develop Skills and Concepts In the practice accompanying this review, students get a chance to choose the strategy or combination of strategies that will help them find solutions. Allow for individual approaches and originality in solutions.

3 Apply

Solution Notes
1. Have students work backward. If the paper were completely covered at one point, it would be $\frac{1}{2}$ covered 3 seconds before.

Name _____

Problem Solving: Review the Four-Step Plan

It's Algebra!

The four-step plan can help us to be better problem solvers.

 SEE
We state what the problem actually is.

We state all the facts we know that will help us to solve that problem.

 PLAN
We think about the important facts and choose a plan or plans to help solve the problem. Among the plans that will help us to solve a problem are:

Draw a picture or diagram
Make a model with manipulatives or on paper
Act the problem out
Make a systematic listing or a table
Make a tally or graph
Look for a pattern
Determine missing data
Collect needed data
Select appropriate notation
Write an open sentence
Use a formula
Test, check, and revise
Work backward
Solve a simpler but related problem
Identify a subgoal
Restate the problem in our own words

★ **DO**
We carry out the plan and reach a solution to the problem.

★ **CHECK**
We check the problem for careless errors.
We see if the solution is reasonable.
We look for another way to solve the problem.

Lesson 16-8 • Problem Solving: Review the Four-Step Plan   365

---

2. Have students act it out by using a toothpick. The first log needs four cuts at 15 s/cut. The second log needs five cuts: $5 \times 15 = 75$ s.

3. Have students make a table.

Score	1	2	3	4	5	6	7	8	9	10
# 3s			1			2			3	1
# 7s							1			1
Not possible	x	x		x	x			x		

4. Have students make a simpler but similar problem.
   1 segment: AB
   3 segments: AB, BC, AC
   6 segments: AB, BC, CD, AC, BD, AD

5. Have students work backward. He gave the last $2,000 to the zoo and an equal amount to the university, $2,000. Added together, $4,000, and added to the sister's $18,000, the amounts total $22,000. The brother received the same amount, $22,000, which was half the total fortune, $44,000.

6. Have students act it out.

Apply

Solve each problem.

1. A bottle of ink spills on your homework paper and the blotch doubles in size every three seconds. How much of the paper will be covered three seconds before it gets completely covered with ink?

    $\frac{1}{2}$ will be covered

2. You are cutting two logs into pieces for your fireplace. One log can be cut into five pieces, and the other into six pieces. If it takes 60 seconds to cut the first log, how long will it take to cut the other log?

    75 seconds

3. Tom and Inga are taking turns throwing bean bags at two targets, one marked 7 points and the other marked 3 points. List all the scores, less than 11, that it would be impossible to get in one turn.

    1, 2, 4, 5, 8

4. How many different segments can be formed using the eight points below as endpoints?

    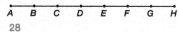

    A   B   C   D   E   F   G   H

    28

5. Mr. Pennypincher gave half of his fortune to his brother, $18,000 to his sister, half of what was left to the university, and the remaining $2,000 to the zoo. How much money did Mr. Pennypincher start with?

    $44,000

Problem-Solving Strategy: Using the Four-Step Plan

★ **SEE** What do you need to find?

★ **PLAN** What do you need to do?

★ **DO** Follow the plan.

★ **CHECK** Does your answer make sense?

---

6. A rectangular sheet of paper can be rolled to create a cylinder. What shape of paper could be rolled to form a slanted cylinder? Hint: First make a cylinder, then dip both ends into water at an angle. Open up the paper and cut along the water line.

    Answer will be a drawing similar to this figure:

7. Does $^+7 + {}^-9$ equal $^-9 + {}^+7$? Support your answer.

    Yes.

8. The sum of two integers is negative. What do you know about the two integers?

    See Solution Notes.

9. There are two different numbers that make the following sentence true even though both n's represent the same number in the sentence. What are the two numbers?

    $$n \times n = {}^+25$$

    $^-5$ and $^+5$

More to Explore • Application

Give students a newspaper or sales brochure. Ask them to find an advertised sale and list 5 items on sale. Have them determine the original price. For example, a dishwasher is on sale for $288. the ad explains that you save $41. Have students determine the price of the dishwasher before the sale. ($288 + $41 = $329)

Higher-Order Thinking Skills

7. **Evaluation:** $^+7 + {}^-9 = {}^-2$ and $^-9 + {}^+7 = {}^-2$.
8. **Analysis:** Either both of the integers are negative, or one is positive and the other is negative where the negative integer is farther from 0 than the positive integer.
9. **Synthesis:** $^-5 \times {}^-5 = {}^+25$, and $^+5 \times {}^+5 = {}^+25$.

4 Assess

Have students determine the strategy they would use to solve the following problem:

There are 8 teams in a soccer tournament. Each team must play each of the other teams once. How many games must be played? (Possible answer: Make a list)

Then have students use the strategy to solve. (28)

T366

Chapter 16
Test

page 367

Items   Objectives

Name _____

Evaluate each algebraic expression for $a = {}^-2$ and $b = 5$.

1. $3a$ ___$^-6$___

2. $a + b$ ___3___

3. $b - a$ ___7___

4. $2a + 3b$ ___11___

5. $b - 5a$ ___15___

6. $12 + a - b$ ___5___

Solve each equation. Check each solution.

7. $n + 7 = 12$
   $n =$ ___5___

8. $\frac{y}{6} = {}^-3$
   $y =$ ___$^-18$___

9. $c - 9 = 5$
   $c =$ ___14___

10. $4k = 24$
    $k =$ ___6___

11. $2x - 1 = 15$
    $x =$ ___8___

12. $18 + z = 13$
    $z =$ ___$^-5$___

13. $\frac{b}{2} = 10$
    $b =$ ___$^-20$___

14. $^-5m = 5$
    $m =$ ___$^-1$___

15. $^-2t + 5 = 25$
    $t =$ ___$^-10$___

16. $^-6a = ^-72$
    $a =$ ___12___

17. $^-5y + 8 = 28$
    $y =$ ___$^-4$___

18. $x - 30 = ^-10$
    $x =$ ___20___

Follow the pattern to complete the table. Write and graph the rule.

19.

x	y
8	5
6	3
4	1
2	$^-1$
0	$^-3$

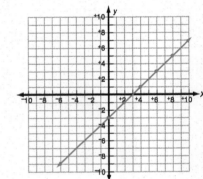

Rule: ___$y = x - 3$___

Alternate Chapter Test

You may wish to use the Alternate
Chapter Test on page 400 of this book
for further review and assessment.

Circle the letter of the correct answer.

1 $7\frac{5}{6}$ $+ 2\frac{3}{8}$

a. $9\frac{5}{24}$
b. $9\frac{4}{7}$
c. $10\frac{5}{24}$
d. NG

2 $6 - 3\frac{2}{3}$

a. $1\frac{2}{3}$
b. $2\frac{1}{3}$
c. $3\frac{1}{3}$
d. NG

3 $4\frac{2}{5} \times 2\frac{1}{2}$

a. $6\frac{3}{7}$
b. $7\frac{1}{10}$
c. $8\frac{1}{5}$
d. NG

4 $2\frac{1}{4} \div 3\frac{3}{4}$

a. $\frac{3}{5}$
b. $1\frac{2}{3}$
c. $2\frac{1}{2}$
d. NG

5 0.15×0.06

a. 0.9
b. 0.09
c. 0.009
d. NG

6 $9 \div 0.03$

a. 3
b. 0.003
c. 300
d. NG

7 5% of 250

a. 12.5
b. 1.25
c. 125
d. NG

8 Both spinners are spun at the same time. What is P(odd, even)?

a. $\frac{1}{5}$
b. $\frac{1}{6}$
c. $\frac{2}{7}$
d. NG

9 $^-8 + {}^+3$

a. $^+11$
b. $^-11$
c. $^+5$
d. NG

10 $^-5 - {}^+4$

a. $^-9$
b. $^-1$
c. $^+9$
d. NG

☐ score

STOP

Items	Objectives
1–2	To add or subtract two mixed numbers with regrouping (see pages 119–120, 127–128)
3–4	To multiply or divide mixed numbers (see pages 141–142, 147–148)
5	To multiply decimals, write zeros in the product to locate the decimal point (see pages 189–190)
6	To divide by decimals with zeros in the dividend (see pages 199–200)
7	To find the percent of a number (see pages 297–298)
8	To find compound probability (see pages 323–324)
9	To add integers with unlike signs (see pages 339–340)
10	To subtract integers (see pages 341–342)

Alternate Cumulative Assessment

Circle the letter of the correct answer.

1. $9\frac{2}{3}$ $+ 2\frac{3}{4}$

a. $12\frac{5}{12}$
b. $11\frac{7}{12}$
c. $11\frac{5}{7}$
d. NG

2. 10 $- 4\frac{3}{5}$

a. $6\frac{2}{5}$
b. $5\frac{3}{5}$
c. $5\frac{2}{5}$
d. NG

3. $5\frac{5}{7} \times 1\frac{3}{4}$

a. $5\frac{15}{28}$
b. 10
c. $10\frac{5}{8}$
d. NG

4. $3\frac{3}{8} \div 3\frac{3}{4}$

a. $\frac{1}{2}$
b. $\frac{9}{10}$
c. $\frac{5}{8}$
d. NG

5. 0.12×0.05

a. 0.006
b. 2.5
c. 0.6
d. NG

6. $28 \div 0.07$

a. 0.04
b. 4
c. 40
d. NG

7. 8% of 180

a. 1.44
b. 14.4
c. 144
d. NG

8. Both spinners are spun at the same time. What is P (even, odd)?

a. $\frac{1}{3}$
b. $\frac{1}{2}$
c. $\frac{1}{6}$
d. NG

9. $^+12 + {}^-7$

a. $^+19$
b. $^-5$
c. $^+5$
d. NG

10. $^+6 - {}^-9$

a. $^+15$
b. $^-3$
c. $^+3$
d. NG

T368

for page 370

Items	Objectives
1–20	To review basic addition and subtraction facts (see pages 1–4)
21–40	To review basic multiplication and division facts (see pages 5–8)
41–44	To write exponential expressions as standard numbers (see pages 9–10)
45–50	To use the Order of Operations to simplify expressions (see pages 11–12)

Add or subtract.

1. $7 + 9 =$ __16__   2. $8 + 5 =$ __13__   3. $4 + 6 =$ __10__   4. $9 + 3 =$ __12__

5. $17 - 9 =$ __8__   6. $15 - 7 =$ __8__   7. $12 - 3 =$ __9__   8. $11 - 8 =$ __3__

9. $\begin{array}{r} 8 \\ +7 \\ \hline 15 \end{array}$
10. $\begin{array}{r} 6 \\ +6 \\ \hline 12 \end{array}$
11. $\begin{array}{r} 5 \\ +0 \\ \hline 5 \end{array}$
12. $\begin{array}{r} 7 \\ +4 \\ \hline 11 \end{array}$
13. $\begin{array}{r} 2 \\ +8 \\ \hline 10 \end{array}$
14. $\begin{array}{r} 1 \\ +9 \\ \hline 10 \end{array}$

15. $\begin{array}{r} 9 \\ -0 \\ \hline 9 \end{array}$
16. $\begin{array}{r} 15 \\ -9 \\ \hline 6 \end{array}$
17. $\begin{array}{r} 8 \\ -1 \\ \hline 7 \end{array}$
18. $\begin{array}{r} 14 \\ -5 \\ \hline 9 \end{array}$
19. $\begin{array}{r} 16 \\ -8 \\ \hline 8 \end{array}$
20. $\begin{array}{r} 13 \\ -6 \\ \hline 7 \end{array}$

Multiply or divide.

21. $2 \times 0 =$ __0__   22. $9 \times 6 =$ __54__   23. $8 \times 8 =$ __64__   24. $5 \times 7 =$ __35__

25. $27 \div 3 =$ __9__   26. $48 \div 6 =$ __8__   27. $81 \div 9 =$ __9__   28. $42 \div 7 =$ __6__

29. $\begin{array}{r} 6 \\ \times 1 \\ \hline 6 \end{array}$
30. $\begin{array}{r} 3 \\ \times 4 \\ \hline 12 \end{array}$
31. $\begin{array}{r} 9 \\ \times 5 \\ \hline 45 \end{array}$
32. $\begin{array}{r} 7 \\ \times 6 \\ \hline 42 \end{array}$
33. $\begin{array}{r} 8 \\ \times 7 \\ \hline 56 \end{array}$
34. $\begin{array}{r} 7 \\ \times 9 \\ \hline 63 \end{array}$

35. $8\overline{)32}$ (4) 36. $4\overline{)20}$ (5) 37. $7\overline{)63}$ (9) 38. $9\overline{)27}$ (3) 39. $8\overline{)40}$ (5) 40. $6\overline{)54}$ (9)

Write as standard numbers.

41. $5^3 =$ __125__   42. $7 \times 10^2 =$ __700__   43. $2^3 =$ __8__   44. $4 \times 10^4 =$ __40,000__

Use the Order of Operations to simplify each expression.

45. $15 - 27 \div 3 =$ __6__   46. $26 + 12 \div 6 =$ __28__

47. $(12 - 7) \times 8 =$ __40__   48. $36 \div (2^3 + 4) =$ __3__

49. $7 + 6 \times 4 - 3 =$ __28__   50. $3^2 + (11 - 8) \times 3 =$ __18__

Add or subtract.

1. $7 + 9 =$ _____
2. $8 + 5 =$ _____
3. $4 + 6 =$ _____
4. $9 + 3 =$ _____

5. $17 - 9 =$ _____
6. $15 - 7 =$ _____
7. $12 - 3 =$ _____
8. $11 - 8 =$ _____

9. $\begin{array}{r} 8 \\ + 7 \\ \hline \end{array}$
10. $\begin{array}{r} 6 \\ + 6 \\ \hline \end{array}$
11. $\begin{array}{r} 5 \\ + 0 \\ \hline \end{array}$
12. $\begin{array}{r} 7 \\ + 4 \\ \hline \end{array}$
13. $\begin{array}{r} 2 \\ + 8 \\ \hline \end{array}$
14. $\begin{array}{r} 1 \\ + 9 \\ \hline \end{array}$

15. $\begin{array}{r} 9 \\ - 0 \\ \hline \end{array}$
16. $\begin{array}{r} 15 \\ - 9 \\ \hline \end{array}$
17. $\begin{array}{r} 8 \\ - 1 \\ \hline \end{array}$
18. $\begin{array}{r} 14 \\ - 5 \\ \hline \end{array}$
19. $\begin{array}{r} 16 \\ - 8 \\ \hline \end{array}$
20. $\begin{array}{r} 13 \\ - 6 \\ \hline \end{array}$

Multiply or divide.

21. $2 \times 0 =$ _____
22. $9 \times 6 =$ _____
23. $8 \times 8 =$ _____
24. $5 \times 7 =$ _____

25. $27 \div 3 =$ _____
26. $48 \div 6 =$ _____
27. $81 \div 9 =$ _____
28. $42 \div 7 =$ _____

29. $\begin{array}{r} 6 \\ \times 1 \\ \hline \end{array}$
30. $\begin{array}{r} 3 \\ \times 4 \\ \hline \end{array}$
31. $\begin{array}{r} 9 \\ \times 5 \\ \hline \end{array}$
32. $\begin{array}{r} 7 \\ \times 6 \\ \hline \end{array}$
33. $\begin{array}{r} 8 \\ \times 7 \\ \hline \end{array}$
34. $\begin{array}{r} 7 \\ \times 9 \\ \hline \end{array}$

35. $8\overline{)32}$
36. $4\overline{)20}$
37. $7\overline{)63}$
38. $9\overline{)27}$
39. $8\overline{)40}$
40. $6\overline{)54}$

Write as standard numbers.

41. $5^3 =$ _____
42. $7 \times 10^2 =$ _____
43. $2^3 =$ _____
44. $4 \times 10^4 =$ _____

Use the Order of Operations to simplify each expression.

45. $15 - 27 \div 3 =$ _____
46. $26 + 12 \div 6 =$ _____

47. $(12 - 7) \times 8 =$ _____
48. $36 \div (2^3 + 4) =$ _____

49. $7 + 6 \times 4 - 3 =$ _____
50. $3^2 + (11 - 8) \times 3 =$ _____

Chapter 2
Alternate Test

for page 372

Items	Objectives
1–4	To identify place value through hundred thousands (see pages 19–20)
5–8	To compare and order numbers through hundred thousands (see pages 21–22)
9–12	To round numbers to the nearest 1,000 and 10,000 (see pages 23–24)
13–16	To identify place value through trillions (see pages 25–26)
17–19	To add two numbers up to 6 digits (see pages 27–28)
20	To add three or more numbers (see pages 29–30)
21–22, 24	To subtract with renaming over zeros (see pages 33–34)
23	To subtract two numbers up to 6 digits (see pages 31–32)
25–30	To use estimation to find sums and differences (see pages 35–36)

Name _____

Write the place value of each green digit.

1. 7,439

   thousands

2. 276,427

   hundreds

3. 381,562

   hundred thousands

4. 59,783

   ten thousands

Write < or > in each circle.

5. 678 $<$ 786

6. 3,871 $>$ 3,781

7. 5,436 $>$ 5,426

8. 29,541 $<$ 29,542

Round to the nearest thousand and to the nearest ten thousand.

9. 15,100

   15,000

   20,000

10. 87,500

    88,000

    90,000

11. 256,603

    257,000

    260,000

12. 724,499

    724,000

    720,000

Write the place value of each green digit.

13. 133,209,645

    millions

14. 6,874,396

    ten thousands

15. 352,175,439

    ten millions

16. 2,468,972,311

    billions

Add.

17.
```
    8,549
 + 28,187
   36,736
```

18.
```
  $249.75
 +  61.08
  $310.83
```

19.
```
   315,755
 + 972,649
 1,288,404
```

20.
```
   229,463
   758,103
 + 653,974
 1,641,540
```

Subtract.

21.
```
   30,754
 -  8,979
   21,775
```

22.
```
   28,002
 -  9,467
   18,535
```

23.
```
  856,273
 - 394,198
  462,075
```

24.
```
  $7,607.52
 - 3,296.94
  $4,310.58
```

Estimate by rounding to the nearest thousand.

25.
```
   8,463     8,000
 + 9,742  + 10,000
           18,000
```

26.
```
  39,000    39,000
 - 21,433  - 21,000
           18,000
```

27.
```
  78,436    78,000
 + 25,841  + 26,000
          104,000
```

28.
```
   6,012     6,000
 - 2,476   - 2,000
            4,000
```

29.
```
  37,242    37,000
 + 18,387  + 18,000
           55,000
```

30.
```
  31,938    32,000
 - 26,099  - 26,000
            6,000
```

Write the place value of each green digit.

1. 7,439

2. 276,427

3. 381,562

4. 59,783

_____   _____   _____   _____

Write < or > in each circle.

5. 678 ◯ 786

6. 3,871 ◯ 3,781

7. 5,436 ◯ 5,426

8. 29,541 ◯ 29,542

Round to the nearest thousand and to the nearest ten thousand.

9. 15,100

10. 87,500

11. 256,603

12. 724,499

_____   _____   _____   _____

_____   _____   _____   _____

Write the place value of each green digit.

13. 133,209,645

14. 6,874,396

15. 352,175,439

16. 2,468,972,311

_____   _____   _____   _____

Add.

17.   8,549
    + 28,187

18. $249.75
    +   61.08

19.   315,755
    + 972,649

20.   229,463
       758,103
    + 653,974

Subtract.

21.   30,754
    − 8,979

22.   28,002
    − 9,467

23.   856,273
    − 394,198

24. $7,607.52
    − 3,296.94

Estimate by rounding to the nearest thousand.

25.   8,463
    + 9,742

26.   39,000
    − 21,433

27.   78,436
    + 25,841

28.   6,012
    − 2,476

29.   37,242
    + 18,387

30.   31,938
    − 26,099

Chapter 3
Alternate Test

for page 374

Items	Objectives
1–8	To multiply 2- or 3-digit factors by 1-digit factors (see pages 45–46)
9, 12–13, 16	To multiply 4-, 5-, or 6-digit factors by 1-digit factors (see pages 47–48)
10–11, 14–15	To multiply money by 1-digit factors (see pages 47–48)
17–24	To multiply whole numbers or money by 2-digit factors (see pages 51–52)
25–32	To multiply whole numbers or money by 3-digit factors (see pages 55–56)

Name _____

Multiply. Use estimation to check your answers.

1. $\begin{array}{r} 76 \\ \times\ 8 \\ \hline 608 \end{array}$

2. $\begin{array}{r} 49 \\ \times\ 5 \\ \hline 245 \end{array}$

3. $\begin{array}{r} 123 \\ \times\ 7 \\ \hline 861 \end{array}$

4. $\begin{array}{r} 908 \\ \times\ 3 \\ \hline 2,724 \end{array}$

5. $\begin{array}{r} 69 \\ \times\ 9 \\ \hline 621 \end{array}$

6. $\begin{array}{r} 54 \\ \times\ 6 \\ \hline 324 \end{array}$

7. $\begin{array}{r} 286 \\ \times\ 5 \\ \hline 1,430 \end{array}$

8. $\begin{array}{r} 405 \\ \times\ 4 \\ \hline 1,620 \end{array}$

9. $\begin{array}{r} 5,162 \\ \times\ 4 \\ \hline 20,648 \end{array}$

10. $\begin{array}{r} \$39.87 \\ \times\ 6 \\ \hline \$239.22 \end{array}$

11. $\begin{array}{r} \$805.46 \\ \times\ 9 \\ \hline \$7,249.14 \end{array}$

12. $\begin{array}{r} 289,597 \\ \times\ 8 \\ \hline 2,316,776 \end{array}$

13. $\begin{array}{r} 7,492 \\ \times\ 2 \\ \hline 14,984 \end{array}$

14. $\begin{array}{r} \$63.59 \\ \times\ 3 \\ \hline \$190.77 \end{array}$

15. $\begin{array}{r} \$602.75 \\ \times\ 7 \\ \hline \$4,219.25 \end{array}$

16. $\begin{array}{r} 125,624 \\ \times\ 2 \\ \hline 251,248 \end{array}$

17. $\begin{array}{r} 84 \\ \times\ 35 \\ \hline 2,940 \end{array}$

18. $\begin{array}{r} \$6.07 \\ \times\ 92 \\ \hline \$558.44 \end{array}$

19. $\begin{array}{r} 415 \\ \times\ 73 \\ \hline 30,295 \end{array}$

20. $\begin{array}{r} 7,936 \\ \times\ 86 \\ \hline 682,496 \end{array}$

21. $\begin{array}{r} 75 \\ \times\ 91 \\ \hline 6,825 \end{array}$

22. $\begin{array}{r} \$6.43 \\ \times\ 95 \\ \hline \$610.85 \end{array}$

23. $\begin{array}{r} 388 \\ \times\ 45 \\ \hline 17,460 \end{array}$

24. $\begin{array}{r} 2,976 \\ \times\ 58 \\ \hline 172,608 \end{array}$

25. $\begin{array}{r} 929 \\ \times\ 452 \\ \hline 419,908 \end{array}$

26. $\begin{array}{r} 5,138 \\ \times\ 865 \\ \hline 4,444,370 \end{array}$

27. $\begin{array}{r} 37,294 \\ \times\ 517 \\ \hline 19,280,998 \end{array}$

28. $\begin{array}{r} \$806.26 \\ \times\ 349 \\ \hline \$281,384.74 \end{array}$

29. $\begin{array}{r} 893 \\ \times\ 415 \\ \hline 370,595 \end{array}$

30. $\begin{array}{r} 6,347 \\ \times\ 624 \\ \hline 3,960,528 \end{array}$

31. $\begin{array}{r} 52,468 \\ \times\ 796 \\ \hline 41,764,528 \end{array}$

32. $\begin{array}{r} \$508.38 \\ \times\ 934 \\ \hline \$474,826.92 \end{array}$

Multiply. Use estimation to check your answers.

1. 76
    × 8

2. 49
    × 5

3. 123
    × 7

4. 908
    × 3

5. 69
    × 9

6. 54
    × 6

7. 286
    × 5

8. 405
    × 4

9. 5,162
    × 4

10. $39.87
    × 6

11. $805.46
    × 9

12. 289,597
    × 8

13. 7,492
    × 2

14. $63.59
    × 3

15. $602.75
    × 7

16. 125,624
    × 2

17. 84
    × 35

18. $6.07
    × 92

19. 415
    × 73

20. 7,936
    × 86

21. 75
    × 91

22. $6.43
    × 95

23. 388
    × 45

24. 2,976
    × 58

25. 929
    × 452

26. 5,138
    × 865

27. 37,294
    × 517

28. $806.26
    × 349

29. 893
    × 415

30. 6,347
    × 624

31. 52,468
    × 796

32. $508.38
    × 934

Items   Objectives

1–4   To divide 2- or 3-digit
numbers by 1-digit
numbers (see pages 63–64)

5–6   To divide 4-, 5-, or 6-digit
numbers by 1-digit
numbers (see pages 65–66)

7–8   To divide by 1-digit
numbers, zeros in the
quotient (see pages 67–68)

9–16   To divide 2- or 3-digit
numbers by 2-digit
numbers (see pages 75–76)

17–19,   To divide 4-, 5-, or 6-digit
23–25   numbers by 2-digit
numbers (see pages 79–80)

20–22   To divide by 3-digit
numbers (see pages 81–82)

Divide and check.

1. $\overset{93\ R2}{4\overline{)374}}$ 2. $\overset{78}{6\overline{)468}}$ 3. $\overset{254\ R1}{3\overline{)763}}$ 4. $\overset{\$1.83}{5\overline{)\$9.15}}$

5. $\overset{567\ R3}{8\overline{)4,539}}$ 6. $\overset{4,189\ R1}{2\overline{)8,379}}$ 7. $\overset{\$13.06}{7\overline{)\$91.42}}$ 8. $\overset{3,058\ R8}{9\overline{)27,530}}$

9. $\overset{5\ R17}{65\overline{)342}}$ 10. $\overset{9}{29\overline{)261}}$ 11. $\overset{8\ R11}{91\overline{)739}}$ 12. $\overset{7\ R36}{57\overline{)435}}$

13. $\overset{21\ R23}{36\overline{)779}}$ 14. $\overset{3\ R70}{74\overline{)292}}$ 15. $\overset{4\ R58}{82\overline{)386}}$ 16. $\overset{8\ R24}{59\overline{)496}}$

17. $\overset{101}{21\overline{)2,121}}$ 18. $\overset{\$156.01}{49\overline{)\$7,644.49}}$ 19. $\overset{384\ R22}{65\overline{)24,982}}$

20. $\overset{8\ R218}{593\overline{)4,962}}$ 21. $\overset{622\ R301}{472\overline{)293,885}}$ 22. $\overset{\$36.75}{914\overline{)\$33,589.50}}$

23. $\overset{128\ R36}{37\overline{)4,772}}$ 24. $\overset{210\ R28}{72\overline{)15,148}}$ 25. $\overset{2,546\ R30}{83\overline{)211,348}}$

Divide and check.

1. $4\overline{)374}$

2. $6\overline{)468}$

3. $3\overline{)763}$

4. $5\overline{)\$9.15}$

5. $8\overline{)4,539}$

6. $2\overline{)8,379}$

7. $7\overline{)\$91.42}$

8. $9\overline{)27,530}$

9. $65\overline{)342}$

10. $29\overline{)261}$

11. $91\overline{)739}$

12. $57\overline{)435}$

13. $36\overline{)779}$

14. $74\overline{)292}$

15. $82\overline{)386}$

16. $59\overline{)496}$

17. $21\overline{)2,121}$

18. $49\overline{)\$7,644.49}$

19. $65\overline{)24,982}$

20. $593\overline{)4,962}$

21. $472\overline{)293,885}$

22. $914\overline{)\$33,589.50}$

23. $37\overline{)4,772}$

24. $72\overline{)15,148}$

25. $83\overline{)211,348}$

for page 378

Items **Objectives**

1–4   To find the least common multiple and the greatest common factor of a pair of numbers (see pages 87–90)

5–12 To write the prime factorization of composite numbers using exponents (see pages 93–94)

13–20 To write equivalent fractions (see pages 97–98)

21–28 To write fractions in lowest terms (see pages 99–100)

29–36 To compare and order fractions (see pages 105–106)

37–44 To change mixed numbers to fractions (see pages 101–102)

45–52 To change improper fractions to whole or mixed numbers (see pages 103–104)

Write the LCM and GCF for each set of numbers.

1. 8, 10

    LCM $\underline{40}$

    GCF $\underline{2}$

2. 6, 24

    LCM $\underline{24}$

    GCF $\underline{6}$

3. 3, 11

    LCM $\underline{33}$

    GCF $\underline{1}$

4. 6, 10

    LCM $\underline{30}$

    GCF $\underline{2}$

Prime factor each number using exponents.

5. 24 $\underline{2^3 \times 3}$

6. 45 $\underline{3^2 \times 5}$

7. 75 $\underline{3 \times 5^2}$

8. 90 $\underline{2 \times 3^2 \times 5}$

9. 35 $\underline{5 \times 7}$

10. 32 $\underline{2^5}$

11. 48 $\underline{2^4 \times 3}$

12. 36 $\underline{2^2 \times 3^2}$

Find the equivalent fraction.

13. $\frac{6}{8} = \frac{18}{24}$

14. $\frac{4}{7} = \frac{36}{63}$

15. $\frac{4}{8} = \frac{16}{32}$

16. $\frac{21}{28} = \frac{3}{4}$

17. $\frac{15}{25} = \frac{3}{5}$

18. $\frac{2}{7} = \frac{12}{42}$

19. $\frac{3}{11} = \frac{12}{44}$

20. $\frac{5}{6} = \frac{30}{36}$

Rename in simplest form.

21. $\frac{14}{63} = \frac{2}{9}$

22. $\frac{28}{32} = \frac{7}{8}$

23. $\frac{24}{42} = \frac{4}{7}$

24. $\frac{18}{36} = \frac{1}{2}$

25. $\frac{14}{35} = \frac{2}{5}$

26. $\frac{12}{42} = \frac{2}{7}$

27. $\frac{27}{45} = \frac{3}{5}$

28. $\frac{24}{56} = \frac{3}{7}$

Write <, =, or > in each circle.

29. $\frac{7}{8}$ ⊘ $\frac{4}{5}$ (>)

30. $\frac{6}{15}$ (=) $\frac{4}{10}$

31. $\frac{12}{16}$ (<) $\frac{7}{9}$

32. $\frac{3}{5}$ (<) $\frac{2}{3}$

33. $\frac{6}{18}$ (<) $\frac{2}{3}$

34. $\frac{3}{7}$ (<) $\frac{4}{9}$

35. $\frac{2}{9}$ (<) $\frac{1}{2}$

36. $\frac{3}{4}$ (>) $\frac{3}{7}$

Rename each mixed number as a fraction.

37. $7\frac{2}{3} = \frac{23}{3}$

38. $6\frac{8}{9} = \frac{62}{9}$

39. $2\frac{3}{10} = \frac{23}{10}$

40. $6\frac{5}{8} = \frac{53}{8}$

41. $9\frac{3}{4} = \frac{39}{4}$

42. $8\frac{2}{7} = \frac{58}{7}$

43. $5\frac{3}{5} = \frac{28}{5}$

44. $4\frac{3}{11} = \frac{47}{11}$

Rename each improper fraction as a whole or mixed number.

45. $\frac{16}{7} = \underline{2\frac{2}{7}}$

46. $\frac{37}{8} = \underline{4\frac{5}{8}}$

47. $\frac{16}{2} = \underline{8}$

48. $\frac{42}{5} = \underline{8\frac{2}{5}}$

49. $\frac{56}{7} = \underline{8}$

50. $\frac{16}{12} = \underline{1\frac{1}{3}}$

51. $\frac{70}{10} = \underline{7}$

52. $\frac{49}{3} = \underline{16\frac{1}{3}}$

Write the LCM and GCF for each set of numbers.

1. 8, 10

   LCM _____

   GCF _____

2. 6, 24

   LCM _____

   GCF _____

3. 3, 11

   LCM _____

   GCF _____

4. 6, 10

   LCM _____

   GCF _____

Prime factor each number using exponents.

5. 24 _____

6. 45 _____

7. 75 _____

8. 90 _____

9. 35 _____

10. 32 _____

11. 48 _____

12. 36 _____

Find the equivalent fraction.

13. $\frac{6}{8} = \frac{}{24}$

14. $\frac{4}{} = \frac{36}{63}$

15. $\frac{}{8} = \frac{16}{32}$

16. $\frac{21}{28} = \frac{}{4}$

17. $\frac{15}{25} = \frac{}{5}$

18. $\frac{2}{7} = \frac{}{42}$

19. $\frac{3}{11} = \frac{12}{}$

20. $\frac{}{6} = \frac{30}{36}$

Rename in simplest form.

21. $\frac{14}{63} = \frac{2}{9}$

22. $\frac{28}{32} =$

23. $\frac{24}{42} =$

24. $\frac{18}{36} =$

25. $\frac{14}{35} =$

26. $\frac{12}{42} =$

27. $\frac{27}{45} =$

28. $\frac{24}{56} =$

Write <, =, or > in each circle.

29. $\frac{7}{8} \bigcirc \frac{4}{5}$

30. $\frac{6}{15} \bigcirc \frac{4}{10}$

31. $\frac{12}{16} \bigcirc \frac{7}{9}$

32. $\frac{3}{5} \bigcirc \frac{2}{3}$

33. $\frac{6}{18} \bigcirc \frac{2}{3}$

34. $\frac{3}{7} \bigcirc \frac{4}{9}$

35. $\frac{2}{9} \bigcirc \frac{1}{2}$

36. $\frac{3}{4} \bigcirc \frac{3}{7}$

Rename each mixed number as a fraction.

37. $7\frac{2}{3} =$

38. $6\frac{8}{9} =$

39. $2\frac{3}{10} =$

40. $6\frac{5}{8} =$

41. $9\frac{3}{4} =$

42. $8\frac{2}{7} =$

43. $5\frac{3}{5} =$

44. $4\frac{3}{11} =$

Rename each improper fraction as a whole or mixed number.

45. $\frac{16}{7} =$ _____

46. $\frac{37}{8} =$ _____

47. $\frac{16}{2} =$ _____

48. $\frac{42}{5} =$ _____

49. $\frac{56}{7} =$ _____

50. $\frac{16}{12} =$ _____

51. $\frac{70}{10} =$ _____

52. $\frac{49}{3} =$ _____

Name _____

for page 380

Items	Objectives
1–4	To add fractions with like denominators (see pages 115–116)
5–8	To add fractions with unlike denominators (see pages 117–118)
9–12	To add mixed numbers with renaming (see pages 119–120)
13–16	To subtract fractions with like denominators (see pages 121–122)
17–20	To subtract fractions with unlike denominators (see pages 123–124)
21	To subtract mixed numbers with regrouping (see pages 125–126)
22–28	To subtract mixed numbers, rename and regroup minuends (see pages 127–128)

Add. Simplify answers if necessary.

1. $\frac{2}{9} + \frac{1}{9} = \frac{1}{3}$

2. $\frac{7}{12} + \frac{11}{12} = 1\frac{1}{2}$

3. $\begin{array}{r} 8\frac{5}{16} \\ + 7\frac{7}{16} \\ \hline 15\frac{3}{4} \end{array}$

4. $\begin{array}{r} 3\frac{13}{24} \\ + 9\frac{8}{24} \\ \hline 12\frac{7}{8} \end{array}$

5. $\frac{2}{3} + \frac{3}{5} = 1\frac{4}{15}$

6. $\frac{2}{7} + \frac{14}{21} = \frac{20}{21}$

7. $\begin{array}{r} 5\frac{5}{12} \\ + 13\frac{3}{8} \\ \hline 18\frac{19}{24} \end{array}$

8. $\begin{array}{r} 6\frac{1}{2} \\ + 6\frac{3}{10} \\ \hline 12\frac{4}{5} \end{array}$

9. $\begin{array}{r} 11\frac{3}{4} \\ + 14\frac{7}{12} \\ \hline 26\frac{1}{3} \end{array}$

10. $\begin{array}{r} 10\frac{5}{9} \\ + 3\frac{8}{18} \\ \hline 14 \end{array}$

11. $\begin{array}{r} 18\frac{2}{8} \\ + 17\frac{5}{6} \\ \hline 36\frac{1}{12} \end{array}$

12. $\begin{array}{r} 12\frac{5}{6} \\ 17\frac{1}{4} \\ + 14\frac{7}{9} \\ \hline 44\frac{31}{36} \end{array}$

Subtract. Simplify answers if necessary.

13. $\frac{9}{10} - \frac{4}{10} = \frac{1}{2}$

14. $\frac{5}{14} - \frac{3}{14} = \frac{1}{7}$

15. $\begin{array}{r} 10\frac{11}{15} \\ - 6\frac{8}{15} \\ \hline 4\frac{1}{5} \end{array}$

16. $\begin{array}{r} 14\frac{5}{6} \\ - 8\frac{1}{6} \\ \hline 6\frac{2}{3} \end{array}$

17. $\frac{3}{4} - \frac{3}{5} = \frac{3}{20}$

18. $\frac{9}{10} - \frac{5}{6} = \frac{1}{15}$

19. $\begin{array}{r} 22\frac{2}{3} \\ - 19 \\ \hline 3\frac{2}{3} \end{array}$

20. $\begin{array}{r} 16\frac{1}{9} \\ - 14\frac{2}{3} \\ \hline 1\frac{4}{9} \end{array}$

21. $\begin{array}{r} 17 \\ - 8\frac{7}{10} \\ \hline 8\frac{3}{10} \end{array}$

22. $\begin{array}{r} 13\frac{1}{4} \\ - 7\frac{5}{6} \\ \hline 5\frac{5}{12} \end{array}$

23. $\begin{array}{r} 14\frac{1}{3} \\ - 6\frac{5}{7} \\ \hline 7\frac{13}{21} \end{array}$

24. $\begin{array}{r} 11\frac{3}{8} \\ - 8\frac{4}{5} \\ \hline 2\frac{23}{40} \end{array}$

25. $\begin{array}{r} 16\frac{2}{7} \\ - 8\frac{2}{3} \\ \hline 7\frac{13}{21} \end{array}$

26. $\begin{array}{r} 14\frac{5}{11} \\ - 6\frac{3}{4} \\ \hline 7\frac{31}{44} \end{array}$

27. $\begin{array}{r} 27\frac{1}{4} \\ - 16\frac{3}{5} \\ \hline 10\frac{13}{20} \end{array}$

28. $\begin{array}{r} 24\frac{1}{2} \\ - 19\frac{4}{7} \\ \hline 4\frac{13}{14} \end{array}$

Add. Simplify answers if necessary.

1. $\frac{2}{9} + \frac{1}{9} =$

2. $\frac{7}{12} + \frac{11}{12} =$

3. $\begin{array}{r} 8\frac{5}{16} \\ + 7\frac{7}{16} \\ \hline \end{array}$

4. $\begin{array}{r} 3\frac{13}{24} \\ + 9\frac{8}{24} \\ \hline \end{array}$

5. $\frac{2}{3} + \frac{3}{5} =$

6. $\frac{2}{7} + \frac{14}{21} =$

7. $\begin{array}{r} 5\frac{5}{12} \\ + 13\frac{3}{8} \\ \hline \end{array}$

8. $\begin{array}{r} 6\frac{1}{2} \\ + 6\frac{3}{10} \\ \hline \end{array}$

9. $\begin{array}{r} 11\frac{3}{4} \\ + 14\frac{7}{12} \\ \hline \end{array}$

10. $\begin{array}{r} 10\frac{5}{9} \\ + 3\frac{8}{18} \\ \hline \end{array}$

11. $\begin{array}{r} 18\frac{2}{8} \\ + 17\frac{5}{6} \\ \hline \end{array}$

12. $\begin{array}{r} 12\frac{5}{6} \\ 17\frac{1}{4} \\ + 14\frac{7}{9} \\ \hline \end{array}$

Subtract. Simplify answers if necessary.

13. $\frac{9}{10} - \frac{4}{10} =$

14. $\frac{5}{14} - \frac{3}{14} =$

15. $\begin{array}{r} 10\frac{11}{15} \\ - 6\frac{8}{15} \\ \hline \end{array}$

16. $\begin{array}{r} 14\frac{5}{6} \\ - 8\frac{1}{6} \\ \hline \end{array}$

17. $\frac{3}{4} - \frac{3}{5} =$

18. $\frac{9}{10} - \frac{5}{6} =$

19. $\begin{array}{r} 22\frac{2}{3} \\ - 19 \\ \hline \end{array}$

20. $\begin{array}{r} 16\frac{1}{9} \\ - 14\frac{2}{3} \\ \hline \end{array}$

21. $\begin{array}{r} 17 \\ - 8\frac{7}{10} \\ \hline \end{array}$

22. $\begin{array}{r} 13\frac{1}{4} \\ - 7\frac{5}{6} \\ \hline \end{array}$

23. $\begin{array}{r} 14\frac{1}{3} \\ - 6\frac{5}{7} \\ \hline \end{array}$

24. $\begin{array}{r} 11\frac{3}{8} \\ - 8\frac{4}{5} \\ \hline \end{array}$

25. $\begin{array}{r} 16\frac{2}{7} \\ - 8\frac{2}{3} \\ \hline \end{array}$

26. $\begin{array}{r} 14\frac{5}{11} \\ - 6\frac{3}{4} \\ \hline \end{array}$

27. $\begin{array}{r} 27\frac{1}{4} \\ - 16\frac{3}{5} \\ \hline \end{array}$

28. $\begin{array}{r} 24\frac{1}{2} \\ - 19\frac{4}{7} \\ \hline \end{array}$

Chapter 7
Alternate Test

for page 382

Items	Objectives
1–4	To multiply fractions (see pages 137–140)
5–8	To multiply whole numbers by fractions (see pages 141–142)
9	To multiply whole numbers by mixed numbers (see pages 141–142)
10–14	To multiply two mixed numbers (see pages 141–142)
15–20	To find reciprocals of fractions and whole and mixed numbers (see pages 143–144)
21–24	To divide two fractions (see pages 145–146)
25–28	To divide whole numbers by fractions or fractions by whole numbers (see pages 145–146)
29–34	To divide two mixed numbers (see pages 147–148)

Name _____

Multiply. Factor wherever possible.

1. $\frac{5}{6} \times \frac{3}{2} = 1\frac{1}{4}$

2. $\frac{9}{10} \times \frac{7}{9} = \frac{7}{10}$

3. $\frac{2}{3} \times \frac{3}{4} = \frac{1}{2}$

4. $\frac{3}{8} \times \frac{4}{7} = \frac{3}{14}$

5. $18 \times \frac{5}{9} = 10$

6. $\frac{5}{6} \times 9 = 7\frac{1}{2}$

7. $64 \times \frac{7}{8} = 56$

8. $\frac{4}{5} \times 30 = 24$

9. $4\frac{1}{6} \times 12 = 50$

10. $3\frac{1}{8} \times 1\frac{3}{10} = 4\frac{1}{16}$

11. $2\frac{1}{4} \times 9\frac{1}{3} = 21$

12. $2\frac{5}{8} \times 1\frac{3}{7} = 3\frac{3}{4}$

13. $4\frac{2}{3} \times 6\frac{1}{2} = 30\frac{1}{3}$

14. $7\frac{1}{5} \times 3\frac{2}{9} = 23\frac{1}{5}$

Write the reciprocal.

15. $\frac{7}{9}$ $\frac{9}{7}$

16. $9\frac{3}{4}$ $\frac{4}{39}$

17. 15 $\frac{1}{15}$

18. $\frac{16}{3}$ $\frac{3}{16}$

19. $4\frac{1}{8}$ $\frac{8}{33}$

20. $5\frac{3}{11}$ $\frac{11}{58}$

Divide.

21. $\frac{5}{8} \div \frac{1}{2} = 1\frac{1}{4}$

22. $\frac{9}{10} \div \frac{3}{5} = 1\frac{1}{2}$

23. $\frac{4}{9} \div \frac{5}{9} = \frac{4}{5}$

24. $\frac{1}{8} \div \frac{2}{3} = \frac{3}{16}$

25. $8 \div \frac{4}{5} = 10$

26. $\frac{6}{11} \div 9 = \frac{2}{33}$

27. $12 \div \frac{3}{7} = 28$

28. $\frac{2}{9} \div 16 = \frac{1}{72}$

29. $6\frac{1}{4} \div 1\frac{2}{5} = 4\frac{13}{28}$

30. $3\frac{2}{3} \div 2\frac{1}{6} = 1\frac{9}{13}$

31. $4\frac{2}{7} \div 1\frac{6}{14} = 3$

32. $5\frac{5}{8} \div 2\frac{1}{4} = 2\frac{1}{2}$

33. $7\frac{2}{3} \div 4\frac{1}{8} = 1\frac{85}{99}$

34. $9\frac{5}{6} \div 3\frac{2}{3} = 2\frac{15}{22}$

Multiply. Factor wherever possible.

1. $\frac{5}{6} \times \frac{3}{2} =$

2. $\frac{9}{10} \times \frac{7}{9} =$

3. $\frac{2}{3} \times \frac{3}{4} =$

4. $\frac{3}{8} \times \frac{4}{7} =$

5. $18 \times \frac{5}{9} =$

6. $\frac{5}{6} \times 9 =$

7. $64 \times \frac{7}{8} =$

8. $\frac{4}{5} \times 30 =$

9. $4\frac{1}{6} \times 12 =$

10. $3\frac{1}{8} \times 1\frac{3}{10} =$

11. $2\frac{1}{4} \times 9\frac{1}{3} =$

12. $2\frac{5}{8} \times 1\frac{3}{7} =$

13. $4\frac{2}{3} \times 6\frac{1}{2} =$

14. $7\frac{1}{5} \times 3\frac{2}{9} =$

Write the reciprocal.

15. $\frac{7}{9}$

16. $9\frac{3}{4}$

17. 15

18. $\frac{16}{3}$

19. $4\frac{1}{8}$

20. $5\frac{3}{11}$

Divide.

21. $\frac{5}{8} \div \frac{1}{2} =$

22. $\frac{9}{10} \div \frac{3}{5} =$

23. $\frac{4}{9} \div \frac{5}{9} =$

24. $\frac{1}{8} \div \frac{2}{3} =$

25. $8 \div \frac{4}{5} =$

26. $\frac{6}{11} \div 9 =$

27. $12 \div \frac{3}{7} =$

28. $\frac{2}{9} \div 16 =$

29. $6\frac{1}{4} \div 1\frac{2}{5} =$

30. $3\frac{2}{3} \div 2\frac{1}{6} =$

31. $4\frac{2}{7} \div 1\frac{6}{14} =$

32. $5\frac{5}{8} \div 2\frac{1}{4} =$

33. $7\frac{2}{3} \div 4\frac{1}{8} =$

34. $9\frac{5}{6} \div 3\frac{2}{3} =$

for page 384

Items	Objectives
1–6	To give the place value for digits in a decimal up to millionths (see pages 163–164)
7–12	To compare and order decimals (see pages 165–166)
13–15	To round decimals to the nearest hundredth (see pages 167–168)
16–21	To round decimals to the nearest thousandth (see pages 167–168)
22–25	To use estimation to find sums and differences by rounding decimals to the nearest whole number (see pages 169–170)
26–29	To add decimals (see pages 171–172)
30–33	To subtract decimals (see pages 173–174)

What is the place value of the 5 in each number?

1. 83.19475 — hundred thousandths
2. 79.54072 — tenths
3. 7.46875 — hundred thousandths

4. 5.92 — ones
5. 153.48 — tens
6. 17.596 — tenths

Write <, =, or > in each circle.

7. 5.478 $<$ 5.487
8. 9.36 $=$ 9.360
9. 6.1 $>$ 6.01

10. 3.411 $>$ 3.141
11. 8.7 $<$ 8.8493
12. 49.743 $>$ 49.599

Round to the nearest hundredth.

13. 9.4397 __9.44__
14. 2.7839 __2.78__
15. 16.1756 __16.18__

Round to the nearest thousandth.

16. 10.5884 __10.588__
17. 3.71263 __3.713__
18. 18.46759 __18.468__

19. 152.14109 __152.141__
20. 3.84921 __3.850__
21. 4.6689 __4.669__

Estimate each sum and difference by first rounding the numbers to the nearest whole number.

22.
```
  65.67    66
+ 49.47  + 49
         ───
          115
```

23.
```
  13.45    13
-  8.459 -  8
         ───
           5
```

24.
```
  16.491    16
- 10.983  - 11
          ───
            5
```

25.
```
  17.4     17
-  8.59  -  9
         ───
           8
```

Add.

26.
```
  26.87
+ 49.09
──────
 75.96
```

27.
```
  18.472
+ 36.75
──────
 55.222
```

28.
```
  53.168
+ 13.9417
───────
 67.1097
```

29.
```
  82.0495
+ 96.3865
───────
 178.4360
```

Subtract.

30.
```
  71.93
- 58.77
──────
 13.16
```

31.
```
  64.258
- 17.96
──────
 46.298
```

32.
```
  43.516
-  9.8724
───────
 33.6436
```

33.
```
  95.3547
- 20.4966
───────
 74.8581
```

What is the place value of the 5 in each number?

1. 83.19475

2. 79.54072

3. 7.46875

_____

4. 5.92

5. 153.48

6. 17.596

_____

Write <, =, or > in each circle.

7. 5.478 ◯ 5.487

8. 9.36 ◯ 9.360

9. 6.1 ◯ 6.01

10. 3.411 ◯ 3.141

11. 8.7 ◯ 8.8493

12. 49.743 ◯ 49.599

Round to the nearest hundredth.

13. 9.4397 _____

14. 2.7839 _____

15. 16.1756 _____

Round to the nearest thousandth.

16. 10.5884 _____

17. 3.71263 _____

18. 18.46759 _____

19. 152.14109 _____

20. 3.84921 _____

21. 4.6689 _____

Estimate each sum and difference by first rounding the numbers to the nearest whole number.

22. 65.67
  + 49.47

23. 13.45
  − 8.459

24. 16.491
  − 10.983

25. 17.4
  − 8.59

Add.

26. 26.87
  + 49.09

27. 18.472
  + 36.75

28. 53.168
  + 13.9417

29. 82.0495
  + 96.3865

Subtract.

30. 71.93
  − 58.77

31. 64.258
  − 17.96

32. 43.516
  − 9.8724

33. 95.3547
  − 20.4966

Chapter 9
Alternate Test

for page 386

Items	Objectives
1–4	To multiply whole numbers and decimals (see pages 185–186)
5–8	To multiply decimals (see pages 187–188)
9–12	To multiply decimals, write zeros in the product to locate the decimal point (see pages 189–190)
13–16	To divide decimals by whole numbers (see pages 191–192)
15	To divide decimals by whole numbers requiring initial zeros in the quotient (see pages 193–194)
17–20	To divide decimals by decimals (see pages 197–198)
21–24	To divide by decimals with fewer decimal places in the dividend than in the divisor (see pages 199–200)
25–28	To change a fraction to a repeating decimal and round to the nearest thousandth (see pages 203–204)

Multiply.

1. $\begin{array}{r} 7.2 \\ \times\ \ 9 \\ \hline 64.8 \end{array}$

2. $\begin{array}{r} 5.63 \\ \times\ \ \ 8 \\ \hline 45.04 \end{array}$

3. $\begin{array}{r} 49 \\ \times\ 1.7 \\ \hline 83.3 \end{array}$

4. $\begin{array}{r} 8.02 \\ \times\ \ 36 \\ \hline 288.72 \end{array}$

5. $\begin{array}{r} 6.47 \\ \times\ 0.5 \\ \hline 3.235 \end{array}$

6. $\begin{array}{r} 3.8 \\ \times\ 0.29 \\ \hline 1.102 \end{array}$

7. $\begin{array}{r} 2.516 \\ \times\ \ \ 4.8 \\ \hline 12.0768 \end{array}$

8. $\begin{array}{r} 7.35 \\ \times\ 9.7 \\ \hline 71.295 \end{array}$

9. $\begin{array}{r} 0.006 \\ \times\ \ .18 \\ \hline 0.00108 \end{array}$

10. $\begin{array}{r} 0.09 \\ \times\ 0.05 \\ \hline 0.0045 \end{array}$

11. $\begin{array}{r} 0.007 \\ \times\ 0.08 \\ \hline 0.00056 \end{array}$

12. $\begin{array}{r} 4.02 \\ \times\ 0.006 \\ \hline 0.02412 \end{array}$

Divide.

13. $5\overline{)23.5}$ = 4.7

14. $2\overline{)1.38}$ = 0.69

15. $9\overline{)0.468}$ = 0.052

16. $6\overline{)51.6}$ = 8.6

17. $0.04\overline{)2.858}$ = 71.45

18. $0.7\overline{)0.042}$ = 0.06

19. $1.2\overline{)7.68}$ = 6.4

20. $5.4\overline{)4.968}$ = 0.92

21. $9.5\overline{)57}$ = 6

22. $0.6\overline{)0.3}$ = 0.5

23. $0.047\overline{)9.4}$ = 200

24. $8.8\overline{)704}$ = 80

Write each fraction as a decimal rounded to the nearest thousandth.

25. $\frac{8}{9} \approx$ __0.889__

26. $\frac{5}{6} \approx$ __0.833__

27. $\frac{3}{11} \approx$ __0.273__

28. $\frac{4}{7} \approx$ __0.571__

Multiply.

1. 7.2
    × 9

2. 5.63
    × 8

3. 49
    × 1.7

4. 8.02
    × 36

5. 6.47
    × 0.5

6. 3.8
    × 0.29

7. 2.516
    × 4.8

8. 7.35
    × 9.7

9. 0.006
    × .18

10. 0.09
    × 0.05

11. 0.007
    × 0.08

12. 4.02
    × 0.006

Divide.

13. $5\overline{)23.5}$

14. $2\overline{)1.38}$

15. $9\overline{)0.468}$

16. $6\overline{)51.6}$

17. $0.04\overline{)2.858}$

18. $0.7\overline{)0.042}$

19. $1.2\overline{)7.68}$

20. $5.4\overline{)4.968}$

21. $9.5\overline{)57}$

22. $0.6\overline{)0.3}$

23. $0.047\overline{)9.4}$

24. $8.8\overline{)704}$

Write each fraction as a decimal rounded to the nearest thousandth.

25. $\frac{8}{9} \approx$ _____

26. $\frac{5}{6} \approx$ _____

27. $\frac{3}{11} \approx$ _____

28. $\frac{4}{7} \approx$ _____

Chapter 10
Alternate Test

for page 388

Items   **Objectives**

1–3,   To add and subtract units
7–8   of time (see pages
     211–212)

4–6, 9   To add and subtract units
     of customary length
     (see pages 215–216)

10–15   To add and subtract units
     of customary capacity and
     weight (see pages
     217–218)

16–30   To change metric units to
     equivalent units (see pages
     219–222)

31–36   To find the change in
     temperature in Fahrenheit
     and Celsius (see pages
     223–224)

Add or subtract.

1.   7 h 36 min
   + 2 h 15 min
   9 h 51 min

2.   43 min 38 s
   + 9 min 27 s
   53 min 5 s

3.   8 h 17 min
   − 3 h 58 min
   4 h 19 min

4.   10 ft 8 in.
   − 5 ft 9 in.
   4 ft 11 in.

5.   7 yd 10 in.
   + 3 yd 27 in.
   11 yd 1 in.

6.   9 ft 11 in.
   + 2 ft 9 in.
   12 ft 8 in.

7.   16 min 52 s
   − 8 min 59 s
   7 min 53 s

8.   5 h 4 min
   − 2 h 20 min
   2 h 44 min

9.   11 yd 15 in.
   + 3 yd 24 in.
   15 yd 3 in.

Add or subtract.

10.   9 gal 2 qt
   + 6 gal 3 qt
   16 gal 1 qt

11.   3 pt 7 fl oz
   + 9 pt 5 fl oz
   12 pt 12 fl oz

12.   15 lb 9 oz
   − 6 lb 15 oz
   8 lb 10 oz

13.   9 lb 6 oz
   + 4 lb 15 oz
   14 lb 5 oz

14.   2 gal 1 qt
   − 1 gal 2 qt
   3 qt

15.   17 pt 1 fl oz
   − 5 pt 7 fl oz
   11 pt 10 fl oz

Find each missing number.

16. 84 mm = __8.4__ cm

17. 0.75 dm = __75__ mm

18. 537 km = __537,000__ m

19. 7 m = __700__ cm

20. 7.5 km = __7,500__ m

21. 55 mm = __0.055__ m

22. 64,500 mm = __0.0645__ km

23. 45 t = __45,000__ kg

24. 0.01 g = __10__ mg

25. 52 kg = __52,000,000__ mg

26. 68 cm = __6.8__ dm

27. 24 kg = __24,000__ g

28. 6.55 km = __6,550__ m

29. 0.01 t = __10,000__ g

30. 0.0011 g = __1.1__ mg

Find each temperature change.

31. 10°C to ⁻2°C __12°__

32. 35°F to 47°F __12°__

33. ⁻5°C to 4°C __9°__

34. 69°F to 54°F __15°__

35. ⁻1°C to ⁻5°C __4°__

36. 28°F to 19°F __9°__

Add or subtract.

1. 7 h 36 min
    + 2 h 15 min

2. 43 min 38 s
    + 9 min 27 s

3. 8 h 17 min
    − 3 h 58 min

4. 10 ft 8 in.
    − 5 ft 9 in.

5. 7 yd 10 in.
    + 3 yd 27 in.

6. 9 ft 11 in.
    + 2 ft 9 in.

7. 16 min 52 s
    − 8 min 59 s

8. 5 h 4 min
    − 2 h 20 min

9. 11 yd 15 in.
    + 3 yd 24 in.

Add or subtract.

10. 9 gal 2 qt
     + 6 gal 3 qt

11. 3 pt 7 fl oz
     + 9 pt 5 fl oz

12. 15 lb 9 oz
     − 6 lb 15 oz

13. 9 lb 6 oz
     + 4 lb 15 oz

14. 2 gal 1 qt
     − 1 gal 2 qt

15. 17 pt 1 fl oz
     − 5 pt 7 fl oz

Find each missing number.

16. 84 mm = _____ cm

17. 0.75 dm = _____ mm

18. 537 km = _____ m

19. 7 m = _____ cm

20. 7.5 km = _____ m

21. 55 mm = _____ m

22. 64,500 mm = _____ km

23. 45 t = _____ kg

24. 0.01 g = _____ mg

25. 52 kg = _____ mg

26. 68 cm = _____ dm

27. 24 kg = _____ g

28. 6.55 km = _____ m

29. 0.01 t = _____ g

30. 0.0011 g = _____ mg

Find each temperature change.

31. 10°C to ⁻2°C _____

32. 35°F to 47°F _____

33. ⁻5°C to 4°C _____

34. 69°F to 54°F _____

35. ⁻1°C to ⁻5°C _____

36. 28°F to 19°F _____

Chapter 11
Alternate Test

for page 390

Items	Objectives
1–3	To identify lines, line segments, and rays (see pages 229–230)
4–6	To identify and name right, obtuse, and acute angles (see pages 231–232)
	To measure angles with a protractor (see pages 233–234)
7	To identify and name perpendicular lines (see pages 235–236)
8	To identify parts of circles (see pages 237–238)
9	To copy an angle with a compass and a straightedge (see pages 241–242)
10	To bisect a line segment with a compass and a straightedge (see pages 239–240)
11	To identify a quadrilateral and draw its diagonals (see pages 247–248)
12	To measure an angle with a protractor (see pages 233–234)
13	To identify congruent polygons and their parts (see pages 249–250)
14	To identify and draw diagonals of a polygon (see pages 243–244)

Name each figure.

1. 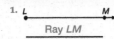 Ray *LM*

2. 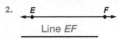 Line *EF*

3.  Line segment *NO*

Use a protractor to measure the angle. Classify the angle according to its size.

4. 60° / acute

5. 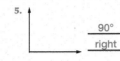 90° / right

6. 120° / obtuse

Complete the statements.

7.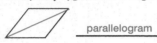
$\overline{FE}$ is perpendicular to ___$\overline{FG}$___ .

8.
The radius is ___$\overline{OY}$___ .

9. Copy angle *R* on $\overrightarrow{MN}$.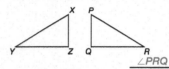

10. Bisect $\overline{AB}$.

11. Identify the polygon. Draw a diagonal.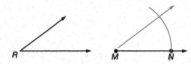
parallelogram

12. Find m∠*MNO*.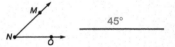
45°

13. The triangles are congruent. Name the angle congruent to ∠XYZ.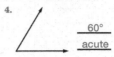
∠PRQ

14. Draw all possible diagonals.

Name each figure.

1.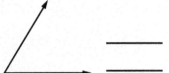
   L _____ M
   _____

2. E _____ F
   _____

3. N _____ O
   _____

Use a protractor to measure the angle. Classify the angle according to its size.

4.
   _____

5.
   _____

6.
   _____

Complete the statements.

7.

   $\overline{FE}$ is perpendicular to _____.

8.
   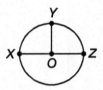
   The radius is _____.

9. Copy angle R on $\overrightarrow{MN}$.

   R _____
   M _____ N

10. Bisect $\overline{AB}$.

    A •———————• B

11. Identify the polygon. Draw a diagonal.

    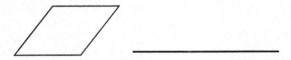
    _____

12. Find m∠MNO.

    _____

13. The triangles are congruent. Name the angle congruent to ∠XYZ.

    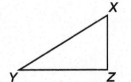
    _____

14. Draw all possible diagonals.

    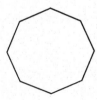

Chapter 12
Alternate Test

for page 392

Items **Objectives**

1–3 To find the perimeter or circumference of a figure (see pages 259–262)

4–6 To find the area of rectangles and squares (see pages 263–264)

7–9 To find the area of parallelograms (see pages 267–268)

10–12 To find the area of triangles (see pages 269–270)

13–15 To find the area of circles (see pages 271–272)

16–17 To find the surface area of rectangular prisms (see pages 277–278)

18 To find the volume of rectangular prisms (see pages 275–276)

Find the perimeter or circumference.

1.  2.5 cm
6.1 cm

$P =$ ___17.2 cm___

2. 8 ft

$C =$ ___25.12 ft___

3. $l = 3.5$ mm
$w = 9.7$ mm

$P =$ ___26.4 mm___

Find the area of each rectangle or square.

4. 3 cm
7 cm

$A =$ ___21 cm²___

5. 17 mm

$A =$ ___289 mm²___

6. $s = 9.9$ mm

$A =$ ___98.01 mm²___

Find the area of each parallelogram.

7. 4 cm
5 cm

$A =$ ___20 cm²___

8. 6.2 cm
6.2 cm

$A =$ ___38.44 cm²___

9. $b = 25.6$ cm
$h = 8.9$ cm

$A =$ ___227.84 cm²___

Find the area of each triangle.

10.  35 mm
45 mm

$A =$ ___787.5 mm²___

11. 0.6 cm
1.5 cm

$A =$ ___0.45 cm²___

12. $b = 9$ cm
$h = 16$ cm

$A =$ ___72 cm²___

Find the area of each circle.

13. $r = 4$ cm

$A =$ ___50.24 cm²___

14. $d = 10$ cm

$A =$ ___78.5 cm²___

15. $r = 6.6$ mm

$A =$ ___136.7784 mm²___

Find the surface area of each box.

16. 4 ft
3 ft   3 ft

Surface Area = ___66 ft²___

17. 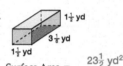 1½ yd
3½ yd
1½ yd

Surface Area = ___23½ yd²___

Find the volume.

18. 2 cm
4 cm
2 cm

$V =$ ___16 cm³___

Chapter 12 • Alternate Test   **392**

Find the perimeter or circumference.

1.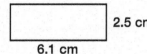

2.5 cm

6.1 cm

$P =$ _____

2.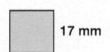

8 ft

$C =$ _____

3. $l = 3.5$ mm
$w = 9.7$ mm

$P =$ _____

Find the area of each rectangle or square.

4.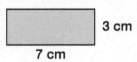

3 cm

7 cm

$A =$ _____

5.

17 mm

$A =$ _____

6. $s = 9.9$ mm

$A =$ _____

Find the area of each parallelogram.

7.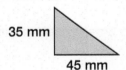

4 cm

5 cm

$A =$ _____

8.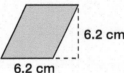

6.2 cm

6.2 cm

$A =$ _____

9. $b = 25.6$ cm
$h = 8.9$ cm

$A =$ _____

Find the area of each triangle.

10.

35 mm

45 mm

$A =$ _____

11.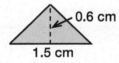

0.6 cm

1.5 cm

$A =$ _____

12. $b = 9$ cm
$h = 16$ cm

$A =$ _____

Find the area of each circle.

13.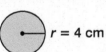

$r = 4$ cm

$A =$ _____

14.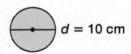

$d = 10$ cm

$A =$ _____

15. $r = 6.6$ mm

$A =$ _____

Find the surface area of each box.

16.

4 ft

3 ft

3 ft

Surface Area = _____

17.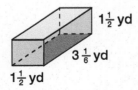

$1\frac{1}{2}$ yd

$3\frac{1}{6}$ yd

$1\frac{1}{2}$ yd

Surface Area = _____

Find the volume.

18.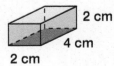

2 cm

4 cm

2 cm

$V =$ _____

Chapter 13
Alternate Test

for page 394

Items	Objectives
1–4	To write the ratio of two quantities (see pages 283–284)
5–8	To solve proportions (see pages 287–288)
9	To find the missing measure of two similar polygons (see pages 289–290)
10	To use scale drawing to find actual length (see pages 291–292)
11–12	To change fractions to percents (see pages 295–296)
13–14	To change percents to fractions (see pages 295–296)
15–18	To find a percent of a number (see pages 297–298)
19–20	To find percent increases or decreases (see pages 299–300)
21–22	To find the total amount of money after one year at a given rate of simple interest (see pages 301–302)

Name _____

Write each ratio as a fraction.

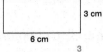
3 cm
6 cm

1. daisies to tulips $\frac{2}{3}$

2. tulips to flowers $\frac{3}{5}$

3. width to length $\frac{3}{6}$

4. length to perimeter $\frac{6}{18}$

Solve each proportion.

5. $\frac{6}{80} = \frac{n}{160}$

$n = \underline{12}$

6. $\frac{6}{14} = \frac{3}{n}$

$n = \underline{7}$

7. $\frac{3}{8} = \frac{18}{n}$

$n = \underline{48}$

8. $\frac{28}{n} = \frac{4}{5}$

$n = \underline{35}$

9. The figures are similar. Solve for n.

12 mm
n
16 mm
24 mm

$n = \underline{18 \text{ mm}}$

10. Use the scale to find the actual distance.

4 cm

D ——— E

Scale: 1 cm = 9 m $\underline{36 \text{ m}}$

Write each fraction as a percent.

11. $\frac{7}{25} = \underline{28\%}$

12. $\frac{3}{5} = \underline{60\%}$

Write the percent as a fraction.

13. $85\% = \underline{\frac{17}{20}}$

14. $58\% = \underline{\frac{29}{50}}$

Find the percent of each number.

15. 60% of 82 = $\underline{49.2}$

16. 20% of 95 = $\underline{19}$

17. $4\frac{1}{2}\%$ of $82 = $\underline{\$3.69}$

18. 7.4% of $820 = $\underline{\$60.68}$

Find the percent increase or decrease.

19. price decrease from $80 to $60
25%

20. rise in temperature from 32° to 48°
50%

Write the total amount after 1 year.

21. $2,500 at 3% $2,575

22. $4,000 at 2.5% $4,100

Chapter 13 • Alternate Test

394

Name _____

Write each ratio as a fraction.

1. daisies to tulips _____

2. tulips to flowers _____

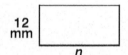

3 cm

6 cm

3. width to length _____

4. length to perimeter _____

Solve each proportion.

5. $\frac{6}{80} = \frac{n}{160}$

6. $\frac{6}{14} = \frac{3}{n}$

7. $\frac{3}{8} = \frac{18}{n}$

8. $\frac{28}{n} = \frac{4}{5}$

$n = $ _____

$n = $ _____

$n = $ _____

$n = $ _____

9. The figures are similar. Solve for n.

12 mm

n

16 mm

24 mm

$n = $ _____

10. Use the scale to find the actual distance.

4 cm

D _____ E

Scale: 1 cm = 9 m _____

Write each fraction as a percent.

11. $\frac{7}{25} = $ _____

12. $\frac{3}{5} = $ _____

Write the percent as a fraction.

13. 85% = _____

14. 58% = _____

Find the percent of each number.

15. 60% of 82 = _____

16. 20% of 95 = _____

17. $4\frac{1}{2}$% of \$82 = _____

18. 7.4% of \$820 = _____

Find the percent increase or decrease.

19. price decrease from \$80 to \$60

20. rise in temperature from 32° to 48°

Write the total amount after 1 year.

21. \$2,500 at 3%

22. \$4,000 at 2.5%

for page 396

Items   Objectives

1–2   To interpret a double-bar graph (see pages 311–312)

3   To construct a double-line graph (see pages 313–314)

4   To interpret a circle graph (see pages 315–316)

5–6   To find the probability of a single event (see pages 319–320)

7–8   To find the probability of compound events (see pages 323–326)

Use the double-bar graph to solve each problem.

1. What type of movie appeals to men and women equally?

   <u>action</u>

2. How many more men than women prefer science-fiction movies?

   <u>2,000</u>

Movie Viewing

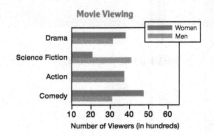

3. Use the following data to draw a double-line graph. Make a key.

Average Monthly Temperature		
Month	**Chicago, IL**	**Richmond, VA**
Jan	20°F	40°F
Feb	25°F	40°F
Mar	40°F	50°F
Apr	50°F	60°F
May	60°F	65°F
June	70°F	70°F

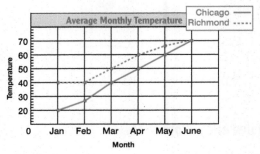

4. Use the circle graph to solve this problem. What percent of people take the train to work?

   <u>20%</u>

Transportation to Work

Use the spinners to find each probability.

5. $P(3)$ $\frac{1}{3}$

6. $P(6)$ $\frac{1}{4}$

7. $P(1, 5)$ $\frac{1}{12}$

8. $P(\text{even, odd})$ $\frac{1}{6}$

Chapter 14 • Alternate Test

396

Use the double-bar graph to solve each problem.

1. What type of movie appeals to men and women equally?

   _____

2. How many more men than women prefer science-fiction movies?

   _____

Movie Viewing

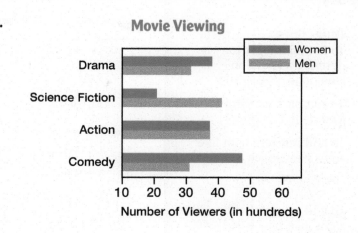

3. Use the following data to draw a double-line graph. Make a key.

Average Monthly Temperature		
Month	**Chicago, IL**	**Richmond, VA**
Jan	20°F	40°F
Feb	25°F	40°F
Mar	40°F	50°F
Apr	50°F	60°F
May	60°F	65°F
June	70°F	70°F

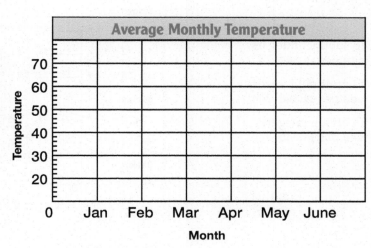

4. Use the circle graph to solve this problem. What percent of people take the train to work?

   _____

Transportation to Work

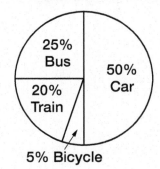

Use the spinners to find each probability.

5. P(3)

6. P(6)

7. P(1, 5)

8. P(even, odd)

Chapter 15
Alternate Test

for page 398

Items	Objectives
1–3	To use an integer to describe a situation (see pages 333–334)
4–9	To compare integers (see pages 335–336)
10–11, 13, 17–19, 22, 25	To add integers with like signs (see pages 337–338)
12, 14–16, 20–21, 23–24	To add integers with unlike signs (see pages 339–340)
26–37	To subtract integers (see pages 341–342)
38–45	To multiply integers (see pages 343–344)
46–53	To divide integers (see pages 345–346)

Name _____

Write an integer to describe each situation.

1. a withdrawal of $25 $^-25$
2. a gain of 10 yards $^+10$
3. 200 feet below sea level $^-200$

Use the number line below to compare the integers. Write >, <, or = in each circle.

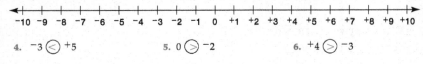

4. $^-3$ ⓒ $^+5$
5. 0 Ⓖ $^-2$
6. $^+4$ Ⓖ $^-3$
7. $^+6$ Ⓖ 0
8. $^-7$ Ⓖ $^-9$
9. $^-1$ Ⓔ $^-4 + ^+3$

Add.

10. $^-4 + ^-8 = $ $^-12$
11. $^+5 + ^+7 = $ $^+12$
12. $^-6 + ^+2 = $ $^-4$
13. $^-9 + ^-1 = $ $^-10$
14. $^-7 + ^+4 = $ $^-3$
15. $^+6 + 0 = $ $^+6$
16. $^+8 + ^-3 = $ $^+5$
17. $^+5 + ^+2 = $ $^+7$
18. $^-3 + ^-1 = $ $^-4$
19. $^-9 + ^-4 = $ $^-13$
20. $^-2 + 0 = $ $^-2$
21. $^+6 + ^-7 = $ $^-1$
22. $^+8 + ^+3 = $ $^+11$
23. $^+5 + ^-4 = $ $^+1$
24. $^-2 + ^+9 = $ $^+7$
25. $^-6 + ^-3 = $ $^-9$

Subtract.

26. $^-5 - ^-9 = $ $^+4$
27. $^+7 - ^+6 = $ $^+1$
28. $^-8 - ^+2 = $ $^-10$
29. $^+4 - ^-3 = $ $^+7$
30. $^-2 - ^+6 = $ $^-8$
31. $^-4 - ^-1 = $ $^-3$
32. $^+3 - ^+5 = $ $^-2$
33. $0 - ^-9 = $ $^+9$
34. $^-8 - ^+4 = $ $^-12$
35. $^-2 - ^-9 = $ $^+7$
36. $^+7 - ^-9 = $ $^+16$
37. $^+6 - ^-4 = $ $^+10$

Multiply.

38. $^-6 \times ^-5 = $ $^+30$
39. $^+2 \times ^-9 = $ $^-18$
40. $^-8 \times ^-4 = $ $^+32$
41. $^+7 \times ^+3 = $ $^+21$
42. $^-4 \times ^+9 = $ $^-36$
43. $^+2 \times ^-5 = $ $^-10$
44. $^+8 \times ^+6 = $ $^+48$
45. $^-7 \times ^-9 = $ $^+63$

Divide.

46. $^+15 \div ^-3 = $ $^-5$
47. $^+64 \div ^-8 = $ $^-8$
48. $^-8 \div ^+2 = $ $^-4$
49. $^+42 \div ^+7 = $ $^+6$
50. $^-35 \div ^+7 = $ $^-5$
51. $^+81 \div ^+9 = $ $^+9$
52. $^+9 \div ^-3 = $ $^-3$
53. $^-12 \div ^-6 = $ $^+2$

Name _____

Write an integer to describe each situation.

1. a withdrawal of $25 _____   2. a gain of 10 yards _____   3. 200 feet below sea level _____

Use the number line below to compare the integers. Write >, <, or = in each circle.

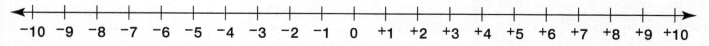

$^-10$ $^-9$ $^-8$ $^-7$ $^-6$ $^-5$ $^-4$ $^-3$ $^-2$ $^-1$ 0 $^+1$ $^+2$ $^+3$ $^+4$ $^+5$ $^+6$ $^+7$ $^+8$ $^+9$ $^+10$

4. $^-3$ ◯ $^+5$ 5. 0 ◯ $^-2$ 6. $^+4$ ◯ $^-3$

7. $^+6$ ◯ 0 8. $^-7$ ◯ $^-9$ 9. $^-1$ ◯ $^-4 + {}^+3$

Add.

10. $^-4 + {}^-8 =$ _____ 11. $^+5 + {}^+7 =$ _____ 12. $^-6 + {}^+2 =$ _____ 13. $^-9 + {}^-1 =$ _____

14. $^-7 + {}^+4 =$ _____ 15. $^+6 + 0 =$ _____ 16. $^+8 + {}^-3 =$ _____ 17. $^+5 + {}^+2 =$ _____

18. $^-3 + {}^-1 =$ _____ 19. $^-9 + {}^-4 =$ _____ 20. $^-2 + 0 =$ _____ 21. $^+6 + {}^-7 =$ _____

22. $^+8 + {}^+3 =$ _____ 23. $^+5 + {}^-4 =$ _____ 24. $^-2 + {}^+9 =$ _____ 25. $^-6 + {}^-3 =$ _____

Subtract.

26. $^-5 - {}^-9 =$ _____ 27. $^+7 - {}^+6 =$ _____ 28. $^-8 - {}^+2 =$ _____ 29. $^+4 - {}^-3 =$ _____

30. $^-2 - {}^+6 =$ _____ 31. $^-4 - {}^-1 =$ _____ 32. $^+3 - {}^+5 =$ _____ 33. $0 - {}^-9 =$ _____

34. $^-8 - {}^+4 =$ _____ 35. $^-2 - {}^-9 =$ _____ 36. $^+7 - {}^-9 =$ _____ 37. $^+6 - {}^-4 =$ _____

Multiply.

38. $^-6 \times {}^-5 =$ _____ 39. $^+2 \times {}^-9 =$ _____ 40. $^-8 \times {}^-4 =$ _____ 41. $^+7 \times {}^+3 =$ _____

42. $^-4 \times {}^+9 =$ _____ 43. $^+2 \times {}^-5 =$ _____ 44. $^+8 \times {}^+6 =$ _____ 45. $^-7 \times {}^-9 =$ _____

Divide.

46. $^+15 \div {}^-3 =$ _____ 47. $^+64 \div {}^-8 =$ _____ 48. $^-8 \div {}^+2 =$ _____ 49. $^+42 \div {}^+7 =$ _____

50. $^-35 \div {}^+7 =$ _____ 51. $^+81 \div {}^+9 =$ _____ 52. $^+9 \div {}^-3 =$ _____ 53. $^-12 \div {}^-6 =$ _____

for page 400

Items **Objectives**

Name _____

Evaluate each algebraic expression for $a = 3$ and $b = {}^-4$.

1. $6b$ ___${}^-24$___

2. ${}^-a + b$ ___${}^-7$___

3. $2b - a$ ___${}^-11$___

4. $2a + b + 10$ ___12___

5. $b + 3a$ ___5___

6. $8 - a - 3b$ ___17___

Solve each equation. Check each solution.

7. $5t = 20$
   $t = $ ___4___

8. $3z + 2 = 11$
   $z = $ ___3___

9. $m - 7 = 7$
   $m = $ ___14___

10. $y + 10 = 30$
    $y = $ ___20___

11. $\frac{c}{-4} = 8$
    $c = $ ___${}^-32$___

12. $9x = {}^-36$
    $x = $ ___${}^-4$___

13. $\frac{a}{2} = 9$
    $a = $ ___18___

14. $11 + n = 25$
    $n = $ ___14___

15. ${}^-3t - 7 = 14$
    $t = $ ___${}^-7$___

16. ${}^-8x = {}^-64$
    $x = $ ___8___

17. $12 - 2y = 32$
    $y = $ ___${}^-10$___

18. $c - 15 = {}^-10$
    $c = $ ___5___

Follow the pattern to complete the table. Write and graph the rule.

19.

x	y
5	9
3	7
1	5
0	4
${}^-1$	3

Rule: $y = x + 4$

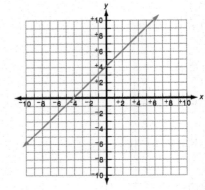

Evaluate each algebraic expression for $a = 3$ and $b = {}^-4$.

1. $6b$ _____

2. ${}^-a + b$ _____

3. $2b - a$ _____

4. $2a + b + 10$ _____

5. $b + 3a$ _____

6. $8 - a - 3b$ _____

Solve each equation. Check each solution.

7. $5t = 20$

   $t =$ _____

8. $3z + 2 = 11$

   $z =$ _____

9. $m - 7 = 7$

   $m =$ _____

10. $y + 10 = 30$

    $y =$ _____

11. $\frac{c}{-4} = 8$

    $c =$ _____

12. $9x = {}^-36$

    $x =$ _____

13. $\frac{a}{2} = 9$

    $a =$ _____

14. $11 + n = 25$

    $n =$ _____

15. ${}^-3t - 7 = 14$

    $t =$ _____

16. ${}^-8x = {}^-64$

    $x =$ _____

17. $12 - 2y = 32$

    $y =$ _____

18. $c - 15 = {}^-10$

    $c =$ _____

Follow the pattern to complete the table. Write and graph the rule.

19.

x	y
5	9
3	7
1	5
0	
⁻1	

Rule: _____

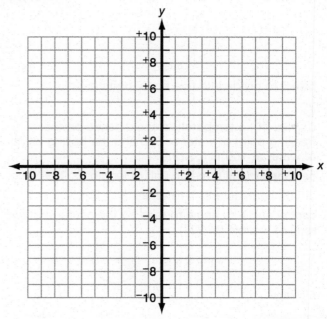

Glossary

A

acute angle an angle that measures less than 90° (p. 231)

addend a number that is added to another number (p. 1)

In 3 + 4 = 7, 3 and 4 are both addends.

algebraic expression a mathematical phrase that consists of variables, numbers, and operations (p. 351)

angle the figure formed by two rays with a common endpoint called the vertex (p. 231)

vertex

arc a part of the circumference of a circle (p. 238)

area the measure of a surface surrounded by a boundary (p. 263)

The shaded part of the square is its area.

Associative Property When the grouping of three or more addends or factors is changed, the sum or product remains the same. (pp. 1, 5)

(2 + 5) + 1 = 2 + (5 + 1)
or
(5 × 3) × 2 = 5 × (3 × 2)

average (mean) the number obtained by adding two or more quantities and dividing by the number of quantities added (pp. 71, 317)

B

bar graph a representation of numerical facts using lengths of bars to show information (p. 311)

base (of a geometric figure) a side of a plane figure or face of a solid figure (p. 273)
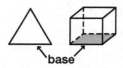
base

bisect to divide into two equal parts (p. 239)

C

Celsius (°C) a temperature scale based on the freezing point of water at 0°C and the boiling point of water at 100°C (p. 223)

centimeter (cm) a metric unit of length (p. 219)

100 centimeters = 1 meter

central angle an angle formed by the center and two radii of a circle (p. 237)
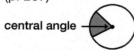
central angle

chord a line segment connecting two points on a circle (p. 237)
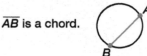
$\overline{AB}$ is a chord.

circle a plane figure bounded by a curved line made up of points all the same distance from the center (p. 237)

circle graph a representation of numerical facts using parts or sections of a circle to show information (p. 315)

circumference the distance around a circle (p. 261)

Commutative Property The order of addends or factors does not change the sum or product. (pp. 1, 5)

5 + 7 = 7 + 5 or 3 × 4 = 4 × 3

composite number a whole number greater than 1 that has more than two factors (p. 91)

cone a solid figure with a circle for its base and a curved surface that forms a point (p. 273)

congruent a word that describes figures, sides, or angles having the same size and shape (pp. 229, 249)

coordinates a pair of numbers that shows the position of a point on the plane (p. 359)

cross products the product of the numerator of the first ratio and the denominator of the second ratio as well as the product of the denominator of the first ratio and the numerator of the second ratio; in a proportion, these products are equal. (p. 109)

cube a solid figure with six equal, square sides (p. 273)

customary units standard measures of length, weight, volume, and capacity (pp. 215, 217)

Inches, miles, pounds, cubic feet, and quarts are examples of customary units.

cylinder a solid figure with two congruent circular bases and a curved rectangular region (p. 273)

D

data numerical information collected for a purpose (p. 309)

decimal a fractional part that uses place value and a decimal point to show tenths, hundredths, and so on (p. 159)

0.6 is the decimal equivalent of the fraction $\frac{3}{5}$.

denominator the number below the line in a fraction (p. 95)

In $\frac{3}{5}$, 5 is the denominator.

dependent events two events in which the outcome of the first event affects the outcome of the second event (p. 325)

diagonal a segment that connects two vertices of a polygon but is not a side (p. 243)

diameter a chord passing through the center of a circle (pp. 237, 261)

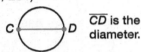

difference the answer in a subtraction problem (p. 3)

discount an amount subtracted from the original price (p. 305)

dividend the number that is being divided in a division problem (p. 7)

In $42 \div 7 = 6$, 42 is the dividend.

divisor the number that is being divided into the dividend (p. 7)

In $42 \div 7 = 6$, 7 is the divisor.

E

edge a segment that is the side of a face on a solid figure (p. 273)

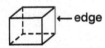

equation a statement that two quantities are equal (p. 353)

equilateral triangle a triangle with three equal sides (p. 245)

equivalent fractions fractions that name the same number (p. 97)

$\frac{3}{4}$ and $\frac{9}{12}$ are equivalent fractions.

evaluate to find the value of a numerical or an algebraic expression (p. 351)

even number a whole number that is divisible by 2 (p. 88)

2, 4, 6, 8, 10, 12, . . . are even numbers.

expanded form a number written as the sum of its place values (p. 19)

426 is $400 + 20 + 6$ or $(4 \times 100) + (2 \times 10) + (6 \times 1)$.

exponent a raised number that tells how many times the base number is used as a factor (p. 9)

$$4^3 = 4 \times 4 \times 4$$

F

face a plane figure making up part of a solid figure (p. 273)

factor a number to be multiplied (p. 5)

In $2 \times 3 = 6$, both 2 and 3 are factors.

factor tree a diagram used to show the prime factors of a number (p. 93)

Fahrenheit (°F) a temperature scale based on the freezing point of water at 32°F and the boiling point of water at 212°F (p. 223)

formula a rule expressed using symbols (p. 133)

fraction a number that represents part of an object or part of a set (p. 95)

G

graphing drawing a picture of relationships among numbers and quantities (pp. 309–317)

greater than a comparison of two numbers with the number of greater value written first (pp. 21, 105)

$10 > 5$, $\frac{7}{8} > \frac{3}{4}$

greatest common factor (GCF) the largest number that is a common factor of two or more numbers (p. 89)

$12 = 3 \times 4$ $15 = 3 \times 5$

3 is the GCF of 12 and 15.

H

height (of a rectangular prism) a side of the prism (p. 275)

height (of a triangle or a parallelogram) the shortest distance between a vertex and the opposite side or the line containing the opposite side (pp. 267, 269)

hexagon a plane figure with six straight sides and six angles (p. 243)

histogram a bar graph that shows the number of times data occurs between continuous intervals (p. 317)

I

Identity Property Any number added to zero is that number; any number multiplied by one is that number. (pp. 1, 5)

$7 + 0 = 7$ or $7 \times 1 = 7$

independent events two events that have no effect on each other (p. 325)

integer a number such as $^-5$, 0, and $^+5$ (p. 333)

intersect to meet and cross over at a point (p. 235)

Line *AB* intersects line *CD* at point *P*.

inverse operations operations that undo each other (pp. 3, 7, 353)

isosceles triangle a triangle with two equal sides (p. 245)
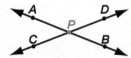

K

kilogram (kg) a metric unit of weight (p. 221)

1 kilogram = 1,000 grams

kilometer (km) a metric unit of length (p. 219)

1 kilometer = 1,000 meters

L

least common denominator (LCD) the least common multiple (LCM) of the denominators of two or more fractions (p. 105)

The LCD of $\frac{1}{2}$ and $\frac{1}{3}$ is 6.

least common multiple (LCM) the smallest number that is a common multiple of two or more numbers (p. 87)

The LCM of 4 and 6 is 12.

less than a comparison of two numbers with the number of lesser value written first (pp. 21, 105)

$3 < 10, \frac{1}{8} < \frac{7}{8}$

line a set of points whose straight path extends indefinitely in opposite directions (p. 229)

line *CD* or $\overleftrightarrow{CD}$

line graph a representation of numerical facts using points and lines on a grid to show information (p. 313)

line segment a part of a line having two endpoints (p. 229)

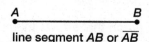
line segment *AB* or $\overline{AB}$

liter (L) a basic metric unit of a liquid measure (p. 221)

1 liter = 1,000 milliliters

M

mean a number representing the average of a group of numbers (p. 317)

median the middle number in a series of numbers that have been arranged in order from least to greatest (p. 317)

$$5, 9, 17, 31, 50$$
$$\uparrow$$
$$\text{median}$$

meter (m) a basic metric unit of length (p. 219)

1 meter = 100 centimeters

metric units measures of length, weight, volume, and capacity based on the decimal system (pp. 219, 221)

Meters, grams, and liters are basic metric units.

midpoint the point halfway between the endpoints of a line segment (p. 239)

milliliter (mL) a metric unit of liquid measure (p. 221)

1 milliliter = $\frac{1}{1000}$ liter

minuend a number or quantity from which another is subtracted (p. 3)

In $18 - 5 = 13$, 18 is the minuend.

mixed number a fractional number greater than 1 that is written as a whole number and a fraction (p. 101)

$5\frac{2}{3}$ is a mixed number.

multiple the product of any given number and a whole number (p. 87)

$10 \times 3 = 30, 10 \times 5 = 50$

30 and 50 are multiples of 10.

N

numerator the number above the line in a fraction (p. 95)

In $\frac{3}{5}$, 3 is the numerator.

O

obtuse angle an angle that measures between 90° and 180° (p. 231)
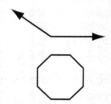

octagon a plane figure with eight sides and eight angles (p. 243)

odd number a whole number that is not divisible by 2 (p. 88)

3, 5, 7, 9, 11, 13, . . . are odd numbers.

opposites numbers that are the same distance away from zero on the number line (p. 333)

Order of Operations the order in which we do operations (p. 11)

Evaluate the powers.

Do the operations inside parentheses.

Multiply and divide in order from left to right.

Add and subtract in order from left to right.

ordered pair two numbers that define one point on a grid; the first number names the distance left or right of the origin, and the second names the distance up or down. (pp. 251, 359)

origin the point on a grid represented by the ordered pair (0, 0) (p. 359)

P

parallel lines lines in the same plane that do not intersect (p. 235)

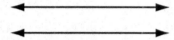

parallelogram a quadrilateral having two pairs of opposite, congruent, parallel sides (p. 247)

pentagon a plane figure with five straight sides and five angles (p. 243)

percent a word meaning "hundredths" (p. 293)

37 percent is written 37% and means 0.37 or $\frac{37}{100}$.

perimeter the distance around a shape that is the sum of the lengths of all of its sides (p. 259)

perpendicular lines lines that form right angles where they intersect (p. 235)

place value The value of a digit depends upon its position in a numeral. (p. 19)

plane a flat surface having infinite length and width (p. 229)

point a location in space (p. 229)

•*B*   point *B*

polygon a closed plane figure having three or more angles or sides (p. 243)

polyhedron a solid figure formed from polygons (p. 273)

prime factor a factor that is a prime number (p. 93)

The prime factors for 60 are 2, 3, and 5.

prime number a counting number greater than 1 whose only factors are itself and 1 (p. 91)

$1 \times 17 = 17$ $1 \times 3 = 3$

17 and 3 are prime numbers.

probability a number that tells how likely it is that a certain event will happen (p. 319)

product the answer to a multiplication problem (pp. 5, 41)

product of primes the prime factors that equal a composite number when multiplied (p. 93)

proportion an equation showing that two ratios are equal (p. 287)

If $\frac{a}{b} = \frac{c}{d}$, then $a \times d = b \times c$.

Q

quadrilateral a plane figure with four straight lines and four angles (p. 247)

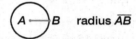

quadrilaterals

quotient the answer to a division problem (p. 7)

R

radius a segment whose endpoints are the center of a circle and a point on the circle (p. 237)

   radius $\overline{AB}$

range the difference between the largest and smallest numbers in a set of data (p. 317)

ratio a comparison of two quantities (p. 283)

The ratio of 3 to 4 can be written $\frac{3}{4}$.

ray a part of a line having one endpoint (p. 229)

ray *EF* or $\overrightarrow{EF}$

E   *F*

reciprocal When the product of two numbers is 1, they are called reciprocals of each other. (p. 143)

rectangle a parallelogram with right angles and opposite sides equal in length (p. 247)

reflection flipping a plane figure (p. 253)

regular polygon a polygon with all sides of equal length and all angles of equal measure (p. 243)

remainder the number left over in division (p. 63)

```
    16 R4
6)100
  − 6
    40
  − 36
     4
```

In the example above, 4 is the remainder.

Glossary

repeating decimal a decimal fraction in which one or more digits are repeated indefinitely (p. 207)

$$\frac{1}{3} = 0.3333\ldots = 0.\overline{3}$$

The bar shows that the digit is repeated.

rhombus a parallelogram with all sides congruent (p. 247)

right angle an angle that measures exactly 90° (p. 231)

In this square, all four angles are right angles.

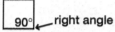 90° → right angle

rotation turning a plane figure (p. 253)

rounding estimating a number's value by raising or lowering any of its place values (p. 23)

S

sale price the cost of an item that has been marked down (p. 305)

sample space possible outcomes written as a set of ordered pairs (p. 321)

scale drawing a representation of an actual object using proportional measurements (p. 291)

scalene triangle a triangle with no sides the same length and no angles the same measure (p. 245)

similar figures plane figures that have the same shape but not necessarily the same size (p. 289)

Figures *A* and *B* are similar.

simplest form a fraction or mixed number whose numerator and denominator cannot be divided by any common factor other than 1 (p. 99)

simplest form   simplest form
   ↓   ↓
$$\frac{12}{36} = \frac{1}{3} \qquad \frac{34}{6} = 5\frac{4}{6} = 5\frac{2}{3}$$

simplify to rename a fraction in simplest form (p. 99)

solid figure a figure that is in more than one plane (p. 273)

cube   pyramid   cylinder

solve to find the value of a variable in an equation (p. 353)

sphere a solid, round figure having a surface equally distant from its center at all points (p. 273)

square a rectangle with four sides equal in length (p. 247)

standard form a number written using the symbols 0 through 9 in place-value form (p. 19)

4,036 is in standard form.

stem-and-leaf plot a graph that groups together all data with the same number of tens (p. 309)

straight angle an angle that measures 180° (p. 231)

substitute to replace a variable with a number (p. 351)

subtrahend the number that is subtracted from the minuend (p. 3)

sum the answer to an addition problem (p. 1)

surface area the total area on the outside of a figure (p. 277)

T

transformation sliding, flipping, or turning a plane figure (p. 253)

translation sliding a plane figure (p. 253)

transversal a line that intersects two other lines (p. 235)

trapezoid a quadrilateral having one pair of parallel sides (p. 247)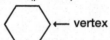

triangle a three-sided polygon (pp. 243, 245)

V

variable a letter used to stand for a number (p. 351)

vertex (pl. vertices) the point at which two sides of an angle, two sides of a plane figure, or three sides of a solid figure meet (p. 231)

 ← vertex

vertical angles the opposite angles formed by two intersecting lines (p. 235)

volume the number of cubic units needed to fill a solid figure (p. 275)

X

***x*-axis** the horizontal number line on the coordinate plane (p. 361)

Y

***y*-axis** the vertical number line on the coordinate plane (p. 361)

Z

Zero Property of Multiplication If a factor is zero, the product will be zero. (p. 5)

TABLES OF MEASURES

Metric Measures

	Unit	Symbol	Relationship
Length	kilometer	km	1 km = 1,000 m
	meter	m	1 m = 100 cm
	centimeter	cm	1 cm = 10 mm
	millimeter	mm	
Mass	metric ton	t	1 t = 1,000 kg
	kilogram	kg	1 kg = 1,000 g
	gram	g	1 g = 1,000 mg
	milligram	mg	
Capacity	liter	L	1 L = 1,000 mL
	milliliter	mL	
Volume	cubic centimeters	cm^3	1 cm^3 = 1 mL = 1 g
Temperature	Celsius	°C	Water freezes at 0°C. Water boils at 100°C.

Time

Unit	Symbol	Relationship
second	s	
minute	min	1 min = 60 s
hour	h	1 h = 60 min
day	d	1 d = 24 h
week	wk	1 wk = 7 d
month	mo	1 mo = approximately 4 wk
year	yr	1 yr = 12 mo = 365 d = 366 d in a leap year
decade		10 yr
century		100 yr

Common Fraction/Decimal Equivalents

$\frac{1}{8} = 0.125$ $\frac{3}{8} = 0.375$ $\frac{5}{8} = 0.625$ $\frac{1}{5} = 0.2$

$\frac{1}{4} = 0.25$ $\frac{1}{2} = 0.5$ $\frac{3}{4} = 0.75$ $\frac{1}{10} = 0.1$

Tables of Measures

TABLES OF MEASURES

Customary Measures

	Unit	Symbol	Relationship
Length	inch	in.	
	foot	ft	1 ft = 12 in.
	yard	yd	1 yd = 3 ft = 36 in.
	mile	mi	1 mi = 5,280 ft = 1,760 yd
Weight	ounce	oz	
	pound	lb	16 oz = 1 lb
	ton	T	1 T = 2,000 lb
Liquid Measure	teaspoon	tsp	
	tablespoon	tbs	1 tbs = 3 tsp
	fluid ounce	fl oz	1 fl oz = 2 tbs
	cup	c	1 c = 8 fl oz
	pint	pt	1 pt = 2 c = 16 fl oz
	quart	qt	1 qt = 2 pt = 32 fl oz
	gallon	gal	1 gal = 4 qt = 8 pt
Dry Measure	pint	pt	
	quart	qt	1 qt = 2 pt
	peck	pk	1 pk = 8 qt
	bushel	bu	1 bu = 4 pk
Temperature	Fahrenheit	°F	Water freezes at 32°F. Water boils at 212°F.

Measurement Formulas

	Figure	Formula	
Area	parallelogram	$A = b \times h$	Area = base × height
	rectangle	$A = l \times w$	Area = length × width
	triangle	$A = \frac{1}{2}b \times h$	Area = $\frac{1}{2}$base × height
Volume	rectangular prism	$V = l \times w \times h$	Volume = length × width × height